Grundlehren der mathematischen Wissenschaften 315

A Series of Comprehensive Studies in Mathematics

T0137819

Editors

S. S. Chern   B. Eckmann   P. de la Harpe
H. Hironaka   F. Hirzebruch   N. Hitchin
L. Hörmander   M.-A. Knus   A. Kupiainen
J. Lannes   G. Lebeau   M. Ratner   D. Serre
Ya. G. Sinai   N. J. A. Sloane   J. Tits
M. Waldschmidt   S. Watanabe

Managing Editors

M. Berger   J. Coates   S. R. S. Varadhan

Springer
Berlin
Heidelberg
New York
Barcelona
Budapest
Hong Kong
London
Milan
Paris
Santa Clara
Singapore
Tokyo

Peter Bürgisser
Michael Clausen
M. Amin Shokrollahi

Algebraic
Complexity Theory

With the Collaboration of Thomas Lickteig

With 21 Figures

 Springer

Peter Bürgisser

Institut für Mathematik
Abt. Angewandte Mathematik
Universität Zürich-Irchel
Winterthurerstraße 190
CH-8057 Zürich, Switzerland
buerg@amath.unizh.ch

Michael Clausen

Institut für Informatik V
Universität Bonn
Römerstraße 164
D-53117 Bonn, Germany
clausen@cs.uni-bonn.de

Mohammad Amin Shokrollahi

International Computer Science
Institute
1947 Center Street, Suite 600
Berkeley, CA 94704-1105, USA
and
Institut für Informatik V
Universität Bonn
Römerstraße 164
D-53117 Bonn, Germany
amin@icsi.berkeley.edu

By courtesy of the publisher, the poem "Der Zweifler" on page VII is reprinted
from volume IV of *Bertolt Brecht: Gesammelte Werke* © Suhrkamp Verlag, Frankfurt am Main 1967.

Cataloging-in-Publication Data applied for

Die Deutsche Bibliothek – CIP-Einheitsaufnahme
Bürgisser, Peter: Algebraic complexity theory / Peter Bürgisser; Michael
Clausen; M. Amin Shokrollahi. With the collab. of Thomas Lickteig. – Berlin;
Heidelberg; New York; Barcelona; Budapest; Hong Kong; London; Milan;
Paris; Santa Clara; Singapore; Tokyo: Springer 1997
(Grundlehren der mathematischen Wissenschaften; 315)

NE: Clausen, Michael:; Shokrollahi, Mohammad Amin:; GT

Mathematics Subject Classification (1991): 68Qxx, 05-xx, 14A10,
14P10, 15-xx, 16A46, 20Cxx, 60C05, 65Fxx, 65T10

ISBN 978-3-642-08228-3

This work is subject to copyright. All rights are reserved, whether the whole or
part of the material is concerned, specifically the rights of translation, reprinting,
reuse of illustrations, recitation, broadcasting, reproduction on microfilm or in
any other way, and storage in data banks. Duplication of this publication or parts
thereof is permitted only under the provisions of the German Copyright Law of
September 9, 1965, in its current version, and permission for use must always be
obtained from Springer-Verlag. Violations are liable for prosecution under the
German Copyright Law.

© Springer-Verlag Berlin Heidelberg 2010
Printed in Germany

Cover design: MetaDesign plus GmbH, Berlin

To

Brigitte
Claudia, Julia, Simone
and Dorothe

DER ZWEIFLER

Immer wenn uns
Die Antwort auf eine Frage gefunden schien
Löste einer von uns an der Wand die Schnur der alten
Aufgerollten chinesischen Leinwand, so daß sie herabfiel und
Sichtbar wurde der Mann auf der Bank, der
So sehr zweifelte.

Ich, sagte er uns
Bin der Zweifler, ich zweifle, ob
Die Arbeit gelungen ist, die eure Tage verschlungen hat.
Ob was ihr gesagt, auch schlechter gesagt, noch für einige Wert hätte.
Ob ihr es aber gut gesagt und euch nicht etwa
Auf die Wahrheit verlassen habt dessen, was ihr gesagt habt.
Ob es nicht vieldeutig ist, für jeden möglichen Irrtum
Tragt ihr die Schuld. Es kann auch eindeutig sein
Und den Widerspruch aus den Dingen entfernen; ist es zu eindeutig?
Dann ist es unbrauchbar, was ihr sagt. Euer Ding ist dann leblos.
Seid ihr wirklich im Fluß des Geschehens? Einverstanden mit
Allem, was wird? Werdet ihr noch? Wer seid ihr? Zu wem
Sprecht ihr? Wem nützt es, was ihr da sagt? Und nebenbei:
Läßt es auch nüchtern? Ist es am Morgen zu lesen?
Ist es auch angeknüpft an Vorhandenes? Sind die Sätze, die
Vor euch gesagt sind, benutzt, wenigstens widerlegt? Ist alles belegbar?
Durch Erfahrung? Durch welche? Aber vor allem
Immer wieder vor allem andern: Wie handelt man
Wenn man euch glaubt, was ihr sagt? Vor allem: Wie handelt man?

Nachdenklich betrachteten wir mit Neugier den zweifelnden
Blauen Mann auf der Leinwand, sahen uns an und
Begannen von vorne.

BERTOLT BRECHT

Preface

The algorithmic solution of problems has always been one of the major concerns of mathematics. For a long time such solutions were based on an intuitive notion of algorithm. It is only in this century that metamathematical problems have led to the intensive search for a precise and sufficiently general formalization of the notions of computability and algorithm.

In the 1930s, a number of quite different concepts for this purpose were proposed, such as Turing machines, WHILE-programs, recursive functions, Markov algorithms, and Thue systems. All these concepts turned out to be equivalent, a fact summarized in Church's thesis, which says that the resulting definitions form an adequate formalization of the intuitive notion of computability. This had and continues to have an enormous effect. First of all, with these notions it has been possible to prove that various problems are algorithmically unsolvable. Among these undecidable problems are the halting problem, the word problem of group theory, the Post correspondence problem, and Hilbert's tenth problem. Secondly, concepts like Turing machines and WHILE-programs had a strong influence on the development of the first computers and programming languages.

In the era of digital computers, the question of finding efficient solutions to algorithmically solvable problems has become increasingly important. In addition, the fact that some problems can be solved very efficiently, while others seem to defy all attempts to find an efficient solution, has called for a deeper understanding of the intrinsic computational difficulty of problems. This has resulted in the development of complexity theory. Complexity theory has since become a very diversified area of research. Each branch uses specific models of computation, like Turing machines, random access machines, Boolean circuits, straight-line programs, computation trees, or VLSI-models. Every computation in such a model induces costs, such as the number of computation steps, the amount of memory required, the number of gates of a circuit, the number of instructions, or the chip area. Accordingly, studies in computational complexity are generally based on some model of computation together with a complexity measure. For an overview, we refer the interested reader to the *Handbook of Theoretical Computer Science* [321], which contains several surveys of various branches of complexity theory.

In this book we focus on *Algebraic Complexity Theory*, the study of the intrinsic algorithmic difficulty of algebraic problems within an algebraic model of computation. Motivated by questions of numerical and symbolic computation, this branch of research originated in 1954 when Ostrowski [403] inquired about the optimality of Horner's rule. Algebraic complexity theory grew rapidly and has since become a well-established area of research. (See the surveys of von zur Gathen [189], Grigoriev [210], Heintz [241], Schönhage [462], and Strassen [506, 510].) However, with the exception of the now classic monograph by Borodin and Munro [65], published in 1975, a systematic treatment of this theory is not available.

This book is intended to be a comprehensive text which presents both traditional material and recent research in algebraic complexity theory in a coherent way. Requiring only some basic algebra and offering over 350 exercises, it should be well-suited as a textbook for beginners at the graduate level. With its extensive bibliographic notes covering nearly 600 research papers, it might also serve as a reference book.

The text provides a uniform treatment of algebraic complexity theory on the basis of the straight-line program and the computation tree models, with special emphasis on *lower complexity bounds*. This also means that this is not a book on Computer Algebra, whose main theme is the design and implementation of efficient algorithms for algebraic problems.

Nonetheless, our book contains numerous algorithms, typically those that are essentially optimal within the specified computation model. Our main goal is to develop methods for proving the optimality of such algorithms.

To emphasize the logical development of the subject, we have divided the book into five parts, with 21 chapters in total. The first chapter consists of an informal introduction to algebraic complexity theory.

The next two chapters form PART I: FUNDAMENTAL ALGORITHMS. Chapter 2 is concerned with efficient algorithms for the symbolic manipulation of polynomials and power series, such as the Schönhage-Strassen algorithm for polynomial multiplication, the Sieveking-Kung algorithm for the inversion of power series, or the Brent-Kung algorithm for the composition of power series. It is followed by a chapter in which the emphasis lies on efficient algorithms within the branching model. In particular, we present the fast Knuth-Schönhage algorithm for computing the greatest common divisor (GCD) of univariate polynomials. This algorithm combined with Huffman coding then yields efficient solutions of algorithmic problems associated with Chinese remaindering. Furthermore the VC-dimension and the theory of epsilon nets are used to show that certain **NP**-complete problems, like the knapsack or the traveling salesman problem, may be solved by "nonuniform polynomial time algorithms" in the computation tree model over the reals. This surprising and important result, due to Meyer auf der Heide, demonstrates that it is not possible to prove exponential lower bounds for the above problems in the model of computation trees. Moreover, it stresses the role of uniformity in the definition of the language class **NP** and, at the same time, puts emphasis on the quality of several lower bounds derived later in Chapter 11.

While the first three chapters rely on the reader's intuitive notion of algorithm, the remaining parts of the book, directed towards lower bounds, call for an exact specification of computation models and complexity measures.

Therefore, in PART II: ELEMENTARY LOWER BOUNDS (Chapters 4–7), we first introduce the models of straight-line programs and computation trees, which we use throughout the rest of the book. We then describe several elementary lower bound techniques. Chapter 5 contains transcendence degree arguments, including results of Motzkin and Belaga as well as the Baur-Rabin theorem. Chapter 6 discusses a unified approach to Pan's substitution method and its extensions. The methods of Chapters 5 and 6 yield lower bounds which are at most linear in the number of input variables. Nonetheless, the methods are strong enough to show the optimality of some basic algorithms, the most prominent being Horner's rule. In Chapter 7 we introduce two fundamental program transformation techniques. The first is Strassen's technique of "avoiding divisions." The second is a method for transforming a program for the computation of a multivariate rational function into one which computes the given function *and* all its first-order partial derivatives. The results of Chapter 7 are of importance in Chapters 8, 14, and 16.

PART III: HIGH DEGREE (Chapters 8–12) shows that concepts from algebraic geometry and algebraic topology, like the degree or Betti numbers, can be applied to prove nonlinear lower complexity bounds. Chapter 8 studies Strassen's degree bound, one of the central tools for obtaining almost sharp lower complexity bounds for a number of problems of high degree, like the computation of the coefficients of a univariate polynomial from its roots. Chapter 9 is devoted to the investigation of specific polynomials that are hard to compute. It may be considered as a counterpart to Chapters 5 and 6 where we study generic polynomials. In Chapter 10 the degree bound is adapted to the computation tree model. With this tool it turns out that the Knuth-Schönhage algorithm is essentially optimal for computing the Euclidean representation. In Chapter 11 Ben-Or's lower complexity bound for semi-algebraic membership problems is deduced from the Milnor-Thom bound. This is applied to several problems of computational geometry. In Chapter 12 the Grigoriev-Risler lower bound for the additive complexity of univariate real polynomials is derived from Khovanskii's theorem on the number of real roots of sparse systems of polynomial equations.

PART IV: LOW DEGREE (Chapters 13–20) is concerned with the problem of computing a finite set of multivariate polynomials of degree at most two. In Chapter 13 we discuss upper and lower complexity bounds for computing a finite set of linear polynomials, which is simply the task of multiplying a generic input vector by a specific matrix. This problem is of great practical interest, as the notable examples of the discrete Fourier transform (DFT), Toeplitz, Hankel and Vandermonde matrices indicate.

The theory of bilinear complexity is concerned with the problem of computing a finite set of bilinear polynomials. Chapters 14–20 contain a thorough treatment of this theory and can be regarded as a book within a book. Chapter 14 introduces the framework of bilinear complexity theory and is meant as a prerequisite for Chapters 15–20. The language introduced in Chapter 14 allows a concise discussion of the matrix multiplication methods in Chapter 15, such as Strassen's original algorithm and the notion of rank, Bini-Capovani-Lotti-Romani's concept of border rank, Schönhage's τ-theorem, as well as Strassen's laser method, and its tricky extension by Coppersmith and Winograd. Chapter 16 shows that several problems in computational linear algebra are about as hard as matrix multiplication, thereby emphasizing the key role of the matrix multiplication problem. Chapter 17 discusses Lafon and Winograd's lower bound for the complexity of matrix multiplication, and its generalization by Alder and Strassen. Moreover, in Chapter 18 we study a relationship, observed by Brockett and Dobkin, between the complexity of bilinear maps over finite fields and a well-known problem of coding theory. Partial solutions to the latter lead to interesting lower bounds, some of which are not known to be valid over infinite fields. This chapter also discusses the Chudnovsky-Chudnovsky interpolation algorithm on algebraic curves which yields a linear upper complexity bound for the multiplication in finite fields.

The bilinear complexity or rank of bilinear problems can be reformulated in terms of tensors, resulting in a generalization of the usual matrix rank. In Chapter 19 tensorial rank is investigated for special classes of tensors, while Chapter 20 is devoted to the study of the rank of "generic" tensors. In the language of algebraic geometry this problem is closely related to computing the dimension of higher secant varieties to Segre varieties.

PART V: COMPLETE PROBLEMS (Chapter 21) presents Valiant's nonuniform algebraic analogue of the P versus NP problem. It builds a bridge both to the theory of NP- and #P-completeness as well as to that part of algebraic complexity theory which is based on the parallel computation model.

A number of topics are not covered in this book; this is due to limitations of time and space, the lack of reasonable lower complexity bounds, as well as the fact that certain problems do not fit into the straight-line program or computation tree model. More specifically, our book treats neither computational number theory nor computational group and representation theory (cf. Cohen [117], Lenstra and Lenstra [326], Sims [484], Atkinson (ed.) [13], Lux and Pahlings [344], Finkelstein and Kantor (eds.) [172]). Also, we have not included a discussion of topics in computational commutative algebra like factorization and Gröbner bases, nor do we speak about the complexity of first-order algebraic theories (cf. Becker and Weispfenning [34], Fitchas et al. [174], Heintz et al. [245], and Kaltofen [284, 286]). We have also omitted a treatment of parallel and randomized algorithms (cf. von zur Gathen [186], Ja'Ja [268]). However, many of these topics have already been discussed in other books or surveys, as the given references indicate.

Clearly, much is left to be done. We hope that our book will serve as a foundation for advanced research and as a starting point for further monographs on algebraic complexity theory.

Zürich, Bonn, and Berkeley   *P. Bürgisser · M. Clausen · M. A. Shokrollahi*
June 1996

Leitfaden

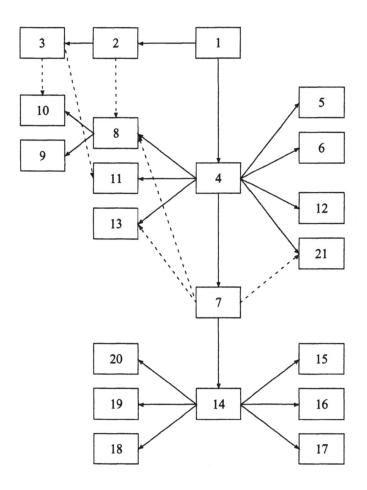

Notes to the Reader

This book is intended as a textbook as well as a reference book. One of the important principal features is the division of the material into the relatively large number of 21 chapters, which are each designed to enable quick acquaintance with a specific topic. Furthermore, we have subdivided each chapter into sections which often make widely differing demands on the reader. Almost every chapter starts at an undergraduate level and ends at a more advanced level. To facilitate the reader's orientation we have marked those sections with asterisks that are of a rather technical nature and may be skipped on a first reading. To provide easy checks on the reader's comprehension of the text, or to challenge her/his proficiency, we have included numerous exercises in each chapter, the harder ones carrying asterisks. Many of the exercises are important results in their own right and are occasionally referred to in later sections. A list of open problems as well as the detailed notes at the end of each chapter should be seen not only as incentives for researchers willing to improve the present knowledge, but also as landmarks pointing to the frontiers of our field.

We believe that the structure of the book facilitates its use in many ways. Generally, all readers interested in lower complexity bounds are expected to study the essential material of Sections 4.1–4.2, where we describe straight-line programs and introduce the notion of complexity. The language developed there will be used throughout the book. Thereafter, those whose primary inclination is to use this book as a reference source can directly traverse to their topic of interest.

The rigorous presentation of many techniques for lower bound proofs in algebraic complexity theory calls not only for the use of tools from different areas of mathematics, but also for technicalities which often obscure the ideas behind those techniques. Whenever we have encountered such a situation, we have tried to familiarize the reader with the underlying ideas by means of examples of increasing difficulty. In so doing, we have designed a textbook for various possible courses. As an example of an introductory course on algebraic complexity theory, one can cover the topics presented in (1) (where (x) means "parts of Chapter x"), 2, 4.1–4.2, 5, 6, 7.2, 8.1. This course could be followed by an advanced course dealing with the content of (1), 4.4–4.5, 3.1–3.2, 8.2–8.5, 10.1–10.2, 11. A special course on bilinear complexity could include (1), 4.1–4.2, 14, 15.1–15.8, 17.1–17.3, 19.1–19.2. A special course on the Degree Bound might consist of (1), (2), (4), 7.2, 8.2–8.4, 3.1–3.2, 10.1–10.2, (11).

Isolated chapters of our book can be used by people from other disciplines as complementary material to courses in their own field of research. Examples of this include courses on **NP**-completeness + (21), coding theory + (18), group representation theory + (13), computational geometry + (11), algebraic number theory + 9.1–9.3, and numerical analysis + (5, 6, 7, 8, 16). Courses in computer algebra can obviously be accompanied by a treatment of several of the lower complexity bounds discussed in this book. In addition, there is also a number of (asymptotically) fast algorithms in Chapters 2, 3, 5, 13, and 15 that are of interest to computer algebraists.

Acknowledgments

Our greatest intellectual debt is to V. Strassen for his many contributions to the field of algebraic complexity theory as well as for his brilliant lectures which introduced the subject to us. Special thanks go to our cooperator Thomas Lickteig who, together with us, first planned this book more than five years ago. His competence in this field has always been of extreme benefit to us. We owe thanks to W. Baur whose clear and concise lecture notes helped us a lot in writing this book.

We are indebted to Ch. Bautz, F. Bigdon, A. Björner, K. Kalorkoti, F. Mauch, M. Nuesken, T. Recio, H.J. Stoß, V. Strassen and Ch. Zengerling for reading parts of the manuscript and their valuable suggestions for improvements. We have benefited from the help of U. Baum, S. Blackburn, J. Buhler, E. Kaltofen, H. Meier-Reinhold, A. McNeil, J. Neubüser, A. Schönhage, and F. Ulmer and would like to express our gratitude to them. We also thank our students at the Universities of Bonn and Zürich for their attention and stimulating questions.

Although this book has been proofread by several people, we take complete responsibility for the errors that may have remained.

Many people, too numerous to mention, have contributed to our project by kindly sending to us a list of their publications relevant for our book. We thank them all very much.

We thank the Schweizerische Nationalfonds for its financial support which allowed the first author to stay at the University of Bonn in the first phase of our project from 1991 until 1993. Thanks go also to the Institute of Applied Mathematics of the University of Zürich for the pleasant working conditions which allowed an efficient continuation of the project after the first author had moved to Zürich.

We have extensively used Email and Internet, mostly after the first and third author had left Bonn for Zürich and Berkeley, respectively. Without these media, communication would have become much harder. Also, we have benefited a lot from the GNU project, in particular from the powerful Emacs-Editor distributed with the GNU-package. We take the opportunity to thank R. Stallman and his team for this public domain software of distinguished quality.

Without the document processing systems TeX and LaTeX we would have had a very hard time. Many thanks to D. Knuth and L. Lamport for providing the community with their wonderful – and free – software. For the camera-ready preparation of this document we have used different style files written by B. Althen, M. Barr, and P. Taylor, whom we would like to thank.

We are especially grateful to the staff at Springer-Verlag Heidelberg for their editorial advise and great patience throughout this enterprise.

Finally, we wish to thank Brigitte, Claudia, and Dorothe for their support, patience, and understanding of the commitment necessary to write such a book.

Table of Contents

Chapter 1. Introduction

Complexity theory investigates the computational resources required to solve algorithmic problems. The goal, at least in simple situations, is to find an optimal procedure among all conceivable ones and to prove its optimality. This optimal cost is called the *complexity* of the problem.

In this book we are concerned with *algebraic complexity theory*, where we investigate algorithmic problems which can and are to be solved by means of algebraic algorithms. We thus restrict the class of all "conceivable" algorithms to those that fit into a given algebraic model of computation. Such models will be introduced and extensively studied in Chap. 4. These describe the scope and limitations of a step-by-step production of intermediate results by means of admissible algebraic operations; the philosophy is that it is allowed to use the intermediate results freely once they have been produced. Such a "computation" is said to be finished if the quantities that the computation is supposed to compute are among the intermediate results; these quantities represent the output. A computation has a certain cost (e.g., its number of steps), and every computation solving a problem yields an upper bound for the complexity of that problem with respect to the given model of computation.

The design and implementation of efficient algorithms for algebraic problems is the main theme of *computer algebra*. Algebraic complexity theory adds to this the search for "lower complexity bounds." Ideally, the lower bound for a problem coincides with an upper bound obtained from an algorithm which solves that problem. This is equivalent to a proof of the optimality of that algorithm within the scope of the given model of computation. It is not surprising that the derivation of nontrivial lower bounds for a computational problem is not a simple task, as all admissible algorithms which solve that problem have to be taken into account. Nevertheless, the past thirty years have witnessed the development of an extensive theory of computational complexity. For reasons of space this book can only reflect a part of the tremendous efforts which have led to this theory.

Lower bounds can often be obtained by establishing a relationship between the complexity of a problem and invariants of appropriate algebraic, topological, geometric, or combinatorial structures. These invariants range from the dimension of vector spaces to the degree of transcendency of field extensions or the degree of algebraic varieties; also the number of connected components or even all the Betti numbers of real manifolds may yield interesting lower bounds.

In this introduction we try to lead the reader to some typical questions, models, and methods of algebraic complexity theory by way of simple examples. In our young field, research has concentrated on problems connected with the manipulation of polynomials and rational functions. This is not a strong limitation, as the important example of matrix multiplication (or inversion of a regular matrix) may indicate: the coefficients of the product of two matrices (the inverse of a regular matrix) are polynomial (rational) functions in the coefficients of the input.

We start our examples with the problem of taking powers. Here we take a ring R and a positive integer n and ask for an optimal algorithm which, on input $r \in R$, computes r^n. The attentive reader will have noticed that we already have used a couple of concepts which still need to be specified: how can we give a precise mathematical model of an input $r \in R$? What is an algorithm? When is an algorithm optimal? The answers to these questions differ according to the point of view taken. An extreme standpoint is the assumption of a concrete computer operated by a certain operating system on which a specific programming language can be compiled. Supplying a coding of an element $r \in R$ (e.g., as a bit string, if R is a suitable ring), the task is to transform this input into a coding for r^n by means of a correct program. Different programs solving this problem can then be compared with respect to their running time or space requirements, to name just two criteria. Possibly one program runs more efficiently than the others while another program requires less memory. Needless to say, the advantage of this point of view is its concreteness and the fact that it describes a real computation rather exactly. This, however, may also be looked upon as a disadvantage since the results obtained depend heavily on the specific computing environment, which restricts the range of validity of the results a lot. This problem may be resolved by studying idealized computer models—such as Turing machines or random access machines—and discussing corresponding computational problems in this new context.

Up to now very few reasonable lower bounds have been derived for specific problems in these "bit-models;" a fact that is probably due to the absence of constraints. A completely different point of view—which we adopt in this book—has turned out to be more fruitful. Loosely speaking, a macroscopic standpoint is taken starting from an idealized computer which, in our case, is capable of multiplying elements of the ring R in one step and store the results in such a way that later optional access to them can be accomplished at no cost. The input $r \in R$, which is not known in advance, can be modeled by an indeterminate X over R. (This will make sense for example in the case of an infinite field R.) The problem of computing the nth power of an element $r \in R$ can then be reformulated as the task of computing X^n from X, where now in every step we are allowed to multiply two intermediate results. Hence, in this case we interpret computation as a step-by-step procedure that starts with the 0th intermediate result X and multiplies in each step two intermediate results already produced. (The first step is thus necessarily the multiplication of X by X yielding the first intermediate result X^2.) We say that the computation is finished if X^n is among the intermediate results. Such a computation is well-modeled by a so-called addition chain (of length r) for n, which is a sequence $(1 = a_0, a_1, \ldots, a_r = n)$ of positive integers satisfying for each $1 \le \rho \le r$, $a_\rho = a_i + a_j$ for some $0 \le i, j < \rho$ (Scholz [453]).

Let $\ell(n)$ denote the minimum number of multiplications sufficient to compute X^n from X with respect to the above notion of computation. Then $\ell(n)$ is the shortest length of an addition chain for n. This number is the complexity (with respect to the above computational model) of the computational problem of raising elements in R to the nth power. Obviously, $\ell(n) \le n - 1$ since one can multiply at each step by X obtaining the following sequence $(X, X^2, X^3, \ldots, X^n)$ of intermediate results. However, the computation can be performed much faster, as the following example indicates:

$$X^{13} = X \cdot (((X \cdot X^2)^2)^2).$$

Generalizing this example we obtain

$$(1.1) \qquad \ell(n) \le \lfloor \log n \rfloor + w_2(n) - 1 \le 2 \log n$$

for $n \ge 2$ where $w_2(n)$ denotes the Hamming weight (i.e., the number of nonzero coefficients) of the binary expansion of n, and $\log := \log_2$. In fact, if $n = 2m$ is even ($n = 2m + 1$ is odd), we first compute X^m (X^{n-1}) and, in a further step, square this result (multiply the result with X) to obtain X^n. This yields the recursion $\ell(2m) \le \ell(m)+1$, $\ell(2m+1) \le \ell(2m)+1$. Noting that $w_2(2m) = w_2(m)$ and $w_2(2m + 1) = w_2(2m) + 1$ we obtain the claim by a simple induction. The above procedure is often referred to as the *binary method*.

A reasonable lower bound can be derived in this context by using a degree argument: we start from the 0th intermediate result, i.e., from a polynomial of degree one. Obviously, the maximum of the degrees obtained can at most be doubled in each step, so the hth intermediate result has degree at most 2^h. We infer that if one can compute X^n from X in h steps, then necessarily $n \le 2^h$. Thus

$$\ell(n) \ge \lceil \log n \rceil.$$

Later in Chap. 8 we shall encounter a generalization of this degree argument which will play an important role in this book.

If n is a power of 2, then the upper bound and the lower bound derived so far coincide. This shows that for powers of 2 the binary method is optimal within the class of algorithms we have considered. For arbitrary n the bounds differ by a factor ≤ 2. Generally, we may in such a case try to derive a better lower bound, or alternatively search for a faster algorithm. Here, a tricky proof due to Schönhage [457] gives the improved lower bound

$$(1.2) \qquad \ell(n) \ge \log n + \log w_2(n) - 2.13.$$

(For weaker versions of this lower bound see Ex. 1.6 and 1.8.) On the other hand, A. Brauer [71] deduced the following better upper bound

$$(1.3) \qquad \ell(n) \leq \log n + \frac{\log n}{\log \log n} + o\left(\frac{\log n}{\log \log n}\right).$$

The proof of Brauer's upper bound proceeds by breaking the binary expansion of n into blocks of appropriate length λ and first computing all X^i, $2 \leq i < 2^\lambda$. More precisely, let λ be a positive integer, $t := 2^\lambda$, and $n = a_0 t^p + a_1 t^{p-1} + \ldots + a_p$ be the t-adic representation of n, where $a_0 \neq 0$ and $0 \leq a_i < t$ for all i. (λ will be chosen later.) Then we have $n \geq t^p = 2^{\lambda p}$, hence $\lambda p \leq \log n$. Now we compute with less than t multiplications all powers $X^0, X^1, \ldots, X^{t-1}$; in particular this will give us all $X^{a_0}, X^{a_1}, \ldots, X^{a_p}$. If we have already computed X^m for some m, then we can obtain X^{mt} by squaring λ times. The resulting *block version* of the binary method computes X^n as follows:

$$\underbrace{X^2, X^3, \ldots, X^t}_{t-1 \text{ mult.}}, \quad \underbrace{X^{a_0 t}}_{\lambda \text{ mult.}}, \quad \underbrace{X^{a_0 t + a_1}}_{1 \text{ mult.}}, \quad \underbrace{X^{(a_0 t + a_1)t}}_{\lambda \text{ mult.}}, \quad \underbrace{X^{(a_0 t + a_1)t + a_2}}_{1 \text{ mult.}}, \ldots$$

Altogether, this shows that $\ell(n) < t + p\lambda + p \leq 2^\lambda + \log n + \lambda^{-1} \log n$. Choosing $\lambda = \lfloor \log \log n - 2 \log \log \log n \rfloor$ gives for large n

$$\ell(n) \leq \frac{\log n}{(\log \log n)^2} + \log n + \frac{\log n}{\lfloor \log \log n - 2 \log \log \log n \rfloor},$$

which yields our assertion.

A comparison of the binary method and its block version reveals that the first is slower than the second by a factor ≤ 2, but that it requires much less memory: the binary method gets along with storing X and the most recent intermediate result; its block version, however, has to store all the values $X^{a_0}, \ldots, X^{a_p}$ in the worst case. Hence, the reduction of the number of multiplication steps is accompanied by an increase of the storage expense. Such a trade-off is typical and can be observed in many other examples as well. *However, in this book we will concentrate solely on the problem of minimizing the number of arithmetic operations for the solution of a computational algebraic problem, and will ignore the amount of storage required.*

For the computation of X^n from X we will now allow division by intermediate results as well as multiplication. Let $d(n)$ denote the minimal number of operations necessary to compute X^n from X in this new model. Obviously, we have $d(n) \leq \ell(n)$. It is a rewarding exercise to show that $d(31) = 6 < 7 = \ell(31)$. Divisions may thus help. On the other hand, the degree argument mentioned above still holds in this new context, which shows that $\lceil \log(n) \rceil \leq d(n)$.

The above explanations illustrate that a computational problem like
"Compute X^n from X."
makes sense only when the algebraic operations that are admitted for an algorithmic solution and the cost of each operation have been agreed. In general, the complexity of a problem heavily depends on these agreements and in particular on the selected cost function.

How realistic are the above complexity claims? If X is regarded as a program variable which can be arbitrarily replaced by elements of R, then the computing time depends not only on the number of multiplications or divisions, but also on the time required to multiply (divide) two elements in R. For instance, when considering algorithms for computing nth powers which work over any finite field, then $\ell(n)$ and $d(n)$ should reflect the real computation times rather exactly (up to a factor depending only on the size of the field). Things differ dramatically in the case of infinite R. For example, if $R = \mathbb{Z}$ is the ring of integers, then the computing time also depends on the length of the input numbers. According to the present state of the art the cost for multiplying two m-bit numbers is of the order $m \log m \log \log m$, [466]. This means that in any reasonable algorithm for computing X^n the last multiplication determines the cost of the whole procedure.

We want to generalize the above problem of computing X^n from X and consider the problem of computing a polynomial

$$p := a_0 X^n + a_1 X^{n-1} + \ldots + a_n.$$

Again we have to provide an exact specification of this task. First, we ask what the inputs should be. For instance, if the coefficients $a_0, \ldots, a_n$ are fixed elements of R, then the task consists of computing the polynomial $p \in R[X]$ from X with as few algebraic operations as possible (like additions, subtractions, scalar and nonscalar multiplications, divisions). On the other hand, if the coefficients $a_0, \ldots, a_n$ are not known in advance, then one can regard these as indeterminates over R and ask for a short computation of the general polynomial of degree n in the algebra $R[X, a_0, \ldots, a_n]$. Alternatively, if R happens to be a field and if divisions are also admissible, then one can ask for as short as possible a computation of p in the rational function field $R(X, a_0, \ldots, a_n)$. To begin with, we adopt this point of view.

Horner's rule shows that it is possible to compute p with n additions and n multiplications. The computation is based on the formula

$$p = (\ldots(((a_0 \cdot X + a_1) \cdot X + a_2) \cdot X + a_3)\ldots) \cdot X + a_n.$$

This means the following: in the first step $g_1 := a_0 \cdot X$ is computed; then a_1 is added to this intermediate result to obtain the second intermediate result $g_2 := g_1 + a_1$. This is then multiplied by X: $g_3 := g_2 \cdot X$, etc. Setting $g_{-n-1} := X$ and $g_{-i} := a_i$, for all $0 \le i \le n$, Horner's rule reads as follows:

$$g_{2i} := g_{2i-1} + g_{-i} \quad \text{and} \quad g_{2i-1} := g_{2i-2} \cdot g_{-n-1},$$

for $1 \le i \le n$. (Note that Horner's rule has been used implicitly in handling the exponents in the problem of computing X^n above.) Ostrowski [403] inquired in 1954 about the optimality of Horner's rule and conjectured that n multiplications or divisions are always required for the computation of the general polynomial of degree n regardless of the number of additions or subtractions used. He also conjectured that the same computational problem needs n additions or subtractions even if one has multiplications and divisions at free disposal. By the substitution method (Pan [405]) and the method of transcendence degree (Motzkin [386], Belaga [35]) these conjectures can be proved. We want to explain these methods at this point without going into technical details.

The substitution method is used to obtain lower bounds in the so-called Ostrowski model of computation. In this model, linear operations like additions, subtractions, or scalar multiplications are free, while nonscalar multiplications and divisions are considered as arithmetic operations of cost one. Starting from a hypothetical computation the first intermediate result of positive cost is trivialized by an appropriate linear substitution of the indeterminates. This reduces the original problem to a simpler one. Now it remains to show that a complete trivialization of the problem requires at least a certain number of trivialization steps. This number, which is typically linear in the number of inputs (=indeterminates), will then represent a lower bound. The substitution method will be discussed extensively in Chap. 6.

The method of transcendence degree associates with a given finite set $F \subseteq K(X_1, \ldots, X_n)$ of rational functions with coefficients in a field extension K of the base field k an intermediate field $k \subseteq \mathrm{Coeff}_k(F) \subseteq K$. This field is obtained by adjoining to k all the coefficients of all the elements of F (after a suitable normalization). In a similar way one can associate with a sequence $(g_1, \ldots, g_r)$ of intermediate results of a computation for F a tower of fields

$$k \subseteq K_1 \subseteq \ldots \subseteq K_r \subseteq K,$$

where by construction $\mathrm{Coeff}_k(F) \subseteq K_r$. The problem is to show that any step of positive cost at most doubles the degree of transcendency, while steps of no cost leave this degree unchanged. This, together with the trivial observation that the transcendence degree of $\mathrm{Coeff}_k(F)$ is at most as big as that of K_r, yields a lower bound. This method will be discussed in detail in Chap. 5.

The example of $p(X) = X^n$ shows that Horner's rule may not be optimal for the computation of specific polynomials in $R[X]$, i.e., of polynomials with known coefficients from R. Surprisingly, if $R = \mathbb{C}$ is the field of complex numbers, one can compute any specific polynomial of degree ≥ 5 faster than by Horner's rule. (The design of an algorithm suitable for a given polynomial may take considerable time, since it involves a "preconditioning of the coefficients.") We give a brief sketch of the basic underlying idea. Let $u \in \mathbb{C}$ be nonzero. Then polynomial division yields unique polynomials $q_u(X)$ of degree $n - 2$ and $r_u(X)$ of degree ≤ 1 such that

$$p(X) = (X^2 - u)q_u(X) + r_u(X).$$

We want to choose u such that $r_u(X)$ becomes a constant polynomial. Let $v^2 = u$. Then $p(v) = r_u(v)$ and $p(-v) = r_u(-v)$. Hence, r_u is constant if and only if $p(v) = p(-v)$, i.e., $v \neq 0$ is a zero of the polynomial $\sum_{i \equiv 1(2)} a_{n-i} X^i$. (An alternative completely rational construction can be found in Chap. 5.) This choice of $u = v^2$ yields

$$p = (X^2 - u)q + r,$$

with $r \in \mathbb{C}$ and $\deg q = n - 2$. Applying this procedure recursively one obtains an algorithm which—if the cost of the preconditioning of the coefficients is neglected—computes any given polynomial p of degree n with at most $n + \lfloor n/2 \rfloor + 2$ arithmetic operations. Details can be found in Chap. 5. There we also prove that the procedure sketched above for the computation of a specific (but sufficiently general) polynomial is optimal for odd n.

What happens if the polynomial p is to be evaluated at many, say n, points? This leads to the computation of $p(X_1), \ldots, p(X_n)$, where $X_1, \ldots, X_n$ are indeterminates. Separate computation by Horner's rule shows that this problem can be solved with an arithmetic cost proportional to n^2. Surprisingly, there exists a faster multiple evaluation algorithm which, in the case say $R = \mathbb{C}$, gets along with $O(n \log n)$ multiplications and $O(n \log^2 n)$ arithmetic operations; see Chap. 3. (In fact, this algorithm even works when the coefficients of the polynomial are indeterminates.)

On the other hand, a degree argument less trivial than that previously encountered shows that this computational task has a lower bound of $n \log n$ in the Ostrowski model. To explain this lower bound we shall work with an alternative concept of degree. In an abstract setting, we start from the rational function field $K := k(X_1, \ldots, X_n)$ over an infinite ground field k. With each sequence $G := (g_1, \ldots, g_r)$ of elements of K we associate a number

$$\deg(G) := \max_b [K : k(b)],$$

where $b = (b_1, \ldots, b_n)$ runs over all bases of transcendency of $K \supset k$ contained in the k-linear hull of $1, X_1, \ldots, X_n, g_1, \ldots, g_r$. We call this number the degree of the sequence G. (For a motivation of this definition note that within the Ostrowski model k-linear operations are free and the inputs are at free disposal.) It is not very difficult to show that in the case of a single polynomial $f \in k[X_1, \ldots, X_n]$ this definition coincides with the usual polynomial degree of f, see Chap. 8. If F is a subsequence of G, then we have trivially

$$\deg(F) \leq \deg(G).$$

Furthermore, it is straightforward that the degree of G remains unchanged if G is extended by a k-linear combination of two intermediate results g_i and g_j:

$$\deg(g_1, \ldots, g_r) = \deg(g_1, \ldots, g_r, \alpha g_i + \beta g_j),$$

where $i, j \leq r$ and $\alpha, \beta \in k$. Finally, some technical effort yields for $g_j \neq 0$

$$\deg(g_1, \ldots, g_r, g_i \circ g_j) \leq 2 \deg(g_1, \ldots, g_r),$$

where $\circ \in \{*, /\}$. This constitutes the heart of the following field theoretic version of *Strassen's degree bound* [497]: If $F \subset k(X_1, \ldots, X_n)$ is a finite set of rational functions, then the Ostrowski complexity $L(F)$ satisfies

$$L(F) \geq \log \deg(F).$$

In Chap. 8 we will discuss the above as well as the original geometric formulation of the degree bound. From this we shall obtain nonlinear lower bounds for some natural computational problems for rational functions. The degree bound will often be used in a weaker form: if $F \subset K$ contains a basis of transcendency of K over k, then $\deg(F) \geq [K:k(F)]$ trivially; hence, under this condition we obtain

$$(1.4) \qquad\qquad L(F) \geq \log[K : k(F)].$$

In the above example of multiple evaluation of a polynomial p over k we have to compute $F := \{p(X_1), \ldots, p(X_n)\}$. It can be easily shown that $[K:k(F)] = n^n$, see the first section of Chap. 8 for details. Hence, (1.4) implies the lower bound

$$L(p(X_1), \ldots, p(X_n)) \geq n \log n.$$

The analysis of the above mentioned fast algorithm for the evaluation of a polynomial at many points assumes, among other techniques, the knowledge of efficient procedures for multiplying polynomials. This will constitute our next topic.

Given two polynomials $a = \sum_{i \leq n} a_i X^i$ and $b = \sum_{i \leq n} b_i X^i$ in $\mathbb{C}[X]$, whose coefficients are regarded as inputs, we are looking for the coefficients of the polynomial $a \cdot b = c = \sum_{\ell} c_{\ell} X^{\ell}$, where $c_{\ell} = \sum_{i+j=\ell} a_i b_j$ for $0 \leq \ell \leq 2n$; the coefficients of c are regarded as the outputs. Direct computation of the coefficients via the defining formula uses $O(n^2)$ arithmetic operations. One can, however, do much better! In fact it is sufficient to find a fast way to multiply polynomials of degree $< N$ modulo $X^N - 1$ if $N > 2n$. Thus we may consider multiplication in the algebra $\mathbb{C}[X]/(X^N - 1)$. If ω denotes a primitive Nth root of unity, then the Chinese remainder theorem shows that the mapping $f \mapsto (f(\omega^i))_{0 \leq i < N}$ constitutes an isomorphism D_N between the algebra $\mathbb{C}[X]/(X^N - 1)$ and the algebra $\mathbb{C}^N$ of diagonal matrices of size N. Using this (cyclic) discrete Fourier transform D_N one can compute the product $a \cdot b$ via the formula

$$a \cdot b = D_N^{-1}(D_N(a) \cdot D_N(b)).$$

Once $D_N(a)$ and $D_N(b)$ are known, the computation of $D_N(a) \cdot D_N(b)$ only uses N multiplications.

The main problem is now the efficient evaluation of the discrete Fourier transform D_N and its inverse D_N^{-1}. This leads to the design of an efficient algorithm for the multiplication of the DFT matrix

$$DFT_N := (\omega^{ij})_{0 \leq i, j < N},$$

resp. its inverse, with a column vector $a = (a_i)_{0 \leq i < N}$. If $N = 2M$ and if we denote by DFT_M the $M \times M$-DFT matrix for the primitive Mth root of unity ω^2, then for $0 \leq i < N$ the ith entry in $DFT_N \cdot a$ can be rewritten as

$$\sum_{0 \leq j < N} a_j \omega^{ij} = \left(\sum_{0 \leq j < M} a_{2j}(\omega^2)^{ij} \right) + \omega^i \left(\sum_{0 \leq j < M} a_{2j+1}(\omega^2)^{ij} \right).$$

This fact together with $\omega^M = -1$ yields

$$DFT_N \cdot a = \begin{pmatrix} DFT_M & \Delta_M DFT_M \\ DFT_M & -\Delta_M DFT_M \end{pmatrix} \cdot \begin{pmatrix} g_a \\ u_a \end{pmatrix}.$$

Here $\Delta_M = \mathrm{diag}(1, \omega, \omega^2, \ldots, \omega^{M-1})$ is an $M \times M$-diagonal matrix (the matrix of twiddle factors) and $g_a = (a_i)_{i \equiv 0(2)}$ and $u_a = (a_i)_{i \equiv 1(2)}$ denote the vectors consisting of the even-indexed and odd-indexed entries of a, respectively. Therefore, in order to compute $DFT_N \cdot a$ it is sufficient to compute $G := DFT_M \cdot g_a$ and $U := DFT_M \cdot u_a$; then, one computes $T := \Delta_M \cdot U$ and finally $G + T$ and $G - T$, which form the first and the second half of the output vector, respectively. This procedure is generally referred to as the fast Fourier transform (FFT). The term "fast" is justified by the following analysis: if $T(N)$ denotes the arithmetic cost of evaluating DFT_N, then the FFT algorithm yields for even N the recursion formula

$$T(N) \le 2T(N/2) + N/2 - 1 + N.$$

Together with $T(1) = 0$ we obtain $T(N) \le 1.5N \log N - N + 1$, for N a power of 2. The inverse of DFT_N equals up to a scalar the DFT matrix with respect to the primitive Nth root of unity ω^{-1}, namely

$$DFT_N^{-1} = \frac{1}{N}(\omega^{-ij})_{0 \le i, j < N}.$$

This demonstrates that the inverse of DFT_N too can be computed by means of the FFT. Altogether we have shown that complex polynomials of degree at most n can be multiplied with $O(n \log n)$ arithmetic operations. (Note that we can choose N as a power of 2 such that $2n < N \le 4n$.) Extensions of this method are discussed in Chap. 2 and Chap. 13.

We now turn our attention to lower complexity bounds for the above problem: let $L(N)$ denote the minimum number of additions, subtractions, and scalar multiplications for evaluating the matrix DFT_N at a generic input column vector of length N. The FFT algorithm shows that $L(N) = O(N \log N)$, for N a power of 2. The question of whether the FFT is optimal up to order of magnitude is one of the big unsolved problems in algebraic complexity theory. Up to now only linear lower bounds for $L(N)$ are known. However, the efforts undertaken to prove a nonlinear lower bound have been very fruitful for a completely different area: to every algorithm for the evaluation of DFT_N one can associate a graph with N input and N output nodes having the "communicative" property that for any $q \le N$ any q input nodes are connected with any q output nodes (in some order) by q node-disjoint paths. Such graphs are called N-superconcentrators. It turns out that the minimum number $s(N)$ of edges of N-superconcentrators yields a lower complexity bound. More precisely, $L(N) \ge s(N)/6 - N/2$. For some time it was conjectured that $s(N)$ is superlinear in N. Surprisingly enough, it turned out that there exist N-superconcentrators with $O(N)$ edges (Pinsker [418], Valiant [525]). This result, though a failure for its original purpose, has led to numerous applications in other areas, see [422].

If we restrict the model of computation in such a way that besides additions and subtractions only scalar multiplications with complex constants of absolute value ≤ 2, say, are allowed, we can prove for the accordingly modified complexity $L_2(N)$ a lower bound which is of the order of magnitude $N \log N$. The starting point of this argument is an alternative characterization of $L_2(N)$. First we define the following model of computation: let $e_1, \ldots, e_N$ denote the standard basis vectors of $\mathbb{C}^N$. We start the computation with the inputs $g_{-N+1} := e_N, \ldots, g_0 := e_1$ considered as the first intermediate results. In every step of the computation we are allowed to add or subtract two previously computed intermediate results, or multiply one previously computed intermediate result by a complex number of absolute value at most 2. In this way we obtain a sequence $G := (g_1, \ldots, g_r)$ of "proper" intermediate results. (This sequence will play the same role as the corresponding sequence G did for the derivation of the degree bound above.) Since the g_i are linear forms, this sequence can be identified with a matrix in $\mathbb{C}^{N \times r}$ in a straightforward way. We say that such a matrix computes DFT_N if the columns of DFT_N are among those of G. Then $L_2(N)$ is the minimum number r such that there exists a computation matrix $G \in \mathbb{C}^{N \times r}$ for DFT_N. We associate with a computation matrix $G = (g_1, \ldots, g_r)$ a volume defined by

$$\mathrm{vol}(G) := \max_b |\det(b)|,$$

where $b = \{b_1, \ldots, b_N\}$ runs over all $\mathbb{C}$-bases of $\mathbb{C}^N$ contained in the set $\{e_1, \ldots, e_N, g_1, \ldots, g_r\}$. (In contrast to the degree bound we do not allow linear combinations of the g_i since we do count $\mathbb{C}$-linear operations.) In exactly the same way as for the degree bound, but with considerably less effort, we can show that

$$\mathrm{vol}(g_1, \ldots, g_r, g_i \pm g_j), \mathrm{vol}(g_1, \ldots, g_r, \alpha g_i) \leq 2\,\mathrm{vol}(g_1, \ldots, g_r),$$

where now $i, j \leq r$ and $\alpha \in \mathbb{C}$ is of absolute value at most 2. Hence, each computation step increases the volume by at most a factor of two. This immediately gives the following lower bound due to Morgenstern [381]

$$L_2(DFT_N) \geq \log \mathrm{vol}(G)$$

for any computation matrix G for DFT_N. If G computes DFT_N, then obviously $\mathrm{vol}(G) \geq |\det(DFT_N)|$. Noting that the latter equals $N^{N/2}$, we finally obtain the lower bound

$$L_2(N) \geq \frac{1}{2} N \log N.$$

As the scalar multiplications involved in the FFT algorithm are multiplications by roots of unity, we see that the FFT algorithm is essentially optimal in this restricted model.

One of the leading problems of algebraic complexity theory is matrix multiplication. Multiplying two $n \times n$-matrices over the field k via the formula

$$(a_{ij}) \cdot (b_{j\ell}) = \left(\sum_j a_{ij} b_{j\ell} \right)$$

uses a number of arithmetic operations which is proportional to n^3. The numerous efforts to prove the optimality of this procedure led Strassen [493] at the end of the

1960's to an asymptotically faster multiplication algorithm (which can be shown to be optimal in case $n = 2$, see Chap. 17). The background of this algorithm is the following pair of bases of $k^{2\times 2}$:

$$A_0 = \begin{pmatrix} 1 & 0 \\ 0 & 1 \end{pmatrix}, \quad A_1 = \begin{pmatrix} 1 & 0 \\ 0 & 0 \end{pmatrix}, \quad A_2 = \begin{pmatrix} 0 & 0 \\ -1 & 1 \end{pmatrix}, \quad A_3 = \begin{pmatrix} 0 & 1 \\ 0 & 1 \end{pmatrix}$$

and

$$B_0 = \begin{pmatrix} 1 & 0 \\ 0 & 1 \end{pmatrix}, \quad B_1 = \begin{pmatrix} 0 & 0 \\ 0 & 1 \end{pmatrix}, \quad B_2 = \begin{pmatrix} 1 & -1 \\ 0 & 0 \end{pmatrix}, \quad B_3 = \begin{pmatrix} 1 & 0 \\ 1 & 0 \end{pmatrix}.$$

This pair of bases has the following property which turns out to be algorithmically advantageous: for all $0 \le i, j, \le 3$ we have $A_i \cdot B_j \in \{A_i, B_j, 0\}$. More precisely, the multiplication table of this pair of bases has the following form:

	B_0	B_1	B_2	B_3
A_0	A_0	B_1	B_2	B_3
A_1	A_1	0	B_2	A_1
A_2	A_2	B_1	A_2	0
A_3	A_3	A_3	0	B_3

To proceed further, we need the Kronecker product of matrices $C = (c_{ij}) \in k^{r\times r}$ and $D \in k^{s\times s}$:

$$C \otimes D := \begin{pmatrix} c_{11}D & c_{12}D & \cdots \\ c_{21}D & c_{22}D & \cdots \\ \vdots & \vdots & \end{pmatrix} \in (k^{s\times s})^{r\times r}.$$

This product satisfies the well-known multiplication property

$$(C_1 \otimes D_1) \cdot (C_2 \otimes D_2) = (C_1 \cdot C_2) \otimes (D_1 \cdot D_2)$$

for matrices $C_1, C_2 \in k^{r\times r}$ and $D_1, D_2 \in k^{s\times s}$. Since (A_0, A_1, A_2, A_3) and (B_0, B_1, B_2, B_3) form bases of $k^{2\times 2}$, we can represent $X, Y \in k^{n\times n} = (k^{m\times m})^{2\times 2}$ for $n = 2m$ in the form

$$X = \sum_{0 \le i \le 3} A_i \otimes \xi_i \quad \text{and} \quad Y = \sum_{0 \le j \le 3} B_j \otimes \eta_j,$$

with uniquely determined matrices $\xi_i, \eta_j \in k^{m\times m}$. The multiplication property of the Kronecker product just mentioned and the above multiplication table imply:

$$
\begin{aligned}
X \cdot Y &= \left(\sum_i A_i \otimes \xi_i \right) \cdot \left(\sum_j B_j \otimes \eta_j \right) = \sum_{i,j} (A_i \cdot B_j) \otimes (\xi_i \cdot \eta_j) \\
&= A_0 \otimes (\xi_0 \cdot \eta_0) \\
&\quad + A_1 \otimes (\xi_1 \cdot (\eta_0 + \eta_3)) \\
(1.5) \qquad &\quad + A_2 \otimes (\xi_2 \cdot (\eta_0 + \eta_2))
\end{aligned}
$$

$$+A_3 \otimes (\xi_3 \cdot (\eta_0 + \eta_1))$$
$$+B_1 \otimes ((\xi_0 + \xi_2) \cdot \eta_1)$$
$$+B_2 \otimes ((\xi_0 + \xi_1) \cdot \eta_2)$$
$$+B_3 \otimes ((\xi_0 + \xi_3) \cdot \eta_3).$$

If $X = (X_{pq})$ and $Y = (Y_{rs})$ with $X_{pq}, Y_{rs} \in k^{m \times m}$, then a simple calculation yields for the coefficients ξ_i and η_j

$$\begin{aligned}
\xi_0 &= -X_{12} + X_{21} + X_{22} \\
\xi_1 &= X_{11} + X_{12} - X_{21} - X_{22} = X_{11} - \xi_0 \\
\xi_2 &= -X_{21} \\
\xi_3 &= X_{12},
\end{aligned}$$

and

$$\begin{aligned}
\eta_0 &= Y_{11} + Y_{12} - Y_{21} \\
\eta_1 &= -Y_{11} - Y_{12} + Y_{21} + Y_{22} = Y_{22} - \eta_0 \\
\eta_2 &= -Y_{12} \\
\eta_3 &= Y_{21}.
\end{aligned}$$

Thus ξ_i and η_j can be computed together with all the $\xi_0 + \xi_i$, $\eta_0 + \eta_i$ ($1 \le i \le 3$) from the $m \times m$-matrices X_{pq}, Y_{rs} with 10 additions or subtractions of $m \times m$-matrices. After the 7 multiplications

$$\begin{aligned}
P_1 &= \xi_0 \cdot \eta_0 \\
P_2 &= \xi_1 \cdot (\eta_0 + \eta_3) \\
P_3 &= \xi_2 \cdot (\eta_0 + \eta_2) \\
P_4 &= \xi_3 \cdot (\eta_0 + \eta_1) \\
P_5 &= (\xi_0 + \xi_2) \cdot \eta_1 \\
P_6 &= (\xi_0 + \xi_1) \cdot \eta_2 \\
P_7 &= (\xi_0 + \xi_3) \cdot \eta_3
\end{aligned}$$

we obtain the "entries" $Z_{pq} \in k^{m \times m}$ of $Z = (Z_{pq}) = X \cdot Y$ from $P_1, \ldots, P_7$ by 8 further additions or subtractions of $m \times m$-matrices, following (1.5):

$$\begin{aligned}
Z_{11} &= P_1 + P_2 + P_6 + P_7 \\
Z_{12} &= P_4 - P_6 \\
Z_{21} &= -P_3 + P_7 \\
Z_{22} &= P_1 + P_3 + P_4 + P_5.
\end{aligned}$$

We infer that it is possible to multiply two $(2m \times 2m)$-matrices with 7 multiplications and 18 additions or subtractions of $m \times m$-matrices. If n is a power of 2, then one recursively applies this procedure to multiply two $n \times n$-matrices.

If $T(n)$ denotes the minimum number of arithmetic operations sufficient to compute the product of two $n \times n$-matrices, we obtain the recursion formula $T(n) \leq 7T(n/2) + 18 \cdot (n/2)^2$. Together with $T(1) = 1$ this gives

$$T(n) \leq 7 \cdot n^{\log 7} - 6n^2.$$

If n is arbitrary, say $2^{s-1} < n \leq 2^s = N$, then filling square matrices of size n with zeros in a straightforward manner to obtain square matrices of size N reduces the multiplication of $n \times n$-matrices to that of $N \times N$-matrices. This gives an upper bound for the *exponent* $\omega := \omega(k)$ of matrix multiplication defined by

$$\omega := \inf\{h \in \mathbb{R} \mid \text{multiplication in } k^{n \times n} \text{ has arithmetic cost } O(n^h)\}$$

in the form $\omega \leq \log 7 < 2.81$.

In the past 20 years a lot of effort has been spent on understanding the nature of the exponent ω. It is now known that $\omega < 2.38$ (Coppersmith and Winograd [135]). The latest developments on this topic can be found in Chap. 15 (after preparations in Chap. 14). We would like to sketch the basic steps that have led to such an improvement of the upper bound for ω. First, it turns out that the exponent does not change if we restrict ourselves to a special type of algorithms, the so-called *bilinear algorithms*. Strassen's algorithm for the multiplication of 2×2-matrices is in fact such a bilinear algorithm: setting $(A_0, \ldots, A_3, B_1, \ldots, B_3) =: (w_1, \ldots, w_7)$ and regarding in the case $m = 1$ the ξ_i and η_j as linear forms in X and Y, then (1.5) shows that the multiplication in the algebra $k^{2 \times 2}$ can alternatively be written in the form $X \cdot Y = \sum_{i=1}^{7} f_i(X)g_i(Y)w_i$ with appropriate linear forms f_i in X and g_i in Y. The multiplication in $A := k^{2 \times 2}$ is a bilinear mapping $A \times A \to A$. More generally, if $\phi: U \times V \to W$ is a bilinear mapping between finite-dimensional k-spaces, then the minimum number r such that there exist $w_1, \ldots, w_r \in W$, $f_1, \ldots, f_r \in \mathrm{Hom}_k(U, k)$, and $g_1, \ldots, g_r \in \mathrm{Hom}_k(V, k)$ satisfying

$$\phi(u, v) = \sum_{i=1}^{r} f_i(u)g_i(v)w_i$$

for all $u \in U, v \in V$ is called the *bilinear complexity* or *rank* $R(\phi)$ of ϕ. The rank $R(A)$ of an algebra A is the rank of the corresponding multiplication map. (The bilinear complexity will be studied extensively in Chap. 14.) Strassen's algorithm implies that $k^{2 \times 2}$ has rank at most 7. Now it is not difficult to show that

$$\omega = \inf\{h \mid R(k^{n \times n}) = O(n^h)\}.$$

Hence, to study the exponent one can work with the rank. This provides many structural advantages; for example

$$R\big(k^{2^m \times 2^m}\big) \leq R\big(k^{2 \times 2}\big)^m,$$

which yields the above estimate $\omega < 2.81$. The next step in the process of obtaining fast matrix multiplication algorithms is the observation that a fast algorithm for multiplication of rectangular matrices of a fixed size also yields a good upper bound for ω. More precisely, one can show that if

$$\langle m, n, p \rangle : k^{m \times n} \times k^{n \times p} \to k^{m \times p}$$

denotes the multiplication of $m \times n$- by $n \times p$-matrices, then

$$R(\langle m, n, p \rangle) \leq r \Rightarrow (mnp)^{\omega/3} \leq r.$$

(In the case of Strassen's algorithm we have $m = n = p = 2$, $r = 7$.)

The next significant progress is related to the fact that there are bilinear maps which can be approximated with arbitrary precision by bilinear maps of smaller rank. This leads to the powerful concepts of *approximative algorithms* and of the *border rank* $\underline{R}(\phi)$ of a bilinear map ϕ due to Bini et al. [49, 45], see Chap. 15. (The concept of approximative complexity was proposed for the first time by Strassen [499] in a different context.) Furthermore, it turns out that fast approximative algorithms for the multiplication of rectangular matrices yield fast *exact* algorithms for the multiplication of large square matrices. More precisely, one can show that

$$\underline{R}(\langle m, n, p \rangle) \leq r \Rightarrow (mnp)^{\omega/3} \leq r.$$

This allows a further significant generalization due to Schönhage [460]: a simultaneous solution of *several independent* matrix multiplication problems by means of an efficient approximative algorithm yields an (asymptotically) fast algorithm for the multiplication of large square matrices:

$$\underline{R}(\langle m_1, n_1, p_1 \rangle \oplus \ldots \oplus \langle m_t, n_t, p_t \rangle) \leq r \Rightarrow (m_1 n_1 p_1)^{\omega/3} + \ldots + (m_t n_t p_t)^{\omega/3} \leq r.$$

This result would be of little use if the following plausible assumption were correct: the approximative complexity of several independent matrix multiplication problems is the sum of the approximative complexities of each of the problems. In other words, separate optimal approximative solutions of several matrix multiplication problems combine to an optimal solution to the problem of simultaneous multiplication of several matrices. It was a big surprise that this seemingly obvious assumption is false! This fact is on the one hand discouraging, since it shows that the nature of the matrix multiplication problem is much more complicated than originally assumed. On the other hand, a counterexample which has proved this assumption false, together with the above inequality has led to a significant improvement on the exponent: $\omega < 2.55$. (Schönhage [460]; details of the constructions leading to this and other upper bounds may be found in Chap. 15.) This is a good point at which to finish discussion of the matrix multiplication problem, since a more thorough explanation of the tools and results calls for much more theory than we intend to present in this introduction.

With regard to the syntax, the algorithms we have considered so far consist of a sequence of instructions of the form

$$(\omega; i, j) \quad \text{resp.} \quad (\lambda; i),$$

where $\omega \in \{+, -, *, /\}$ and $\lambda \in R$. Loosely speaking, the semantics of the first type of instructions is described by the execution of the arithmetic operation ω on the ith and jth intermediate results, while the second type is interpreted as the multiplication of the ith intermediate result with the scalar $\lambda \in R$. Such sequences of instructions are called *straight-line programs*. To a large extent we are concerned in this book with this type of programs. This may seem strange on first sight, since the majority of interesting programs are not at all *straight*, but contain lots of branchings! However, from the point of view of algebraic complexity theory we may restrict ourselves to straight-line programs. The reason for this might best be explained by means of an example. (An exact mathematical argument may be found in Chap. 4.) In what follows we consider the computational problem of the division of complex numbers.

We regard the field $\mathbb{C}$ of complex numbers as a vector space over the field $\mathbb{R}$ of reals. The problem is to construct a program which, on input $(a, b, c, d) \in \mathbb{R}^4$, $cd \neq 0$, computes a pair $(x, y) \in \mathbb{R}^2$, where $x + iy = (a + ib)/(c + id)$, i.e.,

$$x = \frac{ac + bd}{c^2 + d^2} \quad \text{and} \quad y = \frac{bc - ad}{c^2 + d^2}.$$

Following the same line of thought as above we first introduce indeterminates A, B, C, D over $\mathbb{R}$ (regarded as program variables). Next we present (up to linear operations) a sequence $(g_1, \ldots, g_6)$ of intermediate results of a straight-line program which computes the rational functions

$$X := \frac{AC + BD}{C^2 + D^2} \quad \text{and} \quad Y := \frac{BC - AD}{C^2 + D^2}$$

from A, B, C, D with as few as possible nonscalar multiplications and divisions:

$$
\begin{aligned}
g_1 &:= C/D \\
g_2 &:= C \cdot g_1 \\
g_3 &:= B \cdot g_1 \\
g_4 &:= \frac{g_3 - A}{g_2 + D} = Y \\
g_5 &:= A \cdot g_1 \\
g_6 &:= \frac{g_5 + B}{g_2 + D} = X.
\end{aligned}
$$

(For a proof of the optimality of this procedure, consult Lickteig [333].) If we execute this "program" on concrete inputs, then it may happen that the computation is blocked by a division by zero, although the functions to be computed are defined for this input. This is for instance the case if the input for D equals 0 and that of C is nonzero. A remedy for avoiding such abnormalities is to extend the program by means of branchings in such a way that it not only outputs the correct result for all the inputs at which the functions to be computed are defined, but also stops the computation with the error message $\uparrow$ if the input does not belong to this set. For

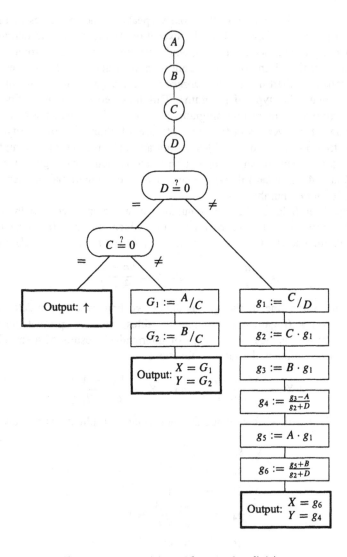

Fig. 1.1. Computation tree for complex division.

reasons of simplicity and in order to show the basic ideas we confine ourselves to branchings of the form

$$\text{if } a = 0 \text{ then goto } i \text{ else goto } j.$$

Here, a denotes a result which is assumed to be already computed.

This extension leads us from straight-line programs to the so-called computation trees. These are finite binary trees whose simple vertices are labeled by arithmetic instructions, while their branching vertices are labeled by zero tests of the above type. In our case, a possible computation tree which solves the problem of complex division for all inputs $(a, b, c, d) \in \mathbb{R}^4$ is described in Fig. 1.1. This computation tree "computes" the pair of rational functions corresponding to complex division. In general, we say that a computation tree computes a set of rational functions $f_1, \ldots, f_p$, if, for every input ξ at which all the f_i are defined, the tree outputs $(f_1(\xi), \ldots, f_p(\xi))$, while inputs ξ at which at least one f_i is not defined lead to the error output $\uparrow$. (For a precise definition the reader is referred to Chap. 4.)

An input ξ of a computation tree uniquely determines a path from the root to one of the leaves of the tree. The cost of the computation tree on input ξ is defined as the sum of the costs of the operations along the path corresponding to ξ. The maximum of these costs taken over all inputs ξ is called the cost of the computation tree. We define the *branching complexity* of a set of rational functions $f_1, \ldots, f_p$ as the minimum cost of a computation tree computing these functions.

We return to our example and consider the sets of inputs that lead to the left, middle, and right leaf, respectively. Among those, the only "thick" set in $\mathbb{R}^4$ is $\{(a, b, c, d) \in \mathbb{R}^4 \mid d \neq 0\}$ corresponding to the right leaf. The other input sets have a dimension smaller than four. The path from the root to the right leaf describes the so-called generic path; it is essentially characterized as the path that always follows the $\neq$-branch for all branching vertices.

Following the generic path in a computation tree for a set $f_1, \ldots, f_p$ of rational functions we obtain a straight-line program for the computation of $f_1, \ldots, f_p$. This observation shows that the straight-line complexity of $f_1, \ldots, f_p$ always represents a lower bound for the branching complexity of these functions. While branching complexity is the minimax cost of a problem, the straight-line complexity is its minimal generic cost. (This will be discussed in greater detail in Chap. 4.) These facts underline the importance of straight-line programs from the point of view of algebraic complexity theory.

The model of computation trees also allows the complexity theoretic specification of computational problems for which the model of straight-line programs makes no sense. A particularly important example for this is that of the continued fraction of a quotient A_1/A_2 of univariate polynomials over a field k, $A_2 \neq 0$. The continued fraction is given by the sequence of quotients in the Euclidean algorithm

$$
\begin{aligned}
A_1 &= Q_1 A_2 + A_3 \\
A_2 &= Q_2 A_3 + A_4 \\
&\cdots \qquad \cdots \\
A_{t-1} &= Q_{t-1} A_t.
\end{aligned}
$$

Up to now we have always been able to give explicit polynomial or rational formulas for the output in terms of the input. In this case, however, we know neither the number t nor the degrees of the quotients Q_i in advance. Hence, the length of the output may vary even if the degrees of the input polynomials A_1

and A_2 are fixed. Such a computation cannot be performed within the straight-line model. In Chap. 3 we shall discuss the computation of the continued fraction of polynomial quotients. Repeated use of a fast polynomial division algorithm typically leads to roughly quadratic cost in the degree n of the input polynomials. A much more efficient algorithm due to Knuth [305] and Schönhage [455] based on early ideas of Lehmer [322] solves this problem with $O(n \log^2 n)$ arithmetic operations and only $O(n \log n)$ multiplications and divisions; this algorithm will be presented in Chap. 3. Strassen [504] has shown that regarding the number of multiplications and divisions this algorithm has an optimal order of magnitude. The lower bound proof uses the degree bound and will be presented in Chap. 10.

Next we would like to discuss briefly the branching complexity of semi-algebraic membership problems. A subset W of $\mathbb{R}^n$ is called *semi-algebraic* if it can be represented as a finite union of sets of all real solutions of finitely many polynomial equalities and inequalities. The problem is to compute the characteristic function of W by means of a computation tree in which we also allow $\leq$-tests besides the zero tests. Thus we search for optimal computation trees which decide for given points in $\mathbb{R}^n$ whether they belong to the semi-algebraic set W. The knapsack problem or the problem of determining the convex hull of a finite set of points in $\mathbb{R}^n$ can be formulated in this abstract setting. The methods that we will describe below allow a satisfactory analysis of these problems.

We start with an optimal computation tree which decides membership of points in $\mathbb{R}^n$ to the semi-algebraic set W in the worst case with r multiplications, divisions or tests. The set D_v of all inputs which lead to the same leaf v of this computation tree turns out to be a semi-algebraic set, which—by a theorem of Milnor and Thom—can have at most 3^{n+r} connected components:

$$b_0(D_v) \leq 3^{n+r}.$$

(We denote by $b_0(X)$ the number of connected components of $X \subseteq \mathbb{R}^n$.) Since W is the union of all D_v corresponding to accepting leaves v, and since there are at most 2^r leaves, one obtains

$$b_0(W) + b_0(\mathbb{R}^n \setminus W) \leq 2^r \cdot 3^{n+r}.$$

This implies the following inequality due to Ben-Or [37] for the branching complexity $C(W) := r$ of W:

$$C(W) \geq (\log 6)^{-1} \left(\log(b_0(W) + b_0(\mathbb{R}^n \setminus W)) - n \log 3 \right).$$

A detailed discussion of these methods and some interesting applications can be found in Chap. 11.

So far we have been concerned with problems for which trivial upper complexity bounds of order n, n^2, or n^3 were known and our goal was to improve these upper bounds or to find matching lower bounds. In the last chapter we will discuss problems for which the known upper and lower complexity bounds differ substantially. Typically, the lower bounds are described by polynomials of small degree, whereas the known upper bounds are exponential. A very similar situation in the discrete setting has led to the development of *structural complexity theory*. That theory aims to classify problems into general complexity classes, to partition the classes via reduction into subclasses (of problems of roughly the same complexity) and to partially order the set of subclasses. Furthermore, structural complexity theory tries to identify hardest problems in each class, the so-called *complete problems*, and compares the complexity classes.

Within this framework, the most fundamental complexity class, denoted by **P**, consists of all languages $A \subseteq \{0, 1\}^* = \cup_{n \geq 0}\{0, 1\}^n$ such that membership of $x \in \{0, 1\}^n$ to A can be decided by a *deterministic* Turing machine in time which is bounded by a polynomial in n. This class is often compared with the (probably much larger) class **NP** of decision problems that can be solved by a *nondeterministic* Turing machine in polynomial time. Such machines are allowed first to "guess" a solution and then to check it. Moreover, the time of incorrect guesses and unsuccessful checks is not taken into account.

Obviously, $\mathbf{P} \subseteq \mathbf{NP}$. With respect to polynomial time reductions, **NP** contains "hardest" problems, the so-called **NP**-complete problems. Among the numerous known **NP**-complete problems are many computational problems that are relevant in practice, like the traveling salesman, or the integer linear programming problem. The fact that these problems have defied all attempts to be solved efficiently, has led to Cook's hypothesis: $\mathbf{P} \neq \mathbf{NP}$, which is one of the most fundamental open problems in complexity theory.

Besides studying decision problems that ask for the existence of a solution, there is also great interest in *counting* all solutions. As a counting analogue of **NP**, Valiant [527] introduced the class #**P** (pronounced as "number P" or "sharp P"). This class consists of all counting functions $\{0, 1\}^* \to \mathbb{N}$ that can be computed by a *counting* Turing machine in polynomial time. (A counting Turing machine C is a nondeterministic Turing machine with an auxiliary output device that (magically) prints in binary notation on a special tape the number of accepting computations induced by an input. The time requirement of C is defined by $t_C(n) := \max_{x \in \{0,1\}^n} \tau_C(x)$, where $\tau_C(x)$ is the maximum number of steps in accepting computations of C on input x.) Consider for example the **NP**-complete problem that asks whether there exists a Hamiltonian cycle in a finite graph. The corresponding counting problem that asks for the *number* of Hamiltonian cycles in that graph is obviously in #**P**. Moreover, this counting problem is #**P**-complete.

Valiant [527, 528] made the surprising discovery that, for numerous decision problems in **P**, the corresponding counting problems are even #**P**-complete! The most prominent example is the perfect matching problem in bipartite graphs (also known as the marriage problem), which according to M. Hall [226] is in **P**. Determining the number of all perfect matchings amounts to computing the permanent of the adjacency matrix of the bipartite graph, and this problem turns out to be #**P**-complete.

In Chap. 21 we introduce Valiant's nonuniform algebraic complexity classes **VP** and **VNP** consisting of *p*-computable and *p*-definable families of multivariate polynomials, respectively. Typical members of **VP** and **VNP** are the families

$DET = (DET_n)$ and $PER = (PER_n)$ of generic determinants and permanents over a fixed field k, defined by

$$DET_n := \sum_{\sigma \in S_n} \text{sgn}(\sigma) \prod_{i=1}^{n} X_{i\sigma(i)} \quad \text{and} \quad PER_n := \sum_{\sigma \in S_n} \prod_{i=1}^{n} X_{i\sigma(i)},$$

respectively. There exists an $O(n^3)$-algorithm for evaluating DET_n. Despite the similarity of determinants and permanents, the best known algorithm for PER_n (in characteristic $\neq 2$) uses $O(n2^n)$ arithmetic operations. A theoretical foundation of this surprising difference is supplied by Valiant's theorem stating that PER is **VNP**-complete (over fields of characteristic $\neq 2$). We will give a detailed proof of this theorem and discuss Valiant's hypothesis, **VP** $\neq$ **VNP**, as well as an extension of it that in purely algebraic terms reads as follows. Over any field k of characteristic $\neq 2$, PER is not a qp-projection of DET (i.e., for fixed constant c there is no way of writing all generic permanents as $PER_n = DET_{m(n)}(a_{ij})$ with $a_{ij} \in k \cup \{X_{\mu\nu}|1 \leq \mu, \nu \leq n\}$ when $m(n) = 2^{O(\log^c n)}$). It is remarkable that no algorithmical concepts appear in the formulation of this so-called extended Valiant hypothesis. Therefore, it might be more accessible to the powerful tools of classical mathematics.

Although the **VNP**-completeness of PER indicates its inherent computational difficulty, the complexity of PER_n is still unknown. This question undoubtedly ranks among the most important unsolved problems in algebraic complexity theory.

In the next two chapters of this book we shall discuss several algorithms, most of them concerning the manipulation of polynomials. One of the central aims of this book is to develop methods for proving the optimality of these (and other, later designed) algorithms in different computational models. The description of these algorithms, which provide upper bounds for the complexity of the computational problems they solve, appeals to the reader's intuitive notion of algorithm; proving lower bounds calls for an accurate specification of the notion of algorithm. So, before our journey through the lower bound methods starts in Chap. 5, we shall introduce in Chap. 4 the necessary concepts and linguistic tools, both by the way of examples, as well as by mathematical definitions.

1.1 Exercises

1.1. Show that the binary method is an application of Horner's rule.

1.2. Use the binary method to design an algorithm for multiplication of integers involving only the simple operations of doubling, halving, and adding.

1.3. Show that $d(31) = 6$ and $\ell(31) = 7$. Furthermore, show that 31 is the smallest n with $d(n) < \ell(n)$.

1.4. Design an algorithm that computes $\ell(n)$, for every positive integer n.

1.5. Show that $\ell(mn) \le \ell(m) + \ell(n)$, for positive integers m and n.

1.6. Prove the following weaker version of equation (1.2):

$$\ell(n) \ge \log n + 0.3 \log w_2(n).$$

(Hint: Let $(a_0, \ldots, a_r)$ be an addition chain for n of optimal length $r = \ell(n)$. W.l.o.g. $a_0 < a_1 < \ldots < a_r$. Consider the set $G := \{\rho \in \underline{r} \mid a_\rho = 2a_{\rho-1}\}$ of giant steps and $B := \{\rho \in \underline{r} \mid a_\rho < 2a_{\rho-1}\}$, the set of baby steps. Define recursively $A_0 = 1$ and $A_\rho := 2A_{\rho-1}$ for $\rho \in G$ and $A_\rho = \gamma A_{\rho-1}$ for $\rho \in B$, where $\gamma = \frac{1}{2}(1 + \sqrt{5}) > 0$. Show that $a_\rho \le A_\rho$, for all $\rho \in \underline{r}$.)

1.7. Let P be a finite subset of $\mathbb{Z}$. A maximal subset Q of P such that

$$\forall u, v \in Q \; \forall v \in \mathbb{Z}: \quad u < v < w \Rightarrow v \in Q$$

is called a *component* of P.

(1) Show that the components of P form a partition of P. (The cardinality of a smallest component of P is called the *width* $\beta(P)$ of P.)
(2) For $d \in \mathbb{N}$ define the d-extension of P by $E_d P := P \cup (P-1) \cup \ldots \cup (P-d)$, where $P + t := \{n + t \mid n \in P\}$. Show that $\beta(P) \ge b$ implies that $\beta(E_d P) \ge b + d$ and $|E_d P|/(b + d) \le |P|/b$.
(3) The 2-support $W_2(x)$ of $x \in \mathbb{N}$ with binary representation $x = \sum_i \xi_i 2^i$ is the set of all i with $\xi_i = 1$. For finite subsets P, R of $\mathbb{Z}$ let $P \nabla R$ denote the minimal set with the property that $W_2(x) \subseteq P$ and $W_2(y) \subseteq R$ implies $W_2(x + y) \subseteq P \nabla R$ for arbitrary x, y. Prove that $\beta(P \nabla R) \ge \beta(P)$, $\beta(R)$ and $|P \nabla R| \le |P| + |R|$.

1.8.* Show the following improvement of Ex. 1.6: $\ell(n) \ge \log n + \log w_2(n) - O(\log \log n)$. (Hint: with the notation of the last two exercises define ranks $r_0 := 0$, $r_i := r_{i-1} + 1$ for $i \in G$, and $r_i := r_{i-1}$ for $i \in B$. Put $d_i := r_{i-1} - r_m$, where $a_i = a_m + a_p$ with $m \le p < i$. Now partition B according to a suitable parameter d into $B = B_1 \cup B_2$, where $B_1 := \{i \in B \mid d_i < d\}$, $B_2 := \{i \in B \mid d_i \ge d\}$. Recursively define numbers $b_0 \le b_1 \le \ldots b_r$ and finite subsets P_i of $\mathbb{Z}$ as follows. Initially set $P_0 = \{0\}$, $b_0 = 1$. For $i \in G$ put $P_i = P_{i-1} + 1$ and $b_i = b_{i-1}$. For $i \in B_1$ put $P_i = E_{d_i+1}(P_{i-1} + 1)$, $b_i = b_{i-1} + 1 + d_i$. Finally, for $i \in B_2$ with $a_i = a_m + a_p$, $m \le p < i$, define $P_i = P_{i-1} \nabla E_\tau P_m$, $b_i = b_{i-1}$, where $\tau = b_{i-1} - b_m$. Now prove that $a_i \in P_i$, $\beta(P_i) \ge b_i$, and $\delta = r_t - r_j$ for some $j \le t$ implies $P_j \subseteq E_\delta P_t$.)

1.9. An addition chain $(1 = a_0, a_1, \ldots, a_r = n)$ is called a *star chain* for n iff for every $\rho \in \underline{r}$ there is a $j < \rho$ such that $a_\rho = a_{\rho-1} + a_j$; $\ell^*(n)$ denotes the minimum length of a star chain for n. Prove that $\ell^*(2^n - 1) \le n - 1 + \ell^*(n)$, for all positive integers n.

1.10. Let $K := k(X_1, \ldots, X_n)$ be the rational function field in n indeterminates over an infinite field k and let $g_1, \ldots, g_r \in K$. Show that for $i, j < r$ and for $\alpha, \beta \in k$ we have

$$\deg(g_1, \ldots, g_r) = \deg(g_1, \ldots, g_r, \alpha g_i + \beta g_j).$$

1.11. Let f be a nonconstant univariate polynomial with coefficients in the field k. Show that $\deg f = [k(X) : k(f)]$. (Hint: use the Gauss lemma.)

1.12. Compute the square of the DFT matrix DFT_N and show that its determinant satisfies $|\det(DFT_N)| = N^{N/2}$.

1.13. The problem of complex multiplication has as inputs four real numbers (a, b, c, d) interpreted as $a + ib$ and $c + id$. It asks for the real and imaginary part of the corresponding product

$$(a + ib) \cdot (c + id) = (ac - bd) + i(ad + bc).$$

The naive algorithm uses 4 real multiplications and 2 real additions or subtractions. Design an algorithm that solves this task with only 3 real multiplications and 5 real additions/subtractions. How could the latter algorithm be applied to the multiplication of complex matrices?

1.14. Prove that for matrices $C_1, C_2 \in K^{r \times r}$ and $D_1, D_2 \in K^{s \times s}$ the following is valid:

$$(C_1 \otimes D_1) \cdot (C_2 \otimes D_2) = (C_1 \cdot C_2) \otimes (D_1 \cdot D_2).$$

1.15. Let K be a field. The symmetric group S_3 of order 6 is isomorphic to a subgroup of $GL(2, K)$. One possible isomorphism (i.e., faithful representation) Δ realizes S_3 as the symmetry group of a regular triangle. If we take its center of origin in 2-space and denote its vertices by e_1, e_2, e_3, then (e_1, e_2) is a basis and $e_3 = -e_1 - e_2$. The natural S_3-action $\pi e_i := e_{\pi(i)}$ yields

$$\Delta((12)) = \begin{pmatrix} 0 & 1 \\ 1 & 0 \end{pmatrix} \quad \text{and} \quad \Delta((123)) = \begin{pmatrix} 0 & -1 \\ 1 & -1 \end{pmatrix}.$$

Thus S_3 acts on $K^{2 \times 2}$ via

$$\pi * X := \Delta(\pi) \cdot X \cdot \Delta(\pi)^{-1}.$$

Consider the S_3-orbits of the matrices $\begin{pmatrix} 0 & 1 \\ 0 & 0 \end{pmatrix}$ and $\begin{pmatrix} 1 & 0 \\ 0 & 1 \end{pmatrix}$ and split the first orbit (of length 6) into two A_3-orbits (each of length 3). Show that this gives the two bases $X_0, \ldots, X_3$ and $Y_0, \ldots, Y_3$ involved in Strassen's algorithm to multiply matrices with $O(n^{\log 7})$ arithmetic steps.

1.16. An n-dimensional K-algebra A is said to have an M-pair iff there exist K-bases $(X_1, X_2, \ldots, X_n)$ and $(Y_1, Y_2, \ldots, Y_n)$ such that $X_i \cdot Y_j = \xi_{ij} X_i + \eta_{ij} Y_j$, for all i, j and suitable $\xi_{ij}, \eta_{ij} \in K$. Show that $K^{n \times n}$ has an M-pair iff $n \leq 2$. (Hint: If $((X_i), (Y_j))$ is an M-pair for A then the K-dimension of both $[X_i, X_j]A$ and $A[Y_i, Y_j]$ is at most 4, here $[X, Y] := XY - YX$ denotes the commutator of $X, Y \in A$.)

1.2 Open Problems

Problem 1.1. The "Scholz-Brauer conjecture" states that $\ell(2^n - 1) \leq n - 1 + \ell(n)$, for all positive integers n. Prove or disprove this conjecture. Does equality always hold?

Problem 1.2. Does $\ell(n) = \ell^0(n)$ hold? (For a definition of $\ell^0(n)$ see the Notes 1.3.)

1.3 Notes

As most topics sketched in this introduction will be studied in detail in later chapters, the present notes report only on addition chains and give references for the exercises.

A. Scholz [453] in 1937 posed the problem of determining the minimum number $\ell(n)$ of additions sufficient to produce a given integer n starting from 1. This has led to the topic of addition chains. For a thorough treatment including additional historical remarks we refer to Sect. 4.6.3. in Knuth [307]. The lower bound on $\ell(n)$ expressed in (1.2) is due to Schönhage [457]. The upper bound in (1.3) was proved by A. Brauer [71]. Erdös [156] showed that equality holds for almost all n. Exact formulas for $\ell(n)$ are known for those n whose binary representations have Hamming weight at most 4, see, e.g., [307]. The most famous open problem about addition chains is the "Scholz-Brauer conjecture" stating that $\ell(2^n - 1) \leq n - 1 + \ell(n)$. As a first step towards its proof, Brauer [71] showed that $\ell^*(2^n - 1) \leq n - 1 + \ell^*(n)$, where $\ell^*(n)$ denotes the minimum length of a *star chain* for n. (Star chains are special addition chains $1 = a_0 < a_1 < \ldots < a_r = n$, where for $\rho \in \underline{r}$, $a_\rho = a_{\rho-1} + a_j$ for some $j < \rho$. Obviously, $\ell(n) \leq \ell^*(n)$.) In [227], Hansen has shown that $\ell^*(n) = e_0 + t$ for those $n = 2^{e_0} + \ldots + 2^{e_t}$, whose exponents $e_0 > \ldots > e_t \geq 0$ are sufficiently far apart in terms of t and n. (For an extension of Hansen's theorem see van Leeuwen [318].) On the basis of this result Hansen [227] was able to prove that there are infinitely many n with $\ell(n) < \ell^*(n)$. Thus Brauer's result does not prove the Scholz-Brauer conjecture. As a further step in this direction, Hansen introduced the concept of ℓ^0-*chains*. In an ℓ^0-chain, certain elements are underlined; the condition is that $a_\rho = a_i + a_j$, where a_i is the largest underlined element less than a_ρ. As ℓ^0-chains in which all elements are underlined are star chains, we have $\ell(n) \leq \ell^0(n) \leq \ell^*(n)$. Hansen showed that $\ell^0(2^n - 1) \leq n - 1 + \ell^0(n)$, but it is an open question, whether always $\ell^0(n) = \ell(n)$.

Knuth [304] (Sect. 4.6.3, Ex. 32) asks for a fast procedure to compute several powers $X^{n_1}, \ldots, X^{n_p}$ of one indeterminate, and in 1976, Yao [563] showed that the minimum length $\ell(n_1, \ldots, n_p)$ of an addition chain containing all n_i is at most $\log N + O(\sum_{i \leq p} \log n_i / \log \log(n_i + 2))$, where $N := \max\{n_1, \ldots, n_p\}$. In 1963, Bellman [36] raised the problem of determining the minimum number $\ell([n_1, \ldots, n_p])$ of multiplications to compute the monomial $X_1^{n_1} \cdots X_p^{n_p}$ from the indeterminates $X_1, \ldots, X_p$. Straus [513] showed that for fixed p, $\ell([n_1, \ldots, n_p]) \sim \log N$, where $N = \max\{n_1, \ldots, n_p\}$. Let $M = (m_{ij})$ be a $p \times q$ matrix of nonnegative integers where no row or column is entirely zero. Define $\ell(M)$ to be the minimum number of multiplications sufficient to compute the set of monomials $\{X_1^{m_{1j}} \ldots X_p^{m_{pj}} \mid 1 \leq j \leq q\}$. Then, by a duality principle, explained in a slightly more general context in Chap. 13 (see Thm. (13.20)), Papadimitriou and Knuth [414] proved that $\ell(M) = \ell(M^\top) + p - q$. This extends a result by Olivos [401] stating that $\ell([n_1, \ldots, n_p]) = \ell(n_1, \ldots, n_p) + (p - 1)$. Pippenger [421] has proved a comprehensive generalization of the above asymptotic results: let $L(p, q, n)$ be the maximum of $\ell(M)$ taken over all $p \times q$ matrices M of nonnegative integers $m_{ij} \leq n$. Then

$$L(p, q, n) = \min\{p, q\} \log n + H / \log H + O(p + q + H (\log \log H)^{1/2} (\log H)^{3/2}),$$

where $H = pq \log(n + 1)$.

Multiplication via the binary method, see Ex. 1.2, is known as the "Russian peasant multiplication," although it was actually used as early as 1800 B.C. by Egyptian mathematicians. Ex. 1.6, 1.7, and 1.8 are from Schönhage [457]. Ex. 1.9 is due to Brauer [71]. Ex. 1.15 and the observation that Strassen's algorithm for multiplying matrices can be derived from certain A_3-orbits of a linear S_3-action, is taken from Clausen [112]. Ex. 1.16 is from Büchi and Clausen [96]. The open problems have been stated in Knuth [307].

Fundamental Algorithms

Chapter 2. Efficient Polynomial Arithmetic

The primary topic of this chapter is the design and analysis of algorithms that solve various problems concerning the symbolic manipulation of polynomials and power series. By *symbolic* computation we mean procedures for computing the coefficients of the resulting polynomials and power series from the coefficients of the inputs. Among these problems are the multiplication, inversion, division, and composition of polynomials and power series. We shall analyse both the total number of arithmetic operations and the number of nonscalar operations of the algorithms. (In the latter case, addition, subtraction, and scalar multiplication are free of charge.) This gives upper bounds for the *total* and *nonscalar complexity* of the problem under investigation. In later chapters we shall see that most of these algorithms are optimal or close to optimal.

It turns out that the problems introduced above have efficient solutions based on fast polynomial multiplication algorithms. After a short informal description of the computation model and the cost measures that will be used in this chapter, we start by discussing the problem of multiplying two univariate polynomials. In Sect. 2.1 we shall see that if the degrees of these polynomials sum up to n, then $n + 1$ is the exact nonscalar complexity of that problem in the *generic case*, i.e., when all the input coefficients are algebraically independent over the groundfield k. Furthermore, we deduce an $O(n \log n)$ upper bound for the total number of arithmetic operations, if k has appropriate roots of unity. For arbitrary fields, a variant of the Schönhage-Strassen integer multiplication algorithm [462] yields an $O(n \log n \log \log n)$ upper bound. This striking result is presented in Sect. 2.2. We will summarize all this by saying that polynomial multiplication has complexity $O(M(n))$ where $M(n)$ denotes the triple $(n, n \log n, n \log n \log \log n)$.

In Sect. 2.3 we discuss the problem of multiplying several polynomials. An obvious divide-and-conquer technique yields the upper bound $O(M(n) \log t)$ for the complexity of the problem of multiplying t polynomials whose degrees sum up to n. A more refined version which uses a Huffman coding technique gives an upper bound of order $M(n)(1 + H)$, where H is the entropy of the probability distribution corresponding to the degree sequence of the polynomials.

In Sect. 2.4 it is shown that the arithmetic complexity of multiplication, inversion, and division of formal power series mod X^{n+1} is also bounded by $O(M(n))$, thus emphasizing the fundamental role of polynomial multiplication. The same upper bound is valid for computing the composition $a \circ b \mod X^{n+1}$ of certain power series, e.g., for $a = \mathrm{LOG}$ or $a = \mathrm{EXP}$. Finally, in Sect. 2.5 the Brent-Kung algorithm [75], which solves the general composition problem in $M(n)\sqrt{n \log n}$ steps is presented.

2.1 Multiplication of Polynomials I

Throughout this section A denotes a commutative algebra over a field k.

A large part of this book is concerned with the computational complexity of manipulating polynomials whose coefficients belong to a commutative k-algebra A. (For later applications we have to include the case, where A is not necessarily an integral domain. However in most applications we can assume that the leading coefficients of the input polynomials are not zero-divisors in A. Under this assumption the degree of the product of two polynomials equals the sum of the degrees of the factors.) A strict definition of computations and algorithms will be given later in Chap. 4; these formal definitions will be important when discussing lower bounds. On the contrary, the algorithms we shall present in this and the next chapter often only require some algorithmic intuition. Therefore, and for the favour of easier access to the results in these two chapters, we deliberately disregard formal aspects. Of course, readers interested in a more formal discussion may use the language introduced in Chap. 4 to make these discussions more rigorous. Let us start with a description of the computation model and cost functions we use in this chapter by means of the problem of polynomial multiplication. The task is to compute from the coefficients of two input polynomials $a = \sum_{i<m} a_i X^i$ and $b = \sum_{j<n} b_j X^j$, $a_i, b_j \in A$, the coefficients of the output polynomial $ab =: c = \sum_\ell c_\ell X^\ell$. We have thus to compute $c_\ell = \sum_{i+j=\ell} a_i b_j$, for $0 \le \ell \le m + n - 2$. As in Chap. 1 we understand by a computation a sequence of instructions (without branchings) which cause an addition, subtraction, multiplication, or division of two intermediate results (including the inputs). A further type of instruction is furnished by scalar multiplication by elements of k. We say that a problem is solved algorithmically if the quantities to be computed are among the intermediate results. What is the cost of such a computation? In the following we shall always consider two different cost measures: the *total arithmetic cost* is the number of all arithmetic operations performed by the computation, while the *nonscalar cost* is the number of multiplications or divisions (additions, subtractions, or scalar multiplications are not counted). The minimum total arithmetic cost of a computation solving a given problem is the *total complexity* of that problem. The *nonscalar complexity* is defined analogously. While the total cost measure might appear to be reasonable, the reader may think of the nonscalar cost measure to be of extreme academic interest only. But the point is that most of the existing methods for lower bound proofs work particularly well in the nonscalar model; obviously these lower bounds remain valid for the total complexity. For most of the algorithms introduced in this chapter we shall

later prove almost matching lower bounds for the nonscalar complexity. Moreover, the total cost of the best known algorithms solving these problems differs from the optimal nonscalar complexity only by subordinate factors. This shows the limitations for the design of more efficient algorithms under the total cost measure.

Addition, subtraction, multiplication, and division with remainder are among the most fundamental problems concerning symbolic polynomial manipulation. Let us start with addition and subtraction of two univariate polynomials a and b of degree less than n. Since $a \pm b = \sum_{i<n}(a_i \pm b_i)X^i$, the nonscalar complexity of this computational problem is 0, since we only need to perform n additions or subtractions at no cost. The total complexity of these problems is at most n. If, for example, we know in advance that some of the coefficients of the input polynomials are 0, then this upper bound can further be improved. But in the *generic case* where all the input coefficients are assumed to be algebraically independent over k, such an improvement is not possible: since we only have the input coefficients and the elements of k at our disposal, the algebraic independence of the input coefficients shows that we have to compute n algebraically independent elements, namely $a_0 \pm b_0, \ldots, a_{n-1} \pm b_{n-1}$. None of these elements is available with free cost. To end the argument, note that at each computation step we can at most compute one of the output coefficients. These considerations are summarized in the following.

(2.1) Remark. Symbolic addition and subtraction of two univariate polynomials of degree less than n has nonscalar complexity 0. The total complexity is at most n, and exactly n in the generic case.   •

For the rest of this and the whole of the next section we will discuss the (symbolic) multiplication of polynomials. In the nonscalar model one has a satisfactory result.

(2.2) Theorem. *Let $|k| \geq n+1$. Then the product of two polynomials $a, b \in A[X]$ with $n = \deg(ab)$ can be computed with $n+1$ nonscalar operations. This is optimal for generic a, b. The statements remain true if $|k| = n$ provided $\deg(ab) = \deg a + \deg b$.*

Proof. Let a and b be of degree α and β, respectively. Then $\alpha + \beta \geq n$, and equality holds in the generic case. To prove the upper bound, let us first assume that $|k| \geq n+1$. Take $n+1$ distinct elements $\lambda_0, \ldots, \lambda_n$ in k. With $c := ab$ we have for all $0 \leq r \leq n$

$$\sum_{\ell=0}^n c_\ell \lambda_r^\ell = \underbrace{\left(\sum_{i=0}^\alpha a_i \lambda_r^i \right)}_{=:u_r} \underbrace{\left(\sum_{j=0}^\beta b_j \lambda_r^j \right)}_{=:v_r} =: g_r.$$

As both u_r and v_r are linear combinations of the input coefficients, the computation of u_r and v_r are free of charge in the nonscalar model, so each g_r can be computed with nonscalar cost 1. Hence the computation of $g_0, \ldots, g_n$ can be performed with $n+1$ nonscalar operations. Now

$$(g_0, \ldots, g_n)^T = (\lambda_r^\ell)_{0 \le r. \ell \le n} \cdot (c_0, \ldots, c_n)^T.$$

As the λ_r are pairwise different, the Vandermonde matrix (λ_r^ℓ) is invertible, hence each c_ℓ is a linear combination of $g_0, \ldots, g_n$ and can thus be computed with no additional nonscalar cost. This proves the upper bound assuming that $|k| \ge n + 1$. Now let $|k| = n, k = \{\lambda_0, \ldots, \lambda_{n-1}\}$. By assumption, $n = \alpha + \beta$. Hence $c_n = a_\alpha b_\beta$. Thus all elements of the vector $(c(\lambda_0), \ldots, c(\lambda_{n-1}), c_n)$ can be computed with $n+1$ nonscalar operations. As

$$
\begin{pmatrix}
\lambda_0^0 & \lambda_0^1 & \cdots & \lambda_0^{n-1} & \lambda_0^n \\
\lambda_1^0 & \lambda_1^1 & \cdots & \lambda_1^{n-1} & \lambda_1^n \\
\vdots & & & & \\
\lambda_{n-1}^0 & \lambda_{n-1}^1 & \cdots & \lambda_{n-1}^{n-1} & \lambda_{n-1}^n \\
0 & 0 & \cdots & 0 & 1
\end{pmatrix}
\begin{pmatrix}
c_0 \\
c_1 \\
\vdots \\
c_{n-1} \\
c_n
\end{pmatrix}
=
\begin{pmatrix}
c(\lambda_0) \\
c(\lambda_1) \\
\vdots \\
c(\lambda_{n-1}) \\
c_n
\end{pmatrix},
$$

and the matrix on the left-hand side is invertible, the c_i can be recovered from the $c(\lambda_j)$ and c_n with k-linear operations only.

To sketch the proof of the lower bound in the generic case we use a linear algebra argument. (The full proof will be presented in Sect. 4.2 after having specified the syntax and semantics of straight-line programs.) In the nonscalar model we have with the input coefficients all elements in the k-linear hull $H_0 := \langle 1, a_0, \ldots, a_\alpha, b_0, \ldots, b_\beta \rangle$ at our disposal. Now suppose we have a computation of the coefficients of c with r nonscalar operations. If H_r denotes the k-linear hull of $1, a_0, \ldots, a_\alpha, b_0, \ldots, b_\beta$ and all intermediate results, then it is easy to see that $\dim_k(H_r) - \dim_k(H_0) \le r$. Now, on the one hand, H_r contains all the coefficients of c. On the other hand, as the input coefficients are assumed to be algebraically independent (generic case), the elements $1, a_0, \ldots, a_\alpha, b_0, \ldots, b_\beta, c_0, \ldots, c_n$ are linearly independent over k. It follows that $r \ge n + 1$.   □

The last theorem can be generalized in different ways. For example, in Sect. 2.3 we shall present an algorithm for multiplying several polynomials. This algorithm will turn out to be essentially optimal in the nonscalar model. The next result discusses the case when one has some knowledge about the coefficients in advance.

(2.3) Corollary. *Let $|k| \ge n + 1$ and $a, b \in A[X]$ with $n = \deg(ab) = \deg a + \deg b$. Suppose one knows ℓ coefficients of their product in advance, for some $0 \le \ell \le n + 1$. Then ab can be computed with nonscalar cost at most $n + 1 - \ell$, and exactly $n + 1 - \ell$ for generic a, b. If a, b are monic or more generally, if their leading coefficients are in k, then the nonscalar cost for computing their product is at most $n - 1$.*

Proof. Let α and β denote the degrees of a and b, respectively, hence, by assumption, $\alpha + \beta = n$. Put $ab =: c = \sum_{j \le n} c_j X^j$. Suppose we know $(c_j)_{j \in J}$ in advance, where J is an ℓ-subset of $\{0, 1 \ldots, n\}$. Let us first consider the generic case. Let $\lambda_0, \ldots, \lambda_n$ be distinct elements in k, and set

$$g_r := \left(\sum_{i=0}^{\alpha} a_i \lambda_r^i\right)\left(\sum_{j=0}^{\beta} b_j \lambda_r^j\right) = \sum_{\ell \leq n} c_\ell \lambda_r^\ell.$$

As $c_0, \ldots, c_n$ and $g_0, \ldots, g_n$ span the same k-space of dimension $n + 1$, there exists a set $I \subseteq \{0, \ldots, n\}$ of cardinality $n + 1 - \ell$ such that

$$\langle g_i \,|\, i \in I\rangle \oplus \langle c_j \,|\, j \in J\rangle = \langle g_0, \ldots, g_n\rangle.$$

Now compute g_i for $i \in I$. This can be performed with $|I| = n + 1 - \ell$ nonscalar operations. As the c_j with $j \in J$ are known in advance, we can then compute all the g_i, and consequently all the c_i with k-linear operations only. This proves the upper bound in the generic case. By the universal mapping property of the polynomial ring $k[a_i, b_j \,|\, i, j]$, the upper bound remains valid for any commutative k-algebra: if the non-generic inputs are $a' = \sum_i a_i' X^i$ and $b' = \sum_j b_j' X^j$, then replace in the original computation each occurrence of a_i and b_j by a_i' and b_j', respectively. (This topic is discussed in more detail in Sect. 4.1.)

We postpone the proof of the matching lower bound in the generic case to Sect. 4.2.

Finally, if a_α and b_β are in k, then $\deg(a - a_\alpha X^\alpha) \leq \alpha - 1$ and $\deg(b - b_\beta X^\beta) \leq \beta - 1$. Hence $(a - a_\alpha X^\alpha)(b - b_\beta X^\beta)$ can be computed with at most $(\alpha - 1) + (\beta - 1) + 1 = n - 1$ nonscalar operations. To finish the proof note that one can recover the coefficients of ab from those of $(a - a_\alpha X^\alpha)(b - b_\beta X^\beta)$ with k-linear operations only. $\qquad\square$

For a variant of this corollary see Ex. 2.4.

We fix some notation for the following. Let $b \in A[X]$ be a nonconstant polynomial whose leading coefficent is a unit in A, and let $a \in A[X]$. Then there is a pair (q, r) of polynomials in $A[X]$, uniquely determined by (a, b), such that $a = bq + r$, $\deg r < \deg b$ provided $r \neq 0$. The polynomial $\lfloor a/b \rfloor := q$ is called the *quotient*, and $a \bmod b := r$ is called the *remainder* of a and b. In particular, if $a = \sum_i a_i X^i$, then $a \bmod X^{n+1} = \sum_{i=0}^n a_i X^i$. In the sequel $a \equiv c \bmod b$ stands for $a \bmod b = c \bmod b$. The following result on squaring will be of use in Sect. 2.4.

(2.4) Corollary. *Let $a \in A[X]$. Given $a \bmod X^{n+1}$ one can compute $a^2 \bmod X^{n+1}$ with at most $n + \lceil\frac{n+1}{2}\rceil \leq \frac{3}{2}n + 1$ nonscalar operations. If $\operatorname{char} k = 2$, then the upper bound can be improved to $\lfloor n/2 \rfloor + 1$, and equality holds in the generic case.*

Proof. Let $a = \sum_j a_j X^j$ and $m := \lceil\frac{n+1}{2}\rceil$. With $\alpha_0 := \sum_{j < m} a_j X^j$ and $\alpha_1 := \sum_{j=m}^n a_j X^{j-m}$ put $\alpha := \alpha_0 + X^m \alpha_1$. Then, as $2m \geq n + 1$, we have

$$a^2 \equiv \alpha^2 \equiv \alpha_0(\alpha_0 + 2\alpha_1 X^m) \bmod X^{n+1}.$$

As α_0 is of degree at most $m - 1$ and $\alpha_0 + 2\alpha_1 X^m$ is of degree at most n, the coefficients of their product can be computed with $(m - 1) + n + 1$ nonscalar steps by Thm. (2.2).

If $\operatorname{char} k = 2$, then $a^2 \equiv \sum_{2j \leq n} a_j^2 X^{2j} \bmod X^{n+1}$. This gives the improved upper bound. Again, we postpone the proof of the lower bound in the generic case to Sect. 4.2. $\qquad\square$

In Chap. 17 we shall see (as a consequence of Alder and Strassen's Thm. (17.14)) that in the generic case the nonscalar complexity of the multiplication mod X^{n+1} is $2n + 1$. Thus squaring mod X^{n+1} is less expensive than multiplying mod X^{n+1}.

When counting all arithmetic operations, polynomial multiplication becomes a delicate problem that is not completely solved. We start our discussion of this topic with some preliminary remarks.

(2.5) Definition. Let N be a positive integer. An element $\omega \in A$ is called a *principal Nth root of unity* iff $\omega^N = 1$ and $1 - \omega^\nu$ is not a zero-divisor in A for all $1 \le \nu < N$.   •

The second condition is in particular satisfied if $1 - \omega^\nu$ is a unit in A for all $1 \le \nu < N$. If $\omega \in A$ is a principal Nth root of unity, then ω is a unit of order N in A, i.e., ω is a *primitive Nth root of unity* in A. For a field A both notions coincide. (However, if for example k has characteristic two and $A := k[X]/(X^2)$, then $X + 1$ is a primitive square root of unity, but it is not principal.) Note that ω is a principal Nth root of unity iff the same is true for ω^{-1}.

(2.6) Theorem. *Let $N \ge 2$ such that $N \cdot 1_A$ is a unit in the commutative k-algebra A and suppose that ω is a principal Nth root of unity in A. Then $\Phi: A[X] \to A^N$, $f \mapsto (f(\omega^i))_{0 \le i < N}$, is a surjective morphism of A-algebras with kernel $(X^N - 1)$. The induced isomorphism $\varphi : A[X]/(X^N - 1) \to A^N$ of A-algebras is called the discrete Fourier transform (DFT) corresponding to ω. With respect to canonical bases the representing matrix of this isomorphism is the DFT-matrix corresponding to ω:*

$$DFT(N, \omega) := (\omega^{pq})_{0 \le p,q < N} \in GL(N, A).$$

Its inverse is given by the formula

$$DFT(N, \omega)^{-1} = \frac{1}{N}(\omega^{-pq})_{0 \le p,q < N} = \frac{1}{N}DFT(N, \omega^{-1}).$$

If in addition $N = 2^n$, then there is a recursive algorithm, the Fast Fourier transform (FFT), *that evaluates $DFT(N, \omega)$ at an arbitrary input vector with at most $1.5N \log N - N + 1$ additions of elements in A or multiplications of elements in A with powers of ω. The inverse DFT-matrix of size N can be evaluated with at most $1.5N \log N + 1$ such arithmetic operations in A.*

Proof. Obviously, Φ is a morphism of A-algebras and its kernel is equal to $\cap_{0 \le i < N}(X - \omega^i)$. We claim that $\ker \Phi = (X^N - 1)$. (Warning: A may have zero-divisors!) To prove this let $f \in A[X]$. Now with the principality of ω an easy induction on $0 \le j < N$ shows that

$$\left(\forall \, 0 \le i \le j : \; f(\omega^i) = 0\right) \Rightarrow \exists u \in A[X] : \; f = u \prod_{i=0}^{j}(X - \omega^i).$$

Hence $\ker \Phi = (X^N - 1)$ and the induced isomorphism $\varphi: A[X]/(X^N - 1) \to \operatorname{im} \Phi$ is described by the DFT-matrix $DFT(N, \omega)$. To show that Φ is surjective, it suffices to prove that $DFT(N, \omega)$ is invertible. To this end note that for all integers ν we have

$$\left(\sum_{q<N} \omega^{\nu q} \right) (1 - \omega^{\nu}) = 1 - \omega^{\nu N} = 0.$$

As $1 - \omega^{\nu}$ is not a zero-divisor in A for all $\nu \not\equiv 0 \bmod N$, it follows that $\sum_{q<N} \omega^{\nu q} = 0$, for those ν. Hence

$$\sum_{q<N} \omega^{pq} \cdot \frac{1}{N} \omega^{-qr} = \frac{1}{N} \sum_{q<N} \omega^{(p-r)q} = \delta_{pr}$$

for all $p, r < N$, thus $DFT(N, \omega)$ is invertible and its inverse has the stated form. Now the FFT-algorithm and its analysis described in Chap. 1 apply to $DFT(2^n, \omega)$. □

We say that A *supports an FFT* of length $N = 2^{\nu}$, if 2 is a unit in A and A contains a principal Nth root of unity. Note that 2 is a unit in A iff char $k \neq 2$.

(2.7) Corollary. *Let A be a commutative k-algebra that supports an FFT of length $N = 2^{\nu}$. Then the total arithmetic complexity of cyclic convolution, i.e., multiplication in $A[X]/(X^N - 1)$, is $O(N \log N)$.*

Proof. Using the DFT-isomorphism $\varphi: A[X]/(X^N - 1) \to A^N$, the cyclic convolution in $A[X]/(X^N - 1)$ can be reduced by the formula

$$a \cdot b = \varphi^{-1}(\varphi(a) \cdot \varphi(b))$$

$(a, b \in A[X]/(X^N - 1))$ to two evaluations of $DFT(N, \omega)$, one multiplication in A^N (componentwise), followed by one evaluation of the inverse of $DFT(N, \omega)$. □

Now we are prepared to give a reasonable upper bound for the total complexity of polynomial multiplication in the case when A supports FFTs.

(2.8) Theorem. *Let N be the smallest power of 2 greater than $n \in \mathbb{N}'$. Let A be a commutative algebra over a field k of characteristic $\neq 2$. If A contains a principal Nth root of unity, then the product of two polynomials $a, b \in A[X]$ with $\deg(ab) = n$ can be computed with $O(n \log n)$ operations.*

Proof. Put $a_i = b_j = 0$, for all i, j with $\deg a < i < N$ and $\deg b < j < N$. For $c := ab$ we have

$$\sum_{\ell=0}^{n} c_{\ell} X^{\ell} \bmod (X^N - 1) = \left(\sum_{i<N} a_i X^i \right) \left(\sum_{j<N} b_j X^j \right) \bmod (X^N - 1).$$

Hence we can apply the fast cyclic convolution of Cor. (2.7) to compute the coefficients $c_0, \ldots, c_n$ with $O(N \log N) = O(n \log n)$ steps. □

For a variant in characteristic 2 see Ex. 2.5.

So far we have derived upper bounds for the nonscalar and the total complexity of polynomial multiplication. However, in the latter case it was a crucial assumption that A supports FFTs of 2-power lengths. In the next section we shall deduce an $O(n \log n \log \log n)$ upper bound for the total number of arithmetic operations when A does not support FFTs. Most of the computational problems studied in later sections involve the polynomial multiplication problem. Thus the complexity of the multiplication problem will be an ingredient in most upper bounds for the problems in question. Each problem will be discussed under three different assumptions or complexity measures. Thus each result typically consists of three claims: Claim 1 is concerned with the nonscalar complexity, Claim 2 is about the total complexity of that problem under the assumption that A supports FFTs of 2-power lengths and $2 \in A^{\times}$, whereas Claim 3 is also about the total complexity but does not assume that A supports FFTs. This motivates to introduce the following function $M: \mathbb{N} \to \mathbb{R}^3$: $M(0) = M(1) := (1, 1, 1)^{\mathsf{T}}$, $M(2) := (2, 2, 2)^{\mathsf{T}}$, and for $n \geq 3$

$$M(n) := \begin{pmatrix} n \\ n \log n \\ n \log n \log \log n \end{pmatrix}.$$

With this vector-valued function, one can rewrite the results of Thm. (2.2), Thm. (2.8) and Thm. (2.13) of the next section in the following compact form.

(2.9) Theorem. *The product of two polynomials $a, b \in A[X]$ with $n = \deg(ab)$ can be computed with $O(M(n))$ operations. The nonscalar cost is at most $n + 1$, and exactly $n + 1$ for generic inputs.*

We summarize some properties of the function M that will be of use in the sequel when analyzing various algorithms.

(2.10) Remark. For $m, n, r \in \mathbb{N}'$, the following inequalities hold componentwise.

(1) $M(m) + M(n) \leq M(m + n)$.
(2) $m \leq n \Rightarrow M(m)/m \leq M(n)/n$; in particular $M(m) \leq M(n)$.
(3) $n = 2^r \Rightarrow \sum_{i=0}^{r} M(n/2^i) < 2M(n)$.
(4) $n = 2^r \Rightarrow \sum_{i=1}^{r} 2^i M(mn/2^i) \leq M(mn) \log n$.   •

2.2* Multiplication of Polynomials II

This section discusses the problem of polynomial multiplication in $A[X]$ when the commutative k-algebra A does not support FFTs. The basic idea is to use for suitable positive integers N algebras of the form

$$A_N := A[X]/(X^N + 1).$$

(Note that the elements of A_N are cosets of the form $a + (X^N + 1)$.) As is shown by the next lemma, such algebras contain principal roots of unity that can be described easily.

(2.11) Lemma. *Let A be a commutative algebra over a field k of characteristic $\neq 2$, $N := 2^n$. Then (the coset of) X is a principal $2N$th root of unity in A_N. In particular, $(X^{pq})_{0 \leq p,q < 2N} \in GL(2N, A_N)$ and X^{2^i} is a principal $(2N/2^i)$th root of unity in A_N for every $0 \leq i \leq n$.*

Proof. In A_N we have $X^{2N} = (-1)^2 = 1$, thus the first condition in Def. (2.5) is satisfied. To prove the second condition it suffices to show that $1 - X^{u2^\nu} \in A_N^\times$, for every odd u and every $0 \leq \nu \leq n$. We proceed by reverse induction on ν. If $\nu = n$, then $1 - X^{u2^n} = 1 - X^{Nu} = 1 - (-1)^u = 2 \in A_N^\times$. For the induction step $\nu + 1 \to \nu$ observe that $(1 + X^{u2^\nu})(1 - X^{u2^\nu}) = 1 - X^{u2^{\nu+1}} \in A_N^\times$, hence $1 - X^{u2^\nu} \in A_N^\times$, which completes the proof. $\square$

(2.12) Remark. Let ψ be a principal $2N$th root of unity in the commutative k-algebra A. Then $\psi^N = -1$ in A and $X \mapsto \psi X$ defines an automorphism Ψ of the A-algebra $A[X]$ mapping $X^N + 1$ onto $-(X^N - 1)$. Hence Ψ induces an isomorphism of A-algebras

$$A[X]/(X^N + 1) \to A[X]/(X^N - 1).$$

W.r.t. canonical bases this isomorphism is described by the diagonal matrix

$$\text{diag}(1, \psi, \psi^2, \ldots, \psi^{N-1}). \qquad \bullet$$

(2.13) Theorem (Schönhage and Strassen). *Let A be a commutative algebra over a field k of characteristic $\neq 2$. Then the product of two polynomials $a, b \in A[X]$ with $n = \deg(ab)$ can be computed with $O(n \log n \log \log n)$ arithmetic operations.*

Proof. To prove our claim, it suffices to find a fast algorithm for the multiplication in the algebra $A_{2n} = A[X]/(X^{2n} + 1)$, where $n = 2^\nu$. First note that for positive integers m and ℓ we have $(A_m)_\ell = A_m[Y]/(Y^\ell + 1)$, where Y is an indeterminate over $A_m = A[X]/(X^m + 1)$. Next, we put

$$L := 2^{\lceil \nu/2 \rceil} \geq \sqrt{n} \geq 2^{\lfloor \nu/2 \rfloor} =: \Lambda.$$

Note that $L\Lambda = n$. We are going to prove the following claim.

(A): *There exist A-linear maps $\alpha : A_{2n} \to (A_{2L})_{2\Lambda}$ and $\beta : (A_{2L})_{2\Lambda} \to A_{2n}$ making the following diagram commutative:*

$$A_{2n} \times A_{2n} \xrightarrow{\quad \text{multipl.} \quad} A_{2n}$$

$$\alpha \times \alpha \qquad\qquad\qquad \beta$$

$$(A_{2L})_{2\Lambda} \times (A_{2L})_{2\Lambda} \xrightarrow{\quad \text{multipl.} \quad} (A_{2L})_{2\Lambda}$$

To define α, write $a \in A[X]$ of degree $< 2n$ as

$$a = \sum_{i<2\Lambda} A_i X^{Li}$$

with $A_i \in A[X]$ of degree $< L$. (Recall that the elements of A_{2n} are cosets.) Then $\bar{a} := a + (X^{2n} + 1) \in A_{2n}$ and $\underline{A_i} := A_i + (X^{2L} + 1) \in A_{2L}$. Now

$$\alpha(\bar{a}) := \sum_{i<2\Lambda} \underline{A_i} Y^i + (Y^{2\Lambda} + 1) \in (A_{2L})_{2\Lambda}.$$

Obviously, α is well-defined and A-linear.

Next we define β. Every $c \in (A_{2L})_{2\Lambda}$ can be written uniquely as

$$c = \sum_{\ell<2\Lambda} \underline{c_\ell} Y^\ell + (Y^{2\Lambda} + 1)$$

with $\underline{c_\ell} = c_\ell + (X^{2L} + 1)$, $c_\ell \in A[X]$ of degree $< 2L$. Put

$$\beta(c) := \sum_{\ell<2\Lambda} c_\ell X^{L\ell} + (X^{2n} + 1) \in A_{2n}.$$

Then β is well-defined and A-linear.

To finish the proof of (A), we have to show that $\bar{a} \cdot \bar{b} = \beta(\alpha(\bar{a}) \cdot \alpha(\bar{b}))$, for all $\bar{a}, \bar{b} \in A_{2n}$. Let $\bar{a}$ and $\bar{b}$ be represented by

$$a = \sum_{i<2\Lambda} A_i X^{Li} \quad \text{and} \quad b = \sum_{j<2\Lambda} B_j X^{Lj},$$

respectively, with $A_i, B_j \in A[X]$ of degree $< L$. Then

$$a \cdot b = \sum_{\ell<4\Lambda} \left(\sum_{i+j=\ell} A_i B_j \right) X^{L\ell},$$

and thus

$$\bar{a} \cdot \bar{b} = \sum_{\ell<2\Lambda} \left(\sum_{i+j=\ell} A_i B_j - \sum_{i+j=\ell+2\Lambda} A_i B_j \right) X^{L\ell} + (X^{2n} + 1).$$

On the other hand,

$$\alpha(\overline{a}) \cdot \alpha(\overline{b}) = \sum_{i<2\Lambda} A_i Y^i \cdot \sum_{j<2\Lambda} B_j Y^j + (Y^{2\Lambda} + 1)$$

$$= \sum_{\ell<4\Lambda} \left(\sum_{i+j=\ell} A_i B_j Y^\ell \right) + (Y^{2\Lambda} + 1)$$

$$= \sum_{\ell<2\Lambda} \left(\sum_{i+j=\ell} A_i B_j - \sum_{i+j=\ell+2\Lambda} A_i B_j \right) Y^\ell + (Y^{2\Lambda} + 1).$$

As the degree of $\sum_{i+j=\ell} A_i B_j - \sum_{i+j=\ell+2\Lambda} A_i B_j$ is less than $2L$, we get $\overline{a} \cdot \overline{b} = \beta(\alpha(\overline{a}) \cdot \alpha(\overline{b}))$, which proves (A).

Now, by Lemma (2.11),

$$\psi := \begin{cases} X & \text{if } L = \Lambda \\ X^2 & \text{if } L = 2\Lambda \end{cases}$$

is a principal 4Λth root of unity in A_{2L}. Hence, by Thm. (2.6) and Rem. (2.12), we have isomorphisms of A-algebras

$$\gamma : (A_{2L})_{2\Lambda} \to A_{2L}[Y]/(Y^{2\Lambda} - 1)$$

defined by $Y \mapsto \psi Y$ and

$$\varphi : A_{2L}[Y]/(Y^{2\Lambda} - 1) \to (A_{2L})^{2\Lambda},$$

the DFT corresponding to the principal 2Λth root of unity $\omega := \psi^2$ in A_{2L}.

Altogether we have the commutative diagram

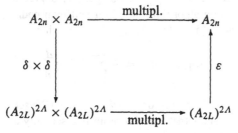

where $\delta := \varphi \circ \gamma \circ \alpha$ and $\varepsilon := \beta \circ \gamma^{-1} \circ \varphi^{-1}$.

Let $T(n)$ denote the cost (= total number of arithmetic operations in A) of the obvious recursive procedure for the multiplication in A_{2n} implied by this last commutative diagram. Furthermore, let $E(\mu)$ denote the total arithmetic cost for evaluating $\mu \in \{\alpha, \beta, \gamma, \gamma^{-1}, \varphi, \varphi^{-1}, \varepsilon, \delta\}$. Then

$$\begin{aligned} T(n) &\leq 2E(\delta) + 2\Lambda \cdot T(L) + E(\varepsilon) \\ &\leq 2\Lambda \cdot T(L) + 2E(\alpha) + 2E(\gamma) + 2E(\varphi) + E(\beta) + E(\gamma^{-1}) + E(\varphi^{-1}). \end{aligned}$$

Adding two elements of A_{2L} or multiplying an element of A_{2L} by a power of X or by an element of $k^\times$ costs at most $2L$ arithmetic operations in A. As no other scalar multiplications occur in the FFT algorithms for φ and its inverse, we obtain by Thm. (2.6)

$$E(\varphi) \leq 2L \cdot (3\Lambda \log(2\Lambda) - 2\Lambda + 1)$$

and $E(\varphi^{-1}) \leq E(\varphi) + 2L \cdot 2\Lambda$. Now $L \cdot \Lambda = n$, thus $O(n \log n)$ is an upper bound for both $E(\varphi)$ and $E(\varphi^{-1})$. Furthermore it is easy to see that $E(\alpha) = 0$ and $E(\mu) = O(n)$, for $\mu \in \{\beta, \gamma, \gamma^{-1}\}$. Altogether, $T(n)$ fulfills the recurrence relation $T(n) \leq 2\Lambda \cdot T(L) + O(n \log n)$, which is easily seen by induction on ν to be satisfied by $T(n) = O(n \log n \log \log n)$ (use $\Lambda \cdot L = n$ and consider the cases $L = \Lambda$ and $\Lambda = 2^{\nu/2 - 1/2}$, $L = 2^{\nu/2 + 1/2}$ separately). This completes the proof of the theorem. $\square$

If the groundfield k has characteristic 2, then a tricky variant of the above proof yields a similar upper bound. This variant, which is due to Schönhage, will be discussed in Ex. 2.6.

2.3* Multiplication of Several Polynomials

Suppose we want to multiply several polynomials $p_1, \ldots, p_t \in A[X]$ of degrees $d_1, \ldots, d_t$, respectively. How does the case $t = 2$, discussed in the first two sections, generalize to larger t? We shall see that an obvious divide-and-conquer technique together with the results of the last section gives an algorithm with cost $O(M(n) \log t)$, where $n := \sum_i d_i$. For general degree sequences, Strassen suggested a multiplication algorithm, based on Huffman coding, which gives the improved upper bound $O(M(n)(1 + \mathcal{H}(d_1, \ldots, d_t)))$, where

$$\mathcal{H}(d_1, \ldots, d_t) := -\sum_{d_i > 0} \frac{d_i}{n} \log \frac{d_i}{n}$$

is the entropy of the probability vector $n^{-1}(d_1, \ldots, d_t)$. (We set $\mathcal{H}(0, \ldots, 0) := 0$.) In Thm. (8.44) we will show that this algorithm is essentially optimal with respect to the nonscalar cost measure.

Let us summarize some well-known properties of the entropy function.

(2.14) Lemma. Let $d = (d_1, \ldots, d_t)$ and $d' = (d_1', \ldots, d_t')$ be t-tuples of nonnegative real numbers, $n = \sum_i d_i$, $n' = \sum_i d_i'$. Then the following holds.

(1) $0 \leq \mathcal{H}(d) \leq \log t$, with both bounds attained: $\mathcal{H}(d) = 0$ iff $n = d_i$ for some i; and $\mathcal{H}(d) = \log t > 0$ iff all d_i are positive and equal.
(2) If $d \leq d'$ componentwise, then $n\mathcal{H}(d) \leq n'\mathcal{H}(d')$.
(3) $(d_1 + \ldots + d_{t-1})\mathcal{H}(d_1, \ldots, d_{t-1}) \leq n\mathcal{H}(d)$.

(4) *Let d be the concatenation of the nonempty sequences d^1 and d^2, i.e., for some $1 < a < t$, $d^1 = (d_1, \ldots, d_a)$ and $d^2 = (d_{a+1}, \ldots, d_t)$; put $u := \sum_{i=1}^{a} d_i$. Then*

$$u\mathcal{H}(d^1) + (n - u)\mathcal{H}(d^2) = n(\mathcal{H}(d) - \mathcal{H}(u, n - u)),$$

and

$$M(u)\mathcal{H}(d^1) + M(n - u)\mathcal{H}(d^2) \leq M(n)(\mathcal{H}(d) - \mathcal{H}(u, n - u)),$$

componentwise.

Proof. See Ex. 2.9.   □

For simplicity, we will assume in the sequel that all p_i are monic polynomials. To warm up, let us briefly discuss the naive divide-and-conquer approach.

(2.15) Corollary. *Given $t = 2^\tau$ monic polynomials in $A[X]$ of positive degrees $d_1, \ldots, d_t$ with $n := \sum_i d_i$, their product can be computed with $O(M(n) \log t)$ operations. The nonscalar cost is at most $n \log t - t/2$.*

Proof. The proof proceeds by induction on τ. Let $T(d_1, \ldots, d_t)$ denote the minimum number of nonscalar operations sufficient to multiply t arbitrary monic polynomials of degrees $d_1, \ldots, d_t$. We know from Cor. (2.3) that two monic polynomials of degrees a and b can be multiplied with at most $a + b - 1$ nonscalar operations. This settles the start $\tau = 1$. For larger $t = 2m$ we multiply the first and the last $t/2$ polynomials separately and multiply these intermediate results together. This yields the recursion

$$T(d_1, \ldots, d_t) \leq T(d_1, \ldots, d_m) + T(d_{m+1}, \ldots, d_t) + (\sum_{i=1}^{m} d_i) + (\sum_{i=m+1}^{t} d_i) - 1,$$

and the upper bound $n \log t - t/2$ follows easily. The other claims can be shown in a similar way.   □

The last result directly applies to the computation of the n-variate elementary symmetric polynomials

$$\sigma_\ell := \sum_{1 \leq i_1 < i_2 < \ldots < i_\ell \leq n} X_{i_1} \cdots X_{i_\ell},$$

$1 \leq \ell \leq n$, $\sigma_0 := 1$. By Vieta's Theorem we have

$$\prod_{i=1}^{n}(X - X_i) = \sum_{\ell=0}^{n}(-1)^\ell \sigma_\ell(X_1, \ldots, X_n) X^{n-\ell}.$$

Thus applying the last corollary to the monic polynomials $p_i := X - X_i$, $1 \leq i \leq n$, we get the following.

(2.16) Corollary. *Given $n = 2^\nu$ indeterminates $X_1, \ldots, X_n$ one can compute the n-variate elementary symmetric polynomials $\sigma_1, \ldots, \sigma_n$ with $O(M(n) \log n)$ operations. The nonscalar cost is at most $n \log n - n/2$.*

For arbitrary n, the last computational task can be solved with nonscalar cost at most $n \log n$, see Ex. 2.8. A similar lower bound in the nonscalar model is deduced in Chap. 8.

For arbitrary degrees an upper bound that is essentially optimal in the nonscalar model will result from an analysis of special computation sequences for $p :=$ $p_1 \cdots p_t$. Every such sequence can be described by a rooted binary tree T with t leaves in which every inner node is of degree 2, together with an enumeration π of its leaves. (Such a pair (T, π) will be called a t-product tree.) Attaching the polynomial p_i to the ith leaf of T and interpreting all the inner vertices of T as multiplication instructions yields a computation of p (at the root of T). For example,

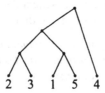

is a 5-product tree (T, π) with $\pi = \binom{12345}{23154}$ that describes the computation of p in "macro-steps" according to the following expression:

$$p = ((p_2 \cdot p_3) \cdot (p_1 \cdot p_5)) \cdot p_4.$$

Now let us come back to the general case. As monic polynomials of degrees a and b can be multiplied with $a + b - 1$ nonscalar operations, an induction shows that

$$\sum_{i=1}^{t} d_i h_i - (t - 1)$$

is an upper bound for the nonscalar cost of computing p via the t-product tree (T, π). Here h_i denotes the depth of the ith leaf in (T, π). The quantity $\sum_{i=1}^{t} d_i h_i$ is called the *weighted external path length* of (T, π) with respect to $d = (d_1, \ldots, d_t)$. As for fixed t there are only finitely many t-product trees, there exists an t-product tree with minimum weighted external path length with respect to d. Such a minimizing (T, π) is called a *Huffman tree* for d and its weighted external path length is denoted by $m(d)$. Let $\mathcal{H} = \mathcal{H}(T, \pi; p_1, \ldots, p_t)$ denote the set of all intermediate results when computing $p = p_1 \cdots p_t$ via the t-product tree (T, π). In the last example $\mathcal{H} = \{p_1, p_2, p_3, p_4, p_5, p_2 p_3, p_1 p_5, p_1 p_2 p_3 p_5, p_1 p_2 p_3 p_4 p_5\}$.

(2.17) Lemma. *Let $d = (d_1, \ldots, d_t)$ be a sequence of positive integers summing up to n, and (T, π) a Huffman tree for d. Then the following holds.*

(1) (T, id) *is a Huffman tree for* $(d_{\pi(1)}, \ldots, d_{\pi(t)})$.

(2) *Let T_L and T_R denote the left and right subtrees of T. If $\pi = $ id and T_L has ℓ leaves, then $(T_L, $ id$)$ and $(T_R, $ id$)$ are Huffman trees for $d_L := (d_1, \ldots, d_\ell)$ and $d_R := (d_{\ell+1}, \ldots, d_t)$, respectively, and $m(d) = m(d_L) + m(d_R) + n$.*

(3) *Let $p_1, \ldots, p_t \in A[X]$ be monic polynomials of degrees $d_1, \ldots, d_t$. Then on input $p_1, \ldots, p_t$ one can compute all elements in $\mathcal{H}(T, \pi; p_1, \ldots, p_t)$, in particular $p_1 \cdots p_t$, with cost $O(\frac{M(n)}{n} m(d))$. The nonscalar cost is at most $m(d) - (t - 1)$.*

Proof. (1) is obvious. To prove (2), note that if the ith leaf of T has depth h_i, then

$$m(d) = \sum_{i=1}^{t} h_i d_i = \sum_{i=1}^{t}(h_i - 1)d_i + n = m(d_L) + m(d_R) + n.$$

(3) After reordering we can assume that $(T, $ id$)$ is a Huffman tree for d. We keep the notation of (2) and observe that $\mathcal{H} = \mathcal{H}_L \cup \mathcal{H}_R \cup \{p_L p_R\}$, where $\mathcal{H}, \mathcal{H}_L$, and $\mathcal{H}_R$ are the sets of polynomials corresponding to T, T_L, and T_R, respectively, and p_L, p_R are the polynomials computed at the root of T_L and T_R, respectively. Hence by induction and by Cor. (2.3) the nonscalar cost is at most

$$m(d_L) - (\ell - 1) + m(d_R) - (t - \ell + 1) + (\deg p_L + \deg p_R - 1) = m(d) - (t - 1).$$

Multiplying this upper bound by $O(\log n)$ (resp. $O(\log n \log \log n)$), we obviously get an upper bound for the total number of arithmetic operations. □

The following provides rather tight upper and lower bounds for $m(d)$.

(2.18) Lemma. *The weighted external path length $m(d)$ of a Huffman tree for the sequence $d = (d_1, \ldots, d_t)$ of positive integral weights satisfies the following bounds:*

$$nH \leq m(d) \leq n(1 + H),$$

where $n := \sum_i d_i$ and $H := \mathcal{H}(d_1, \ldots, d_t)$ is the entropy of the probability vector $n^{-1}(d_1, \ldots, d_t)$.

Proof. The lower bound is proved by induction on t. The start being clear, we only show the induction step. To this end let (T, π) be a Huffman tree for $d = (d_1, \ldots, d_t)$, w.l.o.g. $\pi = 1$. With the notation of the last lemma we obtain by the induction hypothesis $m(d) \geq u\mathcal{H}(d_L) + (n - u)\mathcal{H}(d_R) + n$, where $u := \sum_{i=1}^{\ell} d_i$. With Lemma (2.14)(4), this yields $m(d) \geq n(\mathcal{H}(d) - \mathcal{H}(u, n - u)) + n$, hence $m(d) \geq n\mathcal{H}(d)$.

The proof of the upper bound proceeds in several steps. As

$$m(d) := \min \left\{ \sum_{i=1}^{t} h_i d_i \mid \exists \text{ product tree with } t \text{ leaves of depths } h_1, \ldots, h_t \right\},$$

a first step to solve this minimization problem is a criterion to decide whether or not such a tree with prescribed depths exists. In the sequel, a binary tree T is understood as a rooted tree in which every inner node has degree 1 or 2. If the degree is always 2, then T is called *regular*. (Thus any regular binary tree together with an enumeration of its leaves is a product tree.)

Claim 1. For positive integers $h_1, \ldots, h_t$, the following statements are equivalent:

(1) There exists a binary tree (resp. a regular binary tree) with t leaves of depths $h_1, \ldots, h_t$.

(2) $\sum_{i=1}^{t} 2^{-h_i} \leq 1$ (resp. $= 1$).

To prove (1) $\Rightarrow$ (2), we associate to every edge e (with father node v) in such a tree T the weight $w(e) := \mathrm{degree}(v)^{-1} \in \{1/2, 1\}$. Define the weight $w(p_i)$ of the path p_i from the root to the ith leaf to be the product of the weights of all edges involved in p_i. Then an easy induction (or a probabilistic argument) shows that $\sum_{i=1}^{t} w(p_i) = 1$. As h_i edges are involved in p_i we have $w(p_i) \geq 2^{-h_i}$, hence $\sum_{i=1}^{t} 2^{-h_i} \leq \sum_{i=1}^{t} w(p_i) = 1$. If T is regular, then $w(p_i) = 2^{-h_i}$ for all i.

The proof of (2) $\Rightarrow$ (1) proceeds by induction on t. The start "$t = 1$" is clear. So suppose $t \geq 2$ and $\sum_{i=1}^{t} 2^{-h_i} \leq 1$. We may w.l.o.g. assume that $\sum_i 2^{-h_i} > 1/2$. (Namely, if $2^{-(r+1)} < \sum_i 2^{-h_i} \leq 2^{-r}$, and T_1 is a binary tree with t leaves of depths $h_1 - r, \ldots, h_t - r$, then

$$T = \qquad\qquad (r = 2)$$

satisfies (1).) We assume $h_1 \leq h_2 \leq \ldots \leq h_t$. If we can show that there is some $\tau < t$ such that $\sum_{i=1}^{\tau} 2^{-h_i} = 1/2$, then, using the induction hypothesis, we can find binary trees T_1, T_2 such that T_1 has τ leaves of depths $h_1 - 1, \ldots, h_\tau - 1$, and T_2 has $t - \tau$ leaves of depths $h_{\tau+1} - 1, \ldots, h_t - 1$. Then the binary tree T with T_1 and T_2 as left and right subtree, respectively, satisfies (1) and we are done. The regular case can also be settled in this manner.

It remains to show that for $t \geq 2$ and $h_1 \leq \ldots \leq h_t$, the following holds:

$$\frac{1}{2} < \sum_{i=1}^{t} 2^{-h_i} \leq 1 \Rightarrow \exists \tau < t : \sum_{i=1}^{\tau} 2^{-h_i} = \frac{1}{2}.$$

Again we use induction on t. The case $t = 2$ is easily verified. So assume that $t > 2$ and w.l.o.g. $h_1 > 1$. Then there is some $j < t$ such that $h_j = h_{j+1}$. (Otherwise, $\sum_{i=1}^{t} 2^{-h_i} \leq \sum_{\ell=2}^{\infty} 2^{-\ell} = 1/2$, a contradiction.) Let j be minimal with this property. Then the statement follows by applying the induction hypothesis to the sequence $(h_1, \ldots, h_{j-1}, h_j - 1, h_{j+2}, \ldots, h_t)$. This proves Claim 1.

By Claim 1 we obtain the following characterization of the minimum weighted external path length: for a t-tuple d of positive integers we have

$$m(d) = \min \left\{ \sum_{i=1}^{t} h_i d_i \mid \forall i : h_i \in \mathbb{N}', \sum_i 2^{-h_i} = 1 \right\}.$$

To derive an upper bound for $m(d)$, choose $h_i := \lceil \log \frac{n}{d_i} \rceil$. Then $\sum_{i=1}^n 2^{-h_i} \leq 1$ and

$$\sum_i h_i d_i \leq \sum_i d_i (1 + \log \frac{n}{d_i}) = n(1 + \mathcal{H}(d)).$$

As $\sum_{i=1}^t 2^{-h_i} \leq 1$, there is a binary tree T with t leaves of depths $h_1, \ldots, h_t$, respectively. Deleting every edge in T whose father node has degree one, we obtain a regular tree T' with t leaves of heights $h_i' \leq h_i$, for all i. Hence $\sum_{i=1}^t h_i' d_i \leq \sum_{i=1}^t h_i d_i$, which completes the proof of the lemma. $\qquad\square$

Now we can state the main result of this section.

(2.19) Theorem (Strassen). *Let $p_1, \ldots, p_t$ be $t \geq 2$ monic polynomials in $A[X]$ of positive degrees $d_1, \ldots, d_t$. If (T, π) is a Huffman tree for $d = (d_1, \ldots, d_t)$, then all elements in $\mathcal{H}(T, \pi; p_1, \ldots, p_t)$ can be computed with $O(M(n)(1 + \mathcal{H}(d)))$ operations, where $n = \sum_i d_i$. The nonscalar cost is at most $n(1 + \mathcal{H}(d)) - (t - 1)$. In particular, the product $p_1 \cdots p_t$ can be computed with this cost.*

Proof. Combine Lemma (2.17) and Lemma (2.18). $\qquad\square$

The Huffman coding algorithm is an efficient procedure to construct a Huffman tree for $(d_1, \ldots, d_t)$. The multiplication of t polynomials $p_1, \ldots, p_t \in A[X]$ via Huffman coding goes as follows: start with the list $(p_1, \ldots, p_t)$ and replace two polynomials of smallest degrees in that list by their product until the list consists only of one polynomial.

(2.20) Example. Let $(1, 1, 3, 4, 4, 5)$ be the degree sequence of the polynomials $p_1, \ldots, p_6$. The following table illustrates the steps of the Huffman coding algorithm applied to the above inputs:

$$p_1|p_2|p_3|p_4|p_5|p_6 \quad \underbrace{1\ 1}\ 3\ 4\ 4\ 5$$

$$p_1 p_2|p_3|p_4|p_5|p_6 \quad \underbrace{2\ 3}\ 4\ 4\ 5$$

$$p_4|p_5|p_1 p_2 p_3|p_6 \quad \underbrace{4\ 4}\ 5\ 5$$

$$p_1 p_2 p_3|p_6|p_4 p_5 \quad \underbrace{5\ 5}\ 8$$

$$p_4 p_5|p_1 p_2 p_3 p_6 \quad \underbrace{8\ 10}$$

$$p_4 p_5 p_1 p_2 p_3 p_6 \quad 18\ \bullet$$

For more details on the Huffman coding algorithm see Ex. 2.10.

2.4 Multiplication and Inversion of Power Series

Let us first recall some useful facts on formal power series. A *formal power series* a in indeterminates $X_1, \ldots, X_m$ over A is a formal sum (sequence) $a = \sum_{i \geq 0} a_i$, where $a_i \in A[X_1, \ldots, X_m]$ is a homogeneous polynomial of degree i. The formal power series form a commutative ring $A[[X_1, \ldots, X_m]]$ via

$$\sum_i a_i + \sum_i b_i := \sum_i (a_i + b_i) \text{ and } \left(\sum_i a_i\right)\left(\sum_j b_j\right) := \sum_n \left(\sum_{i+j=n} a_i b_j\right),$$

$0 := \sum_i 0$ is the zero element and $1 := 1 + \sum_{i \geq 1} 0$ is the unit element. The polynomial ring $A[X_1, \ldots, X_m]$ can be viewed in a natural way as a subring of $A[[X_1, \ldots, X_m]]$. A formal power series $a = \sum_i a_i \in A[[X_1, \ldots, X_m]]$ is a unit in $A[[X_1, \ldots, X_m]]$ iff a_0 is a unit in A. Moreover, if $b = \sum_i b_i$ is the inverse of a, then b_i depends only on $a_0, \ldots, a_i$; more precisely, $b_0 = 1/a_0$ and $b_i = -1/a_0 \cdot (\sum_{j=0}^{i-1} a_{i-j} b_j)$, for $i \geq 1$.

In this section we will mainly be concerned with the univariate case, i.e., with power series a over A in one indeterminate X. Such an a can be written uniquely in the form $a = \sum_{i \geq 0} a_i X^i$ with $a_i \in A$. If $c = \sum_n c_n X^n$ is the product of the univariate power series $a = \sum_i a_i X^i$ and $b = \sum_j b_j X^j$, then $c_n = \sum_{i+j=n} a_i b_j$. Thus the computation of $c \bmod X^{n+1} := \sum_{j=0}^{n} c_j X^j$ from $a \bmod X^{n+1}$ and $b \bmod X^{n+1}$ reduces to a variant of the multiplication problem for polynomials.

(2.21) Theorem. *Multiplication mod X^{n+1} in $A[[X]]$ can be done with $O(M(n))$ operations. If $a, b \in A[[X]]$ and ℓ coefficients of ab mod X^{n+1} are known in advance, for some $0 \leq \ell \leq n$, then ab mod X^{n+1} can be computed with nonscalar cost at most*

$$\min\{n, \deg a\} + \min\{n, \deg b\} + 1 - \ell.$$

Here, $\deg a := \infty$, if the power series a is not a polynomial.

Proof. This is a slight modification of Thm. (2.9) and Cor. (2.3). Details are left to the reader. □

In Chap. 17 we will see that in the generic case the nonscalar complexity of multiplication mod X^{n+1} is $2n + 1$.

Next we discuss the problem of inverting a univariate power series. Let $a = \sum_i a_i X^i$ be a unit in $A[[X]]$ with inverse $a^{-1} = b = \sum_i b_i X^i$. As $b_0 = 1/a_0$ and $b_i = -1/a_0 \cdot (\sum_{j=0}^{i-1} a_{i-j} b_j)$, for all $i \geq 1$, the computation of a^{-1} mod X^{n+1} from the input a mod X^{n+1} can be done with $O(n^2)$ arithmetic steps. The proof of the following result will show that Newton-like iterations give much better upper bounds. (Compare also with Ex. 2.13.)

(2.22) Theorem (Sieveking, Kung). *Inversion mod X^{n+1} of univariate power series over a commutative k-algebra A can be performed with $O(M(n))$ operations. The nonscalar cost is at most $3.75n$ if char $k \neq 2$, and at most $3.25n$ if char $k = 2$.*

Proof. We only prove the claim on the nonscalar cost as the other claims can be shown in a similar way. Let $p := \operatorname{char} k$ and put $d_p := 3.75$, if $p \neq 2$, and $d_2 := 3.25$. By L_n we denote the minimum number of nonscalar operations to compute $a^{-1} \bmod X^{n+1}$ from $a \bmod X^{n+1}$. We have to prove that $L_n \leq d_p n$, for all $n \geq 1$.

As $b_0 = 1/a_0$, $b_1 = -b_0 \cdot (a_1 \cdot b_0)$, and $b_2 = -b_0(a_1 \cdot b_1 + a_2 \cdot b_0)$ we see that $L_0 \leq 1$, $L_1 \leq 3$, $L_2 \leq 6$. Below we shall prove that $L_3 \leq 9$. Let $c_p := 3/2$, for $p \neq 2$, and $c_2 := 1/2$. To prove our claim it suffices to show that the following recursion holds for all $n \geq 1$ and all $j \in \{0, 1, 2\}$ with $n + j \geq 2$:

(A)$_j$ $$L_{3n+j} \leq L_n + (6 + c_p)n + (2 + c_p)j - 2c_p + 1.$$

To prove (A)$_j$, suppose we know already $g := \sum_{j=0}^{n} b_j X^j = a^{-1} \bmod X^{n+1}$. Then $1 - ga \equiv 0 \bmod X^{n+1}$. (In general, $u \equiv v \bmod X^\ell$ is the short hand notation for $u \bmod X^\ell = v \bmod X^\ell$.) Cubing yields $1 - 3ga + 3(ga)^2 - (ga)^3 = (1 - ga)^3 \equiv 0 \bmod X^{3n+3}$. Since a is a unit this can be rewritten as $a^{-1} \equiv g \cdot (3 - 3(ga) + (ga)^2) \bmod X^{3n+3}$. In particular, we have for all $j \in \{0, 1, 2\}$:

(B)$_j$ $$a^{-1} \equiv g \cdot (3 - 3(ga) + (ga)^2) \bmod X^{3n+1+j}.$$

Let us describe an efficient way to compute the right-hand side in (B)$_j$. We will proceed in several steps.

Step 1. We compute $ga \bmod X^{3n+1+j}$. Observing that $ga \equiv 1 \bmod X^{n+1}$, we can perform this task by Thm. (2.21) with

(C) $$\min\{3n + j, \deg g\} + \min\{3n + j, \deg a\} + 1 - (n + 1) \leq 3n + j$$

nonscalar operations.

Step 2. We compute $(ga)^2 \bmod X^{3n+1+j}$. Note that

$$ga \equiv 1 + X^{n+1} r_j \bmod X^{3n+1+j},$$

for a suitable polynomial r_j of degree less than $2n + j$, whose coefficients are known by Step 1. But since

$$(ga)^2 \equiv (1 + X^{n+1} r_j)^2 = 1 + 2X^{n+1} r_j + X^{2n+2} r_j^2 \bmod X^{3n+1+j},$$

we have only to compute $r_j^2 \bmod X^{n+j-1}$ (besides linear operations). By Cor. (2.4), we know that in the case $n + j - 2 \geq 0$

(D) $$c_p(n + j - 2) + 1.$$

is an upper bound for the nonscalar cost of computing $r_j^2 \bmod X^{n+j-1}$ from $r_j \bmod X^{n+j-1}$. (In the case $n = 1$, $j = 0$ this step is free of nonscalar costs; this together with (C) and (E) gives $L_3 \leq 9$.)

Step 3. We compute $\Gamma_j := 3 - 3(ga) + (ga)^2 \bmod X^{3n+1+j}$ with no nonscalar operations.

Step 4. We compute $g \cdot \Gamma_j \bmod X^{3n+1+j}$. Since $g \cdot \Gamma_j \equiv a^{-1} \bmod X^{3n+1+j}$, the first $n+1$ coefficients of $g \cdot \Gamma_j$ are already known. Hence, by Thm. (2.21), we can solve this task with at most

(E) $\min\{3n+j, \deg g\} + \min\{3n+j, \deg \Gamma_j\} + 1 - (n+1) \leq 3n+j$

nonscalar operations. Now (A)$_j$ follows from (C), (D), and (E).   □

The inversion of a (general) polynomial can be done a bit faster, see Ex. 2.12. The efficient multiplication and inversion algorithms yield an efficient procedure for the division of power series.

(2.23) Corollary. *Let $a, b \in A[[X]]$, b a unit. Then $a/b \bmod X^{n+1}$ can be computed with $O(M(n))$ operations. The nonscalar cost is at most $5.75n + 1$, if char $k \neq 2$, and at most $5.25n + 1$, if char $k = 2$.*

Proof. $a/b = a \cdot b^{-1}$. Now apply Thm. (2.21) and (2.22).   □

Combining the results on the symbolic multiplication and the inversion of power series with the efficient computation of elementary symmetric polynomials σ_i, we get an efficient way to compute the n-variate *power sums*, defined by $s_\ell := \sum_{i=1}^n X_i^\ell$, for $\ell \in \mathbb{N}$. This result is based on the following.

(2.24) Newton Relation. *Let $K := k(X_1, \ldots, X_n)$, $\sigma := \sum_{i=0}^n (-1)^i \sigma_i T^i \in K[[T]]$, and $s := \sum_{j=0}^\infty s_{j+1} T^j \in K[[T]]$, $\sigma_0 := 1$. If σ' denotes the formal derivative of σ with respect to T, then the Newton relation $s = -\sigma'/\sigma$ holds.*

Proof. Substitute T by T^{-1} in $(T - X_1) \cdot \ldots \cdot (T - X_n) = \sum_{i=0}^n (-1)^i \sigma_i T^{n-i}$ and multiply with T^n to obtain $(1 - X_1 T) \cdot \ldots \cdot (1 - X_n T) = \sum_{i=0}^n (-1)^i \sigma_i T^i = \sigma$. Working in $K[[T]]$ and differentiating w.r.t. T (several useful facts on differentiation of power series are summarized in the next section) we get:

$$\sigma' = \sum_{j=1}^n (-X_j) \cdot \frac{\sigma}{1 - X_j T} = -\sigma \sum_{j=1}^n X_j \sum_{\ell \geq 0} X_j^\ell T^\ell$$

$$= -\sigma \sum_{\ell \geq 0} \left(\sum_{j=1}^n X_j^{\ell+1} \right) T^\ell = -\sigma \sum_{\ell \geq 0} s_{\ell+1} T^\ell = -\sigma \cdot s.$$

As σ is a unit in $K[[T]]$, our claim follows.   □

(2.25) Corollary (Strassen). *The n-variate power sums $s_1, \ldots, s_n$ can be computed from the given indeterminates $X_1, \ldots, X_n$ with $O(M(n) \log n)$ operations. The nonscalar cost is at most $n \log n + 5.75n + 1$. If char $k = 2$, then the constant 5.75 can be replaced by 5.25.*

Proof. We prove the statement concerning the nonscalar cost. According to Cor. (2.16) and Ex. 2.8 one can compute σ from the input with at most $n \log n$ nonscalar operations. In the second stage we compute σ' from σ at no nonscalar cost. Finally, we use the Newton relation $s = -\sigma'/\sigma$ to compute $s \bmod T^{n+1}$ in additional $5.75n + 1$ nonscalar steps, see Cor. (2.23). Altogether our claim follows. The other claims can be shown in a similar way.   □

Using the concept of reciprocal polynomials one can reduce the problem of polynomial division with remainder to that of polynomial multiplication and inversion.

(2.26) Corollary (Strassen). *Let* $a = \sum_{i=0}^{n} a_i X^i$ *and* $b = \sum_{j=0}^{m} b_j X^j$ *be polynomials in* $A[X]$ *with* $\deg a = n \geq m = \deg b \geq 1$ *and* $b_m \in A^\times$. *Then the quotient and the remainder of* (a, b) *can be computed with* $O(M(n-m)+M(m))$ *arithmetic operations. The nonscalar cost is at most* $5.75(n-m)+m+1$. *If char* $k = 2$, *then the constant* 5.75 *can be replaced by* 5.25.

Proof. Let $q = \lfloor a/b \rfloor$ and $r = a \bmod b$. Replace in $a(X) = q(X)b(X) + r(X)$ the indeterminate X by X^{-1} and multiply by X^n. This gives

$$\sum_{i=0}^{n} a_i X^{n-i} = \left(\sum_{i=0}^{n-m} q_i X^{n-m-i}\right)\left(\sum_{i=0}^{m} b_i X^{m-i}\right) + \sum_{i=0}^{m-1} r_i X^{n-i}.$$

Since $\sum_{i=0}^{m-1} r_i X^{n-i}$ is a multiple of X^{n-m+1} and $b_m \in A^\times$ we get

$$\left(\sum_{i=0}^{n} a_i X^{n-i}\right)\left(\sum_{i=0}^{m} b_i X^{m-i}\right)^{-1} \equiv \sum_{i=0}^{n-m} q_i X^{n-m-i} \bmod X^{n-m+1}.$$

By Cor. (2.23), q can be computed with $O(M(n-m))$ operations, and as $r \equiv a - qb \bmod X^{m+1}$ one can compute r in $O(M(m))$ steps, see Thm. (2.21). This proves our first claim. Now let us estimate the nonscalar cost. By Cor. (2.23) q can be obtained with nonscalar cost at most $5.75(n-m)+1$ (resp. $5.25(n-m)+1$ in characteristic 2). Computing r needs by Thm. (2.21) at most m further nonscalar operations, since the $n-m+1$ highest coefficients of a and $q \cdot b$ are equal. $\square$

2.5* Composition of Power Series

Throughout this section A denotes a commutative algebra over a field k of characteristic zero.

Let us first specify the computational problem, which will be discussed in the present section. If $b = \sum_{i \geq 1} b_i X^i$ and $a = \sum_{j \geq 0} a_j X^j$ are univariate power series over A, then, as $b_0 = 0$, their *composition* $c = a \circ b = a(b)$,

$$c = \sum_{n \geq 0} c_n X^n := \sum_{j \geq 0} a_j \left(\sum_{i \geq 1} b_i X^i\right)^j,$$

is well-defined. We have $c_0 = a_0$, and for $n \geq 1$ the coefficient c_n is a polynomial expression in $a_1, \ldots, a_n, b_1, \ldots, b_n$:

$$c_n = \sum_{j=1}^{n} a_j \sum_{\ell_1 + \cdots + \ell_j = n} b_{\ell_1} \cdots b_{\ell_j}.$$

Let $a, b \in A[[X]]$ with $b_0 = 0$. The composition problem is to compute $a(b) \bmod X^{n+1}$ from $a \bmod X^{n+1}$ and $b \bmod X^{n+1}$. To investigate this problem we need some preliminaries. The (formal) *derivative* of $f = \sum_{j \geq 0} f_j X^j \in A[[X]]$ is defined by $f' := D(f) := \sum_{i \geq 1} i f_i X^{i-1}$. One defines the higher derivatives recursively by $D^0(f) := f$ and $D^i(f) := D^{i-1}(D(f))$, for $i \geq 1$.

(2.27) Theorem. *Let $D: A[[X]] \to A[[X]]$, $f \mapsto f'$, denote the derivation operator. Then the following holds.*

(1) $D(af + bg) = aD(f) + bD(g)$, *for all $a, b \in A$ and $f, g \in A[[X]]$ (A-linearity).*
(2) $D(fg) = D(f)g + fD(g)$, *for all $f, g \in A[[X]]$ (product rule).*
(3) $D(f^n) = nf^{n-1}D(f)$, *for all $f \in A[[X]]$ and all $n \in \mathbb{N}'$ (power rule).*
(4) $D(f/g) = (D(f)g - fD(g))/g^2$, *for all $f, g \in A[[X]]$, $g_0 \in A^\times$ (quotient rule).*
(5) $D(f \circ g) = (D(f) \circ g) \cdot D(g)$, *for all $f, g \in A[[X]]$ with $g_0 = 0$ (chain rule).*
(6) $D(f) = D(g)$ *and $f_0 = g_0$ implies $f = g$.*

Proof. See, e.g., [572]. □

Formal power series admit a Taylor expansion.

(2.28) Taylor Expansion. *Let $f, g, h \in A[[X]]$ with $g_0 = h_0 = 0$. Then*

$$f(g + h) = f(g) + \sum_{i \geq 1} \frac{(D^i f)(g)}{i!} h^i.$$

Proof. (Sketch) First of all we remark that the composition $a \circ b$ of power series a and b generalizes in an obvious way to the case when b is a multivariate power series with vanishing constant term. Let $f = \sum_{j \geq 0} f_j X^j$ and Y be a new indeterminate. Then

$$\begin{aligned}
f(X + Y) &= \sum_{j \geq 0} f_j (X + Y)^j = \sum_{j \geq 0} f_j \left(\sum_{i=0}^{j} \binom{j}{i} X^{j-i} Y^i \right) \\
&= \sum_{i \geq 0} Y^i \left(\sum_{j \geq i} f_j \binom{j}{i} X^{j-i} \right) = \sum_{i \geq 0} Y^i \left(\frac{(D^i f)(X)}{i!} \right).
\end{aligned}$$

Substituting X by g and Y by h yields our claim. □

(2.29) Example. The formal power series corresponding to the logarithm function $(x \mapsto \ln(1 + x))$ and the exponential function $(x \mapsto e^x)$ of classical analysis are given by

$$\text{LOG} := \sum_{i \geq 1} \frac{(-1)^{i-1}}{i} X^i \quad \text{and} \quad \text{EXP} := \sum_{j \geq 0} \frac{1}{j!} X^j,$$

respectively. A straightforward computation shows that

$$\text{LOG}' = \sum_{j \geq 0} (-1)^j X^j = (1 + X)^{-1} \quad \text{and} \quad \text{EXP}' = \text{EXP}.$$

Furthermore, LOG and EXP $- 1$ are inverse to each other w.r.t. composition,

$$\text{LOG} \circ (\text{EXP} - 1) = X \quad \text{and} \quad (\text{EXP} - 1) \circ \text{LOG} = X.$$

(Use Thm. (2.27)(5+6) to prove the first claim; apply Ex. 2.19(1) to deduce the second claim.) Let $a, b \in A[[X]]$ with $a_0 = b_0 = 0$. By Taylor expansion and the fact that $\text{EXP}' = \text{EXP}$, we obtain

$$\text{EXP}(a + b) = \text{EXP}(a) \cdot \text{EXP}(b). \qquad \bullet$$

We start with some special cases of the composition problem.

(2.30) Corollary (Brent). *Let* $b \in A[[X]]$, $b_0 = 0$. *Given* $b \bmod X^{n+1}$ *one can compute* $\text{LOG} \circ b \bmod X^{n+1}$ *with* $O(M(n))$ *operations. The nonscalar cost is at most* $5.75n - 4.75$.

Proof. We prove only the last statement, the other claim follows similarly. Let $c = \text{LOG} \circ b$. By the chain rule, $c' = \frac{b'}{1+b}$. As the computation of $b' \bmod X^n$ from $b \bmod X^{n+1}$ is free of charge in the nonscalar model, one can compute $c' \bmod X^n = \sum_{i=1}^{n} i c_i X^{i-1}$ with nonscalar cost at most $5.75(n - 1) + 1$, see Cor. (2.23). Thus with $c_0 = 0$ we get $c \bmod X^{n+1}$ with no additional nonscalar operations. $\qquad \square$

A similar result holds for the complexity of computing $\text{EXP} \circ b \bmod X^{n+1}$, when $b \in A[[X]]$, $b_0 = 0$. As $\text{EXP} \circ b = 1 + (\text{EXP} - 1) \circ b$, we are left with the problem to compute $(\text{EXP} - 1) \circ b \bmod X^{n+1}$ efficiently. Let $Y = (\text{EXP} - 1) \circ b$. As LOG and EXP $- 1$ are inverse to each other w.r.t. composition, we obtain $\text{LOG}(Y) = b$. Thus we are looking for the (unique) zero of the function $S \mapsto \text{LOG}(S) - b$. Here, the following Newton iteration technique for formal power series applies.

(2.31) Theorem. *Let* $F = \sum_{j \geq 0} f_j Y^j$ *be a power series in* Y *with coefficients* $f_j = \sum_{i \geq 0} f_{ji} X^i \in A[[X]]$. *Suppose that* $f_{00} = 0$ *and* $f_{10} \in A^{\times}$. *Then there is a unique power series* $S = \sum_{\ell \geq 1} s_\ell X^\ell \in A[[X]]$ *with* $F(S) = 0$. *For* $m \geq 0$ *let* $Y_m := S \bmod X^{2^m}$. *If* F' *denotes the derivative of* F *w.r.t.* Y, *then the sequence* $(Y_m)_{m \geq 0}$ *satisfies the Newton iteration* $Y_0 = 0$ *and, for* $m \geq 0$,

$$Y_{m+1} = Y_m - \frac{F(Y_m)}{F'(Y_m)} \bmod X^{2^{m+1}}.$$

Proof. Comparing in $\sum_{j\geq 0}(\sum_{i\geq 0} f_{ji} X^i)(\sum_{\ell\geq 1} s_\ell X^\ell)^j = 0$ the coefficients of X^m yields a system of equations: $f_{00} = 0$, for $m = 0$, $f_{10}s_1 + f_{01} = 0$, for $m = 1$, and $f_{10}s_m + p_m = 0$, for $m \geq 2$, where p_m is a polynomial expression in $s_1, \ldots, s_{m-1}$, and some of the f_{ji} with $i + j \leq m$. As according to our assumptions, $f_{00} = 0$ and $f_{10} \in A^\times$, this system has a unique solution S.

We now show by induction on m that the sequence $(Y_m)_{m\geq 0}$ satisfies the Newton iteration. The start being clear, we prove the induction step $m \to m + 1$. By Taylor expansion of $F \in A[[X]][[Y]]$, we obtain

$$
\begin{aligned}
0 = F(S) &= F(Y_m + (S - Y_m)) \\
&= F(Y_m) + \sum_{i\geq 1} \frac{(D^i F)(Y_m)}{i!}(S - Y_m)^i \\
&= F(Y_m) + F'(Y_m)(S - Y_m) + (S - Y_m)^2 G(Y_m, S - Y_m),
\end{aligned}
$$

for some $G \in A[[X, Y]]$. As $F'(Y_m) \bmod X = f_{10} \in A^\times$, $F'(Y_m)$ is invertible in $A[[X]]$, hence

$$
0 = \frac{F(S)}{F'(Y_m)} = \frac{F(Y_m)}{F'(Y_m)} + (S - Y_m) + (S - Y_m)^2 \frac{G(Y_m, S - Y_m)}{F'(Y_m)}.
$$

Together with $0 = (S - Y_m)^2 \bmod X^{2^{m+1}}$, we obtain

$$
Y_m - \frac{F(Y_m)}{F'(Y_m)} \bmod X^{2^{m+1}} = S \bmod X^{2^{m+1}} = Y_{m+1}. \qquad \square
$$

As each iteration step doubles the number of known coefficients of S, the sequence $(Y_m)_{m\geq 0}$ is said to *converge quadratically* to the solution S.

(2.32) Corollary (Brent). *Let* $b \in A[[X]]$, $b_0 = 0$. *Given* $b \bmod X^{n+1}$ *one can compute* $\mathrm{EXP} \circ b \bmod X^{n+1}$ *with* $O(M(n))$ *operations. The nonscalar cost is at most* $27n - 11.5 \log n - 13.5$.

Proof. We only prove the last statement as the other claims can be shown in a similar way. Let $F(Y) = \sum_{j\geq 0} f_j Y^j := \mathrm{LOG}(Y) - b$. Then $f_0 = b$ and $\sum_{j\geq 1} f_j Y^j = \mathrm{LOG}(Y)$; in particular, $f_{00} = 0$ and $1 = f_1 = f_{10} \in A^\times$. Thus Thm. (2.31) applies, and the unique zero S of F satisfies $S = (\mathrm{EXP} - 1) \circ b$.

Next we deduce an upper bound for the nonscalar cost of the mth iteration step, i.e., computing Y_{m+1} from Y_m. As $F'(Y) = (1 + Y)^{-1}$, we have

$$
Y_{m+1} = Y_m - \left[(\mathrm{LOG}(Y_m) - b) \cdot (1 + Y_m) \bmod X^{2^{m+1}}\right].
$$

By Cor. (2.30), the computation of $\mathrm{LOG}(Y_m) - b \bmod X^{2^{m+1}}$ from Y_m and $b \bmod X^{2^{m+1}}$ costs at most $5.75 \cdot (2^{m+1} - 1) - 4.75$ nonscalar operations, and the computation of the product has nonscalar cost at most $(2^{m+1}-1)+(2^m-1)+1-2^m$. (To see this, use Thm. (2.21) and the fact that $[\ldots] \bmod X^{2^m} = 0$.) Altogether, the mth iteration has nonscalar cost at most $6.75 \cdot 2^{m+1} - 11.5$. Now let $2^m \leq n < 2^{m+1}$. As $S \bmod X^{n+1} = Y_{m+1} \bmod X^{n+1}$, it suffices to compute Y_{m+1}. The overall nonscalar cost is thus at most $\sum_{\mu=1}^{m+1}(6.75 \cdot 2^\mu - 11.5) = 6.75(2^{m+2} - 2) - 11.5(m + 1)$, and by a straightforward computation our last claim follows. $\qquad \square$

As a nice application, we discuss the problem of computing the mth power mod X^{n+1} of a power series $a \in A[[X]]$. By the binary method, see Chap. 1, this can be done in $O(M(n) \log m)$ operations. Surprisingly, one can improve this upper bound as follows.

(2.33) Corollary (Brent). *Let* $a \in A[[X]]$, $m \geq 2$. *Given* a *mod* X^{n+1} *one can compute* a^m *mod* X^{n+1} *with* $O(M(n) + \log m)$ *operations, where* $M(n) + \log m$ *means that* $\log m$ *has to be added to each component of* $M(n)$.

Proof. Write $a = \alpha X^{\ell} b$ with $0 \neq \alpha \in A$, $\ell \geq 0$, and $b \in A[[X]]$ with $b_0 = 1$. First use the binary method to compute α^m with at most $2 \log m$ operations. In the worst case ($\ell = 0$), we next compute b^m mod X^{n+1} from b mod X^{n+1}. Here, we proceed as follows. By formal differentiation one easily checks that $\mathrm{LOG}((X + 1)^m - 1) = m\mathrm{LOG}(X)$. As $b_0 = 1$, we can replace X by $b - 1$ and obtain $\mathrm{LOG}(b^m - 1) = m\mathrm{LOG}(b - 1)$. Thus $b^m - 1 = (\mathrm{EXP} - 1) \circ (m\mathrm{LOG}(b - 1))$, and by Cor. (2.30) and (2.32), b^m mod X^{n+1} can be computed with $O(M(n))$ operations. Finally, we compute $\alpha^m \cdot (b^m \mod X^{n+1})$ with at most $n + 1$ additional operations. $\qquad\square$

For the general composition problem the asymptotically best known upper bound is the following.

(2.34) Theorem (Brent and Kung). *Let* $a, b \in A[[X]]$, $b_0 = 0$, $b_1 \in A^{\times}$. *Given* a *mod* X^{n+1} *and* b *mod* X^{n+1} *the computation of* $a \circ b$ *mod* X^{n+1} *can be performed with* $O(M(n)\sqrt{n \log n})$ *operations.*

Proof. We prove all three claims simultaneously. W.l.o.g. we may assume that n is a power of 2. Let $m < n$ and $\ell := \lceil n/m \rceil$. (The value of m will be specified later.) Let $b = \beta_m + \gamma_m$ with $\beta_m := \sum_{i=1}^{m} b_i X^i = b \mod X^{m+1}$. By Taylor expansion, see Thm. (2.28), we have

$$a(b) = a(\beta_m + \gamma_m) = a(\beta_m) + \sum_{i=1}^{\infty} \frac{(D^i a)(\beta_m)}{i!}(\gamma_m)^i.$$

As for $i > 0$ the degree of any term in $(\gamma_m)^{\ell+i}$ is at least $n + 1$, we obtain

$$a(b) \equiv a(\beta_m) + \sum_{i=1}^{\ell} \frac{(D^i a)(\beta_m)}{i!}(\gamma_m)^i \mod X^{n+1}.$$

This equality suggests the following procedure for computing $a(b)$ mod X^{n+1}.

(I)   Compute $a(\beta_m)$ mod X^{n+1}.
(II)   Compute $(D^i a)(\beta_m)$ mod X^{n+1}, for every $i \in \ell$.
(III)   Compute $a(b)$ mod X^{n+1}.

A more detailed discussion of each step follows.

(I) We claim that the computation of $a(\beta_m)$ mod X^{n+1} can be performed with $O(M(n)m \log n)$ operations.

The idea is to use $a(\beta_m) = \alpha_1(\beta_m) + \beta_m^{n/2}\alpha_2(\beta_m) \bmod X^{n+1}$ (with polynomials α_1, α_2 of degree at most $n/2$) as the basis of a recursive procedure for computing $a(\beta_m) \bmod X^{n+1}$. Our claim easily follows from the following.

(A)   Let j and n be powers of 2, and $\alpha := a_0 + a_1 X + \cdots + a_j X^j$ and $\beta := b_1 X + \cdots + b_m X^m$ in $A[X]$ with $j, m \le n$. Given α and β, one can compute $\alpha \circ \beta \bmod X^{n+1}$ with $O(M(n) \cdot \frac{jm}{n} \cdot \log n)$ operations.

Proof of (A). Write $\alpha(\beta) = \alpha_1(\beta) + \beta^{j/2}\alpha_2(\beta)$ with polynomials α_1, α_2 of degree at most $j/2$. This gives us a recursive procedure for computing $\alpha(\beta) \bmod X^{n+1}$. (During the computation we always truncate terms of degree higher than n.) Assume that $T(j)$ is the triple describing the minimal number of arithmetic operations sufficient to compute both $\beta^{j/2} \bmod X^{n+1}$ and $\alpha(\beta) \bmod X^{n+1}$ for arbitrary α with $\deg \alpha = j$. As $\deg \beta^{j/2} \le mj/2$ and $\deg \alpha(\beta) \le jm$, the recursive procedure yields the recursion

$$T(j) \le 2T(j/2) + O(M(\min(jm, n))).$$

Let $r \in \mathbb{Z}$ such that

$$\frac{jm}{2^{r+1}} < n \le \frac{jm}{2^r}.$$

If $r < 0$, then $jm < n$, and with $j =: 2^h$ we get by Rem. (2.10)

$$
\begin{aligned}
T(j) &\le c \sum_{s=0}^{h-1} 2^s M(jm/2^s) + j \cdot T(1) \\
&\le cM(jm) \log j \\
&\le cM(n)\frac{jm}{n} \log n,
\end{aligned}
$$

for some constant $c > 0$. If $r \ge 0$, we have

$$
\begin{aligned}
T(j) &\le c(M(n) + 2M(n) + \cdots + 2^r M(n)) + 2^{r+1}T(j/2^{r+1}) \\
&\le c(2^{r+1} - 1)M(n) + 2^{r+1}T(j/2^{r+1}) \\
&\le 2c\frac{jm}{n}\left(M(n) + T(j/2^{r+1})\right).
\end{aligned}
$$

Since $jm/2^{r+1} < n$ and $T(j) \le cM(n)\frac{jm}{n} \log n$ we obtain

$$T(j/2^{r+1}) \le cM(n)\frac{jm}{2^{r+1}n} \log n \le cM(n) \log n.$$

Hence $T(j) = O(M(n) \cdot \frac{jm}{n} \cdot \log n)$. This proves (A) and shows that Step (I) can be done with $T(n) = O(M(n)m \log n)$ operations.

(II) By the chain rule, we have $(a \circ \beta_m)' = (a' \circ \beta_m) \cdot \beta_m'$. As $b_1 \in A^\times$, β_m' is a unit in $A[[X]]$, and we get $a' \circ \beta_m = (a \circ \beta_m)'/\beta_m'$. Thus

$$(D^i a)(\beta_m) = ((D^{i-1}a) \circ \beta_m)'/\beta_m' \bmod X^{n+1},$$

for all $i \leq \ell$. Consequently, by Cor. (2.23), Step (II) can be performed with $O(\ell M(n)) = O(M(n)n/m)$ steps.

(III) Using Horner's rule, the computation of $a \circ b \bmod X^{n+1}$ can now be done with $O(\ell M(n))$ additional operations. Choosing $m \sim \sqrt{n/\log n}$, our claim follows. $\square$

2.6 Exercises

2.1. Suppose T satisfies the recursion $T(0) = O(1)$ and $T(n) = T(\lfloor n/2 \rfloor) + O(M(n))$. Prove that $T(n) = O(M(n) \log n)$.

2.2. This exercise introduces the Karatsuba algorithm for polynomial multiplication. Let A be a commutative k-algebra and $a, b \in A[X]$ of degree less than $n = 2m$. Write a, b as $a = \alpha_0 + \alpha_1 X^m$ and $b = \beta_0 + \beta_1 X^m$ with polynomials α_i and β_j of degree less than m.

(1) Put $\gamma_0 := \alpha_0 \beta_0$, $\gamma_1 := \alpha_1 \beta_1$, $\gamma_2 := (\alpha_0 - \alpha_1)(\beta_0 - \beta_1)$, and $\gamma_3 := \gamma_0 + \gamma_1 - \gamma_2$. Show that $ab = \gamma_0 + \gamma_3 X^m + \gamma_1 X^{2m}$.
(2) Let $n = 2^r$. On the basis of (1) design an algorithm that performs multiplication of polynomials of degree less than n with only 3^r multiplications in A (instead of 4^r multiplications with the naive algorithm). What is the total number of arithmetic operations?

2.3. Let A be a commutative algebra over an infinite field k, and let $a, b \in A[X]$ be nonconstant and monic such that all input coefficients (except the leading ones) are algebraically independent. Prove that the nonscalar complexity of computing ab is exactly $\deg(ab) - 1$.

2.4. Let a, b be univariate polynomials over a commutative k-algebra A such that $n = \deg(ab) = \deg a + \deg b$. Let $ab =: c = \sum_{j \leq n} c_j X^j$.

(1) Suppose one knows $c_n, c_{n-1}, \ldots, c_{n+1-\ell}$ in advance, for some $0 \leq \ell \leq n+1$. Then c can be computed with $n + 1 - \ell$ nonscalar operations provided that $|k| \geq n + 1 - \ell$.
(2) Let $a = \sum_{i \leq \alpha} a_i X^i$ and $b = \sum_{j \leq \beta} b_j X^j$ with $\alpha + \beta = n$. Suppose that $0 \leq \alpha' < \alpha$ and $0 \leq \beta' < \beta$ exist such that $a_i, b_j \in k$ for all $\alpha' < i \leq \alpha$ and $\beta' < j \leq \beta$. Prove that c can be computed with $\alpha' + \beta' + 1$ nonscalar operations provided that $|k| \geq \alpha' + \beta' + 1$.

2.5. Let k be a field containing a primitive 3^rth root of unity, for every $r \geq 1$.

(1) Show that the corresponding 3×3 DFT-matrix can be multiplied by a generic vector with only 8 arithmetic operations (additions, subtractions, scalar multiplications).

(2) Design an (inverse) FFT for all $N = 3^r$.

(3) If A is a commutative k-algebra, then multiplication of $a, b \in A[X]$ with $\deg(ab) = n$ can be performed with $O(n \log n)$ arithmetic operations. (In characteristic two, this result complements Thm. (2.8).)

2.6.* Let A be a commutative algebra over a field k of characteristic 2. Prove that the product of two polynomials $a, b \in A[X]$ with $n = \deg(ab)$ can be computed with $O(n \log n \log \log n)$ arithmetic operations in A. (Hints: for $N = 3^n$ consider the k-algebra $A_{3,N} := A[X]/(1 + X^N + X^{2N})$. Define $L := 3^{\lceil n/2 \rceil}$ and $\Lambda := 3^{\lfloor n/2 \rfloor}$. Use an FFT of length 3Λ over $A_{3,L}$ to reduce one multiplication in $A_{3,N}$ to several multiplications in $A_{3,L}$. Further hints may be found in the proofs of Lemma (2.11) and Thm. (2.13).)

2.7. Let $(\ell_i)_{i \geq 0}$ be a sequence of real numbers with $\ell_1 = 0$, $\ell_2 \leq 1$, and $\ell_{2n} \leq 2\ell_n + 2n$ as well as $\ell_{2n+1} \leq \ell_n + \ell_{n+1} + 2n$, for all $n \geq 1$. Prove that $\ell_n \leq n \log n$, for all $n \geq 1$. (Hint: show that $n^n \geq 2(n-1)^n$, for $n \geq 1$, and work with $\lambda_n := 2^{\ell_n}$ instead of ℓ_n.)

2.8. Show that for arbitrary n the nonscalar cost to compute all n-variate elementary symmetric polynomials is at most $n \log n$. (Hint: let $\sigma_j = \sigma_j^{(n)}$ denote the jth n-variate elementary symmetric polynomial. Recursively define sets Σ_n as follows:
$\Sigma_1 := \Sigma_1(X_1) := \{X_1\}$,

$$\Sigma_{2n} := \Sigma_n(X_1, \ldots, X_n) \cup \Sigma_n(X_{n+1}, \ldots, X_{2n}) \cup \{\sigma_1^{(2n)}, \ldots, \sigma_{2n}^{(2n)}\},$$

$$\Sigma_{2n+1} := \Sigma_n(X_1, \ldots, X_n) \cup \Sigma_{n+1}(X_{n+1}, \ldots, X_{2n+1}) \cup \{\sigma_1^{(2n+1)}, \ldots, \sigma_{2n+1}^{(2n+1)}\}.$$

Define L_n as the nonscalar complexity to compute from $X_1, \ldots, X_n$ all elements in Σ_n. Use Cor. (2.3) to prove that $L_{2n} \leq 2L_n + 2n - 1$ and $L_{2n+1} \leq L_n + L_{n+1} + 2n$. Now apply the last exercise.)

2.9.* This exercise discusses properties of the (non-normalized) entropy function. Let $a, b \in \mathbb{R}$. A function $f : [a, b] \to \mathbb{R}$ is called *strongly convex* if for all $x_1, x_2 \in [a, b]$ we have $f(\frac{x_1 + x_2}{2}) \leq \frac{1}{2}(f(x_1) + f(x_2))$ with equality holding iff $x_1 = x_2$. In the following we assume that $f \in C^2[a, b]$, i.e., the second derivative f'' of f exists and is continuous on $[a, b]$.

(1) Suppose that $f'' > 0$ on $[a, b]$. Prove that f is strongly convex. (Hint: use the Taylor expansion of f.)

(2) (Jensen's inequality) Let $f'' > 0$ on $[a, b]$. Show that for all $k \geq 2$ and all $x_1, \ldots, x_k \in [a, b]$ we have

$$f\left(\frac{x_1 + \ldots + x_k}{k}\right) \leq \frac{1}{k}(f(x_1) + \ldots + f(x_k)),$$

with equality holding iff $x_1 = \ldots = x_k$. (Hint: first show the assertion for powers of 2, then prove the "induction step" $k \to k - 1$.)

(3) Let $x = (x_1, \ldots, x_n)$ be a probability distribution. Show that $0 \leq \mathcal{H}(x) \leq \log n$, $\mathcal{H}(x) = 0$ iff there exists some i such that $x_i = 1$, and $\mathcal{H}(x) = \log n$ iff x is the uniform distribution, i.e., $x_i = \frac{1}{n}$, for all i.

(4) Prove the properties of the entropy function stated in Lemma (2.14). (Hint: for part (2) of the lemma show that for fixed $\alpha \in \mathbb{R}_{>0}$ the function $f_\alpha(x) := (x + \alpha) \log(x + \alpha) - x \log x$ is monotonically increasing. Note that if $\alpha = d_1 + \ldots + d_{t-1}$, then $n\mathcal{H}(d) = f_\alpha(d_t) + C$, for some constant C.)

2.10. In this exercise we introduce the Huffman coding algorithm. The objects to be processed are quadruples (T, w, m, ℓ), where T is a regular binary tree with positive root weight w, ℓ is a sequence of positive integers describing an enumeration of the leaves of T, and $m \in \mathbb{N}$ is the weighted external path length of (T, ℓ) with respect to a suitable weight sequence. Besides obvious Delete and Insert operations on sequences of such quadruples, there is one further basic operation Connect. This operation maps a pair $((T_1, w_1, m_1, \ell_1), (T_2, w_2, m_2, \ell_2))$ of such quadruples to the quadruple (T, w, m, ℓ), where T is the tree with left and right subtree T_1 and T_2, respectively, $w = w_1 + w_2$, $m = m_1 + m_2 + w$, and ℓ is the concatenation of the sequences ℓ_1 and ℓ_2. The main subroutine of Huffman's algorithm is as follows. Given a list $((T_1, w_1, m_1, \ell_1), \ldots, (T_t, w_t, m_t, \ell_t))$ of $t > 1$ such quadruples with root weights $w_1 \leq \ldots \leq w_t$, this subroutine deletes the first two quadruples from that list, connects the deleted quadruples, and inserts the result of Connect into the list in such a way that the root weights of the trees are weakly increasing. (Compare with Example (2.20).) Now let $d_1 \leq \ldots \leq d_t$ be positive integers, $t > 1$. On input $(d_1, \ldots, d_t)$ the Huffman coding algorithm starts with the list $((T_1, d_1, 0, (1)), (T_2, d_2, 0, (2)), \ldots, (T_t, d_t, 0, (t)))$, where each T_i consists of exactly one node (=root) and runs the above subroutine $t - 1$ times. The final list contains only one entry (T, w, m, ℓ), say.

Prove that (T, ℓ) is a Huffman tree for $(d_1, \ldots, d_t)$ and $m = m(d_1, \ldots, d_t)$. (Interpret ℓ as a permutation π with $\ell = (\pi(1), \ldots, \pi(t))$.)

2.11. Prove Thm. (2.21).

2.12.* Let $a^{-1} = \sum_{i \geq 0} b_i X^i$ be the inverse of the generic polynomial $a = \sum_{i=0}^m a_i X^i$ in $A[[X]]$, where $A := k(a_0, \ldots, a_m)$. By $L(m, n)$ we denote the nonscalar complexity of computing $a^{-1} \bmod X^{n+1}$ from a. Prove the following statements.

(1) For $s \leq n + 1$ the nonscalar complexity of computing $a^{-1} \bmod X^{n+s+1}$ from $a^{-1} \bmod X^{n+1}$ and a is at most $L(m, n) + m + s + m - 1$.

(2) $L(m, n) \leq n + (2m - 1)\lceil \log n \rceil + 2$.

(Hint: first show that

$$\sum_{i=n+1}^{n+s} b_i X^i \equiv \left(1 - \sum_{i=0}^n b_i X^i \sum_{i=0}^m a_i X^i\right) \sum_{i=0}^{s-1} b_i X^i \bmod X^{n+s+1}.$$

Then apply Thm. (2.21) to the paranthesized expression.)

2.13. The Newton iteration of second order for a differentiable function $f: \mathbb{R} \to \mathbb{R}$ is defined by $\xi_{n+1} := \xi_n - \frac{f(\xi_n)}{f'(\xi_n)}$. This can be applied to invert a positive real number a.

(1) Show that the Newton iteration applied to $f := (\xi \mapsto \xi^{-1} - a)$ yields the following division-free iteration $\xi_{n+1} = 2\xi_n - \xi_n^2 a$.
(2) Show that for $\xi_0 \in \mathbb{R}_{>0}$ with $|1 - \xi_0 a| < 1$, the sequence $(\xi_n)_{n \geq 0}$ converges *quadratically* to a^{-1}, i.e., $|1/a - \xi_{n+1}| \leq c \cdot |1/a - \xi_n|^2$, for some constant c and all n.
(3) Generalize this to the following third order Newton iteration:

$$\xi_{n+1} := \xi_n - \frac{f(\xi_n)}{f'(\xi_n)} - \frac{1}{2} \frac{f''(\xi_n) \cdot f(\xi_n)^2}{f'(\xi_n)^3},$$

and compare your result with the division-free iteration used in the proof of Thm. (2.22).

2.14. Let $x = (x_0, x_1, x_2, \ldots)$ be a sequence of real numbers satisfying the recursion $x_s = a x_{s-1} + c b^s$, for some real numbers a, b, c. Show that for $s \geq 0$

$$x_s = \begin{cases} a^s x_0 + (a^s - b^s) \frac{bc}{a-b} & \text{if } a \neq b, \\ (cs + x_0) a^s & \text{if } a = b. \end{cases}$$

2.15. For $0 \neq a \in A = k[[X_1, \ldots, X_n]]$ let $v(a) := \min\{i \mid a_i \neq 0\}$ and define $|a| := 2^{-v(a)}$, $|0| := 0$. Prove the following.

(1) $|a| \geq 0$, $|a \cdot b| = |a| \cdot |b|$, $|a + b| \leq \max\{|a|, |b|\}$, for all $a, b \in A$.
    (Thus $| \ |: A \to \mathbb{R}_{\geq 0}$ is a valuation.)
(2) (A, d) is a complete metric space, where $d(a, b) := |a - b|$.
(3) Addition and multiplication are continuous operations in (A, d).
(4) $k[X_1, \ldots, X_n]$ is dense in A.
(5) For every sequence $(f_i)_{i \in \mathbb{N}}$ in A with $\lim_{i \to \infty} f_i = 0$ there exists $\sum_{i \geq 0} f_i := \lim_{m \to \infty} \sum_{i=0}^m f_i$.
(6) For $a \in A$ with $a_0 = 0$, the formal power series $\sum_{i \geq 0} a^i$ is well-defined and $(1 - a) \cdot \sum_{i \geq 0} a^i = 1$.

In the remaining exercises, A denotes a commutative algebra over a field k of characteristic zero.

2.16. Let $\text{ARCTG} := \sum_{n \geq 0} (-1)^n \frac{1}{2n+1} X^{2n+1} \in k[[X]]$, and $b \in A[[X]]$ with $b_0 = 0$. Prove that on input $b \mod X^{n+1}$ one can compute $\text{ARCTG} \circ b \mod X^{n+1}$ with $O(M(n))$ arithmetic operations in A.

2.17. The *integral operator* $I: k[[X]] \to k[[X]]$ is defined by

$$I(a) := \sum_{i \geq 0} \frac{a_i}{i+1} X^{i+1},$$

for $a = \sum_{i \geq 0} a_i X^i \in k[[X]]$. Prove the following statements.

(1) I is k-linear and $D(I(a)) = a$, for all $a \in k[[X]]$.
(2) For $f, g \in k[[X]]$ and $\lambda \in k$ the first-order linear differential equation $y' = y \cdot f + g$ has a unique solution $y = \sum_{i \geq 0} y_i X^i$ in $k[[X]]$ with $y_0 = \lambda$. Show that this solution is given by the formula

$$y = (\text{EXP}(I(f))) \cdot [\lambda + I(g \cdot \text{EXP}(-I(f)))].$$

2.18. Let $a, b \in k[[X]]$. If power series f, g are known such that $a \circ b = y$ is a solution to the first-order linear differential equation $y' = y \cdot f + g$, then on input $f \bmod X^{n+1}$ and $g \bmod X^{n+1}$ one can compute $a \circ b \bmod X^{n+1}$ with $O(M(n))$ operations.

2.19.

(1) Show that under composition the set of all power series $a = \sum a_j X^j \in A[[X]]$ with $a_0 = 0$ and $a_1 \in A^{\times}$ is a group G_A with unit element X. (The inverse of $a \in G_A$, denoted by $\text{rev}(a)$, is called the *reverse* of a.)
(2) Let $a = \sum_{i \geq 1} a_j X^j \in G_A$ and $\text{rev}(a) = \sum_{i \geq 1} b_j X^j$. Show that b_n is a rational expression in $a_1, \ldots, a_n$.
(3) The *reversion problem* for a power series $a \in G_A$ is to compute $\text{rev}(a) \bmod X^{n+1}$ from $a \bmod X^{n+1}$. Prove that on input $a \bmod X^{n+1}$ one can compute $\text{rev}(a) \bmod X^{n+1}$ with $O(M(n)\sqrt{n \log n})$ operations. (Hint: $\text{rev}(a)$ is the zero of the function $f_a := (G_A \ni b \mapsto a \circ b - X \in A[[X]])$. Now use Newton iteration to reduce the reversion problem to the composition problem.)

2.7 Open Problems

Problem 2.1. Is the total complexity of polynomial multiplication nonlinear?

Problem 2.2. What is the exact nonscalar complexity for squaring power series in characteristic different from two?

Problem 2.3. What is the exact nonscalar cost of the inversion of power series?

Problem 2.4. Can Brent and Kung's algorithm for the general composition problem be improved substantially?

2.8 Notes

The multiplication of polynomials via evaluation-interpolation, as used in the proofs of Thm. (2.2) and Cor. (2.3), goes back to Newton [390], Waring [544], Lagrange [312], and Gauss [193], among others. It was not until 1962 that a faster than the classical $O(n^2)$ algorithm for polynomial multiplication under the total cost measure appeared. This was an order $n^{1.57}$ algorithm due to Karatsuba and Ofman [296]. (In fact, Karatsuba and Ofman discussed the multiplication of integers. However, their algorithm can be readily adapted to the polynomial setting, see also Ex. 2.2.) Next, Toom [520] presented an $O(n^{1+\varepsilon})$-algorithm for the same problem. By 1966 there had been discovered two asymptotically faster algorithms (for integer multiplication), one due to Cook [123], the other to Schönhage [454]. For a comprehensive account of these algorithms see Knuth [304].

In their celebrated 1965 paper, Cooley and Tukey [127] (re-)discovered the FFT and obtained an algorithm of total cost $O(n \log n)$ for the multiplication of polynomials of degree at most n over the complex field. (Actually, the FFT has a long and involved history predating the Cooley-Tukey paper, and the basic idea can be traced back to Gauss, see Cooley et al. [126], Goldstine [199], and Heideman et al. [238].) A major breakthrough was the 1971 paper by Schönhage and Strassen [466]. The authors established an FFT-based algorithm for the multiplication of integers that readily translates to an $O(n \log n \log \log n)$ algorithm for polynomial multiplication over fields of characteristic $\neq 2$, see also Nussbaumer [394]. This algorithm is the content of Section 2.2. Schönhage [459] presented a variant of this which works in characteristic 2, see Ex. 2.6. Kaminski [293], as well as Cantor and Kaltofen [105] exhibited different methods solving this task in a similarly efficient way. The advantage of the Cantor-Kaltofen algorithm is that it works uniformly for polynomials over arbitrary, not necessarily commutative, not necessarily associative algebras with $O(n \log n)$ algebra multiplications and $O(n \log n \log \log n)$ additions/subtractions. The problem of polynomial multiplication over (small) finite fields is discussed in Chap. 18.

In [504], Strassen used Huffman coding [257] for multiplying several polynomials and proved by his degree bound that this technique is essentially optimal in the nonscalar model (see Chap. 8; for an upper bound see Thm. (2.19)). In the same paper, he introduced a variant of Huffman coding that works for the multiplication of several (noncommuting) matrices with polynomial entries, see Ex. 3.6. For further details on Huffman trees we refer the reader to Knuth [306]. Priority queues are an appropriate data structure to implement Huffman's algorithm, see, e.g., Kingston [302]. Cor. (2.16) is due to Horowitz [254].

Let L_n denote the nonscalar complexity of inverting a generic power series mod X^{n+1}. Sieveking [482] showed that $L_n \leq 5n - 2$. Kung [309] used Newton iteration techniques to improve this to $L_n \leq 4n - \log n$, and Schönhage observed that a closer look at Kung's proof even gives the results stated in Thm. (2.22), see Kalorkoti [280]. In the same paper, Kalorkoti investigated the problem of inverting a general polynomial of degree d mod X^{n+1}, see also Ex. 2.12. Cor. (2.26) is due to Strassen [497].

Brent [73] has designed fast algorithms for the composition of special power series like those in Cor. (2.30) and (2.32). Thm. (2.34) is due to Brent and Kung [75]. Ritzmann [440] proved that the problem of computing the composition mod $\prod_{i=0}^{n}(X - \varepsilon_i)$, where $\varepsilon_0, \ldots, \varepsilon_n$ are pairwise different elements in $k^{\times}$, has nonscalar complexity of order $n \log n$. Brent and Traub [76] studied the complexity of computing the iterated composition mod X^{n+1} of power series. Thm. (2.31) on Newton iteration in formal power series rings is a special case of Hensel and Newton methods in valuation rings, see von zur Gathen [183].

Ex. 2.6 is due to Schönhage [459]. Ex. 2.7 and Ex. 2.8 follow Strassen [497]. Ex. 2.16 is from Brent [73], and Ex. 2.18 and 2.19 follow Brent and Kung [75].

There is an extensive literature on fast *parallel* algorithms for polynomial arithmetic. We confine ourselves to mentioning von zur Gathen [184], Eberly [152], Ja'Ja [268], and Bini and Pan [51].

Chapter 3. Efficient Algorithms with Branching

The classical Euclidean algorithm typically amounts to order n^2 nonscalar operations to construct the greatest common divisor G and the continued fraction expansion $A_1/A_2 = Q_1 + 1/(Q_2 + 1/(Q_3 \ldots))$ of two univariate polynomials A_1, A_2 with $n = \deg A_1 \geq \deg A_2 \geq 1$. The first section discusses the Knuth-Schönhage algorithm [301, 451] that performs this task with only $O(n \log n)$ nonscalar operations. Strassen's local analysis [500], presented in the subsequent section, yields the improved upper bound $O(n(H + 1))$, where H is the entropy of the probability vector $n^{-1}(\deg G, \deg Q_1, \deg Q_2, \ldots)$. On the other hand, Strassen proved a lower bound of the same order of magnitude for uniformly almost all inputs, see Chap. 10. Thus under the nonscalar cost measure, the Knuth-Schönhage algorithm is uniformly most powerful for computing continued fraction expansions together with greatest common divisors. (For the total number of arithmetic operations one obtains the upper bound $O(n(H + 1) \log n)$, if the groundfield supports FFTs.) In Sect. 3.3 we combine efficient greatest common divisor computations and Huffman coding to solve algorithmic problems around the Chinese remainder theorem, such as evaluation and interpolation.

Sect. 3.4–3.5 have no connection with the first three sections of this chapter apart from the fact that we describe efficient algorithms with branching. We study the basic problem of computational geometry of locating a query point in an arrangement of hyperplanes in Euclidean space, and prove the existence of fast algorithms solving this problem. A consequence is the existence of "nonuniform polynomial time algorithms" for the real knapsack problem, a discovery which is due to Meyer auf der Heide [356]. We present a proof relying on the concepts of the Vapnik-Chervonenkis-dimension of a range space and of epsilon-nets. Haussler and Welzl's result [233] stating the existence of small epsilon-nets in range spaces of finite VC-dimension is here crucial.

3.1 Polynomial Greatest Common Divisors

Elements a, b of a unique factorization domain R are called *associated* iff $a = bu$ for some unit u in R. If $\mathbb{P}$ is a complete set of representatives of the classes of associated primes in R, then every nonzero $a \in R$ can be written uniquely as $a = \alpha \prod_{p \in \mathbb{P}} p^{\alpha_p}$ with $\alpha_p \in \mathbb{N}$ and α a unit in R. If $0 \neq b \in R$ has the prime factorization $b = \beta \prod_{p \in \mathbb{P}} p^{\beta_p}$, then $\mathrm{GCD}_{\mathbb{P}}(a, b) := \prod_{p \in \mathbb{P}} p^{\min(\alpha_p, \beta_p)}$ is *the greatest common divisor* of a and b with respect to $\mathbb{P}$. (By convention, the empty product is put to 1.) Let $\mathbb{P}'$ be another such complete set. As $\mathrm{GCD}_{\mathbb{P}}(a, b)$ and $\mathrm{GCD}_{\mathbb{P}'}(a, b)$ differ only by a unit in R, we shall write (after having fixed such $\mathbb{P}$) simply $\mathrm{GCD}(a, b)$. For $R = \mathbb{Z}$ we shall take as $\mathbb{P}$ the set of positive prime numbers, and for $R = K[X]$, K a field, $\mathbb{P}$ will be the set of all monic irreducible polynomials in $K[X]$. The above discussion shows that the GCD of a and b can be easily deduced from the corresponding prime factorizations. However, the factorization problem in unique factorization domains like $\mathbb{Z}$ and $K[X]$ seems to be much harder than the corresponding GCD problem. Fortunately, the classical Euclid algorithm solves the GCD problem for the integers and for univariate polynomials over a field in a rather efficient way. Nonetheless, as GCD computations occur so frequently (for examples see below), it is important to look for nearly optimal GCD algorithms.

After a short analysis of the classical Euclidean algorithm we shall present a GCD algorithm for $R = K[X]$, which will turn out to be essentially optimal. We will concentrate in this section on the problem of computing GCDs of univariate polynomials with coefficients from a field, but pointers to the literature, where GCD computations in other unique factorization domains are discussed, may be found in the notes to this section.

The efficient computation of GCDs of univariate polynomials is of fundamental importance for the solution of several algorithmic problems. Among these problems are the simplification of quotients of univariate polynomials, and the problem of squarefree as well as distinct-degree factorization of polynomials, see Ex. 3.3, 3.4, and 3.5. Moreover, the complete factorization of polynomials over finite fields via the Berlekamp algorithm or via the Cantor-Zassenhaus algorithm calls for a number of GCD computations. A further example is provided by the computation of inverses in simple algebraic field extensions (in particular inversion in finite fields). The close relationship between GCDs and the computation of Padé approximants is discussed in Ex. 3.8. Polynomial GCD computations play also a prominent role in symbolic integration, see, e.g., Chapters 11 and 12 in Geddes et al. [194]. To mention further applications we introduce some terminology, which will be used throughout this section.

Let $k \subset K$ denote a field extension. (In the context of the nonscalar cost measure, k-linear operations will be free of charge.) Let $A_1, A_2 \in K[X]$ be nonzero univariate polynomials over K with $n = \deg A_1 \geq m = \deg A_2$. The Euclidean algorithm consists of successively applying the division with remainder algorithm. This results in the *Euclidean remainder scheme*:

$$
\begin{aligned}
A_1 &= Q_1 A_2 + A_3, \\
A_2 &= Q_2 A_3 + A_4, \\
&\ \ \vdots \\
A_{t-2} &= Q_{t-2} A_{t-1} + A_t, \\
A_{t-1} &= Q_{t-1} A_t,
\end{aligned}
$$

where $A_i \neq 0$, $\deg A_i < \deg A_{i-1}$ for $3 \le i \le t$. The polynomial A_t is (up to a unit) the GCD of A_1 and A_2. The sequence $(Q_1, \dots, Q_{t-1})$ depends only on the rational function A_1/A_2 and constitutes the continued fraction expansion of A_1/A_2. In fact, we have

$$\frac{A_1}{A_2} \;=\; Q_1 + \cfrac{1}{Q_2 + \cfrac{1}{Q_3 + \cfrac{\ddots}{\quad + \cfrac{1}{Q_{t-1}}}}}$$

The extended sequence $(Q_1, \dots, Q_{t-1}, A_t)$ represents the pair (A_1, A_2) uniquely and is called its *Euclidean representation*. The sequence

$$(d_1, \dots, d_{t-1}, d_t) := (\deg Q_1, \dots, \deg Q_{t-1}, \deg A_t)$$

will be called the corresponding *degree pattern*.

Obviously, for $1 \le i \le t$, we have $\deg A_i = \sum_{j=i}^t d_j$.

(3.1) Remark. A sequence $(d_1, \dots, d_t)$ of integers is the degree pattern of the Euclidean representation of some pair of polynomials (A_1, A_2) with $\deg A_1 = n \ge m = \deg A_2$ iff $t \ge 2$, $d_1 \ge 0$, $d_t \ge 0$, $d_i \ge 1$ for all $2 \le i < t$, and $n = \sum_{j=1}^t d_j$ as well as $m = \sum_{j=2}^t d_j$. Such a sequence will sometimes be called a *degree pattern* of *type* (n, m). •

The Euclidean representation is rather informative. As we have seen, it contains the GCD and the continued fraction expansion of the given polynomials. Moreover, the resultant of A_1, A_2 can be written as a power product of A_t and the leading coefficients of the Q_i (cf. Ex. 3.11). When A_2 is the formal derivative of A_1, the discriminant of A_1 is obtained in this way. If $K = \mathbb{R}$, one can obtain from the Euclidean representation the number of zeros of A_1 in any real interval in linear time $O(n)$ by using Sturm's theorem (cf. Ex. 3.10).

In this section we are going to design an efficient algorithm to compute the coefficients of the Euclidean representation $(Q_1, \dots, Q_{t-1}, A_t)$ from those of A_1, A_2. Since the length of the output (number of coefficients of the Q_i and of A_t) depends on the input (A_1, A_2) even if we keep the degrees of the inputs fixed, it is clear that our model of computation has to be extended by allowing branching instructions. Thus we are now working with computation trees. (The only type of branching instruction we shall allow in this context are equality tests. For a formal specification of our problem we refer the reader to Chap. 4.)

Repeated application of Cor. (2.26) together with $t \le n+2$ gives the following result.

(3.2) Corollary. *(Analysis of the Euclidean algorithm) Let $A_1, A_2 \in K[X]$ be univariate polynomials with $n = \deg A_1 \ge \deg A_2 = m \ge 1$. Then the coefficients of the Euclidean representation of (A_1, A_2) with degree pattern $(d_1, \dots, d_t)$ can be computed with $O(\sum_{i=1}^{t-1} M(\sum_{j=i}^t d_j))$ operations. In the worst (and typical) case,* *when the degree pattern of type (n, m) is equal to $(n - m, 1, \ldots, 1, 0)$, the above upper bound is of size $O(M(n) + M(m) \cdot m)$. For $n = m$ this gives an upper bound of size $O(M(n)n)$.*

For a proof that $(n - m, 1, \ldots, 1, 0)$ is the "typical" degree pattern of type (n, m) see Ex. 3.1 (for the case of finite fields) and Ex. 5.11 (for infinite fields). It is worthwhile to mention that with respect to nonscalar cost the Euclidean algorithm is essentially optimal if one insists on computing the A_i in addition to the Q_i, see Prop. (10.9).

The rest of this section is concerned with an improved algorithm that computes the Euclidean representation with $O(M(n)(1 + \mathcal{H}(d_1, \ldots, d_t)))$ operations, which is of size $O(M(n) \cdot \log n)$ in the worst case. In Sect. 10.2 we will prove that this algorithm is essentially optimal with respect to nonscalar cost.

Throughout this section we will keep the following.

(3.3) Conventions.

(1) (A_1, A_2) denotes a pair of nonzero polynomials in $K[X]$ with $n = \deg A_1 \geq \deg A_2 = m \geq 0$. $(Q_1, \ldots, Q_{t-1}, A_t)$ is its Euclidean representation and $d = (d_1, \ldots, d_t)$ is the corresponding degree pattern; $A_i = Q_i A_{i+1} + A_{i+2}$, for $1 \leq i < t$, with $A_{t+1} = 0$, is the Euclidean remainder scheme of (A_1, A_2).
(2) (A'_1, A'_2) denotes another such pair with corresponding data n', m', t' as well as $(Q'_1, \ldots, Q'_{t'-1}, A'_{t'})$ and $d' = (d'_1, \ldots, d'_{t'})$.
(3) $\ell \geq 0$ is a fixed integer, if not otherwise specified. The triple (A_1, A_2, ℓ) uniquely determines an integer $s = s(A_1, A_2, \ell)$, $1 \leq s \leq t$, such that

$$\sum_{i < s} d_i \leq \ell \;\wedge\; \left(s = t \;\vee\; \sum_{i \leq s} d_i > \ell \right).$$

Obviously, s only depends on ℓ and the degree pattern $(d_1, \ldots, d_t)$ of (A_1, A_2). So we sometimes write $s = s(d_1, \ldots, d_t; \ell)$. Analogously, s' denotes the integer corresponding to (A'_1, A'_2, ℓ). ●

For a fixed degree pattern $d = (d_1, \ldots, d_t)$ and integers ℓ, ℓ' we have

$$\ell \leq \ell' \Rightarrow s(d; \ell) \leq s(d; \ell').$$

Recall that a pair of polynomials can be recovered from its Euclidean representation. So two different pairs will have different Euclidean representations. Nevertheless, it will turn out that the Euclidean representations corresponding to two different pairs of polynomials whose corresponding most significant coefficients are equal, coincide to a certain extent. In the sequel we will make this more precise.

Given a polynomial $A = \sum_{i=0}^{q} a_i X^i \in K[X]$ of degree $q \geq 0$ and an integer ℓ, we set

$$A|\ell := \pi(A \cdot X^{\ell - q}),$$

where $\pi : K[X, X^{-1}] \to K[X]$ is the K-linear map defined by $X^i \mapsto X^i$ for $i \geq 0$, and $X^i \mapsto 0$ for $i < 0$. Thus $A|\ell = \sum_{i=0}^{\min(\ell, q)} a_{q-i} X^{\ell-i}$. By convention, $0|\ell := 0$. Altogether we have

$$(3.4) \qquad A|\ell = \begin{cases} 0 & \text{if } \ell < 0 \text{ or } A = 0 \\ a_q X^\ell + \ldots + a_{q-\ell} & \text{if } 0 \le \ell \le q \\ a_q X^\ell + \ldots + a_0 X^{\ell-q} & \text{if } q < \ell. \end{cases}$$

Roughly speaking, $A|\ell$ describes the most significant part of length $\ell + 1$ of A normalized to degree ℓ. If $A \ne 0$ and $\ell \ge 0$, then $\deg(A|\ell) = \ell$. From the definition one easily deduces the following facts.

(3.5) Remark. Let $A, B \in K[X]$. Then for all $j \in \mathbb{N}$ the following holds.

(1) $(A \cdot X^j)|\ell = A|\ell$.
(2) $A|(\deg A + j) = A \cdot X^j$.
(3) $A|\ell = B|\ell$ implies $A|(\ell - j) = B|(\ell - j)$.
(4) $(A|\ell)|(\ell - j) = A|(\ell - j)$; in particular, $(A|\ell)|\ell = A|\ell$. •

For a fixed integer ℓ we define the relation *coincidence up to ℓ* on the set of all pairs of polynomials (A, B) with $\deg A \ge \deg B \ge 0$ as follows:

$$(A, B) \equiv_\ell (A', B')$$
$$:\Longleftrightarrow \begin{cases} A|\ell = A'|\ell \text{ and} \\ B|(\ell - (\deg A - \deg B)) = B'|(\ell - (\deg A' - \deg B')). \end{cases}$$

(3.6) Remark.

(1) Coincidence up to ℓ is an equivalence relation.
(2) $(A, B) \equiv_\ell (A \cdot X^j, B \cdot X^j)$, for all $j \ge 0$.
(3) If $(A, B) \equiv_\ell (A', B')$ and $\ell \ge \deg A - \deg B$, then $\deg A - \deg B = \deg A' - \deg B'$.
(4) If $(A, B) \equiv_\ell (A', B')$ and $j \ge 0$, then $(A, B) \equiv_{\ell-j} (A', B')$.
(5) $(A, B) \equiv_\ell (A|\ell, B|(\ell-d))$ for all integers ℓ, where $d := \deg A - \deg B \ge 0$. •

The following two fundamental lemmas compare the Euclidean representations of pairs of polynomials which coincide to a certain extent. These lemmas are the main ingredients of the Knuth-Schönhage GCD algorithm.

(3.7) Lemma. *Let $(A, B) \equiv_{2\lambda} (A', B')$ with $\lambda \ge \deg A - \deg B$. If $A = QB + C$ with $\deg C < \deg B$ and $A' = Q'B' + C'$ with $\deg C' < \deg B'$, then $Q = Q'$ and exactly one of the following three cases occurs:*

(1) $C = 0$.
(2) $C \ne 0$ and $\lambda - \deg Q < \deg B - \deg C$.
(3) $C \ne 0$ and $\lambda - \deg Q \ge \deg B - \deg C$. *Moreover, in this case, (B, C) and (B', C') coincide up to $2(\lambda - \deg Q)$.*

Proof. For convenience we use the convention $\deg 0 := -\infty$. By multiplying (A, B) and (A', B') with appropriate powers of X, we can assume w.l.o.g. that

$$\deg A = \deg A' > 2\lambda$$

(use Rem. (3.6)). Therefore, with Rem. (3.6)(3) we obtain

$$\deg(A - A') < \deg A - 2\lambda, \ \deg B = \deg B',$$

$$\deg(B - B') < \deg B - (2\lambda - (\deg A - \deg B)) = \deg A - 2\lambda.$$

From this, our assumption $2\lambda \geq \deg A - \deg B = \deg Q$, and

(A) $$A - A' = Q(B - B') + (Q - Q')B' + C - C'.$$

we see that the degrees of the polynomials $A - A'$, $Q(B - B')$ and $C - C'$ are all smaller than $\deg B$. Hence $\deg(Q - Q')B' < \deg B$, which implies $Q = Q'$. Moreover, we get from (A)

$$\deg(C - C') \leq \max\{\deg(A - A'), \deg(Q(B - B'))\}$$
(B) $$< \deg Q + \deg A - 2\lambda.$$

Now assume $C \neq 0$ and $\lambda - \deg Q \geq \deg B - \deg C$. Then (B) and $\lambda \geq \deg A - \deg B$ imply $\deg(C - C') < \deg C$, hence $\deg C = \deg C'$ (in particular $C' \neq 0$). Again, by (B), we get

$$\begin{aligned}
\deg C - \deg(C - C') &> \deg C - \deg Q - \deg A + 2\lambda \\
&= 2(\lambda - \deg Q) + \deg Q - \deg A + \deg C \\
&= 2(\lambda - \deg Q) - (\deg B - \deg C).
\end{aligned}$$

Together with $b := \deg B = \deg B' =: b'$, $c := \deg C = \deg C' =: c'$, this gives

(C) $$C|(2(\lambda - \deg Q) - (b - c)) = C'|(2(\lambda - \deg Q) - (b' - c')).$$

Finally, the assumption $(A, B) \equiv_{2\lambda} (A', B')$ implies

$$B|(2\lambda - (\deg A - \deg B)) = B'|(2\lambda - (\deg A' - \deg B')).$$

As $\deg A - \deg B = \deg Q = \deg Q' = \deg A' - \deg B'$, we get with Rem. (3.5)(3)

(D) $$B|(2(\lambda - \deg Q)) = B'|(2(\lambda - \deg Q)).$$

(C) and (D) state that (B, C) and (B', C') coincide up to $2(\lambda - \deg Q)$ which proves the lemma. $\qquad \square$

The last lemma is crucial for the following result.

(3.8) Lemma. *With Convention (3.3), in particular with $s = s(d; \ell)$ and $s' = s(d'; \ell)$, the following holds.*

$$(A_1, A_2) \equiv_{2\ell} (A'_1, A'_2) \ \Rightarrow \ s = s' \wedge \forall 1 \leq i < s : Q_i = Q'_i.$$

Proof. By induction on j, $1 \leq j \leq s$, we will show the following statements:

$(1)_j$: $j \leq s'$, and $Q_i = Q'_i$ for all $i < j$,

$(2)_j$: $j = s$ or (A_j, A_{j+1}) and (A'_j, A'_{j+1}) coincide up to $2(\ell - \sum_{i<j} d_i)$.

(Note that $(1)_s$ then yields $s \leq s'$ as well as $Q_i = Q'_i$ for all $i < s$, hence the lemma follows by symmetry.) The start $j = 1$ is easy: We have $s' \geq 1 = j$. Furthermore, there exists no integer i satisfying $1 \leq i < j = 1$. Hence $(1)_1$ is valid, and $(2)_1$ follows from our assumption $(A_1, A_2) \equiv_{2\ell} (A'_1, A'_2)$. Next we prove the induction step "$j \Rightarrow j+1$." Let $j < s$ and suppose that both $(1)_j$ and $(2)_j$ are valid. We apply the last lemma to $(A, B) := (A_j, A_{j+1})$, $(A', B') := (A'_j, A'_{j+1})$, and $\lambda := \ell - \sum_{i<j} d_i$. Then $Q = Q_j$, $Q' = Q'_j$, $C = A_{j+2}$, $C' = A'_{j+2}$. By $(1)_j$ we have $j \leq s'$. Now $j < s$ together with $(2)_j$ gives $(A_j, A_{j+1}) \equiv_{2\lambda} (A'_j, A'_{j+1})$. Since $j < s$ and $\sum_{i<s} d_i \leq \ell$ we have $\deg A_j - \deg A_{j+1} = \deg Q_j = d_j \leq \lambda$. Hence all assumptions of the last lemma are satisfied. Consequently, $Q_j = Q'_j$ and the fact that $(d_1, \ldots, d_j) = (d'_1, \ldots, d'_j)$ together with $d_j \leq \lambda$ yields $j + 1 \leq s'$. This proves $(1)_{j+1}$.

Next we show that $(2)_{j+1}$ is valid. According to the last lemma we consider several cases. If $C = 0$, i.e., $A_{j+2} = 0$, then $j + 1 = t \geq s > j$; thus $j + 1 = s$ and $(2)_{j+1}$ follows in this case. Now let $A_{j+2} \neq 0$. In the second case we have $\lambda - \deg Q < \deg B - \deg C$, i.e., $\lambda - d_j < \deg A_{j+1} - \deg A_{j+2} = d_{j+1}$. In other words, $\ell < \sum_{i \leq j+1} d_i$, hence $j + 1 = s$. Thus $(2)_{j+1}$ is valid in this case. In the last case we know that (A_{j+1}, A_{j+2}) and (A'_{j+1}, A'_{j+2}) coincide up to $2(\lambda - d_j) = 2(\ell - \sum_{i \leq j} d_i)$, and $(2)_{j+1}$ follows. This completes the proof. $\square$

It will be of help to rewrite the equations $A_j = Q_j A_{j+1} + A_{j+2}$ in terms of matrices. To this end let $A_{t+1} := 0$ and $M_i := \begin{pmatrix} 0 & 1 \\ 1 & -Q_i \end{pmatrix}$, for $1 \leq i < t$. Then

$$(3.9) \qquad \begin{pmatrix} A_j \\ A_{j+1} \end{pmatrix} = M_{j-1} \cdots M_1 \begin{pmatrix} A_1 \\ A_2 \end{pmatrix},$$

for all $1 \leq j \leq t$. (By convention, the empty product of matrices is the unit matrix.) An easy calculation shows that for $1 \leq a < b \leq t$ the entries in $M_b M_{b-1} \cdots M_{a+1} M_a$ are polynomials whose degrees satisfy (componentwise)

$$(3.10) \qquad \deg(M_b \cdots M_a) \leq \begin{pmatrix} \sum_{a<i<b} d_i & \sum_{a \leq i<b} d_i \\ \sum_{a<i \leq b} d_i & \sum_{a \leq i \leq b} d_i \end{pmatrix}.$$

While the Euclidean algorithm is based on the relations $\mathrm{GCD}(a, b) = \mathrm{GCD}(b, a)$, $\mathrm{GCD}(a, 0) = a$, and $\mathrm{GCD}(a, b) = \mathrm{GCD}(b, a \bmod b)$, the Knuth-Schönhage algorithm makes intimate use of certain properties of a function Γ, which is closely related to the Euclidean representation.

(3.11) Definition. The function Γ is defined for all triples (A, B, ℓ), where $\ell \in \mathbb{N}$ and $A, B \in K[X]$ are nonzero polynomials satisfying $\deg A \geq \deg B \geq 0$. The image $\Gamma(A, B, \ell)$ is the triple (z, Q, M), where $z = s(A, B; \ell)$, see Convention

(3.3), $Q = (Q_1, \ldots, Q_{z-1})$ is a prefix of the Euclidean representation of (A, B), and $M = M_{z-1} \cdots M_1$ with $M_i = \begin{pmatrix} 0 & 1 \\ 1 & -Q_i \end{pmatrix}$.   •

The next lemma summarizes crucial properties of the function Γ.

(3.12) Lemma. *Let Γ be defined on (A, B, ℓ), (A', B', ℓ), and (A_1, A_2, ℓ), and let $g := \deg A - \deg B$. Then the following holds.*

(1) $\Gamma(A, B, \ell) = \Gamma(A, B, \min(\ell, \deg A))$.

(2) *If $g = 0$, then* $\Gamma(A, B, 0) = (2, (\lfloor A/B \rfloor), \begin{pmatrix} 0 & 1 \\ 1 & -\lfloor A/B \rfloor \end{pmatrix})$.

(3) *If $0 \leq \ell < g$, then* $\Gamma(A, B, \ell) = (1, (), \begin{pmatrix} 1 & 0 \\ 0 & 1 \end{pmatrix})$.

(4) *If $(A, B) \equiv_{2\ell} (A', B')$, then $\Gamma(A, B, \ell) = \Gamma(A', B', \ell)$.*

(5) *If $\Gamma(A, B, \ell) = \Gamma(A', B', \ell)$, then $\Gamma(A, B, v) = \Gamma(A', B', v)$, for all $0 \leq v \leq \ell$.*

(6) *Let $\Gamma(A_1, A_2, \ell) = (z, Q, M)$ with $Q = (Q_1, \ldots, Q_{z-1})$ and $M = M_{z-1} \cdots M_1$. Let $0 \leq r \leq z-2$. If $\Gamma(A_{r+1}, A_{r+2}, \ell - \sum_{i=1}^{r} d_i) = (s, Q'', M'')$, then $s = z - r$, $Q'' = (Q_{r+1}, \ldots, Q_{z-1})$, and $M'' = M_{z-1} \cdots M_{r+1}$.*

Proof. (1), (2), and (3) follow easily from the definition of Γ, and (4) is a direct consequence of Lemma (3.8). To prove (5) use $\Gamma(A, B, \ell) = \Gamma(A', B', \ell)$ and the fact that $s(A, B; v) = s(A', B'; v)$, for all $0 \leq v \leq \ell$. Finally, (6) follows by observing that $(Q_{r+1}, \ldots, Q_{t-1}, A_t)$ is the Euclidean representation of (A_{r+1}, A_{r+2}) and $s := s(A_{r+1}, A_{r+2}; \ell - \sum_{i \leq r} d_i) = s(A_1, A_2; \ell) - r = z - r$.   □

Now we are well prepared to discuss the main subroutine of the Knuth-Schönhage GCD algorithm, which computes the function Γ. (As usual, when speaking of computing polynomials from polynomials, we always think of computing their coefficients from the coefficients of the given polynomials. Similar remarks apply to sequences or matrices built up from polynomials.) The efficient computation of Γ is based on a *Controlled Euclidean Descent*. As the computation of Γ is rather tricky and involved, we prefer to present this as a procedure *CED*, written down in pseudocode, see Fig. 3.1. The following theorem shows that the procedure *CED* correctly computes the function Γ, and gives a worst case upper bound for the number of operations used.

(3.13) Theorem (Knuth, Schönhage, Strassen). *Let $A, B \in K[X]$ with $n = \deg A \geq \deg B \geq 0$ and ℓ be a nonnegative integer. Then on input (A, B, ℓ) the procedure CED outputs $\Gamma(A, B, \ell) = (z, Q, M)$, where $z = s(A, B; \ell)$, $Q = (Q_1, \ldots, Q_{z-1})$ is a prefix of the Euclidean representation of (A, B), and $M = M_{z-1} \cdots M_1$ with $M_i = \begin{pmatrix} 0 & 1 \\ 1 & -Q_i \end{pmatrix}$. In the worst case this computation costs $O(M(\lambda) \log \lambda)$ operations, where $\lambda := \min(n, \ell)$.*

---

procedure $CED(A, B, \ell; z, Q, M)$

0    $\ell := \min(\ell, \deg A)$;

1    $d := \deg A - \deg B$;

2   **if** $\ell = d = 0$ **then**

3        $(z, Q, M) := (2, (\lfloor A/B \rfloor), \begin{pmatrix} 0 & 1 \\ 1 & -\lfloor A/B \rfloor \end{pmatrix})$;

4   **else**

5   **if** $\ell < d$ **then**

6            $(z, Q, M) := (1, (), \begin{pmatrix} 1 & 0 \\ 0 & 1 \end{pmatrix}))$;

7   **else**

8            $A_1 := A|2\ell$; $A_2 := B|(2\ell - d)$;

9            $u := \lfloor \ell/2 \rfloor$;

10           $CED(A_1, A_2, u; r, Q', M')$;

11           $\begin{pmatrix} C \\ D \end{pmatrix} := M'\begin{pmatrix} A_1 \\ A_2 \end{pmatrix}$;

12   **if** $D = 0$ **or** $(\deg C - \deg D + \sum_{i=1}^{r-1} \deg Q_i') > \ell$ **then**

13               $(z, Q, M) := (r, Q', M')$;

14   **else**

15               $Q' := \text{append}(Q', \lfloor C/D \rfloor)$;

16               $M' := \begin{pmatrix} 0 & 1 \\ 1 & -\lfloor C/D \rfloor \end{pmatrix} M'$;

17               $E := C \bmod D$;

18   **if** $E = 0$ **then**

19                   $(z, Q, M) := (r + 1, Q', M')$;

20   **else**

21                   $v := \ell - \sum_{i=1}^{r} \deg Q_i'$;

22                   $CED(D, E, v; s, Q'', M'')$;

23                   $(z, Q, M) := (r + s, \text{concat}(Q', Q''), M'' \cdot M')$;

24   **fi**;

25   **fi**;

26   **fi**;

27   **fi**;

  end.

Fig. 3.1. Procedure *CED*.

Proof. As by Lemma (3.12)(1) $\Gamma(A, B, \ell) = \Gamma(A, B, \min(n, \ell))$, we can replace ℓ by $\min(n, \ell)$. After this reduction is performed in line 0 of *CED*, it remains to prove the claim for all $0 \le \ell \le n$, and this will be done by induction on ℓ. Let $g := \deg A - \deg B$.

According to Lemma (3.12)(2) and (3), the start "$\ell = 0$" is correctly handled by lines 2 to 6 of *CED*.

Now let $\ell \geq 1$. The cases where $1 \leq \ell < g$ are correctly handled by lines 5 and 6, see Lemma (3.12)(3). Next let $0 \leq g \leq \ell \leq n$, $\ell \geq 1$. Then (A, B) and $(A_1, A_2) := (A|2\ell, B|(2\ell - g))$ coincide up to 2ℓ, and by Lemma (3.12)(4), $\Gamma(A, B, \ell) = \Gamma(A_1, A_2, \ell)$. Thus in the sequel we can work with (A_1, A_2) instead of (A, B). Let (A_1, A_2) have Euclidean representation $(Q_1, \ldots, Q_{t-1}, A_t)$ with degree sequence $(d_1, \ldots, d_t)$, and let $A_i = Q_i A_{i+1} + A_{i+2}$, for all $i < t$, $A_{t+1} = 0$. (The reader should note that $Q_1, \ldots, Q_{z-1}$ are also the first $z - 1$ quotients of the Euclidean representation of (A, B), as $\Gamma(A, B, \ell) = \Gamma(A_1, A_2, \ell)$.) As a first approximation to $\Gamma(A_1, A_2, \ell)$, line 10 computes by the induction hypothesis $\Gamma(A_1, A_2, \lfloor \ell/2 \rfloor) =: (r, Q', M')$. Thus $r = s(A_1, A_2; \lfloor \ell/2 \rfloor)$, $Q' = (Q_1, \ldots, Q_{r-1})$, and $M' = M_{r-1} \cdots M_1$.

Now one has to distinguish two cases: $r = z$ and $r < z$.

The case $r = z$ happens iff $A_{r+1} = 0$ or $\sum_{i=1}^{r} d_i > \ell$. (Note that $A_{r+1} = 0$ means $r = z = t$.) According to Eq. (3.9), the vector (C, D) computed in line 11 equals (A_r, A_{r+1}). This together with $d_r = \deg C - \deg D$ shows that line 12 performs the test "$r = z$?". If $r = z$, then line 13 returns the correct output.

If however $r < z$, line 15 appends $\lfloor C/D \rfloor = Q_r$ to Q', thus after line 15 the content of Q' is $(Q_1, \ldots, Q_r)$, and after line 16, $M' = M_r \cdots M_1$. After line 17, $E = A_{r+2}$. If $E = 0$, then $z = r+1$. In this case, line 19 returns the correct output. Finally, if $E \neq 0$, then $r+2 \leq z$. In particular, $\sum_{i=1}^{r-1} d_i \leq \lfloor \ell/2 \rfloor < \sum_{i=1}^{r} d_i$. In this case, $v := \ell - \sum_{i=1}^{r} \deg Q_i' = \ell - \sum_{i=1}^{r} d_i \leq \lfloor \ell/2 \rfloor$. By the induction hypothesis, line 22 computes $\Gamma(A_{r+1}, A_{r+2}, v) =: (s, Q'', M'')$, which by Lemma (3.12)(6) equals $(z - r, (Q_{r+1}, \ldots, Q_{z-1}), M_{z-1} \cdots M_{r+1})$. Thus line 23 outputs the correct triple in this case. As the procedure obviously terminates, the first claim of the theorem is proved.

For the worst case analysis let $T(\ell)$ denote the supremum of the number of operations on inputs (A, B, ℓ') with $\ell' \leq \ell$. Then $T(0) = O(1)$ regardless of how large $\deg A$ is, and a straightforward analysis of *CED* yields the recursion

$$T(\ell) = 2T(\lfloor \ell/2 \rfloor) + O(M(\ell)).$$

Hence $T(\ell) = O(M(\ell) \log \ell)$, see Ex. 2.1. As after line 0 we have $\ell \leq \deg A$, our last claim follows. $\qquad \square$

By Eq. (3.9), the first row of the matrix $M_{t-1} \cdots M_1$ consists of two polynomials a_1, a_2 satisfying $a_1 A_1 + a_2 A_2 = A_t$. By Eq. (3.10), we know that $\deg a_1 \leq \deg A_2 - \deg A_t - \deg Q_{t-1}$ and $\deg a_2 \leq \deg A_1 - \deg A_t - \deg Q_{t-1}$. Thus if $t \geq 3$, we have $\deg Q_{t-1} \geq 1$ and consequently

$$\deg a_1 < \deg A_2 - \deg A_t$$
$$\deg a_2 < \deg A_1 - \deg A_t.$$

In case $t \geq 3$, there is a unique pair (a_1, a_2) satisfying both $a_1 A_1 + a_2 A_2 = A_t$ and the last degree restrictions, see Ex. 3.2. The sequence $(Q_1, \ldots, Q_{t-1}, A_t, a_1, a_2)$

will be called the *extended Euclidean representation* of (A_1, A_2). The last theorem enables us to give the following upper complexity bound for the computation of the extended Euclidean representation.

(3.14) Corollary. *Let* $(A_1, A_2) \in K[X]^2$ *with* $n = \deg A_1 \geq \deg A_2 \geq 0$. *Then its extended Euclidean representation* $(Q_1, \ldots, Q_{t-1}, A_t, a_1, a_2)$ *is computable with* $O(M(n) \log n)$ *operations.*

Proof. The last theorem with $\ell = n$ shows that

$$(A_1, A_2) \mapsto (Q_1, \ldots, Q_{t-1}, M_{t-1} \cdots M_1)$$

is computable with $O(M(n) \log n)$ operations. As the first row of the matrix $M_{t-1} \cdots M_1$ contains a_1 and a_2, the computation of $A_t = a_1 A_1 + a_2 A_2$ costs by Thm.(2.9) $O(M(n))$ additional operations. $\square$

3.2* Local Analysis of the Knuth-Schönhage Algorithm

This section discusses Strassen's local analysis of the Knuth-Schönhage algorithm. The resulting upper bound not only depends on the maximum degree n of the input polynomials A_1, A_2, but also on the degree sequence $(d_1, \ldots, d_t)$ of the corresponding Euclidean representation $(Q_1, \ldots, Q_{t-1}, A_t)$. This will be complemented in Sect. 10.2 by a similar lower bound, which is also due to Strassen. Altogether we will obtain the striking result that the Knuth-Schönhage algorithm is optimal up to a constant factor with respect to nonscalar cost.

Let us have a look at Eq. (3.9). This equation indicates a rough analogy between the polynomial GCD problem and the problem of multiplying several polynomials as discussed in Sect. 2.3. Solving the GCD problem on the basis of Eq. (3.9) leads to two complications that were not present when we discussed the multiplication of several polynomials: first of all, one does not know the matrices M_j in advance, and secondly, matrix multiplication is not commutative. Nonetheless, Strassen was able to prove the following upper bound for the computation of the Euclidean representation that is surprisingly similar to the result on the multiplication of several polynomials. In the sequel, $D(d_1, \ldots, d_t)$ denotes the set of all pairs (A_1, A_2) of polynomials in $K[X]$ whose Euclidean representation has degree pattern $(d_1, \ldots, d_t)$.

(3.15) Theorem (Strassen). *Let* $(d_1, \ldots, d_t)$ *be a degree pattern of type* (n, m), *and* $(A, B, \ell) \in D(d_1, \ldots, d_t) \times \mathbb{N}$. *Then on input* (A, B, ℓ) *the procedure CED computes* $\Gamma(A, B, \ell) =: (z, Q, M)$ *with*

$$O(M(\ell)\big(1 + \mathcal{H}(d_1, \ldots, d_{z-1}, \ell - \textstyle\sum_{i<z} d_i)\big))$$

operations. For the number of nonscalar operations we have more precisely (with $\gamma = 30, \delta = 5.4, \varepsilon = 16$) *the upper bound*

$$\gamma \ell \left[\mathcal{H}(d_1, \ldots, d_{z-1}, \ell - \textstyle\sum_{i<z} d_i) + \delta \right] + \varepsilon(z-1) + 1.$$

Proof. We shall stick to the notation of Thm. (3.13) and its proof. As usual, we only give the proof for the claim concerning the nonscalar complexity. In the sequel, *time* refers to the number of nonscalar operations. Note that only lines 3, 10, 11, 15, 16, 17, 22, and 23 of the procedure *CED* give rise to nonscalar operations. Let $T(A, B, \ell)$ denote the minimum number of nonscalar operations sufficient to compute $\Gamma(A, B, \ell)$ with the procedure *CED*, and let t_i denote the minimum number of nonscalar operations sufficient to perform line i of *CED* on input (A, B, ℓ).

The proof of the theorem proceeds by induction on ℓ. Let $\ell = 0$. If $g := \deg A - \deg B = 0$, then $z = 2$ and $T(A, B, 0) \leq t_3 \leq 1$. If however $g > 0$, then $T(A, B, 0) = 0$. Altogether, this settles the start.

Now let $\ell \geq 1$. If $\ell < g$, then $T(A, B, \ell) = 0$. So it remains to consider the case $0 \leq g \leq \ell \leq n$. Here,

$$T(A, B, \ell) = \begin{cases} T(A_1, A_2, \lfloor \ell/2 \rfloor) + t_{11} =: T_1, & \text{if } r = z \\[2ex] T(A_1, A_2, \lfloor \ell/2 \rfloor) + t_{11} + t_{15} + t_{16} + t_{17} \\ \hspace{4cm} =: T_2, & \text{if } r+1 = z \\[2ex] T(A_1, A_2, \lfloor \ell/2 \rfloor) + t_{11} + t_{15} + t_{16} + t_{17} \\ \hspace{2cm} + T(D, E, v) + t_{23} =: T_3, & \text{if } r+2 \leq z. \end{cases}$$

Let us next estimate the t_i. We claim that

$$\begin{aligned} t_{11} &\leq 10\ell + 4 \\ t_{15} + t_{17} &\leq 6.75\ell + 1 \\ t_{16} &\leq 3\ell + 2 \\ t_{23} &\leq 7(\ell + 1). \end{aligned}$$

In fact, the entries in $M' = M_{r-1} \cdots M_1$ in line 11 are polynomials of degree at most $\sum_{i=1}^{r-1} d_i \leq \ell/2$, see Eq. (3.10), and A_1, A_2 are of degree at most 2ℓ. Hence, by Thm. (2.2), $t_{11} \leq 4[(\sum_{i=1}^{r-1} d_i) + 2\ell + 1] \leq 10\ell + 4$. To prove the second claim note that lines 15 and 17 are concerned with the division with remainder problem on the instance (A_r, A_{r+1}). According to Cor. (2.26) this can be solved in time

$$5.75(\deg A_r - \deg A_{r+1}) + \deg A_{r+1} + 1 = 5.75d_r + d_{r+1} + \ldots + d_t + 1 \leq 6.75\ell + 1.$$

(For the last inequality use $r < z$, $\sum_{i<z} d_i \leq \ell$, and $\sum_{i \leq t} d_i = \deg A_1 \leq 2\ell$.) To prove the estimate for t_{16} we use Eq. (3.10), apply Thm. (2.9) and obtain $t_{16} \leq 2(\deg \lfloor C/D \rfloor + \ell/2 + 1) = 2(d_r + \ell/2 + 1)$. As $d_r \leq \ell$, it follows that $t_{16} \leq 3\ell + 2$. To show the estimate on t_{23} note that $M'' = M_{z-1} \cdots M_{r+1}$ and $M' = M_r \cdots M_1$. Using Strassen's algorithm for the multiplication of 2×2 matrices, see Chap. 1, we get with Eq. (3.10) and Thm. (2.9) $t_{23} \leq 7(\ell + 1)$.

Now we can estimate T_1, T_2, and T_3. By Lemma (2.14)(2) and the induction hypothesis we obtain in the first case

$$T_1 \leq \gamma \ell/2 \left(\mathcal{H}(d_1, \ldots, d_{z-1}, \frac{\ell}{2} - \sum_{i<z} d_i) + \delta \right) + \varepsilon(z-1) + 10\ell + 5$$

$$\leq \gamma \ell \left(\mathcal{H}(d_1, \ldots, d_{z-1}, \ell - \sum_{i<z} d_i) + \delta \right) - \gamma \delta \frac{\ell}{2} + \varepsilon(z-1) + 10\ell + 5,$$

which easily implies our main claim for $\gamma = 30$ and $\delta = 5.4$. In the third case we obtain

$$T_3 \leq \gamma \ell/2 \left(\mathcal{H}(d_1, \ldots, d_{r-1}, \frac{\ell}{2} - \sum_{i<r} d_i) + \delta \right)$$
$$+ \varepsilon(r-1) + 1 + (10\ell + 4) + (6.75\ell + 1) + (3\ell + 2) + (7\ell + 7)$$
$$+ \gamma(\ell - \sum_{i \leq r} d_i) \left(\mathcal{H}(d_{r+1}, \ldots, d_{z-1}, \ell - \sum_{i<z} d_i) + \delta \right)$$
$$+ \varepsilon(z - 1 - r) + 1.$$

Let T_{31} and T_{32} denote the first and the remaining rows of the right-hand side of the last inequality, respectively. Then $T_3 \leq T_{31} + T_{32}$. With $\varepsilon = 16$ we obtain

$$T_{32} = \gamma(\ell - \sum_{i \leq r} d_i) \left(\mathcal{H}(d_{r+1}, \ldots, d_{z-1}, \ell - \sum_{i<z} d_i) + \delta \right) + \varepsilon(z-1) + 26.75\ell.$$

To estimate T_3 we choose $0 < \eta < 1/2$ and distinguish two cases.

Case $\sum_{i \leq r} d_i < (1 - \eta)\ell$.
As $\ell/2 - \sum_{i<r} d_i \leq d_r$ we get by Lemma (2.14)(2)

$$\frac{\ell}{2} \mathcal{H}(d_1, \ldots, d_{r-1}, \ell/2 - \sum_{i<r} d_i) \leq (\sum_{i \leq r} d_i) \mathcal{H}(d_1, \ldots, d_r).$$

Hence together with Lemma (2.14)(4) we obtain

$$T_3 \leq \gamma \cdot (\sum_{i \leq r} d_i)(\mathcal{H}(d_1, \ldots, d_r) + \delta) + T_{32}$$

$$\leq \gamma \cdot \ell(\mathcal{H}(d_1, \ldots, d_{z-1}, \ell - \sum_{i<z} d_i) + \delta) + \varepsilon(z-1) + 1 + \Lambda_1,$$

where

$$\Lambda_1 := \ell \cdot \left(26.75 - \gamma \cdot \mathcal{H}(\sum_{i \leq r} d_i, \ell - \sum_{i \leq r} d_i) \right) - 1.$$

As $\ell/2 < \sum_{i \leq r} d_i < (1 - \eta)\ell$ the convexity of the entropy function implies that $\mathcal{H}(\sum_{i \leq r} d_i, \ell - \sum_{i \leq r} d_i) \geq \mathcal{H}(\eta, 1 - \eta)$. Hence

$$\Lambda_1 \leq \ell \cdot (26.75 - \gamma \cdot \mathcal{H}(\eta, 1 - \eta)) - 1.$$

Thus in this case our main claim is a consequence of the condition that the right-hand side of the last inequality is ≤ 0. This in turn is satisfied if

$$(A) \qquad\qquad 26.75 \leq \gamma \cdot \mathcal{H}(\eta, 1 - \eta).$$

Case $\sum_{i \leq r} d_i \geq (1 - \eta)\ell$.
Here we have $\ell/2 < (1 - \eta)\ell \leq \sum_{i=1}^{r} d_i$, hence the first term T_{31} of T_3 can be estimated as follows (use $d_r \geq \ell/2 - \sum_{i < r} d_i$ and apply Lemma (2.14)(2))

$$
\begin{aligned}
T_{31} &\leq \gamma \cdot (\sum_{i \leq r} d_i)\,(\mathcal{H}(d_1, \ldots, d_r) + \delta) + \gamma \delta \frac{\ell}{2} - \gamma \delta (\sum_{i \leq r} d_i) \\
&\leq \gamma \cdot (\sum_{i \leq r} d_i)\,(\mathcal{H}(d_1, \ldots, d_r) + \delta) + \gamma \delta \ell (\eta - \frac{1}{2}) =: T'_{31}.
\end{aligned}
$$

Thus

$$
\begin{aligned}
T_3 &\leq T'_{31} + T_{32} \\
&\leq \gamma \cdot \ell \left(\mathcal{H}(d_1, \ldots, d_{z-1}, \ell - \sum_{i < z} d_i) + \delta \right) + \varepsilon(z - 1) + 1 + \Lambda_2,
\end{aligned}
$$

where

$$\Lambda_2 := \ell\,(26.75 + \gamma \delta(\eta - 0.5)) - 1.$$

Thus in this case our main claim is a consequence of the condition

$$(B) \qquad\qquad 26.75 \leq (0.5 - \eta)\gamma \delta.$$

Now both conditions (A) and (B) are satisfied for $\gamma = 30$, $\delta = 5.4$ and $\eta = 1/3$. Finally, the estimation of T_2, which is similar to that of T_3, is left to the reader. $\square$

The last theorem has an important corollary.

(3.16) Corollary. *Let* $d = (d_1, \ldots, d_t)$ *be a degree pattern,* $n := d_1 + \ldots + d_t$. *Then the function that assigns to every* $(A_1, A_2) \in D(d)$ *its extended Euclidean representation is computable with* $O(M(n)(1 + \mathcal{H}(d)))$ *operations. More precisely, the number of nonscalar operations is at most*

$$30 \cdot n \cdot (\mathcal{H}(d_1, \ldots, d_t) + 6.1).$$

Proof. Let $(Q_1, \ldots, Q_{t-1}, A_t, a_1, a_2)$ be the extended Euclidean representation of (A_1, A_2). The last theorem with $\ell = n$ shows that

$$D(d) \ni (A_1, A_2) \mapsto (Q_1, \ldots, Q_{t-1}, M_{t-1} \cdots M_1)$$

is computable with $30n \cdot (\mathcal{H}(d) + 5.4) + 16(t - 1) + 1$ nonscalar operations. By Eq. (3.9) and Thm. (2.9), $(A_1, A_2, M_{t-1} \cdots M_1) \mapsto A_t$ is computable with at most $4n$ nonscalar operations. Using $t - 1 \leq n + 1$, our last claim follows. The other claims can be shown in a similar way. $\square$

3.3 Evaluation and Interpolation

This section discusses algorithmic problems closely related to the well-known Chinese remainder theorem.

(3.17) Chinese Remainder Theorem. *Let R be a commutative ring, $I_1, \ldots, I_t$ ideals in R, which are pairwise coprime, i.e., $I_i + I_j = R$, for all $i \neq j$. Then the ring morphism $R \ni x \mapsto (x + I_1, \ldots, x + I_t) \in \prod_j R/I_j$ is surjective with kernel $I := \cap_{j=1}^t I_j$. This induces a ring isomorphism $R/I \to R/I_1 \times \ldots \times R/I_t$.*

Proof. See, e.g., Lang [316, p. 69].   □

In our discussion, $R = K[X]$ is the ring of univariate polynomials over a field K which is an extension field of k. As $K[X]$ is a principal ideal domain, every nonzero ideal in $K[X]$ is generated by a unique monic polynomial, and ideals $I_1 = (p_1), \ldots, I_t = (p_t)$ of $K[X]$ are pairwise coprime iff the polynomials $p_1, \ldots, p_t$ are pairwise coprime, i.e., $\mathrm{GCD}(p_i, p_j) = 1$, for all $i \neq j$. Identifying the algebra $K[X]/(p)$, p a polynomial of degree $n \geq 1$, with the algebra of all polynomials in $K[X]$ which are of degree strictly less than n, equipped with multiplication modulo p, we see that the mapping $K[X] \ni f \mapsto (f \bmod p_1, \ldots, f \bmod p_t)$ induces an isomorphism of K-algebras

$$\phi : K[X]/(p) \to K[X]/(p_1) \times \ldots \times K[X]/(p_t),$$

where $p := p_1 \cdots p_t$. These types of isomorphisms (and their analogues for $R = \mathbb{Z}$) are the basis of modular arithmetic. Of particular importance is the case when all $p_i = X - \xi_i$ are linear polynomials corresponding to pairwise different elements $\xi_1, \ldots, \xi_t$ in K. Here $t = n$, and the target algebra $\prod_j K[X]/(p_j)$ can be identified with the algebra K^n (equipped with componentwise multiplication), and the problem to compute ϕ is equivalent to the problem of evaluating a polynomial f of degree strictly less than n at the n points $\xi_1, \ldots, \xi_n$. Computing the inverse of ϕ reduces to the interpolation problem: given $\eta_1, \ldots, \eta_n \in K$, find the unique polynomial f of degree smaller than n in $K[X]$ such that $f(\xi_i) = \eta_i$, for all $1 \leq i \leq n$. In the special case when $\xi_i = \omega^{i-1}$, $\omega \in K$ a primitive nth root of unity, we have $p = \prod_{i=0}^{n-1}(X - \omega^i) = X^n - 1$ and $\phi: K[X]/(X^n - 1) \to K^n$ is the classical discrete Fourier transform. (For a detailed discussion of Fourier transforms the reader is referred to Chap. 13.)

Our aim is to show that in general both ϕ and its inverse can be computed efficiently. The algorithmic solutions of both problems involve Huffman coding. To avoid tedious repetitions we fix for the rest of this section the following terminology (compare Sect. 2.3).

(3.18) Convention. Let $p_1, \ldots, p_t$ denote pairwise coprime monic polynomials in $K[X]$ of positive degrees $d_1, \ldots, d_t$, respectively; $p := p_1 \cdots p_t$, and $n := \deg p$. (T, π) denotes a Huffman tree for $d = (d_1, \ldots, d_t)$, w.l.o.g. $\pi = \mathrm{id}$. $\mathcal{H} = \mathcal{H}(T, \pi; p_1, \ldots, p_t)$ is the set of all intermediate results obtained by computing $p_1 \cdots p_t$ via (T, π), whereas $\mathcal{H}_L$ and $\mathcal{H}_R$ denote the portions of $\mathcal{H}$ corresponding to the left subtree T_L and the right subtree T_R of T, respectively. (Thus if $p_L = p_1 \cdots p_\ell$ and $p_R = p_{\ell+1} \cdots p_t$ denote the polynomials computed at the roots of T_L and T_R, respectively, then $\mathcal{H} = \mathcal{H}_L \cup \mathcal{H}_R \cup \{p_L p_R\}$.)   •

An upper complexity bound for the computation of ϕ is given by the following theorem. We remark that an almost matching lower bound will be proven later on in Thm. (8.45).

(3.19) Theorem. *With the above convention, let u be a univariate polynomial over K such that $\deg u < n$. Then on input $(p_1, \ldots, p_t, u)$ one can compute $u \bmod q$ for all $q \in \mathcal{H}$ with $O(M(n)(1 + \mathcal{H}(d)))$ operations. The nonscalar cost is at most $7.75n(1 + \mathcal{H}(d)) - 10.5(t - 1)$.*

Proof. Let us prove the statement concerning the nonscalar cost. We first compute all elements in $\mathcal{H}$ with at most $m(d) - (t - 1)$ nonscalar operations, see Lemma (2.17). Let $\mathcal{L}(\mathcal{H}, u)$ denote the minimal number of nonscalar operations to compute $u \bmod q$ for all $q \in \mathcal{H}$ from $\mathcal{H}$ and u. Then it suffices to show that

$$\mathcal{L}(\mathcal{H}, u) \leq 6.75m(d) - 9.5(t - 1).$$

We are going to prove this by induction on t. As the case $t = 1$ is clear let us check the induction step. In order to do this we first compute $u \bmod p_L$ and $u \bmod p_R$. In the worst case, when $n - 1 = \deg u \geq \max\{\deg p_L, \deg p_R\}$, this can be performed according to Cor. (2.26) with at most

$$5.75(\deg u - \deg p_L) + \deg p_L + 1 + 5.75(\deg u - \deg p_R) + \deg p_R + 1 = 6.75n - 9.5$$

nonscalar operations. The crucial observation is now

$$u \bmod q = (u \bmod p_L) \bmod q,$$

for all $q \in \mathcal{H}_L$, and

$$u \bmod q = (u \bmod p_R) \bmod q,$$

for all $q \in \mathcal{H}_R$. Based on this we obtain by induction

$$
\begin{aligned}
\mathcal{L}(\mathcal{H}, u) &\leq \mathcal{L}(\mathcal{H}_L, u \bmod p_L) + \mathcal{L}(\mathcal{H}_R, u \bmod p_R) + 6.75n - 9.5 \\
&\leq 6.75m(d_L) - 9.5(\ell - 1) + 6.75m(d_R) - 9.5(t - \ell - 1) \\
&\quad + 6.75n - 9.5 \\
&= 6.75m(d) - 9.5(t - 1).
\end{aligned}
$$

(For the last equality use Lemma (2.17)(2).) This proves our claim concerning the nonscalar cost measure. Multiplying this upper complexity bound with $O(\log n)$ (resp. $O(\log n \log \log n)$), we obviously get an upper bound for the total number of arithmetic operations. This proves the theorem.   □

As an immediate consequence we get upper complexity bounds for the evaluation of a polynomial at many points.

(3.20) Corollary. *Let $\xi_1, \ldots, \xi_n \in K$ and $u \in K[X]$ with $\deg u < n$. Then one can compute $u(\xi_1), \ldots, u(\xi_n)$ from u and $\xi_1, \ldots, \xi_n$ with $O(M(n)\log n)$ operations. The nonscalar cost is at most $7.75n \log n - 2.75n + 10.5$.*

Proof. W.l.o.g. we may assume that $\xi_1, \ldots \xi_n$ are pairwise different. Now apply the last theorem to $p_i := X - \xi_i$ and note that $u \bmod p_i = u(\xi_i)$. $\qquad\square$

If $u = \sum_{j=0}^{n-1} u_j X^j$, then $(u(\xi_i))_{1 \le i \le n} = (\xi_i^{j-1})_{1 \le i,j \le n} \cdot (u_j)_{0 \le j < n}$. Hence the last corollary shows that an $n \times n$ Vandermonde matrix can be multiplied with a vector in $O(M(n)\log n)$ steps. In Chap. 13 the problem of multiplying a known matrix by an input vector is studied in detail.

Let us turn to the interpolation problem, i.e., the computation of ϕ^{-1}. As GCDs will be involved in the course of the computation, branching instructions will occur (in contrast to the above solution of the evaluation problem that could be formalized by a straight-line program). For a lower bound see Thm. (8.45).

(3.21) Theorem. *With Convention (3.18) let $u_1, \ldots, u_t \in K[X]$ such that $\deg u_i < d_i$ for all i. Then on input $p_1, \ldots, p_t, u_1, \ldots, u_t$, one can compute the unique $u \in K[X]$ with $\deg u < n$ satisfying $u \bmod p_i = u_i$, for all i, with*

$$O\left(M(n)(H+1) + \sum_{i=1}^{t} M(d_i)\log(d_i + 1)\right)$$

operations, where $H := \mathcal{H}(d)$. Furthermore, the nonscalar cost is at most $13.75n(H + 15.2) + 30\sum_i d_i \log(d_i + 1)$.

Proof. We prove the claim concerning the nonscalar cost. The other claims follow similarly. According to the *Lagrange interpolation formula* we have

$$(A) \qquad\qquad u = \sum_{i=1}^{t} u_i a_i b_i \bmod p,$$

where $a_i := p/p_i$, and $b_i := a_i^{-1} \bmod p_i$. (Note that a_i and p_i are coprime.) Define $\hat{p} := a_1 + \cdots + a_t$. As $GCD(a_i, p_i) = 1$ and p_i divides a_j for $i \ne j$, we have for all i

$$(B) \qquad\qquad e_i := a_i \bmod p_i = \hat{p} \bmod p_i.$$

The computation of u from $p_1, \ldots, p_t, u_1, \ldots, u_t$ proceeds in several steps.

Step 1. We compute $\mathcal{H} = \mathcal{H}(T, \pi; p_1, \ldots, p_t)$. By Lemma (2.17) this can be done with nonscalar cost at most $m(d) - (t - 1)$. In particular, we then know $p = p_1 \cdots p_t$.

Step 2. We compute $\hat{p}$ by a Huffman coding technique based on the formula:

(C) $\quad \hat{p} = \left(\sum_{i=1}^{\ell} \frac{p_1 \cdots p_\ell}{p_i} \right) \cdot \underbrace{p_{\ell+1} \cdots p_t}_{=p_R \in \mathcal{H}} + \underbrace{p_1 \cdots p_\ell}_{=p_L \in \mathcal{H}} \cdot \left(\sum_{i=\ell+1}^{t} \frac{p_{\ell+1} \cdots p_t}{p_i} \right).$

Let $\tau(\mathcal{H})$ denote the complexity of computing $\hat{p}$ from $\mathcal{H}$. As two polynomials of degree a and b whose leading coefficients belong to k can be multiplied with $a+b-1$ nonscalar operations, formula (C) gives the recursion $\tau(\mathcal{H}) \le \tau(\mathcal{H}_L) + \tau(\mathcal{H}_R) + 2(n-2)$. With Lemma (2.17)(2) an easy induction shows that $\tau(\mathcal{H}) \le 2m(d) - 2t$, for $t > 1$.

Step 3. We compute $e_i = \hat{p} \bmod p_i$, for all $i \in \underline{t}$. As $\deg \hat{p} < n$ this can be done with nonscalar cost at most $6.75m(D) - 9.5(n-1)$. (To see this recall the proof of Thm. (3.19) and note that we already know $\mathcal{H}$.)

Step 4. We compute $e_i^{-1} \bmod p_i = b_i$, for all $i \in \underline{t}$. As b_i is part of the extended Euclidean representation of (p_i, e_i), Cor. (3.16) shows that each b_i can be computed in at most $30d_i(\log d_i + 6.1)$ nonscalar steps. Altogether, the computation of $b_1, \ldots, b_t$ costs at most $30 \sum_{i=1}^{t} d_i(\log d_i + 6.1)$ nonscalar operations.

Step 5. We compute $\hat{u} := \sum_{i=1}^{t} u_i b_i a_i$ using again Huffman coding:

$$\hat{u} = \left(\sum_{i=1}^{\ell} u_i b_i \frac{p_1 \cdots p_\ell}{p_i} \right) \cdot \underbrace{p_{\ell+1} \cdots p_t}_{=p_R \in \mathcal{H}} + \underbrace{p_1 \cdots p_\ell}_{=p_L \in \mathcal{H}} \cdot \left(\sum_{i=\ell+1}^{t} u_i b_i \frac{p_{\ell+1} \cdots p_t}{p_i} \right).$$

Let $\sigma(\mathcal{H})$ denote the minimal number of nonscalar operations to compute $\hat{u}$ from $\mathcal{H} \cup \{u_1, \ldots, u_t, b_1, \ldots, b_t\}$. The last formula gives the recursion $\sigma(\mathcal{H}) \le \sigma(\mathcal{H}_L) + \sigma(P_R) + 3n$, and by Lemma (2.17)(2) we obtain $\sigma(\mathcal{H}) \le 3m(d)$.

Step 6. We compute $u = \hat{u} \bmod p$. As $\deg \hat{u} \le n + \delta - 2$, where $\delta := \max\{d_1, \ldots, d_t\} - 2$, this is possible in additional $5.75(\delta - 2) + n + 1$ nonscalar steps, see Cor. (2.26).

Altogether, the nonscalar cost is at most

$$12.75m(d) + 6.75n - 12.5t + 30 \sum_i d_i(\log d_i + 6.1).$$

Finally, use $m(d) \le n(H+1)$ to complete the proof of the theorem. $\qquad \square$

It is worthwhile to mention that the first four steps in the above proof only depend on $p_1, \ldots, p_t$. Thus for fixed $p_1, \ldots, p_t$ one can precompute the polynomials specified in Step 1 to Step 4. Thus for each input $u_1, \ldots, u_t$ only Step 5 and Step 6 are to be performed. This is a form of preconditioning, which will be discussed further in Chap. 5.

As a special case, the last theorem contains upper complexity bounds for the classical interpolation problem. A closer look at the proof yields in this special situation an improved bound for the nonscalar cost.

(3.22) Corollary (Horowitz). *Given $\xi, \eta \in K^n$, where $\xi_1, \ldots, \xi_n$ are pairwise different, one can compute the unique polynomial $u \in K[X]$ of degree strictly less* *than n satisfying $u(\xi_i) = \eta_i$, for all i, with $O(M(n)\log n)$ operations. The non-scalar cost is at most $10.75n\log n - 1.75n + 10.5$.*

Proof. Only the last claim remains to be proved. This is done in Ex. 3.13. $\square$

The Hermite interpolation problem is closely related to the computation of ϕ^{-1} when $p_1 = (X - \xi_1)^{d_1}, \ldots, p_t = (X - \xi_t)^{d_t}$ with pairwise different $\xi_1, \ldots, \xi_t \in K$. In this case one can avoid GCD computations which results in the improved upper bound $O(M(n)\mathcal{H}(d))$, see Ex. 3.14.

3.4* Fast Point Location in Arrangements of Hyperplanes

The topic of this section is a problem from computational geometry. To describe it, we introduce first some notation. By a hyperplane in $\mathbb{R}^n$ we will always understand an affine one, so it is not assumed that it contains the origin. A hyperplane H in $\mathbb{R}^n$ defines two open halfspaces H^-, H^+. By distinguishing one of them as the positive halfspace H^+ we give H an orientation. Let a finite set $\mathcal{H} = \{H_1, \ldots, H_m\}$ of hyperplanes in $\mathbb{R}^n$ be given and choose orientations for each H_i. Moreover set $H_i^0 := H_i$. Then we can assign to each point $\xi \in \mathbb{R}^n$ its *position vector* $\mathrm{pv}(\xi) \in \{-, 0, +\}^n$ with respect to $\mathcal{H}$ defined by $\mathrm{pv}_i(\xi) := \sigma$ iff $\xi \in H_i^\sigma$. We call two points in $\mathbb{R}^n$ equivalent iff they have the same position vector. The *arrangement* $\mathcal{A}(\mathcal{H})$ of $\mathcal{H}$ is defined as the partition of $\mathbb{R}^n$ into equivalence classes. We call the equivalence classes the *faces* of $\mathcal{A}(\mathcal{H})$. A *cell* of $\mathcal{A}(\mathcal{H})$ is a face F corresponding to a position vector in $\{-, +\}^n$ (this means that F has dimension n). (Cf. Fig. 3.2.)

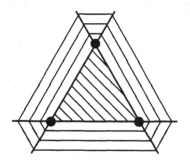

Fig. 3.2. Three lines in the plane, 19 faces, 7 cells.

For a fixed arrangement $\mathcal{A}(\mathcal{H})$ we are going to study the computational problem of finding the face in which a query point $\xi \in \mathbb{R}^n$ lies. In other words, our goal is to compute the position vector of a given ξ with respect to $\mathcal{H}$ as fast as possible.

Let H_i be the zeroset of an affine linear polynomial f_i in n indeterminates. Of course, the position vector, and thus the face in which an input ξ lies, can be found by simply evaluating all the f_i at ξ and determining the sign of $f_i(\xi)$. This trivial algorithm runs in time $O(mn)$. Our main result is the *existence* of an algorithm solving this problem in time $O(n^4 \log^2 n \log m)$. (We will not study the complexity of actually constructing it.) All the algorithms we are going to describe in this section take as input n real numbers, perform the four basic arithmetic operations, and branch according to tests for equality and inequality. (Note that n is fixed and not considered to be part of the input.) By the running time of the algorithm we understand the maximal number of arithmetic operations and tests to be performed on an input. These algorithms can be formalized as computation trees over $\mathbb{R}$, a notion which is defined in Sect. 4.4.

In the sequel we assume some familiarity with convex polytopes. For the reader's convenience we recall some definitions and facts (cf. Grünbaum [223]). A *polytope* $P \subset \mathbb{R}^n$ is a bounded subset which is the intersection of finitely many closed halfspaces. One can show that the polytopes are exactly the convex hulls of finitely many points. The *dimension* of P is the dimension of its affine hull (which is the smallest affine subspace of $\mathbb{R}^n$ containing P). A hyperplane H *supports* P iff P is contained in one of the closed halfspaces of H. (Note that we do not exclude the cases $P \subseteq H$ or $P \cap H = \emptyset$.) We say that a hyperplane H *cuts* P iff it does not support P. A polytope F is a *face* of P iff there is some hyperplane H supporting P such that $F = P \cap H$. A *vertex* of P is a zero-dimensional face of P. We denote by vert(P) the set of vertices of P. A *facet* of P is a face of dimension dim $P - 1$. The *relative interior* relint(P) of P is its interior in its affine hull. Note that if H suports P then either $P \subseteq H$, or relint(P) is contained in one of the open halfspaces of H. A *d-simplex* is a d-dimensional polytope with $d + 1$ vertices.

(3.23) Definition. A *triangulation* of a d-dimensional polytope P is a collection $\mathcal{T}$ of d-simplices such that

$$\bigcup_{S \in \mathcal{T}} S = P, \quad \bigcup_{S \in \mathcal{T}} \text{vert}(S) = \text{vert}(P)$$

and any nonempty $S \cap S'$ is a face of S and S', for all $S, S' \in \mathcal{T}$. (Cf. Fig. 3.3.) •

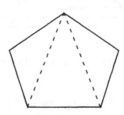

Fig. 3.3. A triangulated polytope.

A triangulation of P can be obtained by the following recursive procedure. If $\dim P \leq 1$, then $\{P\}$ is a triangulation of P. If $\dim P > 1$, then choose a vertex p of P as a reference point. For each facet F of P not containing p find a triangulation T_F of F by induction. Then the union T of all sets

$$\{\text{conv}(\{p\} \cup S) \mid S \in T_F\}$$

taken over all such facets F is a triangulation of P. (The proof of this intuitive fact is left to the reader; $\text{conv}(M)$ denotes the convex hull of M.)

(3.24) Lemma. *Let $P \subset \mathbb{R}^n$ be a polytope having m facets and T be a triangulation obtained by the above procedure. There is an algorithm which for a point $\xi \in P$ finds some simplex $S \in T$ containing ξ in time $O(mn^2)$.*

Proof. Let p be the reference point used in the triangulation T. We may w.l.o.g. assume that the input $\xi \in P$ is different from p. Let η be the rightmost point in P on the ray emanating from p in direction to ξ. Among all intersections of this ray with the affine hulls of the facets of P not containing p, the point η is closest to p. As the intersection of a ray with a hyperplane can be computed in time $O(n)$, we can compute η and find a facet F of P such that $p \notin F$ in time $O(mn)$. By induction, we can find a simplex S in the triangulation T_F of F such that $\eta \in S$. Then ξ lies in the simplex $\text{conv}(\{p\} \cup S) \in T$. As the recursion depth is n, the total running time of this algorithm is $O(mn^2)$. $\qquad\square$

In order to quickly locate a point ξ in a face of an arrangement $\mathcal{A}(\mathcal{H})$ we use a divide and conquer strategy which is based on the following crucial theorem.

(3.25) Theorem (Clarkson). *Let $\mathcal{H}$ be a finite set of affine hyperplanes in $\mathbb{R}^n$ and $0 < \varepsilon < 1$. Then there is a subset $\mathcal{N} \subseteq \mathcal{H}$ of cardinality $|\mathcal{N}| = O(n^2 \varepsilon^{-1} \log^2 n\varepsilon^{-1})$ such that for every simplex S in $\mathbb{R}^n$ the following is true: if the number of hyperplanes in $\mathcal{H}$ cutting S is strictly larger than $\varepsilon|\mathcal{H}|$, then there exists some hyperplane in $\mathcal{N}$ which cuts S. (An explicit bound is $|\mathcal{N}| \leq \lceil a \log a \rceil$, where $a := 32\varepsilon^{-1} n(n+1) \log 2n(n+1)$.)*

We will deduce this theorem in the next section from general principles. There we will also see that $\mathcal{N}$ can actually be constructed by a random sampling. Note that $|\mathcal{N}|$ is bounded by a polynomial in n and ε^{-1} which does not depend on the cardinality of $\mathcal{H}$.

We turn now to the informal description of our point location algorithm. Let $\mathcal{H}$ be a finite set of hyperplanes in $\mathbb{R}^n$ and ξ be a query point from which we first assume that it lies in the cube $[-1, 1]^n$. We denote by $\mathcal{C}$ the set of $2n$ bounding hyperplanes of $[-1, 1]^n$. Let $\mathcal{N} \subseteq \mathcal{H}$, $|\mathcal{N}| = O(n^2 \log^2 n)$, satisfy the conditions of Thm. (3.25) with $\varepsilon = 1/2$. In a first step we find the face F of the arrangement $\mathcal{A}(\mathcal{N} \cup \mathcal{C})$ containing ξ. This can be done in a trivial way in time $O(|\mathcal{N} \cup \mathcal{C}|n)$. The face F must be bounded, thus its closure $\overline{F}$ is a polytope of dimension $\leq n$. Note that every $H \in \mathcal{N} \cup \mathcal{C}$ supports $\overline{F}$. In a second step we find a simplex S containing ξ in a triangulation of $\overline{F}$ as described by Lemma (3.24). This can be done in time $O(|\mathcal{N} \cup \mathcal{C}|n^2)$. In a third step we find a face S' of S which contains ξ in its relative interior just by identifying the facets of S which contain ξ. This can be clearly done in time $O(n^2)$. Let

$$\mathcal{H}_{S'} := \{H \in \mathcal{H} \mid H \text{ cuts } S'\}.$$

Since every $H \in \mathcal{N}$ supports S', we must have $|\mathcal{H}_{S'}| \leq |\mathcal{H}| \cdot 2^{-1}$ by Thm. (3.25). Every $H \in \mathcal{H} \setminus \mathcal{H}_{S'}$ supports S', therefore either $S' \subseteq H$, or relint(S') is contained in one of the open halfspaces of H. This implies that all elements of relint(S') have the same position with respect to H. As this position can be determined in beforehand for all $H \in \mathcal{H} \setminus \mathcal{H}_{S'}$, it remains to compute the position vector of ξ with respect to the hyperplanes in $\mathcal{H}_{S'}$. That is, we need to find the face of the arrangement $\mathcal{A}(\mathcal{H}_{S'})$ which contains ξ. Now we proceed recursively until we arrive at an arrangement $\mathcal{A}(\tilde{\mathcal{H}})$ with $|\tilde{\mathcal{H}}| \leq 1$. Let m denote the cardinality of $\mathcal{H}$. As the recursion depth of this algorithm is $O(\log m)$, the total running time is $O(n^4 \log^2 n \log m)$.

If the query point $\xi = (\xi_1, \ldots, \xi_n)$ is not contained in the cube $[-1, 1]^n$ we do the following. Let H_i be the zeroset of $\alpha_{i0} + \alpha_{i1}X_1 + \ldots + \alpha_{in}X_n$. Set $\xi_0 := 1$ and find j such that $|\xi_j| = \max\{|\xi_i| \mid 0 \leq i \leq n\}$. For ease of notation we assume $j = n$. Compute $\tilde{\xi}_i := \xi_i/|\xi_n|$, $\varepsilon := \xi_n/|\xi_n| = \pm 1$, and locate the point $\tilde{\xi} := (\tilde{\xi}_0, \ldots, \tilde{\xi}_{n-1}) \in [-1, 1]^n$ in the arrangement of $\{\tilde{H}_1, \ldots, \tilde{H}_m\}$, where $\tilde{H}_i$ is the zeroset of $\alpha_{i0}X_0 + \ldots + \alpha_{in-1}X_{n-1} + \alpha_{in}\varepsilon$. Then we have clearly located ξ in the arrangement $\mathcal{H}$.

We have thus proved the following theorem.

(3.26) Theorem (Clarkson, Meiser). *Let $\mathcal{H}$ be a finite set of m affine hyperplanes in $\mathbb{R}^n$. There is an algorithm which finds the face of the arrangement $\mathcal{A}(\mathcal{H})$ containing a query point in time $O(n^4 \log^2 n \log m)$.*

We emphasize that the determination of a subset $\mathcal{N} \subseteq \mathcal{H}$ satisfying the conditions of Thm. (3.25) as well as the computation of triangulations of all bounded faces of the arrangement $\mathcal{A}(\mathcal{N} \cup \mathcal{C})$ may be thought of as a preprocessing which is needed to construct the above algorithm from a given $\mathcal{H}$. Therefore this does *not* contribute to the running time of the above algorithm. For a more explicit description of the above algorithm in terms of a suitable data structure, and a discussion of the preprocessing phase we refer to the Notes. We remark that in the terminology introduced in Sect. 4.4 the statement of the above theorem reads as follows: the total complexity $C^{\text{tot}}(\mathcal{A}(\mathcal{H}))$ of the partition $\mathcal{A}(\mathcal{H})$ of $\mathbb{R}^n$ into faces is of order of magnitude $O(n^4 \log^2 n \log m)$.

Let $\mathcal{H}$ be a finite set of hyperplanes defined by linear polynomials with rational coefficients, and let $\xi \in \mathbb{Q}^n$. We think now of a rational number as being represented by its nominator and denominator encoded in binary. Let us argue that the algorithm of Thm. (3.26) performs well in the bit-model. In fact, in this algorithm, arithmetic operations are only performed to determine the position of a point with respect to a hyperplane and to intersect a ray with a hyperplane. From this it easily follows that the binary length of all intermediate results is polynomially bounded in the binary length of the input ξ.

We are going to discuss some applications. The n-dimensional *real knapsack problem* is the problem to decide for a given point in $\mathbb{R}^n$ whether it lies in the union

$$KS_n := \left\{ x \in \mathbb{R}^n \mid \exists S \subseteq \underline{n} : \textstyle\sum_{i \in S} x_i = 1 \right\}$$

of 2^n affine hyperplanes. Its restriction to rational inputs in $\mathbb{Q}^n$ is called the n-dimensional *(classical) knapsack problem*. It is well-known that the knapsack problem is NP-complete. Therefore, it is very unlikely that this problem can be solved by a polynomial time algorithm (in the bit-model). However, the following is true.

(3.27) Corollary (Meyer auf der Heide). *There is an algorithm solving the real n-dimensional knapsack problem in time $O(n^5 \log^2 n)$.*

Proof. KS_n is the union of a set $\mathcal{H}$ of 2^n hyperplanes. Thus the complement $\mathbb{R}^n \setminus KS_n$ is the union of the cells of the arrangement $\mathcal{A}(\mathcal{H})$. The assertion follows immediately from Thm. (3.26). □

At first glance it might appear that this implies that the classical knapsack problem can be solved by a polynomial time algorithm. This is not the case for the following reason. For any n we have proved the existence of an algorithm A_n deciding for an input $\xi \in \mathbb{Q}^n$ whether it lies in $KS_n \cap \mathbb{Q}^n$, with running time bounded by a polynomial in n and the binary length of ξ. However, the parameter n is not considered as part of the input, and we do not say anything about how A_n can be quickly "computed" from n. (In the literature such a sequence $(A_n)_n$ is called a "nonuniform polynomial time algorithm.")

Before discussing two further applications we proceed with a general remark whose proof is left as an exercise to the reader.

(3.28) Remark. Let $f_1, \ldots, f_m$ be nonzero affine linear polynomials in n variables and let $\mathcal{B}$ denote the boolean algebra generated by the open halfspaces $H_i^+ := \{x \in \mathbb{R}^n \mid f_i(x) > 0\}$. (This means that a set in $\mathcal{B}$ is obtained from $H_1^+, \ldots, H_m^+$ by finitely many union, intersection and complement operations.) Then the sets in $\mathcal{B}$ are exactly the unions of faces of the arrangement $\mathcal{A}(\{f_1, \ldots, f_m\})$. •

In conjunction with Thm. (3.26), this remark implies that membership to any set in the above boolean algebra $\mathcal{B}$ can be tested by an algorithm in time $O(n^4 \log^2 n \log m)$.

The (n-dimensional) *real traveling salesman problem* is the following. There are given n towns, an $n \times n$-matrix of distances (x_{ij}) between all pairs of towns, and a number y. Is there a roundtrip visiting each town exactly once with total length at most y? Thus we need to decide membership of a given point of $\mathbb{R}^{\binom{n}{2}} \times \mathbb{R}$ to the subset

$$TS_n := \left\{ x_{ij} \geq 0 \mid 1 \leq i < j \leq n \right\} \cap \bigcup_{\pi \in S_n} \left\{ \sum_{i=1}^{n-1} x_{\pi(i)\pi(i+1)} + x_{\pi(n)\pi(1)} \leq y \right\}.$$

Let $\mathcal{H}$ be the set of hyperplanes corresponding to the inequalities occuring in the definition of TS_n. (Note that $|\mathcal{H}| = \binom{n}{2} + n!$.) By Rem. (3.28) the set TS_n is the union of some faces of the arrangement $\mathcal{A}(\mathcal{H})$. Thus membership to TS_n can be decided in time $O(n^9 \log^3 n)$.

The *integer linear programming problem with bounded solutions* is defined as follows. Let $m, n, t \in \mathbb{N}'$ and let m inequalities in n variables be given. Is there an integer vector $\xi \in \{0, \ldots, t\}^n$ which satisfies all these inequalities? This can be formulated as the problem to decide for a given point $(a, b) \in \mathbb{R}^{m \times n} \times \mathbb{R}^m$ whether it lies in the subset

$$\mathrm{IP}_{m,n,t} := \bigcup \left\{ (a, b) \in \mathbb{R}^{m \times n} \times \mathbb{R}^m \mid \forall i \in \underline{m} : \sum_{j=1}^n a_{ij}\xi_j \le b_i \right\}$$

where the union is over all $\xi \in \{0, 1, \ldots, t\}^n$. The set $\mathrm{IP}_{m,n,t}$ is by Rem. (3.28) the union of some faces of an arrangement of a set of $m(t+1)^n$ hyperplanes. Thus, by Thm. (3.26), membership to $\mathrm{IP}_{m,n,t}$ can be decided in time at most $O\big((\log m + n \log t)m^4 n^4 \log^2(mn)\big)$. Summarizing we have proved the following.

(3.29) Corollary (Meyer auf der Heide). *The n-dimensional real traveling salesman problem can be algorithmically solved in time $O(n^9 \log^3 n)$. The integer linear programming problem with m inequalities in n variables and solutions bounded by t can be algorithmically solved in time $c_t m^4 n^5 \log^3(mn)$ where the constant c_t may depend on t.*

3.5* Vapnik-Chervonenkis Dimension and Epsilon-Nets

We introduce the abstract concepts of range spaces, their VC-dimensions, epsilon-nets, and prove a general result stating the existence of small epsilon-nets in range spaces of finite VC-dimensions. The crucial Thm. (3.25) in the last section is then obtained as a consequence.

(3.30) Definition. A *range space* S is a pair (X, R) where X is a set and R a set of subsets of X. Elements of X are called *points*, and elements of R are called *ranges* of S. For a finite subset $A \subseteq X$ we define

$$\Pi_R(A) := \{A \cap r \mid r \in R\}.$$

We say that A *is shattered by R* iff every subset of A can be obtained by intersecting A with a range, that is, $\Pi_R(A) = 2^A$. The *Vapnik-Chervonenkis-dimension* (VC-dimension) $\mathrm{VCdim}(S)$ of S is the largest integer d such that there exists a subset $A \subseteq X$ of cardinality d that is shattered by R. If no such d exists, we say that the VC-dimension of S is infinite.   •

(3.31) Examples.

(1) The VC-dimension of (X, R) is zero iff $|R| \leq 1$.
(2) The range space having as points all real numbers and as ranges all closed rays unbounded to the right has VC-dimension one. If we allow also for rays which are unbounded to the left, then the resulting range space has VC-dimension two.
(3) The range space whose set of points is $\mathbb{R}^2$ and whose ranges are all closed halfplanes has VC-dimension three.
(4) Consider the range space $S = (\mathbb{R}^2, R)$ where R consists of all closed rectangles $[a, b] \times [c, d]$ which are parallel to the coordinate axis. It is easy to see that $\text{VCdim}(S) = 4$.
(5) The range space $(\mathbb{R}^2, R)$, where R is the set of all convex polygons, has infinite VC-dimension. (For example, there exist finite sets of every cardinality on the unit circle which are shattered by R.)   •

In a range space (X, R) of finite VC-dimension d the size of any $\Pi_R(A)$ can be explicitly bounded by a function depending only on d and the size of A. For $d, m \in \mathbb{N}$ we define $\Phi_d(m)$ as follows: $\Phi_d(0) := 1$ for all $d \geq 0$, $\Phi_0(m) := 1$ for all $m \geq 0$, and

$$\Phi_d(m) := \Phi_d(m - 1) + \Phi_{d-1}(m - 1) \text{ for } d, m \geq 1.$$

(3.32) Lemma (Vapnik and Chervonenkis). (1) $\Phi_d(m) = \sum_{i=0}^{d} \binom{m}{i}$ *if* $d < m$, *otherwise* $\Phi_d(m) = 2^m$.
(2) *The estimate* $\Phi_d(m) \leq m^d$ *is true for* $d, m \geq 2$.
(3) *Let* (X, R) *be a range space of finite VC-dimension* d. *Then for every finite subset* $A \subseteq X$ *of size* m *we have* $|\Pi_R(A)| \leq \Phi_d(m)$.

Proof. (1) This is an easy induction using the identity $\binom{m-1}{i-1} + \binom{m-1}{i} = \binom{m}{i}$. Statement (2) is straightforward to verify.

(3) Note that $(A, \Pi_R(A))$ is a range space of VC-dimension $\leq d$. It therefore suffices to prove the following statement: if (X, R) is a range space of VC-dimension d with $|X| =: m < \infty$, then $|R| \leq \Phi_d(m)$.

For $d = 0$ or $m = 0$ the assertion is trivially true. We proceed now by induction on $d + m$. Let $S = (X, R)$ be a range space of VC-dimension $d \geq 1$ with $|X| = m \geq 1$ and let $x \in X$. We define $R_1 := \{r \setminus \{x\} \mid r \in R\}$ and consider the surjective map $R \to R_1, r \mapsto r \setminus \{x\}$. Its fibers consist of one or two points, and exactly the elements of $R_2 := \{r \in R \mid x \notin r, r \cup \{x\} \in R\}$ have two point fibers. Therefore, $|R| = |R_1| + |R_2|$. We define now the two range spaces $S_i := (X \setminus \{x\}, R_i)$ for $i = 1, 2$. Obviously, $\text{VCdim}(S_1) \leq d$, hence by the induction hypothesis $|R_1| \leq \Phi_d(m - 1)$. Let A be a subset of $X \setminus \{x\}$ that is shattered by R_2. Then it is easy to see that $A \cup \{x\}$ is shattered by R. Therefore $\text{VCdim}(S_2) \leq d - 1$ and by the induction hypothesis $|R_2| \leq \Phi_{d-1}(m-1)$. Summarizing we obtain $|R| \leq \Phi_d(m - 1) + \Phi_{d-1}(m - 1) = \Phi_d(m)$ which proves the assertion.   □

The following range space $(\mathcal{H}_n^+, \mathcal{R}_n)$ will be important for us. Its points consist of the open halfspaces in $\mathbb{R}^n$ and its ranges are the sets

$$\xi^+ := \{\text{open halfspaces in } \mathbb{R}^n \text{ containing } \xi\},$$

where $\xi \in \mathbb{R}^n$. What is the Vapnik-Chervonenkis-dimension of this range space? Let $A = \{H_1^+, \ldots, H_m^+\}$ be a finite set of open halfspaces $H_i^+ = \{f_i > 0\} \subset \mathbb{R}^n$ defined by polynomials f_i of degree one. For a subset $I \subseteq \underline{m}$ let us call

$$D_I := \bigcap_{i \in I} \{f_i > 0\} \cap \bigcap_{i \notin I} \{f_i \leq 0\}$$

a *domain* of A if it is nonempty. The domains form a partition of $\mathbb{R}^n$ which is exactly the partition into fibers of the map $\mathbb{R}^n \ni \xi \mapsto \xi^+ \cap A$. Therefore, $|\Pi_{\mathcal{R}_n}(A)|$ equals the number of domains of A. In Ex. 3.19 we show that the number of domains of A is at most $\Phi_n(m)$. This implies VCdim$(\mathcal{H}_n^+, \mathcal{R}_n) \leq n$, as $\Phi_n(m) < 2^m$ for $n < m$. On the other hand, if we choose $H_i^+ := \{X_i > 0\}$ for $1 \leq i \leq n$, then the number of domains of $A = \{H_1^+, \ldots, H_n^+\}$ is clearly 2^n, hence A is shattered by $\mathcal{R}_n$. Thus we have proved the following.

(3.33) Lemma. *The range space $(\mathcal{H}_n^+, \mathcal{R}_n)$ has VC-dimension n.*

Starting from $(\mathcal{H}_n^+, \mathcal{R}_n)$ we can get estimates for the VC-dimension of more complicated range spaces.

(3.34) Lemma. *Consider the range space whose points are the affine hyperplanes in $\mathbb{R}^n$ and whose ranges are the sets*

$$\{H \subset \mathbb{R}^n \mid \text{hyperplane } H \text{ cuts the simplex } S\},$$

where S is a simplex in $\mathbb{R}^n$ (not necessarily n-dimensional). This range space has VC-dimension at most $4n(n+1)\log(2n(n+1)) = O(n^2 \log n)$.

Proof. W.l.o.g. $n \geq 2$. We modify the range space under consideration a little bit: let its points be *oriented* hyperplanes H in $\mathbb{R}^n$. H^+ denotes the positive open halfspace of H and H^- its negative open halfspace. Obviously, the VC-dimension of the range space is not affected by this modification. If $S \subset \mathbb{R}^n$ is the t-dimensional simplex with vertices $s_0, \ldots, s_t$, $t \leq n$, then we may express the range associated with S as

$$r_S := \{H \mid H \text{ cuts } S\} = \bigcup_{i \neq j} (\{H \mid s_i \in H^+\} \cap \{H \mid s_j \in H^-\}).$$

Observe that $s_i^+ = \{H \mid s_i \in H^+\}$ is just a range of $(\mathcal{H}_n^+, \mathcal{R}_n)$ if we identify oriented hyperplanes with their positive halfspaces.

Now let A be a finite set of $m > 1$ oriented hyperplanes in $\mathbb{R}^n$. We want to estimate the number of sets $A \cap r_S$, where S varies over all simplices in $\mathbb{R}^n$. By Lemma (3.33) the VC-dimension of $(\mathcal{H}_n^+, \mathcal{R}_n)$ equals n, hence the sets

$$\Pi_+ := \{\{H \in A \mid s \in H^+\} \mid s \in \mathbb{R}^n\}, \quad \Pi_- := \{\{H \in A \mid s \in H^-\} \mid s \in \mathbb{R}^n\}$$

both have cardinality at most $\Phi_n(m)$. Let $1 \le t \le n$ and S be a t-dimensional simplex. The set $A \cap r_S$ is obtained as follows: corresponding to the vertices of S we pick $t + 1$ sets in Π_+, pick $t + 1$ sets in Π_- and take a boolean combination of them. This implies the following rough estimate

$$|\{A \cap r_S \mid S \text{ simplex in } \mathbb{R}^n\}| \le \sum_{t=1}^{n} \Phi_n(m)^{2(t+1)} \le nm^{2n(n+1)}.$$

As $nm^{2n(n+1)} < 2^m$ for $m := \lceil 4n(n + 1) \log(2n(n + 1)) \rceil$, the VC-dimension of the range space under consideration is at most $4n(n + 1) \log(2n(n + 1))$. $\square$

We introduce now the concept of epsilon-nets.

(3.35) Definition. Let (X, R) be a range space, A a finite subset of X and let $0 < \varepsilon < 1$. We denote by $R_{A,\varepsilon}$ the set of all $r \in R$ which contain a fraction of the points in A of size greater than ε, that is,

$$R_{A,\varepsilon} := \{r \in R \mid |A \cap r|/|A| > \varepsilon\}.$$

An ε-*net* of A for the range space (X, R) is a subset $N \subseteq A$ containing a point in each $r \in R_{A,\varepsilon}$. $\bullet$

Consider for example the range space whose set of points is $\mathbb{R}^2$ and whose ranges are all closed halfplanes. Let A be a set of m points on a circle and $0 < \varepsilon < 1$. A subset $N \subseteq A$ which contains at least one point among any $\lfloor \varepsilon m \rfloor + 1$ consecutive points of A is an ε-net of A for this range space. Clearly, we can find such an N with $|N| \le 1 + \varepsilon^{-1}$. Note that the size of this ε-net does not depend on the size of A. The main result of this section states that this is not coincidental. A finite subset A of an arbitrary range space (X, R) of finite VC-dimension d contains for any $\varepsilon \in (0, 1)$ an ε-net whose size depends only on d and ε, but not on the size of A. We will prove this by the so-called probabilistic method in combinatorics: we show that a randomly selected subset N of A of sufficiently large size is an ε-net of A with high probability. The probabilistic method is a powerful and general technique to show the existence of combinatorial structures. A systematic exposition of this method with numerous applications can be found in Alon et al. [7].

(3.36) Theorem (Haussler and Welzl). *Let (X, R) be a range space of finite VC-dimension $d > 0$, $0 < \varepsilon, \delta < 1$, and let A be a finite subset of X. If N is the set of distinct elements obtained by*

$$m \ge \max\{4\varepsilon^{-1} \log 2\delta^{-1}, 8d\varepsilon^{-1} \log 8d\varepsilon^{-1}\}$$

independent random selections of an element of A (according to uniform distribution), then N is an ε-net of A for the range space (X, R) with probability at least $1 - \delta$.

In particular, there exist ε-nets N of A of size $|N| \le \lceil 8d\varepsilon^{-1} \log 8d\varepsilon^{-1} \rceil$.

(3.37) Example. The above theorem implies that for any finite set A of points in the plane $\mathbb{R}^2$ there is a subset N of A of size at most 26950 such that every closed halfplane containing at least 1% of the points of A contains at least one point in N. •

Proof of Thm. (3.36). Let the range space (X, R), the finite subset $A \subseteq X$, and $m \geq 1, 0 < \varepsilon < 1$ be fixed. Let Q be the set of all vectors in A^m whose elements $x = (x_1, \ldots, x_m)$ do not form an ε-net of A, that is,

$$Q := \{x \in A^m \mid \exists\, r \in R_{A,\varepsilon} : \{x_1, \ldots, x_m\} \cap r = \emptyset\}.$$

If we choose a vector in A^m uniformly at random, then it lies in Q with probability $\Pr(Q) := |Q|/|A|^m$. Our goal is to show that $\Pr(Q) \leq \delta$ if m is as large as required in the statement of the theorem.

For a range $r \in R_{A,\varepsilon}$ consider the random variable

$$U_r \colon A^m \to \mathbb{N}, \quad y \mapsto |\{i \mid y_i \in r\}|$$

which has a binomial distribution with parameters m and $p := |r \cap A|/|A| \geq \varepsilon$, so $\Pr(U_r = j) = \binom{m}{j} p^j (1 - p)^j$. It is well-known that the expectation $E(U_r)$ of U_r equals pm, while its variance satisfies $\operatorname{Var}(U_r) = p(1 - p)m$. Recall also Chebyshev's inequality which for an arbitrary random variable (with existing second moment) states that for a positive δ

$$\Pr(|U_r - E(U_r)| \geq \delta) \leq \delta^{-2} \operatorname{Var}(U_r)$$

(see for instance Feller [166, Chap. IX, §6]).

Claim 1. $\Pr(U_r \leq \varepsilon m/2) < 1/2$ provided $m \geq 8\varepsilon^{-1}$.

It is easily verified that this probability is a monotonically decreasing function of $p \in [\varepsilon, 1]$ and thus maximized for $p = \varepsilon$. If $p = \varepsilon$, the expectation of U_r equals εm and its variance is $\varepsilon(1 - \varepsilon)m$. Thus, by Chebyshev's inequality, we get

$$\Pr(U_r \leq \varepsilon m/2) \leq \left(\frac{2}{\varepsilon m}\right)^2 \varepsilon(1 - \varepsilon)m < \frac{4}{\varepsilon m} \leq 1/2$$

which proves Claim 1.

Consider now the following event

$$J := \{(x, y) \in A^m \times A^m \mid \exists\, r \in R : \{x_1, \ldots, x_m\} \cap r = \emptyset \text{ and } U_r(y) > \varepsilon m/2\}.$$

From Claim 1 we immediately get for $m \geq 8\varepsilon^{-1}$

$$\Pr(Q) \cdot \frac{1}{2} \leq \Pr(J).$$

It therefore suffices to bound $\Pr(J)$ from above.

The symmetric group S_{2m} acts on A^{2m} in the natural way. Set $M := \{(\pi', z') \in S_{2m} \times A^{2m} \mid \pi'z' \in J\}$ and define for $\pi \in S_{2m}, z \in A^{2m}$

$$M_z := \{\pi' \in S_{2m} \mid \pi' z \in J\}, \quad M^\pi := \{z' \in A^{2m} \mid \pi z' \in J\} = \pi^{-1}(J).$$

Then we obtain by changing the order of summation

$$|S_{2m}||J| = \sum_\pi |M^\pi| = |M| = \sum_z |M_z| \le |A^{2m}| \max_z |M_z|,$$

hence $\Pr(J) \le \max_z |M_z|/(2m)!$.

Claim 2. We have $|M_z|/(2m)! \le \Phi_d(2m)2^{-\varepsilon m/2}$ for any $z \in A^{2m}$.

Let $z = (z_1, \ldots, z_{2m}) \in A^{2m}$ be fixed and $S \subseteq A$ be the set of distinct components of z. We know by Lemma (3.32) that $\Pi_R(S) = \{S \cap r \mid r \in R\}$ has at most $\Phi_d(2m)$ elements. If $\pi \in M_z$, then there exists $r \in \Pi_R(S)$ such that $\pi z = (x, y) \in A^m \times A^m$ and $\{x_1, \ldots, x_m\} \cap r = \emptyset$, $U_r(y) > \varepsilon m/2$. Therefore, it suffices to show that for a fixed $r \in \Pi_R(S)$ the set

$$T := \{\pi \in S_{2m} \mid \pi z = (x, y), \ \{x_1, \ldots, x_m\} \cap r = \emptyset, \ U_r(y) > \varepsilon m/2\}$$

has at most $(2m)! \, 2^{-\varepsilon m/2}$ elements. Let $I := \{i \in \underline{2m} \mid z_i \in r\}$. We may assume that $\varepsilon m/2 < \ell := |I| \le m$, since otherwise T is empty. A permutation $\pi \in S_{2m}$ lies in T iff it maps I into the index set $\{m+1, \ldots, 2m\}$. Therefore, $|T| = \binom{m}{\ell}\ell!(2m-\ell)!$, and we obtain

$$\frac{1}{(2m)!}|T| = \frac{\binom{m}{\ell}}{\binom{2m}{\ell}} = \frac{m(m-1)\cdots(m-\ell+1)}{2m(2m-1)\cdots(2m-\ell+1)} \le 2^{-\ell} < 2^{-\varepsilon m/2}$$

which proves Claim 2.

Summarizing, we have shown that if $m \ge 8\varepsilon^{-1}$,

(A) $$\Pr(Q) \le 2\Phi_d(2m)2^{-\varepsilon m/2} \le 2(2m)^d 2^{-\varepsilon m/2},$$

the inequality on the right-hand side being valid for $d > 1$. A calculation shows that if $m \ge 8d\varepsilon^{-1} \log 8d\varepsilon^{-1}$, then $4d\varepsilon^{-1} \log 2m \le m$ (use the general inequality $a \log n \le n$ valid for $a \ge 4$, $n \ge 2a \log a$). If additionally $m \ge 4\varepsilon^{-1} \log 2\delta^{-1}$, then

$$4d\varepsilon^{-1} \log 2m \le 2m - m \le 2m - 4\varepsilon^{-1} \log 2\delta^{-1}.$$

This implies $\Pr(Q) \le \delta$ by (A) and proves the theorem. (Check the case $d = 1$ separately.) $\qquad\square$

If we apply this theorem to the range space of Lemma (3.34), then Thm. (3.25) in the previous section immediately follows.

3.6 Exercises

3.1. Let q be a prime power and $(d_1, \ldots, d_t)$ a degree pattern of type (n, m). Show that there are exactly $\prod_{i=1}^{t}(q^{d_i} \cdot (q-1)) = q^n(q-1)^t$ pairs (A_1, A_2) of polynomials over $\mathbb{F}_q$, whose Euclidean representation has this degree pattern. (Thus the most typical degree pattern of type (n, m) is the $(m+2)$-tuple $(n-m, 1, \ldots, 1, 0)$.)

3.2. Let $(d_1, \ldots, d_t)$ of type (n, m) be the degree pattern corresponding to the Euclidean representation of $(A_1, A_2) \in K[X]^2$. Prove that in case $t \geq 3$ there is exactly one pair of polynomials (a_1, a_2) with $\deg a_1 < \deg A_2 - d_t$ and $\deg a_2 < \deg A_1 - d_t$ satisfying $a_1 A_1 + a_2 A_2 = \text{GCD}(A_1, A_2)$.

3.3. A monic polynomial $a \in K[X]$ is called *squarefree* iff there is no polynomial $b \in K[X]$ of degree at least 1 such that b^2 divides a.

(1) Prove that every monic $a \in K[X]$ of degree n has a unique factorization $a = \prod_{i=1}^{n} a_i^i$, with pairwise coprime, monic, squarefree polynomials $a_1, \ldots, a_n$. (Typically, most of the a_i are equal to 1.) This factorization is called the *squarefree factorization* of a.

(2) With the above notation show that $g := \prod_{i=2}^{n} a_i^{i-1}$ is a common divisor of a and its formal derivative a'; furthermore, $a/g = a_1 \cdots a_n$. In particular, if $\text{GCD}(a, a') = 1$, then a is squarefree.
For the remaining parts of this exercise we assume that char $K = 0$.

(3) With the above notation we have $g = \text{GCD}(a, a')$, in particular, a is squarefree iff $\text{GCD}(a, a') = 1$.

(4) Put $w_1 := a/g$, $y_1 := a'/g$, and $z_1 := y_1 - w_1'$. For $2 \leq i \leq n$ let $w_i := w_{i-1}/a_{i-1}$, $y_i := z_{i-1}/a_{i-1}$, and $z_i := y_i - w_i'$. With $A_j := \prod_{i \neq j} a_i$ show that for all $1 \leq i < n$ the following identities hold:

$$w_i = \prod_{j=i}^{n} a_j$$

$$y_i = \sum_{j=i}^{n}(j+1-i)a_j' \frac{A_j}{a_1 \cdots a_{i-1}}$$

$$z_i = \sum_{j=i+1}^{n}(j-i)a_j' \frac{A_j}{a_1 \cdots a_{i-1}}$$

$$a_i = \text{GCD}(w_i, z_i).$$

Furthermore, $w_n = a_n$.

(5) On the basis of (4) design and analyze an algorithm that computes the coefficients of $a_1, \ldots, a_n$ from those of a.

3.4.* This exercise discusses the squarefree factorization of monic polynomials a of degree n over $\mathbb{F}_q$, where $q = p^m$, p prime.

(1) If $a' = 0$, then $a = b^p$, for some polynomial $b \in \mathbb{F}_q[X]$. Design an algorithm to compute the coefficients of b from those of a.
(2) On the basis of the last exercise and (1) design an efficient algorithm that computes the coefficients of $a_1, \ldots, a_n$ from those of a.

3.5. In this exercise we study the *distinct degree (partial) factorization* of square-free polynomials $a \in \mathbb{F}_q[X]$, where $q = p^m$. This is a factorization $a = \prod_i \alpha_i$, where α_i is the product of all monic irreducible polynomials of degree i that divide a.

(1) Let $\mathrm{Irr}(q, d)$ denote the set of all monic irreducible polynomials of degree d in $\mathbb{F}_q[X]$. Prove that for all positive integers r

$$X^{q^r} - X = \prod_{d \mid r} \prod_{b \in \mathrm{Irr}(q,d)} b.$$

(2) Let $b_1 := a$, and $g_1 := \mathrm{GCD}(b_1, X^q - X)$. For $i \geq 2$ let $b_i := b_{i-1}/g_{i-1}$ and $g_i := \mathrm{GCD}(b_i, X^{q^i} - X)$. Prove that $\alpha_i = g_i$, for all $i \geq 1$.
(3) On the basis of (2) design and analyze an algorithm that computes the distinct degree factorization of squarefree polynomials over $\mathbb{F}_q$.

3.6. The goal of this exercise is to describe a fast algorithm for converting the Euclidean representation $(Q_1, \ldots, Q_{t-1}, A_t)$ with degree pattern $d = (d_1, \ldots, d_t)$ of type (n, m) into the corresponding rational fraction represented by the pair $(A_1, A_2) \in K[X]^2$. (The optimality of this algorithm with respect to nonscalar cost will be proved in Ex. 10.1.)

(1) For $1 \leq i < t$ let $G_i := \begin{pmatrix} Q_i & 1 \\ 1 & 0 \end{pmatrix}$ and $G_t := \begin{pmatrix} A_t & 0 \\ 0 & 0 \end{pmatrix}$. Prove that for all $1 \leq i < t$
$$\begin{pmatrix} A_i & 0 \\ A_{i+1} & 0 \end{pmatrix} = G_i G_{i+1} \cdots G_t.$$
(2) Prove that one can compute $G_1 \cdots G_t$ with $O(M(n)(H+1))$ operations from $G_1, \ldots, G_t$, where $H := \mathcal{H}(d)$. The nonscalar cost is at most $8n(H+5) + 7(t-1)$. (Hints: prove the last claim by induction on t. Let $s \in \underline{t}$ with $pn := \sum_{i=1}^{s-1} d_i \leq n/2 < \sum_{i=1}^{s} d_i =: p'n$, and put $\varepsilon := 0.325$. Use Lemma (2.14) and Strassen's algorithm for the multiplication of 2×2 matrices (see Chap. 1) and consider the following three cases.
Case $p' \leq 1 - \varepsilon$: Compute $A := G_1 \cdots G_s$, $B := G_{s+1} \cdots G_t$, and finally $A \cdot B$.
Case $p' > 1 - \varepsilon$, $p \geq \varepsilon$: Compute $A := G_1 \cdots G_{s-1}$, $B := G_s \cdots G_t$, and finally $A \cdot B$.
Case $p' > 1 - \varepsilon$, $p < \varepsilon$: Compute $A := G_1 \cdots G_{s-1}$, $B := G_{s+1} \cdots G_t$, and finally $A \cdot G_s \cdot B$.)

3.7. This exercise prepares our subsequent discussion of Padé approximation.

Let $(A_1, A_2) \in K[X]^2$ have Euclidean remainder scheme $A_i = Q_i A_{i+1} + A_{i+2}$, for $i \in \underline{t}$, and $A_{t+1} := 0$, with corresponding degree pattern $(d_1, \ldots, d_t)$ of type (n, m), $n \geq m \geq 1$. Then, for $i \in \underline{t}$

$$\begin{pmatrix} S_i & T_i \\ S_{i+1} & T_{i+1} \end{pmatrix} := M_{i-1} \cdots M_1,$$

where $M_j := \begin{pmatrix} 0 & 1 \\ 1 & -Q_j \end{pmatrix}$, is well-defined. Prove the following:

(1) For all $i \in \underline{t}$ we have

$$\begin{pmatrix} A_1 \\ A_2 \end{pmatrix} = (-1)^{i-1} \begin{pmatrix} T_{i+1} & -T_i \\ -S_{i+1} & S_i \end{pmatrix} \begin{pmatrix} A_i \\ A_{i+1} \end{pmatrix}.$$

(2) For all $i \in \underline{t}$ we have $\deg T_{i+1} + \deg A_i = n$.
(3) Let μ, ν be non-negative integers with $\mu + \nu = n - 1$ and $\mu \geq d_t$. Then there exists a unique $j \in \underline{t}$ with $\deg A_j \leq \mu$ and $\deg T_j \leq \nu$.
(4) Suppose that there exists some j, $2 \leq j \leq t$, and polynomials $T, S, G \in K[X]$ such that $\deg T < \deg A_1 - \deg A_j$, $\deg G < \deg A_{j-1}$, and $SA_1 + TA_2 = G$. Show that there exists $f \in K[X]$ such that $S = fS_j$, $T = fT_j$, and $G = fA_j$. (Hint: use induction on $\deg G$. For the induction step choose ℓ such that $\deg A_\ell \leq \deg G < \deg A_{\ell-1}$, and show that $S = gS_\ell$, $T = gT_\ell$, and $G = gA_\ell$ for some $g \in K[X]$. In the last step prove that $\ell = j$.)

3.8. This exercise discusses rational function approximations to formal power series. Given a power series $a = \sum_{j \geq 0} a_j X^j \in K[[X]]$, $a_0 \neq 0$, and non-negative integers μ, ν, a *Padé approximant* of type (μ, ν) to a is a rational function p/q represented by a pair (p, q) of nonzero polynomials satisfying $\deg p \leq \mu$, $\deg q \leq \nu$, and $qa - p = X^{\mu+\nu+1}b$ for some power series b.

(1) Show that Padé approximants of type (μ, ν) to $a \in K[[X]]$ with $a_0 \neq 0$ always exist. Moreover, if (p, q) and (P, Q) represent two Padé approximants of type (μ, ν) to a, then $p/q = P/Q$.
(2) Let $n := \mu + \nu + 1$. Prove that the Padé approximant of type (μ, ν) to a can be computed by applying a modified extended GCD algorithm to $A_1 := X^n$ and $A_2 := \sum_{j<n} a_j X^j$. (Hint: use the last exercise.)

3.9. A sequence $a = (a_0, a_1, \ldots) \in K^{\mathbb{N}}$ is called *linearly generated* iff there exists some nonzero $f = f_0 + f_1 X + \cdots + f_n X^n \in K[X]$ such that for all $j \geq 0$ we have $\sum_{i=0}^n f_i a_{i+j} = 0$; f is called a *generating polynomial* of a. A generating polynomial of least degree is called a *minimum polynomial* of a. We associate to a the power series $A := \sum_{i=0}^{\infty} a_i X^i \in K[[X]]$.

(1) Let $f = \sum_{i=0}^n f_i X^i \in K[X]$ be a generating polynomial of a with $n = \deg f$, and let $G := \mathrm{rec}(f) := X^n f(1/X) = \sum_i f_i X^{n-i}$ denote the reciprocal of f. Show that $H := GA$ is a polynomial of degree at most $n - 1$.

(2) Let $N \geq n$, $A_1 := X^{2N}$, $A_2 := \sum_{i=0}^{2N-1} a_i X^i$. Let T_j, S_j be as defined in Ex. 3.7 and j be such that $\deg A_j < N \leq \deg A_{j-1}$. Show that there exists $\alpha \in k$ such that $G := \alpha S_j$. Hence there exists $\nu \in \mathbb{N}$ such that $f = \alpha X^\nu \mathrm{rec}(S_j)$. (Hint: first show that G and the polynomial H from (1) are coprime, then use Ex. 3.7(4).)

(3) Given $(a_0, \ldots, a_{2n-1})$, design an algorithm of cost $O(M(n) \log n)$ that computes a polynomial f of degree at most n satisfying $\sum_{i=0}^n f_i a_{i+j} = 0$ for $0 \leq j \leq n-1$.

3.10. (Sturm's algorithm) This exercise discusses the problem of computing the number of roots of a squarefree polynomial $u \in \mathbb{R}[X]$ in a real half-open interval $(a, b]$. As u is squarefree, we have $\mathrm{GCD}(u, u') = 1$, see Ex. 3.3. Sturm suggested to consider the polynomial remainder sequence corresponding to $A_1 = u$ and $A_2 = u'$, but with negated remainders:

$$
\begin{aligned}
A_1 &= Q_1 A_2 - A_3 \\
A_2 &= Q_2 A_3 - A_4
\end{aligned}
$$

$$\cdots$$

$$
\begin{aligned}
A_{t-2} &= Q_{t-2} A_{t-1} - A_t \\
A_{t-1} &= Q_{t-1} A_t.
\end{aligned}
$$

As $\mathrm{GCD}(u, u') = 1$, we have $\deg A_t = 0$. The *variation* $V(u, a)$ of u at $a \in \mathbb{R}$ is the number of changes of sign in the sequence $A_1(a), \ldots, A_t(a)$, not counting zeros.

(1) Prove Sturm's theorem stating that the number of real roots of u in the interval $(a, b]$ is equal to $V(u, a) - V(u, b)$. (Hint: fix a and consider the function $v := ([a, b] \ni x \mapsto V(u, a) - V(u, x))$. Taking into account that some sign changes are impossible one can show that v is a monotonically increasing step function.)

(2) Based on (1) show that this number can be computed in $O(M(n) \log n)$ steps.

(3) Given $(Q_1, \ldots, Q_{t-1}, A_t)$ show that the number of roots of u in the interval $(a, b]$ can be computed in $O(n)$ steps.

3.11. Let A, B be univariate polynomials of degree $n \geq m \geq 0$.

(1) Show that the resultant $\mathrm{Res}(A, B)$ of A and B can be computed from the Euclidean representation of (A, B) with $O(n)$ arithmetic operations.

(2) Let $(Q_1, \ldots, Q_{t-1}, A_t)$ be the Euclidean representation of (A, B) having degree pattern $(d_1, \ldots, d_t)$ with $d_t = 0$. Set $n_i := \sum_{j=i}^t d_j$ for $i \in \underline{t}$ and denote by $\mathrm{lc}(A)$ the leading coefficient of a nonzero polynomial A. Prove the following formula:

$$\mathrm{Res}(A, B) = (-1)^\sigma A_t^{m+n} \prod_{i=2}^{t-1} \mathrm{lc}(Q_j)^{s_i},$$

where $s_i = d_1 + d_i + 2 \sum_{j=2}^{i-1} d_j$ and $\sigma = \sum_{j=1}^{t-2} n_j n_{j+1}$.

(Hint: use the subresultant theorem (10.14) which gives

$$\text{Res}(A_i, A_{i+1}) = (-1)^{n_i n_{i+1}} \text{lc}(A_{i+1})^{n_i - n_{i+2}} \text{Res}(A_{i+1}, A_{i+2})$$

for the sequence of remainders $A_1 = A, A_2 = B, A_3, \ldots, A_t$, produced by the Euclidean algorithm.)

3.12. Let K be a field of characteristic 0.

(1) Design an algorithm that computes all derivatives $(D^j f)(\xi)$, $0 \leq j \leq n$, of an input polynomial $f \in K[X]$ of degree n at an input $\xi \in K$ with $O(M(n))$ operations.

(2) Let n be a positive integer, $\xi \in K$, and $a \in K[X]$ of degree less than n with $a(\xi) \neq 0$. Then a can be inverted mod $(X - \xi)^n$ with $O(M(n))$ operations.

3.13. Prove the following statement: given $\xi_1, \ldots, \xi_n, \eta_1, \ldots, \eta_n \in K$ with pairwise different $\xi_1, \ldots, \xi_n$, one can compute the unique polynomial $u \in K[X]$ of degree at most $n - 1$ satisfying $u(\xi_i) = \eta_i$ for all $i \in \underline{n}$ with nonscalar cost at most $10.75n \log n - 1.75n + 10.5$. (Hint: by Langrange interpolation we have $u = \sum_{i<n} u_i X^i = \sum_{i \leq n} \alpha_i \prod_{j \leq n, j \neq i}(X - \xi_j)$, where $\alpha_i = \eta_i / (\prod_{j \leq n, j \neq i}(\xi_i - \xi_j))$. In a first step compute all the α_i. Use the fact that $\alpha_i = \eta_i / g'(\xi_i)$, where $g(X) := \prod_{j \leq n}(X - \xi_j)$. Use Ex. 2.8 to compute the coefficients u_i recursively and analyze the nonscalar cost with Ex. 2.7.)

3.14. This exercise discusses the *Hermite interpolation problem* over a field K of characteristic 0. Let $d = (d_1, \ldots, d_t)$ be a sequence of positive integers summing up to n, and let $H_i = (\xi_i, \eta_{i0}, \ldots, \eta_{i,d_i-1}) \in K^{d_i+1}$ for $i \in \underline{t}$ with pairwise different $\xi_1, \ldots, \xi_t$.

(1) Show that there is a unique polynomial $h \in K[X]$ of degree less than n such that $h(\xi_i) = \eta_{i0}$ and $(D^j h)(\xi_i) = \eta_{ij}$, for all $1 \leq i \leq t$ and all $1 \leq j < d_i$. (This polynomial h is called the Hermite interpolation polynomial corresponding to $(H_1, \ldots, H_t)$.)

(2) With the notation of (1) let $h_i := \sum_{j<d_i} \frac{\eta_{ij}}{j!}(X - \xi_i)^j$, and let ϕ denote the isomorphism of the Chinese remainder theorem corresponding to $p_i := (X - \xi_i)^{d_i}$ for $i \in \underline{t}$. Show that the Hermite interpolation polynomial h corresponding to $(H_1, \ldots, H_t)$ satisfies $h = \phi^{-1}(h_1, \ldots, h_t)$.

(3) Design an algorithm that on input $(H_1, \ldots, H_t)$ computes the corresponding Hermite interpolation polynomial with $O(M(n)(H + 1))$ operations, where $H = \mathcal{H}(d)$. (Hint: modify the proof of Thm. (3.21) by means of Ex. 3.12.)

3.15. (Linear decision trees) Let $n \in \mathbb{N}'$. A *linear decision tree* is a regular binary tree together with a function that assigns to each of its inner nodes v a real affine linear function f_v in the indeterminates $X_1, \ldots, X_n$, and to each of its leaves a label "accept" or "reject." The semantics of linear decision trees is defined as follows. To any $\xi \in \mathbb{R}^n$ we assign a unique path in the binary tree from the root to a leaf by continuing with the right son of a node v if $f_v(\xi) \geq 0$, and with the left son otherwise. We say that the linear decision tree accepts the set of those $\xi \in \mathbb{R}^n$, whose path ends up with an accepting leaf. The number of steps of a linear decision tree is defined as the depth of the underlying tree. Show that a linear decision tree which accepts a subset $W \subseteq \mathbb{R}^n$ with $b_0(W)$ connected components needs at least $\log b_0(W)$ steps. (Compare this obvious lower bound with Thm. (11.9).)

3.16. We denote by $s(m, n)$ the minimal number of steps needed by a linear decision tree to accept a union of m hyperplanes in $\mathbb{R}^n$. Show that $s(m, 1) \leq \log m + 1$ and $s(m, n) \leq s(\binom{m}{2}, n-1) + s(m, 1)$ for $n \geq 1$. Conclude that $s(m, n) = O(2^n \log m)$. (Hint: project the pairwise intersections of the given hyperplanes onto a hyperplane in "general position.")

3.17. Prove Rem. (3.28).

3.18.

(1) Let $W \subseteq \mathbb{R}^n$ be a union of m affine subspaces. Show that there is an algorithm which decides membership to W for a query point in $\mathbb{R}^n$ in time $O(n^5 \log^2 n \log m)$.

(2) Let $\Sigma_n \simeq \mathbb{R}^{\binom{n}{2}}$ be the set of symmetric real $n \times n$-matrices with zero diagonal. (A point in $\Sigma_n \cap \{0, 1\}^{\binom{n}{2}}$ may be interpreted as a graph on n vertices.) Illustrate the result of part one by considering the following examples inspired by the clique, resp. Hamilton cycle problem:

$$W_1 = \{(x_{ij}) \in \Sigma_n \mid \exists S \subseteq \underline{n}, |S| = \lfloor n/2 \rfloor, \forall i, j \in S : i < j \Rightarrow x_{ij} = 1)\},$$
$$W_2 = \{(x_{ij}) \in \Sigma_n \mid \exists \pi \in S_n, \pi \text{ cycle of length } n, \forall i \in \underline{n} : x_{i,\pi(i)} = 1\}.$$

3.19. Let $A = \{H_1^+, \dots, H_m^+\}$ be a finite set of open halfspaces in $\mathbb{R}^n$, $\mathcal{H} := \{H_1, \dots, H_m\}$.

(1) Show that the number of domains of A is at most $\Phi_n(m)$.

(2) Give an example where the number of cells of $\mathcal{H}$ is strictly less than the number of domains of A.

(3) We say that the hyperplanes $H_1, \dots, H_m \subset \mathbb{R}^n$ are in *general position* iff $\dim \cap_{i \in I} H_i = n - |I|$ for all subsets $I \subseteq \underline{n}$ with $1 \leq |I| \leq n$. Prove that under this assumption

number of cells of $\mathcal{H}$ = number of domains of A = $\Phi_n(m)$.

This shows that the inequality $|\Pi_R(A)| \leq \Phi_n(m)$ in Lemma (3.32) is sharp.

3.20. Prove that the range space $(\mathbb{R}^n, \{\text{closed halfspaces in } \mathbb{R}^n\})$ has VC-dimension $n + 1$.

3.21. Consider the range space $S = (\mathbb{R}^n, R)$ where R consists of the axis parallel rectangles $[a_1, b_1] \times \cdots \times [a_n, b_n]$. Show that $\text{VCdim}(S) = 2n$.

3.22. Let (X, R) be a range space of VC-dimension $d \geq 2$, let $t \in \mathbb{N}'$, and let R_t denote the set of all Boolean combinations of at most t ranges in R. Prove that $\text{VCdim}(X, R_t) \leq 2dt \log dt$. Conclude that the range space whose set of points is $\mathbb{R}^n$ and whose ranges are the polytopes with at most t facets, has VC-dimension at most $2(n + 1)t \log(n + 1)t$. (Hint: compare the proof of Lemma (3.33).)

3.23. Let (X, R) be a range space with the property that there is a function $f:[0, 1] \to \mathbb{N}$ such that every finite subset A of X contains for every $\varepsilon \in (0, 1)$ an ε-net N of A of size $|N| \leq f(\varepsilon)$. Prove that the VC-dimension of (X, R) must be necessarily finite. Thus the assumption $\text{VCdim}(X, R) < \infty$ in Thm. (3.36) cannot be avoided.

3.24. This exercise shows that Thm. (3.36) in fact holds in much more generality. Let $(X, \mathcal{A}, P)$ be a probability space and $R \subseteq \mathcal{A}$. For $0 < \varepsilon < 1$ let $R_{P,\varepsilon}$ denote the set of ranges $r \in R$ with probability $P(r) > \varepsilon$. We will call a subset $N \subseteq X$ which contains a point in each $r \in R_{P,\varepsilon}$ an ε-net of the range space (X, R) with respect to the probability measure P.

Prove the following generalization of Thm. (3.36). Let (X, R) be of finite VC-dimension $d > 0$, $0 < \varepsilon, \delta < 1$, and let N be the set of distinct elements obtained by $m \geq \max\{4\varepsilon^{-1} \log 2\delta^{-1}, 8d\varepsilon^{-1} \log 8d\varepsilon^{-1}\}$ independent selections of an element of X according to the probability measure P. Then N is an ε-net of (X, R) w.r.t. P with probability at least $1 - \delta$ (if this probability is defined).

It is remarkable that m is independent of the probability measure P. In the special case where P is the uniform distribution on a finite subset A of X, the original statement of Thm. (3.36) follows.

3.25.* In this exercise we present Vapnik and Chervonenkis original result. Its deduction is very similar to the proof of Thm. (3.36). Let $(X, \mathcal{A}, P)$ be a probability space and $R \subseteq \mathcal{A}$ such that the range space (X, R) has finite VC-dimension $d > 0$. (Readers not familiar with measure theory may assume that $\mathcal{A} = 2^X$ and that P is the uniform distribution on a finite subset $A \subseteq X$.) If m elements $x_1, \ldots, x_m$ are picked independently at random from X, then

$$v_r^m(x_1, \ldots, x_m) := m^{-1}|\{i \in \underline{m} \mid x_i \in r\}|$$

denotes the relative frequency of elements lying in the range $r \in R$. Assume that $\pi^m := \sup\{|v_r^m - P(r)| \mid r \in R\}$ is measurable, thus $\pi^m : X^m \to \mathbb{R}$ is a random variable. We want to show that for $\varepsilon > 0$ the probability $P(\pi^m > \varepsilon)$ tends to zero very rapidly as m goes to infinity.

(1) For $(x, y) \in X^m \times X^m$ define

$$\rho^m(x, y) := \sup\{|v_r^m(x) - v_r^m(y)| \mid r \in R\}$$

and assume that $\rho^m: X^{2m} \to \mathbb{R}$ is a random variable. Consider the events

$$Q := \{\pi^m > \varepsilon\}, \quad C := \{\rho^m \geq \varepsilon/2\}.$$

Show that $P(Q)/2 \leq P(C)$ if $m > 2\varepsilon^{-2}$.

(2) For $z \in X^{2m}$ let $M_z := \{\pi \in S_{2m} \mid \pi z \in C\}$. Prove that

$$P(C) \leq (2m)!^{-1} \sup \{|M_z| \mid z \in X^{2m}\}.$$

(3) Let $z \in X^{2m}$, $r \in R$, and $\ell := 2m v_r^{2m}(z)$. Show that the set

$$\{\pi \in S_{2m} \mid \pi z = (x, y), \ |v_r^m(x) - v_r^m(y)| \geq \varepsilon/2\}$$

has at most

$$(2m)! \sum \binom{\ell}{t} \binom{2m - \ell}{m - t} \binom{2m}{m}^{-1} =: (2m)! \cdot S$$

elements, where the sum is over all $t \in \mathbb{N}$ satisfying $|2t - \ell| \geq \varepsilon m/2$.

(4) Estimate the sum S appearing in (3) by $S \leq 4\varepsilon m^{3/2} \exp(-\varepsilon^2 m/4)$. (Hint: use the estimates ($\lambda := t/n$, H is the entropy function)

$$\binom{n}{t} \leq \frac{n^n}{t^t (n - t)^{n-t}} = 2^{nH(\lambda)}, \quad H(\lambda) \leq H(1/2) + \frac{1}{2} H''(1/2)(\lambda - 1/2)^2,$$

which imply $\binom{n}{t} \leq 2^n \exp(-n(2\lambda - 1)^2/2)$.)

(5) Conclude that $P(Q) \leq 8\Phi_d(2m)\varepsilon m^{3/2} \exp(-\varepsilon^2 m/4)$ provided $m > 2\varepsilon^{-2}$.

(6) Prove that for all $0 < \varepsilon, \delta < 1$ we have $P(\pi^m > \varepsilon) \leq \delta$ if

$$m \geq \max \{16\varepsilon^{-2}(d + 3/2) \ln 8\varepsilon^{-2}(d + 3/2), \ 8\varepsilon^{-2}(\ln 8\delta^{-1} + d \ln 2)\}.$$

3.26. We discuss a consequence of Ex. 3.25. Let (X, R) be a range space of finite dimension $d > 0$, $0 < \varepsilon < 1$, and $A \subseteq X$ be a finite set. A subset $V \subseteq X$ is called an *ε-approximation* of A for (X, R) iff for all $r \in R$

$$\left| \frac{|r \cap A|}{|A|} - \frac{|r \cap V|}{|V|} \right| \leq \varepsilon.$$

Let $m := \lceil 16\varepsilon^{-2}(d + 3/2) \ln 8\varepsilon^{-2}(d + 3/2) \rceil$. Prove that any finite subset $A \subset X$ of size $\geq m^2$ contains an ε-approximation of A of size m.

3.7 Open Problems

Problem 3.1. Can the upper complexity bound for the general interpolation problem in Thm. (3.21) be improved to $O(M(n)(H + 1))$?

Problem 3.2. What is the complexity of computing the GCD of multivariate polynomials?

3.8 Notes

Using standard integer arithmetic, the classical Euclidean algorithm needs about n^2 bit operations in the worst case to compute the GCD of two integers $< 2^n$. Substantial computational improvements are due to Lehmer [322] and Knuth [305]. Knuth's method allows to compute the greatest common divisor of two n-bit integers in $O(n \log^5 n \log \log n)$ bit operations. Schönhage [455] proposed a variant of Knuth's method that solves this task in $O(n \log^2 n \log \log n)$ bit operations. This algorithm has been translated to the case of univariate polynomials over a field by Moenck [374]. A rigorous verification of the Knuth-Schönhage algorithm in this case, as well as an improved analysis of its arithmetic cost are due to Strassen [502, 504]. Sect. 3.1 and 3.2 follow his presentation rather closely, in particular, Thm. (3.15) and Cor. (3.16) are due to him. The qualitative idea of Lemma (3.8) goes back to Lehmer [322]. For a variant of the controlled Euclidean descent for integers see Schönhage et al. [465, p. 239].

For univariate polynomials over (small) finite fields Strassen's analysis rather realistically reflects the running time of the Knuth-Schönhage algorithm, when implemented in a reasonable way. This is however not the case for univariate *integer* polynomials, since one has to take into account the growth of the size of the coefficients. Brent et al. [74] and Schönhage [463] reduce GCD computations in $\mathbb{Z}[X]$ to GCD computations on long integers. With this technique, Schönhage [463] designed a probabilistic algorithm for the computation of the GCD of two univariate integer polynomials of degrees at most n with their ℓ^1-norms being bounded by 2^h, and estimated its expected running time by a worst-case bound of $O(n(n+h)^{1+o(1)})$ bit operations.

There is a vast literature on GCD computations for multivariate polynomials over a field or over the integers. The proposed algorithms are based on the theory of subresultants, Chinese remaindering, or Hensel lifting, see Collins [119, 120], Brown [82, 83], Brown and Traub [84], Moses and Yun [385], Zippel [573], Hearn [236], Wang [542, 543], Knuth [304, 307], Buchberger et al. [94], Stoutemeyer [492], Akritas [3, 4], Geddes et al. [194], among others. A major breakthrough is due to Kaltofen [283]. He has designed a probabilistic algorithm that takes as input two multivariate polynomials of degree at most d given by straight-line programs of length at most ℓ, and a failure allowance ε. The algorithm either outputs a straight-line program of length $O(\ell d + d^2)$ computing the GCD of the given polynomials, or outputs "failure," the latter happening with probability $\leq \varepsilon$. It requires a polynomial number of arithmetic steps in d and ℓ (on a probabilistic algebraic random access machine over K). Compared to other data structures for multivariate polynomials (like dense or sparse representations, or formulas), straight-line programs behave favourably with respect to the various algebraic manipulations on polynomials, see Kaltofen [283].

Thm. (3.19) and Thm. (3.21) combine well-known evaluation and interpolation techniques (see, e.g., Lipson [338], Borodin and Moenck [64], Yun [570]) with Huffman coding. Cor. (3.20) goes back to Fiduccia [170], Moenck and

Borodin [373], Strassen [497], and Borodin and Moenck [64], and Cor. (3.22) is due to Horowitz [254].

GCD computations play also a fundamental role for the problem of factoring polynomials. This topic, which is beyond the scope of this book, is surveyed in Kaltofen [281, 284, 286], see also Kaltofen and Shoup [291]. Ex. 3.3, 3.4, and 3.5 discuss important steps in the process of completely factoring polynomials. Ex. 3.6 is due to Strassen [504]. The problem of computing Padé approximants, sketched in Ex. 3.8, has been discussed for example in Brent et al. [74], Czabor and Geddes [141], Cabay and Choi [104], see also Gragg [202]. Finding minimum polynomials of linearly generated sequences over the field $\mathbb{F}_2$ is especially useful for cryptography in the context of linear feed back shift register sequences, see, e.g., Rueppel [444] or van Tilborg [517]. Massey [355] was the first to notice that an algorithm due to Berlekamp [41] for decoding BCH-codes could be modified to find minimum polynomials of linearly generated sequences. The resulting algorithm is often called the *Berlekamp-Massey algorithm*. Its cost is $O(nd)$, where d is the degree of the generating polynomial of the sequence. Several authors (McEliece [357], Welch and Scholtz [547], Cheng [107], Dornstetter [148]) have noticed the relationship between the partial solutions produced by this algorithm and the extended Euclidean algorithm, and that the latter can also be used to solve the problem of finding a minimum polynomial, as presented in Ex. 3.9. The lecture notes of Kaltofen [285] have been helpful in designing this exercise.

For computational solutions to the problem of isolating real zeros of polynomials, briefly sketched in Ex. 3.10, see Collins and Loos [121], Akritas [4], and in particular Schönhage [461, 462, 465].

The study of the combinatorial and topological properties of arrangements of real and complex hyperplanes has a long tradition. The reader interested in this beautiful theory is referred to the books by Björner et al. [56] and by Orlik and Terao [402]. We remark that the word arrangement there just denotes a family of hyperplanes, and not a partition into faces. For recent results on arrangements of subspaces of arbitrary dimension the reader may consult Björner [52].

The investigation of the computational complexity of geometric problems is the topic of the young field of computational geometry, for a survey see F.F. Yao [569]. Point location in general subdivisions of Euclidean space is in fact one of the basic problems in this area. For locating a point in an arrangement of hyperplanes it is quite natural to restrict oneself to linear search trees. There the basic operation, counted as one step, is a test whether a query point lies to the left of, to the right of, or on a hyperplane. (For a formal definition see Ex. 3.15.) Dobkin and Lipton [145] described a linear search tree which solves the point location problem for an arrangement of m hyperplanes in $\mathbb{R}^n$ with $O(2^n \log m)$ steps (cf. Ex. 3.16). Meyer auf der Heide [360] was the first who showed that a number of steps polynomial in $\log m$ *and* n is sufficient. He designed a linear search tree solving the point location problem for an arrangement of hyperplanes in $\mathbb{R}^n$ defined by integer polynomials with coefficients of absolute value $\leq q$ with at most $O(n^4 \log nq)$ steps. From this he drew the astonishing conclusion that the NP-complete real knapsack problem can be solved in "nonuniform" polynomial time. (For the notion of NP-completeness see Garey and Johnson [181].) In contrast to the method explained in the text, Meyer auf der Heide's strategy was based on the metric notion of a "coarseness" of an arrangement which is the minimum of the inner radii of its cells. Cor. (3.27) and (3.29) are due to him [360, 362], see also [363]. For lower bounds we refer to Chap. 11.

The concept of the VC-dimension goes back to Vapnik and Chervonenkis [532]. They proved that the relative frequency of an event r tends to the probability of r, uniformly over all events r in a class R, provided the VC-dimension of R is finite (cf. Ex. 3.25). Haussler and Welzl [235] extended Vapnik and Chervonenkis' ideas by inventing the notion of epsilon-nets and proving Thm. (3.36) which states the existence of small epsilon-nets. This theorem has found various applications in computational geometry [235], and in computational learning theory (see Blumer et al. [60]). At the same time, Clarkson [111] used random sampling techniques in order to design efficient algorithms in computational geometry without explicitly using the notion of VC-dimension. Thm. (3.25) and the algorithm of Thm. (3.26) are essentially due to him. However, in the latter theorem, Clarkson only obtained a running time $O(\log m)$ with a constant factor which might depend exponentially on the dimension n. The polynomial dependence of the running time in $\log m$ *and* n is due to Meiser [359], who gave a detailed proof of these results based on Thm. (3.36) and the estimation of the VC-dimension in Lemma (3.33). His work also contains a rather explicit description of the data structure to be used by the fast point location algorithm.

Ex. 3.16 is due to Dobkin and Lipton [145]. The statement of Ex. 3.22 is contained in Dudley [150]. In Ex. 3.25 and 3.26 we describe Vapnik and Chervonenkis' [532] result.

Part II

Elementary Lower Bounds

Chapter 4. Models of Computation

The standard models for investigating issues in computational complexity in the discrete setting are the Turing machine and the random access machine. However, these models are not well-suited for a discussion of complexity questions in a general algebraic framework, where one assumes that arithmetic operations (over the reals, say) can be performed with infinite-precision at unit cost. For the search of lower bounds in such an algebraic framework, two computational models have proved to be particularly useful: the straight-line program, also called arithmetic circuit, and the computation tree.

In this chapter we define and analyze the notion of straight-line programs which formalize step-by-step computations without branching. Starting from an input set I which is a subset of a k-algebra A (e.g., a polynomial ring or a field of rational functions), such programs compute a finite subset F of A by means of the four basic arithmetic operations and scalar multiplications by elements of k. By assigning costs to the different operations we arrive at the notion of complexity of a set F (with respect to I and A), which is the minimum cost of a straight-line program computing F in A from I. If we neglect scalar multiplications, additions, and subtractions we obtain the nonscalar complexity which was first considered by Ostrowski. It will play an important role in this book, as most of the known lower bounds hold for this complexity measure. For instance, we shall prove the dimension bound, the first lower bound result for the nonscalar complexity in this book. In Sect. 4.3 we discuss the question to what extent computing in a field extension may decrease the complexity. In the last two sections of this chapter we formalize branching algorithms by computation trees. We then continue by analyzing the objects a computation tree computes, as well as defining their complexity, and show that the latter has a lower bound in terms of the straight-line complexity of certain rational functions.

4.1 Straight-Line Programs and Complexity

Throughout this book we shall be mainly concerned with the computation of multivariate rational functions. The notion of computation will become precise in the course of this section; for the moment, however, we want to appeal to the reader's intuition. Let us start with a simple example. Suppose we want to compute the univariate rational function $f = (X^5 - 1)/(X - 1)$, i.e, we want to design a step-by-step procedure, which, on input X, computes the value of f at X. We assume that the procedure does not contain any branchings; for people familiar with programming languages, this means that we avoid using constructs such as if ... then ... else. A sequence of operations using the binary method for computing X^5 could look like this:

$$X_0 := X,$$
$$X_1 := X_0 * X_0, \quad X_2 := X_1 * X_1, \quad X_3 := X_2 * X_0,$$
$$X_4 := X_3 - 1, \quad X_5 := X_0 - 1, \quad X_6 := X_4/X_5.$$

Let us call this program $\Gamma 1$. (Γ, the third letter in the greek alphabet, is an acronym for the word Computation.) Given X, $X \neq 1$, this program computes the value of f at X. Suppose now that we have at hand an intelligent computer algebra system capable of performing the four basic arithmetic operations in any k-algebra A. Suppose further, that the system stops the current computation with an error message whenever it encounters an operation a/b where $b \notin A^\times$, i.e., b is not a unit in A. Which inputs to the above program provoke an error message? Of course, we have now to specify the algebra to which X belongs, so that it becomes convenient to regard the *pair* $(A; X)$ as an input. We say that $\Gamma 1$ is *executable* on input $(A; X)$ if it does not stop with an error message. Otherwise $\Gamma 1$ is said to be *not executable* on this input. It is clear that $\Gamma 1$ is executable on $(A; X)$ iff $X - 1 \in A^\times$ is a unit. For instance, $\Gamma 1$ is executable on $(k(X, Y); XY - 1)$ while it is not executable on $(k[X, Y]; XY - 1)$.

Let us now turn to a more abstract setting by specifying the set Ω of operations our system is capable of performing. It is clear that Ω should contain the set $\{+, -, *, /\}$ of the four (binary) basic arithmetic operations. Furthermore, since we should be able to perform computations in a k-algebra, it is necessary to have the (unary) scalar multiplication with elements of k at our disposal. Identifying the scalar multiplication with $\lambda \in k$ with λ and the entity of all these operations with k, we see that Ω should contain the set $k \cup \{+, -, *, /\}$. How can we perform the computation step $X_4 := X_3 - 1$? This is done by means of another operation, denoted by 1^c: the step $X_4 := X_3 - 1$ is replaced by the sequence $X_4 := 1$, $X_5 := X_3 - X_4$. More generally, for $\lambda \in k$ we denote by λ^c the (0-ary) operation of taking the constant λ and denote the entity of all these operations by k^c. Decomposing $\Gamma 1$ with the aid of the set $\Omega := k^c \cup k \cup \{+, -, *, /\}$ of operations we obtain the program Γ given as

$$X_0 := X,$$
$$X_1 := X_0 * X_0, \quad X_2 := X_1 * X_1, \quad X_3 := X_2 * X_0,$$
$$X_4 := 1, \quad X_5 := X_3 - X_4, \quad X_6 := X_0 - X_4,$$
$$X_7 := X_5/X_6.$$

We could also have encoded the program Γ as $\Gamma = (\Gamma_1, \ldots, \Gamma_7)$, where the Γ_i are instructions given as

(4.1)
$$\Gamma_1 = (*; 0, 0), \quad \Gamma_2 = (*; 1, 1), \quad \Gamma_3 = (*; 2, 0), \quad \Gamma_4 = (1^c),$$
$$\Gamma_5 = (-; 3, 4), \quad \Gamma_6 = (-; 0, 4), \quad \Gamma_7 = (/; 5, 6).$$

This gives rise to the following formal definition.

(4.2) Definition. Let $\Omega = k^c \cup k \cup \{+, -, *, /\}$, $n \in \mathbb{N}$, A be a k-algebra, and $a \in A^n$. The pair $(A; a)$ is called an *input of length n*. The arity of an operation $\omega \in \Omega$ is denoted by $\mathrm{ar}(\omega)$.

(1) (Syntax of straight-line programs) A *straight-line program* Γ (over k, expecting inputs of length n) is a sequence $(\Gamma_1, \ldots, \Gamma_r)$ of *instructions*

$$\Gamma_i = (\omega_i; u_{i1}, \ldots, u_{i\,\mathrm{ar}(\omega_i)}),$$

where $\omega_i \in \Omega$ and the $u_{i\ell}$ are integers satisfying $-n < u_{i1}, \ldots, u_{i\,\mathrm{ar}(\omega_i)} < i$.

(2) (Semantics of straight-line programs) Let $\Gamma = (\Gamma_1, \ldots, \Gamma_r)$ be a straight-line program expecting inputs of length $n \in \mathbb{N}$, Γ_i as in (1). Γ is said to be *executable on* $(A; a)$ (or *executable in A on input a*) with *result sequence* $b = (b_{-n+1}, \ldots, b_r) \in A^{n+r}$, if $b_i = a_{n+i}$ for $-n + 1 \leq i \leq 0$ and $b_i = \omega_i(b_{u_{i1}}, \ldots, b_{u_{i\,\mathrm{ar}(\omega_i)}})$ for $1 \leq i \leq r$. (b is unique if it exists.) $(A; b)$ is called the *output* corresponding to the input $(A; a)$.

(3) Let the straight-line program Γ expecting inputs of length n be executable on $(A; a)$ with result sequence $(b_{-n+1}, \ldots, b_r)$. Γ is said to *compute* the set $F \subseteq A$ on input a, if $F \subseteq \{b_{-n+1}, \ldots, b_r\}$.   ●

In the sense of the above definition we see that the program Γ in (4.1) computes every subset of $\{X, X^2, X^4, X^5, 1, X^5-1, X-1, (X^5-1)/(X-1)\}$. If we regard all the operations as being of equal interest, we may represent straight-line programs by means of directed acyclic multigraphs. (A multigraph is a graph in which more than one edge is allowed between a pair of vertices.) For instance, we can associate to the program Γ in (4.1) the following multigraph:

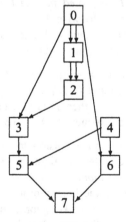

More generally, any straight-line program Γ defines a directed acyclic multigraph: keeping the notation of the above definition, the nodes of this graph are the elements in $\{i \in \mathbb{Z} \mid -n < i \leq r\}$, and it has the edges $u_{i1} \to i, \ldots, u_{i\,\mathrm{ar}(\omega_i)} \to i$ for $1 \leq i \leq r$. The length (= number of edges) of the longest directed path in this graph is called the *depth* $D(\Gamma)$ of Γ. (This graph theoretic viewpoint will be helpful in Chap. 13 and in Chap. 21.) If every operation involved in Γ can be performed in unit time, then $D(\Gamma)$ is the parallel execution time of the program. For example, our program in (4.1) has parallel execution time 5.

Very often we shall be dealing with techniques for transforming one straight-line program to another. For example, it is rather easy to transform the above program Γ to another program $\overline{\Gamma}$ computing the set $\{(X^{10}-1)/(X^2-1)\}$ on input $(A; X)$: we first compute X^2 and then start the program Γ with the new inputs X, X^2 instead of X. In other words, we concatenate two straight-line programs, one for X^2, the other for $(X^5-1)/(X-1)$. The following definition makes this concept precise.

(4.3) Definition. Let $\Gamma = (\Gamma_1, \ldots, \Gamma_r)$ be a straight-line program expecting inputs of length n and $\Gamma' = (\Gamma'_1, \ldots, \Gamma'_t)$ be a straight-line program expecting inputs of length $n + r$. For $1 \leq \tau \leq t$ let $\Gamma'_\tau = (\omega_\tau; u_{\tau 1}, \ldots, u_{\tau \operatorname{ar}(\omega_\tau)})$ and put $\tilde{\Gamma}_\tau :=$ $(\omega_\tau; u_{\tau 1} + r, \ldots, u_{\tau \operatorname{ar}(\omega_\tau)} + r)$. Then

$$\Gamma\Gamma' = (\Gamma_1, \ldots, \Gamma_r, \tilde{\Gamma}_1, \ldots, \tilde{\Gamma}_t)$$

is a straight-line program, the *concatenation* of Γ and Γ'.   ●

Let us now turn to an important topic, that of studying those inputs on which a given straight-line program is executable. We start with some trivial cases. By convention, the empty straight-line program (having no instructions) is executable on any input of length n. At the other extreme, there are inputs of any length n on which any straight-line program expecting inputs of this length is executable: let $\mathbf{0} = \{0\}$ denote the zero k-algebra. Then, as $0 = 1$ in $\mathbf{0}$, every straight-line program expecting inputs of length n is executable on the zero input $(\mathbf{0}; 0, \ldots, 0)$ of length n. The straight-line program Γ is called *inconsistent* iff it is only executable on this zero input, and *consistent* otherwise. In particular, the empty straight-line program over n is consistent. We continue with one more trivial observation: since morphisms of k-algebras map units to units the following result follows.

(4.4) Remark. Let $\varphi: A \to A'$ be a morphism of k-algebras. If a straight-line program Γ is executable on an input $(A; a_1, \ldots, a_n)$ of length n with result sequence $(A; b_1, \ldots, b_m)$ of length m, then Γ is also executable on $(A'; \varphi \circ a) = (A'; \varphi(a_1), \ldots, \varphi(a_n))$ with result sequence $(A'; \varphi \circ b)$.   ●

In the following we shall derive an extension to this trivial remark. Namely, we shall show that for any consistent straight-line program Γ there exists an input $(U; u)$, called the *universal input for* Γ, such that any other input $(A; a)$ on which Γ is executable is a unique homomorphic image of $(U; u)$, i.e., there exists a unique morphism $\varphi: U \to A$ of k-algebras such that $a = \varphi \circ u$. Calling two inputs $(A; a)$ and $(A'; a')$ of the same length *isomorphic* iff there is a k-algebra isomorphism $A \to A'$ mapping a to a', we see that an input $(U; u)$ with the above properties is unique (up to isomorphism) if it exists. Let us examine the existence for the case of the computation Γ in (4.1). We have already seen that Γ is executable on $(A; a)$ iff $a - 1$ is a unit in A. The program is thus not executable on $(k[X]; X)$, but it is executable on $(U; X)$, where U is the subalgebra $U := \{g/(X-1)^m \mid g \in k[X], m \in \mathbb{N}\}$ of $K(X)$. It is easy to verify that $(U; X)$ is in fact a universal input for Γ. More generally, let $d \in k[X] := k[X_1, \ldots, X_n]$ be a nonzero polynomial. The subalgebra

$$k[X]_d := \{g/d^m \mid g \in k[X], m \in \mathbb{N}\}$$

of $k(X)$ is called the *localization* of $k[X]$ with respect to the denominator d. It is the smallest k-algebra between $k[X]$ and $k(X)$ in which d is invertible. The following *universal mapping property* is easy to show: if $\varphi: k[X] \to A$ is a k-algebra morphism such that $\varphi(d) \in A^\times$, then there is a unique extension of φ to a k-algebra morphism $k[X]_d \to A$.

We can now state the general theorem on the existence and uniqueness of universal inputs.

(4.5) Theorem. *Let Γ be a consistent straight-line program expecting inputs of length n. Then, up to isomorphism, there is a unique universal input $(U; u)$ for Γ, i.e., an input $(U; u)$ satisfying*

- *Γ is executable on $(U; u)$,*
- *if Γ is executable on an input $(A; a)$ then there is a unique k-algebra morphism $\varphi: U \to A$ such that $\varphi \circ u = a$.*

Moreover, U is the localization $k[X_1, \ldots, X_n]_d$ of the polynomial ring with respect to a suitable polynomial d, $u = (X_1, \ldots, X_n)$, and Γ is executable on $(A; a)$ iff $d(a) \in A^\times$.

Proof. The uniqueness of the universal input is clear, once we have proved its existence. Let $\Gamma = (\Gamma_1, \ldots, \Gamma_r)$ be a consistent straight-line program. We proceed by induction on r. If $r = 0$, then Γ is empty and by convention, Γ is executable on any input of length n. We have $k[X_1, \ldots, X_n]_1 = k[X_1, \ldots, X_n] =: k[X]$, and by the universal mapping property for the polynomial algebra $(k[X]; X)$ is a universal input for the empty straight-line program expecting inputs of length n. Now let $r > 0$ and assume that we already know a polynomial δ such that $(k[X]_\delta; X)$ is a universal input for $\Gamma' := (\Gamma_1, \ldots, \Gamma_{r-1})$. Let $b_{-n+1}, \ldots, b_{r-1}$ be the corresponding result sequence. If Γ_r is not a division instruction, then the input $(k[X]_\delta; X)$ is also universal for Γ. If however $\Gamma_r = (/; i, j)$ with $i, j < r$ then, as Γ is consistent, b_j is a non-zero element in $k[X]_\delta$. Hence $b_j = \beta/\delta^\ell$, with some non-zero polynomial β and some non-negative integer ℓ. Let $d := \delta \cdot \beta$. Then Γ is executable on $(k[X]_d; X)$. We claim that $(k[X]_d; X)$ is a universal input for Γ. To see this, let Γ be executable on $(A; a)$. Then Γ' is executable on $(A; a)$ as well. But $(k[X]_\delta; X)$ is universal for Γ'. Hence there is a unique k-algebra morphism $\phi: k[X]_\delta \to A$ mapping X_i onto a_i. As Γ is executable on $(A; a)$, we have $\phi(b_j) \in A^\times$. By the universal mapping property of localizations there is a unique k-algebra morphism $\varphi: (k[X]_\delta)_{b_j} \to A$ extending ϕ. Finally notice that $(k[X]_\delta)_{b_j}$ equals $k[X]_d$. The last assertion of the theorem is obvious. $\square$

We next define the length of a straight-line program and are then able to compare different straight-line programs that solve the same problem. The length of the shortest such program defines the complexity of the problem.

(4.6) Definition. Let A be a k-algebra and $\Gamma = (\Gamma_1, \ldots, \Gamma_r)$ be a straight-line program, and let ω_i denote the operational symbol corresponding to Γ_i. Finally, let $c: \Omega \rightarrow \mathbb{N}$ be a function (called *cost function* in the sequel).

(1) c-length$(\Gamma) := \sum_{i=1}^{r} c(\omega_i)$ is called the c-*length* of Γ.
(2) The *complexity*
$$L_A^c(F|I) \in \mathbb{N} \cup \{\infty\}$$
of the finite subset F of A *with respect to the input set* $I \subseteq A$ *and the cost function* c is defined as the minimal c-length of a straight-line program expecting inputs of length n for some $n \in \mathbb{N}$ which computes F on some input $a \in I^n$.
(3) Replacing in (2) "c-length" by "depth" defines the *depth* $D_A^c(F|I)$ of F with respect to I.   •

The depth will play a subordinate role in the sequel. However, a fundamental relationship between complexity and depth will be deduced in Chap. 21. In part (2) of the foregoing definition we allow for infinite complexities as well. Note that F can be computed by a straight-line program from inputs in I iff each element of F is of the form ab^{-1}, where a, b are elements of the k-subalgebra of A generated by I.

Some cost functions and complexity measures are studied so frequently that it is convenient to give them separate names. In the sequel, 1_S denotes the indicator function of a subset $S \subseteq \Omega$.

(4.7) Definition. Let $c_* := 1_{\{*,/\}}$, $c_+ := 1_{\{+,-\}}$, tot $:= 1_{k \cup \{+,-,*,/\}}$, and F, I be subsets of a k-algebra A, F finite. We call

$$
\begin{array}{ll}
L_A^{\text{tot}}(F|I) & \text{the \textit{total complexity},} \\
L_A^+(F|I) := L_A^{c_+}(F|I) & \text{the \textit{additive complexity}, and} \\
L_A^*(F|I) := L_A^{c_*}(F|I) & \text{the \textit{multiplicative complexity}}
\end{array}
$$

of F modulo I. Instead of c_*-length (c_+-length, tot-length) of a straight-line program we say *multiplicative length* (*additive length, total length*).   •

Note first that none of the above cost functions c charges for taking a constant in k, that is, $c(\lambda^c) = 0$ for all $\lambda \in k$. *So we always think of all constants in k as being freely at our disposal*. A very important point is that c_* *does not charge for multiplications with scalars in k!* The multiplicative complexity $L_A^*(F|I)$ is thus the minimum number of nonscalar multiplications or divisions sufficient to compute F from elements of I. For this reason it is also called the *nonscalar complexity*. The nonscalar complexity plays an important role in this book since many of the interesting lower bounds in this book are proved in this model. It is therefore convenient to have a more streamlined notation for this.

(4.8) Notation. We often write $L_A(F|I)$ instead of $L_A^*(F|I)$. Sometimes we write $L_{k \to A}(F|I)$ for $L_A(F|I)$ to stress the role of k. If $I = \{X_1, \ldots, X_n\}$ is the standard input, we write $L_A(F)$ or even $L(F)$ when A is clear from the context. •

As an example, let us reformulate Thm. (2.9) of Sect. 2.1. For a polynomial a we denote by $C(a)$ the set of the coefficients of a. Let a and b be polynomials whose coefficients are algebraically independent over a field k. Further let K denote the field generated over k by the coefficients of a and b. Then, if $n = \deg(ab)$, we have

$$
\begin{aligned}
L_K(C(ab) \mid C(a) \cup C(b)) &= n + 1, \\
L_K^{tot}(C(ab) \mid C(a) \cup C(b)) &= O(n \log n) \quad \text{if } k \text{ supports FFTs,} \\
L_K^{tot}(C(ab) \mid C(a) \cup C(b)) &= O(n \log n \log \log n) \quad \text{for arbitrary } k.
\end{aligned}
$$

It is a useful exercise to state the results in Chap. 2 in this new terminology.

We finish this section by some trivial but important observations.

(4.9) Remarks. Let A be a k-algebra, $c: \Omega \to \mathbb{N}$ a cost function, and I, F, E subsets of A with F, E finite.

(1) If $F_1 \subseteq F$ and $I \subseteq I_1$, then $L_A^c(F_1|I_1) \leq L_A^c(F|I)$.

(2) For a field extension $k \subseteq K$ and a K-algebra A we have $L_{K \to A}(F|I) \leq L_{k \to A}(F|I)$.

(3) If $\varphi: A \to A'$ is a morphism of k-algebras, then $L_{A'}^c(\varphi(F)|\varphi(I)) \leq L_A^c(F|I)$. (Check Rem. (4.4).)

(4) $L_A^c(F|I) \leq L_A^c(E|I) + L_A^c(F|E \cup I)$. (Use concatenation of straight-line programs.)   •

4.2 Computation Sequences

In this section we are going to discuss computation sequences, which are in a sense equivalent to straight-line programs in the context of multiplicative complexity. Computation sequences combine both the syntactic and the semantic aspects of straight-line programs, and it is sometimes more convenient to use this alternative concept.

(4.10) Definition. Let I be a subset of a k-algebra A. We call a sequence $(g_1, \ldots, g_r) \in A^r$ a *computation sequence of length r modulo I* (in A over k) if for all $\rho \in \underline{r}$ there exist $u_\rho, v_\rho \in k + \langle I \rangle_k + \langle g_1, \ldots, g_{\rho-1} \rangle_k$ such that

$$
g_\rho = u_\rho v_\rho \text{ or } (v_\rho \in A^\times, g_\rho = \frac{u_\rho}{v_\rho}).
$$

A computation sequence $(g_1, \ldots, g_r)$ is said to *compute* $F \subseteq A$ modulo I, if $F \subseteq k + \langle I \rangle_k + \langle g_1, \ldots, g_r \rangle_k$.   •

The following result characterizes the multiplicative complexity in terms of the length of computation sequences.

(4.11) Theorem. *Let A be a k-algebra and F, I ⊆ A, F finite. Then for all r ∈ ℕ, $L_A(F|I) \le r$ if and only if there exists a computation sequence of length r modulo I (in A over k) which computes F.*

Proof. "⟹" Let $\Gamma = (\Gamma_1, \ldots, \Gamma_s)$ be a straight-line program of multiplicative length r executable in A on input $a = (a_1, \ldots, a_n) \in I^n$ and computing F. Let $(b_{-n+1}, \ldots, b_s)$ be the corresponding result sequence. We will associate to the pair (Γ, a) a computation sequence $(g_1, \ldots, g_r)$ such that for all $i \le s$

$$b_i \in k + \langle I \rangle_k + \langle g_1, \ldots, g_r \rangle_k.$$

We proceed by induction on s. The start "s = 0" is clear. So let s > 0 and let $\Gamma' = (\Gamma_1, \ldots, \Gamma_{s-1})$ be of multiplicative length r'. Then $r - 1 \le r' \le r$. Let $(g_1, \ldots, g_{r'})$ be the computation sequence associated to (Γ', a). If $r' = r$, then Γ_s is a linear operation (i.e., $\omega_s \in k^c \cup k \cup \{+, -\}$) and our claim follows. If, however $r' = r - 1$, then $\Gamma_s = (\omega; i, j)$ with $\omega \in \{*, /\}$. Define $g_r := b_i \omega b_j$. Applying the induction hypothesis to Γ' we see that b_i and b_j both belong to $k + \langle I \rangle_k + \langle g_1, \ldots, g_{r-1} \rangle_k$. Hence $(g_1, \ldots, g_r)$ is a computation sequence computing F modulo I.

"⟸" Let $(g_1, \ldots, g_r)$ be a computation sequence computing F modulo I. To prove $L_A(F|I) \le r$ we proceed by induction on r. The start "r = 0" being clear, let r > 0. Every $f \in F$ can be written in the form

$$f = \hat{f} + \lambda_f g_r$$

with $\hat{f} \in k + \langle I \rangle_k + \langle g_1, \ldots, g_{r-1} \rangle_k$ and $\lambda_f \in k$. Put $\hat{F} := \{\hat{f} \mid f \in F\}$. Let $\omega \in \{*, /\}$ and suppose that $g_r = u_r \omega v_r$ with $u_r, v_r \in k + \langle I \rangle_k + \langle g_1, \ldots, g_{r-1} \rangle_k$. By induction, $L_A(\hat{F} \cup \{u_r, v_r\} \mid I) \le r - 1$; furthermore $L_A(F \mid \hat{F} \cup \{u_r, v_r\}) \le 1$. Hence, by Rem. (4.9),

$$L_A(F \mid I) \le L_A(\hat{F} \cup \{u_r, v_r\} \mid I) + L_A(F \mid \hat{F} \cup \{u_r, v_r\} \cup I) \le r - 1 + 1 = r. \quad \square$$

It is a useful exercise for the reader to explicitly write down computation sequences for the algorithms presented in Sect. 2.1.

The last theorem shows in particular that $L_A(F|I) = L_A(F'|I')$ if $\langle F \rangle_k = \langle F' \rangle_k$ and $\langle I \rangle_k = \langle I' \rangle_k$. As an immediate consequence we can derive our first lower bound result.

(4.12) Dimension Bound. *Let A be a k-algebra, I and F be subsets of A, F finite. Then*

$$L_A(F \mid I) \ge \dim_k(k + \langle I \rangle_k + \langle F \rangle_k) - \dim_k(k + \langle I \rangle_k).$$

Proof. Let $(g_1, \ldots, g_r)$ be a computation sequence in A over k of length $r = L_A(F|I)$ computing F modulo I. By the last theorem we have

$$k + \langle I \rangle_k + \langle F \rangle_k \subseteq k + \langle I \rangle_k + \langle g_1, \ldots, g_r \rangle_k.$$

Now our claim follows easily. $\square$

As a first application of the dimension bound we discuss some of the lower bounds of Sect. 2.1.

(4.13) Theorem. (1) *Let* $k[a, b] = k[a_0, \ldots, a_m, b_0, \ldots, b_n]$ *be a polynomial ring in* $m + 1 + n + 1$ *indeterminates. For* $0 \leq \ell \leq m + n$ *put* $c_\ell := \sum_{i+j=\ell} a_i b_j$, *and let* $J \subseteq \{0, \ldots, m + n\}$. *The we have*

$$L_{k(a,b)}(c_0, \ldots, c_{m+n}|I) = L_{k[a,b]}(c_0, \ldots, c_{m+n}|I) = m + n + 1 - |J|,$$

where $I := \{a_0, \ldots, a_m, b_0, \ldots, b_n\} \cup \{c_j \mid j \in J\}$, *provided* $|k| \geq m + n + 1$.
(2) *Let* $a_0, \ldots, a_n$ *be indeterminates over* k, $k[a] := k[a_0, \ldots, a_n]$, *and* $m \leq n$. *With* $I := \{a_0, \ldots, a_n\}$ *we have*

$$L_{k(a)}(a_0^2, \ldots, a_m^2|I) = L_{k[a]}(a_0^2, \ldots, a_m^2|I) = m + 1.$$

Proof. (1) The upper bound has been proved in Sect. 2.1. So let us proceed with the lower bound. We only need to prove this for $L_{k(a,b)}$, by virtue of Rem. (4.4). We write L for $L_{k(a,b)}$. It is clear that $L(c_0, \ldots, c_{m+n}|I) = L(\{c_\ell \mid \ell \notin J\}|I) =: L(F|I)$. Since the polynomials $c_0, , \ldots, c_{m+n}, a_0, \ldots, a_m,$ $b_0, \ldots, b_n, 1$ are linearly independent over k, we have by the dimension bound $L(F|I) \geq \dim(k + \langle I \rangle_k + \langle F \rangle_k) - \dim(k + \langle I \rangle_k) = (1 + |I| + |F|) - (1 + |I|) = |F| = m + n + 1 - |J|$.

(2) The upper bound is clear. As the polynomials $1, a_0, \ldots, a_n, a_0^2, \ldots, a_m^2$ are linearly independent over k, the dimension bound gives the desired lower bound. $\square$

4.3* Autarky

Let $k \subseteq K$ be an extension of fields. Suppose we want to compute a finite set $F \subset k(X) := k(X_1, \ldots, X_n)$ of rational functions over k. Can it be helpful to compute in $K(X_1, \ldots, X_n)$ and to use scalar multiplications with elements of K or to use constants from K? In other words, is it useful to take straight-line programs over K instead of straight-line programs over k? For making the discussion less technical, we only consider here the multiplicative cost function c_*. Let us first look at an example.

(4.14) Example. Let X_1, X_2 be indeterminates over $\mathbb{C}$, $F := \{X_1^2 + X_2^2\}$ and $I := \{X_1, X_2\}$. Then $L_{\mathbb{C} \to \mathbb{C}(X)}(F|I) = 1 < 2 = L_{\mathbb{R} \to \mathbb{R}(X)}(F|I)$ (check this!). Thus use of constants from $\mathbb{C}$ may decrease the multiplicative complexity of a problem. On the other hand, if $Y_1, \ldots, Y_m$ are additional indeterminates, then $L_{\mathbb{R} \to \mathbb{R}(X)}(F|I) = L_{\mathbb{R}(Y) \to \mathbb{R}(Y)(X)}(F|I) = 2$ (check this!). Thus in this case additional indeterminates are of no use. $\bullet$

By Rem. (4.9)(2+3) we have $L_{K \to K(X)}(F|I) \leq L_{k \to k(X)}(F|I)$. In which case does equality hold?

(4.15) Definition. We call a field extension $k \subseteq K$ *autarkical* iff for any indeterminates $X_1, \ldots, X_\ell$ and subsets $F, I \subseteq k(X) := k(X_1, \ldots, X_\ell)$, F finite, the following is true:

$$L_{K \to K(X)}(F|I) = L_{k \to k(X)}(F|I). \qquad \bullet$$

To proceed further we have to fix some notation. Let $k \subseteq R \subseteq K$ be an intermediate ring, and $\Omega^R := R^c \cup R \cup \{+, -, *, /\}$. We have the inclusions $\Omega^k \subseteq \Omega^R \subseteq \Omega^K$. A straight-line program whose operations belong to Ω^R will be called a straight-line program over R in the sequel.

Before entering into technical details, let us informally describe the basic ideas which show that suitable extensions of fields $k \subseteq K$ are autarkical. Let Γ be a straight-line program computing $F \subset k(X)$ on some input $(K(X); a)$ with result sequence b. Let $R = k[T]$, where T denotes the finite set of all non-zero elements of K that are involved in Γ. In other words, T consists of all λ such that scalar multiplication with λ or λ^c is an operation in Γ. Then Γ can be viewed as a straight-line program over R. Now comes the crucial point of the reasoning: assume that one can find a morphism $\rho: R \to k$ of k-algebras mapping T into $k^\times$. This morphism transforms Γ into a straight-line program over k computing F on input $(A; a)$, where A is a suitable localization of $R[X]$ depending on b. These considerations easily imply that $k \subseteq K$ is autarkical.

(4.16) Proposition. *Let the field extension $k \subseteq K$ have the property that for every finite subset T of $K^\times$ there exists a morphism $\rho: k[T] \to k$ of k-algebras such that $\rho(t) \neq 0$ for all $t \in T$. Then the extension $k \subseteq K$ is autarkical.*

Proof. We have to show that

$$L_{K \to K(X)}(F|I) \geq L_{k \to k(X)}(F|I),$$

for all subsets F, I of $k(X)$ with finite F. Let $\Gamma = (\Gamma_1, \ldots, \Gamma_s)$ be a straight-line program over K of multiplicative length $r = L_{K(X)}(F|I)$ which computes F on input $a \in I^n$. Let $b = (b_{-n+1}, \ldots, b_s)$ be the corresponding result sequence. Then we can write each b_i as $b_i = u_i/v_i$ with $u_i, v_i \in K[X]$. We define T as the set of all non-zero coefficients of all the u_i, v_i and all those operational symbols ω_j (of Γ_j) such that $\omega_j = \lambda$ or $\omega_j = \lambda^c$ for some $\lambda \in K^\times$. The set T is obviously finite. Now let R be the k-subalgebra of K generated by T. The assumptions imply the existence of a morphism $\rho: R \to k$ of k-algebras mapping T into $k^\times$. Note that Γ can be viewed as a straight-line program over R of multiplicative length r. Furthermore, all the u_i, v_i are in $R[X]$. If A denotes the localization of $R[X]$ by the product of all the non-zero u_i and v_i, then $a \in I^n \cap A^n$, and the straight-line program Γ over R computes on input $(A; a)$ the set F. As A is a subalgebra of $K(X)$ we get

$$L_{R \to A}(F|\{a_1, \ldots, a_n\}) = L_{K \to K(X)}(F|I).$$

Now $\rho: R \to k$ extends to a morphism $R[X] \to k[X]$ which maps the element d to a non-zero element of $k[X]$. By the universal mapping property of localizations we can extend this to a morphism $\rho_A: A \to k(X)$. As, in addition, ρ_A is the identity on $A \cap k(X)$, Rem. (4.9) yields

$$L_{k \to k(X)}(F|I) \leq L_{k \to k(X)}(F|\{a_1, \ldots, a_n\})$$
$$\leq L_{R \to A}(F|\{a_1, \ldots, a_n\}) = L_{K \to K(X)}(F|I),$$

and our claim follows. □

The next result describes three types of field extensions to which the last proposition applies.

(4.17) Theorem. *In the following cases the field extension $k \subseteq K$ is autarkical:*

(1) K *is a purely transcendental extension of the infinite field k.*
(2) k *is algebraically closed.*
(3) k *is real closed and K is formally real.*

Proof. It suffices to show that in all three cases the property required in Prop. (4.16) is satisfied.

(1) W.l.o.g. we can assume that $K = k(Y_1, \ldots, Y_m)$ with indeterminates Y_i. Let T be a finite subset of $K^\times$ and $R := k[T]$. Let D denote the product of all numerators and denominators of all elements in T. Then $D \neq 0$. As k is infinite, $D(\eta) \neq 0$, for some $\eta \in k^m$. R is a subalgebra of the local ring

$$\mathcal{R} := \{a/b \mid a, b \in k[Y], b(\eta) \neq 0\},$$

thus the substitution morphism $\mathcal{R} \to k$, $r \mapsto r(\eta)$ restricted to R proves (1).

(2) Now let k be algebraically closed. In this case, the desired property is a version of Hilbert's Nullstellensatz (see, e.g., Lang [316, p. 375]).

(3) If k is real closed, the required property is a version of the real Nullstellensatz or the Artin-Lang Theorem (see, e.g., Lang [316, p. 399, Cor. 3.2]). □

The following corollary states that the use of additional indeterminates for computing a finite subset of $k(X)$ does not help. This result will be used tacitly throughout the book.

(4.18) Corollary. *Let k be infinite, $X_1, \ldots, X_\ell, Y_1, \ldots, Y_m$ be indeterminates over k and $F, I \subseteq k(X)$, F finite. Then we have*

$$L_{k(X,Y)}(F|I) = L_{k(X)}(F|I).$$

4.4* Computation Trees

The model of straight-line programs is inadequate for an algorithmic solution of many important problems. For instance, the problem of sorting a given sequence of real numbers naturally calls for instructions of the form if $a < b$ then ... else ... Allowing such branchings in the course of a computation leads to computation trees, the topic of this section. It should be intuitively clear that computation trees are more powerful than straight-line programs: straight-line programs can be viewed as "degenerated" computation trees. However, as we shall see below, a theoretical investigation of computation trees requires some more technical reasoning.

For motivating the concepts to be introduced later, we start with a simple example: consider the problem of computing a basis for the kernel of a 2×2-matrix $\begin{pmatrix} a & b \\ c & d \end{pmatrix}$ over a field k. The input of our procedure will consist of the entries a, b, c, d of the given matrix. The output will either be zero, or else will consist of one or two linearly independent vectors in k^2. One possible procedure for solving this problem is given by the computation tree $T1$ in Fig. 4.1. Let us first investigate the nodes of this tree. Any (binary) tree carries a natural partial order $\prec$ on the set of its nodes, in which the root is the smallest element. ($u \prec v$ iff u is predecessor of v). An *initial segment* of *length* n of a binary tree is a set of nodes $\mathcal{I} = \{v_1, \ldots, v_n\}$ such that $v_1 \prec v_2 \prec \ldots \prec v_n$, and such that there is

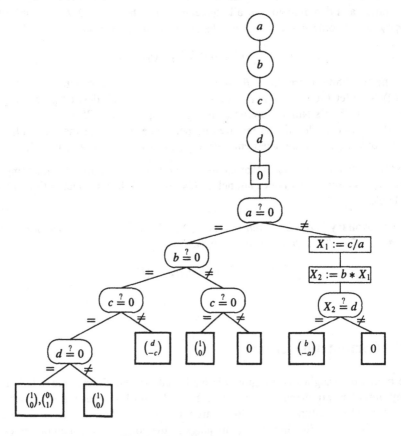

Fig. 4.1. Computation tree $T1$.

a node u which is successor of v_n, but predecessor of all the nodes not lying in $\mathcal{I} \cup \{u\}$. $T1$ has an initial segment of length 4 consisting of the input nodes which are represented by circles. Further, it has branching nodes represented by ovals; these nodes contain test instructions. The rectangles stand for computations nodes. The rectangle containing (the 0-ary operation) zero occurs in order to enable us to compare intermediate results with the constant zero (compare our formal definition below). Finally, the leaves of the tree, marked with doubly framed rectangles, contain the outputs. Here we encounter an interesting point: different output nodes do not necessarily contain outputs of the same length. We have three classes of output nodes: those in the first class correspond to a two-dimensional kernel, those in the second class to a one-dimensional kernel, while those in the third correspond to a zero-dimensional kernel. We have already observed this behavior in Chap. 3: the Euclidean representation $(Q_1, \ldots, Q_{t-1}, A_t)$ of two univariate polynomials A_1 and A_2 has a length depending on the pair (A_1, A_2); hence, if we write the program given in Sect. 3.1 as a computation tree, the output nodes will contain outputs of different lengths; these lengths are not known in advance.

Before turning to a formal definition of computation trees we have to specify the test instructions we will be using. In the example of the tree $T1$ we only have one type of test instructions, namely testing on equality. If k is the field of real numbers (or any other ordered field), we also want to test whether $a \leq b$; without such instructions problems such as sorting would become algorithmically inaccessible. In both cases, $=$ and $\leq$ represent relations in $k \times k$ of arity two. In the following, we denote by P the set of the relations the tree is allowed to use. Usually, the computation tree tests only for equality, thus P $= \{=\}$, but if k is an ordered field, we also allow for inequality testing, thus P $= \{=, \leq\}$. In what follows, it is often not important which choice of P we are considering. In those occasions where this becomes important, however, we shall explicitly say which choice we have in mind. In the computation nodes of a computation tree we allow operations belonging to $\Omega = k^c \cup k \cup \{+, -, *, /\}$. Here is now the formal definition of a computation tree.

(4.19) Definition. (Syntax of computation trees) Let $\Omega = k^c \cup k \cup \{+, -, *, /\}$, $n \in \mathbb{N}$, and P $= \{=\}$ or P $= \{=, \leq\}$ (in case k is ordered). Let a binary tree be given which has an initial segment $\mathcal{I}$ of length n. The elements of $\mathcal{I}$ are called *input nodes*. We partition the set of nodes not lying in $\mathcal{I}$ according to their outdegree d into the sets of *output nodes* $\mathcal{O}$ ($d = 0$), *computation nodes* $\mathcal{C}$ ($d = 1$), and *branching nodes* $\mathcal{B}$ ($d = 2$). A *computation tree* T (over k, expecting inputs of length n) is such a binary tree together with

(1) a function (the *instruction function*) that assigns to any
   - computation node v an operational instruction of the form

$$(\omega; u_1, \ldots, u_m),$$

   where $\omega \in \Omega$ is of arity m and $u_1, \ldots, u_m \in \mathcal{I} \cup \mathcal{C}$ are predecessors of v,

- branching node v a test instruction of the form

$$(\rho; u_1, u_2),$$

where $\rho \in P$ and $u_1, u_2 \in \mathcal{I} \cup C$ are predecessors of v,
- output node v an output instruction of the form

$$(u_1, \ldots, u_m),$$

where $m \in \mathbb{N}$ and $u_1, \ldots, u_m \in \mathcal{I} \cup C$ are predecessors of v,
(2) a partition σ of the set of output nodes such that the length m of the assigned output instructions is constant on σ-classes.   ●

The reader will have no difficulty in identifying the tree $T1$ as a computation tree.

We define an *input* to the computation tree T to be an element $a \in k^n$. More generally, we could allow for inputs pairs $(A; a)$ where $a \in A^n$ and A is a k-algebra which additionally carries a binary relation $\leq$ in the case $P = \{=, \leq\}$. But we will not need this more general viewpoint.

We are now going to describe the semantics of computation trees. Let us first discuss how an input $a \in k^n$ is processed by the tree T. To this end, we successively construct a path T_a in T starting at the root, as well as functions

$$\alpha : \mathcal{I} \cup \{\text{computation nodes on } T_a\} \to A,$$

$$\beta : \{\text{branching nodes on } T_a\} \to \{\text{false, true}\}.$$

We start by setting $T_a := (v_1, \ldots, v_n)$ and $\alpha : \mathcal{I} \to A, \alpha(v_i) := a_i$ for $i \in \underline{n}$, where $v_1 \prec \ldots \prec v_n$ are the input nodes of T. Suppose the path constructed so far ends with the node w. If w is an output node, then the path cannot be extended further and we are done. So assume that w is not an output node. If w is a branching node, then we define the successor of w as being its left (resp. right) son if $\beta(w) = $ true (resp. false). If w is a computation node or $w = v_n$, then its successor is defined to be its son. Let v be the successor of w. We distinguish now three cases. If v is an output node, then we extend the path by v. If v is a branching node with test instruction $(\rho; u_1, u_2)$, then we extend the path by v and set $\beta(v) := \rho(\alpha(u_1), \alpha(u_2))$. Finally, if v is a computation node with instruction $(\omega; u_1, \ldots, u_m)$, then we extend the path to v if $\omega(\alpha(u_1), \ldots, \alpha(u_m))$ is defined and set $\alpha(v) := \omega(\alpha(u_1), \ldots, \alpha(u_m))$. If $\omega(\alpha(u_1), \ldots, \alpha(u_m))$ is undefined, then the path ends at w. If T_a ends up with an output node v, then we call the computation tree T *executable on input* a. Let $(u_1, \ldots, u_m)$ be the output instruction of v. Then $(\alpha(u_1), \ldots, \alpha(u_m))$ is called the *output of T on input a*. The σ-class in which v lies is called the *output class of a*.

Let us now state precisely what a computation tree computes. Let T be a computation tree expecting inputs of length n and let $J \subseteq k^n$. We say that T is *executable on J* iff T is executable on every element of J. Let T have the output partition $\sigma = \{\sigma_1, \ldots, \sigma_t\}$ and let m_i be the common output length of the nodes in σ_i. The set J is partitioned into the subsets $(1 \leq i \leq t)$

$$J_i := \{a \in J \mid \text{output class of } a \text{ is } \sigma_i\}$$

(discard the empty ones). We call this the *partition π of inputs of T on J*. For each i we have a mapping $\varphi_i: J_i \to k^{m_i}$ assigning to $a \in J_i$ its output. The unique extension φ of the φ_i to a map defined on J is called the *computation map of T on J*. On a given input $a \in J$ the tree T decides in which of the classes J_i the element a lies and computes the output $\varphi(a)$ of T on a. This motivates the following definition.

(4.20) Definition. (Collection) A *collection for a subset* $J \subseteq k^n$ consists of a partition $\pi = \{J_1, \ldots, J_t\}$ of J and of a family of functions $\varphi_i: J_i \to k^{m_i}$ for $i \in \underline{t}$. We will denote a collection by (φ, π) where φ is the unique map defined on J which extends the φ_i.   •

The pair (φ, π) consisting of the computation map and the partition of inputs of a computation tree T is clearly a collection for J. Thus computation trees compute collections.

Let us discuss some special cases. A collection (φ, π) for J where φ maps every $a \in A$ to an output of length zero describes a *decision problem*. If $|\pi| = 2$ then we have a *membership problem*. In these cases we omit φ and regard the partition π as a special collection. On the other hand, a collection (φ, π) for J with $\pi = \{J\}$ describes a "pure computational problem." In this case we omit π and speak of the collection φ.

(4.21) Example. For $n \geq m \geq 0$ let $J(n, m)$ be the set of pairs of polynomials (A_1, A_2) of degree n, m, respectively. (We view $J(n, m)$ as a subset of k^{m+n+2}.) For $(d_1, \ldots, d_t) \in \mathbb{N}^t$ with $t \geq 2$, $d_i > 0$ for $1 < i < t$, and $n = \sum_{i=1}^{t} d_i$, $m = \sum_{i=2}^{t} d_i$ we define $D(d_1, \ldots, d_t)$ as the set of all $(A_1, A_2) \in J(n, m)$ whose Euclidean representation $(Q_1, \ldots, Q_{t-1}, A_t)$ has degree sequence $(d_1, \ldots, d_t)$, i.e., $d_i = \deg Q_i$, for $i < t$, and $d_t = \deg A_t$. The sets $D(d_1, \ldots, d_t)$ form a partition of $J(n, m)$. This partition, together with the functions

$$D(d_1, \ldots, d_t) \ni (A_1, A_2) \mapsto (Q_1, \ldots, Q_{t-1}, A_t) \in \prod_{i=1}^{t} k^{d_i+1}$$

that assign to (A_1, A_2) its Euclidean representation define a collection for $J(n, m)$ which formalizes the problem of computing the Euclidean representation.   •

We proceed with introducing the notion of complexity. Let $c: \Omega \cup P \to \mathbb{N}$ be a cost function and T be a computation tree. By adding the costs of the various instructions encountered when going from the root of T to an output node, we may define the cost of any output node of T. If T is executable on J, this gives us a function $t: J \to \mathbb{N}$, the *cost of T on J*, where $t(a)$ is the cost of the output node in which the path T_a ends.

(4.22) Definition. Let (φ, π) be a collection for J, $c: \Omega \cup P \to \mathbb{N}$ be a cost function and $\tau: J \to [0, \infty)$. We say that (φ, π) is *computable in time τ* (with respect to c) iff there is a computation tree T which computes (φ, π) and has cost $\leq \tau$ on J. The *(worst-case) complexity* $C^c(\varphi, \pi)$ of the collection (φ, π) (with respect to c)

is defined as the minimum natural number $r \in \mathbb{N}$ such that (φ, π) is computable in constant time $J \to [0, \infty), a \mapsto r$.   •

Sometimes we also say that a computation tree computes a collection (φ, π) in time $\tau: J \to [0, \infty)$ iff it has cost $\leq \tau$ on J.

Only two cost functions will be of relevance later on. In Chap. 10 we will prove lower bounds on the *multiplicative complexity* $C^*(\varphi, \pi) := C^{c_*}(\varphi, \pi)$ defined with respect to $c_* = 1_{\{*,/\}}$ which charges for the nonscalar multiplications and divisions. As for straight-line programs, we sometimes write C instead of C^*. In Chap. 11 we will succeed in establishing lower bounds on the *multiplicative branching complexity* $C^{*,\leq}(\pi) := C^c(\pi)$ defined with respect to $c = 1_{\{*,/,=,\leq\}}$, which takes nonscalar multiplications, divisions and comparisons into account. Also, we denote by $C^{tot}(\pi)$ the *total complexity* which is defined with respect to $1_{k \cup \{+,-,*,/\} \cup P}$.

(4.23) Example. It is straightforward (though cumbersome) to formulate Gaussian elimination and the Euclidean algorithm as computation trees. Charging any arithmetic operation and test by one, this yields upper bounds on the complexity of the appropriate collections of order of magnitude n^3 and n^2, respectively.   •

4.5* Computation Trees and Straight-line Programs

Throughout this section k is assumed to be an infinite field.

We continue our discussion of the semantics of computation trees with the goal of relating the complexity of collections to straight-line complexity. For this, let us come back to our example of the computation tree $T1$ in Fig. 4.1. If at each test instruction we follow the right path, we obtain the following straight-line program Γ expecting inputs of length 4:

$$\Gamma_1 = (0;), \quad \Gamma_2 = (/; -1, -3), \quad \Gamma_3 = (*; -2, 2).$$

Note that among all the inputs $(a, b, c, d) \in k^4$ only those do not take this path for which $a = 0$, or $ad - bc = 0$; geometrically, these inputs lie on the union of two hypersurfaces in k^4. Hence a *generic* input, that is an input from some Zariski open subset of k^4, follows the path given by the straight-line program Γ. (For some general facts on the Zariski topology see the Exercises of this chapter.) This straight-line program computes some rational functions and the straight-line complexity of these functions is a lower bound for the worst-case cost of the tree in question. In the following we will work out this idea in more generality (compare also Ex. 4.21).

Let T be a computation tree over k which is executable on $J \subseteq K^n$. As usual, we assume $P = \{=\}$ for arbitrary k, and $P = \{=, \leq\}$ if k is ordered. To an output node v of T we assign the *input set belonging to node v*

$$D_v := \{a \in J \mid \text{path } T_a \text{ ends up with } v\}$$

consisting of the inputs in J whose path ends up with the node v. It is obvious that the family $\{D_v\}_v$ is a refinement of the input partition of T, namely any set $J_i = \{a \in J \mid \text{output class of } a \text{ is } \sigma_i\}$ of the input partition of T is the disjoint union of the input sets D_v belonging to the nodes in the class σ_i. Let an output node v be fixed in the following. To v we may assign a unique straight-line program $\Gamma := \Gamma_v$ expecting inputs of length n by forgetting the test and the output instructions along the path from the root to v. We assume that Γ is consistent. Let $(k[X]_d; X)$ be the universal input for Γ. The result sequence of Γ on its universal input defines a map

$$\alpha : \{\text{input nodes}\} \cup \{\text{computation nodes on path to } v\} \to k[X]_d.$$

Let $\mathcal{B}_v$ denote the set of branching nodes of T on the path to v and $\mathcal{B}_{v,\ell} \subseteq \mathcal{B}_v$ be the set of branching nodes where the path to v takes the left branch. To $w \in \mathcal{B}_v$ we assign

$$g_w := \alpha(u_1), \quad h_w := \alpha(u_2) \in k[X]_d$$

if $(\rho_w; u_1, u_2)$ is the test instruction of w. We leave the straightforward proof of the following fact to the reader:

(4.24) $$D_v = \Big\{ a \in J \mid d(a) \neq 0 \text{ and for all } w \in \mathcal{B}_v:$$

$$\rho_w(g_w(a), h_w(a)) \text{ true } \Leftrightarrow w \in \mathcal{B}_{v,\ell} \Big\}.$$

(Recall from Thm. (4.5) that Γ is executable on a iff $d(a) \in k^\times$.) The reader should note that the condition expressed in (4.24) amounts to a system of equalities, non-equalities, and inequalities of the form

$$g_{w_p}(a) = h_{w_p}(a), g_{w_q}(a) \neq h_{w_q}(a), g_{w_r}(a) \leq h_{w_r}(a), \text{ or } g_{w_s}(a) > h_{w_s}(a),$$

where the latter two may only occur if $P = \{=, \leq\}$. If $(u_1, \ldots, u_m)$ is the output instruction of v, we define

$$f_1 := \alpha(u_1), \ldots, f_m := \alpha(u_m) \in k[X]_d.$$

It is easy to see that $(f_1(a), \ldots, f_m(a))$ is the output of $a \in D_v$. As Γ computes the g_w, h_w and f_i on its universal input, we conclude that

(4.25) $$L^c_{k[X]_d}(\{f_1, \ldots, f_m\} \cup \{g_w, h_w \mid w \in \mathcal{B}_v\}) \leq c\text{-length}(\Gamma)$$

for an arbitrary cost function $c: \Omega \to \mathbb{N}$.

The facts expressed in (4.24) and (4.25) will be the basis for all our lower bound proofs on the complexity of collections in Chap. 10. Let us draw a first, very simple conclusion.

(4.26) Proposition. *Consider a collection φ consisting of the rational map*

$$\varphi = (\varphi^1, \ldots, \varphi^m) : k^n \supseteq \text{Def } \varphi \to k^m.$$

Then we have for any cost function $c \colon \Omega \to \mathbb{N}$

$$L^c_{k(X)}(\varphi^1, \ldots, \varphi^m) \leq C^c(\varphi).$$

Proof. Let $J := \text{Def } \varphi$ and let T be a computation tree which computes φ in time $\tau \colon J \to \mathbb{N}$. Then J is a finite disjoint union of the sets

$$D_v = \{a \in J \mid \text{path } T_a \text{ ends with } v\},$$

where v runs over all output nodes of T. Since J is irreducible in the Zariski topology, there is some v such that D_v is dense in J (cf. Ex. 4.12). By the preceding considerations there are $f_1, \ldots, f_m \in k[X]_d$ such that for all $a \in D_v$ we have $(f_1(a), \ldots, f_m(a)) = \varphi(a)$. Therefore, $f_i = \varphi_i$ for all i and since

$$L^c_{k[X]_d}(f_1, \ldots, f_m) \leq c\text{-length}(\Gamma_v) \leq \tau(a),$$

the assertion follows.   □

This proposition justifies a thorough study of the straight-line complexity of rational maps, since every lower bound on this implies a lower bound on the complexity of φ viewed as a collection.

Let us draw another conclusion and assume $\mathrm{P} = \{=\}$. A subset of k^n is called *constructible* iff it is a Boolean combination of *Zariski closed* subsets, i.e., of subsets of the form

$$\{a \in k^n \mid f(a) = 0\}, \quad f \in k[X_1, \ldots, X_n].$$

A subset of k^n described by some polynomial equalities and nonequalities

$$\{a \in k^n \mid f_1(a) = 0, \ldots, f_r(a) = 0, g(a) \neq 0\}, \quad f_i, g \in k[X_1, \ldots, X_n]$$

is called *locally closed*. These subsets are just the intersections of an open and a closed subset of k^n in the Zariski topology. A locally closed set is clearly constructible. Assume that the computation tree T (with $\mathrm{P} = \{=\}$) is executable on a constructible subset $J \subseteq k^n$. Then, by the above description of the input set D_v in (4.24), all the sets D_v are locally closed and the restriction of the computation map of T to the D_v's are restrictions of rational maps. From this observation immediately follows one direction of the following proposition. The proof of the other direction is left to the reader.

(4.27) Proposition. *Let k be any algebraically closed field and $\mathrm{P} = \{=\}$. Further, let (φ, π) be a collection for a constructible subset $J \subseteq k^n$. Then (φ, π) can be computed by a computation tree iff the partition π has a refinement $\{D_1, \ldots, D_s\}$ into locally closed subsets such that all $\varphi_{|D_i} \colon D_i \to k^{m_i}$ are restrictions of rational maps. In this case all the sets in the partition π of inputs are constructible.*

For P $= \{=, \leq\}$ we have a similar statement. A subset of $\mathbb{R}^n$ is called *semi-algebraic* if it is a Boolean combination of subsets of the form

$$\{a \in \mathbb{R}^n \mid f(a) \geq 0\}, \quad f \in \mathbb{R}[X_1, \ldots, X_n].$$

The proof of the next proposition is similar to that of the previous one.

(4.28) Proposition. *Prop. (4.27) holds for the field $k = \mathbb{R}$ with P $= \{=, \leq\}$ when we replace "locally closed" and "constructible" by "semi-algebraic."*

4.6 Exercises

4.1. Horner's rule to evaluate a polynomial $f_n = \sum_{i=0}^n a_i X^{n-i}$ is as follows:

$$f_0 := a_0$$
$$f_{i+1} := f_i * X + a_{i+1} \quad (0 \leq i < n).$$

Describe this formally by a straight-line program of multiplicative and additive length n.

4.2. Consider the straight-line programs to invert a generic triangular, resp. generic 4×4-matrix, which are described in Prop. (16.3), resp. Ex. 16.3. What is the universal input of these straight-line programs?

4.3. Let $(g_1, \ldots, g_r)$ be a computation sequence in $k(X) := k(X_1, \ldots, X_n)$ modulo a subset $I \subseteq k(X)$. Show that there is some nonzero polynomial $d \in k[X]$ such that $(g_1, \ldots, g_r)$ is a computation sequence in the localization $k[X]_d$.

4.4. Suppose k is an infinite field and let $\mathcal{O}_\lambda \subseteq k(X) := k(X_1, \ldots, X_n)$ denote the ring consisting of the rational functions which are defined at $\lambda \in k^n$. ($\mathcal{O}_\lambda$ is called the local ring of λ.) Let $F, I \subset k(X)$, F finite. Show that $L_{k(X)}^c(F|I) = L_{\mathcal{O}_\lambda}^c(F|I)$ for Zariski almost all $\lambda \in k^n$, where c denotes any cost function.

4.5. Generalize the autarky result of Thm. (4.17) to the additive and to the total complexity.

4.6. Given a sequence of $n = 2^m$ real numbers we can sort them as follows: we divide the sequence into two sequences of length $n/2$, sort these sequences recursively, and then merge the sorted sequences. (This algorithm is called *mergesort*.)

(1) Show that this algorithm can be modeled by a computation tree over $\mathbb{R}$ without computation nodes. (Such trees are also called *decision trees*.) Describe the collection π_n this tree computes.
(2) Let $C^\leq(\pi_n)$ be the minimum r such that the collection π_n is computable in constant time r by a decision tree. Using mergesort show that $C^\leq(\pi_n) \leq 2C^\leq(\pi_{n/2}) + (n-1)$. Deduce that $C^\leq(\pi_{2^m}) \leq (m-1)2^m + 1$.
(3) Show that $C^\leq(\pi_n) \geq \log n! \geq n(\log n - 2)$ for all $n \geq 2$. (Hint: what ist $|\pi_n|$?)

4.7. What about universal inputs for computation trees?

4.8. Prove the description of the input set D_v expressed in (4.24).

4.9. Describe the input sets D_v of the computation tree $T1$ in Fig. 4.1.

4.10. Complete the proof of Prop. (4.27).

4.11. Prove Prop. (4.28).

The goal of the following exercises is to recall notions and fundamental facts from general topology which are relevant for the Zariski topology. (Compare Hartshorne [232, Chap. 1, §1].) These will be needed explicitly in Chap. 8 and Chap. 10.

4.12. A topological space X is called *irreducible* iff it cannot be written as the union $X = X_1 \cup X_2$ of two closed proper subsets. Prove the following facts.

(1) X is irreducible iff any nonempty open subset of X is dense in X.
(2) Any dense subset of an irreducible topological space is irreducible in its induced topology.
(3) If Y is a subset of a topological space X, which is irreducible in its induced topology, then its closure $\overline{Y}$ is irreducible as well.
(4) If the irreducible topological space X is written as a finite union of subsets X_i, $X = X_1 \cup \ldots \cup X_r$, then at least one of the X_i is dense in X.

4.13. Prove that the following conditions on a topological space X are equivalent.

(1) X satisfies the *descending chain condition* for closed subsets, i.e., for any sequence $Y_1 \supseteq Y_2 \ldots$ of closed subsets, there is an integer r such that $Y_r = Y_{r+1} = \ldots$.
(2) X satisfies the *ascending chain condition* for open subsets, i.e., for any descending sequence $Y_1 \subseteq Y_2 \subseteq \ldots$ of open subsets, there is an integer r such that $Y_r = Y_{r+1} = \ldots$.
(3) Every nonempty family of open subsets of X has a maximal element.

A topological space satisfying one (and hence all) of the above conditions is called *Noetherian*.

4.14. Show that any subset of a Noetherian topological space is Noetherian in its induced topology.

4.15. Show that a Noetherian topological space X can be expressed as a finite union $X = X_1 \cup \ldots \cup X_r$ of irreducible closed subsets X_i. If we require that $X_i \not\supseteq X_j$ for $i \neq j$, then the X_i are uniquely determined. They are called the *irreducible components* of X.

4.16. Let k be an infinite field. The *Zariski topology* of k^n is defined by considering the zerosets of families of polynomials as the closed sets. For short we call such sets *Zariski closed*. Hilbert's Basissatz states that every ideal of $k[X_1, \ldots, X_n]$ is finitely generated (cf. Lang [316, Chap. V, §2]). Use this to prove that every Zariski closed subset of k^n is the zeroset of finitely many polynomials. Conclude that k^n with the Zariski topology is Noetherian. Prove also that k^n is irreducible.

4.17. Prove that for a subset Y of a topological space X the following statements are equivalent:

(1) Y is the intersection of an open set with a closed set.
(2) Y is open in its closure.

Such a subset Y is called *locally closed*.

4.18. Let X be a topological space. The Boolean algebra generated by the closed (open) subsets of X consists of the so-called *constructible* subsets of X. The constructible subsets of k^n encountered in the text are those with respect to the Zariski topology. Show that every constructible subset of X is a finite union of locally closed subsets. If X is Noetherian, then we may even assume that these locally closed subsets are irreducible.

4.19. Let Y be a constructible subset of an irreducible topological space X. Show that Y contains an open and dense subset of X.

4.20. Show that every semi-algebraic set in $\mathbb{R}^n$ can be written as a finite union of semi-algebraic sets of the form

$$\{\xi \in \mathbb{R}^n \mid f_1(\xi) = \ldots = f_p(\xi) = 0, g_1(\xi) > 0, \ldots, g_q(\xi) > 0\},$$

where $p, q \in \mathbb{N}'$, $f_1, \ldots, f_p, g_1, \ldots, g_q \in \mathbb{R}[X_1, \ldots, X_n]$.

4.21. (Generic path) Let T be a computation tree which is executable on an irreducible constructible subset $J \subseteq k^n$, k being an infinite field. We call a path from the root to an output node *generic* for J iff all inputs in a dense subset of J (with respect to the Zariski topology) take this path. (This means that the input set D_v is dense in J.)

(1) Show that T possesses at least one generic path for J.
(2) If P $= \{=\}$, then T has exactly one generic path.
(3) If P $= \{=, \leq\}$ and $k = \mathbb{R}$, then T may have many generic paths.

(Hint: use Ex. 4.12 and 4.19.)

4.22. What is the generic path of the computation tree $T1$ of Fig. 4.1? What about the generic paths of the computation trees discussed in Ex. 4.6?

4.7 Notes

The notion of a straight-line program as a model for algebraic computations without branching goes back to Ostrowski's investigation [403] of the optimality of Horner's rule in 1954. Also in that paper, computation sequences and the nonscalar complexity measure were considered for the first time. The model of computation trees was designed to capture algebraic computations in fixed dimensions with branchings. The first evidences for this model seem to be Strassen's general definition of a "program" in [494] as well as Rabin's paper [429] where the model of algebraic decision trees was introduced (for a definition see Ex. 11.4).

The Def. (4.2) of a straight-line program can essentially be found in Belaga [35], Winograd [552], and Strassen [494]. The notion of a universal input for a straight-line program is from Bürgisser and Lickteig [102]. The dimension bound (4.12) is due to Fiduccia [169]. The notion of autarky as well as Thm. (4.17) is from Strassen [494].

The works of Strassen [502, 504], Ben-Or [37], and Smale [486] contain slightly different definitions of computation trees along with lower bounds. Our Def. (4.19) of computation trees closely follows Strassen [504], from where we have also taken the notion of a collection. A version of Prop. (4.27) is already in Strassen [496].

We finish these notes by mentioning some further computational models which have been studied in the algebraic setting. Von zur Gathen [186] defines arithmetic networks which are in a sense equivalent to algebraic computation trees. In computational geometry, one usually considers the model of a "real random access machine," which is similar to the ordinary RAM, introduced by Cook and Rackoff [122], but with the assumption of unit cost and infinite-precision for arithmetic operations with reals. (Cf. Preparata and Shamos [425].) All the models mentioned so far do not take explicitly into account the aspect of uniformity: the algorithms only treat problems of a fixed input dimension, which is not considered as part of the input. In 1989 Blum, Shub and Smale [59] have defined a model of computation over the reals (or over any ring) which allows to speak about uniformity in a precise way. Their model (BSS-model for short) may be viewed as sort of a Turing machine [523] working over any ring. For fixed input dimension the BSS-model is equivalent to the model of computation tree. Based on their model, Blum, Shub and Smale have developed a theory of recursive functions, defined complexity classes, and identified complete problems over the reals. In particular, an analogue of Cook's hypothesis $\mathbf{P} \neq \mathbf{NP}$ reappears as a major open problem in that setting. The problem to decide whether a given multivariate polynomial of degree four with real coefficients has a real zero turns out to be an $\mathbf{NP}$-complete problem in that framework. For details and further references the reader is referred to the forthcoming book by Blum et al. [58].

Chapter 5. Preconditioning and Transcendence Degree

In 1955 Motzkin discovered that Horner's rule is not the fastest way to evaluate a polynomial when allowing for preconditioning of the coefficients. He found that the minimum number of multiplications needed to evaluate almost all real univariate polynomials of degree $n \notin \{3, 5, 7, 9\}$ equals $\lfloor n/2 \rfloor + 1$ when starting from the variable, real constants and allowing arbitrarily many additions. In the cases $n \in \{3, 5, 7, 9\}$ he claimed that the optimal number of multiplications equals $\lfloor n/2 \rfloor + 2$. Motzkin also determined the multiplicative complexity of a "typical" univariate rational function. Unfortunately, only an abstract of his results without proofs is published [386]. Belaga [35] exhibited algorithms to evaluate any complex univariate polynomial of degree n with $\lfloor (n + 3)/2 \rfloor$ multiplications and $n + 1$ additions. Furthermore, he proved that n additions are necessary in general. Pan [404] obtained similar results for real polynomials.

In this chapter we present a uniform treatment of these results. We exhibit several variants of straight-line programs which compute a polynomial $f \in K[X]$ of degree n from X and (precomputed) elements of K with about $n/2$ multiplications and n additions. Then we assign to a finite set of rational functions $F \subseteq K(X_1, \ldots, X_n)$ its coefficient field over a subfield $k \subseteq K$ and prove that its transcendence degree yields a lower bound on both the multiplicative and additive complexity of F. For almost all polynomials this implies the near optimality of the straight-line programs considered before. Finally, we discuss an extension of the above transcendence degree bound to the context of linearly disjoint fields, due to Baur and Rabin [31].

5.1 Preconditioning

In this section all straight-line programs use only operations in $\Omega = \{+, -, *\}$.

We want to study here the fundamental computational problem of evaluating a univariate polynomial. To begin with, let us consider a fourth degree polynomial $f = \sum_{i=0}^{4} \alpha_i X^i \in \mathbb{C}[X]$ with complex coefficients ($\alpha_4 \neq 0$). In order to evaluate f at some complex number we can use Horner's rule; alternatively we may proceed as follows. If there exist $\beta_0, \ldots, \beta_4 \in \mathbb{C}$ such that

$$f = ((g + X + \beta_2)g + \beta_3)\beta_4,$$

where
$$g := (X + \beta_0)X + \beta_1,$$
then we can compute f from X and $\beta_0, \ldots, \beta_4$ using only three multiplications and five additions. However, a straightforward calculation shows that

$$\beta_0 := (\alpha_3/\alpha_4 - 1)/2, \quad \gamma := \alpha_2/\alpha_4 - \beta_0(\beta_0 + 1), \quad \beta_1 := \alpha_1/\alpha_4 - \beta_0\gamma,$$
$$\beta_2 := \gamma - 2\beta_1, \qquad\quad \beta_3 := \alpha_0/\alpha_4 - \beta_1(\beta_1 + \beta_2), \quad \beta_4 := \alpha_4$$

satisfy the above conditions. Thus, in comparison with Horner's rule, we have traded one multiplication for one addition. Of course, the computation of the β_i from the α_i also needs time, but this preliminary calculation must be done only once in order to be able to evaluate f many times with this method.

More generally, let a polynomial $f = \sum_{i=0}^n \alpha_i X^i \in \mathbb{C}[X]$ be given. In a first phase, the so called *preconditioning phase*, we compute complex numbers $\beta_0, \ldots, \beta_m$ from the coefficients $\alpha_0, \ldots, \alpha_n$ of f. In a second phase, we compute $f(\xi)$ from a given $\xi \in \mathbb{C}$ and from $\beta_0, \ldots, \beta_m$ by a straight-line program. The preconditioning phase might consist of rational operations, as in the previously discussed example, or it might involve more complex tasks as finding roots of polynomials. *In the following, we will completely ignore the complexity of the preconditioning phase and only focus on the complexity of the second phase.* This is justified if we intend to evaluate the fixed polynomial f many times since the preconditioning must be done only once. We try to determine the minimum of the complexities of f with respect to the input sets $\{X, \beta_0, \ldots, \beta_m\}$ taken over all $m \in \mathbb{N}$, $\beta_0, \ldots, \beta_m \in \mathbb{C}$. This quantity is of course the complexity of f with respect to the input set $\mathbb{C} \cup \{X\}$.

We will work in little more generality. Let K be a fixed field. In this section we will describe straight-line programs using operations in $\Omega = \{+, -, *\}$ which compute polynomials in $K[X]$ from X and elements of K. The reader should note that scalar multiplications in a subfield of K (as well as taking constants from there) are not considered here! With respect to such straight-line programs we define for a finite set of polynomials $F \subseteq K[X]$ the *complexities with preconditioning*

$$L^*[F] \quad := \quad L_{K[X]}^{c_*}(F|K \cup \{X\}),$$
$$L^+[F] \quad := \quad L_{K[X]}^{c_+}(F|K \cup \{X\}),$$
$$L^{\text{tot}}[F] \quad := \quad L_{K[X]}^{\text{tot}}(F|K \cup \{X\}),$$

where $c_* = 1_{\{*\}}$, $c_+ = 1_{\{+,-\}}$, $\text{tot} = 1_{\{+,-,*\}}$. Note that $L^*[F]$ counts *all* multiplications, so multiplications with scalars in K are not free of charge.

(5.1) Theorem. *Let K be real or algebraically closed and $f \in K[X]$ of degree n. Then $L^*[f] \leq \lfloor n/2 \rfloor + 2$. More precisely, there is a straight-line program computing f in $K[X]$ from X and some elements of K using at most $\lfloor n/2 \rfloor + 2$ multiplications and $n + 1$ additions or subtractions, hence*

$$L^{\text{tot}}[f] \leq n + \lfloor n/2 \rfloor + 3.$$

We first prove the following lemma which immediately implies the statement of the above theorem in the case of algebraically closed fields K.

(5.2) Lemma. *Let $f = \sum_{i=0}^{n} \alpha_i X^i \in K[X]$ be a polynomial of degree n, K an arbitrary field, and assume that $h := \sum_{2\ell+1 \le n} \alpha_{2\ell+1} X^\ell$ is either constant or a product of linear factors in $K[X]$. Then there is a straight-line program computing f and X^2 in $K[X]$ from X and some elements of K using at most $\lfloor n/2 \rfloor + 2$ multiplications, and n additions or subtractions.*

Proof. We proceed by induction on n. The start "$n = 1$" being clear let us assume that $n > 1$. Putting $g := \sum_{2\ell \le n} \alpha_{2\ell} X^\ell$ we can write $f(X) = g(X^2) + Xh(X^2)$. If $h \in K$, then the assertion follows by computing $g(X^2)$ using Horner's rule. So assume $h \notin K$ and let $\xi \in K$ be a root of h. We have

$$h(X) = (X - \xi)h_1(X), \quad g(X) = (X - \xi)g_1(X) + \eta$$

with $g_1, h_1 \in K[X]$, $\eta \in K$. Therefore

$$f(X) = (X^2 - \xi)f_1(X) + \eta,$$

where $f_1(X) := g_1(X^2) + Xh_1(X^2)$. We can compute $f(X)$ from X^2, $f_1(X)$ and ξ with one multiplication, one addition and one subtraction. As h_1 is again constant or a product of linear factors, the assertion follows by the induction hypothesis. $\square$

(5.3) Remark. The preconditioning phase described in the above lemma involves finding roots of polynomials. $\bullet$

For simplicity we supply a proof of Thm. (5.1) only for $K = \mathbb{R}$, the reader familiar with the theory of real closed fields can easily settle the general case. We need the following lemma.

(5.4) Lemma. *Let $p, q \in \mathbb{R}[X]$ and $f = p + iq \in \mathbb{C}[X]$ be of positive degree. If all complex roots of f have a positive imaginary part, then all the roots of p and q are reals.*

Proof. Let $f = \alpha_n \prod_{j=1}^{n}(X - \xi_j)$ where $\alpha_n, \xi_1, \ldots, \xi_n \in \mathbb{C}$, $\alpha_n \neq 0$, $n \ge 1$. Suppose that $\zeta \in \mathbb{C}$ is a root of p and let $\bar{\zeta}$ be its complex conjugate. Then

$$f(\zeta) = i\,q(\zeta), \quad f(\bar{\zeta}) = iq(\bar{\zeta}) = i\,\overline{q(\zeta)},$$

hence $|f(\zeta)| = |f(\bar{\zeta})|$. This means

$$\prod_{j=1}^{n} |\zeta - \xi_j| = \prod_{j=1}^{n} |\bar{\zeta} - \xi_j|.$$

Since all the ξ_j have positive imaginary part this is only possible when $\zeta \in \mathbb{R}$ (cf. Fig. 5.1). Analogously one shows that all the roots of q are reals. $\square$

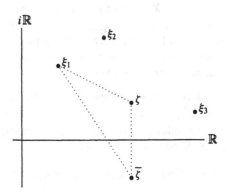

Fig. 5.1. Proof of Lemma (5.4).

Proof of Thm. (5.1). According to Lemma (5.2) it remains to consider the case $K = \mathbb{R}$. Let $f \in \mathbb{R}[X]$ be of positive degree. Choose $c \in \mathbb{R}$ such that all the roots of $f_1(X) := f(X - c)$ have a positive real part. Let $f_1(X) = g(X^2) + Xh(X^2)$ with $g, h \in \mathbb{R}[X]$. Then all the roots of

$$f_2(X) := f_1(-iX) = g(-X^2) - iXh(-X^2)$$

have a positive imaginary part. Lemma (5.4) implies therefore that all the roots of $-Xh(-X^2)$, and in particular all the roots of $h(X)$, are reals. The assertion follows now from Lemma (5.2) (the shift $X \mapsto X - c$ accounts for one subtraction). $\square$

(5.5) Remark. We have actually shown the better bound $L^{tot}[f] \le n + \lfloor n/2 \rfloor + 2$ for an algebraically closed K. This improvement by one can also be made in the case of a real closed K; for the more involved proof we refer to Ex. 5.7.   •

The preconditioning process just considered involves finding roots of polynomials. We are now going to discuss an algorithm which uses only rational operations in the preconditioning phase, and which works over any field of characteristic zero.

(5.6) Theorem (Paterson and Stockmeyer). *Let K be a field of characteristic zero and $f \in K[X]$ with $\deg f < n$. Then*

$$L^*[f] \le n/2 + 2 \log n.$$

More precisely, there is a straight-line program computing f in $K[X]$ from X and some elements of K using at most $n/2 + 2 \log n$ multiplications and $3n/2 - 1$ additions or subtractions.

Proof. For $\ell \geq 1$ let μ_ℓ denote the maximum of the complexities

$$L^*_{K[X]}\left(g \mid K \cup \{X, X^2, X^4, \ldots, X^{2^{\ell-1}}\}\right)$$

taken over all *monic* polynomials $g \in K[X]$ of degree $2^\ell - 1$. Let $m = 2^\ell$ and

$$f = X^{2m-1} + \alpha_{2m-2}X^{2m-2} + \ldots + \alpha_1 X + \alpha_0 \in K[X].$$

We can write

$$f = (X^m + \delta)f_1 + f_2,$$

where

$$\begin{aligned}
f_1 &:= X^{m-1} + \alpha_{2m-2}X^{m-2} + \ldots + \alpha_{m+1}X + \alpha_m, & \delta &:= \alpha_{m-1} - 1, \\
f_2 &:= X^{m-1} + \beta_{m-2}X^{m-2} + \ldots + \beta_1 X + \beta_0, & \beta_i &:= \alpha_i - \alpha_{m+i}\delta.
\end{aligned}$$

This implies $\mu_{\ell+1} \leq 2\mu_\ell + 1$. Since $\mu_1 = 0$ we obtain $\mu_\ell \leq 2^{\ell-1} - 1$ for $\ell \geq 1$. So we have shown that for all $\ell \in \mathbb{N}$ and all (not necessarily monic) polynomials $f \in K[X]$ of degree $2^\ell - 1$ that

$$L^*_{K[X]}\left(f \mid K \cup \{X, X^2, X^4, \ldots, X^{2^{\ell-1}}\}\right) \leq 2^{\ell-1}.$$

Now let $f \in K[X]$ be any nonzero polynomial with $\deg f < n$, w.l.o.g. $\deg f = n - 1$. We may write

$$n = 2^{\ell_0} + 2^{\ell_1} + \ldots + 2^{\ell_s}, \quad \ell_0 > \ell_1 > \ldots > \ell_s,$$

$$f = f_0 + f_1 X^{2^{\ell_0}} + f_2 X^{2^{\ell_0}+2^{\ell_1}} + \ldots + f_s X^{2^{\ell_0}+2^{\ell_1}+\ldots+2^{\ell_{s-1}}},$$

where $f_i \in K[X]$ with $\deg f_i \leq 2^{\ell_i} - 1$. Replacing f by $f(X+c)$ for some $c \in K$ satisfying $f(c)f^{(1)}(c) \cdots f^{(n-1)}(c) \neq 0$, we may even assume that $\deg f_i = 2^{\ell_i} - 1$ for all i. (Here the assumption char $K = 0$ is used.) We compute now f as follows. First we calculate $X^2, X^4, X^8, \ldots, X^{2^{\ell_0}}$ with ℓ_0 multiplications. Then we compute $f_0, f_1, \ldots, f_s$ as above with at most $\sum_{j=0}^s 2^{\ell_j-1} = n/2$ multiplications. Finally, we obtain

$$f = \left(\ldots(f_s X^{2^{\ell_{s-1}}} + f_{s-1})X^{2^{\ell_{s-2}}} + \ldots\right)X^{2^{\ell_0}} + f_0$$

using s further multiplications. Since $s \leq \ell_0 \leq \log n$ we get

$$L^*[f] \leq \ell_0 + n/2 + s \leq n/2 + 2\log n.$$

It is easy to check that the previously described straight-line program computing a monic polynomial $f \in K[X]$ of degree $2^\ell - 1$ from $K \cup \{X, X^2, X^4, \ldots, X^{2^{\ell-1}}\}$ uses $3 \cdot 2^{\ell-1} - 2$ additions. From this we see that the above straight-line program for computing an arbitrary $f \in K[X]$ of degree $n - 1$ uses at most

$$1 + \sum_{j=0}^s (3 \cdot 2^{\ell_j-1} - 2) + s \leq 3n/2 - 1$$

additions (the shift $X \mapsto X + c$ accounts for one addition). $\qquad\square$

5.2 Transcendence Degree

For the whole section $k \subseteq K$ denotes a fixed field extension.

Our goal is here to prove that the algorithms presented in the last section for evaluating polynomials with preconditioning are essentially optimal.

(5.7) Definition. Let $K(X_1, \ldots, X_m)$ be a rational function field in the indeterminates $X_1, \ldots, X_m$ over K and $F = \{f_1, \ldots, f_r\} \subseteq K(X_1, \ldots, X_m)$ a finite subset. Let $f_i = u_i/v_i$ where $u_i, v_i \in K[X]$ are relatively prime and v_i has a nonzero coefficient lying in the subfield k. We define the *coefficient field* Coeff(F) of F over k as the subfield of K generated by k and all the coefficients of $u_1, \ldots, u_r, v_1, \ldots, v_r$. •

This notion is well-defined. The coefficient field over k of a finite subset $F \subseteq K[X_1, \ldots, X_m]$ of polynomials is obviously the subfield of K generated by k and all the coefficients of all $f \in F$. For shortness we write Coeff($f_1, \ldots, f_r$) instead of Coeff($\{f_1, \ldots, f_r\}$).

(5.8) Lemma. *Assume* $f_i = g_i/h_i$, $g_i, h_i \in K[X_1, \ldots, X_m]$, $h_i \neq 0$ *for* $i \in \underline{r}$. *Then*

$$\text{Coeff}(f_1, \ldots, f_r) \subseteq \text{Coeff}(g_1, \ldots, g_r, h_1, \ldots, h_r).$$

Proof. We denote the coefficient field on the right-hand side by $\tilde{K}$. Let $f_i = u_i/v_i$ with relatively prime $u_i, v_i \in K[X]$ and assume that the coefficient of $X_1^{\mu_{i1}} \cdots X_m^{\mu_{im}}$ in v_i equals one. Then we have for $i \in \underline{r}$

$$u_i h_i = v_i g_i \text{ and coefficient of } X_1^{\mu_{i1}} \cdots X_m^{\mu_{im}} \text{ in } v_i \text{ equals one.}$$

This condition can be written as a system of linear equations over the field $\tilde{K}$ in the coefficients of $u_1, \ldots, u_r, v_1, \ldots, v_r$. Because this system has a unique solution we conclude that

$$\text{Coeff}(f_1, \ldots, f_r) = \text{Coeff}(u_1, \ldots, u_r, v_1, \ldots, v_r) \subseteq \tilde{K}. \qquad \square$$

In the following we view $K(X) := K(X_1, \ldots, X_m)$ as an algebra over the subfield k. We are going to study the multiplicative complexity $L^*_{K(X)}(F|I)$, additive complexity $L^+_{K(X)}(F|I)$, and total complexity $L^{tot}_{K(X)}(F|I)$ of a finite subset $F \subseteq K(X)$ with respect to an input set $I \subseteq K(X)$. There is a danger of confusion here. The multiplicative complexity $L^*_{K(X)}$ counts all the multiplications and divisions except the multiplications with scalars in the *subfield* k, because we view $K(X)$ as a k-algebra. If we viewed $K(X)$ as a K-algebra, then $L^*_{K(X)}$ had a different meaning, namely it would only count the multiplications which are not multiplications by scalars in K. In order to distinguish these two situations we often write $k \to K(X)$ and $K \to K(X)$, respectively. The reader might wonder about the special treatment of the subfield k. In $L^*_{k \to K(X)}$ multiplications with scalars in k are not counted. Why do we not count just all the multiplications? It turns out that the method of this section yields lower bounds not only on the total number of multiplications, but even on $L^*_{k \to K(X)}$. For this reason we focus on this quantity. However, the reader should note that a distinction of multiplications with scalars in k from other multiplications is not relevant for the additive and total complexity.

For a finite set of rational functions $F \subseteq K(X_1, \ldots, X_m)$ it is useful to introduce the following abbreviations

$$
\begin{aligned}
L^*(F) &:= L^*_{k \to K(X)}(F \mid K \cup k(X)), \\
L^+(F) &:= L^+_{k \to K(X)}(F \mid K \cup k(X)), \\
L^{tot}(F) &:= L^{tot}_{k \to K(X)}(F \mid K \cup k(X)).
\end{aligned}
$$

Apart from the special treatment of the subfield k, the difference to the complexities considered in the previous section is that we allow here the computations to take place in the rational function field, and that the input set is $K \cup k(X_1, \ldots, X_m)$. We have $L^*(F) \leq L^*[F]$ for a set of polynomials F and similar relations hold for L^+ and L^{tot}. The above abbreviations are of course tailored for the subsequent lower bounds in terms of the transcendence degree of the field extension $k \subseteq \text{Coeff}(F)$. It is a remarkable fact that these lower bounds even hold for the input set $K \cup k(X_1, \ldots, X_m)$ and not only for $K \cup \{X_1, \ldots, X_m\}$. (Thus precomputing rational functions in the $X_1, \ldots, X_m$ over k does not help in the examples considered below.)

(5.9) Transcendence Degree Bound. *Let $F \subseteq K(X_1, \ldots, X_m)$ be a finite subset of a rational function field in the indeterminates $X_1, \ldots, X_m$ over K. Then*

$$
\begin{aligned}
L^*(F) &\geq \frac{1}{2}\left(\text{tr.deg}_k \, \text{Coeff}(F) - |F| \right), \\
L^+(F) &\geq \text{tr.deg}_k \, \text{Coeff}(F) - |F|.
\end{aligned}
$$

Proof. To prove the statement on the multiplicative complexity we show the following.

> Let $\Gamma = (\Gamma_1, \ldots, \Gamma_t)$ be a straight-line program over k (expecting inputs of length N) which is executable in $K(X)$ on $(b_{-N+1}, \ldots, b_0) \in (K \cup k(X))^N$, and denote the corresponding result sequence by $(b_{-N+1}, \ldots, b_t)$. Then there exist $u_i, v_i \in K[X]$, $c_i \in K$ for $-N < i \leq t$ such that $v_i \neq 0$, $b_i = c_i + u_i/v_i$, and
>
> $$\text{tr.deg}_k \, \text{Coeff}(u_{-N+1}, \ldots, u_t, v_{-N+1}, \ldots, v_t) \leq 2 \, c^*\text{-length of } \Gamma.$$

This statement implies the asserted lower bound for the multiplicative complexity. Namely, let $F = \{b_i \mid i \in I\}$, $|I| = |F|$. Then, by Lemma (5.8), we have $\text{Coeff}(F) \subseteq \text{Coeff}\left(\bigcup_{i \in I} \{u_i, v_i, c_i\} \right)$. Hence

$$
\begin{aligned}
\text{tr.deg}_k \, \text{Coeff}(F) &\leq \text{tr.deg}_k \, \text{Coeff}\left(\bigcup_{i \in I} \{u_i, v_i\} \right) + |F| \\
&\leq 2L^*(F) + |F|.
\end{aligned}
$$

We are going to prove the above statement by induction on t. Write $K_i :=$ Coeff$(u_{-N+1}, \ldots, u_i, v_{-N+1}, \ldots, v_i)$. To verify the start "$t = 0$" take $c_i := b_i$, $u_i := 0$, $v_i := 1$ if $b_i \in K$ and choose $c_i := 0$, $u_i, v_i \in k[X]$ such that $b_i = u_i/v_i$ if $b_i \in k(X)$. Let us assume now that $t > 0$.

Case 1. $\Gamma_t = (+; \mu, \nu)$. We have

$$b_t = b_\mu + b_\nu = (c_\mu + c_\nu) + \frac{u_\mu v_\nu + u_\nu v_\mu}{v_\mu v_\nu}$$

and set $c_t := c_\mu + c_\nu$, $u_t := u_\mu v_\nu + u_\nu v_\mu$, $v_t := v_\mu v_\nu$. Obviously $K_t = K_{t-1}$.

Case 2. $\Gamma_t = (*; \mu, \nu)$. We have

$$b_t = b_\mu b_\nu = c_\mu c_\nu + \frac{c_\mu u_\nu v_\mu + c_\nu u_\mu v_\nu + u_\mu u_\nu}{v_\mu v_\nu}$$

and set $c_t := c_\mu c_\nu$, $u_t := c_\mu u_\nu v_\mu + c_\nu u_\mu v_\nu + u_\mu u_\nu$, $v_t := v_\mu v_\nu$. Since $K_t \subseteq K_{t-1}(c_\mu, c_\nu)$, the transcendence degree increases at most by two.

Case 3. $\Gamma_t = (\lambda; \mu)$ where $\lambda \in k$. We have

$$b_t = \lambda c_\mu + \frac{\lambda u_\mu}{v_\mu}$$

and set $c_t := \lambda c_\mu$, $u_t := \lambda u_\mu$, $v_t := v_\mu$. Here $K_t = K_{t-1}$.

The operations subtraction, division and of taking a constant in k can be treated analogously.

In order to show the statement on the additive complexity we prove the following.

> Let $\Gamma = (\Gamma_1, \ldots, \Gamma_t)$ be a straight-line program over k which is executable in $K(X)$ on $(b_{-N+1}, \ldots, b_0) \in (K \cup k(X))^N$ and denote the corresponding result sequence by $(b_{-N+1}, \ldots, b_t)$. Then there exist $u_i, v_i \in K[X]$, $c_i \in K$ for $-N < i \leq t$ such that $v_i \neq 0$, $b_i = c_i u_i/v_i$, and
>
> $$\text{tr.deg}_k \text{ Coeff}(u_{-N+1}, \ldots, u_t, v_{-N+1}, \ldots, v_t) \leq c^+\text{-length of } \Gamma.$$

On can see as before that this implies the assertion.

Again we proceed by induction on t. The start "$t = 0$" is as before. Assume now $t > 0$, e.g., $\Gamma_t = (+; \mu, \nu)$. Then we have

$$b_t = b_\mu + b_\nu = c_\mu \frac{u_\mu v_\nu + d u_\nu v_\mu}{v_\mu v_\nu}$$

where $d := c_\nu/c_\mu$, and we set $c_t := c_\mu$, $u_t := u_\mu v_\nu + d u_\nu v_\mu$, $v_t := v_\mu v_\nu$. Using the notation K_i from before we have $K_t \subseteq K_{t-1}(d)$, thus the transcendence degree increases at most by one. The remaining cases can be settled in an analogous manner. $\square$

(5.10) Corollary. *If* $F \subseteq K[X_1, \ldots, X_m]$ *consists of polynomials without constant term, then*

$$L^*(F) \geq \frac{1}{2} \text{tr.deg}_k \text{Coeff}(F).$$

Proof. Let $F = \{f_1, \ldots, f_t\}$. The assertion follows by applying Thm. (5.9) to the polynomials $f_1 + Y_1, \ldots, f_t + Y_t$, where $Y_1, \ldots, Y_t$ denote further indeterminates. $\square$

From this corollary we immediately get lower bounds on the complexity of evaluating a univariate polynomial with preconditioning.

(5.11) Corollary (Motzkin, Belaga). *Let* $f = \sum_{i=0}^n \alpha_i X^i \in K[X]$ *be a polynomial with coefficients* $\alpha_0, \ldots, \alpha_n$ *which are algebraically independent over the subfield k. Then* $L^*(f) \geq n/2$ *and* $L^+(f) = n$.

(5.12) Remark. By Thm. (5.6) the above lower bound on the multiplicative complexity is asymptotically optimal for any field extension $k \subseteq K$ in characteristic zero. (It is also asymptotically optimal for positive characteristic, cf. Ex. 5.2.)

Assume now K being real or algebraically closed and char $K = 0$. In Ex. 5.5 the upper bound of Thm. (5.1) for the multiplicative complexity is improved by one to $L^*[f] \leq n/2 + 1$ for all polynomials $f \in K[X]$ of even degree n. In Ex. 5.9 we show that this bound is sharp. For odd n one can prove that

$$L^*[f] = \begin{cases} \lfloor n/2 \rfloor + 1 & \text{if } n \geq 9, \\ \lfloor n/2 \rfloor + 2 & \text{if } n \in \{3, 5, 7\} \end{cases}$$

is optimal (cf. [437, 407]). Furthermore, we know from Thm. (5.1), Rem. (5.5) and the above corollary that

$$3n/2 \leq L^{\text{tot}}[f] \leq n + \lfloor n/2 \rfloor + 2$$

for $f \in K[X]$ of degree n having algebraically independent coefficients over k. In Ex. 5.9 we prove that the right-hand side inequality is actually an equality if n is odd. $\bullet$

As another application we study the problem of evaluating a rational function.

(5.13) Corollary (Motzkin). *Let* f *be the rational function*

$$f = \frac{\alpha_n X^n + \alpha_{n-1} X^{n-1} + \ldots + \alpha_0}{X^n + \beta_{n-1} X^{n-1} + \ldots + \beta_0},$$

where $\alpha_0, \ldots, \alpha_n, \beta_0, \ldots, \beta_{n-1} \in K$ *are algebraically independent over the subfield k. Then we have* $L^*(n) = n$, $L^+(f) = 2n$, *and* $L^{\text{tot}}(f) = 3n$.

Proof. The lower bounds are a consequence of Thm. (5.9). To prove the upper bound we remark that for any $(\alpha, \beta) \in K^{2n+1}$ not being a zero of some nonzero polynomial with integer coefficients, the Euclidean algorithm produces on the pair of polynomials $(\sum_{i=0}^n \alpha_i X^i, X^n + \sum_{i=0}^{n-1} \beta_i X^i)$ a remainder sequence with degrees decreasing in each step exactly by one, except for the first step (see Ex. 5.11). For those (α, β) we can write f as a continued fraction of the following form:

$$f \;=\; \gamma_0 + \cfrac{\delta_0}{X + \gamma_1 + \cfrac{\delta_1}{X + \gamma_2 + \cfrac{}{\ddots \cfrac{\delta_{n-2}}{X + \gamma_{n-1} + \cfrac{\delta_{n-1}}{X + \gamma_n}}}}}$$

where $\gamma_0, \ldots, \gamma_n, \delta_0, \ldots, \delta_{n-1} \in K$. These considerations apply in particular to a vector (α, β) having components which are algebraically independent over k. The above representation of f as a continued fraction immediately gives a straight-line program for computing f from $K \cup \{X\}$ with n divisions and $2n$ additions.   □

Thus for the evaluation of rational functions with numerator and denominator having the same degrees, the evaluation according to the above continued fraction expansion is optimal with respect to the number of additions or subtractions, the number of multiplications or divisions, and the total number of operations counted, even when preconditioning is allowed.

5.3* Extension to Linearly Disjoint Fields

This section may be skipped in a first reading. We assume some familiarity with tensor products of fields and linear disjointness. Our goal is to show that the lower bounds proved in the previous section actually hold in a much more general context.

Let $\alpha_1, \ldots, \alpha_m \in \mathbb{C}$ be linearly independent over $\mathbb{Q}$ and consider the linear form $f = \alpha_1 X_1 + \ldots + \alpha_m X_m$. If the α_i are algebraic over $\mathbb{Q}$, then Thm. (5.9) yields no information for the complexities $L^*(f), L^+(f)$. Nevertheless, we will be able to prove that, e.g., $L^*_{\mathbb{Q} \to \mathbb{C}(X)}(f|\mathbb{C} \cup \mathbb{Q}(X)) \geq (m-1)/2$. This is done by interchanging the roles of $\mathbb{C}$ and $\mathbb{Q}(X)$, and by assigning to subsets $F \subseteq \mathbb{C}(X_1, \ldots, X_m)$ a coefficient field which is a subfield of $\mathbb{Q}(X_1, \ldots, X_m)$ and whose transcendence degree over $\mathbb{Q}$ yields also lower bounds. It turns out that the general concept of linearly disjoint fields is the proper framework for discussing these questions.

Two fields K, E containing a field k are called *linearly disjoint* over k iff their tensor product $K \otimes_k E$ over k is an integral domain. Denoting its field of fractions by the symbol KE and interpreting K, E as subfields of KE we have the following situation:

By the definition of the tensor product, for all k-linearly independent sequences $(\kappa_1, \ldots, \kappa_m) \in K^m$ and $(\varepsilon_1, \ldots, \varepsilon_n) \in E^n$ the sequence $(\kappa_i \otimes \varepsilon_j)_{i,j}$ is also linearly independent over k.

A proof of the statements below can be found in Jacobson [264, Sections 8.18, 8.19].

(5.14) Fact.
(1) If $E = k(X_1, \ldots, X_m)$ is a rational function field in the indeterminates X_i, then K, E are linearly disjoint over k for any K and $KE \simeq K(X_1, \ldots, X_m)$ is a rational function field over K in m indeterminates. Moreover,

$$K \otimes_k k(X) \simeq \{f/s \mid f \in K[X], s \in k[X] \setminus 0\}.$$

(2) If k is algebraically closed, then all fields K, E containing k are linearly disjoint over k. Moreover, the canonical morphism $K \otimes_k \overline{k(X)} \to \overline{K(X)}$ is injective, if X is an indeterminate over K. ($\overline{E}$ denotes the algebraic closure of a field E.) $\bullet$

(5.15) Definition. Let the fields K, E be linearly disjoint over k. The *coefficient field* $\mathrm{Coeff}_{K/k}(F)$ of a subset $F \subseteq K \otimes_k E$ with respect to the extension K/k is defined as the smallest intermediate field $k \subseteq \tilde{K} \subseteq K$ such that $F \subseteq \tilde{K} \otimes_k E$. $\bullet$

The subsequent remark explains how coefficient fields may be determined.

(5.16) Remark. Let $(e_j)_{j \in J}$ be a k-basis of E, F a finite subset of $K \otimes_k E$. Write $f \in F$ as $f = \sum_{j \in J} f_j \otimes e_j$ with $f_j \in K$. Then the coefficent field of F satisfies $\mathrm{Coeff}_{K/k}(F) = k(f_j \mid f \in F, j \in J)$. $\bullet$

This remark implies that the above definition is consistent with the previous one (5.7) in the situation where $E = k(X_1, \ldots, X_m)$ is a rational function field: $\mathrm{Coeff}_k(F) = \mathrm{Coeff}_{K/k}(F)$ for finite $F \subseteq K \otimes_k k(X)$.

Analogously to the proof of Thm. (5.9) one can show the following generalization. (For a detailed proof we refer the reader to Ex. 5.15.)

(5.17) Theorem (Baur and Rabin). *Let $F \subseteq K \otimes_k E$ be finite, K, E being linearly disjoint over k. Then*

$$L^*_{k \to KE}(F \mid K \cup E) \geq \frac{1}{2}\left(\mathrm{tr.deg}_k \, \mathrm{Coeff}_{K/k}(F) - |F|\right),$$

$$L^+_{k \to KE}(F \mid K \cup E) \geq \mathrm{tr.deg}_k \, \mathrm{Coeff}_{K/k}(F) - |F|.$$

Note that this theorem implies the statement of Thm. (5.9) only for subsets $F \subseteq K \otimes_k k(X)$. In the general context of linearly disjoint fields it is not possible to define a coefficient field for arbitrary subsets $F \subseteq KE$ as before, since $K \otimes_k E$ may not be a unique factorization domain.

Now the lower bound for the example at the beginning of this section follows.

(5.18) Corollary. *Let* $f := \sum_{j=1}^{m} \alpha_j X_j$, *where the* X_j *are indeterminates over* K *and* $\alpha_j \in K$. *Let* d *be the dimension of the* k-*linear hull of* $\alpha_1, \ldots, \alpha_m$. *Then* $L^+(f) = d - 1$, $L^*(f) \geq (d-1)/2$.

Proof. By Fact (5.14) the fields $K, E = k(X)$ are linearly disjoint over k. We may assume w.l.o.g. that $\alpha_1, \ldots, \alpha_d$ are linearly independent over k and $\alpha_\ell = \sum_{j=1}^{d} \lambda_{\ell j} \alpha_j$ for $\ell > d$, $\lambda_{\ell j} \in k$. Then $f = \sum_{j=1}^{d} \alpha_j Y_j$, where $Y_j := X_j + \sum_{\ell > d} \lambda_{\ell j} X_\ell$, and by Rem. (5.16) we have $\text{Coeff}_{E/k}(f) = k(Y_1, \ldots, Y_d)$. (Think of $\alpha_1, \ldots, \alpha_d$ as being extended to a k-basis of K.) Obviously $\text{tr.deg}_k k(Y_1, \ldots, Y_d) = d$, hence the lower bounds follow from Thm. (5.17). □

An interesting consequence of the above theorem is that for evaluating a sufficiently general polynomial $f := \sum_{i=0}^{n} \alpha_i X^i \in K[X]$ a precomputation not only of rational functions in $k(X)$, but also of any *algebraic functions* in $\overline{k(X)}$ does not help.

(5.19) Corollary. *Let* $f := \sum_{i=0}^{n} \alpha_i X^i \in K[X]$ *where the* $\alpha_0, \ldots, \alpha_n$ *are algebraically independent over the subfield* k. *Then*

$$L^*_{k \to \overline{K(X)}}\left(f \mid K \cup \overline{k(X)}\right) \geq n/2, \quad L^+_{k \to \overline{K(X)}}\left(f \mid K \cup \overline{k(X)}\right) = n.$$

Proof. By replacing k, K by their algebraic closures we may assume w.l.o.g. that k is algebraically closed. By Fact (5.14)(2) the fields K and $\overline{k(X)}$ are linearly disjoint over k and we have an embedding $K \otimes_k \overline{k(X)} \to \overline{K(X)}$. The assertion follows from Thm. (5.17) when we note that a computation in $\overline{K(X)}$ starting with inputs in $K \cup \overline{k(X)}$ can only compute elements in $K\overline{k(X)}$. □

5.4 Exercises

5.1.

(1) Describe a straight-line program computing the complex polynomial

$$\sqrt{2}X^6 + iX^5 + X^4 + (1 - \sqrt{2})X^2 - iX - 2$$

from X and some complex numbers using only 5 multiplications and 6 additions.

(2) Find a straight-line program which computes the integer polynomial

$$3X^6 + 2X^5 + X^4 - 2X^2 - 2X - 2$$

from X and some integers using only 3 nonscalar multiplications, 2 multiplications by integers and 6 additions.

(3) Find a straight-line program computing the rational function

$$\frac{X^4 - 3X^2 + 19X - 5}{X^4 - X^3 - 4X^2 + 24X - 28}$$

from X and some integers using only 4 divisions and 8 multiplications or subtractions.

5.2. (This exercise supplements Thm. (5.6).) Let $f \in K[X]$, $\deg f < n$, K a field of arbitrary characteristic. Describe a straight-line program computing f in $K(X)$ from X and some elements of K using at most $n/2 + 3\log n + 1$ multiplications or divisions, and $3n/2 - 1$ additions or subtractions.

5.3. Let $F := \{\sum_{j=1}^{n} \alpha_{ij} X_j \mid 1 \leq i \leq r\} \subseteq K[X_1, \ldots, X_n]$, where all the α_{ij} are algebraically independent over the subfield k of K. Show that $L^+(F) = r(n-1)$ and $rn/2 \leq L^*(F) \leq (r+1)\lceil n/2 \rceil$.

5.4. Let $L(f) := L^*_{K \to K(X)}(f|X)$ denote the multiplicative complexity of $f \in K(X)$ with respect to the field of scalars K (multiplications with scalars $\lambda \in K$ are free). Show that for a subfield $k \subseteq K$

$$L^+(f) \leq (1 + L(f))^2, \quad L^*(f) \leq (1 + L(f))^2, \quad L^{tot}(f) \leq 2(1 + L(f))^2.$$

Deduce that $L(f) \geq \sqrt{n} - 1$ for a polynomial f of degree n having algebraically independent coefficients over k. (Compare Sect. 9.1.)

The next five exercises form a logical unit. The goal is the exact complexity determination stated in Ex. 5.9. Although we only improve estimates by one, everything becomes considerably more complicated. (Compare Rem. (5.12).)

5.5.* (Improvement by one of the upper bound on L^* given in Thm. (5.1) for polynomials of even degree.) Let K be a real or algebraically closed field, and assume that $\operatorname{char} K = 0$, $m \geq 1$.

(1) Prove that for a given $f = X^{2m} + u_{2m-1} X^{2m-1} + \ldots + u_0 \in K[X]$ and $a \in K$ there exist $g = X^{2m-2} + v_{2m-3} X^{2m-3} + \ldots + v_0 \in K[X]$ and $b, \beta \in K$ such that

$$f = g(X^2 + aX + b) + \beta,$$

provided $u_{2m-1} - ma \neq 0$. (Hint: put $u_{2m} = 1$, $u_i = 0$ for $i > 2m$ and show

$$v_r = \sum_{i,j \geq 0} (-1)^{i+j} \binom{i+j}{j} u_{r+2+i+2j} a^i b^j .)$$

(2) Prove that for any monic polynomial $f \in K[X]$ of degree $2m$ there exist polynomials $f_1, f_2, \ldots, f_m = f$ and $\alpha_i, \beta_i \in K$ such that

$$
\begin{aligned}
f_1 &= X(X + \alpha_1) + \beta_1, \\
f_2 &= f_1(f_1 + X + \alpha_2) + \beta_2, \\
f_i &= f_{i-1}(f_1 + \alpha_i) + \beta_i, \quad 2 < i \leq m.
\end{aligned}
$$

Therefore $L^*[f] \leq m$. (Hint: determine first α_1.)

5.6.* (Preparation for Ex. 5.7.) Let $f(X) = g(X^2) + Xh(X^2)$ be a real polynomial of degree n. From the simple Lemma (5.4) we concluded that all the roots of g and h must be reals if all the roots of f have a positive real part. In this exercise we want to generalize this. Assume that f has at most one root with negative real part. We claim that we can still conclude that all the roots of g and h are reals. Prove this by proceeding as follows.

W.l.o.g. f has no roots on the imaginary axis. The integral $\frac{1}{2\pi i} \oint_{\gamma_R} \frac{f'(\zeta)}{f(\zeta)} d\zeta$ equals the number of roots of f inside the region bounded by the path γ_R, by assumption it is at most one (see Fig. 5.2). The contribution to the above integral

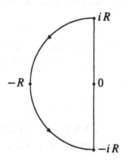

Fig. 5.2. The path γ_R.

of the path $[\pi/2, 3\pi/2] \to \mathbb{C}, t \mapsto Re^{it}$ is about $n/2$. Now let n be even. Then $f(\pm iR) \approx a_n R^n i^n$ for large $R > 0$. Hence the path $f(it)$ for $-R \leq t \leq R$ must go clockwise around the origin at least $n/2 - 1$ times. Conclude that $f(it)$ crosses the imaginary axis at least $n - 2$ times and the real axis at least $n - 3$ times. Thus both g and h have at least $n/2 - 1$ different real roots. As $\deg g, \deg h \leq n/2$ the claim follows for even n. Argue similarly for odd n.

5.7. (Improvement by one of the upper bound on L^{tot} given in Thm. (5.1) over $K = \mathbb{R}$. See also Rem. (5.5).) Prove that $L^{\text{tot}}[f] \leq n + \lfloor n/2 \rfloor + 2$ for $f \in \mathbb{R}[X]$ of degree n. (Hint: Show the existence of some $c \in \mathbb{R}$ such that at most one root of $f(X - c)$ has negative real part, and such that $f(X - c)$ has a divisor of the form $X^2 - \xi$. Then proceed as in the proof of Thm. (5.1), but use Ex. 5.6 instead of Lemma (5.4).)

5.8. (Improvement upon the transcendence degree bound for univariate polynomials.) Let $\Gamma = (\Gamma_1, \ldots, \Gamma_t)$ be a straight-line program over k which is executable in $K[X]$ on $(b_{-N+1}, \ldots, b_0) \in K^{N-1} \times \{X\}$ and denote the corresponding result sequence by $(b_{-N+1}, \ldots, b_t)$. We call an instruction $\gamma_s = (\omega; i, j)$ *parametric* iff $i < 0$ or $j < 0$. So b_s is the result of an arithmetic operation where one of the operands is an element of K. We denote the number of (parametric) multiplication or division instructions by c^* (c_p^*), whereas c_p^+ denotes the number of parametric addition or subtraction instructions. We may assume w.l.o.g. that there are no parametric scalar multiplications, i.e., instructions $\gamma_s = (\lambda; i)$ with $\lambda \in k$, $i < 0$. Prove:

(1) $\operatorname{tr.deg}_k \operatorname{Coeff}(b_0, \ldots, b_t) \le c_p^+ + c_p^*$,
(2) there are $u_i \in K[X]$, $c_i \in K$ for $-N < i \le t$ such that $b_i = c_i + u_i$ and

$$\operatorname{tr.deg}_k \operatorname{Coeff}(u_{-N+1}, \ldots, u_t) \le 2c^* - c_p^*.$$

Moreover, this inequality is strict if $c_p^* < c^*$. (Hint: save one parameter in the first nonparametric multiplication step.)

5.9. (Exact complexity determination in some cases.) Let K be real or algebraically closed, char $K = 0$, and $f \in K[X]$ be of degree n having algebraically independent coefficients over k.

(1) $L^*[f] = n/2 + 1$ if $n \ge 2$ is even,
(2) $L^{\text{tot}}[f] = n + \lfloor n/2 \rfloor + 2$ if $n \ge 3$ is odd.

(Hint: combine the statements of the previous four Exercises.)

5.10. Let K be algebraically closed. Show that for all $(\alpha, \beta) \in K^{2n+1}$ not being a zero of some nonzero polynomial with integer coefficients the rational function

$$f = \frac{\alpha_n X^n + \alpha_{n-1} X^{n-1} + \ldots + \alpha_0}{X^n + \beta_{n-1} X^{n-1} + \ldots + \beta_0}$$

is reduced and can be written as a partial fraction in the form

$$f = \gamma_0 + \frac{\gamma_1}{X - \delta_1} + \ldots + \frac{\gamma_n}{X - \delta_n},$$

where $\gamma_i, \delta_i \in K$. Using this prove that $L^*(f) = n$, $L^+(f) = 2n$, and $L^{\text{tot}}(f) = 3n$ if $\alpha_0, \ldots, \beta_{n-1}$ are algebraically independent over k. (We obtained this already in Cor. (5.13) by writing f as a continued fraction, which works over any field K.) (Hint: resultant and discriminant.)

5.11. Let $n \ge m$. Convince yourself that there exists a nonzero polynomial P with integer coefficients such that for any pair of univariate polynomials $A_1 = \sum_{i=0}^n \alpha_i X^i$, $A_2 = \sum_{j=0}^m \beta_j X^j$ over a field k satisfying $P(\alpha, \beta) \ne 0$ the Euclidean algorithm

$$
\begin{aligned}
A_1 &= Q_1 A_2 + A_3, \\
A_2 &= Q_2 A_3 + A_4, \\
&\vdots \\
A_m &= Q_m A_{m+1} + A_{m+2}, \\
A_{m+1} &= Q_{m+1} A_{m+2},
\end{aligned}
$$

produces remainders $Q_1, Q_2, \ldots, Q_m, Q_{m+1}$ of degrees $n - m, 1, \ldots, 1$, respectively. (You only need to show the existence of such P. Explicitly writing it down is complicated, compare Sect. 10.3.)

5.12. A *coding of rational functions* of degree (n, n) is some rational function $F \in \mathbb{C}(U_0, \ldots, U_{2n}, X)$ having the following property: for Zariski almost all $(\alpha, \beta) \in \mathbb{C}^{2n+1}$ there exists $u \in \mathbb{C}^{2n+1}$ such that $F(u, X)$ is defined and

$$
\frac{\alpha_n X^n + \alpha_{n-1} X^{n-1} + \ldots + \alpha_0}{X^n + \beta_{n-1} X^{n-1} + \ldots + \beta_0} = F(u, X).
$$

(1) Show that the continued fraction

$$
F = U_0 + \cfrac{U_{n+1}}{X + U_1 + \cfrac{U_{n+2}}{X + U_2 + \cfrac{}{\ddots \cfrac{U_{2n-1}}{X + U_{n-1} + \cfrac{U_{2n}}{X + U_n}}}}}
$$

is a coding of rational functions of degree (n, n) satisfying $L^*(F) \leq n$, $L^+(F) \leq 2n$.

(2) Prove that for any coding F of rational functions of degree (n, n) we have

$$
L^*(F) \geq n, \quad L^+(F) \geq 2n.
$$

(The complexities are taken with respect to $k = \mathbb{C}$, $K = \mathbb{C}(U)$. Hint for (2): replace $\mathbb{C}$ by a subfield k which is finitely generated over $\mathbb{Q}$. Make a substitution $U_i \mapsto u_i \in \mathbb{C}$.)

5.13. A *coding of polynomials* of degree n is some rational function $F \in \mathbb{C}(U, X)$, $U = (U_0, \ldots, U_n)$, such that for Zariski almost all $\alpha \in \mathbb{C}^{n+1}$ there exists $u \in \mathbb{C}^{n+1}$ satisfying

$$
\alpha_n X^n + \ldots + \alpha_0 = F(u, X).
$$

(1) Describe such a coding F satisfying $L^*(F) \leq \lfloor n/2 \rfloor + 2$, $L^+(F) \leq n$.
(2) Prove that for any such coding $L^*(F) \geq n/2$, $L^+(F) \geq n$.

(The complexities are taken with respect to $k = \mathbb{C}$, $K = \mathbb{C}(U)$. Hint for (1): See the proof of Lemma (5.2). Show that Zariski almost all polynomials f of degree $n \geq 3$ can be written in the form $f = (X^2 - \xi)f_1 + \eta$ for some $\xi, \eta \in \mathbb{C}$, $f_1 \in \mathbb{C}[X]$.)

5.14. Verify Fact (5.14)(1) and Rem. (5.16). Conclude that the Definitions (5.7) and (5.15) are consistent in the situation where E is a rational function field over k.

5.15. Let K, E be linearly disjoint over a common subfield k, $F \subseteq K \otimes_k E$.

(1) Show that adding an element of K to F can increase the transcendence degree of $\operatorname{Coeff}_{K/k}(F)$ over k at most by one.
(2) Let $u, v, w \in K \otimes_k E$, $v \neq 0$ and $u = vw$. Show that $\operatorname{Coeff}_{K/k}(w) \subseteq \operatorname{Coeff}_{K/k}(u, v)$. (Compare Lemma (5.8).)
(3) Give a detailed proof of Thm. (5.17).

5.16. The following is a useful tool for calculating transcendence degrees of coefficient fields. Let k be algebraically closed, $f_1, \ldots, f_s \in k(a_1, \ldots, a_m, X)$ be rational functions and $U \subset k^m$ be a nonempty Zariski open subset such that the map

$$U \to k(X)^s, \alpha \mapsto (f_1(\alpha, X), \ldots, f_s(\alpha, X))$$

is well defined and has only finite fibers. Prove that the transcendence degree of $\operatorname{Coeff}_{k(a)/k}(f_1, \ldots, f_s)$ equals m. (Hint: Use the fact that the dimension of the image of a polynomial map $g : k^m \to k^n$ equals the transcendence degree of $k(g_1, \ldots, g_n)$ over k.)

5.17. (Partial fraction formed by general polynomials.) Let $f \in k(a, X)$ be the partial fraction

$$f := \sum_{j=0}^{n_0} a_{0j} X^j + \sum_{i=1}^{r} \sum_{j=1}^{n_i} \frac{a_{ij}}{(x - a_{i0})^j}.$$

Determine $\operatorname{Coeff}_{k(a)/k}(f)$. Which lower bounds do you obtain from this? (Hint: W.l.o.g. assume that k is algebraically closed. Apply Ex. 5.16 and use the uniqueness of the partial fraction expansion (see, e.g., Lang [316, V,§5]).)

5.18. (Continued fraction formed by general polynomials, compare Thm. (6.15).) To a sequence of general polynomials

$$Y_0 := \sum_{j=0}^{n_0} a_{0j} X^j, Y_1 := \sum_{j=0}^{n_1} a_{1j} X^j, \ldots$$

we assign the continued fraction defined recursively by

$$f_0 := Y_0, \quad f_i := Y_i + f_{i-1}^{-1} \quad i > 0.$$

Determine the transcendence degree of $\operatorname{Coeff}_{k(a)/k}(f_r)$ over k. Which lower bounds do you obtain from this? (Hint: W.l.o.g. assume that k is algebraically closed. Prove the uniqueness of the continued fraction expansion and apply Ex. 5.16.)

5.5 Open Problems

Problem 5.1. Is it possible to evaluate polynomials of degree n with $n/2 + O(1)$ multiplications using only rational preconditioning? (That is, can the term $2 \log n$ in Thm. (5.6) be replaced by a constant?)

5.6 Notes

The proof of Thm. (5.1) over the complex numbers goes back to a method by Knuth [303]. This method has been modified by Eve [157] who set it to work over the reals (cf. Exercises 5.6, 5.7). The elegant way of proving Thm. (5.1) over the reals by invoking Lemma (5.4) is due to Baur [29] who attributes this lemma to Hermite. For an exact determination of the minimum number of multiplications necessary to evaluate almost all polynomials of a certain degree n we refer to Revah [437] (see also Pan [407]). (In fact, for $n = 9$, this number equals 5 and not 6 as claimed by Motzkin.) Further examples of straight-line programs computing polynomials using preconditioning may be found in Knuth [307, Sect. 4.6.4].

Thm. (5.6) on evaluating polynomials using only rational preconditioning is due to Paterson and Stockmeyer [415]. Rabin and Winograd [430] have also designed several algorithms for evaluating polynomials which use only rational operations in the preconditioning phase. One of their methods requires only $n/2 + O(\sqrt{n})$ multiplications and $n + O(\sqrt{n})$ additions or subtractions to evaluate a polynomial of degree n. Another of their algorithms uses only $n/2 + O(\log n)$ multiplications and $n + o(n)$ additions or subtractions. In their paper they conjecture that Problem 5.1 has a negative answer.

The lower bound in Thm. (5.9) for the multiplicative complexity goes back to Motzkin [386], whereas the lower bound on the additive complexity in this theorem is a result by Belaga [35]. (Proofs for these lower bounds can be found also in Winograd [552], and Reingold and Stocks [434].) Cor. (5.13) goes back to Motzkin [386]. The generalization in Sect. 5.3 to linearly disjoint fields is due to Baur and Rabin [31].

Ex. 5.5 is taken from Knuth [307, 4.6.4 Ex. 39] who gives the credit for it to Motzkin (unpublished notes). Exercises 5.6, 5.7 are in Knuth [307, 4.6.4, Thm. E and Ex. 23] and due to Eve [157]. Exercises 5.8, 5.9 are also taken from Knuth [307, 4.6.4, Ex. 30,33] and essentially due to Pan [405]. Exercises 5.10, 5.12 and 5.13 are taken from Strassen [503].

Chapter 6. The Substitution Method

In 1954 Ostrowski published his note [403] "On two problems in abstract algebra connected with Horner's rule" which led to numerous investigations on the complexity of algebraic problems. In this paper he inquired about the optimality of the so-called Horner's rule, a rule which was however already known to Newton (see [548, p. 222]). Ostrowski conjectured that there is no general evaluation procedure to evaluate a polynomial of degree n (in the polynomial ring) which requires less than n multiplications. In the cases $n \leq 4$ he succeeded to prove his conjecture. It is remarkable that Ostrowski, guided by good mathematical intuition, already suggested the nonscalar counting. In 1966 Pan [405] invented the substitution method and proved Ostrowski's conjecture (even when divisions are allowed). We remark that the optimality of Horner's rule with respect to the number of additions and subtractions had already been settled before by Belaga [35] (compare Chap. 5). The substitution method was further developed by Winograd [552] and Strassen [495].

In our presentation we do not put the emphasis on the substitution method in its original form. Rather, we employ the idea of reducing the multiplicative complexity by adjoining intermediate results to the field of scalars. This elegant method was first used by Lickteig [333] in his determination of the complexity of the complex division. Taking up this idea, we associate to a finite set F of rational functions in n indeterminates its degree of linearization $\delta(F)$, which is defined as the maximum rank of an affine linear substitution that maps all functions in F to elements of complexity zero. The main theorem of this chapter states that the multiplicative complexity of F is bounded from below by $n - \delta(F)$. This theorem allows both a sharp determination of the multiplicative complexity for various basic computational problems and a unified treatment of results by Pan, Winograd, Strassen, Hartmann and Schuster, among others.

Throughout this chapter we denote the multiplicative complexity L^* by L, thus we count only the multiplications and divisions. Unless otherwise specified, the input set is always the set of indeterminates of the field of fractions under consideration.

6.1 Discussion of Ideas

We start by considering two examples.

(6.1) Example. Let us study the problem of computing the sum of squares $f = \sum_{j=1}^{n} a_j^2$ from the indeterminates $a_1, \ldots, a_n$ in the polynomial ring $k[a] = k[a_1, \ldots, a_n]$ over the field k. If k is the field of complex numbers, $i^2 = -1$, and n is even, then

$$\sum_{j=1}^{n} a_j^2 = \sum_{j=1}^{n/2} (a_j + ia_{n/2+j})(a_j - ia_{n/2+j}),$$

hence $n/2$ multiplications are sufficient. Over the reals, such an improvement upon the trivial algorithm is not possible; we claim that $L_{\mathbb{R}[a]}(a_1^2 + \ldots + a_n^2) = n$. To see this, let $(g_1, \ldots, g_r)$ be a computation sequence for $f = a_1^2 + \ldots + a_n^2$ in $\mathbb{R}[a]$ modulo $\{a_1, \ldots, a_n\}$. This means that $f \in \langle 1, a_1, \ldots, a_n, g_1, \ldots, g_r \rangle$ and that for all $\rho \in \underline{r}$ there exist $u_\rho, v_\rho \in \langle 1, a_1, \ldots, a_n, g_1, \ldots, g_{\rho-1} \rangle$ such that $g_\rho = u_\rho v_\rho$. We assume that r is minimal, i.e., the above computation sequence is optimal. Then there are nonconstant affine linear polynomials u_1, v_1 such that $g_1 = u_1 v_1$. After permuting the $a_1, \ldots, a_n$ and some scaling we may assume that $u_1 = a_n - \sum_{j=1}^{n-1} \gamma_j a_j - \kappa$ where $\gamma_j, \kappa \in \mathbb{R}$. Consider now the $\mathbb{R}$-algebra morphism $\sigma: \mathbb{R}[a_1, \ldots, a_n] \to \mathbb{R}[a_1, \ldots, a_{n-1}]$ sending a_n to $\sum_{j=1}^{n-1} \gamma_j a_j + \kappa$ and leaving the other a_i fixed. Then $(\sigma(g_1), \sigma(g_2), \ldots, \sigma(g_r))$ is a computation sequence for $\sigma(f)$, where the first step $\sigma(g_1) = 0$ can be omitted. Therefore, the complexity of $a_1^2 + \ldots + a_{n-1}^2 + (\sum_{j=1}^{n-1} \gamma_j a_j + \kappa)^2$ in $\mathbb{R}[a_1, \ldots, a_{n-1}]$ is at most $r - 1$. By performing $m \leq r$ substitutions of the above type, we obtain, after permuting the a_i, some polynomial of complexity zero having the form

$$h := a_1^2 + \ldots + a_{n-m}^2 + A_{n-m+1}^2 + \ldots + A_n^2,$$

where the A_i are affine linear polynomials in $a_1, \ldots, a_{n-m}$, say $A_i = \sum_j \alpha_{ij} a_j + \beta_i$. A polynomial of multiplicative complexity zero is by definition affine linear, hence we conclude that the coefficient $1 + \sum_{i>n-m} \alpha_{ij}^2$ of a_j^2 in h must vanish for all $j \leq n-m$. Since we work over the field of reals, this is only possible when $m = n$ and $\alpha_{ij} = 0$ for all i, j. Thus we have proved that $L_{\mathbb{R}[a]}(a_1^2 + \ldots + a_n^2) = n$.   ●

(6.2) Example. Let us investigate the problem of computing a general polynomial $f = \sum_{i=0}^{n} a_i X^i$ from indeterminates $a_0, \ldots, a_n, X$ over a field k. Horner's rule

$$f = (\ldots((a_n X + a_{n-1})X + a_{n-2})X + \ldots + a_1)X + a_0$$

shows that n multiplications are sufficient for performing this task. On the other hand, we know from Chap. 5 that the multiplicative complexity to compute $\sum_{i=0}^{n} a_i X^i$ from $k(a_0, \ldots, a_n) \cup \{X\}$ is $n/2 + o(n)$ for $n \to \infty$ (cf. Thm. (5.6) and Cor. (5.11)). We claim now that n multiplications or divisions are necessary if we do not allow preconditioning of the coefficients. In fact, we are going to show the stronger statement that n multiplications or divisions are necessary even if we allow preconditioning of rational functions in the variable X, thus if we take $k(X) \cup \{a_0, \ldots, a_n\}$ as input set. To avoid technical difficulties we make here the simplifying assumption that all our computations take place in the k-algebra $k(X)[a_0, \ldots, a_n]$, so only divisions by elements of $k(X)[a_0, \ldots, a_n]^\times = k(X)^\times$ may occur.

Let $(g_1, \ldots, g_r)$ be an optimal computation sequence for $f = \sum_{i=0}^n a_i X^i$ in $k(X)[a]$ modulo $k(X) \cup \{a_0, \ldots, a_n\}$. Then either $g_1 = uv$ or $g_1 = u/v$ with elements $u, v \in \langle a_0, \ldots, a_n \rangle_k + k(X)$. The optimality of the computation sequence implies that not both u and v lie in $k(X)$. If $g_1 = uv$, we may assume w.l.o.g. that $u \notin k(X)$. In case $g_1 = u/v$ we have $v \in k(X)^\times$ and therefore also $u \notin k(X)$. After scaling, we may assume that there exist $\ell \in \{0, 1, \ldots, n\}$, $\gamma_j \in k$ for $j \neq \ell$ and $\kappa \in k(X)$ such that

$$u = a_\ell - \sum_{j \neq \ell} \gamma_j a_j - \kappa.$$

We apply now the $k(X)$-algebra morphism

$$\sigma : k(X)[a_0, \ldots, a_n] \to k(X)[a_0, \ldots, a_{\ell-1}, a_{\ell+1}, \ldots, a_n]$$

sending a_ℓ to $\sum_{j \neq \ell} \gamma_j a_j + \kappa$ and leaving the other a_i fixed. Since $\sigma(g_1) = 0$, it is clear that the complexity of $\sigma(f)$ is at most $r - 1$. By repeating this argument $m \leq r$ times we obtain a subset $I \subseteq \underline{n}$ of cardinality $n + 1 - m$ and a polynomial h of complexity zero having the form $h = \sum_{j=0}^n (A_j + \beta_j) X^j$ where $\beta_j \in k(X)$ for all j and $A_j = a_j$ for $j \in I$, $A_j \in \langle a_i \mid i \in I \rangle$ for $j \notin I$. Let $A_j = \sum_{i \in I} \alpha_{ji} a_i$ for $j \notin I$. Then we have

$$h = \sum_{i \in I} \left(X^i + \sum_{j \notin I} \alpha_{ji} X^j \right) a_i + \sum_{j=0}^{n-1} \beta_j X^j \in \langle a_0, \ldots, a_n \rangle_k + k(X).$$

The coefficients of a_i in h must all lie in k, which is only possible when $I \subseteq \{0\}$. Hence $r \geq m = n + 1 - |I| \geq n$. Thus we have proved that

$$L_{k(X)[a]}\left(\sum_{i=0}^n a_i X^i \mid k(X) \cup \{a_0, \ldots, a_n\} \right) = n. \qquad \bullet$$

We are now going to discuss how the lower bound results of the previous two examples can be generalized to computations in the corresponding *rational function fields*. Before doing this we proceed with two remarks.

(6.3) Remark. A morphism $\sigma : k[X_1, \ldots, X_n] \to A$ from a polynomial ring over k to a k-algebra can be extended to the subalgebra

$$\operatorname{def} \sigma := \{g/h \mid g, h \in k[X], \sigma(h) \in A^\times\}$$

of $k(X)$ in a natural way. ($\operatorname{def} \sigma$ is just the localization of $k[X]$ with respect to the system $\sigma^{-1}(A^\times)$ of denominators.) We will often think of σ as being extended to $\operatorname{def} \sigma$. We call σ the *substitution defined by* $(a_1, \ldots, a_n)$, where $a_i := \sigma(X_i)$. For a rational function $f \in k(X_1, \ldots, X_n)$ we say that $f(a_1, \ldots, a_n)$ *is defined* if $f \in \operatorname{def} \sigma$. $\qquad \bullet$

(6.4) Remark. The affine substitution

$$\sigma: k[X_1, \ldots, X_n] \to k(X_1, \ldots, X_n) =: k(X), \quad X_i \mapsto \sum_{j=1}^{n} \alpha_{ij} X_j + \beta_i,$$

where $(\alpha_{ij}) \in \mathrm{GL}(n, k)$, $(\beta_i) \in k^n$, defines a k-algebra automorphism of $k(X)$. Since the vector space of polynomials of complexity zero is stabilized by σ, we have

$$L_{k(X)}(F) = L_{k(X)}(\sigma(F))$$

for every finite subset $F \subseteq k(X)$.   ●

Let us return to our discussion in Example (6.1). We assume that $(g_1, \ldots, g_r)$ is an optimal computation sequence for $f = a_1^2 + \ldots + a_n^2$ in $\mathbb{R}(a)$ modulo $\{a_1, \ldots, a_n\}$. Suppose that $g_1 = uv$, $u = a_n - \sum_{j<n} \gamma_j a_j - \kappa$. The problem is that the $\mathbb{R}$-algebra morphism $\sigma: \mathbb{R}[a_1, \ldots, a_n] \to \mathbb{R}(a_1, \ldots, a_{n-1})$ sending u to zero and fixing $a_1, \ldots, a_{n-1}$ is not injective, hence cannot be extended to the field $\mathbb{R}(a_1, \ldots, a_n)$. However, the computation sequence $(g_1, \ldots, g_r)$ computes f already in a subalgebra

$$\mathbb{R}[a]_d := \{g/d^s \mid g \in \mathbb{R}[a], s \in \mathbb{N}\}$$

for some nonzero $d \in \mathbb{R}[a]$ (compare Ex. 4.3). The idea is now to consider a *family* of $\mathbb{R}$-algebra morphisms $(\sigma_\lambda)_{\lambda \in \mathbb{R}}$, where $\sigma_\lambda: \mathbb{R}[a_1, \ldots, a_n] \to \mathbb{R}(a_1, \ldots, a_n)$ sends a_n to $\sum_{j<n} \gamma_j a_j + \kappa + \lambda$ and leaves the other a_i fixed. Note that $\sigma_0 = \sigma$ and $\sigma_\lambda(u) = \lambda$. The domain of definition

$$\mathrm{def}\,\sigma_\lambda = \{g/h \mid g, h \in \mathbb{R}[a], \sigma_\lambda(h) \neq 0\}$$

of σ_λ contains $\mathbb{R}[a]_d$, provided $\sigma_\lambda(d) = d(a_1, \ldots, a_{n-1}, \sum_{j<n} \gamma_j a_j + \kappa + \lambda) \neq 0$. Therefore, as $\mathbb{R}$ is infinite, we have $\mathbb{R}[a]_d \subseteq \mathrm{def}\,\sigma_\lambda$ for all but finitely many $\lambda \in \mathbb{R}$. For such λ the sequence $(\sigma_\lambda(g_1), \ldots, \sigma_\lambda(g_r))$ computes $\sigma_\lambda(f)$ in $\mathbb{R}(a_1, \ldots, a_{n-1})$ modulo $\{a_1, \ldots, a_{n-1}\}$, and the first step $\sigma_\lambda(g_1) = \sigma_\lambda(u)\sigma_\lambda(v) = \lambda\sigma_\lambda(v)$ can be omitted. Similarly as in Example (6.1) one can prove that $L_{\mathbb{R}(a)}(a_1^2 + \ldots + a_n^2) = n$.

Let us turn to the second Example (6.2). Suppose we have an optimal sequence $(g_1, \ldots, g_r)$ computing $f = \sum_{i=0}^{n} a_i X^i$ in $k(a, X)$ modulo $k(X) \cup \{a_0, \ldots, a_n\}$ such that $g_1 = u/v$ for some $v \in k(X)^\times$, $u \in \langle a_0, \ldots, a_n \rangle_k + k(X)$, $u \notin k(X)$. W.l.o.g. we assume $u = a_n - \sum_{j<n} \gamma_j a_j - \kappa$ where $\gamma_j \in k$, $\kappa \in k(X)$. We apply the $k(X)$-algebra morphisms $\sigma_\lambda: k(X)[a_0, \ldots, a_n] \to k(a_0, \ldots, a_{n-1}, X)$ sending a_n to $\sum_{j<n} \gamma_j a_j + \kappa + \lambda v$ and leaving the other a_i fixed. By construction, we have $\sigma_\lambda(u/v) = \lambda v/v = \lambda$, and we can eliminate the first computation step. Using these ideas it is now quite easy to write down a complete proof for $L_{k(a,X)}(\sum_{i=0}^{n} a_i X^i) = n$. (Cf. Ex. 6.8.) However, in order to avoid to consider families of substitutions σ_λ and somewhat clumsy statements of the form "for all but finitely many $\lambda \in k$ we have ..." we suggest a different, more elegant approach. We are going to eliminate in a hypothetical optimal computation sequence $(g_1, \ldots, g_r)$ the first multiplication or division step $g_1 = uv$ or $g_1 = u/v$ by viewing the k-algebra

$k(a_0, \ldots, a_n, X)$ as a $k(v)$-algebra. So we have *enlarged the field of scalars by adjoining an intermediate result*, the first step $g_1 = vu$ or $g_1 = v^{-1}u$ has become a scalar multiplication step in the new algebra and can therefore be omitted. Let us explain this method in detail by proving the optimality of Horner's rule with respect to the number of multiplications and divisions in $k(a, X)$ (instead of in $k(X)[a]$ as before). Since the change of the field of scalars is crucial here, we will often notationally stress it by denoting a k-algebra A by $k \to A$.

(6.5) Theorem (Pan). *Let $a_0, \ldots, a_n, X$ be indeterminates over a field k. Then*

$$L_{k \to k(a,X)}\left(\sum_{i=0}^{n} a_i X^i\right) = n.$$

Proof. The upper bound is a consequence of Horner's rule. To show the lower bound, we will prove by induction on $r \in \mathbb{N}$ the following, more general statement.

$$\begin{cases} \text{For all } k, \text{ all } n \text{ and all } A_0, \ldots, A_n \in \langle a_0, \ldots, a_n \rangle_k, \ \beta_0, \ldots, \beta_n \in k(X): \\[2mm] L_{k \to k(a,X)}\left(\sum_{i=0}^{n}(A_i + \beta_i)X^i \mid k(X) \cup \{a_0, \ldots, a_n\}\right) \leq r \\[2mm] \text{implies } \dim\langle A_0, \ldots, A_n \rangle_k \leq r + 1. \end{cases}$$

When putting $A_i = a_i$, $\beta_i = 0$, we obtain the desired lower bound, even with the larger input set $k(X) \cup \{a_0, \ldots, a_n\}$.

The induction start "$r = 0$" is easily verified, namely

$$\sum_{i=0}^{n}(A_i + \beta_i)X^i \in \langle a_0, \ldots, a_n \rangle_k + k(X) \implies A_1 = \ldots = A_n = 0.$$

So let us suppose that $r > 0$. Let $(g_1, \ldots, g_r)$ be an optimal computation sequence for $\sum_i (A_i + \beta_i)X^i$ in $k(a, X)$ modulo $k(X) \cup \{a_0, \ldots, a_n\}$. We have $g_1 = uv$ or $g_1 = u/v$ with some $u, v \in \langle a_0, \ldots, a_n \rangle_k + k(X)$. By optimality $u \notin k(X)$ or $v \notin k(X)$.

Case 1. $v \notin k(X)$. For ease of notation we assume that the coefficient of a_n in v is not zero, so after scaling we have $v = a_n - \sum_{j<n} \gamma_j a_j - \kappa$ with some $\gamma_j \in k$, $\kappa \in k(X)$. By construction, the complexity of $\sum_i (A_i + \beta_i)X^i$ in the algebra $k(v) \to k(a, X)$ is at most $r - 1$. Putting $A_i = \sum_{j=0}^{n} \alpha_{ij} a_j$, where $\alpha_{ij} \in k$, we have

$$A_i + \beta_i = \sum_{j<n} \alpha_{ij} a_j + \alpha_{in}\left(\sum_{j<n} \gamma_j a_j + v + \kappa\right) + \beta_i$$

$$\text{(6.6)} \qquad = \underbrace{\sum_{j<n}\left(\alpha_{ij} + \alpha_{in}\gamma_j\right)a_j}_{=:A_i'} + \underbrace{\alpha_{in}(v + \kappa) + \beta_i}_{=:\beta_i'}.$$

We may apply the induction hypothesis to $k(a, X)$, viewed as a rational function field in the indeterminates $a_0, \ldots, a_{n-1}, X$ over $k(v)$, the linear forms $A_i' \in \langle a_0, \ldots, a_{n-1} \rangle_k \subseteq \langle a_0, \ldots, a_{n-1} \rangle_{k(v)}$ and the elements $\beta_i' \in k(v)(X)$. (It does no harm that the number of indeterminates $a_0, \ldots, a_{n-1}$ is different from the number n of A_i', since we may adjoin a further indeterminate to achieve this.) We conclude therefore that

$$\dim \langle A_0', \ldots, A_n' \rangle_k = \dim \langle A_0', \ldots, A_n' \rangle_{k(v)} \leq r.$$

On the other hand, we have $A_i' = A_i + \alpha_{in}(\sum_{j<n} \gamma_j a_j - a_n)$ for all i, hence the dimensions of $\langle A_0, \ldots, A_n \rangle_k$ and $\langle A_0', \ldots, A_n' \rangle_k$ can differ at most by one. Summarizing, we conclude that $\dim \langle A_0, \ldots, A_n \rangle_k \leq r + 1$, which was to be shown.

Case 2. $v \in k(X)$. Then $u \notin k(X)$. If $g_1 = uv$ we can proceed as in the first case. The remaining situation where $g_1 = u/v$, $u \in \langle a_0, \ldots, a_n \rangle_k + k(X)$, $v \in k(X)^\times$ is settled analogously by going over to the algebra $k(u/v) \to k(a, X)$. $\square$

6.2 Lower Bounds by the Degree of Linearization

We summarize the ideas developed in the previous section in a theorem giving a general lower bound on the multiplicative complexity of rational functions. All the results obtained in the foregoing section are immediate consequences of this theorem. In order to be able to formulate it, we need to define the following quantity.

(6.7) Definition. Let $k \subseteq K$ be a field extension, $K(a_1, \ldots, a_n)$ a rational function field in the indeterminates $a_1, \ldots, a_n$ over K, $F \subseteq K(a_1, \ldots, a_n)$ a finite subset and $p \in K[a_1, \ldots, a_n]$ a nonzero polynomial. The *degree of linearization* $\delta_{k \to K(a)}^p(F)$ of F with respect to p is the maximum $d \in \mathbb{N}$ such that there exists

$$A_1, \ldots, A_n \in \langle a_1, \ldots, a_n \rangle_k, \quad \beta_1, \ldots, \beta_n \in K,$$

having the properties

(1) $f(A_1 + \beta_1, \ldots, A_n + \beta_n)$ is defined and is contained in $\langle a_1, \ldots, a_n \rangle_k + K$ for all $f \in F$,
(2) $p(\beta_1, \ldots, \beta_n) \neq 0$,
(3) $\dim \langle A_1, \ldots, A_n \rangle_k = d$.   •

Thus, if $p = 1$, the degree of linearization is the maximal rank of an affine linear substitution of the type $a_i \mapsto \sum_j \alpha_{ij} a_j + \beta_i$, the α_{ij} being in k, which maps all $f \in F$ to elements of complexity zero. The meaning of the technical assumption on the polynomial p will soon become clear. To ease the notation we write $\delta(f)$ instead of $\delta(\{f\})$.

One proves immediately the following (compare Examples (6.1) and (6.2) in the previous section).

(6.8) Examples.
(1) We have $\delta^1_{\mathbb{R} \to \mathbb{R}(a)}(\sum_{i=1}^n a_i^2) = 0$, the a_i being indeterminates over $\mathbb{R}$.
(2) We have $\delta^1_{k \to K(a)}(\sum_{i=0}^n a_i X^i) = 1$, where $a_0, \ldots, a_n, X$ are indeterminates over a field k and $K := k(X)$. •

We will see in a moment that small degrees of linearization yield better lower bounds, so we will try to choose the polynomial p such as to minimize $\delta^p_{k \to K(a)}(F)$. The choice $p = 1$ is not always the best possible. For instance, it is easy to see that $\delta^1_{k \to k(a)}(a_1 \cdots a_n) = n - 1$. (Take $A_i = a_i$ for $i < n$ and $A_n = 0 = \beta_n$.) On the other hand, the degree of linearization of $a_1 \cdots a_n$ with respect to the polynomial $p := a_1 \cdots a_n$ is one. To see this, note that $\deg((A_1 + \beta_1) \cdots (A_n + \beta_n)) \leq 1$ and $\beta_1 \cdots \beta_n \neq 0$ imply that at most one of the A_i may be different from zero. Thus $\dim\langle A_1, \ldots, A_n\rangle_k \leq 1$. But for the choice $A_1 = a_1, A_2 = \ldots = A_n = 0$ and $\beta_i = 1$ we have $\dim\langle A_1, \ldots, A_n\rangle_k = 1$.

We can now state the announced theorem. Its proof will be given later.

(6.9) Theorem. *Let $k \subseteq K$ be a field extension, k infinite, $K(a_1, \ldots, a_n)$ be a rational function field in the indeterminates $a_1, \ldots, a_n$ over K, $F \subseteq K(a_1, \ldots, a_n)$ be a finite subset and $p \in K[a_1, \ldots, a_n]$ be a nonzero polynomial. Then we have*

$$n - \delta^p_{k \to K(a)}(F) \leq L_{k \to K(a)}\big(F \mid K \cup \{a_1, \ldots, a_n\}\big).$$

(6.10) Remarks.
(1) In our applications K is usually a rational function field in indeterminates $X_1, \ldots, X_m$. We would like to emphasize that in those examples the theorem yields lower bounds valid not only for the input set $\{a_1, \ldots, a_n, X_1, \ldots, X_m\}$, but also for the much larger input set $k(X_1, \ldots, X_m) \cup \{a_1, \ldots, a_n\}$.
(2) The above theorem yields only lower bounds which are not larger than the number n of variables. In particular, nonlinear lower bounds are outside the range of its applications.
(3) If $\delta^p_{k \to K(a)}(F) = \delta^p_{k(t) \to K(t)(a)}(F)$ for some indeterminate t over $K(a)$, then the statement of the theorem is also true for a finite field k. (This condition on δ is obviously satisfied in Example (6.8)(2), and it will turn out to be true for all other examples considered in this chapter.) •

Thm. (6.5) is clearly a consequence of the foregoing theorem and Examples (6.8). As further applications we obtain the following corollaries.

(6.11) Corollary. *For indeterminates $a_1, \ldots, a_n$ we have*

$$L_{k(a)}(a_1 \cdots a_n) = n - 1, \quad L_{\mathbb{R}(a)}(a_1^2 + \ldots + a_n^2) = n.$$

(6.12) Corollary (Winograd). *Let a_{ij}, X_j be indeterminates over a field k for $i \in \underline{m}, j \in \underline{n}$. Then*

$$L_{k(a,X)}\Big(\Big\{\sum_{j=1}^n a_{ij} X_j \mid i \in \underline{m}\Big\}\Big) = mn.$$

The proof that the degree of linearization vanishes for the family of quadratic forms considered above with respect to $p = 1$ is left as an exercise to the reader (cf. Ex. 6.1).

In the proof of Thm. (6.9) we need the invariance of the degree of linearization with respect to purely transcendental extensions.

(6.13) Lemma. *We make the same assumptions as in Def. (6.7) and require additionally k to be infinite. Then, if v denotes an indeterminate over $K(a)$, we have*

$$\delta^p_{k \to K(a)}(F) = \delta^p_{k(v) \to K(v)(a)}(F).$$

Proof. The inequality "$\leq$" is clear. To show the reverse inequality, let $A_1, \ldots, A_n \in \langle a_1, \ldots, a_n \rangle_{k(v)}, \beta_1, \ldots, \beta_n \in K(v)$ satisfy the assumptions (1)–(3) of Def. (6.7) with respect to the field extension $k(v) \subseteq K(v)$ and $d := \delta^p_{k(v) \to K(v)(a)}(F)$. We consider the k-family of K-algebra morphisms $(\sigma_v)_{v \in k}$ defined by

$$\sigma_v \colon K[a_1, \ldots, a_n, v] \to K(a_1, \ldots, a_n), a_i \mapsto a_i, v \mapsto v.$$

It is straightforward to show that there is a nonzero polynomial $D \in K[v]$ such that for all $v \in k$ satisfying $D(v) \neq 0$ the elements $\sigma_v(A_1), \ldots, \sigma_v(A_n)$, resp. $\sigma_v(\beta_1), \ldots, \sigma_v(\beta_n)$ are defined, contained in $\langle a_1, \ldots, a_n \rangle_k$, resp. in K, and such that conditions (1)–(3) of Def. (6.7) are valid with respect to $k \subseteq K$. This implies the assertion. We leave the details as an exercise to the reader (cf. Ex. 6.11). $\square$

Proof of Thm. (6.9). We are going to show the following, more general statement, by induction on $r \in \mathbb{N}$.

> We make the same assumptions as in the statement of (6.9). Moreover, let $b_1, \ldots, b_m$ be $m \leq n$ indeterminates over K and let
>
> $$A_1, \ldots, A_n \in \langle b_1, \ldots, b_m \rangle_k, \ \beta_1, \ldots, \beta_n \in K$$
>
> be such that $f(A + \beta) := f(A_1 + \beta_1, \ldots, A_n + \beta_n)$ is defined for all $f \in F$, $p(\beta_1, \ldots, \beta_n) \neq 0$ and
>
> $$L_{k \to K(b)}(\{f(A + \beta) \mid f \in F\} \mid K \cup \{b_1, \ldots, b_m\}) \leq r.$$
>
> Then we have $\dim \langle A_1, \ldots A_n \rangle_k \leq r + \delta^p_{k \to K(a)}(F).$

(Introducing a new set of indeterminates helps to clarify the proof.) We remark first that the assertion of the theorem follows from the above claim. To see this, we choose $m = n$, $A_i = b_i = a_i$ and $\beta \in K^n$ such that $p(\beta) \neq 0$, and use the invariance of L under affine automorphisms, cf. Rem. (6.4).

The definition of the degree of linearization has been made just to guarantee the induction start "$r = 0$."

So let us suppose that $r > 0$. We proceed as in the proof of Thm. (6.5). Let $(g_1, \ldots, g_r)$ be an optimal computation sequence for $\{f(A + \beta) \mid f \in F\}$ in $k \to K(b)$ modulo $K \cup \{b_1, \ldots, b_m\}$. We have $g_1 = uv$ or $g_1 = u/v$ with some $u, v \in \langle b_1, \ldots, b_m \rangle_k + K$. By optimality not both u and v are contained in K.

Case 1. $v \notin K$. W.l.o.g. we suppose that the coefficient of b_m in v is not zero, say $v = b_m - \sum_{j<m} \gamma_j b_j - \kappa$, $\gamma_j, \kappa \in K$ after some scaling. By construction, the complexity of $\{f(A + \beta) \mid f \in F\}$ in $k(v) \to K(b)$ is at most $r - 1$. We view now $K(b_1, \ldots, b_m)$ as a rational function field in $b_1, \ldots, b_{m-1}$ over $K(v)$ and note that $A_i + \beta_i = A_i' + \beta_i'$ for some linear forms $A_i' \in \langle b_1, \ldots b_{m-1} \rangle_k$ and elements $\beta_i' = \alpha_{im}(v + \kappa) + \beta_i \in K(v)$, where α_{im} is the coefficient of b_m in A_i (compare (6.6)). Moreover, the dimensions of $\langle A_1, \ldots, A_n \rangle_k$ and $\langle A_1', \ldots, A_n' \rangle_k$ differ at most by one. We may now apply the induction hypothesis to the situation described in the following.

We focus on the field extension $k(v) \subseteq K(v)$ and take the same F and p, but F interpreted as a subset of the rational function field $K(v)(a_1, \ldots, a_n)$ and p as an element of the polynomial ring $K(v)[a_1, \ldots, a_n]$. Further, we consider the linear forms $A_1', \ldots, A_n'$ in the indeterminates $b_1, \ldots, b_{m-1}$ over $K(v)(a)$ and the elements $\beta_1', \ldots, \beta_n' \in K(v)$. We have $p(\beta') \neq 0$ since the substitution $v \mapsto -\kappa$ maps β' to β. Because $A_i' + \beta_i' = A_i + \beta_i$, the $f(A' + \beta')$ are certainly defined, and we know that their complexity in $k(v) \to K(v)(b_1, \ldots, b_{m-1})$ with respect to the input set $K(v) \cup \{b_1, \ldots, b_{m-1}\}$ is strictly less than r. Therefore, by the induction hypothesis, we conclude that

$$\dim\langle A_1', \ldots, A_n' \rangle_{k(v)} \leq r - 1 + \delta^p_{k(v) \to K(v)(a)}(F).$$

However, by Lemma (6.13), the degree of linearization of F in $k \to K(a)$ does not change when we adjoin a transcendental element v. The induction step is thus settled in this case.

Case 2. $v \in K$. Then $u \notin K$. If $g_1 = uv$ we can proceed as in Case 1. The remaining situation where $g_1 = u/v$ is settled analogously by going over to the algebra $k(u/v) \to K(b)$.

The proof of the theorem is now complete. $\square$

6.3* Continued Fractions, Quotients, and Composition

This section is devoted to a discussion of three further, more complicated applications of the substitution method, which yield optimal lower bounds for the multiplicative complexity. The reader less interested in the elaboration of this method may omit this section.

We start with an application of Thm. (6.9) to the problem of computing continued fractions. Let us proceed with some preliminary considerations. To a sequence of indeterminates $Y_0, Y_1, \ldots$ we assign a sequence of continued fractions, recursively defined by $F_0 := Y_0$ and $F_i := Y_i + F_{i-1}^{-1}$ for $i > 0$. In order to write the rational functions F_i as a quotient of two polynomials, we define the *Euler brackets* $[Y_i, Y_{i-1}, \ldots, Y_0]$ recursively by setting $[Y_0] := Y_0$, $[Y_1, Y_0] := Y_1 Y_0 + 1$ and for $i \geq 2$

$$[Y_i, \ldots, Y_0] := Y_i [Y_{i-1}, \ldots, Y_0] + [Y_{i-2}, \ldots, Y_0].$$

We leave the proof of the following lemma as an exercise to the reader. (Cf. Ex. 6.12.)

(6.14) Lemma. (1) *We have $F_i = [Y_i, \ldots, Y_0]/[Y_{i-1}, \ldots, Y_0]$ for $i > 0$.*

(2) *$[Y_i, \ldots, Y_0]$ is the sum of the product $Y_i \cdots Y_0$ and of all subproducts obtained from this by deleting disjoint pairs of adjacent factors. Hereby, the empty product $[\,] = 1$ is one of these subproducts if i is odd. For instance*

$$[Y_3, \ldots, Y_0] = Y_3 Y_2 Y_1 Y_0 + Y_1 Y_0 + Y_3 Y_0 + Y_3 Y_2 + 1.$$

(3) *Let $\kappa_i, \ldots, \kappa_0$ be field elements and $i \geq \ell_s > \ell_{s-1} > \ldots > \ell_1 \geq 0$. Then the polynomial h obtained from $[Y_i + \kappa_i, \ldots, Y_0 + \kappa_0]$ by substituting the $Y_{\ell_s}, \ldots, Y_{\ell_1}$ by zero is of the form*

$$h = [\kappa_{\ell_s}, \kappa_{\ell_{s-1}}, \ldots, \kappa_{\ell_1}] \cdot \prod Y_j + \text{ terms of lower degree,}$$

where the product is over all j satisfying $0 \leq j \leq i$ and $j \notin \{\ell_1, \ldots, \ell_s\}$. In particular, the degree of h is $i + 1 - s$ if the bracket $[\kappa_{\ell_s}, \ldots, \kappa_{\ell_1}]$ does not vanish.

We are going to prove now that the continued fraction formed by general polynomials has the expected multiplicative complexity.

(6.15) Theorem (Strassen, Hartmann and Schuster). *To a sequence of general polynomials*

$$\sum_{j=0}^{n_0} a_{0j} X^j, \sum_{j=0}^{n_1} a_{1j} X^j, \ldots$$

(the a_{ij} being indeterminates over k) we assign the continued fractions $f_0, f_1, \ldots$, recursively defined by $f_0 := \sum_{j=0}^{n_0} a_{0j} X^j$ and for $i > 0$

$$f_i := \sum_{j=0}^{n_i} a_{ij} X^j + f_{i-1}^{-1}.$$

Then we have for all r

$$L_{k(a,X)}(f_r) = \sum_{i=0}^{r} n_i + r.$$

Proof. The upper bound follows by computing the general polynomials with Horner's rule. To settle the lower bound, we will prove that

$$L_{k(a,X)}(f_r^{-1}) \geq \sum_{i=0}^{r}(n_i + 1) = \text{ number of } a\text{-variables occurring in } f_r^{-1}$$

by showing that the degree of linearization $\delta_{k \to k(X)(a)}^p (f_r^{-1})$ vanishes for a suitable polynomial p.

Let $b_1, \ldots, b_m$ be further indeterminates, $m := \sum_{i=0}^{r}(n_i + 1)$, and let $A_{ij} \in \langle b_1, \ldots, b_m \rangle_k$, $\beta_{ij} \in k(X)$ for $0 \le i \le r$, $0 \le j \le n_i$. We denote the substitution $a_{ij} \mapsto A_{ij} + \beta_{ij}$ by σ. (We have chosen another set of indeterminates in order to simplify the notation.) We define

$$N_i := \left[\sum_{j=0}^{n_i} a_{ij}X^j, \ldots, \sum_{j=0}^{n_0} a_{0j}X^j\right].$$

By the previous lemma we have $f_i^{-1}N_i = N_{i-1}$, hence

(A) $$\sigma(f_i^{-1})\sigma(N_i) = \sigma(N_{i-1}).$$

Suppose now that the following condition on $\sigma(N_0), \ldots, \sigma(N_r)$ is satisfied:

(B) $\begin{cases} \sigma(N_0), \ldots, \sigma(N_r) \text{ are nonzero polynomials in } b_1, \ldots, b_m \text{ of weakly} \\ \text{increasing degrees with the additional property} \\ \\ \qquad \deg_b \sigma(N_{i-1}) = \deg_b \sigma(N_i) \implies A_{i0} = \ldots A_{in_i} = 0. \end{cases}$

Then, starting from the assumption $\sigma(f_r^{-1}) \in \langle b_1, \ldots, b_m \rangle_k + k(X)$, we conclude from this and (A) that

$$\deg_b \sigma(N_{r-1}) = \deg_b \sigma(N_r), \quad \sigma(f_r^{-1}) \in k(X)^\times \text{ and } A_{r0} = \ldots = A_{rn_r} = 0.$$

By definition of the brackets we have $N_r = (\sum_{j=0}^{n_r} a_{rj}X^j)N_{r-1} + N_{r-2}$. Hence

$$\sigma(N_r) = \left(\underbrace{\sum_{j=0}^{n_r} A_{rj}X^j + \sum_{j=0}^{n_r} \beta_{rj}X^j}_{=0}\right)\sigma(N_{r-1}) + \sigma(N_{r-2}).$$

From this and (A) we conclude that

$$\sigma(f_{r-1}^{-1}) = \frac{\sigma(N_{r-2})}{\sigma(N_{r-1})} = \sigma(f_r) - \sum_{j=0}^{n_r} \beta_{rj}X^j \in k(X)^\times.$$

Therefore, $\sigma(f_{r-1}^{-1})$ is also contained in $k(X)$. By continuing this argument inductively, we see that all the A_{ij} for $0 \le i \le r$, $0 \le j \le n_i$ must vanish.

What remains to show is the existence of some polynomial p over $k(X)$ such that the nonvanishing of p at a sequence $(\beta_{ij})_{0 \le i \le r, 0 \le j \le n_i}$ is sufficient to guarantee the above condition (B) on $\sigma(N_0), \ldots, \sigma(N_r)$. Let $\ell_s > \ell_{s-1} > \ldots > \ell_1$ be exactly those indices $i \in \{r, r-1, \ldots, 0\}$ for which $A_{i0} = \ldots = A_{in_i} = 0$. Lemma (6.14)(3) implies

$$\sigma(N_r) = \left[\sum_{j=0}^{n_{\ell_s}} \beta_{\ell_s j}X^j, \ldots, \sum_{j=0}^{n_{\ell_1}} \beta_{\ell_1 j}X^j\right] \cdot \prod_i \left(\sum_{j=0}^{n_i} A_{ij}X^j\right)$$

$$+ \text{ terms of lower degree in } b_1, \ldots, b_m,$$

where the product is over all j satisfying $0 \le j \le i$ and $j \notin \{\ell_1, \dots, \ell_s\}$. Therefore, when we require the first factor to be nonzero, $\sigma(N_r)$ is not zero and $\deg \sigma(N_r) \ge \deg \sigma(N_{r-1})$, where equality is only possible when $A_{r0} = \dots = A_{rn_r} = 0$. Hence it is sufficient to choose for p the product of the brackets

$$\left[\sum_{j=0}^{n_{\ell_s}} B_{\ell_s j} X^j, \dots, \sum_{j=0}^{n_{\ell_1}} B_{\ell_1 j} X^j \right]$$

taken over all subsequences $\ell_s > \dots > \ell_1$ of $r, r-1, \dots, 0$. (The B_{ij} are indeterminates over $k(X)$ thought to be substituted by β_{ij}.)   □

We turn now to the problem of evaluating a quotient of general polynomials.

(6.16) Theorem (Hartmann and Schuster). *Let a_{ij}, X be indeterminates over a field k, for $0 \le i \le 1$ and $0 \le j \le n_i$. Then we have*

$$L_{k(a,X)}\left(\sum_{j=0}^{n_1} a_{1j} X^j \Big/ \sum_{j=0}^{n_0} a_{0j} X^j \right) = n_0 + n_1 + 1.$$

Proof. The upper bound is clear. Let $f := \sum_{j=0}^{n_1} a_{1j} X^j / \sum_{j=0}^{n_0} a_{0j} X^j$. The most obvious thing to do now is to try to show that the degree of linearization $\delta^p_{k \to k(X)(a)}(f)$ is one with respect to some nonzero polynomial $p \in k(X)[a]$. Unfortunately, this quantity exceeds one for all nonzero polynomials p if k is infinite and $n_0, n_1 > 0$. (Cf. Ex. 6.13.) Therefore, we can not use Thm. (6.9). Instead, we give a direct induction proof which differs from those of the mentioned theorem by a slight modification of the induction claim. In fact, we will replace the condition $p(\beta) \ne 0$ by $\sigma(f) \notin k(X)$.

We are going to prove the following statement by induction on $r \in \mathbb{N}$.

Let $b_1, \dots, b_m$ be indeterminates over $k(X)$, let $A_{ij} \in \langle b_1, \dots, b_m \rangle_k$, $\beta_{ij} \in k(X)$ for $0 \le i \le 1, 0 \le j \le n_i$ and denote the substitution $a_{ij} \mapsto A_{ij} + \beta_{ij}$ by σ. Assume that

$$\sigma\left(\sum_{j=0}^{n_0} a_{0j} X^j \right) \ne 0, \quad \sigma(f) \notin k(X),$$

and

$$L_{k \to k(X)(b)}\left(\sigma(f) \,\Big|\, k(X) \cup \{b_1, \dots, b_m\} \right) \le r.$$

Then

$$\dim \sum_{i,j} k A_{ij} \le r + 1.$$

It is obvious that this claim implies the desired assertion.

Let us verify the induction start "$r = 0$." Assume

$$\sigma(f) = \sum_{j=0}^{n_1} (A_{1j} + \beta_{1j}) X^j \Big/ \sum_{j=0}^{n_0} (A_{0j} + \beta_{0j}) X^j = B + \kappa,$$

where $B \in \langle b_1, \ldots, b_m \rangle_k, B \neq 0$ and $\kappa \in k(X)$. Then a comparison of the degrees in the b-variables yields $A_{00} = \ldots = A_{0n_0} = 0$. Therefore $\sum_j A_{1j} X^j = B \sum_j \beta_{0j} X^j$, and we see that the rational function $\sum_j \beta_{0j} X^j$ must be a polynomial, say $\sum_j \gamma_j X^j, \gamma_j \in k$. Hence $A_{1j} = \gamma_j B$ for all j and the dimension of $\sum k A_{ij}$ is indeed at most one. (Note that without the assumption $B \neq 0$ we could not have drawn this conclusion!)

Now let us check the induction step. We proceed as in the proof of Thm. (6.9), our exposition will thus be brief. We may assume that there is some $v = b_m - \sum_{j<m} \gamma_j b_j - \kappa, \gamma_j \in k, \kappa \in k(X)$ such that the complexity of $\sigma(f)$ in $k(v) \to k(X, b_1, \ldots, b_m) = k(v)(X)(b_1, \ldots, b_{m-1})$ is at most $r - 1$. Furthermore, we have linear forms $A'_{ij} \in \langle b_1, \ldots, b_{m-1} \rangle_k$ and elements $\beta'_{ij} \in k(v, X)$ such that $A_{ij} + \beta_{ij} = A'_{ij} + \beta'_{ij}$, and the dimensions of $\sum k A_{ij}$ and $\sum k A'_{ij}$ differ at most by one. If we can show that $\sigma(f) \notin k(v, X)$, then we can apply the induction hypothesis to the ground field $k(v)$, the indeterminates $b_1, \ldots, b_{m-1}$, the A'_{ij} and β'_{ij} and we are done.

If $\sigma(f) \in k(v, X)$, then

$$\sigma(f) \sum_j (A'_{0j} + \beta'_{0j}) X^j = \sum_j (A'_{1j} + \beta'_{1j}) X^j$$

implies by a comparison of coefficients that

(A) $$\sigma(f) \sum_j A'_{0j} X^j = \sum_j A'_{1j} X^j.$$

If it happens to be the case that $\sum_j A'_{0j} X^j = 0$, then $A'_{ij} = 0$ for all i, j and therefore trivially $\dim \sum k A_{ij} \le 1 \le r + 1$. On the other hand, if $A'_{0j} \neq 0$ for some j, then (A) implies $\sigma(f) \in k(X)$, a contradiction. $\square$

Our final application deals with the problem of computing the composition of polynomials.

(6.17) Theorem (Hartmann and Schuster). *Let $n_0, n_1 \ge 1$ and a_{ij}, X be indeterminates over a field k, for $0 \le i \le i$ and $0 \le j \le n_i$. Then we have*

$$L_{k(a,X)} \left(\sum_{i=0}^{n_1} a_{1i} \left(\sum_{j=0}^{n_0} a_{0j} X^j \right)^i \right) = n_0 + n_1.$$

Proof. The upper bound is clear. Let f be the polynomial under consideration. To show the lower bound, we actually prove a stronger statement. Let us rename the variable a_{00} by calling it t and put $k_0 := k(t)$. We claim that

$$L_{k_0 \to k_0(X)(a)}(f \mid k_0(X) \cup \{a_{01}, \ldots, a_{1n_1}\}) \ge n_0 + n_1$$

$$= \text{number of } a\text{-variables} - 1.$$

As in the proof of Thm. (6.16), we do not refer to Thm. (6.9), but give a direct proof. The hardest step is to find a substitute for the condition "$p(\beta) \neq 0$" in the induction claim, which on the one hand makes the induction work and on the other hand yields a large enough lower bound.

In this proof we will call $\deg q := \deg N - \deg D$ the *degree* of a nonzero rational function $q = N/D$, N, D being polynomials in one variable over a field.

We are going to prove the following statement by induction on $r \in \mathbb{N}$.

Let $b_1, \ldots, b_m$ be indeterminates over $k_0(X)$, $A_{ij} \in \langle b_1, \ldots, b_m \rangle_{k_0}$, $\beta_{ij} \in k_0(X)$ and denote the substitution $a_{ij} \mapsto A_{ij} + \beta_{ij}$ by σ. Assume that

$$\sum_{j=1}^{n_0} \beta_{0j} X^j \notin k_0, \quad \beta_{1n_1} \neq 0, \quad \deg_X \beta_{1n_1} \geq 0$$

and

$$L_{k_0 \to k_0(X)(b)}\left(\sigma(f) \mid k_0(X) \cup \{b_1, \ldots, b_m\}\right) \leq r.$$

Then

$$\dim \sum_{i,j} kA_{ij} \leq r+1.$$

This statement implies the desired assertion. (Consider the affine automorphism $a_{ij} \mapsto a_{ij} + 1$.)

Let us check the start "$r = 0$." We assume $n_1 > 1$ and leave the case $n_1 = 1$ as an exercise to the reader. Let

(A) $$\sigma(f) = \sum_{i=0}^{n_1} (A_{1i} + \beta_{1i})g^i \in \langle b_1, \ldots, b_m \rangle_{k_0} + k_0(X),$$

where we have set $g := t + \sum_{j=1}^{n_0} (A_{0j} + \beta_{0j})X^j$.

If $g \notin k_0(X)$, then by considering the degrees in the b-variables, we conclude first that $A_{1n_1} = 0$ and then, since $n_1 > 1$,

$$A_{1,n_1-1}g^{n_1-1} + \beta_{1n_1}g^{n_1} = (A_{1,n_1-1} + \beta_{1n_1}g)g^{n_1-1} = 0.$$

Therefore, $A_{1,n_1-1} + \beta_{1n_1} \sum_{j=1}^{n_0} A_{0j}X^j = 0$, which implies either $\beta_{1n_1} = 0$ or $\deg_X \beta_{1n_1} < 0$, which are both contradictions.

Assume that $g \in k_0(X)$, that is $A_{01} = \ldots = A_{0n_0} = 0$. From (A) we obtain $\sum_{i=0}^{n_1} A_{1i}g^i \in \langle b_1, \ldots, b_m \rangle_{k_0}$. If $A_{1i} \neq 0$ for some $i > 0$, then g must be algebraic over k_0. Since k_0 is algebraically closed in $k_0(X)$, we conclude that $g \in k_0$, which is contradicting our assumption $\sum_j \beta_{0j}X^j \notin k_0$. Thus we see that only A_{10} can be nonzero, hence $\dim \sum kA_{ij} \leq 1$, which was to be shown.

We proceed now with the verification of the induction step. As in the proof of Thm. (6.9) we may assume that there is some $v = b_m - \sum_{j<m} \gamma_j b_j - \kappa$, $\gamma_j \in k_0$, $\kappa \in k_0(X)$, such that the complexity of $\sigma(f)$ in $k_0(v) \to k_0(v)(X)(b_1, \ldots, b_{m-1})$ is at most $r-1$. We have linear forms A'_{ij} contained in $\langle b_1, \ldots, b_{m-1} \rangle_{k_0}$, and elements $\beta'_{ij} = \alpha_{ij}(v+\kappa) + \beta_{ij} \in k_0(v, X)$ for some $\alpha_{ij} \in k_0$ such that $A_{ij} + \beta_{ij} = A'_{ij} + \beta'_{ij}$, and such that the dimensions of $\sum kA_{ij}$ and $\sum kA'_{ij}$ differ at most by one. If we can show that

$$\sum_{j=1}^{n_0} \beta'_{0j} X^j \notin k_0(v), \quad \beta'_{1n_1} \neq 0, \quad \deg_X \beta'_{1n_1} \geq 0,$$

then we can apply the induction hypothesis and we are done. So let us verify the above conditions. We have $\beta'_{1n_1} = \alpha_{1n_1}(v+\kappa)+\beta_{1n_1}$. W.l.o.g. $\alpha_{1n_1} \neq 0$, since β_{1n_1} satisfies corresponding conditions. Clearly $\beta'_{1n_1} \neq 0$. If we write $\alpha_{1n_1}\kappa + \beta_{1n_1} = N/D$ with $N, D \in k_0[X]$, then $\beta'_{1n_1} = (\alpha_{1n_1}vD + N)/D$, hence

$$\deg_X \beta'_{1n_1} = \max\{\deg_X D, \deg_X N\} - \deg_X D \geq 0.$$

It remains to show that

$$\sum_{j=1}^{n_0} \beta'_{0j} X^j = \sum_{j=1}^{n_0} \beta_{0j} X^j + (v+\kappa) \sum_{j=1}^{n_0} \alpha_{0j} X^j \notin k_0(v).$$

Let us assume the contrary. Then $\sum_j \alpha_{0j} X^j \neq 0$, and we have

(B) $$\sum_{j=1}^{n_0} \beta'_{0j} X^j = v \sum_{j=1}^{n_0} \alpha_{0j} X^j + h \in k_0(v)$$

for some $h \in k_0(X)$, which therefore also lies in $k_0(v)[X]$. Hence $h \in k_0[X]$. We have now arrived at a contradiction, since some monomial X^j for $j > 0$ occurs on the left-hand side of (B).   □

6.4 Exercises

The X-, Y-, a-, and b-variables denote indeterminates over a field k.

6.1. Verify that $\delta^1_{k \to k(a,X)}\left(\{\sum_{j=1}^n a_{ij} X_j \mid i \in \underline{m}\}\right) = 0$.

6.2. (Evaluation of a general polynomial in several variables) Show that

$$L_{k(a,X)}\left(\sum a_{ij} X^i Y^j\right) = (m+1)(n+1) - 1,$$

where the sum is over all pairs (i, j) satisfying $0 \leq i \leq m$ and $0 \leq j \leq n$.

6.3. (Evaluation of several general univariate polynomials) Determine

$$L_{k(a,X)}\left(\{\sum_{j=0}^{n_i} a_{ij} X^j \mid i \in \underline{p}\}\right).$$

6.4. (Inner product) Prove that $L_{k(a,b)}(\sum_{i=1}^{n} a_i b_i) = n$.

(Hint: show the stronger statement

$$L_{k \to k(b)(a)}\left(\sum_{i=1}^{n} a_i b_i \mid k(b) \cup \{a_1, \ldots, a_n\}\right) = n.)$$

6.5. Prove that $L_{k(a)}(\sum_{i=1}^{n} 1/a_i) = n$.

6.6. (Evaluation of a product of general polynomials) Prove

$$L_{k(a,b,X)}\left(\left(\sum_{i=0}^{m} a_i X^i\right)\left(\sum_{i=0}^{n} b_j X^j\right)\right) = m + n + 1.$$

6.7. Let $k \subseteq K$ be a field extension and $g_{ij}, g_i \in K$ for $i \in \underline{m}, j \in \underline{n}$. We define the *column-rank* cr(G) of the matrix $G = (G_{ij}) \in K^{m \times n}$ as the dimension of the k-space generated by the columns of this matrix modulo k^m. For instance, the matrix $[1, X, \ldots, X^n] \in K^{1 \times (n+1)}$ has column-rank n, where $K = k(X)$. Let F denote the set of polynomials $\{\sum_{j=1}^{n} g_{ij} a_j + g_i \mid i \in \underline{m}\}$.

(1) Prove that the degree of linearization $\delta_{k \to K(a)}^1(F)$ of F equals $n - $ cr(G).
(2) Conclude that $L_{k \to K(a)}(F) \geq$ cr(G).
(3) Deduce Thm. (6.5) and Cor. (6.12) from (2).

6.8. Give a detailed proof of $L_{k(a,X)}(\sum_{i=0}^{n} a_i X^i) = n$ based on families of substitutions σ_λ.

6.9. Show that $L_{k(a)}(a_1^2 + \ldots + a_n^2) \geq n/2$ if chark $\neq 2$. (Hint: replace a_i by $a_i + X_i$ where X_i is a new indeterminate and use the transcendence degree bound (5.9).)

6.10. Show that for even n

$$mn/2 \leq L_{k(a,X)}\left(\left\{\sum_{j=1}^{n} a_{ij} X_j \mid i \in \underline{m}\right\} \mid k(a) \cup \{X_1, \ldots, X_n\}\right) \leq (m+1)n/2.$$

Thus a precomputation of rational functions in the a-variable may help (whereas a precomputation of rational functions in the X-variables does not).

6.11. Give a detailed proof of Lemma (6.13).

6.12. Prove Lemma (6.14).

6.13. Let $f = \sum_{j=0}^{n_1} a_{1j} X^j / \sum_{j=0}^{n_0} a_{0j} X^j$ and assume k to be infinite. Prove that for all nonzero $p \in k(X)[a]$ we have $\delta_{k \to k(X)(a)}^p(f) \geq \min\{n_0, n_1\} + 1$. (Hint: show that for all $p \neq 0$ there is some $\beta \in k(X)^{n_0+n_1+2}$ such that $p(\beta) \neq 0$ and $\sum_j \beta_{0j} X^j, \sum_j \beta_{1j} X^j \in k^\times$.)

6.14. Let $f_1, \ldots, f_q \in k(X_1, \ldots, X_n)$. Then there exist $p \in \underline{q}$ and $\alpha_i \in k$ such that

$$L_{k(X)}(\tilde{f}_1, \ldots, \tilde{f}_{p-1}, \tilde{f}_{p+1}, \ldots, \tilde{f}_q) \leq L_{k(X)}(f_1, \ldots, f_q) - 1,$$

where $\tilde{f}_i := f_i - \alpha_i f_p$.

6.15. (Multiplication of complex numbers) Show that

$$L_{\mathbb{R}(X,Y,U,V)}(XU - YV, XV + YU) = 3.$$

(Hint: Use Ex. 6.14. Compare also Prop. (17.7).)

6.16. (Inversion of complex numbers) Prove that

$$L_{\mathbb{R}(X,Y)}\left(\frac{X}{X^2 + Y^2}, -\frac{Y}{X^2 + Y^2}\right) = 4.$$

(Hint: First substitute, and then use Ex. 6.14.)

6.17. (Inversion of 2×2-matrices) Let $d := a_{11}a_{22} - a_{12}a_{21}$. Show that

$$L_{k(a)}(a_{11}/d, -a_{12}/d, -a_{21}/d, a_{22}/d) = 6.$$

(Hint: Use Ex. 6.14 three times, and then substitute.)

6.5 Open Problems

Problem 6.1. Determine the complexity of one Newton step for the approximation of a root of a general polynomial:

$$L_{k(a,X)}\left(X - \frac{\sum_{i=0}^{n} a_i X^i}{\sum_{i=1}^{n} i a_i X^{i-1}}\right) = ?$$

6.6 Notes

The substitution technique introduced by Pan [405] was further developed by Winograd in his influential paper [552]. There Winograd proved a lower bound on the multiplicative complexity of a family of polynomials $\sum_{j=1}^{n} g_{ij}a_j + g_i$, $i \in \underline{m}$, $g_{ij}, g_i \in k(X)$, in terms of the so-called column-rank of the matrix (g_{ij}). (Winograd's result is a special case of Thm. (6.9), see Ex. 6.7.) As a consequence, he obtained Pan's result and the optimality of the natural algorithm for multiplying a general matrix by a general vector (Cor. (6.12)).

Strassen [495] extended Winograd's result to an arbitrary set of rational functions. Thm. (6.9) is a reformulation of his result. His paper contains further applications of the substitution technique, among them are Cor. (6.11) (which has been independently obtained by Vari [533]), and a special case of Thm. (6.15) on the complexity of continued fractions. All the results of Sect. 6.3 are due to Hartmann and Schuster [231].

Borodin [67] has proved that Horner's rule is essentially the only way to compute a general polynomial of degree n in $2n$ operations.

The viewpoint of reducing the multiplicative complexity by adjoining intermediate results to the field of scalars is also taken in Baur's lectures [29] which have been particularly helpful in writing this chapter.

We remark that the techniques developed in Chap. 17 for proving lower bounds on the complexity of bilinear maps are nothing but refined variants of the substitution method. For comments on this we refer to the notes of Chap. 17.

Most of the exercises of this chapter are taken from Strassen's lectures [503]. Ex. 6.17 on the inversion of complex numbers is due to Alt and van Leeuwen [8]. We note that the multiplicative complexity of complex division (as well as that of solving a linear system of size 2) equals 6 as was shown by Lickteig [332, 333]. Problem 6.1 is formulated in Strassen [510].

Chapter 7. Differential Methods

We will investigate two problems which will prove to be of particular importance for the rest of the book. The first question is how much divisions may help for the computation of a set of polynomials. The example of the univariate polynomial $f = X^{31}$ shows that divisions indeed can help. Strassen discovered in 1973 [498] a technique for transforming a straight-line program for a set of rational functions to a division free straight-line program for the "coefficients" of the Taylor series of these functions (Thm. (7.1)). In particular, he showed that divisions do not help for the computation of a set of quadratic forms. This result, which marks the beginning of the theory of bilinear complexity, will be exploited later in Chap. 14. Following Baur and Strassen [32] we will show in the second part of the present chapter how to transform a straight-line program for computing a multivariate rational function into one that computes this function *and* its gradient. Combined with other lower bound techniques, such as Strassen's degree bound introduced in the next chapter, this so-called derivative inequality (7.7) allows us to derive sharp lower bounds for the nonscalar complexity of numerous computational problems. The results of this chapter will also be used extensively in Chap. 16 where we will show that several computational problems in linear algebra are about as hard as matrix multiplication.

Unless otherwise stated, L denotes the multiplicative complexity throughout this chapter, and $X_1, \ldots, X_n$ are indeterminates over the field k; $k[X]$ and $k(X)$ will denote the ring $k[X_1, \ldots, X_n]$ and the field $k(X_1, \ldots, X_n)$, respectively.

7.1 Complexity of Truncated Taylor Series

Throughout this section k is an infinite field.

Let $n \in \mathbb{N}'$, $\lambda \in k^n$, and let $k[[X - \lambda]]$ denote the ring of formal power series in the indeterminates $X_1 - \lambda_1, \ldots, X_n - \lambda_n$. Further, let $\mathcal{O}_\lambda \subset k(X)$ be the local ring of λ; hence, $\mathcal{O}_\lambda$ consists of all f/g, $f, g \in k[X]$, $g \neq 0$, such that $g(\lambda) \neq 0$. There is a canonical embedding $\iota: \mathcal{O}_\lambda \to k[[X - \lambda]]$, $f \mapsto \iota(f) := \sum_{j \geq 0} f^{(j)}$, where $f^{(j)}$ is a homogeneous polynomial of degree j in the $X_1 - \lambda_1, \ldots, X_n - \lambda_n$, see Chap. 2. If $\lambda = 0$ and $f \in k[X]$, then $f^{(j)}$ is the homogeneous part of degree j of f. If k is of characteristic zero, then $f^{(j)} = f^{[j]}/j!$, where

$$f^{[j]} := \sum_{\mu_1,\ldots,\mu_j=1}^{n} \frac{\partial^j f}{\partial X_{\mu_1} \cdots \partial X_{\mu_j}}\bigg|_{X=\lambda} (X_{\mu_1} - \lambda_{\mu_1}) \cdots (X_{\mu_j} - \lambda_{\mu_j});$$

hence, in this case, $\iota(f)$ is the formal Taylor series of f at λ. Having this in mind, we will refer to $\iota(f)$ as the *Taylor series of f at λ* in the sequel.

For $d \in \mathbb{N}$ and $f \in \mathcal{O}_\lambda$ let $C_d(f) := \{f^{(0)}, \ldots, f^{(d)}\}$. We extend C_d to finite subsets $F \subset \mathcal{O}_\lambda$ by setting $C_d(F) := \cup_{f \in F} C_d(f)$. Note that the elements of $C_d(f)$ belong to $k[X]$, whereas those of F may not. The aim of this section is to establish a relationship between $L_{\mathcal{O}_\lambda}(F)$ and $L_{k[X]}(C_d(F))$ (Thm. (7.1) below).

Before going into technical details, let us extract the basic idea of our considerations: by a standard argument we may suppose that $\lambda = 0$. We replace computations in $\mathcal{O}_\lambda$ by computations in the ring $k[[X]]$. Recall that $f \in k[[X]]$ is invertible iff $f^{(0)} \neq 0$. Furthermore, if $f \in k[[X]]$ is such that $f^{(0)} = 0$, then

$$(1 - f)^{-1} = \sum_{i \geq 0} f^i.$$

(Note that the right-hand side of the above equality is a well-defined element of $k[[X]]$ since $f^{(0)} = 0$.) This shows that a division by a unit in $k[[X]]$ can be replaced by (in general infinitely many) multiplications and additions. But since the results of our computations are polynomials, it is enough to consider instead of full Taylor series only their truncated parts up to a sufficiently high degree. In this way divisions can be reduced to a finite number of multiplications and additions.

(7.1) Theorem. *Let $\lambda \in k^n$, $F = \{f_1, \ldots, f_m\} \subset \mathcal{O}_\lambda$, and $d \in \mathbb{N}'$. Then the multiplicative complexity of the $f_i^{(j)}$ satisfy the following inequality:*

$$L_{k[X]}\Big(\{f_i^{(j)} \mid 1 \leq i \leq m, 0 \leq j \leq d\}\Big) \leq \binom{d}{2} L_{\mathcal{O}_\lambda}(f_1, \ldots, f_m),$$

where $\sum_{j=0}^{\infty} f_i^{(j)}$ is the Taylor series of f_i at λ.

Proof. By Rem. (4.9)(3) we may assume that $\lambda = 0$. (Note that we do not count linear operations.) We denote by $\mathcal{O}$ the local ring of 0. The imbedding $\mathcal{O} \to k[[X]]$ which associates to f its Taylor series shows that

$$\ell := L_{\mathcal{O}}(F) \geq L_{k[[X]]}(F),$$

see Rem. (4.9)(3). Let $(u_1, \ldots, u_\ell)$ be a computation sequence for F in $k[[X]]$. We extend C_d to finite subsets of $k[[X]]$ by setting $C_d(f) := \{f^{(0)}, \ldots, f^{(d)}\}$ for $f = \sum_{j \geq 0} f^{(j)} \in k[[X]]$ and $C_d(F) := \cup_{f \in F} C_d(f)$ for finite subsets $F \subset k[[X]]$. If we know how to compute $C_d(u_i)$ from $C_d(a_i)$ and $C_d(b_i)$, then we can replace the instruction $u_i = a_i \omega b_i$ by a whole bunch of instructions in $k[X]$. In this way we can transform $(u_1, \ldots, u_\ell)$ into a computation sequence for $C_d(F)$ in $k[X]$. The proof of the theorem is thus reduced to that of the following lemma.

(7.2) Lemma. *Let $f, g \in k[[X]]$ and $d \in \mathbb{N}'$. Then*

$$L_{k[X]}\big(C_d(fg) \mid C_d(f, g)\big) \le \binom{d}{2}.$$

Further, if $g^{(0)} \ne 0$ we have

$$L_{k[X]}\big(C_d(f/g) \mid C_d(f, g)\big) \le \binom{d}{2}.$$

Proof. Let $u := fg$. Noting that $f^{(0)}, g^{(0)} \in k$, we have

$$
\begin{aligned}
u^{(0)} &= f^{(0)}g^{(0)} \\
u^{(1)} &= f^{(1)}g^{(0)} + f^{(0)}g^{(1)} \\
u^{(2)} &= f^{(2)}g^{(0)} + f^{(1)}g^{(1)} + f^{(0)}g^{(2)} \qquad\qquad \text{1 mult.} \\
&\;\;\vdots \\
u^{(d)} &= \textstyle\sum_{i=0}^{d} f^{(i)}g^{(d-i)} \qquad\qquad\qquad\qquad (d-1) \text{ mult.}
\end{aligned}
$$

This scheme gives rise to a computation sequence and an easy induction argument proves the first inequality. To prove the second, let $u = f/g$. Since $g^{(0)} \ne 0$, g is invertible in $k[[X]]$. We have $gu = f$ and we can compute the components $u^{(i)}$ of u by the following scheme:

$$
\begin{aligned}
u^{(0)} &= f^{(0)}/g^{(0)} \\
u^{(1)} &= (f^{(1)} - u^{(0)}g^{(1)})/g^{(0)} \\
u^{(2)} &= (f^{(2)} - u^{(1)}g^{(1)} - u^{(0)}g^{(2)})/g^{(0)} \qquad\qquad \text{1 mult.} \\
&\;\;\vdots \\
u^{(d)} &= (f^{(d)} - \textstyle\sum_{i=0}^{d-1} u^{(i)}g^{(d-i)})/g^{(0)} \qquad\qquad (d-1) \text{ mult.}
\end{aligned}
$$

Again, this scheme gives rise to a computation sequence and the second inequality follows by an induction argument. $\qquad\qquad\qquad\qquad\qquad\qquad\qquad\square$

An immediate consequence of the foregoing theorem is the following.

(7.3) Corollary. *Let F be a finite subset of $k[X]$ consisting of polynomials of degree $\le d$. Then*

$$L_{k[X]}(F) \le \binom{d}{2} L_{k(X)}(F).$$

The bounds given in Thm. (7.1) or in the last corollary are not optimal: one can employ the techniques of Chap. 2 for manipulating power series to obtain better upper bounds. In particular we have (with the notation of Thm. (7.1))

$$(7.4) \quad L_{k[X]}\big(\{f_i^{(j)} \mid 1 \le i \le m, 0 \le j \le d\}\big) \le 5.75\, d\, L_{O_\lambda}(f_1, \ldots, f_m),$$

see Ex. 7.1. However, Cor. (7.3) yields an important consequence which we shall be dealing with in Chap. 14: *divisions do not help for the computation of a set of quadratic polynomials.*

(7.5) Corollary (Strassen). *Let F be a finite subset of $k[X]$ consisting of polynomials of degree at most two. Then*

$$L_{k[X]}(F) = L_{k(X)}(F).$$

Finally, we remark that the technique of the proof of Thm. (7.1) can be used to generalize Thm. (7.1) to any cost function, see Ex. 7.2.

7.2 Complexity of Partial Derivatives

In this section we relate the complexity of a rational function $f \in k(X)$ with that of f *and* its gradient. We confine ourselves to the study of the *multiplicative* complexity of f; the method introduced here works as well for other complexity measures, in particular for the *total* complexity.

We first state a result on straight-line programs in general. It is in a sense an inverse autarky result (see Sect. 4.3) . Given a morphism $\varphi \colon A \to B$ of k-algebras it says that (under mild conditions) complexity results in B can be lifted via φ to A. Recall that $\Omega = k^c \cup k \cup \{+, -, *, /\}$ is the set of operations.

(7.6) Proposition. *Let $\varphi \colon A \to B$ be a morphism of k-algebras, and let I be a subset of A. Assume that $\varphi^{-1}(B^\times) = A^\times$, and let $G \subseteq B$ be finite. Then there exists a finite subset F of A with $\varphi(F) = G$ and $L_B(G|\varphi(I)) = L_A(F|I)$.*

Proof. Let $b_{-n+1}, \ldots, b_0 \in \varphi(I)$ and $\Gamma = (\Gamma_1, \ldots, \Gamma_r)$ be a straight-line program expecting inputs of length n, which on input $(B; b_{-n+1}, \ldots, b_0)$ computes G and produces the result sequence $(b_{-n+1}, \ldots, b_r)$. As $b_i \in \varphi(I)$, for $-n < i \le 0$, we can find for every such i an element a_i in I such that $\varphi(a_i) = b_i$. Next we define $a_\rho \in \varphi^{-1}(b_\rho)$, $1 \le \rho \le r$, inductively along Γ. If $\Gamma_\rho = (\lambda^c)$, $\lambda \in k$, then $a_\rho := \lambda$. If $\Gamma_\rho = (\lambda; j), \lambda \in k$, then $a_\rho := \lambda a_j$. If $\Gamma_\rho = (\omega; i, j)$ with $\omega \in \{+, -, *\}$ then $a_\rho := a_i \omega a_j$. Finally, if $\Gamma_\rho = (/; i, j)$ then $b_\rho = b_i/b_j$, thus $b_j \in B^\times$ and consequently, $a_j \in \varphi^{-1}(b_j) \subseteq \varphi^{-1}(B^\times) = A^\times$. Hence $a_\rho := a_i/a_j$ is well-defined and $\varphi(a_\rho) = b_\rho$. Altogether, this shows that Γ is executable on $(A; a_{-n+1}, \ldots, a_0)$ with result sequence $(a_{-n+1}, \ldots, a_r)$ and

$$\varphi \circ (a_{-n+1}, \ldots, a_r) = (b_{-n+1}, \ldots, b_r).$$

In particular, if c_*-length$(\Gamma) = L_B(G|\varphi(I))$ and $F := \varphi^{-1}(G) \cap \{a_i \mid -n < i \le r\}$ then

$$L_A(F|I) \le L_B(G|\varphi(I)) = L_B(\varphi(F)|\varphi(I)) \le L_A(F|I),$$

and our second claim follows.   □

Let us fix some notation for the rest of this section. We denote by ∂_i the differential operator $\partial/\partial X_i$ of $k(X)$. A k-subalgebra of $k(X)$ is called ∂_i-closed, if $\partial_i f \in A$ whenever $f \in A$. We call A ∂-closed, if A is ∂_i-closed for all $i = 1, \ldots, n$.

In the sequel we will be dealing with *localizations* of the k-algebra $k[X]$. Let us briefly recall this concept. Let $S \subset k[X]$ be a multiplicative subset, i.e., $1 \in S$

and for all $x, y \in S$ we have $xy \in S$. The localization of A at S denoted by $k[X]_S$ is the subalgebra of $k(X)$ defined as $k[X]_S := \{f/g \mid f \in k[X], g \in S\}$. Note that $k[X]_S$ is ∂-closed.

(7.7) Derivative Inequality (Baur and Strassen). *Let A be a localization of $k[X]$ and $f \in A$. Then*

$$L_A(f, \partial_1 f, \ldots, \partial_n f) \leq 3L_A(f).$$

Proof. Let A be such a k-algebra and $\Gamma = (\Gamma_1, \ldots, \Gamma_t)$ be a straight-line program expecting inputs of length n which is executable on the input $(A; X)$ with the result sequence $(X_1, \ldots, X_n, r_1, \ldots, r_t)$. We show by induction on t the following: if Γ computes $f \in A$ on input $(A; X)$, then Γ induces a straight-line program $\partial \Gamma$ expecting inputs of length n which computes $f, \partial_1 f, \ldots, \partial_n f$ on input $(A; X)$, and

$$c_*\text{-length}(\partial \Gamma) \leq 3 \cdot c_*\text{-length}(\Gamma).$$

This is true for $t = 0$ with $\partial \Gamma = \Gamma$. For $t \geq 1$ let $\Gamma' = (\Gamma'_2, \ldots, \Gamma'_t)$ denote the straight-line program expecting inputs of length $n+1$ derived from Γ by deleting the first instruction Γ_1 and replacing in all successor instructions calls to the result of the first instruction by calls to an additional new input component. In other words $(\omega; -n, j)$ (or $(\omega; j, -n)$) is an instruction of Γ' iff $(\omega; 1, j)$ (or $(\omega; j, 1)$) is an instruction of Γ. Let X_0 be an indeterminate over $k(X)$ and let S denote the preimage of $A^\times$ under the substitution

$$\sigma : k[X_0, \ldots, X_n] \to A, \quad X_0 \mapsto r_1, \quad \forall i \in \underline{n} : X_i \mapsto X_i.$$

Note that σ is uniquely extendable to the localization $R := S^{-1}k[X_0, \ldots, X_n]$ and R is ∂-closed. By construction, Γ' is executable on input $(R; X_0, \ldots, X_n)$ and computes some $F \in R$ with $\sigma(F) = f$, see the proof of Prop. (7.6). By the induction hypothesis there exists a straight-line program $\partial \Gamma'$ expecting inputs of length $n + 1$ which computes $F, \partial_0 F, \ldots, \partial_n F$ on input $(R; X_0, \ldots, X_n)$. Moreover, $c_*\text{-length}(\partial \Gamma') \leq 3 \cdot c_*\text{-length}(\Gamma')$. Applying σ we see that $\partial \Gamma'$ is executable on $(A; r_1, X_1, \ldots, X_n)$ and computes $f = \sigma(F), \sigma(\partial_0 F), \ldots, \sigma(\partial_n F)$. Inserting Γ_1 as the first instruction into $\partial \Gamma'$ we obtain a straight-line program $\widetilde{\partial \Gamma}$ expecting inputs of length n which on input $(A; X)$ computes $f, \sigma(\partial_0 F), \ldots, \sigma(\partial_n F)$; moreover,

$$c_*\text{-length}(\widetilde{\partial \Gamma}) \leq c_*\text{-length}(\partial \Gamma') + c_*(\Gamma_1) \leq 3 \cdot c_*\text{-length}(\Gamma) - 2c_*(\Gamma_1).$$

By the chain rule we have for all $i \in \underline{n}$

$$\partial_i f = \sigma(\partial_i F) + \sigma(\partial_0 F)(\partial_i r_1).$$

Suppose that $\Gamma_1 = (/; -m + \mu, -n + \nu)$. Then $r_1 = X_\mu/X_\nu$, $\partial_\mu r_1 = 1/X_\nu$, and $\partial_\nu r_1 = -X_\mu/X_\nu^2$. We contend that

$$\sigma(\partial_0 F)(\partial_\mu r_1) = \sigma(\partial_0 F)/X_\nu, \quad \sigma(\partial_0 F)(\partial_\nu r_1) = -r_1 \cdot (\sigma(\partial_0 F)/X_\nu).$$

Hence there exists a straight-line program $\tilde{\Gamma}$ expecting inputs of length $2n + 2$ of c_*-length 2 which on input $(A; X_1, \ldots, X_n, r_1, f, \sigma(\partial_0 F), \ldots, \sigma(\partial_n F))$ computes $\sigma(\partial_0 F)(\partial_i r_1)$ for all $i \in \underline{n}$. Then, $\partial \Gamma$ defined by concatenating $\tilde{\Gamma}$ to $\partial \overline{\Gamma}$ satisfies the properties required.

The other cases for Γ_1 can be handled similarly.   □

(7.8) Remarks.

(1) The above argument yields a similar assertion for the *total* complexity with the factor 3 replaced by the factor 4; see the Notes.

(2) It is unknown whether there exists a useful generalization of Thm. (7.7) to higher derivatives. Ex. 7.5 shows that we cannot expect a result similar to Thm. (7.7) with 3 replaced by some other constant.   •

An interesting application of the derivative inequality is the following, which asserts that matrix inversion has roughly the same complexity as computing the determinant. Later in Chap. 16 we will see that the latter problem is, in a sense, as hard as matrix multiplication. Hence, the complexity of matrix multiplication is a measure for that of matrix inversion.

(7.9) Corollary. *Let a_{ij} $(i, j \in \underline{n})$ be indeterminates over the field k, $(b_{ij}) = (a_{ij})^{-1}$ as matrices. Further let $K = k(a_{ij} \mid i, j \in \underline{n})$. Then*

$$L_K(\{b_{ij} \mid i, j \in \underline{n}\}) \leq 3 \cdot L_K(\det(a_{ij})) + n^2,$$

where $L_K(F) := L_K(F \mid \{a_{ij}, i, j \in \underline{n}\})$.

Proof. Let $a := (a_{ij})$. By Cramer's rule we have

$$\frac{\partial}{\partial a_{ij}} \det(a) = b_{ji} \det(a).$$

Hence,

$$
\begin{aligned}
L_K(\{b_{ij} \mid i, j \in \underline{n}\}) &\leq L_K\left(\{\det(a)\} \cup \left\{\frac{\partial}{\partial a_{ij}} \det(a) \,\middle|\, i, j \in \underline{n}\right\}\right) + n^2 \\
&\leq 3 \cdot L_K(\det(a_{ij})) + n^2. \quad \square
\end{aligned}
$$

Another application of the above theorem is in connection with computing power sums. Its proof is straightforward.

(7.10) Corollary. *Let $m \in \mathbb{N}'$. Then*

$$L_K(X_1^{m-1}, \ldots, X_n^{m-1}) \leq 3 \cdot L_K\left(\sum_{i=1}^{n} X_i^m\right).$$

To obtain a lower bound for the left-hand side in Cor. (7.10) and also for further applications of Thm. (7.7) we need the so-called degree bound. This will be the subject of Chap. 8.

7.3 Exercises

7.1. Prove the formula in (7.4).

7.2. During this exercise we keep the notation of Sect. 7.1. We assume that $f, g \in k[[X]]$ and that $d \in \mathbb{N}'$. Further, $c \colon \Omega \to \mathbb{N}$ denotes an arbitrary cost function such that $c(\mathrm{sc}) := \max_{\alpha \in k}(c(\alpha))$ exists and $c(\pm) := \max\{c(+), c(-)\}$.

(1) Prove that

$$L^c_{k[X]}(C_d(fg) \mid C_d(f, g)) \leq \frac{d(d+1)}{2} c(+) + \frac{d(d-1)}{2} c(*) + (2d+1)c(\mathrm{sc}),$$

and if $g^{(0)} \neq 0$ then

$$L^c_{k[X]}(C_d(f/g) \mid C_d(f, g)) \leq \frac{d(d+1)}{2} c(-) + \frac{d(d-1)}{2} c(*) + (2d+1)c(\mathrm{sc}).$$

(2) Show that

$$L^c_{k[X]}(C_d(f \pm g) \mid C_d(f, g)) \leq c(\pm)(d+1)$$

$$\forall \alpha \in k \colon \quad L^c_{k[X]}(C_d(\alpha f) \mid C_d(f)) \leq c(\mathrm{sc})(d+1).$$

7.3. Let $\lambda \in k^n$, $F = \{f_1, \ldots, f_m\} \subset \mathcal{O}_\lambda$, and Γ be a straight-line program which computes F on input $(\mathcal{O}_\lambda; X_1, \ldots, X_n)$. Assume c is a cost function as in Ex. 7.2 satisfying $c(\lambda^c) = 0$ for all $\lambda \in k$. For $\omega \in \{+, -, *, /\}$ denote by $\ell^\omega(\Gamma)$ the $1_{\{\omega\}}$-length of Γ, and let $\ell^{\mathrm{sc}}(\Gamma)$ denote the total number of scalar multiplications in Γ. Prove that

$$L^c_{k[X]}\left(\{f_i^{(j)} \mid 1 \leq i \leq m, 0 \leq j \leq d\}\right) \leq$$
$$\frac{d(d+1)}{2}\left[\ell^*(\Gamma) + \frac{2}{d}\ell^+(\Gamma)\right]c(+) + \frac{d(d+1)}{2}\left[\ell^/(\Gamma) + \frac{2}{d}\ell^-(\Gamma)\right]c(-)$$
$$+ \frac{d(d-1)}{2}(\ell^*(\Gamma) + \ell^/(\Gamma))c(*)$$
$$+ \left[(2d+1)[\ell^*(\Gamma) + \ell^/(\Gamma)] + (d+1)\ell^{\mathrm{sc}}(\Gamma)\right]c(\mathrm{sc}).$$

7.4. Let $f = X_1 \cdots X_n$. In Chap. 6 it was proved that $L(f) = n - 1$. Starting from a division free computation sequence of length $n - 1$ for f, derive another one of length $3(n - 1)$ for $\{f, \partial_1 f, \ldots, \partial_n f\}$.

7.5. Let f be as in the previous exercise and assume that $n \geq 4$. Using the Dimension Bound (4.12) show that

$$L\left(\{\partial_{ij}f \mid 1 \leq i, j \leq n\}\right) \geq \binom{n}{2},$$

where $\partial_{ij}f := \partial^2 f / \partial X_i \partial X_j$. Hence, the naive generalization of the derivative inequality to higher derivatives is not true.

7.4 Open Problems

Problem 7.1. Find a useful generalization of the derivative inequality to several functions or higher (e.g., second) derivatives.

7.5 Notes

Thm. (7.1) and Cor. (7.5) are due to Strassen [498]. The method leading to the derivative inequality (7.7) has been first used by Linnainmaa [336] in the context of error analysis. It was rediscovered by Baur and Strassen [32] who realized its significance for reductions as well as lower bounds in complexity theory. The proof we have presented here has been modeled after Morgenstern's proof [383] and has been taken from Lickteig [334]; he attributes it to Morgenstern, as well as to Schönhage. The following variant of the derivative inequality is due to Baur and Strassen [32, Thm. 2]. We keep the notation of Sect. 7.2: Let $f \in k(X)$ be computable from $\{X_1, \ldots, X_n\} \cup k$ using A additions/subtractions, S scalar multiplications, and M further multiplications/divisions. Then $\{f, \partial_1 f, \ldots, \partial_n f\}$ can be computed from $\{X_1, \ldots, X_n\} \cup k$ using $2A + M$ additions/subtractions, $2S$ scalar multiplications, and $3M$ further multiplications/divisions. In particular,

$$L^{\mathrm{tot}}(f, \partial_1 f, \ldots, \partial_n f) \leq 4L^{\mathrm{tot}}(f).$$

The applications of the derivative inequality presented in Sect. 7.2 are from Baur and Strassen [32]. Problem 7.1 has been taken from Strassen's survey article [510].

Kaltofen [282] has proved the following generalization of Thm. (7.1) (we keep the notation of Sect. 7.1): let $g, h \in k[X]$ be relatively prime multivariate polynomials of degree at most d and $f = g/h \in k(X)$ be a rational function. If f can be computed by a straight-line program of length ℓ, then g and h can be computed by a program without divisions of length polynomial in d and ℓ. For a connection between the derivative inequality, Euler derivations, and lower bounds methods involving the degree of transcendency, see Lickteig [334].

Part III

High Degree

Chapter 8. The Degree Bound

Strassen's degree bound [497], one of the fundamental tools for proving non-linear lower bounds in algebraic complexity theory, states that the multiplicative complexity of a finite set of rational functions is bounded from below by the logarithm of the geometric degree of the graph of the associated rational map. Before discussing this bound in Sect. 8.3, we start with one of its field theoretic versions which can be derived in an elementary way and which will suffice to prove that most of the algorithms derived in Chap. 2 are essentially optimal. In Sect. 8.2 we proceed by defining the geometric degree and deduce a special version of the Bézout inequality. Our intention has been to give a detailed account for non-specialists in algebraic geometry by keeping the prerequisites at a minimum. Applications not easily implied by the field theoretic version of the degree bound will be discussed in Sect. 8.4, before developing methods for estimating the degree in Sect. 8.5. In the last section of this chapter we show how the degree bound can be employed to derive lower bounds for the complexity of rational functions over finite fields.

Unless otherwise specified, k denotes throughout this chapter an algebraically closed field, $X_1, \ldots, X_n$ are indeterminates over k, $K := k(X) := k(X_1, \ldots, X_n)$; L stands for the multiplicative complexity.

8.1 A Field Theoretic Version of the Degree Bound

In this section k is only assumed to be infinite.

Let $f \in k[T]$ be a polynomial and $(g_1, \ldots, g_r)$ be a division free computation sequence for f. Induction on i yields for the polynomials g_i the inequality $\deg g_i \leq 2^i$, and hence $L_{k[T]}(f) \geq \log \deg f$. For proving a similar assertion for the nonscalar complexity of several functions we have to find a suitable generalization of the degree. Our first step in this direction will be an alternative characterization of polynomial degree. We start with a version of the so-called Gauss lemma, which will be frequently used in the sequel. Recall that the *content* a univariate polynomial over a unique factorization domain R is defined as the greatest common divisor of its coefficients. A polynomial is called *primitive* if its content is 1.

(8.1) Gauss Lemma. *Let R be a unique factorization domain and Q be the field of fractions of R. Further, let T be an indeterminate over Q and $f \in R[T]$ be primitive. Then f is irreducible in $R[T]$ if and only if it is irreducible in $Q[T]$.*

(For a proof see, e.g., Lang [316, p. 198, Thm. 6.1].)

(8.2) Proposition. *Let $f \in k[T] \setminus k$. Then $\deg f = [k(T) : k(f)]$.*

Proof. Let Z be an indeterminate over $k(f)$ and $\tilde{f}(Z) := f(Z) - f \in k[f][Z]$. Then $\tilde{f}(T) = 0$. It thus remains to show that $\tilde{f}$ is irreducible in $k(f)[Z]$. This is a consequence of the Gauss lemma: the primitive polynomial $\tilde{f}$ is irreducible in $k[f, Z]$ since it is of degree one in the variable f. Hence, by the Gauss lemma, it is also irreducible in $k(f)[Z]$. $\qquad\square$

We define the *degree* $\deg Y$ of a finite set of rational functions $Y \subseteq K$ as

$$\deg Y := \max_{\Lambda} [K : k(\Lambda)],$$

where Λ runs over all n-subsets of $\sum_{y \in Y} ky + \sum_{j=1}^{n} kX_j + k$ consisting of algebraically independent elements. If the maximum should not exist, we set $\deg Y := \infty$. For $f_1, \ldots, f_m \in K$ we write $\deg(f_1, \ldots, f_m)$ for the degree of $\{f_1, \ldots, f_m\}$. In particular, $\deg(f)$ denotes the degree of the set $\{f\}$ in the above sense. We shall see below that it coincides with the polynomial degree if f is a non-constant polynomial. The following remarks are obvious.

(8.3) Remarks.
(1) For any $A = (a_{ij}) \in \mathrm{GL}(n, k)$ and any $b = (b_i) \in k^n$ the affine linear map $X_i \mapsto \sum_{j=1}^{n} a_{ij} X_j + b_i$ gives an automorphism φ of K. We have $\deg Y = \deg \varphi(Y)$ for all finite subsets $Y \subseteq K$.
(2) $\deg(\emptyset) = 1$.
(3) If $Y_1 \subseteq \sum_{y \in Y_2} ky + \sum_{j=1}^{n} kX_j + k$, then $\deg(Y_1) \leq \deg(Y_2)$.   ●

The next proposition shows that deg is indeed a generalization of the usual degree. Its proof is left as an exercise to the reader. (Cf. Ex. 8.1.)

(8.4) Proposition. *Let $f \in k[X_1, \ldots, X_n] \setminus k$. Then $\deg(f)$ is finite and equals $\deg f$.*

Let us now state the main theorem of this section.

(8.5) Theorem (Baur). *Let $S \subset K$, $\deg S < \infty$, $f_1, \ldots, f_m \in K$. Then*

$$L(f_1, \ldots, f_m \mid S) \geq \log \deg(S \cup \{f_1, \ldots, f_m\}) - \log \deg S.$$

(In particular, $\deg(S \cup \{f_1, \ldots, f_m\})$ is finite.)

(8.6) Remarks.
(1) In many cases $S = \{X_1, \ldots, X_n\}$. Then we obtain

$$L(f_1, \ldots, f_m) \geq \log \deg(f_1, \ldots, f_m).$$

In particular, if $f_1, \ldots, f_n$ form a transcendence basis of K, then

$$L(f_1, \ldots, f_n) \geq \log[K : k(f_1, \ldots, f_n)].$$

(2) If log is replaced by $\log_3$ in the assertion of Thm. (8.5), then the proof becomes considerably easier; see Ex. 8.4 and Ex. 8.8.　　　　　　　　　●

Before proving the theorem, we mention some interesting applications.

(8.7) Corollary (Strassen). *Let $\sigma_1, \ldots, \sigma_n$ be the elementary symmetric polynomials in the indeterminates $X_1, \ldots, X_n$. Then*

$$n(\log n - 2) \leq L(\sigma_1, \ldots, \sigma_n) \leq n \log n.$$

Proof. For the upper bound see Cor. (2.16) and Ex. 2.8. As for the lower bound, note that

$$\deg(\sigma_1, \ldots, \sigma_n) \geq [K : k(\sigma_1, \ldots, \sigma_n)] = n!$$

(for the equality note that the Galois group of the general polynomial of degree n is S_n; see, e.g., Lang [316]). Moreover, $\log n! \geq n(\log n - 2)$.　　□

We can generalize the lower bound of the last corollary in the following way.

(8.8) Corollary (Strassen). *Let $F \subset K$ be a finite set of symmetric rational functions with $\text{tr.deg}_k \, k(F) = r$. Then we have*

$$L(F) \geq r(\log r - 2).$$

Proof. We may w.l.o.g. assume that $F = \{f_1, \ldots, f_r\}$ for some rational functions $f_1, \ldots, f_r$ such that $f_1, \ldots, f_r, X_{r+1}, \ldots, X_n$ are algebraically independent. It suffices to show that

$$[K : k(f_1, \ldots, f_r, X_{r+1}, \ldots, X_n)] \geq r!.$$

Let $\Omega := k(X_{r+1}, \ldots, X_n)$. Then we have

$$[K : k(f_1, \ldots, f_r, X_{r+1}, \ldots, X_n)] = [\Omega(X_1, \ldots, X_r) : \Omega(f_1, \ldots, f_r)] \geq r!,$$

since the assumptions imply that $\Omega(f_1, \ldots, f_r) \subseteq \Omega(\tilde{\sigma}_1, \ldots, \tilde{\sigma}_r)$, where $\tilde{\sigma}_1, \ldots, \tilde{\sigma}_r$ are the elementary symmetric polynomials in the indeterminates $X_1, \ldots, X_r$.　　□

The next result discusses the problem of multiple evaluation.

(8.9) Corollary (Strassen). *Let T be an indeterminate over k, $f(T) \in k[T]$ be a polynomial of degree $m > 0$. Then*

$$
\begin{aligned}
n \log m \; &\leq \; L(f(X_1), \ldots, f(X_n) \mid X_1, \ldots, X_n) \\
&\leq \; 9 \left\lceil \frac{n}{m} \right\rceil m \log m \leq 9(n + m) \log m.
\end{aligned}
$$

Proof. The upper bound has already been proved in Cor. (3.20). We remark that in the upper bound we can replace 9 by 7.75 if $m \geq 4$.

For the lower bound we use the following estimates:

$$
\begin{aligned}
\deg(f(X_1), \ldots, f(X_n)) &\geq [K:k_0] \\
&= \prod_{i=1}^{n}[k_i:k_{i-1}] \\
&= m^n,
\end{aligned}
$$

where $k_i := k(X_1, \ldots, X_i, f(X_{i+1}), \ldots, f(X_n))$ for $0 \leq i \leq n$. The first inequality is valid since $f(X_1), \ldots, f(X_n)$ are algebraically independent over k and the last equality follows from Prop. (8.2). $\square$

For $m \in \underline{n}$ the polynomial $s_m := \sum_{i=1}^{n} X_i^m$ is called the mth power sum. The next result discusses the multiplicative complexity of these polynomials.

(8.10) Corollary (Baur and Strassen). *Let $m \geq 2$ be an integer not divisible by char k. Then*

$$
\frac{1}{3}n \log(m-1) \leq L(s_m) \leq 2n \log m.
$$

Proof. The upper bound is clear. (Compute each X_i^m separately using the binary method.) For the lower bound we use Cor. (7.10) and the previous corollary. $\square$

The multiplicative complexity of the set of all power sums is discussed in Ex. 8.6. The last application of the above theorem is concerned with the evaluation of the resultant. For this, however, we need some preliminaries.

(8.11) Lemma. *Let $f, g \in k[T]$ be relatively prime univariate polynomials and $f/g \notin k$. Then $[k(T) : k(f/g)] = \max\{\deg f, \deg g\}$.*

Proof. Let $u := f/g$ and Z be transcendental over $k(u)$. Consider the polynomial $G(Z) := f(Z) - ug(Z) \in k(u)[Z]$. Clearly, $G(T) = 0$, and G has degree $\max\{\deg f, \deg g\}$ as a polynomial in $k(u)[Z]$. It remains to prove that G is irreducible in $k(u)[Z]$. For this we first note that G is irreducible in $k(Z)[u]$ since it has degree one in the variable u. Hence, since $\mathrm{GCD}(f, g) = 1$, G is also irreducible in $k[u, Z]$ by the Gauss lemma. Now the Gauss lemma applies again and gives the result. $\square$

(8.12) Remark. The analogy between Lemma (8.11) and Prop. (8.2) shows that a possible extension of the polynomial degree in the univariate case to rational functions could be to set $\deg f := \max\{\deg g, \deg h\}$, if $f = g/h$ for coprime polynomials g, h. However this would not give $\deg f = \deg(f)$ in this case, see Ex. 8.2.   •

(8.13) Corollary (Baur and Strassen). *Let $Y_1, \ldots, Y_n$ be indeterminates over K. Then*

$$
\frac{1}{3}n(\log n - 1) \leq L\left(\prod_{i,j}(X_i - Y_j)\right) \leq n(9 \log n + 1).
$$

Proof. In this proof we set $\Omega := k(Y_1, \ldots, Y_n)$ and $\tilde{\Omega} := \Omega(X_1, \ldots, X_n)$.

We first prove the upper bound. Let $g(T) = \prod_{j=1}^n (T - Y_j) \in k(Y)[T]$, and $R := \prod_i g(X_i)$. We first compute the coefficients of g, i.e., the elementary symmetric polynomials in the Y_i, with cost $n \log n$; then we compute $g(X_1), \ldots, g(X_n)$ with cost $8n \log n$, and finally R with additional $n - 1$ multiplications. This gives the upper bound.

For the lower bound we first apply the derivative inequality (7.7). We have $\partial R / \partial X_i = R g'(X_i)/g(X_i)$ for $1 \leq i \leq n$. Hence,

$$
\begin{aligned}
3L(R) \quad &\geq \quad L\left(R, R\frac{g'(X_1)}{g(X_1)}, \ldots, R\frac{g'(X_n)}{g(X_n)} \right) \\
&\geq \quad L\left(R, \frac{g'(X_1)}{g(X_1)}, \ldots, \frac{g'(X_n)}{g(X_n)} \right) - n \\
&=: \quad r - n.
\end{aligned}
$$

Note that for a finite set $F \subset \tilde{\Omega}$ we have $L_{k \to \tilde{\Omega}}(F) \geq L_{\Omega \to \tilde{\Omega}}(F)$. Hence, since $g'(X_1)/g(X_1), \ldots, g'(X_n)/g(X_n)$ are algebraically independent over $k(Y)$, we can apply Thm. (8.5) to obtain

$$
r \geq \log \left[\tilde{\Omega} : \Omega\left(\frac{g'(X_1)}{g(X_1)}, \ldots, \frac{g'(X_{n-1})}{g(X_n)} \right) \right].
$$

Consider the tower of fields

$$
\begin{aligned}
&\Omega(g'(X_1)/g(X_1), \ldots, g'(X_n)/g(X_n)) \qquad &(=: K_0) \\
\subseteq \quad &\Omega(X_1, g'(X_2)/g(X_2), \ldots, g'(X_n)/g(X_n)) \qquad &(=: K_1) \\
&\quad \vdots \\
\subseteq \quad &\Omega(X_1, \ldots, X_{n-1}, g'(X_n)/g(X_n)) \qquad &(=: K_{n-1}) \\
\subseteq \quad &\tilde{\Omega} \qquad &(=: K_n).
\end{aligned}
$$

Applying Lemma (8.11) we obtain $[K_i : K_{i-1}] = n$ for all $1 \leq i \leq n$. This implies the result. $\square$

Let us now prepare for the proof of Thm. (8.5). Let $M \subseteq K$ be an extension of (arbitrary) fields. The subfield

$$
N := \{\alpha \in K \mid \alpha \text{ algebraic over } M\}
$$

of K is called the *algebraic closure* of M in K. M is said to be *algebraically closed* in K if $M = N$.

(8.14) Lemma. *Let $M \subseteq K$ be an extension of fields, N be the algebraic closure of M in K, and $x \in K \setminus N$ be such that K over $M(x)$ is a finite extension. Then $[K : M(x)] = [K : N(x)][N : M]$.*

Proof. Since x is transcendental over N, N and $M(x)$ are linearly disjoint over M (see Lang [316, X, Prop. 5.3]). Hence, $[N(x) : M(x)] = [N : M]$. $\square$

The following lemma will be central in the proof of Thm. (8.5).

(8.15) Lemma. *Let* $N \subseteq K$ *be an extension of fields,* N *be algebraically closed in* K, *and* tr.deg $K/N = 1$. *Further, let* $p \in \mathbb{Q}$ *and* $z_1, z_2 \in K \setminus N$ *be such that* $[K : N(z_i)] \leq p$ *for* $i = 1, 2$. *Then the following assertions hold:*

(1) *For all* $b \in N$ *such that* $z_1 + bz_2 \notin N$ *we have* $[K : N(z_1 + bz_2)] \leq 2p$.
(2) *If* $z_1/z_2 \notin N$ *and there exist infinitely many* $b \in N$ *such that* $[K : N(z_1+bz_2)] \leq p$, *then* $[K : N(z_1/z_2)] \leq p$.

Proof. We may suppose that $K = N(z_1, z_2)$. The main part of the proof consists of showing that certain polynomials are irreducible. Before proving the two assertions of the lemma, we summarize the main lines of our reasoning.

Let T_1, T_2 be indeterminates over K and $0 \neq h(T_1, T_2) \in N[T_1, T_2]$ be a polynomial of minimum degree such that $h(z_1, z_2) = 0$; h is irreducible. Moreover,

(A) $\quad \forall \begin{pmatrix} a & b \\ c & d \end{pmatrix} \in GL(2, k): \quad h(aT_1 + bT_2, cT_1 + dT_2)$ is irreducible.

In particular, since N is assumed to be algebraically closed in K, any $x \in K \setminus N$ is transcendental over N and for any such x and any $b \in N$ we have

(B) $\qquad\qquad h(x - bT_2, T_2)$ is irreducible in $N[x, T_2]$.

Equation (A) together with the Gauss lemma imply that $h(z_1, T_2)$ is irreducible in $N(z_1)[T_2]$ and $h(T_1, z_2)$ is irreducible in $N(z_2)[T_1]$. Thus we obtain for $i = 1, 2$

(C) $\qquad \deg_{T_i} h(T_1, T_2) = [N(z_1, z_2) : N(z_i)] = [K : N(z_i)] \leq p$.

For $b \in N$ we define $h_b(T_2) := h((z_1 + bz_2) - bT_2, T_2) \in N(z_1 + bz_2)[T_2]$. If $z_1 + bz_2 \notin N$, then the Gauss lemma and (B) imply that h_b is irreducible.

(1) (C) implies

$$2p \geq \deg h_b = [N(z_1 + bz_2, z_2) : N(z_1 + bz_2)] = [K : N(z_1 + bz_2)].$$

(2) If $b \in N$ is such that $[K : N(z_1 + bz_2)] \leq p$, then $z_1 + bz_2 \notin N$. Thus for such b the polynomial h_b is irreducible and of degree $\leq p$. Let $m := \deg h$ and $\sum_{i+j=m} a_{ij} T_1^i T_2^j$ be the homogeneous part of degree m of h. The coefficient of T_2^m of h_b equals $\sum_{i+j=m} a_{ij}(-b)^i$. Thus, there exist at most finitely many $b \in N$ such that h_b has degree $< m$. Thus the assumptions imply $m \leq p$. Further, $\tilde{h}(T_2) := h(T_2 z_1/z_2, T_2) \in N(z_1/z_2)[T_2]$ is nonzero (z_1/z_2 is transcendental over N), has degree $\leq p$, and $\tilde{h}(z_2) = 0$. This gives the assertion. $\qquad\square$

Proof of Thm. (8.5). We may w.l.o.g. assume that $X_1, \ldots, X_n \in S$. We will show by induction on r:

If $(g_1, \ldots, g_r)$ is a computation sequence mod S in K, then

$$\deg(S \cup \{g_1, \ldots, g_r\}) \leq 2^r \deg S.$$

The assertion of the theorem follows then with Rem. (8.3)(3).

For $r = 0$ there is nothing to prove. Suppose that $r > 0$, set

$$V := \sum_{y \in S} ky + \sum_{i=1}^{r-1} kg_i + k,$$

and let $y_1, \ldots, y_n \in V + kg_r$ be algebraically independent over k. We have to show that $[K : k(y_1, \ldots, y_n)] \leq 2^r \deg S$. By the induction hypothesis we have $[K : k(y_1, \ldots, y_n)] \leq 2^{r-1} \deg S$ if $y_1, \ldots, y_n \in V$. After a linear transformation of the y_i we may therefore assume that $y_1 = x + g_r$ and $x, y_2, \ldots, y_n \in V$.

Let $M = k(y_2, \ldots, y_n)$ and N be the algebraic closure of M in K. By the induction hypothesis we have for all $x \in V \setminus N$: $[K : M(x)] \leq 2^{r-1} \deg S$. Lemma (8.14) therefore implies

(A) $\forall x \in V \setminus N$: $[K : N(x)] \leq \dfrac{2^{r-1} \deg S}{[N : M]} =: p.$

Note that $g_r = uv$ or $g_r = u/v$ for some $u, v \in V$, v nonzero in the latter case. Hence, in view of Lemma (8.14), it suffices to prove the following two assertions:

(B) $\forall x, u, v \in V, v \neq 0, x + \dfrac{u}{v} \notin N$: $[K : N(x + u/v)] \leq 2p,$

(C) $\forall x, u, v \in V, x + uv \notin N$: $[K : N(x + uv)] \leq 2p.$

Let us prove (B). We may w.l.o.g. assume that $u/v \notin N$. (Otherwise the assertion becomes trivial by (A).) Hence, $u + bv \in N$ holds for at most one value $b \in k \subseteq N$ which shows that for infinitely many $b \in k$ we have $u + bv \in V \setminus N$. Therefore, by (A), we have for infinitely many $b \in k$

$$[K : N(u + bv)] \leq p.$$

Lemma (8.15)(2) applies and we get $[K : N(u/v)] \leq p$; Lemma (8.15)(1) implies $[K : N(x + u/v)] \leq 2p$.

We proceed by proving (C). We may w.l.o.g. assume that $v \notin N$. (Otherwise we apply Lemma (8.15)(1) with $b = v$, $z_1 = x$, and $z_2 = u$.) Therefore, $u + x/v + b/v \notin N$ for all but at most one $b \in k$. By (B) we have

$$[K : N(u + x/v)] \leq 2p$$

provided $u + x/v \notin N$. Moreover, $[K : N(1/v)] = [K : N(v)] \leq p \leq 2p$. We may now apply Lemma (8.15)(2) to $z_1 := 1/v$, $z_2 := u + x/v$ and conclude that

$$[K : N(x + uv)] = [K : N\left(\dfrac{u + x/v}{1/v}\right)] \leq 2p. \qquad \square$$

(8.16) Remark. The above proof can be canonically embedded into the theory of algebraic function fields of one variable; see Ex. 8.8.   ●

8.2 Geometric Degree and a Bézout Inequality

We assume familiarity with the contents of Hartshorne [232, Chap. I]. For basic facts about the Zariski topology compare Ex. 4.12 –4.19.

In this section we will introduce the concept of the *degree* of a locally closed subset of $\mathbb{P}^n$. Our main aim is to give a self-contained elementary proof of a special version of Bézout's inequality (see Thm. (8.28)) which will be used in the next section for deriving nontrivial lower bounds.

Let us first recall some basic definitions. By a projective variety (affine variety) we mean a Zariski closed subset of $\mathbb{P}^n$ (of k^n). (Here our notation differs from that of [232, Chap. I] since we do not assume irreducibility.) By a *variety in* $\mathbb{P}^n$ (*variety in* k^n) we mean a projective (affine) variety. Since $\mathbb{P}^n$ (or k^n) is a Noetherian topological space, any variety has a decomposition into a finite number of irreducible components. The dimension of a variety is the maximum of the dimensions of its irreducible components. We call a variety *equidimensional* if all its irreducible components have the same dimension. The definition of the degree of a variety presented below coincides with the one used in algebraic geometry only for equidimensional varieties. The deviation from the classical notion of degree allows us to prove a general Bézout inequality used later to prove nonlinear lower bounds.

In the sequel we follow closely the discussion in [232, Chap. I.7] and begin by defining the degree of a homogeneous ideal I of the polynomial ring $S = k[X_0, \ldots, X_n]$. The ring S is graded via $S = \oplus_{t \geq 0} S^{(t)}$, where $S^{(t)}$ is the k-space of homogeneous polynomials of degree t in S. The S-module I is also graded via $I^{(t)} := I \cap S^{(t)}$ and so is S/I, where $(S/I)^{(t)} := S^{(t)}/I^{(t)}$. For $t \in \mathbb{N}$ we set

$$H(I; t) := \dim_k (S/I)^{(t)}.$$

The function $H(I; \cdot) : \mathbb{N} \to \mathbb{N}$, $t \mapsto H(I; t)$ is called the *Hilbert function* of the ideal I.

(8.17) Hilbert-Serre Theorem. *Let I be a homogeneous ideal of the polynomial ring $S = k[X_0, \ldots, X_n]$, and assume that the zeroset of I in $\mathbb{P}^n$ is nonempty and of dimension $d \geq 0$. Then there exist uniquely determined integers $h_0(I), h_1(I), \ldots, h_d(I) \in \mathbb{Z}$, $h_0(I) > 0$, such that the polynomial*

$$h(I; T) := \sum_{j=0}^{d} h_j(I) \binom{T}{d - j}$$

satisfies $h(I; t) = H(I; t)$ for all sufficiently large $t \in \mathbb{N}$.

Proof. See Hartshorne [232, I.7, Prop. 7.3, Thm. 7.5]. ☐

(8.18) Remark. It is easy to show that a proper homogeneous ideal I of S has empty zeroset if and only if $\sqrt{I} = \oplus_{t > 0} S^{(t)}$. Such ideals are called *irrelevant*. •

The polynomial $h(I; T)$ of the above theorem is called the *Hilbert polynomial* of the homogeneous ideal I; $h_0(I)$ is called the *(geometric) degree* of I and is denoted by DEG I. Thus, the degree of I is $d!$ times the leading coefficient of $h(I; T)$. The *(geometric) degree* DEG V of a variety $V \subseteq \mathbb{P}^n$ is defined as the geometric degree of its vanishing ideal $I(V)$.

(8.19) Remark. In textbooks on algebraic geometry and commutative algebra the geometric degree of an ideal I is denoted by deg I and is simply called the degree of I. Later we shall be interested in a modified definition of the degree for which we have reserved the notation deg. •

The proof of the following remarks are left to the reader. (See Ex. 8.12.)

(8.20) Remarks.
(1) If $V \subseteq \mathbb{P}^n$ is finite, then DEG $V = |V|$.
(2) The geometric degree is invariant under nonsingular projective linear transformations of $\mathbb{P}^n$.
(3) DEG $\mathbb{P}^n = 1$.
(4) Let f be a homogeneous polynomial and I be the principal ideal generated by f. Then DEG $I = \deg f$. •

Next we focus on the problem of determining the degree of the intersection of a projective variety and a hypersurface.

(8.21) Bézout's Theorem. *Let V be an irreducible variety in $\mathbb{P}^n$ and let H be an irreducible hypersurface not containing V such that $V \cap H \neq \emptyset$. Further let $Z_1, \ldots, Z_s$ be the irreducible components of $V \cap H$. Then there exist positive integers $i(V, H; Z_j)$ such that*

$$\sum_{j=1}^{s} i(V, H; Z_j) \, \text{DEG} \, Z_j = \text{DEG} \, V \cdot \text{DEG} \, H.$$

Proof. See Hartshorne [232, I.7, Thm. 7.7]. □

The numbers $i(V, H; Z_j)$ are called the *intersection multiplicities* of V and H along Z_j. For the sequel it is sufficient to know that they are positive integers.

(8.22) Example. An m-dimensional *projective linear subspace* $U_{a,b} \subseteq \mathbb{P}^n$ is the zeroset in $\mathbb{P}^n$ of the set of linear forms $\{b_i X_0 + \sum_{j=1}^{n} a_{ij} X_j \mid i = 1, \ldots, n - m\}$, where $a = (a_{ij}) \in k^{(n-m) \times n}$ is a matrix of full rank and $b = (b_1, \ldots, b_{n-m})^T \in k^{n-m}$. Obviously, projective linear subspaces of $\mathbb{P}^n$ are irreducible. By induction on the codimension it is easily proved, using Bézout's theorem above, that projective subspaces have degree 1. In particular, a single point also has degree 1. •

Let us now introduce the notion of degree, as used in complexity theory.

(8.23) Definition. Let $V = V_1 \cup \ldots \cup V_s$ be the decomposition of the nonempty locally closed subset $V \subseteq \mathbb{P}^n$ into its irreducible components. We define the *degree* deg V of V as

$$\deg V := \sum_{i=1}^{s} \mathrm{DEG}\, \overline{V}_i,$$

where $\overline{V}_i$ is the closure of V_i in $\mathbb{P}^n$. For technical reasons we set $\deg \emptyset := -1$. •

(8.24) Remarks.
(1) The degree is also defined for locally closed subsets of k^n, since there is an open embedding $k^n \subseteq \mathbb{P}^n$.
(2) $\deg \mathbb{P}^n = \deg k^n = 1$.
(3) If $V \subset \mathbb{P}^n$ is finite, then $\deg V = |V|$.
(4) If $V \subseteq \mathbb{P}^n$ is locally closed and irreducible, then $\deg V = \mathrm{DEG}\, \overline{V}$.
(5) If $V \subseteq \mathbb{P}^n$ is locally closed and $V_1, \ldots, V_s$ are the irreducible components of V, then $\deg V = \sum_{i=1}^{s} \deg V_i$. •

Our notion of degree, as defined in (8.23), differs from that of algebraic geometry. Trivially, $\deg V = \mathrm{DEG}\, V$ for an irreducible variety $V \subseteq \mathbb{P}^n$. More generally, one can show that for varieties $V \subseteq \mathbb{P}^n$ we have $\deg V = \mathrm{DEG}\, V$ if and only if V is equidimensional, see Ex. 8.17. If $V \subseteq \mathbb{P}^n$ is a hypersurface, the assertion $\deg V = \mathrm{DEG}\, V$ can quickly be verified as follows.

(8.25) Example. Let H be a hypersurface in $\mathbb{P}^n$ whose ideal is generated by the homogeneous polynomial $f \in S = k[X_0, \ldots, X_n]$ of degree d. Since the ideal of H is radical, f is square-free. Let $f = p_1 \cdots p_s$ with distinct irreducible polynomials $p_1, \ldots, p_s$ and let H_i be the zeroset of $p_i S$ for $i = 1, \ldots, s$. Since $p_i S$ is radical, we have $\mathrm{DEG}\, H_i = \deg p_i$, see Example (8.20)(4). Hence, we have $\deg H = \sum_{j=1}^{s} \mathrm{DEG}\, H_j = \sum_{j=1}^{s} \deg p_j = \deg f$. •

For the proof of the Bézout inequality below we still need an auxiliary lemma.

(8.26) Lemma. *Let $V \subseteq \mathbb{P}^n$ be a locally closed subset and $U \subseteq V$ be open in V. Then $\deg U \leq \deg V$. Moreover, if U is dense in V, then $\deg U = \deg V$.*

Proof. W.l.o.g. $U \neq \emptyset$. Let $U_1, \ldots, U_r$ be the irreducible components of U. Let C_i denote the closure of U_i in V. It suffices to prove that

(A) $C_1, \ldots, C_r$ are distinct irreducible components of V.

Namely, we have by Rem. (8.24)(5) $\deg U = \sum_{i=1}^{r} \deg U_i$ and (A) implies $\deg V \geq \sum_{i=1}^{r} \deg C_i$. Now note that $\deg U_i = \deg C_i$, since the closures in $\mathbb{P}^n$ of U_i and C_i coincide. Further, if U is dense in V, then $C_1, \ldots, C_r$ are all the irreducible components of V and the the equality of the degrees follows. To prove (A) let $V_1, \ldots, V_t$ denote the irreducible components of V. Suppose that $U \cap V_i \neq \emptyset$ for $i = 1, \ldots, s$ and $U \cap V_i = \emptyset$ for $i = s+1, \ldots, t$. It suffices to show that $U \cap V_i$, $i = 1, \ldots, s$, are the irreducible components of U. For this we note that $U \cap V_i$ is an open subset of V_i and thus irreducible. Moreover, $U \cap V_i \not\subseteq U \cap V_j$ for $1 \leq i < j \leq s$ as $V_i \neq V_j$. □

An immediate consequence of the foregoing lemma is the following.

(8.27) Corollary. *If $V \subseteq \mathbb{P}^n$ ($V \subseteq k^n$) is locally closed and W is the closure of V in $\mathbb{P}^n$ (W is the closure of V in k^n), then $\deg V = \deg W$.*

Proof. If $V \subseteq \mathbb{P}^n$ is locally closed, then V is open and dense in W and the assertion follows by Lemma (8.26). This also implies the claim for locally closed $V \subseteq k^n$ since the Zariski closures of V and W in $\mathbb{P}^n$ coincide.   □

Now we can state and prove the following special version of Bézout's inequality. For the general version we refer the reader to Ex. 8.18.

(8.28) Bézout Inequality. *Let $V \subseteq \mathbb{P}^n$ ($V \subseteq k^n$) be a nonempty locally closed subset, and let $H_1, \ldots, H_r$ be hypersurfaces in $\mathbb{P}^n$ (in k^n). Then we have*

$$\deg(V \cap H_1 \cap \ldots \cap H_r) \leq \deg V \cdot \deg H_1 \cdots \deg H_r.$$

Proof. We may w.l.o.g. assume that $r = 1$ and set $H := H_1$. We first prove the assertion for $V \subseteq \mathbb{P}^n$. The first step consists of reducing to closed subsets of $\mathbb{P}^n$. Namely, if $V \subseteq \mathbb{P}^n$ is locally closed, then V is open in $\overline{V}$, hence $V \cap H$ is open in $\overline{V} \cap H$ and Lemma (8.26) and its Cor. (8.27) imply the assertion of the theorem. So let $V \subseteq \mathbb{P}^n$ be a closed subset such that $V \cap H \neq \emptyset$. Suppose that V and H are irreducible. If $V \subseteq H$, then $\deg V \cap H = \deg V \leq \deg V \deg H$. Otherwise, let $C_1, \ldots, C_s$ be the irreducible components of $V \cap H$. Then we have by Bézout's theorem (8.21)

$$\begin{aligned}
\deg V \cap H &= \sum_{j=1}^{s} \mathrm{DEG}\, C_j \leq \sum_{j=1}^{s} i(V, H; C_j)\, \mathrm{DEG}\, C_j \\
&= \mathrm{DEG}\, V \cdot \mathrm{DEG}\, H = \deg V \cdot \deg H.
\end{aligned}$$

Now assume that V has the irreducible components $V_1, \ldots, V_t$ and H is irreducible. Then each irreducible component of $V \cap H$ is an irreducible component of some of the $V_i \cap H$. (Note that an irreducible component of $V_i \cap H$ may be contained in an irreducible component of $V_j \cap H$ for $i \neq j$.) Hence, the above inequality implies

$$\deg V \cap H \leq \sum_{j: V_j \cap H \neq \emptyset} \deg V_j \cap H \leq \sum_{j=1}^{t} \deg V_j \deg H = \deg V \deg H.$$

The case where H has several irreducible components can be settled in a similar way.

Suppose now that $V \subseteq k^n$ is locally closed. Let $\mathcal{H}$ be the closure of H in $\mathbb{P}^n$. Since $V \cap H = V \cap \mathcal{H} \cap k^n$ is open in $V \cap \mathcal{H}$, we have $\deg V \cap H \leq \deg V \cap \mathcal{H}$ by Lemma (8.26). This implies

$$\deg V \cap H \leq \deg V \cap \mathcal{H} \leq \deg V \deg \mathcal{H} = \deg V \deg H.   \qquad \square$$

(8.29) Remark. In Def. (8.23) we could have extended the notion of degree to arbitrary constructible subsets $V \subseteq \mathbb{P}^n$. However, the Bézout inequality would become false for such subsets. Here is an example: let $E \subseteq k^3$ be a plane, L be a line on E, and $P_1, \ldots, P_n$ be distinct points on L. The closure of the irreducible constructible set $V := (E \setminus L) \cup \{P_1, \ldots, P_n\}$ in k^3 equals E, hence $\deg V = 1$. Let H be a hyperplane intersecting E in L. Then $H \cap V$ consists of n points, hence has degree n, while $\deg H = \deg V = 1$.   ●

We finish this section by stating—without proof—an alternative geometric characterization of the degree of an irreducible affine variety. For $a \in k^{r \times n}$ and $b \in k^r$ we denote by $L_{a,b}$ the affine linear subspace of $\mathbb{P}^n$ consisting of those $x \in k^n$ satisfying $ax + b = 0$.

(8.30) Theorem. *Let $r \geq 1$ and $V \subseteq k^n$ be an r-dimensional irreducible locally closed subset of degree d. Then there exists an open set $O \subseteq k^{r \times n} \times k^r$ such that for all $(a, b) \in O$ we have $|L_{a,b} \cap V| = d$. Moreover, for all $(a, b) \in k^{r \times n} \times k^r$ such that $|L_{a,b} \cap V|$ is finite, we have $|L_{a,b} \cap V| \leq d$.*

Note that the second part follows from the first by the Bézout inequality (8.28). The heart of the proof of the above theorem is contained in Harris [228, Lect. 13, Lect. 18]. For more details the reader is referred to Ex. 8.19.

8.3 The Degree Bound

The basic idea of the degree bound is simple. Let Γ be a straight-line program executable on, say, $(k(X); X_1, \ldots, X_n)$. Consider the result sequence $(b_{-n+1}, \ldots, b_t)$ of Γ on this input. We would like to associate to this sequence a locally closed subset V_Γ of k^{t+n} such that $\deg V_\Gamma \leq 2^{c_*(\Gamma)}$. This would then give a lower bound for $c_*(\Gamma)$. However, we only know some of the results of a hypothetical straight-line program for computing a set of rational functions; namely, we just know the inputs and the functions to be computed. We have thus to ensure that the construction of V_Γ is done in such a way, that it is possible to give a lower bound for $\deg V_\Gamma$ in terms of the known results only.

A straight-forward way to construct a locally closed set from the result sequence of Γ, or more generally, from a sequence $f := (f_1, \ldots, f_r)$ of rational functions in $k(X)$ is to study the rational map associated to f; by abuse of notation it will also be denoted by f:

$$f : k^n \supseteq \operatorname{def} f \to k^r, \quad f(x) := (f_1(x), \ldots, f_r(x)).$$

Note that graph f is a locally closed and irreducible subset of k^{r+n} (Ex. 8.9). The set V_Γ we are looking for will be graph b, where $b = (b_{-n+1}, \ldots, b_t)$ is the result sequence of Γ. In the sequel we will discuss why V_Γ has the properties we are interested in. Because of its extensive use, it is convenient to introduce a short-hand notation for the degree of the graph of a rational map.

(8.31) Definition. For $f_1, \ldots, f_r \in K$ we set

$$\deg(f_1, \ldots, f_r) := \deg \operatorname{graph}(f_1, \ldots, f_r). \qquad \bullet$$

We shall see later that we can majorize the degree of V_Γ by 2^ℓ where ℓ is the c_*-length of Γ. However, as the only known b_i are the functions to be computed, we will need a tool to compare for a rational map $f = (f_1, \ldots, f_r)$ the degrees $\deg f$ and $\deg f'$, where $f' = (f_{i_1}, \ldots, f_{i_\ell})$ for some $\ell < r$ and some $i_1, \ldots, i_\ell \in \underline{r}$. This will be done in the following two lemmas.

(8.32) Lemma. Let $\pi : k^r \to k^{r-1}$, $(z_1, \ldots, z_r) \mapsto (z_1, \ldots, z_{r-1})$ be the projection onto the first $r-1$ components, and $X \subseteq k^r$, $Y \subseteq k^{r-1}$ be irreducible locally closed subsets. Then we have:

(1) $\pi^{-1}(Y) = Y \times k$ is irreducible, $\dim \pi^{-1}(Y) = \dim Y + 1$, and $\deg \pi^{-1}(Y) = \deg Y$.

(2) If $\pi(X)$ is locally closed, then $\dim X - 1 \leq \dim \pi(X) \leq \dim X$, and $\deg \pi(X) \leq \deg X$.

Proof. We may assume that X and Y are closed since $\overline{\pi(X)} = \overline{\pi(\overline{X})}$ by the continuity of π and $\pi^{-1}(\overline{Y}) = \overline{Y} \times k = \overline{Y \times k} = \overline{\pi^{-1}(Y)}$. (The bar means closure in k^n.) The irreducibility of $Y \times k$ is obvious, so we may focus on the other assertions. Their verification is accomplished by translating the claims into algebraic terms. Let $Z_0, \ldots, Z_r$ be indeterminates over k. For $W \subseteq k^r$ we denote by $I(W) \subseteq S := k[Z_0, \ldots, Z_r]$ the vanishing ideal of the closure of W in $\mathbb{P}^r$. It is easy to check that $I(\pi^{-1}(Y)) = I(Y)S$, the ideal of S generated by $I(Y) \subseteq S' := k[Z_0, \ldots, Z_{r-1}]$. Further, $I(\pi(X)) = I(X) \cap S'$. More generally, let $I \subseteq S$ and $J \subseteq S'$ be homogeneous prime ideals and set $I^c := I \cap S'$, $J^e := JS$. We prove a couple of claims which will ultimately yield the assertions of the lemma. We will make use of the notion of (Krull) dimension of a ring; if $W \subseteq \mathbb{P}^r$ is a closed subset, then the dimension of $S/I(W)$ equals $\dim W + 1$ (see Hartshorne [232]).

(a) *The grading* $((S'/J)[Z_r])^{(t)} := \oplus_{j=0}^{t} (S'/J)^{(j)} Z_r^{t-j}$ *for* $t \in \mathbb{N}$ *turns the polynomial ring* $(S'/J)[Z_n]$ *into a graded k-algebra and there exists a canonical isomorphism*

$$S/J^e \simeq (S'/J)[Z_r]$$

of graded k-algebras. (Obvious!)

(b) *We have* $\dim S'/J = \dim S/J - 1$ *and* DEG $J^e =$ DEG J. (This proves part (1) of the lemma.) We use Hilbert polynomials. Let $d := \dim S'/J - 1$. By (a) we have for large t

$$H(J^e; t) = \sum_{j=0}^{t} H(J; j) = \sum_{j=0}^{t} h(J; j) + O(1).$$

We use the definition of the Hilbert polynomial h to obtain for large t

$$h(J^e; t) = H(J^e; t) = \sum_{j=0}^{t}\sum_{i=0}^{d} h_i(J)\binom{j}{d-i} + O(1)$$

$$= \sum_{i=0}^{d} h_i(J)\sum_{j=0}^{t}\binom{j}{d-i} + O(1)$$

$$= \sum_{i=0}^{d} h_i(J)\binom{t+1}{d-i+1} + O(1)$$

$$= h_0(J)\frac{t^{d+1}}{(d+1)!} + O(t^d).$$

This implies the assertions.

(c) *We have* $\dim S'/I^c \le \dim S/I \le \dim S'/I^c + 1$ *and* $\mathrm{DEG}\, I^c \le \mathrm{DEG}\, I$. (This implies part (2) of the lemma.) The inclusion $S'/I^c \hookrightarrow S/I$ gives $\dim S'/I^c \le \dim S/I$. Further, since $I^{ce} \subseteq I$, we have $\dim S/I \le \dim S/I^{ce} = \dim S'/I^c + 1$, the equality being a consequence of (b). Since the inclusion $S'/I^c \hookrightarrow S/I$ respects the grading, we have $H(I^c; t) \le H(I; t)$ for all $t \in \mathbb{N}$. Hence, we are done if $\dim S/I^c = \dim S/I$. Otherwise, $\dim S'/I^c = \dim S/I - 1 \ge 0$. Comparing dimensions we see that $I^{ce} = I$, thus, using (b) we obtain $\mathrm{DEG}\, I^c = \mathrm{DEG}\, I^{ce} = \mathrm{DEG}\, I$. □

(8.33) Remarks.
(1) Statement (2) of the previous lemma also holds if $\pi(X)$ is not locally closed, when we replace $\pi(X)$ by its closure.
(2) Statement (2) of the previous lemma becomes false if we replace deg by DEG; see Ex. 8.13 and 8.17. ●

(8.34) Lemma. *Let* $f_1, \ldots, f_{r-1}, f_r \in K$ *be rational functions. For the rational maps* $f := (f_1, \ldots, f_r)$ *and* $f' := (f_1, \ldots, f_{r-1})$ *we have:*

(1) $\deg(\mathrm{graph}\, f' \times k) = \deg f' \le \deg f$.
(2) *If* $f_r = f_i f_j$, *or* $f_j \ne 0$ *and* $f_r = f_i/f_j$ *for some* $i, j \le r - 1$, *then we have* $\deg f \le 2 \deg f'$.
(3) *If* $f_r = f_i \pm f_j$, *or* $f_r = cf_j$ *for some* $c \in k$, $i, j \le r - 1$, *or* $f_r \in k$, *then* $\deg f = \deg f'$.

Proof. (1) The equality follows from Lemma (8.32)(1), the inequality from Lemma (8.32)(2) by noting that graph f and $\pi(\mathrm{graph}\, f) = \mathrm{graph}\, f'$ are locally closed and irreducible.

(2) In case $f_r = f_i f_j$ let H be the quadratic hypersurface in k^{n+r} defined as the zeroset of $Y_r - Y_i Y_j$. (We identify the coordinate ring of k^{n+r} with the polynomial ring $k[X_1, \ldots, X_n, Y_1, \ldots, Y_r]$.) Then graph $f = (\mathrm{graph}\, f' \times k) \cap H$. Since $\deg H = 2$ (its defining equation has degree 2; see Example (8.25)), the Bézout inequality (8.28) gives the result. If $f_r = f_i/f_j$, let H be the quadric defined as the zeroset of $Y_r Y_j - Y_i$ in k^{n+r}. Then we have

$$\{Y_j \neq 0\} \cap \text{graph } f = ((\text{graph } f' \times k) \cap \{Y_j \neq 0\}) \cap H =: X \cap H.$$

The left-hand side is nonempty since $f_j \neq 0$, and open in the irreducible set graph f. It follows that its degree equals deg f by Cor. (8.27). X is open in the irreducible subset graph $f' \times k$ of k^{n+r}. Hence its degree equals deg f' by part (1) of this lemma and Cor. (8.27). Since H has degree 2, the Bézout inequality (8.28) gives deg $f \leq 2 \deg f'$.

(3) In all these cases we obtain graph $f = (\text{graph } f' \times k) \cap E$, where E is an appropriate hyperplane of k^{n+r}. The Bézout inequality (8.28) and part (1) of this lemma imply deg $f = \deg f'$. $\qquad \square$

The foregoing lemmas constitute the heart of the proof of the main theorem of this section.

(8.35) Relative Degree Bound (Strassen). *Let $u_1, \ldots, u_n$ and $f_1, \ldots, f_r$ belong to K, $f := (f_1, \ldots, f_r)$, $u := (u_1, \ldots, u_n)$, and $uf := (u, f)$. Then we have*

$$L(f \mid u) \geq \log \deg uf - \log \deg u.$$

Proof. Let $\Gamma = (\Gamma_1, \ldots, \Gamma_t)$ be a straight-line program executable on $(K; u)$ for f. We use induction on t to show that if $b = (u_1, \ldots, u_n, b_1, \ldots, b_t)$ is the result sequence of Γ on input $(K; u)$, then $\deg b / \deg u \leq 2^{c_*(\Gamma)}$. For $t = 0$ the assertion is obvious. For the induction step "$t - 1 \to t$" let Γ have length $t \geq 1$ and $\tilde{b} := (u_1, \ldots, u_n, b_1, \ldots, b_{t-1})$. Consider the last operation Γ_t of Γ. If $\Gamma_t = (*; i, j)$ for some $i, j < t$, then $b_t = b_i b_j$ and by Prop. (8.34)(2) we have $\deg b \leq 2 \deg \tilde{b}$. The same is true if $\Gamma_t = (/; i, j)$. If Γ_t is any other operation, we obtain by (8.34)(3) $\deg b = \deg \tilde{b}$. The assertion follows by the induction hypothesis.

By multiple use of (8.34)(1) and the fact that $\deg b$ is independent of the ordering of the b_i (see Rem. (8.20)(2)), we obtain $\deg uf \leq \deg b$, which yields the assertion of the theorem. $\qquad \square$

Obviously, $\deg(X_1, \ldots, X_n) = 1$. Hence we obtain the following corollary.

(8.36) Degree Bound (Strassen). *Let $f_1, \ldots, f_r$ be functions in K and $f = (f_1, \ldots, f_r)$. Then we have*

$$L(f) \geq \log \deg f.$$

(8.37) Remark. In the situation of the last corollary we obtain the following geometric description of the locally closed set V_Γ corresponding to the result sequence of a straight line program Γ for f (see the discussion at the beginning of this section): if Γ has length t and $\ell := c_*(\Gamma)$, then V_Γ is a an open subset of a component of the intersection of ℓ quadrics with $t - \ell$ hyperplanes in k^{n+t}. $\quad \bullet$

In Sect. 8.1 we discussed the algebraic degree of a finite set of rational functions. We are going to show (in characteristic zero) that the geometric and the algebraic notions of degree coincide. We shall need the following classical result which is a special case of Thm. 6 in Chap. II, §5 of Shafarevich [471].

(8.38) Theorem on Fibers. *Let k be an algebraically closed field of characteristic zero, $f_1, \ldots, f_n \in K$ be algebraically independent, and $f = (f_1, \ldots, f_n) \colon k^n \supseteq \mathrm{def}\, f \to k^n$ be the corresponding rational map. Further set $d := [K \colon k(f_1, \ldots, f_n)]$. Then there exists an open subset $O \subseteq k^n$ such that for all $y \in O$ we have $|f^{-1}(y)| = d$. Moreover, any $y \in k^n$ with finite fiber $f^{-1}(y)$ satisfies $|f^{-1}(y)| \le d$.*

(8.39) Proposition. *Let k be of characteristic zero, $f_1, \ldots, f_r \in K$, and $f := (f_1, \ldots, f_r)$. Then*

$$\deg f = \max_{\Lambda} [K \colon k(\Lambda)],$$

where Λ runs over all algebraically independent n-subsets of $k + \sum_i k X_i + \sum_j k f_j$.

Proof. For $a \in k^{n \times (n+r)}$ and $b \in k^n$ we let $L_{a,b}$ be the affine linear subspace of k^{n+r} given as the zeroset of $\{\sum_j c_{ij} X_j + \sum_\ell d_{i\ell} Y_\ell + b_i \mid i = 1, \ldots, n\}$, where $(c_{ij}) \in k^{n \times n}$ and $(d_{i\ell}) \in k^{n \times r}$ denote the matrices consisting of the first n and the last r columns of a, respectively. Similarly, we define the rational map $h_{a,b} = (h_1, \ldots, h_n) \colon \mathrm{def}\, f \to k^n$ by $h_i(x) := \sum_j c_{ij} x_j + \sum_\ell d_{i\ell} f_\ell(x) + b_i$. It is easily seen that there is a bijection between $\mathrm{graph}\, f \cap L_{a,b}$ and $h_{a,b}^{-1}(0)$. Noting that $h_{a,b}(x) = c$ iff $h_{a,b-c}(x) = 0$, we obtain by the characterization of the degree given in Thm. (8.30)

$$\text{(A)} \qquad \deg f = \max \big\{ |h_{a,b}^{-1}(c)| \mid a \in k^{n \times (n+r)}, b, c \in k^n, |h_{a,b}^{-1}(c)| \text{ finite} \big\}.$$

Let $h_1, \ldots, h_n \in k + \sum_i k X_i + \sum_j k f_j$ be algebraically independent, and $h := (h_1, \ldots, h_n) \colon k^n \supseteq \mathrm{def}\, f \to k^n$. For appropriate $a \in k^{n \times (n+r)}$, $b \in k^n$ we have $h = h_{a,b}$. By Thm. (8.38) we have for for Zariski almost all $\xi \in k^n$

$$[K \colon k(h_1, \ldots, h_n)] = |h^{-1}(\xi)| \le \deg f,$$

the inequality being a consequence of (A). Conversely, let a, b, c be such that the maximum in (A) is attained for $h = h_{a,b}$, i.e., $\deg f = |h^{-1}(c)| < \infty$. Lemma (8.38) shows that for Zariski almost $\xi \in k^n$ the fiber $h^{-1}(\xi)$ is finite and $\deg f = |h^{-1}(c)| \le |h^{-1}(\xi)| = [K \colon k(h_1, \ldots, h_n)]$. □

8.4 Applications

In this section we shall derive a technique for estimating the degree of the graph of a rational map and use this to give some applications of the degree bound not discussed in Sect. 8.1.

(8.40) Proposition. *Let $f_1, \ldots, f_r \in k(X)$ and $\lambda_1, \ldots, \lambda_r \in k$ be such that there are exactly $N < \infty$ different $\xi \in k^n$ satisfying $\xi \in \mathrm{def}\, f_i$ for all i and*

$$f_1(\xi) = \lambda_1, \ldots, f_r(\xi) = \lambda_r.$$

Then $\deg(f_1, \ldots, f_r) \ge N$ and hence

$$L(f_1, \ldots, f_r) \ge \log N.$$

Proof. Let $\xi^{(1)}, \ldots, \xi^{(N)}$ be the different solutions of the above system of equations and consider graph f, where $f := (f_1, \ldots, f_r)$. Let us denote by $(x_1, \ldots, x_n, y_1, \ldots, y_r)$ the coordinate functions of k^{n+r} and let $H_1, \ldots, H_r$ be the hyperplanes given by $y_i = \lambda_i$, $i = 1, \ldots, r$. Then the $\xi^{(i)}$ form the set U of intersection points of graph f and $H_1 \cap \ldots \cap H_r$. Hence, $|U| = \deg U \leq \deg f$ by Bézout's inequality since the degree of the H_i is 1. □

The above proposition is thus a special case of Strassen's degree bound. In the exercises we shall derive an elementary proof of it. (See Ex. 8.21.) Our first application will be another proof of Cor. (8.7).

(8.41) Corollary. *Let $\sigma_1, \ldots, \sigma_n \in k[X_1, \ldots, X_n]$ be the elementary symmetric polynomials, and $\sigma := (\sigma_1, \ldots, \sigma_n)$. Then we have $\deg \sigma \geq n!$. Hence, $L(\sigma) \geq n(\log n - 2)$.*

Proof. Let $a_0, \ldots, a_{n-1} \in k$ be such that $T^n + \sum_{j=0}^{n-1}(-1)^{n-j} a_j T^j$ has n different roots in k. Then the system of equations $\sigma_j(x) = a_j$, $j = 1, \ldots, n$, has exactly $n!$ different solutions and we are done by Prop. (8.40). □

The next application is to the problem of interpolation: for indeterminates X_i, Y_j, $0 \leq i, j \leq n - 1$, over k we ask for the multiplicative complexity $L(a_0, \ldots, a_{n-1})$ of the uniquely determined rational functions $a_0, \ldots, a_{n-1} \in k(X, Y)$ satisfying $\sum_{i=0}^{n-1} a_i X_j^i = Y_j$ for all $j = 0, \ldots, n - 1$. In Cor. (3.22) we have already derived an upper bound for this complexity which is of the order of magnitude $n \log n$. Using the degree bound we can prove a matching lower bound.

(8.42) Theorem (Strassen). *Let $X_0, \ldots, X_{n-1}, Y_0, \ldots, Y_{n-1}$ be indeterminates over k and let $a_0 = a_0(X, Y), \ldots, a_{n-1} = a_{n-1}(X, Y) \in k(X, Y)$ satisfy*

$$\forall j = 0, \ldots, n-1: \quad \sum_{i=0}^{n-1} a_i X_j^i = Y_j.$$

Then we have

$$L(a_0, \ldots, a_{n-1} \mid X, Y) \geq n \log(n - 1).$$

Proof. Let $\alpha_0, \ldots, \alpha_{n-1}, \zeta_0, \ldots, \zeta_{n-1} \in k$, $\alpha_{n-1} \neq 0$, be such that for all $j = 0, \ldots, n - 1$ the equation $\sum_{i=0}^{n-1} \alpha_i T^i = \zeta_j$ has $(n - 1)$ different zeros in k. Then the system of equations

$$a_0 = \alpha_0, \ldots, a_{n-1} = \alpha_{n-1}, Y_0 = \zeta_0, \ldots, Y_{n-1} = \zeta_{n-1}$$

has exactly $(n - 1)^n$ different solutions. Therefore, Prop. (8.40) yields

$$\begin{aligned} L(a_0, \ldots, a_{n-1} \mid X, Y) &= L(a_0, \ldots, a_{n-1}, Y_0, \ldots, Y_{n-1} \mid X, Y) \\ &\geq n \log(n - 1), \end{aligned}$$

which proves the theorem. □

We are also able to derive a lower bound for the problem of the multiplication of several polynomials which shows that the algorithm discussed in Sect. 2.3 is essentially optimal.

Let $d_1, \ldots, d_t$ be nonnegative integers that sum up to n. Recall that the entropy of the probability distribution $n^{-1}(d_1, \ldots, d_t)$ is defined by

$$\mathcal{H}(d_1, \ldots, d_t) := \frac{1}{n} \sum_{d_i > 0} d_i \log \frac{n}{d_i},$$

see Sect. 2.3. An elementary calculation using the estimate $n! \geq (n/e)^n \geq (n/4)^n$ shows the following.

(8.43) Remark. Let $d_1, \ldots, d_t \in \mathbb{N}$, $n := \sum_{i=1}^{t} d_i$. Then we have

$$\log \left(\frac{n!}{d_1! \cdots d_t!} \right) \geq n(\mathcal{H}(d_1, \ldots, d_t) - 2). \qquad \bullet$$

In the sequel we denote by $C(p_1, \ldots, p_t)$ the set of coefficients of the polynomials $p_1, \ldots, p_t$.

(8.44) Theorem (Strassen). *Let $p_1, \ldots, p_t$ be univariate polynomials of degrees $d_1, \ldots, d_t$ whose coefficients are indeterminates over k. Let $p := p_1 \cdots p_t$ and $n := \sum_{i=1}^{t} d_i$. Then we have*

$$L_K\big(C(p) \mid C(p_1, \ldots, p_t)\big) \geq n\big(\mathcal{H}(d_1, \ldots, d_t) - 2\big)$$

for the multiplicative complexity in the rational function field K generated by the coefficients of the p_i over k.

Proof. We identify a polynomial of degree n over k with an element of the irreducible open subset $k^\times \times k^n$ of k^{n+1}. Consider the polynomial map

$$\psi: \prod_{i=1}^{t} (k^\times \times k^{d_i}) \to k^\times \times k^n, \quad (p_1, \ldots, p_t) \mapsto p_1 \cdots p_t$$

which associates to polynomials $p_1, \ldots, p_t$ their product. We have to estimate the degree of graph ψ from below. Let f be a monic polynomial of degree n with set of roots M such that $|M| = n$. We intersect graph ψ with an affine linear subspace by requiring that all p_i be monic and their product equal f. Then this intersection is finite and consists of the $n!/(d_1! \cdots d_t!)$ sequences $(p_1, \ldots, p_t, f)$ satisfying

$$\forall i \in \underline{t}: \quad p_i(T) = \prod_{\theta \in M_i} (T - \theta)$$

for some partition $\{M_1, \ldots, M_t\}$ of M into subsets of cardinality $d_1, \ldots, d_t$. Therefore, Bézout's inequality yields

$$\deg \text{graph } \psi \geq \frac{n!}{d_1! \cdots d_t!}.$$

Using Rem. (8.43) we obtain the assertion.   □

With the same method one can prove lower bounds for the Chinese remainder morphism and its inverse (compare Sect. 3.3). Let us recall the problem setting. In the following let $d_1, \ldots, d_t$ be fixed positive natural numbers and $n := \sum_{i=1}^{t} d_i$. We denote by V the set of all t-tuples $(p_1, \ldots, p_t)$ of monic univariate polynomials $p_i \in k[X]$ of degree d_i which are pairwise coprime. By identifying a polynomial with its coefficient sequence we may consider V as a subset of $\prod_{i=1}^{t} k^{d_i}$. The set V is Zariski open, as the condition $GCD(p_i, p_j) = 1$ may be expressed by the nonvanishing of the resultant of p_i and p_j (cf. Ex. 10.8). Let U_m denote the set of polynomials in $k[X]$ of degree $< m$. U_m may be identified with k^m. By the Chinese remainder theorem (3.17) we have a bijection

$$\varphi: \quad V \times U_n \;\to\; V \times U_{d_1} \times \ldots \times U_{d_t}$$
$$(p_1, \ldots, p_t, u) \;\mapsto\; (p_1, \ldots, p_t, u \bmod p_1, \ldots, u \bmod p_t).$$

In Sect. 3.3 we have designed fast algorithms for computing φ and its inverse. We may now complement this by the following lower bounds.

(8.45) Theorem. *Let $p_1, \ldots, p_t$ be monic univariate polynomials of positive degrees $d_1, \ldots, d_t$, and $u_1, \ldots, u_t, u$ be univariate polynomials of degrees less than $d_1, \ldots, d_t$, $n := \sum_{i=1}^{t} d_i$, respectively. Assume that all but the highest coefficients of the p_i and the coefficients of $u_1, \ldots, u_t, u$ are distinct indeterminates over k, and let K denote the field generated over k by all these coefficients. Moreover, let $\tilde{u}$ be the unique polynomial of degree $< n$ with coefficients in K such that $\tilde{u} \bmod p_i = u_i$. Then we have*

$$L_K(C(u \bmod p_1, \ldots, u \bmod p_t) \mid C(u, p_1, \ldots, p_t)) \;\geq\; n(\mathcal{H}(d_1, \ldots, d_t) - 2),$$
$$L_K(C(\tilde{u}) \mid C(u_1, \ldots, u_t, p_1, \ldots, p_t)) \;\geq\; n(\mathcal{H}(d_1, \ldots, d_t) - 2).$$

Proof. The coefficients of $p_1, \ldots, p_t, u \bmod p_1, \ldots, u \bmod p_t$ are elements of K, i.e., rational functions in the coefficients of $p_1, \ldots, p_t, u$ and define a rational map

$$\left(\prod_{i=1}^{t} k^{d_i}\right) \times U_n \supseteq \operatorname{def} \varphi' \xrightarrow{\varphi'} \left(\prod_{i=1}^{t} k^{d_i}\right) \times U_1 \times \cdots \times U_{d_t}.$$

It is easy to see that in fact $\operatorname{def} \varphi' = (\prod_i k^{d_i}) \times U_n$ and that φ' coincides with φ on $V \times U_n$. Similarly, $(p_1, \ldots, p_t, \tilde{u})$ corresponds to a rational map defined on an open subset of $(\prod_i k^{d_i}) \times U_1 \times \cdots \times U_{d_t}$. Ex. 8.25 shows that this map is defined on $V \times U_1 \times \cdots \times U_{d_t}$ and equals φ^{-1}.

By the degree bound it is therefore sufficient to show that the logarithm of the degree of graph φ is at least $n(\mathcal{H}(d_1, \ldots, d_t) - 2)$. Note that this is trivial for $t \leq 4$, since $\mathcal{H}(d_1, \ldots, d_t) \leq \log t$. So we may assume $t \geq 5$. Let $f \in k[X]$ be of degree $n - 1$ and $\xi_1, \ldots, \xi_t$ be pairwise distinct elements of k such that $f - \xi_i$ has $n - 1$ distinct zeros in k for all i. Then, by intersecting graph φ with an appropriate affine subspace and taking a projection, we see that

$$\deg \operatorname{graph} \varphi \geq \deg \left\{ (p_1, \ldots, p_t) \in V \mid \forall i \in \underline{t} : f - \xi_i \equiv 0 \bmod p_i \right\}.$$

The set on the right-hand side is finite and has $\prod_{i=1}^{t} \binom{n-1}{d_i}$ elements. By Rem. (8.43) it is sufficient to show the estimate

(A)
$$\prod_{i=1}^{t} \binom{n-1}{d_i} \geq \frac{n!}{d_1! \cdots d_t!}.$$

To do this we may w.l.o.g. assume that not all d_i are equal to one. Then our assumption $t \geq 5$ implies that there is some i such that $2 \leq d_i \leq n-3$, say $i = t$. Now a combinatorial interpretation of the occurring quantities yields

$$\prod_{i=1}^{t} \binom{n-1}{d_i} \geq \binom{n-1}{d_t} \binom{n-1}{n-d_t} \frac{(n-d_t)!}{d_1! \cdots d_{t-1}!} = \binom{n-1}{d_t} \frac{d_t}{n} \cdot \frac{n!}{d_1! \cdots d_t!}.$$

As $\binom{n-1}{d_t} \geq \binom{n-1}{2} \geq n/2 \geq n/d_t$, the estimate (A) follows.   □

The preceding theorem together with Prop. (4.26) implies that any computation tree over k which computes the map φ (respectively its inverse φ^{-1}) needs nonscalar time at least $n(\mathcal{H}(d_1, \ldots, d_t) - 2)$ on all inputs in a Zariski dense subset.

Another application is concerned with the evaluation of a single elementary symmetric polynomial.

(8.46) Theorem (Baur and Strassen). *Let $\sigma_1, \ldots, \sigma_n \in k[X_1, \ldots, X_n]$ be the elementary symmetric polynomials in $X_1, \ldots, X_n$.*

(1) *For all $1 \leq q \leq n/2$ we have*

$$\frac{1}{3}(n - q + 1) \log(q - 1) \leq L(\sigma_q) \leq n \log q + 2n.$$

(2) *For all $q \in \underline{n}$ we have*

$$|L(\sigma_q) - L(\sigma_{n-q})| \leq 2n.$$

Proof. We may suppose that k is algebraically closed and of infinite degree of transcendency over its prime field k_0.

The second assertion follows from the equality

$$\sigma_{n-q} = \sigma_q \left(\frac{1}{X_1}, \ldots, \frac{1}{X_n} \right) X_1 \cdots X_n.$$

Lower bound in (1): Here we will use the derivative inequality (7.7) and Strassen's degree bound. Let $\partial_i \sigma_q$ denote the partial derivative $\partial \sigma_q / \partial X_i$. Our first aim is to prove for $i = 1, \ldots, n$

(A) $\partial_i \sigma_q = \sigma_{q-1} - X_i \sigma_{q-2} + \cdots + (-1)^{q-2} X_i^{q-2} \sigma_1 + (-1)^{q-1} X_i^{q-1}.$

In fact, let T be an indeterminate over K. Then

$$f(T) := \prod_{i=1}^{n} (1 + X_i T) = 1 + \sigma_1 T + \cdots + \sigma_n T^n.$$

Differentiating both sides with respect to X_i and dividing by T shows that

$$\frac{f(T)}{(1 + X_i T)} = \frac{\sum_{j=0}^n \sigma_j T^j}{(1 + X_i T)} = \partial_i \sigma_1 + \cdots + \partial_i \sigma_n T^{n-1}.$$

By comparing coefficients in $K[[T]]$ we obtain Statement (A). The derivative inequality (7.7) and the degree bound imply

$$L(\sigma_q) \geq \frac{1}{3} L(\partial_1 \sigma_q, \ldots, \partial_n \sigma_q) \geq \frac{1}{3} \log \deg G,$$

where $G := \mathrm{graph}(\partial_1 \sigma_q, \ldots, \partial_n \sigma_q) \subseteq k^{2n}$. Consider

$$V := \{(x, y) \in k^{2n} \mid \sigma_1(x) = \lambda_1, \ldots, \sigma_{q-1}(x) = \lambda_{q-1}, y_q = \mu_q, \ldots, y_n = \mu_n\},$$

where $\lambda_1, \ldots, \lambda_{q-1}, \mu_q, \ldots, \mu_n \in k$ are fixed elements algebraically independent over k_0. V is the intersection of hypersurfaces and a double use of Bézout's inequality yields $\deg G \cap V \leq (q-1)! \deg G$. Hence it suffices to show that

(B) $$|G \cap V| = (q-1)^{n-q+1}(q-1)!.$$

Let $\xi \in k^n$ be fixed. There exists $\eta \in k^n$ such that $(\xi, \eta) \in G \cap V$ if and only if there exists $\eta \in k^n$ such that

$$\partial_1 \sigma_q(\xi) = \eta_1, \ldots, \partial_n \sigma_q(\xi) = \eta_n,$$

$$\sigma_1(\xi) = \lambda_1, \ldots, \sigma_{q-1}(\xi) = \lambda_{q-1}, \eta_q = \mu_q, \ldots, \eta_n = \mu_n.$$

This is equivalent to

$$\partial_q \sigma_q(\xi) = \mu_q, \ldots, \partial_n \sigma_q(\xi) = \mu_n, \sigma_1(\xi) = \lambda_1, \ldots, \sigma_{q-1}(\xi) = \lambda_{q-1}.$$

Using (A) we see that the latter condition is equivalent to

(C) $$\forall \, j \geq q: \quad \mu_j = \lambda_{q-1} - \lambda_{q-2} \xi_j + \cdots + (-1)^{q-1} \xi_j^{q-1},$$

and

(D) $$(1 + \xi_1 T) \cdots (1 + \xi_n T) \equiv 1 + \lambda_1 T + \cdots + \lambda_{q-1} T^{q-1} \bmod T^q,$$

where T is an indeterminate over k. By the algebraic independence of the λ_i, μ_j, there exist exactly $(q-1)^{n-q+1}$ solutions $(\xi_q, \ldots, \xi_n) \in k^{n-q+1}$ of (C). (Recall that k is algebraically closed.) Choose one such solution $(\xi_q, \ldots, \xi_n)$. Then the element $(\xi_1, \ldots, \xi_{q-1})$ in k^{q-1} is such that $(\xi_1, \ldots, \xi_n)$ satisfies (D) if and only if $(1 + \xi_1 T) \cdots (1 + \xi_{q-1} T) = F(T)$ where $F(T)$ is the polynomial

$$F(T) := (1 + \lambda_1 T + \cdots + \lambda_{q-1} T^{q-1})/(1 + \xi_q T) \cdots (1 + \xi_n T) \bmod T^q.$$

Since $\xi = (\xi_1, \ldots, \xi_n)$ satisfies (C) and (D), $\xi_1, \ldots, \xi_n$ are algebraically independent over k_0. (Note that $\lambda_1, \ldots, \lambda_{q-1}, \mu_q, \ldots, \mu_n \in k_0(\xi_1, \ldots, \xi_n)$.) In particular, $\xi_1, \ldots, \xi_{q-1}$ are pairwise different. This implies that to every solution $(\xi_q, \ldots, \xi_n)$ of (C) there exist exactly $(q-1)!$ solutions of (D) and completes the proof of Statement (B).

Upper bound in (1): see Ex. 8.22.   □

(8.47) Remark. The above theorem shows for example that the complexity of $\sigma_{\lfloor n/2 \rfloor}$ is of the same order of magnitude as the complexity of all of the σ_i.   •

8.5* Estimates for the Degree

For the purpose of obtaining lower bounds via the degree bound it is necessary to develop techniques for estimating the degree of a rational map in terms of its coordinate functions. One result in this direction is the following.

(8.48) Theorem. Let $f_1, \ldots, f_r \in k[X_1, \ldots, X_n]$ be polynomials of degree at most d for some $d \geq 1$. Then
$$\deg \overline{\mathrm{im}\, f} \leq d^{\dim \mathrm{im}\, f},$$
where the bar means closure in k^r.

Note that im f does not need to be locally closed, as is shown by the example $f = (X, XY)$ with im $f = (k^\times \times k) \cup \{(0,0)\}$.

   For the proof of Thm. (8.48) we need some preliminaries. Let $d \geq 1, N := \binom{n+d}{n} - 1$. The mapping
$$v_{n,d} \colon \mathbb{P}^n \to \mathbb{P}^N, \quad (a_0 : \ldots : a_n) \mapsto (M_0(a) : \ldots : M_N(a)),$$
where $M_0, \ldots, M_N$ are the monomials in $X_0, \ldots, X_n$ of degree d in some order, is called the *Veronese mapping*. Its affine version is given by
$$\tilde{v}_{n,d} \colon k^n \to \{1\} \times k^N, \quad (a_1, \ldots, a_n) \mapsto (M_0'(a), \ldots, M_N'(a)),$$
where $1 = M_0', \ldots, M_N'$ are the monomials of degree $\leq d$ in $X_1, \ldots, X_n$. The image of $\tilde{v}_{n,d}$ is the graph of a polynomial map and therefore closed. The image of $v_{n,d}$ is also closed (cf. Harris [228, p. 23]), but we will not need this for our argument. As $\overline{\mathrm{im}\, \tilde{v}_{n,d}}$ is dense in im $v_{n,d}$ by the natural embedding, we have $\deg \mathrm{im}\, \tilde{v}_{n,d} = \deg \overline{\mathrm{im}\, v_{n,d}}$ by Cor. (8.27), the bar meaning closure in $\mathbb{P}^n$. The latter degree can be computed by means of the Hilbert polynomial.

(8.49) Lemma. For $n, d \geq 1$ the degree of im $\tilde{v}_{n,d}$ equals d^n.

Proof. See Ex. 8.26.   □

This lemma shows that the estimate in Thm. (8.48) is sharp.

Proof of Thm. (8.48). The proof consists of two steps. First we show that

(A) $$\deg \overline{\mathrm{im}\, f} \leq d^n.$$

Indeed, Lemma (8.49) and repeated use of Lemma (8.34)(1),(3) gives
$$\deg \overline{\mathrm{im}\, f} \leq \deg \mathrm{im}(\tilde{v}_{n,d} f) = \deg \mathrm{im}\, \tilde{v}_{n,d} = d^n,$$

where $\tilde{v}_{n,d} f$ is the mapping $(M_0', \ldots, M_N', f_1, \ldots, f_r)$. (Note that we may apply Lemma (8.32)(1) by Rem. (8.33)(1).)

We proceed with the second step of the proof. Let $t := \text{tr.deg}_k k(f_1, \ldots, f_r)$ $= \dim \text{im } f$. It suffices to show that there exists a linear map $\alpha: k^t \to k^n$ such that $\text{im } f \circ \alpha$ is dense in $\text{im } f$. (For then $f \circ \alpha: k^t \to k^r$ is a polynomial map and we are done by (A).) W.l.o.g. we may assume that $f_1, \ldots, f_t, X_1, \ldots, X_{n-t}$ are algebraically independent over k. Let $X' := (X_1, \ldots, X_{n-t})$ and $X'' :=$ $(X_{n-t+1}, \ldots, X_n)$, and $f' := (f_1, \ldots, f_t)$. Since X_j is algebraic over $k(f', X')$ for $n - t < j \le n$, there exist $b_{ij}, c \in k[f', X']$, $c \ne 0$, such that

$$(B) \qquad X_j^{e_j} + \sum_{i=0}^{e_j-1} \frac{b_{ij}}{c} X_j^i = 0$$

for some $e_j \ge 1$. As k is infinite, there exists $\xi = (\xi_1, \ldots, \xi_n) \in k^n$ such that $c(\xi) \ne 0$. Set $\xi' := (\xi_1, \ldots, \xi_{n-t})$. Then (B) implies that X_j is algebraic over $F := k(f'(\xi', X''), X')$. Hence, $F \subseteq k(X_1, \ldots, X_n)$ is an algebraic extension and $f_1'(\xi, X''), \ldots, f_t'(\xi, X'')$ are algebraically independent. Now define $\alpha: k^t \to k^n$ by $x := (x_1, \ldots, x_t) \mapsto (\xi, x)$. $\qquad \square$

The subsequent theorem is another result to be used in Chap. 9 and Chap. 10.

(8.50) Theorem (Heintz and Sieveking). *Let $V \subseteq W \subseteq k^n$ be nonempty and closed in k^n, assume that all the irreducible components of V have the same dimension and that there are polynomials $f_1, \ldots, f_r \in k[X_1, \ldots, X_n]$ of degree $\le d$ with $d \ge 1$, such that each component of V is also a component of the zeroset of $f_1, \ldots, f_r$ in W. Then we have*

$$\deg W \ge \frac{\deg V}{d^{\dim W - \dim V}}.$$

Proof. We use induction on $\dim W$. The start $\dim W = \dim V$ being clear, let $\dim W > \dim V$. We may assume that every component of W contains a component of V. Let $C_1, \ldots, C_t$ be the components of W of highest dimension. If all f_j vanished on a C_i, then a component of V contained in C_i would not be a component of the zeroset of $f_1, \ldots, f_r$ in W. Thus, for each $i \le t$ there exists $j_i \le r$ and $c_i \in C_i$ such that $f_{j_i}(c_i) \ne 0$. Since k is infinite, there exist $\lambda_1, \ldots, \lambda_t \in k$ such that $\sum_{i \le t} \lambda_i f_{j_i}$ does not vanish on any C_i. Let

$$U := W \cap \left\{ x \in k^n \mid \sum_{i \le t} \lambda_i f_{j_i}(x) = 0 \right\}.$$

Then $\dim U \le \dim W - 1$ and $V \subseteq U$. The induction hypothesis applied to U and the Bézout inequality give

$$d \cdot \deg W \ge \deg U \ge \deg V \cdot d^{\dim V - \dim U}. \qquad \square$$

The next result gives a relative lower bound for the degree of a polynomial map. We shall use this result later in Chap. 10.

(8.51) Theorem. *Let $f_1, \ldots, f_r \in k[X_1, \ldots, X_n]$ be polynomials of degree $\leq d$. Let $f_i^{(d)}$ denote the homogeneous part of degree d of f_i. Then*

$$\deg(f_1, \ldots, f_r) \geq \deg(f_1^{(d)}, \ldots, f_r^{(d)}).$$

We shall prove this result by studying the Hilbert polynomials of the corresponding varieties.

Let $\mathcal{P} := k[X_1, \ldots, X_n]$. We denote by $\mathcal{P}^{\leq t}$ the set of polynomials of degree $\leq t$ in $\mathcal{P}$. If $S \subseteq \mathcal{P}$, then $S^{\leq t} := S \cap \mathcal{P}^{\leq t}$. The following remark is obvious.

(8.52) Remark. Let $I \subseteq S = k[X_0, \ldots, X_n]$ be a homogeneous ideal and $J \subseteq \mathcal{P}$ be its dehomogenization. Then $H(I; t) = \dim_k \mathcal{P}^{\leq t} - \dim_k J^{\leq t}$.   •

For an ideal $J \subseteq \mathcal{P}$ we define DEG J as the degree of its homogenization. Hence, if the ideals J_1, J_2 of $\mathcal{P}$ have the same dimension and $\dim_k J_1^{\leq t} \leq \dim_k J_2^{\leq t}$ for sufficiently large t, then DEG $J_1 \geq$ DEG J_2. Next, we ask for the vanishing ideal of the graph of a polynomial map $f := (f_1, \ldots, f_r): k^n \to k^r$. It is an ideal in $k[X_1, \ldots, X_n, Y_1, \ldots, Y_r] =: k[X, Y]$.

(8.53) Lemma. *Let R be any subring of k, $f_1, \ldots, f_r \in R[X]$ and $I \subseteq R[X, Y]$ be the vanishing ideal of graph$(f_1, \ldots, f_r)$. Then the ideal $I \cap R[X, Y]$ of $R[X, Y]$ is generated by $Y_i - f_i(X)$, $1 \leq i \leq r$.*

Proof. Let $g \in I \cap R[X, Y]$. We can find $h_1 \in R[X, Y]$ and $g_1 \in R[X, Y_2, \ldots, Y_r]$ such that $g = (Y_1 - f_1(X))h + g_1$. Repeating this shows that g may be written as $g = \sum_{i=1}^{r} (Y_i - f_i(X))h_i + g_{r+1}$, where $h_i \in R[X, Y]$ and $g_{r+1} \in R[X]$. We obtain

$$\forall x \in k^n: \quad 0 = g(x, f(x)) = g_{r+1}(x).$$

This shows that $g_{r+1} = 0$ and we are done.   □

For the rest of this section ε denotes an indeterminate over k, $R := k[[\varepsilon]]$ is the ring of formal power series in ε, and $K := k((\varepsilon))$ is the quotient field of R. There exists a k-algebra morphism $R^n \to k^n$, $v = (v_1, \ldots, v_n) \mapsto (v_1(0), \ldots, v_n(0)) =: v_{\varepsilon=0}$. If $W \subseteq K^n$ is a K-subspace, then $W_{\varepsilon=0} := (W \cap R^n)_{\varepsilon=0}$ is a k-subspace of k^n. We have the following result.

(8.54) Proposition. *Let $W \subseteq K^n$ be a K-subspace. Then*

$$\dim_k W_{\varepsilon=0} = \dim_K W.$$

Proof. The proof consists of two parts. We introduce the notation $W_R := W \cap R^n$.

(i) We use induction on r to prove the following assertion: for all K-linearly independent $u_1, \ldots, u_r \in W_R$ there exist $v_1, \ldots, v_r \in W_R$ such that the K-spaces generated by $u_1, \ldots, u_r$ and $v_1, \ldots, v_r$ are equal and $(v_1)_{\varepsilon=0}, \ldots, (v_r)_{\varepsilon=0}$ are k-linearly independent. The start $r = 1$ is trivial. For the induction step we assume that we have constructed $v_1, \ldots, v_{r-1}$ from $u_1, \ldots, u_{r-1}$ and give an algorithm to compute v_r. We set $w := u_r$.

(1) If $w_{\varepsilon=0}, (v_1)_{\varepsilon=0}, \ldots, (v_{r-1})_{\varepsilon=0}$ are linearly independent over k, then set $v_r := w$ and stop.

(2) Otherwise there exist $\lambda \in k^{r-1}$ such that $\left(w - \sum_{i=1}^{r-1} \lambda_i v_i\right)_{\varepsilon=0} = 0$. Now change w to $\varepsilon^{-1}\left(w - \sum_{i=1}^{r-1} \lambda_i v_i\right)$ and goto (1).

It is sufficient to prove that this algorithm terminates. Otherwise, for any $i \leq r-1$ there would exist a sequence $(\lambda_i^{(0)}, \lambda_i^{(1)}, \ldots)$ in k such that

$$\forall n \in \mathbb{N} \, \exists \, w_n \in W_R: \quad u_r - \sum_{i=1}^{r-1} (\lambda_i^{(0)} + \varepsilon \lambda_i^{(1)} + \cdots + \varepsilon^n \lambda_i^{(n)}) v_i = \varepsilon^{n+1} w_n.$$

Setting $\rho_i := \sum_{j=0}^{\infty} \varepsilon^j \lambda_i^{(j)} \in R$, this implies $u_r = \sum_{i=1}^{r-1} \rho_i u_i$, a contradiction to the linear independence of the u_i.

(ii) Let $\dim_K W = r$ and $u_1, \ldots, u_r \in W_R$ be a K-basis of W. Construct $v_1, \ldots, v_r$ as in the last step. Then $(v_1)_{\varepsilon=0}, \ldots, (v_r)_{\varepsilon=0}$ form a k-basis of $W_{\varepsilon=0}$. (Note that the conditions $\sum \rho_i v_i \in W_R$, $\rho_i \in K$ imply that $\rho_i \in R$ for all i.) Hence, $\dim_k W_{\varepsilon=0} = \dim_K W$. $\qquad \square$

Proof of Thm. (8.51). Let $I_k := \sum_{i=1}^{r} (Y_i - f_i(X))k[X, Y]$ be the vanishing ideal of graph f in $k[X, Y]$. The scalar extension of I_k by K with respect to $k \to K$ gives an ideal I of $K[X, Y]$, which obviously satisfies

$$\text{DEG } I = \text{DEG } I_k.$$

The ideal

$$\tilde{I} := \sum_{j=1}^{r} \left(Y_j - \varepsilon^d f_j(X/\varepsilon)\right) K[X, Y]$$

satisfies $\text{DEG } \tilde{I} = \text{DEG } I$, since the affine transformation $Y_i \mapsto Y_i/\varepsilon^d$, $X_j \mapsto X_j/\varepsilon$ does not change the degree. Note that

$$\varepsilon^d f_i(X/\varepsilon) = f_i^{(d)} + \varepsilon f_i^{(d-1)} + \cdots + \varepsilon^d f_i^{(0)}.$$

Since $\tilde{I} \cap R[X, Y]$ is generated by $Y_j - \varepsilon^d f_j(X/\varepsilon)$, $1 \leq j \leq r$, see Lemma (8.53), we have

$$\tilde{I}_{\varepsilon=0} \subseteq J := \text{ vanishing ideal of graph } f^{(d)}.$$

This immediately implies $(\tilde{I}^{\leq t})_{\varepsilon=0} \subseteq J^{\leq t}$ for all $t \geq 0$. Now Prop. (8.54) gives

$$\dim_K \tilde{I}^{\leq t} = \dim_k (\tilde{I}^{\leq t})_{\varepsilon=0} \leq \dim_k J^{\leq t}$$

for all $t \geq 0$. Hence, $\text{DEG } \tilde{I} \geq \text{DEG } J$. $\qquad \square$

8.6* The Case of a Finite Field

In this section, k denotes a finite field and $\bar{k}$ denotes its algebraic closure.

Strassen's degree bound can be employed to give nontrivial lower bounds for the complexity of the evaluation of functions on finite fields. More precisely, let k be a finite field, $n \geq 1$ be a positive integer, and $D \subseteq k^n$ be a nonempty subset. The set $A = A_D$ of all functions from D to k is a k-algebra under pointwise addition and multiplication of functions. Note that the units in A are exactly those functions which do not vanish anywhere on D. Let $x_1, \ldots, x_n$ denote the projection functions of k^n. In this section we are interested in lower bounds for the nonscalar complexity

$$L_A(f_1, \ldots, f_r) := L_A(f_1, \ldots, f_r \mid x_1, \ldots, x_n)$$

of elements $f_1, \ldots, f_r \in A$ to be computed from $x_1, \ldots, x_n$ in the k-algebra A.

Since k is finite, the canonical morphism $\pi : k[X] := k[X_1, \ldots, X_n] \to A$ of k-algebras sending the indeterminate X_i to x_i is surjective; hence, every element of A can be viewed as a polynomial map in many ways. The morphism π extends to the localization $B := \{F/G \mid F, G \in k[X], \pi(G) \in A^\times\}$. By Prop. (7.6) there are elements $F_1, \ldots, F_r \in B$ such that $\pi(F_i) = f_i$ and

$$(8.55) \qquad L_A(f_1, \ldots, f_r) \geq L_B(F_1, \ldots, F_r) \geq L_{\bar{k}(X)}(F_1, \ldots, F_r).$$

Now we can use the degree bound to estimate the right-hand side from below. There is, however, a problem: the F_i are not uniquely determined by the f_i. We only know that $S := \mathrm{graph}(f_1, \ldots, f_r) \subset k^{n+r}$ is contained in $\mathrm{graph}(F_1, \ldots, F_r) \subset \bar{k}^{n+r}$. We thus need an effective condition on S to ensure that the graph of any rational map from $\bar{k}^n$ to $\bar{k}^r$ containing S has large degree. This is given by the following lemma. (Note that Thm. (8.50) serves a similar purpose.)

(8.56) Lemma. *Let $S \subseteq \bar{k}^n$ be a finite set such that there exist $t \in \mathbb{N}'$ and linear forms $\ell_1, \ldots, \ell_q$ on $\bar{k}^n$ with the following properties:*

(1) *For all $1 \leq i \leq q$ and for any $c_1, \ldots, c_{i-1} \in \bar{k}$ the linear form ℓ_i restricted to the set $S \cap \{x \in \bar{k}^n \mid \ell_1(x) = c_1, \ldots, \ell_{i-1}(x) = c_{i-1}\}$ assumes at most t different values.*

(2) *$\ell_1, \ldots, \ell_q$ separate the points of S, i.e., for any $c_1, \ldots, c_q \in \bar{k}$ there exists at most one $x \in S$ such that $\ell_1(x) = c_1, \ldots, \ell_q(x) = c_q$.*

Then, for any $m \geq 1$ and any Zariski closed set $W \subseteq \bar{k}^n$ of dimension d containing S we have

$$\deg W \geq \frac{|S|}{t^d}.$$

Proof. We may assume that W is irreducible. We use induction on q. If $q = 1$ the assumptions imply $|S| \leq t$ and hence the assertion. Now suppose that $q > 1$ and let the assertion of the lemma be true for $q - 1$. By condition (1) the linear form ℓ_1 assumes on S at most t values, say $\{b_1, \ldots, b_m\}$, $m \leq t$. Define for $i = 1, \ldots, m$ the hyperplanes $H_i := Z(\ell_1 - b_i)$ and partition S into the sets $S_i := S \cap H_i$. The induction hypothesis implies for all i

$$\deg W \cap H_i \geq \frac{|S_i|}{t^{\dim W \cap H_i}}.$$

Since W is irreducible, we have $W \subseteq H_i$ or $\dim W \cap H_i = d - 1$ by Krull's principal ideal theorem. If $W \subseteq H_i$ for some i, then $S \subseteq H_i$ and the induction hypothesis applies to S. Otherwise, summing over all i we get

$$t \deg W \geq \sum_{i=1}^{m} \deg W \cap H_i \geq \sum_{i=1}^{m} \frac{|S_i|}{t^{d-1}} = \frac{|S|}{t^{d-1}}. \qquad \square$$

(8.57) Theorem (Strassen). *Let $D \subseteq k^n$ be nonempty, $f_1, \ldots, f_r$ be functions from D to k, and let $S \subseteq k^{n+r}$ denote the graph of the map $(f_1, \ldots, f_r): D \to k^r$. Suppose that there exists $t \in \mathbb{N}'$ and k-linear forms $\ell_1, \ldots, \ell_q$ on k^{r+n} with the following properties:*

(1) *For any $1 \leq i \leq q$ and any $c_1, \ldots, c_{i-1} \in k$ the form ℓ_i assumes at most t different values on the set*

$$S \cap \{x \in k^{r+n} \mid \ell_1(x) = c_1, \ldots, \ell_{i-1}(x) = c_{i-1}\}.$$

(2) $\ell_1, \ldots, \ell_q$ *separate the points of S.*

Then we have

$$L_{A_D}(f_1, \ldots, f_r) \geq \log \frac{|D|}{t^n}.$$

Proof. By the discussion at the beginning of this section there are rational functions $F_1, \ldots, F_r \in \bar{k}(X_1, \ldots, X_n)$ extending $f_1, \ldots, f_r$ such that $L_{A_D}(f_1, \ldots, f_r) \geq L_{\bar{k}(X)}(F_1, \ldots, F_r)$. The degree bound implies $L_{\bar{k}(X)}(F_1, \ldots, F_r) \geq \log \deg W$, where W is the graph of $(F_1, \ldots, F_r)$. The finite set S satisfies the properties in Lemma (8.56) and is contained in W. Hence, $\deg W \geq |S|/t^n$. Since $|S| = |D|$, the assertion follows. $\qquad \square$

Let us now turn to an application concerning the evaluation of a polynomial function at several points.

(8.58) Corollary (Strassen). *Let $n \geq 2$ and A be the ring of all functions from k^{2n+2} to k. Define $f_0, \ldots, f_n \in A$ by*

$$f_0(\alpha, \xi) := \alpha_0 \xi_0^n + \cdots + \alpha_n, \ldots, f_n(\alpha, \xi) := \alpha_0 \xi_n^n + \cdots + \alpha_n$$

for any $(\alpha, \xi) = (\alpha_0, \ldots, \alpha_n, \xi_0, \ldots, \xi_n) \in k^{2n+2}$. Then

$$L_A(f_0, \ldots, f_n) \geq \frac{n+1}{2} \log m - n,$$

where $m = \min\{n + 1, |k|\}$.

Proof. We choose pairwise different $\lambda_1, \ldots, \lambda_m \in k$ and put $t := \lceil \sqrt{m} \rceil$. Next we choose $\alpha_0, \ldots, \alpha_n \in k$ such that on the set $\{\lambda_1, \ldots, \lambda_m\}$ the polynomial function

$$p: k \to k, \quad p(\xi) := \alpha_0 \xi^n + \ldots + \alpha_n$$

assumes only the values $\lambda_1, \ldots, \lambda_t$ and each value at most t times. (Since $m \leq n + 1$, this can be done by interpolation.) Now let $\tilde{D} := \{\lambda_1, \ldots, \lambda_m\}^{n+1} \subseteq k^{n+1}$, and define functions $\tilde{f}_i$ from $\tilde{D}$ to k as

$$\tilde{f}_i(\xi_0, \ldots, \xi_n) := p(\xi_i) \quad \text{for } (\xi_0, \ldots, \xi_n) \in \tilde{D}.$$

Since there is a k-algebra morphism from A to the ring $\tilde{A}$ of functions from $\tilde{D}$ to k which carries f_i to $\tilde{f}_i$, we have $L_A(f_0, \ldots, f_n) \geq L_{\tilde{A}}(\tilde{f}_0, \ldots, \tilde{f}_n)$. Now we apply Thm. (8.57) to $\tilde{f}_0, \ldots, \tilde{f}_n$. Let $\tilde{S}$ be the graph of the map

$$(\tilde{f}_0, \ldots, \tilde{f}_n): \tilde{D} \to k^{n+1},$$

i.e., $\tilde{S}$ is the set of all $(\xi_0, \ldots, \xi_n, \eta_0, \ldots, \eta_n) \in k^{2n+2}$ with $(\xi_0, \ldots, \xi_n) \in \tilde{D}$ and $p(\xi_i) = \eta_i$ for all $0 \leq i \leq n$. Let $x_0, \ldots, x_n, y_0, \ldots, y_n$ be the coordinate projections of k^{2n+2} and put

$$\ell_1 := y_0, \ldots, \ell_{n+1} := y_n, \quad \ell_{n+2} := x_0, \ldots, \ell_{2n+2} := x_n.$$

Then condition (2) of Thm. (8.57) is trivially satisfied. As to condition (1), note that the projection y_i assumes on all of $\tilde{S}$ at most the values $\lambda_1, \ldots, \lambda_t$ by the definition of p. Similarly, the projection x_i assumes on $\tilde{S} \cap \{y_i = c\}$ at most t values. So both conditions of Thm. (8.57) hold. As a consequence we have

$$\begin{aligned} L_{\tilde{A}}(\tilde{f}_0, \ldots, \tilde{f}_n) &\geq \log(|\tilde{D}|/t^{n+1}) = (n+1)\log(m/\lceil\sqrt{m}\rceil) \\ &\geq (n+1)\log(2\sqrt{m}/3) \geq \frac{n+1}{2}\log m - n. \quad \square \end{aligned}$$

8.7 Exercises

$k[X]$ denotes the ring $k[X_1, \ldots, X_n]$ throughout this section.

8.1. Prove Prop. (8.4).

8.2. Let g and h be univariate coprime polynomials, $h \neq 0$, and $f = g/h$. Show that $\deg(f) = \max\{\deg g, 1 + \deg h\}$.

8.3. In this exercise we will see that for some of the computational problems we discussed in this chapter the degree bound will not yield a better lower bound than our estimates.

(1) Let $f_1, \ldots, f_p \in k[X]$ be multivariate polynomials of degree ≥ 1. Show that $\deg(f_1, \ldots, f_p) \leq \prod_{i=1}^p \deg f_i$.
(2) Let $\sigma_1, \ldots, \sigma_n$ be the elementary symmetric polynomials in $X_1, \ldots, X_n$. Show that $\deg(\sigma_1, \ldots, \sigma_n) = n!$.
(3) Let $f \in k[T]$ be a univariate polynomial of degree d, and $f_i(X) := f(X_i)$. Show that $\deg(f_1, \ldots, f_n) = d^n$.

8.4. The aim of this exercise is to give a simplified proof of Thm. (8.5) if log is replaced by $\log_3$.

(1) Let $N \subseteq K$ be an extension of fields, N be algebraically closed in K, and tr.deg $K/N = 1$. Further, let $p \in \mathbb{N}$ and $z_1, z_2 \in K \setminus N$ be such that $[K:N(z_i)] \leq p$ for $i = 1, 2$. Prove that $[K:N(z_1/z_2)] \leq 2p$ if $z_1/z_2 \notin N$. (Note that we are not imposing further conditions on z_1, z_2; compare Lemma (8.15)(2).)

(2) Prove Thm. (8.5) with log replaced by $\log_3$. (Hint: take the original proof as a model. Consider the claims (B) and (C). Using Lemma (8.15)(1) and part (1) of this exercise give a short proof of these claims in case where $2p$ is replaced by $3p$.)

8.5. Let $f_1, \ldots, f_s \in k(X)$. Suppose that the s by n matrix $(\partial f_i/\partial X_j)_{i,j}$ has rank s. Show that $f_1, \ldots, f_s$ are algebraically independent over k. (Hint: Let $P \in k[Y_1, \ldots, Y_s] \setminus \{0\}$ be of minimal degree subject to $P(f_1, \ldots, f_s) = 0$. Show that $\partial P/\partial Y_i \neq 0$ for some i.)

8.6. (Evaluation of power sums) Show that

$$n(1 - c)\big(\log(n(1 - c)) - 2\big) \leq L(s_1, \ldots, s_n) \leq n \log n + 5.75\, n + 1,$$

where $c = 1/p$ if k is of characteristic $p > 0$ and $c = 0$ otherwise. (Hint: use the Newton relations (2.24), the proof of Cor. (2.25), Cor. (8.8) and Ex. 8.5.)

8.7. Let $f \in k[T]$ be a univariate polynomial of degree n, and denote by F the product $F := f(X_1) \cdots f(X_n)$. Show that $L(F)$ if of order $n \log n$.

8.8. The aim of this exercise is an alternative proof of Thm. (8.5) based on the theory of algebraic function fields of one variable. We assume familiarity with this theory. (A good reference is, e.g., Stichtenoth [488].) Throughout this exercise k will denote an arbitrary field and K/k is an algebraic function field of one variable, i.e., tr.deg$_k K = 1$ and there exists $x \in K$ such that K is a finite extension of $k(x)$. We assume that k is algebraically closed in K. An element $f \in K \setminus k$ is called a function, elements of k are called constants. For a function $f \in K$ we let N_f and D_f denote the numerator and the denominator of the divisor (f) of f, respectively. We write $\mathbb{P}(K)$ and $\mathbb{D}(K)$ for the set of prime divisors and the multiplicative group of divisors of K, respectively. If A is a divisor of K, we let $\mathcal{L}(A)$ denote the linear space of A.

(1) Let $f \in K$ be a function. Show that $[K:k(f)] = \deg N_f = \deg D_f$. (This generalizes Prop. (8.2).)

(2) Let $A \in \mathbb{D}(K)$, and $f \in \mathcal{L}(A)$ be a function. If A is integral, then we have $[K:k(f)] \leq \deg A$.

(3) If $A, B \in \mathbb{D}(K)$, $f \in \mathcal{L}(A)$, $g \in \mathcal{L}(B)$, then $f + g \in \mathcal{L}(\mathrm{lcm}(A, B))$ and $fg \in \mathcal{L}(AB)$. Hence $[K:k(f + g)]$, $[K:k(fg)] \leq [K:k(f)] + [K:k(g)]$.

(4) Use (3) to show that with the notation of Thm. (8.5) we have

$$\mathcal{L}(f_1, \ldots, f_m \mid S) \geq \log_3 \deg(S \cup \{f_1, \ldots, f_m\}) - \log_3 \deg S.$$

(5) If f and g are functions in K, then $[K:k(f/g)] \leq [K:k(f)] + [K:k(g)]$.

(6) Let f, g be functions in K, $B := \mathrm{lcm}(D_f, D_g)$, and m be an integer $\geq \max\{\deg D_f, \deg D_g\}$. Suppose that $\deg B > m$. Show that for every $\alpha \in k$ satisfying $[K:k(f + \alpha g)] \leq m$ there exists a prime divisor p_α such that $p_\alpha | B$ and $f + \alpha g \in L(B/p_\alpha)$.

(7) Use (6) to show that if the set $\{\alpha \in k \mid [K:k(f + \alpha g)] \leq m\}$ has cardinality $\geq 2m + 1$, then $\deg B \leq m$.

(8) Suppose $h \in K$ satisfies $[K:k(h)] \leq m$ and that the cardinality of the set $\{\alpha \in k \mid [K:k(f + \alpha h)] \leq m\}$ is at least $2m + 1$. Under the assumptions of (7) show that $[K:k(f + hg)]$, $[K:k(h + f/g)] \leq 2m$.

(9) Prove Thm. (8.5).

8.9. Let $f_1, \ldots, f_r \in k(X)$, and let $f = (f_1, \ldots, f_r): k^n \supseteq \mathrm{def}\, f \to k^r$ be the corresponding rational map. Show that graph f is a locally closed and irreducible subset of k^{n+r}.

8.10. Prove Rem. (8.18).

8.11. Let $V \subseteq \mathbb{P}^n$ be finite. Using Hilbert polynomials show that DEG $V = |V|$.

8.12. Prove Rem. (8.20).

8.13. Let Z be the affine variety in k^2 consisting of the line $\{X_1 = 0\}$ and the point $(1, 1)$. Compute $H(Z; t)$ and deduce that Z has degree 1. Now consider the first projection π of Z and show that DEG $\pi(Z) = 2$.

Throughout the next five exercises we use the following notation: X_0 is an indeterminate over $k(X)$ and $S := k[X_0, X_1, \ldots, X_n]$.

8.14. Let $I, J \subset S$ be homogeneous ideals. Prove the following:

(1) If $I \subseteq J$, then $H(I; t) \geq H(J; t)$ for all $t \in \mathbb{N}$.
(2) $H(I + J; t) = H(I; t) + H(J; t) - H(I \cap J; t)$ for all $t \in \mathbb{N}$.

8.15. Let $I \subset S$ be a homogeneous ideal, furthermore let $f \in S$ be homogeneous of degree r. Let $(I:f) := \{g \in S \mid gf \in I\}$. Prove the following:

(1) $(I:f)$ is a homogeneous ideal of S.
(2) $0 \to S/(I:f) \to S/I \to S/(I + fS) \to 0$ is an exact sequence of S-modules, where $S/(I:f) \to S/I$ is the multiplication by f.
(3) $H(I + fS; t) = H(I; t) - H((I:f); t - r)$.
(4) $H(I \cap fS; t) = H(fS; t) + H((I:f); t - r)$.

8.16. Let $I \subset S$ be a homogeneous ideal such that $\dim S/I \geq 2$. Further let $f \in S$ be a homogeneous polynomial of degree ≥ 1 such that $(I: f) = I$. Show that $\mathrm{DEG}(I + fS) = \mathrm{DEG}\, I \cdot \deg f$. (Hint: Use Krull's principal ideal theorem to show that $\dim S/(I + fS) = \dim S/I - 1$. Now use part (3) of the previous exercise.)

8.17. The aim of this exercise is to show that $\mathrm{DEG}\, V = \deg V$ iff V is equidimensional.

(1) Let $I, J \subset S$ be homogeneous ideals such that $\dim S/I > \dim S/J$. Show that $\mathrm{DEG}(I \cap J) = \mathrm{DEG}\, I$. Hence, if V and W are projective varieties, $\dim W < \dim V$, then $\mathrm{DEG}(V \cup W) = \mathrm{DEG}\, V$.

(2) Let $I, J \subset S$ be proper homogeneous ideals such that $\dim S/(I + J) < \dim S/I = \dim S/J$. Show that $\mathrm{DEG}(I \cap J) = \mathrm{DEG}\, I + \mathrm{DEG}\, J$.

(3) Let $P_1, \ldots, P_t$ be prime ideals of S such that for all j we have $\cap_{i \neq j} P_i \neq \cap_i P_i$ and $\dim S/P_i = \dim S/P_j$ for $i \neq j$. Show that $\mathrm{DEG}(P_1 \cap \cdots \cap P_t) = \sum_{i=1}^{t} \mathrm{DEG}\, P_i$.

(4) Let $V \subseteq \mathbb{P}^n$ be a projective variety of dimension n, and $V_1, \ldots, V_s$ be the n-dimensional components of V. Prove that $\mathrm{DEG}\, V = \sum_{i=1}^{s} \mathrm{DEG}\, V_i$. Hence show that if $V \subseteq \mathbb{P}^n$ is a projective variety, then $\deg V \geq \mathrm{DEG}\, V$ with equality holding iff V is equidimensional.

(5) Let $\pi: \mathbb{P}^n \to \mathbb{P}^{n-1}$ be the projection onto the first $n-1$ components. Give further examples of projective varieties $V \subseteq \mathbb{P}^n$ such that $\mathrm{DEG}\, V < \mathrm{DEG}\, \pi(V)$. (See also Ex. 8.13.)

8.18.* (General Bézout Inequality) Let $I, J \subseteq S$ be homogeneous ideals, and $T := k[X_0, \ldots, X_n, Y_0, \ldots, Y_n]$. We define the mappings $\varphi_1: S \to T$, $f \mapsto f(X_0, \ldots, X_n)$ and $\varphi_2: S \to T$, $g \mapsto g(Y_0, \ldots, Y_n)$. The *join* $\mathcal{J}(I, J)$ of I and J is defined as the ideal in T generated by $\varphi_1(I)$ and $\varphi_2(J)$. The *join* $J(X, Y)$ of two projective varieties $X, Y \subseteq \mathbb{P}^n$ is the zeroset of $\mathcal{J}(I(X), I(Y))$ in $\mathbb{P}^{2n+1}$, where $I(X), I(Y) \subseteq S$ are the vanishing ideals of X and Y, respectively.

(1) Show that $H(\mathcal{J}(I, J); t) = \sum_{j=0}^{t} H(I; j) H(J; t - j)$.

(2) Show that $\mathrm{DEG}\, \mathcal{J}(I, J) = \mathrm{DEG}\, I \cdot \mathrm{DEG}\, J$. (Hint: proceed in a similar way as in the proof of Lemma (8.32), claim (b).)

(3) For locally closed subsets $X, Y \subseteq \mathbb{P}^n$ show that $\deg X \cap Y \leq \deg X \deg Y$. (Hint: w.l.o.g. X and Y are irreducible and closed. Show that $J(X, Y)$ is an irreducible projective variety, and $X \cap Y$ can be obtained as a projection of $J(X, Y) \cap H$, where $H \subseteq \mathbb{P}^{2n+1}$ is an appropriate projective subspace. Now use the Bézout inequality, Lemma (8.32)(2).)

8.19. We use the same notation as in Example (8.22). The set $F_{r,n}$ of all those $(a, b) \in k^{r \times n} \times k^r$ such that $U_{a,b}$ is an $(n - r)$-dimensional subspace of $\mathbb{P}^n$ is open in $k^{r \times n} \times k^r$. The goal of this exercise is to derive Thm. (8.30) from the following assertion contained in Harris [228, Lect. 13, Lect. 18]: *Let $V \subseteq \mathbb{P}^n$ be* an r-dimensional irreducible variety of degree d. Then there exists an open subset O of $F_{r,n}$ such that for all $(a, b) \in O$ we have $|V \cap U_{a,b}| = d$.

(1) Let $B \subseteq k^N$ be an affine variety. A closed subset $V \subseteq B \times \mathbb{P}^n$ is called a *family of projective varieties* with base B. For $b \in B$ the fibers $\pi_1^{-1}(b) =: V_b$ are called the *members* of V. Note that $(b, p) \in V$ iff $p \in \pi_2(V_b)$, where π_2 is the second projection. Let $X \subseteq \mathbb{P}^n$ be a projective variety. Show that $\{b \in B \mid X \cap \pi_2(V_b) \neq \emptyset\}$ is closed in B. (Hint: use the fact that π_1 maps closed sets to closed sets, see Harris [228, Thm. 3.12].)

(2) Let $V \subseteq \mathbb{P}^n$ be an r-dimensional irreducible projective variety, $O \subseteq \mathbb{P}^n$ be an open subset. Show that the set $\{(a, b) \in k^{r \times n} \times k^r \mid U_{a,b} \cap V \subseteq O\}$ is open in $k^{r \times n} \times k^r$.

(3) Prove Thm. (8.30).

8.20. Let $f : k^m \supseteq \mathrm{def}\, f \to k^n$ be a rational map. Show that for any $\xi \in k^n$ the number of components of $f^{-1}(\xi)$ is a lower bound for $\deg f$.

8.21.* In this exercise we give an elementary proof of Prop. (8.40).

(1) Let $f_0, f_1, \ldots, f_n \in k[X]$, and $d_i := \deg f_i$, $i = 0, \ldots, n$. Show that there exists a polynomial $R \in k[Y_0, \ldots, Y_n]$ such that $R(f_0, \ldots, f_n) = 0$ and $\deg_{Y_0} R \leq \prod_{j=1}^n d_j$. (Hint: let $d \in \mathbb{N}$. We may assume that $D := \prod_{j=1}^n d_j \geq 1$. Let $R = \sum_{j=(j_0, \ldots, j_n)} R_j Y_0^{j_0} \cdots Y_n^{j_n}$, subject to $j_0 \leq D$ and $\sum_{i=0}^n j_i d_i \leq d$. Then $R(f_0, \ldots, f_n) = \sum_{m=(m_1, \ldots, m_n)} C_m X_1^{m_1} \cdots X_n^{m_n}$. Prove that for sufficiently large d the set of homogeneous linear equations $C_m = 0$ in the R_j's has a nontrivial solution.)

(2) Let $f_0 \in k[X]$, $f_1, \ldots, f_n \in k(X)$. Let $\ell := L(f_1, \ldots, f_n)$. Show that there exists a polynomial $R \in k[Y_0, \ldots, Y_n]$ such that $R(f_0, \ldots, f_n) = 0$ and $\deg_{Y_0} R \leq 2^\ell$. (Hint: Let $\Gamma := (\Gamma_1, \ldots, \Gamma_t)$ be straight-line program over k expecting inputs of length n, of c_*-length ℓ executable on input $(k(X); X)$ and yielding the result sequence $(r_{-n+1} = X_n, \ldots, r_0 = X_1, r_1, \ldots, r_t)$, where $r_{i_1} = f_1, \ldots, r_{i_n} = f_n$. Passing along Γ define polynomials $F_1, \ldots, F_t \in k[Z_{-n+1}, \ldots, Z_t]$ of degrees one or two, where according to whether the computation step Γ_m is of the form

$$\begin{aligned}
s_m &:= s_i \circ s_j & (\circ \in \{+, -, *\}), \text{ or} \\
s_m &:= s_i / s_j, \text{ or} \\
s_m &:= \lambda_m \cdot s_i & (\lambda_m \in k), \text{ or} \\
s_m &:= c, & (c \in k)
\end{aligned}$$

F_m is going to be defined as

$$\begin{aligned}
F_m &:= Z_m - Z_i \circ Z_j, \text{ or} \\
F_m &:= Z_m \cdot Z_j - Z_i, \text{ or} \\
F_m &:= Z_m - \lambda_m \cdot Z_i, \text{ or} \\
F_m &:= Z_m - c,
\end{aligned}$$

respectively. Consider for $m = 0, \ldots, t$ the substitutions

$$\sigma_m : k[Z_{-n+1}, \ldots, Z_t] \longrightarrow k(X)[Z_{m+1}, \ldots, Z_t]$$

sending Z_i to r_i for $i \leq m$. Argue by induction on m that there is a nonzero polynomial $R_m \in k[Y_0, Y_1, \ldots, Y_n, Y_{n+1+m}, \ldots, Y_{n+t}]$ with $\deg_{Y_0} R_m \leq 2^{\ell}$ such that

$$R_m(f_0, \sigma_m Z_{i_1}, \ldots, \sigma_m Z_{i_n}, \sigma_m F_{m+1}, \ldots, \sigma_m F_t) = 0.)$$

(3) Prove Prop. (8.40). (Hint: let $\xi_1, \ldots, \xi_N \in k^n$ be the different solutions of $f_1(\xi) = \lambda_1, \ldots, f_n(\xi) = \lambda_n$. Show that there exists $f_0 \in k[X]$ such that $f_0(\xi_i) \neq f_0(\xi_j)$ for $i \neq j$. Now apply part (2).)

8.22. Prove the upper bound in Thm. (8.46). (Hint: let $m = \lfloor n/q \rfloor$, $p = n - mq$. For any $i = 0, \ldots, m-1$ compute the elementary symmetric polynomials $\sigma_{i,1}, \ldots, \sigma_{i,q}$ in $X_{iq+1}, \ldots, X_{(i+1)q}$ and all elementary symmetric polynomials $\sigma_{m,1}, \ldots, \sigma_{m,p}$ in $X_{mq+1}, \ldots, X_n$. Let

$$Q_i := 1 - \sigma_{i,1}t + \cdots + (-1)^q \sigma_{i,q} t^q, \quad i = 0, \ldots, m-1,$$
$$Q_m := 1 - \sigma_{m,1}t + \cdots + (-1)^p \sigma_{m,p} t^p.$$

Then $\prod_{i=0}^m Q_i \equiv 1 - \sigma_1 t + \cdots + (-1)^q \sigma_q t^q \mod t^{q+1}$.)

8.23.* (Evaluation of the discriminant) The polynomial $D := \prod_{i \neq j}(X_i - X_j)$ is called the *discriminant* in n variables..

(1) Show that $L(D) < 9n(\log n + 1)$. (Hint: first compute the coefficients of $A(T) := \sum_{j=1}^n (T - X_j)$ and evaluate dA/dT at $X_1, \ldots, X_n$.)

(2) Prove that $L(D) \geq \frac{1}{6} n \log n - \frac{2}{3} n$. (Hint: let $p = \lceil n/2 \rceil$, $q = \lfloor n/2 \rfloor$, and replace $X_1, \ldots, X_n$ by $X_1, \ldots, X_p, Y_1, \ldots, Y_q$. Then, for some polynomials g and h, we have $D(X, Y) = g(X) \prod_{i,j}(X_i - Y_j)^2 h(Y)$. First show that

$$\frac{\partial D}{\partial Y_j} = D\left(-\sum_{\ell=1}^p \frac{2}{X_\ell - Y_j} + \frac{\partial h}{\partial Y_j}(Y) \cdot \frac{1}{h(Y)}\right).$$

Let K be the algebraic closure of $k(Y_1, \ldots, Y_q)$. Use the derivative inequality (7.7) to prove that

$$L_K\left(\frac{\prod_{i=1}^p(X_i - Y_1)}{\sum_{\ell=1}^p \prod_{i \neq \ell}(X_i - Y_1)}, \ldots, \frac{\prod_{i=1}^p(X_i - Y_q)}{\sum_{\ell=1}^p \prod_{i \neq \ell}(X_i - Y_q)}\right) \leq 3L(D) + q.$$

Let $\phi: K^p \supseteq \operatorname{def}\phi \to K^q$ be the rational map defined by the above q rational functions. Consider $\phi^{-1}(0)$ and use Ex. 8.20 to show that $\deg \operatorname{graph}\phi \geq p!$.)

8.24. * In Cor. (8.13) we determined the complexity of the resultant of two polynomials as a function of their roots. What can we say about the resultant as a function of the coefficients of the given polynomials? Let $R = R(a_1, \ldots, a_n, b_1, \ldots, b_n)$ be the resultant of the monic polynomials $X^n + \sum_{i=1}^n a_i X^{i-1}$, $X^n + \sum_{i=1}^n b_i X^{i-1}$ with indeterminate coefficients (cf. Def. (10.11)). By Ex. 3.11 we know that $L(R|a, b) = O(n \log n)$. The degree bound (8.36) combined with the derivative inequality (7.7) implies that

$$\frac{1}{3} \log \deg(R, \partial R/\partial a_1, \ldots, \partial R/\partial a_n, \partial R/\partial b_1, \ldots, \partial R/\partial b_n) \leq L(R|a, b).$$

Prove that the left-hand side has order of magnitude $O(n)$. Thus no nonlinear lower bound can be derived in this way! (See Problem 8.2.) (Hint: Appeal to Thm. (8.30) to express the degree as the number of intersection points of the graph of $(R, \partial R/\partial a_1, \ldots, \partial R/\partial b_n)$ with a sufficiently general affine subspace. In the corresponding system of equations substitute $a_i \mapsto \sigma_i(x)$, $b_i \mapsto \sigma_i(y)$ and use $R(\sigma(x), \sigma(y)) = \prod_{1 \leq i, j \leq n}(x_i - y_i)$ (cf. Lang [316, V, §10]). Moreover, use the formula (A) in the proof of Thm. (8.46) which expresses $\partial_i \sigma_q$ in terms of $\sigma_{q-1}, \ldots, \sigma_1$.)

8.25.
(1) Let $p \in k[X]$ be monic of degree $n \geq 1$ and $a \in k[X]$, $\deg a < n$, be coprime to p. The inverse of a modulo p is the unique polynomial $b \in k[X]$ of degree $< n$ such that $ab \equiv 1 \mod p$. Show that the coefficients of b may be expressed in the form $h\Delta^{-m}$, where h is a polynomial in the coefficients of p and a, $m \in \mathbb{N}$, and Δ is the resultant of p and a. (Hint: write the relation $ba + cp = 1$ with $c \in k[X]$, $\deg c < n$, as a linear system of equations and use Cramer's rule.)
(2) Show that the inverse of the Chinese remainder map φ defined just before Thm. (8.45) is a rational map with domain of definition $V \times U_1 \times \cdots \times U_{d_t}$. (Hint: look at the proof of Thm. (3.21).)

8.26. Prove Lemma (8.49) using Hilbert polynomials.

8.27. Assumptions being as in Cor. (8.58), show that there exists a universal constant c such that
$$L_A(f_0, \ldots, f_n) \leq cn \log(m).$$

(Hint: Use the upper bound in Cor. (8.9).)

8.28. * Let the finite field k have at least $n + 1$ elements and let

$$D := \{(\xi_0, \ldots, \xi_n, \eta_0, \ldots, \eta_n) \in k^{2n+2} \mid \xi_i \neq \xi_j \text{ for } i \neq j\}.$$

Let A be the ring of all functions from D to k. Define $g_0, \ldots, g_n \in A$ by

$$\eta_0 = g_0(\xi, \eta)\xi_0^n + \cdots + g_n(\xi, \eta),$$
$$\eta_n = g_0(\xi, \eta)\xi_n^n + \cdots + g_n(\xi, \eta)$$

for any $(\xi, \eta) \in D$. Prove that

$$L_A(g_0, \ldots, g_n) \geq (n/2) \log n - 3n.$$

(Hint: Choose $\lambda_0, \ldots, \lambda_n \in k$ pairwise different. Set $t := \lceil \sqrt{n+1} \rceil$ and let $\tilde{\lambda}_i :=$ $\lambda_{\lfloor i/t \rfloor}$. There are at most t different $\tilde{\lambda}_i$. Let

$$\tilde{D} := \{(\xi_0, \ldots, \xi_n) \in \{\lambda_0, \ldots, \lambda_n\}^{n+1} \mid \xi_i \neq \xi_j \text{ for } i \neq j\},$$

and $\tilde{g}_i(\xi_0, \ldots, \xi_n) := g_i(\xi_0, \ldots, \xi_n, \tilde{\lambda}_0, \ldots, \tilde{\lambda}_n)$ for $(\xi_0, \ldots, \xi_n) \in \tilde{D}$. Denote by $\tilde{A}$ the ring of all functions from $\tilde{D}$ to k. Then $L_A(g_0, \ldots, g_n) \geq L_{\tilde{A}}(\tilde{g}_0, \ldots, \tilde{g}_n)$. Now proceed as in the proof of Cor. (8.58) with the linear forms

$$\ell_1 := a_0\lambda_0^n + \cdots + a_n, \ldots, \ell_{n+1} := a_0\lambda_n^n + \cdots + a_n, \ell_{n+2} := x_0, \ldots, \ell_{2n+2} := x_n,$$

where $x_0, \ldots, x_n, a_0, \ldots, a_n$ are the coordinate projections of k^{2n+2}.)

8.8 Open Problems

Problem 8.1. Is the algorithm for computing the Hermite interpolation polynomial which was described in Ex. 3.14 optimal up to order of magnitude?

Problem 8.2. What is the complexity of the resultant of two univariate polynomials of degree at most n as a function of the polynomial coefficients? How about the discriminant? (By Ex. 3.11 we know upper bounds $O(n \log n)$ for the multiplicative complexity of these problems.)

8.9 Notes

The degree bound was discovered in 1973 by Strassen [497] and used by him in that paper to prove most of the lower bounds of Sect. 8.1. The degree bound can also be used for the derivation of lower bounds for the cost of computation trees (see Strassen [504] or Chap. 10), or the study of the complexity of deciding membership to semi-algebraic sets (see Lickteig [334, 335] and the Notes to Chap. 11). In conjunction with other methods, Strassen's degree bound even yields nontrivial lower bounds for the multiplicative complexity of a single rational function. The first such method was given by Schnorr [450]. Here we use the derivative inequality (7.7), whose applications, all due to Baur and Strassen [32], are given by Corollaries (8.10), (8.13) in Sect. 8.1, or Thm. (8.46) in Sect. 8.4.

The first *elementary* proof of a version of the degree bound was given by Schönhage [458] in 1976. We have included the statement as well as the proof as Ex. 8.21. The field theoretic version of Strassen's degree bound presented in Sect. 8.1 is due to Baur [29]. Of course, these two elementary variants of the degree bound hide the geometric concepts underlying this bound, which are best understood by using the degree in its geometric setting.

The geometric tool for proving Strassen's degree bound is Bézout's inequality (8.28). This topic is discussed in several books at different levels. As it has been our intention to keep the material at an elementary level, we have confined ourselves to those aspects of the theory relevant for the proof of the degree bound; in particular, we have avoided the discussion of the so-called intersection multiplicities, which form the central part of the general theorem of Bézout. The reader interested in these topics may wish to consult the books of Fulton [177], or Vogel [538]. Among other books which discuss the intersection theory of algebraic varieties we would like to mention the classic books of van der Waerden [539], Weil [545], Shafarevich [471], Mumford [387], as well as Hartshorne [232], or Harris [228]. For the few results we have used from commutative algebra we refer the reader to Atiyah and Macdonald [12] or Matsumura [356]. Our definition of the degree of a locally closed subset of $\mathbb{P}^n$ follows Heintz [239], from where we have also taken the general Bézout inequality, included here as Ex. 8.18. For a more thorough discussion of the affine Bézout inequality we refer the reader to Rojas [442].

Theorems (8.42) and (8.44) in Sect. 8.4 are due to Strassen [497, 504]. There are several other applications which are not included due to lack of space. For instance, one can show that the computation of the derivatives, a single coefficient, or the value of the interpolation polynomial at a new point all have complexity of order $n \log n$ (Strassen [500], Stoß [489]). Thm. (8.50) in Sect. 8.5 is due to Heintz and Sieveking [247]. Our formulation follows Strassen [504]. Prop. (8.54) is from Bürgisser [98].

In Sect. 8.6 we have followed Strassen [501] almost word by word. The methods described in this section give no lower bounds better than linear for the computation of elementary symmetric functions. This is, however, not astonishing since these functions can be computed in linear time over finite fields, as has been shown by Mihaĭljuk [389].

Ex. 8.6 is from Baur and Strassen [497]. The authors give the credit for this algorithm to Stoß. For Ex. 8.11 consult Harris [228, Lect. 13]. Ex. 8.19 is from Harris [228, Ex. 4.2]. Ex. 8.21 is from Schönhage [458]. Ex. 8.22 and 8.23 are from Baur and Strassen [32]. For a solution to Ex. 8.26 see Harris [228, Example 13.4]. Ex. 8.27 and 8.28 are from Strassen [501]. Problem 8.2 is from Strassen [510].

Ritzmann [439] has employed Strassen's degree bound for the investigation of the problem of approximating the zeros of a complex polynomial by iteration procedures. For a general discussion of iteration procedures from the viewpoint of complexity theory the reader is referred to Baur [28] and the references therein.

Chapter 9. Specific Polynomials which Are Hard to Compute

We discuss various techniques for exhibiting specific polynomials of provably high complexity which originate in a landmark paper by Strassen [499] written in 1974. In the first section we restrict ourselves to computations in the polynomial algebra (no divisions) and derive a lower bound in an elementary way. In the next section we study computations in the ring of formal power series, define complexity classes, and prove a representation theorem for them. Based on this, a general lower bound theorem on the multiplicative complexity follows. As a result we exhibit in Sect. 9.3 various specific polynomials with algebraic coefficients which are hard to compute. Some tools from algebraic number theory are required for computing the degrees of the occurring field extensions. In Sect. 9.4 we proceed with rather technical lower bound proofs for polynomials with rapidly growing integer coefficients. Finally, in Sect. 9.5, we extend most of what we have done so far to other complexity measures, in particular to additive complexity.

Except for the first section we assume familiarity with the notion of degree of an affine variety, in particular with Sect. 8.2 and 8.5.

K denotes an algebraically closed field throughout this chapter.

9.1 A Generic Computation

In Chap. 5 we have established lower bounds on the complexity of polynomials $f \in \mathbb{C}[X]$ in terms of the transcendence degree of their coefficient field over $\mathbb{Q}$. However, if the coefficients of f are algebraic or even rational, then the transcendence degree bounds are useless. Our goal is to develop a technique which allows to prove lower bounds in those cases.

For simplicity we will study the multiplicative complexity

$$L_{K[X]}(f) := L^*_{K \to K[X]}(f|X)$$

of $f \in K[X]$ in the *polynomial algebra*. Thus divisions are not allowed. (Note that multiplications with scalars $\lambda \in K$ are free of charge.) Let us first prove an upper bound.

(9.1) Proposition (Paterson and Stockmeyer). *We have $L_{K[X]}(f) \leq 2\sqrt{n}$ for any polynomial $f \in K[X]$ of degree n.*

Proof. Let $p, q \in \mathbb{N}$ with $pq > n$ and write f in the form $f = \sum_{i=0}^{q-1} f_i X^{pi}$ with coefficient polynomials f_i of degree $\leq p - 1$. First we compute $X^2, X^3, \ldots, X^p$ with $p - 1$ multiplications. Then we have all the f_i for free at our disposal since we do not count multiplications with scalars in K. In a second step we compute f from X^p and the f_i with $q - 1$ multiplications by Horner's rule. Therefore $L_{K[X]}(f) \leq p + q - 2$. By choosing $p := q := \lfloor 1 + \sqrt{n} \rfloor > \sqrt{n}$ we conclude that $L_{K[X]}(f) \leq 2\sqrt{n}$. □

We remark that the algorithm of this proposition is for instance useful to evaluate a polynomial at a matrix. As a multiplication of a matrix by a scalar is cheaper than multiplying two matrices, it is natural to try to minimize the number of nonscalar multiplications (compare Ex. 16.11).

We continue with the investigation of lower bounds. By Thm. (4.11) we have $L_{K[X]}(f) \leq r$ iff there exists a computation sequence $(g_1, \ldots, g_r)$ in $K[X]$ of length r which computes f modulo X. This means that there exist $u_\rho, v_\rho \in \langle 1, X, g_1, \ldots, g_{\rho-1} \rangle$ for all $\rho \in \underline{r}$ such that $g_\rho = u_\rho v_\rho$ and $f \in \langle 1, X, g_1, \ldots, g_r \rangle$. W.l.o.g. we may assume that $u_\rho, v_\rho \in \langle X, g_1, \ldots, g_{\rho-1} \rangle$. (Indeed, put $u'_\rho := u_\rho - u_\rho(0)$, $v'_\rho := v_\rho - v_\rho(0)$, $g'_\rho := u'_\rho v'_\rho$ and show that $\langle 1, X, g_1, \ldots, g_\rho \rangle = \langle 1, X, g'_1, \ldots, g'_\rho \rangle$ by induction on ρ.) The idea is to investigate the dependence of the coefficients of the polynomials to be computed on the parameters appearing in a computation. Let $a_{\rho j}, b_{\rho j}$ $(1 \leq \rho \leq r, 0 \leq j < \rho)$, $c_0, \ldots, c_{r+1}$, and X be indeterminates. Setting $G_0 := X$ we define a *generic computation* $(G_1, \ldots, G_r)$ inductively as follows

$$G_\rho := \Big(\sum_{j=0}^{\rho-1} a_{\rho j} G_j \Big) \Big(\sum_{j=0}^{\rho-1} b_{\rho j} G_j \Big).$$

Moreover, we set $F_r := c_0 + c_1 X + \sum_{\rho=1}^r c_{\rho+1} G_\rho$. By the preceding discussion it is clear that a polynomial $f \in K[X]$ satisfies $L_{K[X]}(f) \leq r$ iff there exist values $a'_{\rho j}, b'_{\rho j}, c'_\rho \in K$ such that $f = F_r(a', b', c', X)$. We can express this in the following way. For $r, n \in \mathbb{N}$ we define the *complexity class*

$$C'_{r,n} := \{ f \in K[X] \mid \deg f \leq n, \ L_{K[X]}(f) \leq r \}.$$

If we write $F_r =: \sum_{\nu \geq 0} F_{r\nu} X^\nu$ with $F_{r\nu} \in \mathbb{Z}[a, b, c]$, then we obtain

$$C'_{r,n} = \Big\{ \sum_{\nu=0}^n F_{r\nu}(a', b', c') X^\nu \ \Big| \ (a', b', c') \in K^{(r+1)^2 + 1} \Big\}.$$

(Note that $(r + 1)^2 + 1$ equals the number of indeterminates $a_{\rho j}, b_{\rho j}, c_j$.) Let us draw a first conclusion from this. Let $n > 1$ and take $r = \lceil \sqrt{n} - 2 \rceil$. Then $(r + 1)^2 + 1 < n + 1$, hence the polynomials $F_{r0}, \ldots, F_{rn}$ must be algebraically dependent over the prime field k of K. Thus there exists some nonzero polynomial

$H \in k[Y_0, \ldots, Y_n]$ such that $H(F_{r0}, \ldots, F_{rn}) = 0$, i.e., H vanishes on $C'_{r,n}$. Thus any polynomial $f = \sum_{j=0}^n \alpha_j X^j$ with $H(\alpha) \neq 0$ has complexity $L_{K[X]}(f) > r$. So we have proved the following lower bound, which is optimal up to a constant factor by Prop. (9.1).

(9.2) Proposition (Paterson and Stockmeyer). *Zariski almost all polynomials $f \in K[X]$ of degree n have multiplicative complexity $L_{K[X]}(f) \geq \sqrt{n} - 1$. More specifically, there exists some nonzero $H \in k[Y_0, \ldots, Y_n]$ over the prime field k of K such that $L_{K[X]}(f) \geq \sqrt{n} - 1$ for all $f = \sum_{j=0}^n \alpha_j X^j$ satisfying $H(\alpha) \neq 0$.*

How can we show for a specific polynomial f that it is not contained in $C'_{r,n}$? We first extract more information about the polynomials $F_{r\nu}$. Up to now, we only used that the $F_{r\nu}$ depend on at most $(r+1)^2 + 1$ variables. (In other words, we used that the dimension of the variety $C'_{r,n}$ is less than or equal to $(r+1)^2 + 1$.) We now take a closer look at the degrees of the $F_{r\nu}$. We set $G_\rho =: \sum_{\nu \geq 0} G_{\rho\nu} X^\nu$ with $G_{\rho\nu} \in \mathbb{Z}[a, b]$. Obviously $G_{\rho 0} = 0$ for all ρ. From the definition we get

$$\sum_{\pi \geq 1} G_{\rho\pi} X^\pi = \left(\sum_{\mu \geq 1} \left(\sum_{j=0}^{\rho-1} a_{\rho j} G_{j\mu} \right) X^\mu \right) \left(\sum_{\nu \geq 1} \left(\sum_{j=0}^{\rho-1} b_{\rho j} G_{j\nu} \right) X^\nu \right).$$

By induction on ρ we easily conclude that $\deg G_{\rho\nu} \leq \rho\nu$ (we set $\deg 0 := -\infty$). Therefore $\deg F_{r\nu} \leq 1 + r\nu$.

(9.3) Theorem. *Let $f = \sum_{\nu=0}^n \alpha_\nu X^\nu \in K[X]$ be of degree $n \geq 1$, and $d_0, \ldots, d_n$ be positive natural numbers not all equal to one such that the power products*

$$\alpha_0^{j_0} \alpha_1^{j_1} \cdots \alpha_n^{j_n}, \quad 0 \leq j_\nu < d_\nu \text{ for } 0 \leq \nu \leq n,$$

are linearly independent over the prime field k of K. Let

$$N := \prod_{\nu=0}^n d_\nu, \quad M := \sum_{\nu=0}^n (d_\nu - 1).$$

Then we have

$$L_{K[X]}(f) \geq \sqrt{\frac{\log N}{\log 4Mn^{3/2}}} - 2.$$

Proof. W.l.o.g. $r := L_{K[X]}(f) > 0$. Consider the k-space

$$\mathcal{H} := \{H \in k[Y_0, \ldots, Y_n] \mid \forall \nu : \deg_{Y_\nu} H < d_\nu\}$$

of dimension N. Our assumption says that there is no nonzero $H \in \mathcal{H}$ such that $H(\alpha_0, \ldots, \alpha_n) = 0$. Hence, as $f \in C'_{r,n}$, there is no nonzero $H \in \mathcal{H}$ such that

(A) $\qquad\qquad\qquad H(F_{r0}, \ldots, F_{rn}) = 0.$

Recall that the $F_{r\nu}$ are integer polynomials in $m := (r+1)^2 + 1$ indeterminates of degree at most $c := 1 + rn$. Hence for all $H \in \mathcal{H}$ the degree of $H(F_{r0}, \ldots, F_{rn})$

is at most cM. Thus condition (A) amounts to a homogeneous k-linear system of $\binom{m+cM}{m}$ equations for the N coefficients of H. As this system has only the trivial solution by our assumption, we must have $N \leq \binom{m+cM}{m}$. By using the general inequality $\binom{m+cM}{m} \leq (cM)^m$ (compare Rem. (9.27)), and taking into account that $r \leq 2\sqrt{n}$ by Prop. (9.2), we conclude that

$$N \leq (cM)^m \leq (2rnM)^{(r+2)^2} \leq (4n^{3/2}M)^{(r+2)^2},$$

which implies the asserted lower bound.   □

We proceed with some applications. First we note that the above theorem implies $L_{K[X]}(f) \geq \sqrt{n} - 2$ if the coefficients of f are algebraically independent over k (let $d_1 = \ldots = d_n \to \infty$). Of course this is essentially the content of Prop. (9.2).

(9.4) Corollary. *Let* $p_1, \ldots, p_n \in \mathbb{N}$ *be pairwise distinct primes. Then*

$$L_{\mathbb{C}[X]}\left(\sum_{v=1}^n \sqrt{p_v}X^v\right) \geq \sqrt{\frac{2n}{5\log n + 4}} - 2.$$

Proof. We apply Thm. (9.3) with $d_0 := 1, d_1 := 2, \ldots, d_n := 2$ and observe that $\sqrt{p_1}^{j_1} \cdots \sqrt{p_n}^{j_n}$, $0 \leq j_v \leq 1$, are linearly idependent over $\mathbb{Q}$ since the degree of the field extension $\mathbb{Q}(\sqrt{p_1}, \ldots, \sqrt{p_n})/\mathbb{Q}$ equals 2^n. (This follows, e.g., from Lemma (9.20) proved in Sect. 9.3.)   □

For the next application we need some tools from number theory. Let $\pi(n)$ denote the number of primes $\leq n$ for $n \in \mathbb{N}$. The prime number theorem (see, e.g., Apostol [10]) says that

(9.5) $$\pi(n) \sim \frac{n}{\ln n} \quad (n \to \infty).$$

(An earlier result of Chebyshev stating only $1/6 \leq \frac{1}{n}\pi(n)\ln n \leq 6$ for all $n > 1$ is proved in Ex. 9.19-9.21.) The Euler totient function φ assigns to a natural number n the order of the group of units of $\mathbb{Z}/n\mathbb{Z}$. We have (cf. Lang [316, Chap. II, §2])

$$\varphi(n) := |(\mathbb{Z}/n\mathbb{Z})^{\times}| = n \prod_{v=1}^r (1 - p_v^{-1}),$$

where $p_1, \ldots, p_r$ are the different primes dividing n. Moreover, we remark that for $n \in \mathbb{N}'$ (cf. Lang [316, Chap. VIII, §3])

$$[\mathbb{Q}(e^{2\pi i/n}) : \mathbb{Q}] = \varphi(n).$$

(9.6) Corollary. *We have* $L_{\mathbb{C}[X]}\left(\sum_{\nu=1}^{n} e^{2\pi i/\nu} X^{\nu}\right) \gtrsim \sqrt{\frac{2n}{7 \ln n}}.$

Proof. Set $\alpha_{\nu} := e^{2\pi i/\nu}$. We apply Thm. (9.3) with the numbers d_{ν} defined by $d_{\nu} := \nu - 1$ if ν is prime, and $d_{\nu} := 1$ otherwise. Let $p_1, \ldots, p_s$ be the different primes $\leq n$, $s = \pi(n)$, and set $\ell := \prod_{\sigma=1}^{s} p_{\sigma}$. We have

$$\mathbb{Q}(\alpha_{p_1}, \ldots, \alpha_{p_s}) = \mathbb{Q}(e^{2\pi i/\ell}), \quad [\mathbb{Q}(e^{2\pi i/\ell}) : \mathbb{Q}] = \varphi(\ell) = \prod_{\sigma=1}^{s} (p_{\sigma} - 1)$$

by the remarks preceding the corollary. As $a_{p_{\sigma}}^{p_{\sigma}-1} = 1$, we conclude that the power products $\alpha_{p_1}^{j_1} \cdots \alpha_{p_s}^{j_s}$, $0 \leq j_{\sigma} < p_{\sigma} - 1$, are linearly independent over $\mathbb{Q}$. We use the following estimates

$$N := \prod_{\sigma=1}^{s} (p_{\sigma} - 1) \geq s! \geq (s/e)^{s}, \quad M := \sum_{\sigma=1}^{s} (p_{\sigma} - 1) \leq n^2.$$

The prime number theorem (9.5) implies that $\ln N \geq s(\ln s - 1) \gtrsim n$. Putting this together, the assertion follows. $\qquad\square$

The lower bounds we have obtained so far hold for the multiplicative complexity $L_{K[X]}$ in the polynomial algebra, i.e., when divisions by nonconstant polynomials are excluded. The goal of the next section is to extend our method, so that it will work also for the multiplicative complexity in the rational function field $K(X)$.

9.2 Polynomials with Algebraic Coefficients

In the following $L_A(f) := L_{K \to A}^{*}(f|X)$ denotes the the multiplicative complexity of computing an element f in a K-algebra A from a distinguished element $X \in A$. We note that the finite dimensional K-space $K[X]/(X^{n+1})$ can be considered as an affine space.

(9.7) Definition. Let $n, r \in \mathbb{N}$ and $A := K[X]/(X^{n+1})$. We call the subset

$$C_{r,n} := \{f \in A \mid L_A(f) \leq r\}$$

a *complexity class* (in A). Its Zariski closure in A, denoted by $\overline{C}_{r,n}$, will be called a *closed complexity class* in A. $\qquad\bullet$

The subsequent lemma, closely related to the discussions in Sect. 7.1, shows that we may work in $K[X]/(X^{n+1})$ instead of in $K(X)$.

(9.8) Lemma. *Let $n \in \mathbb{N}$. The image of any polynomial $f \in K[X]$ of complexity $L_{K(X)}(f) \leq r$ under the canonical map $\pi: K[X] \to K[X]/(X^{n+1})$ lies in $\overline{C}_{r,n}$.*

Proof. For all but finitely many $\lambda \in K$ we have $L_{K(X)}(f) = L_{\mathcal{O}_\lambda}(f)$, where $\mathcal{O}_\lambda$ denotes the subring of $K(X)$ consisting of the rational functions defined at λ. Therefore $L_{\mathcal{O}_0}(f(X + \lambda)) \leq r$ for such λ. Applying the K-algebra morphism

$$\mathcal{O}_0 \to K[[X]] \to K[[X]]/(X^{n+1}) \simeq K[X]/(X^{n+1})$$

we conclude that $\pi(f(X + \lambda)) \in C_{r,n}$ for all but finitely many $\lambda \in K$. Hence $\pi(f) \in \bar{C}_{r,n}$ (cf. Ex. 9.1). $\qquad\square$

The *height* $\mathrm{ht}(f)$ of a polynomial $f \in \mathbb{Z}[Y_1, \ldots, Y_n]$ is defined as the maximum of the absolute values of its coefficients; the *weight* $\mathrm{wt}(f)$ of f is the sum of the absolute values of its coefficients. Note that the weight is subadditive and submultiplicative.

Our first goal is to prove the following theorem which describes the complexity classes $C_{r,n}$.

(9.9) Representation Theorem (Strassen, Schnorr). *For $r, v \in \mathbb{N}$ there are polynomials $F_{rv} \in \mathbb{Z}[Y_1, \ldots, Y_r, Z_1, \ldots, Z_{(r+1)^2+1}]$ such that for all $n \in \mathbb{N}$*

$$C_{r,n} = \left\{ \sum_{v=0}^{n} F_{rv}(\eta, \zeta) X^v \bmod X^{n+1} \;\middle|\; \eta \in \{0, 1\}^r, \zeta \in K^{(r+1)^2+1} \right\},$$

and such that for $r \geq 1$, $v \geq 0$, $n > 1$,

$$\deg F_{rv} \leq (2r - 1)v + 2, \quad \sum_{v=1}^{n} \mathrm{wt}(F_{rv}) \leq 2^{n^{2r}}.$$

We note that the bound on the weight given in this theorem will be used only in Sect. 9.4, so its proof may be skipped at a first reading.

The following elementary lemma will be useful for the proof. (Recall that $\deg 0 := -\infty$.)

(9.10) Lemma. *Let $X, Z_1, Z_2, \ldots$ be indeterminates over K, $R := K[Z_i \mid i \geq 1]$. For positive integers s, a define the K-subspaces*

$$B_{s,a} := \left\{ \sum_{v \geq 0} f_v X^v \in R[[X]] \;\middle|\; \forall v \geq 0 \colon \deg f_v \leq vs + a \right\}.$$

Then we have:

(1) $B_{s,a} \cdot B_{t,b} \subseteq B_{\max(s,t),a+b}$.
(2) $\iota(B_{s,a}) \subseteq B_{s+a,0}$, where for $f \in B_{s,a}$ we define $\iota(f) := 1/(1 + (f - f(0)))$.

Proof. (1) Let $f = \sum_{v \geq 0} f_v X^v \in B_{s,a}$, $g = \sum_{v \geq 0} g_v X^v \in B_{t,b}$. Then we have for $f \cdot g = \sum_{v \geq 0} r_v X^v$ the identity $r_v = \sum_{\mu=0}^{v} f_\mu g_{v-\mu}$ for all $v \geq 0$. Hence, $\deg r_v \leq \max_{\mu=0}^{v}(\deg f_\mu + \deg g_{v-\mu}) = v \cdot \max\{s, t\} + a + b$.

(2) Let $f = \sum_{v \geq 0} f_v X^v \in B_{s,a}$. We may suppose that $f(0) = f_0 = 0$. Let $\iota(f) = \sum_{v \geq 0} g_v X^v$. Then $g_0 = 1$ and for all $v \geq 1$ we have $g_v = -\sum_{\mu=0}^{v-1} g_\mu f_{v-\mu}$. We use induction on v. Note that $\deg g_0 = 0 \cdot (s + a)$. For $v \geq 1$ we have by the induction hypothesis

$$\deg g_v \leq \max_{\mu=0}^{v-1}(\deg g_\mu + \deg f_{v-\mu}) \leq \max_{\mu=0}^{v-1}(v \cdot s + (\mu - 1) \cdot a) \leq v \cdot (s + a). \qquad\square$$

Before turning to the next auxiliary lemma we note the following: if $f \in K[[X]]$ is computed by a computation sequence modulo X, then we may w.l.o.g. assume that this computation sequence has the form $(1 + g_1, \ldots, 1 + g_r)$ modulo X, where $g_\rho(0) = 0$ for all ρ. (Let $(g_1, \ldots, g_r)$ be a computation sequence modulo X computing f. If $g_\rho = u_\rho v_\rho$, then put $u_\rho' := u_\rho - u_\rho(0) + 1$, $v_\rho' := v_\rho - v_\rho(0) + 1$, $1 + g_\rho' := u_\rho' v_\rho'$. If $g_\rho = u_\rho v_\rho^{-1}$, then $v_\rho(0) \neq 0$. In this case put $v_\rho' := v_\rho^{-1}(0) v_\rho$, $u_\rho' := v_\rho^{-1}(0)(u_\rho + [1 - u_\rho(0) v_\rho^{-1}(0)] v_\rho)$, $1 + g_\rho' := u_\rho'(v_\rho')^{-1}$. By induction on ρ show that $\langle 1, X, g_1', \ldots, g_\rho' \rangle = \langle 1, X, g_1, \ldots, g_\rho \rangle$.)

(9.11) Proposition. *Let $X, Y_1, Y_2, \ldots, Z_1, Z_2, \ldots$ be indeterminates and let $R := K[Y_j, Z_j \mid j \geq 1]$. There is a sequence $(G_r)_{r \in \mathbb{N}}$ of formal power series in $R[[X]]$ without constant term having the following properties:*

(1) *For sequences $\eta = (\eta_r)_{r \geq 1}$ in $\{0, 1\}$, $\zeta = (\zeta_r)_{r \geq 1}$ in K, and $r \in \mathbb{N}$,*

$$(1 + G_1(\eta, \zeta, X), \ldots, 1 + G_r(\eta, \zeta, X))$$

is a computation sequence in $K[[X]]$ modulo X. Moreover, any computation sequence $(1 + g_1, \ldots, 1 + g_r)$ in $K[[X]]$ modulo X with $g_\rho(0) = 0$ for all ρ is of this form.

(2) *$G_0 = X$, and $G_r \in \mathbb{Z}[Y_1, \ldots, Y_r, Z_1, \ldots, Z_{r^2 + r}][[X]]$ for all $r \in \mathbb{N}$.*

(3) *Let $G_r = \sum_{\nu=1}^{\infty} G_{r\nu} X^\nu$ with $G_{r\nu} \in R$. Then $\deg G_{r\nu} \leq (2r - 1)\nu + 1$ for all $r, \nu \geq 1$.*

(4) *$\sum_{\rho=0}^{r} \sum_{\nu=1}^{n} \mathrm{wt}(G_{\rho\nu}) \leq 2^{n^{2r}}$ for all $r \geq 1$, $n > 1$.*

Proof. We define the G_r recursively with respect to r. Set $G_0 := X$. Assume that $r \geq 1$ and let $a_0, \ldots, a_{r-1}, b_0, \ldots, b_{r-1}, c$ be further indeterminates. We define G_r by the following equation

$$
1 + G_r = \left(1 + \sum_{\rho=0}^{r-1} a_\rho G_\rho\right)\left(c\left(1 + \sum_{\rho=0}^{r-1} b_\rho G_\rho\right)\right.
$$
(A)
$$
\left. + (1 - c)\left(1 + \sum_{\rho=0}^{r-1} b_\rho G_\rho\right)^{-1}\right).
$$

holding in the ring

$$\mathbb{Z}[Y_1, \ldots, Y_{r-1}, Z_1, \ldots, Z_{r^2-r}, a_0, \ldots, a_{r-1}, b_0, \ldots, b_{r-1}, c][[X]]$$

of formal power series. (Later we will substitute for the variable c the values zero or one in order to model a division or multiplication.) At the end of the proof we may rename the new indeterminates as follows: $Y_r := c$, $Z_{r^2-r+1+\rho} := a_\rho$, $Z_{r^2+1+\rho} := b_\rho$ for $0 \leq \rho < r$. Statement (2) is now clear.

We show statement (1) by induction on r. The start "$r = 0$" is trivial, so let us assume that $r > 0$. Let $(1 + g_1, \ldots, 1 + g_r)$ be a computation sequence in $K[[X]]$ modulo X such that $g_\rho(0) = 0$ for all $\rho \in \underline{r}$ and put $g_0 := X$. Let us consider first the case where $1 + g_r = u_r v_r^{-1}$. We may assume w.l.o.g. that $u_\rho(0) = v_\rho(0) = 1$ for all $\rho \in \underline{r}$ and

$$u_r = 1 + \sum_{\rho=0}^{r-1} \alpha_\rho g_\rho, \quad v_r = 1 + \sum_{\rho=0}^{r-1} \beta_\rho g_\rho$$

for $\alpha_\rho, \beta_\rho \in K$. By the induction hypothesis there are $\eta \in \{0, 1\}^{r-1}$, $\zeta \in K^{r^2-r}$ such that $g_\rho = G_\rho(\eta, \zeta, X)$ for all $0 \leq \rho < r$. Note that by statement (2) the coefficients of the power series G_ρ, $\rho < r$, depend only on $Y_1, \ldots, Y_{r-1}, Z_1, \ldots, Z_{r^2-r}$. From this and (A) we conclude that

$$
\begin{aligned}
1 + g_r &= \left(1 + \sum_{\rho=0}^{r-1} \alpha_\rho G_\rho(\eta, \zeta, X)\right)\left(1 + \sum_{\rho=0}^{r-1} \beta_\rho G_\rho(\eta, \zeta, X)\right)^{-1} \\
&= 1 + G_r(\eta, 0, \zeta, \alpha_0, \ldots, \alpha_{r-1}, \beta_0, \ldots, \beta_{r-1}, X).
\end{aligned}
$$

If we define

$$\eta' := (\eta_1, \ldots, \eta_{r-1}, 0), \quad \zeta' := (\zeta_1, \ldots, \zeta_{r^2-r}, \alpha_0, \ldots, \alpha_{r-1}, \beta_0, \ldots, \beta_{r-1}),$$

then in fact $g_r = G_r(\eta', \zeta', X)$. The case where $1 + g_r = u_r v_r$ is similar. (There we have to substitute $c \mapsto \eta_r := 1$.) The reverse direction of statement (1) is clear. The reader should note that $\eta \in \{0, 1\}^r$ describes the "$*$, /-structure" of the computation sequence $(g_1, \ldots, g_r)$: $\eta_\rho = 1$ iff we have a division in the ρth step.

We prove the degree estimate (3) by induction on r. The start "$r = 1$" can be proved as the induction step, so let us assume that $r > 1$. We are going to apply Lemma (9.10) and use the notation introduced there. The induction hypothesis says that $G_\rho \in B_{2r-3,1}$ for $\rho < r$, in particular $1 + \sum_{\rho=0}^{r-1} a_\rho G_\rho \in B_{2r-1,0}$. The definition of G_r in (A) implies

$$1 + G_r \in B_{2r-1,0} \cdot \left(B_{0,1} \cdot B_{2r-1,0} + B_{0,1} \cdot B_{2r-1,0}\right) \subseteq B_{2r-1,1},$$

which was to be shown.

We turn to the demonstration of (4). Assume that $n > 1$. We set

$$G_r^n := \sum_{v=1}^{n} G_{rv} X^v, \quad S_r^n := 1 + \sum_{\rho=0}^{r} \text{wt}(G_\rho^n) = 1 + \sum_{\rho=0}^{r}\sum_{v=1}^{n} \text{wt}(G_{\rho v}).$$

Note that $S_0^n = 2$ and $\text{wt}(1 + \sum_{\rho=0}^{r-1} a_\rho G_\rho^n) = S_{r-1}^n$. Let $\mathcal{R}$ be the ring $\mathbb{Z}[Y, c, Z, a, b]$. We apply the canonical ring morphism $\mathcal{R}[[X]] \to \mathcal{R}[[X]]/(X^{n+1})$ to (A), interprete the resulting equation back in $\mathcal{R}[[X]]$, and obtain for $r \geq 1$

$$
\begin{aligned}
1 + G_r^n + X^{n+1} P = \ &\left(1 + \sum_{\rho=0}^{r-1} a_\rho G_\rho^n\right)\left(c\left(1 + \sum_{\rho=0}^{r-1} b_\rho G_\rho^n\right)\right. \\
&+ (1 - c)\sum_{\sigma=0}^{n-1}(-1)^\sigma\left(\sum_{\rho=0}^{r-1} b_\rho G_\rho^n\right)^\sigma \bigg) \\
&+ (1 - c)(-1)^n\left(\sum_{\rho=0}^{r-1} b_\rho G_\rho^n\right)^n
\end{aligned}
$$

for some $P \in \mathcal{R}[[X]]$. Using the subadditivity and the submultiplicativity of the weight we conclude from this for $r \geq 1$

$$1 + \text{wt}(G_r^n) \leq S_{r-1}^n \left(S_{r-1}^n + 2 \sum_{\sigma=0}^{n-1} (S_{r-1}^n)^\sigma \right) + 2(S_{r-1}^n)^n < 7(S_{r-1}^n)^n$$

(for the second inequality we have used $n \geq 2$ and $S_{r-1}^n \geq S_0^n = 2$). This implies for $r \geq 1$

$$\sum_{\rho=0}^r \text{wt}(G_\rho^n) \leq r + \sum_{\rho=1}^r \text{wt}(G_\rho^n) < 7r(S_{r-1}^n)^n,$$

hence $S_r^n \leq 7r(S_{r-1}^n)^n$. From this recursion and $S_0^n = 2$ we obtain for $r \geq 1$

(B) $\qquad S_r^n \leq r(r-1)^{n^1}(r-2)^{n^2} \cdots 2^{n^{r-2}} 7^{n^{r-1}+\ldots+n+1} 2^{n^r},$

therefore

$$S_r^n \leq (14r)^{n^r}.$$

The rough estimate $14r \leq 2^{n^r}$ holds for all $r \geq 1, n > 1$ except for the values $(r,n) \in \{(1,2),(1,3),(2,2)\}$. So for proving statement (3) it remains to consider these exceptional values. A direct calculation shows:

(C) $\qquad \begin{aligned} G_{11} &= a_0 - b_0 + 2b_0 c, \\ G_{12} &= -a_0 b_0 + 2a_0 b_0 c + b_0^2 - b_0^2 c, \\ G_{1\nu} &= (-1)^{\nu-1}(a_0 b_0^{\nu-1} - a_0 b_0^{\nu-1} c - b_0^\nu + b_0^\nu c) \text{ for } \nu > 2. \end{aligned}$

From this we see that $S_1^n = 4n + 3 \leq 2^{n^2}$ if $n > 1$, which in particular settles the cases $(r,n) \in \{(1,2),(1,3)\}$. For the value $(r,n) = (2,2)$ we note that $S_2^2 \leq 2 \cdot 7^3 \cdot 2^4 \leq 2^{16}$ by (B). $\qquad\qquad\square$

Proof of the Representation Theorem (9.9). Let $r \in \mathbb{N}$ and $W_0, \ldots, W_{r+1}$ be further indeterminates. With the help of the power series G_ρ introduced in Prop. (9.11) we define

$$\sum_{\nu=0}^\infty F_{r\nu} X^\nu := W_0 + \sum_{\rho=0}^r W_{\rho+1} G_\rho,$$

where the $F_{r\nu}$ are polynomials in $\mathbb{Z}[Y_1, \ldots, Y_r, Z_1, \ldots, Z_{r^2+r}, W_0, \ldots, W_{r+1}]$. Now rename the variables: $Z_{r^2+r+1+\rho} := W_\rho$ for $0 \leq \rho \leq r+1$. Since for a computation sequence $(g_1, \ldots, g_r)$ in $K[[X]]$ (or $K[X]/(X^{n+1})$) we may w.l.o.g. assume that $g_\rho(0) = 1$ for all ρ, the description

$$C_{r,n} = \left\{ \sum_{\nu=0}^n F_{r\nu}(\eta, \zeta) X^\nu \bmod X^{n+1} \;\middle|\; \eta \in \{0,1\}^r, \zeta \in K^{(r+1)^2+1} \right\}$$

follows easily from part one of Prop. (9.11). The bounds on the degree and the sum of weights of $F_{r\nu}$ are a consequence of part three and four of that lemma. $\square$

Based on the representation theorem (9.9) one can show that Thm. (9.3) also holds for the multiplicative complexity $L_{K(X)}$ in the rational function field. Thus the lower bounds exhibited in Cor. (9.4) and (9.6) are also true for $L_{K(X)}$ (compare Ex. 9.2). However, we follow here a less elementary but more powerful approach based on the notion of the degree of a variety as introduced in Sect. 8.2. For this we need some preparations.

For a subfield $k \subseteq K$ we denote by $\mathrm{Aut}(K/k)$ the group of field automorphisms of K which fix the elements of k. (So this is the Galois group if $k \subseteq K$ is a Galois extension.) We remark that the group $\mathrm{Aut}(K/k)$ acts on K^m componentwise, and it acts on the polynomial ring $K[Y_1, \ldots, Y_m]$ via $\sigma(\sum_j \alpha_j Y^j) := \sum_j \sigma(\alpha_j) Y^j$. A Zariski closed subset $V \subseteq K^m$ which is stable under $\mathrm{Aut}(K/k)$ is said to be defined over k (if $\mathrm{char}\, k = 0$).

(9.12) Lemma. *Let $k \subseteq K$ be a subfield and $V \subseteq K^m$ be a Zariski closed subset.*

(1) *V is stable under $\mathrm{Aut}(K/k)$ if and only if its vanishing ideal $I(V)$ is stable under $\mathrm{Aut}(K/k)$.*

(2) *Let $\alpha_1, \ldots, \alpha_m \in K$ be algebraic over k. Then the cardinality of the orbit of $\alpha := (\alpha_1, \ldots, \alpha_m) \in K^m$ under the action of $\mathrm{Aut}(K/k)$ equals the separable degree of the field extension $k(\alpha)/k$, i.e.,*

$$|\mathrm{Aut}(K/k)\alpha| = [k(\alpha) : k]_s.$$

(3) *The Zariski closure of the image of a morphism $\psi \colon K^m \to K^n$ given by polynomials in $k[Y_1, \ldots, Y_m]$ is stable under $\mathrm{Aut}(K/k)$.*

Proof. (1) Note that $(\sigma f)(\sigma \eta) = \sigma(f(\eta))$ for $\sigma \in \mathrm{Aut}(K/k)$, $f \in K[Y]$, $\eta \in K^m$. Hence $I(\sigma(V)) = \sigma(I(V))$.

(2) Let E/k be the normal closure of $k(\alpha)/k$. Then the orbits of α under $\mathrm{Aut}(K/k)$ and $\mathrm{Aut}(E/k)$ coincide. Moreover, as $\mathrm{Aut}(E/k(\alpha))$ is the stabilizer of α in $\mathrm{Aut}(E/k)$, we have

$$|\mathrm{Aut}(E/k)\alpha| = \big[\mathrm{Aut}(E/k) : \mathrm{Aut}(E/k(\alpha))\big].$$

We can express the number of automorphisms of a finite normal field extension by the separable degree (cf. Lang [316, Chap. VII, §4]):

$$|\mathrm{Aut}(E/k)| = [E : k]_s, \quad |\mathrm{Aut}(E/k(\alpha))| = [E : k(\alpha)]_s.$$

Therefore, by the multiplicativity of the separable degree,

$$|\mathrm{Aut}(K/k)\alpha| = |\mathrm{Aut}(E/k)\alpha| = [k(\alpha) : k]_s.$$

(3) The image of ψ is clearly stable under $\mathrm{Aut}(K/k)$. Moreover, the maps $K^m \to K^m, \eta \mapsto \sigma\eta$ are continuous for all $\sigma \in \mathrm{Aut}(K/k)$. $\square$

The representation theorem (9.9) combined with Lemma (9.12) and Thm. (8.48) lead to the following description of the closure $\overline{C}_{r,n}$ of the complexity class $C_{r,n}$.

(9.13) Theorem. *Let $r, n \geq 1$. The afffine variety $\overline{C}_{r,n}$ has at most 2^r irreducible components. Any of its components W satisfies*

$$\dim W \leq (r+1)^2 + 1, \quad \deg W \leq (2rn)^{(r+1)^2+1},$$

and is stable under $\mathrm{Aut}(K/k)$, k denoting the prime field of K.

Proof. W.l.o.g. assume that $n > 1$ and put $m := (r+1)^2 + 1$. For $\eta \in \{0, 1\}^r$ let W_η denote the closure of the image of the morphism

$$K^m \to K[X]/(X^{n+1}), \zeta \mapsto \sum_{\nu=0}^{n} F_{r\nu}(\eta, \zeta) X^\nu \bmod X^{n+1}$$

defined by the polynomials $F_{r\nu}$, see the representation theorem (9.9). Clearly $\dim W_\eta \leq m$. Thm. (8.48) implies that $\deg W_\eta \leq (2rn)^{\dim W_\eta} \leq (2rn)^m$, as $\deg F_{r\nu} \leq (2r-1)\nu + 2 \leq 2rn$ for $0 \leq \nu \leq n$. Lemma (9.12)(3) shows that $\dim W_\eta$ is stable under $\mathrm{Aut}(K/k)$. As $C_{r,n}$ is the union of the images of the above morphisms, taken over all $\eta \in \{0, 1\}^r$, we conclude that $\overline{C}_{r,n} = \cup_{\eta \in \{0,1\}^r} W_\eta$. Hence every irreducible component of $\overline{C}_{r,n}$ is one of the W_η, which finishes the proof. $\square$

To exploit this description of $\overline{C}_{r,n}$ we use the following special case of Thm. (8.50) (which is based on Bézout's inequality).

(9.14) Lemma. *Let $W \subseteq K^n$ be a nonempty Zariski closed subset. If $g_1, \ldots, g_r$ are non-constant polynomials of degree at most d in $K[Y_1, \ldots, Y_n]$ such that the intersection $W \cap Z(g_1, \ldots, g_t)$ is finite, then*

$$|W \cap Z(g_1, \ldots, g_t)| \leq d^{\dim W} \deg W.$$

The following theorem is a refinement of Thm. (9.3). It is one of the main results of this chapter and gives a general lower bound on the multiplicative complexity of a polynomial $f \in K[X]$ with coefficients that are algebraic over a subfield k. Numerous applications of this theorem will be given in the next section. The reader should note that the quantities M, N have here a different meaning than in Thm. (9.3).

(9.15) Theorem (Heintz and Sieveking). *Let $f = \sum_{\nu=0}^{n} \alpha_\nu X^\nu \in K[X]$ be a polynomial of degree $n \geq 1$ whose coefficients are algebraic over a subfield k of K. Let N be the separable degree of $k(\alpha_0, \ldots, \alpha_n)/k$. We denote by M the minimal natural number with the property that there exist polynomials $g_1, \ldots, g_t \in k[T_0, \ldots, T_n]$ of degree at most M which have a finite zeroset $Z(g_1, \ldots, g_t) \subseteq K^{n+1}$ containing $(\alpha_0, \ldots, \alpha_n)$. Then the multiplicative complexity of f satisfies*

$$L_{K(X)}(f) \geq \sqrt{\frac{\log N}{\log 4Mn^{3/2}}} - 2.$$

Proof. We set $r := L_{K(X)}(f)$, $m := (r+1)^2 + 1$ and assume w.l.o.g. that $r > 0$. By Lemma (9.8) $f \bmod X^{n+1}$ is contained in $\overline{C}_{r,n}$, say in its irreducible component W. Thm. (9.13) implies that $\dim W \leq m$, $\deg W \leq (2rn)^m$. We identify points $\sum_{\nu=0}^{n} \beta_\nu X^\nu \bmod X^{n+1}$ in $K[X]/(X^{n+1})$ with their coefficient vectors $\beta := (\beta_0, \ldots, \beta_n)$. Let $g_1, \ldots, g_t \in k[T_0, \ldots, T_n]$ be of degree $\leq M$ such that their zeroset $Z(g_1, \ldots, g_t) \subseteq K^{n+1}$ is finite and contains $\alpha := (\alpha_0, \ldots, \alpha_n)$. The set $W \cap Z(g_1, \ldots, g_t)$ is stable under the action of $\mathrm{Aut}(K/k)$, hence contains the orbit α. By Lemma (9.12)(2) the cardinality of this orbit equals the separable degree N of $k(\alpha_0, \ldots, \alpha_n)/k$. Lemma (9.14) thus implies $N \leq M^m (2rn)^m$. Using the estimates $r \leq 2\sqrt{n}$ and $m \leq (r+2)^2$ we obtain the assertion. $\square$

9.3 Applications

In this section we give various applications of Thm. (9.15) starting with a simple example. Let $g \in \mathbb{Q}[X]$ be a polynomial of degree n having as Galois group the full symmetric group S_n and denote by $\theta_0, \ldots, \theta_{n-1} \in \mathbb{C}$ its roots. By applying Thm. (9.15) to the polynomial $f = X^n + \sum_{j=0}^{n-1} \theta_j X^j$ we obtain with $N = n!$ and $M \geq n$ the lower bound

$$L_{\mathbb{C}(X)}(f) \geq \sqrt{\frac{\log n!}{\log 4n^{5/2}}} - 2 \gtrsim \sqrt{\frac{2n}{5}}.$$

For more interesting examples we will need the tools from number theory mentioned at the end of Sect. 9.1.

(9.16) Corollary. *For $n \to \infty$ the following asymptotic lower bounds hold:*

(1) $L_{\mathbb{C}(X)}\left(\sum_{j=1}^{n} e^{2\pi i/j} X^j\right) \gtrsim \sqrt{\frac{2n}{5 \ln n}}$,

(2) $L_{\mathbb{C}(X)}\left(\sum_{j=1}^{n} e^{2\pi i/2^j} X^j\right) \gtrsim \sqrt{\frac{2n}{3 \log n}}$,

(3) $L_{\mathbb{C}(X)}\left(\sum_{j=1}^{n} e^{2\pi i/p_j} X^j\right) \gtrsim \sqrt{n/5}$, *where p_j denotes the jth prime number.*

Proof. We apply Thm. (9.15) to the polynomial $f := \sum_{j=1}^{n} e^{2\pi i/d_j} X^j$ where $d_1, \ldots, d_n$ are positive natural numbers. Let us first note that

$$\mathbb{Q}(e^{2\pi i/d_1}, \ldots, e^{2\pi i/d_n}) = \mathbb{Q}(e^{2\pi i/\ell}),$$

where ℓ is the least common multiple of $d_1, \ldots, d_n$ and $N := [\mathbb{Q}(e^{2\pi i/\ell}) : \mathbb{Q}] = \varphi(\ell)$. From Thm. (9.15) we conclude therefore

(A) $$L_{\mathbb{C}(X)}(f) \geq \sqrt{\frac{\log \varphi(\ell)}{\log 4M n^{3/2}}} - 2,$$

where M is the minimal natural number such that there exist polynomials $g_1, \ldots, g_t \in \mathbb{Q}[T_1, \ldots, T_n]$ with $\deg g_j \leq M$ and having a finite zeroset containing the coefficient vector of f.

(1) Let $f_n := \sum_{j=1}^{n} e^{2\pi i/j} X^j$ and ℓ_n be the least common multiple of $1, 2, \ldots, n$. We need to estimate $\varphi(\ell_n)$. If $p_1, \ldots, p_{\pi(n)}$ denote the different primes smaller than or equal to n we have $\ell_n = p_1^{e_1} \cdots p_{\pi(n)}^{e_{\pi(n)}}$ for suitable $e_i > 0$, hence

$$n^{\pi(n)} \geq \ell_n \geq \varphi(\ell_n) = \prod_{j=1}^{\pi(n)} p_j^{e_j-1}(p_j - 1) \geq \prod_{j=1}^{\pi(n)} (p_j - 1) \geq \pi(n)! \geq (\pi(n)/e)^{\pi(n)}.$$

By the prime number theorem (9.5) we conclude that $\ln \varphi(\ell_n) \sim n$ for $n \to \infty$. Noting that $M \leq n$, we get the desired statement from (A).

(2) Let $f_n := \sum_{j=1}^{n} e^{2\pi i/2^j} X^j$. Here obviously $N = \varphi(2^n) = 2^{n-1}$. Considering the polynomials

$$g_1 = T_1^2 - 1, g_2 = T_2^2 - T_1, \ldots, g_n := T_n^2 - T_{n-1}$$

we see that $M \leq 2$. Hence the assertion follows from (A).

(3) From the fact $\pi(p_n) = n$ and (9.5) it follows easily that $p_n \sim n \ln n$ for $n \to \infty$. Here $M \leq p_n$ and $N = (p_1 - 1) \cdots (p_n - 1)$. For $n \to \infty$ we therefore have the asymptotic estimate

$$\ln N \geq \sum_{p_n/2 < p_j} \ln(p_j - 1) \geq (n - \pi(p_n/2)) \ln(p_n/2 - 1) \sim \frac{n}{2} \ln n$$

which implies the assertion by (A). □

We give now an application of Thm. (9.15) of a somewhat different type.

(9.17) Corollary. *Let k be a subfield of the algebraically closed field K. Let $f \in k[X]$, $n := \deg f \geq 8$, and assume that the Galois group of f is isomorphic to S_n. Then for all monic factors $h \in K[X]$ of f with $\deg h \geq 1$ we have*

$$L_{K(X)}(h) \geq \sqrt{\frac{2}{5} \deg h \left(1 - \frac{\log \deg h}{\log n}\right)} - 2.$$

In particular, if $\deg h = n/2$, then $L_{K(X)}(h) \geq \sqrt{\frac{n}{5 \log n}} - 2$.

Proof. Write $f = \sum_{j=0}^{n} f_j X^j \in k[X]$ and let $h = X^m + \sum_{j=0}^{m-1} h_j X^j \in K[X]$ be a monic factor of f of degree m, $1 \leq m \leq n$. We have

$$f = f_n \prod_{\theta \in S} (X - \theta), \quad h = \prod_{\theta \in S'} (X - \theta),$$

where $S' \subseteq S \subseteq K$ with $|S'| = m$, $|S| = n$. The automorphisms of $k(S)$ which fix the elements of $k(h_0, \ldots, h_{m-1})$ are given by the permutations of the set S of roots which map S' into itself. Hence

$$[k(S) : k(h_0, \ldots, h_{m-1})] = |\mathrm{Aut}(k(S)/k(h_0, \ldots, h_{m-1}))| = m!(n - m)!,$$

which implies

$$N := [k(h_0, \ldots .h_{m-1}) : k] = \binom{n}{m}.$$

Let $r = \sum_{j=0}^{n-m} r_j X^j \in K[X]$ be such that $f = hr$. Then we have

(A) $$\forall\ m \le j \le n: \quad f_j = r_{j-m} + \sum h_\ell r_s,$$

where the sum is taken over all pairs (ℓ, s) satisfying $\ell < m$, $s \le n-m$, $\ell + s = j$. Replacing $h_0, \ldots, h_{m-1}$ by indeterminates $y_0, \ldots, y_{m-1}$ in (A) and solving the resulting system of linear equations we see that there exist polynomials

$$R_0, \ldots, R_{n-m} \in k[Y_0, \ldots, Y_{m-1}]$$

such that for all $0 \le j \le n - m$

$$r_j = R_j(h_0, \ldots, h_{m-1}), \quad \deg R_j \le n - m - j \le n - m.$$

We define now for $0 \le j \le n$

$$g_j := -f_j + R_{j-m} + \sum Y_\ell R_s \in k[Y_0, \ldots, Y_{m-1}],$$

where the sum is taken over the same pairs (ℓ, s) as in (A). It is easy to see that $(h'_0, \ldots, h'_{m-1}) \in K^m$ is a common zero of $g_0, \ldots, g_n$ if and only if $X^m + \sum_{j=0}^{m-1} h'_j X^j$ is a monic factor of f. Moreover, $\max_j \deg g_j \le n - m + 1$.

We can now apply Thm. (9.15) to h with $N = \binom{n}{m}$ and $M \le n - m + 1$. Therefore

$$L_{K(X)}(h) \ge \sqrt{\frac{\log \binom{n}{m}}{\log(4m^{3/2}(n - m + 1))}} - 2.$$

Using the estimates

$$\binom{n}{m} = \frac{n}{m} \cdot \frac{n-1}{m-1} \cdot \ldots \cdot \frac{n-m+1}{1} \ge \left(\frac{n}{m}\right)^m$$

and

$$4m^{3/2}(n + 1 - m) \le \frac{24\sqrt{3}}{25\sqrt{5}}(n + 1)^{5/2} \le n^{5/2} \text{ for } n \ge 8$$

we conclude that

$$L_{K(X)}(h) \ge \sqrt{\frac{2m}{5}\left(1 - \frac{\log m}{\log n}\right)} - 2,$$

which was to be shown.   □

Not all polynomials $f \in k[X]$ with Galois group S_n are hard to compute. We state the following proposition without proof. A field k is called *Hilbertian* iff any irreducible polynomial $f \in k[T_1, \ldots, T_r, X_1, \ldots, X_s]$ remains irreducible for Zariski almost all substitutions of $(T_1, \ldots, T_r)$ by an element of k^r. For instance, one can show that all number fields are Hilbertian. (See Lang [314, Chap. VIII].)

(9.18) Proposition (Heintz). *Let k be a Hilbertian field. Then for any $n \geq 1$ there is a polynomial $f_n \in k[X]$ of degree n having the Galois group S_n such that*

$$L_{k(X)}^{\text{tot}}(f_n | X) = O(\log n).$$

This and the previous proposition show that there exist polynomials over k of arbitrary degree n, which are themselves easy to compute, but which have the property that each factor of them, of degree not too close to n, is hard to compute. In Ex. 9.8 we will discuss a different approach for obtaining results of this kind.

We give another application showing that most polynomials with coefficients in $\{0, 1\}$ are hard to compute. However, this is purely a statement about the existence of such polynomials (see Problem 9.3).

(9.19) Corollary. *For $n \geq 2$ let $\rho_n := \sqrt{(3n \log n)/2}$. Then*

$$\left| \left\{ \alpha \in \{0, 1\}^n \,\middle|\, L_{K(X)}\left(\sum_{j=1}^{n} \alpha_j X^j \right) \leq \sqrt{2n/(3 \log n)} - 2 \right\} \right| \leq 2^{n - \rho_n}.$$

Proof. We set $r := \lfloor \sqrt{2n/(3 \log n)} \rfloor - 2$. We can assume w.l.o.g. that $r \geq 0$. If $r = 0$ then $6 \leq n/\log n$, hence $30 \leq n$ and the assertion is true since $\rho_n \leq n-1$. So let us assume that $r \geq 1$. By Lemma (9.8) it is sufficient to show that (identifying $K[X]/(X^{n+1})$ with K^{n+1})

(A) $|\overline{C}_{r,n} \cap \{0, 1\}^{n+1}| \leq 2^{n - \rho_n}.$

Thm. (9.13) tells us

$$\dim \overline{C}_{r,n} \leq (r + 1)^2 + 1, \quad \deg \overline{C}_{r,n} \leq 2^r (2rn)^{(r+1)^2+1} \leq (2rn)^{(r+1)(r+2)}.$$

(Note that $\overline{C}_{r,n}$ has at most 2^r irreducible components.) $\{0, 1\}^{n+1}$ is the intersection of $n + 1$ hypersurfaces in K^{n+1} defined by the equations $Y_j(Y_j - 1) = 0$ for $0 \leq j \leq n$ (Y_j denoting the coordinate functions). Therefore, by Lemma (9.14), we have

$$|\overline{C}_{r,n} \cap \{0, 1\}^{n+1}| \leq 2^{\dim \overline{C}_{r,n}} \deg \overline{C}_{r,n} \leq (4rn)^{(r+1)(r+2)}.$$

A routine calculation yields

$$(r + 1)(r + 2) \log(4rn)$$

$$\leq \left(\sqrt{\frac{2n}{3 \log n}} - 1 \right) \sqrt{\frac{2n}{3 \log n}} \log\left[4n\left(\sqrt{\frac{2n}{3 \log n}} - 2 \right) \right]$$

$$\leq \left(\frac{2n}{3 \log n} - \sqrt{\frac{2n}{3 \log n}} \right) \left[\frac{3}{2} \log n + \log\left(\sqrt{\frac{32}{3 \log n}} - \frac{8}{\sqrt{n}} \right) \right],$$

and since $\sqrt{\frac{32}{3 \log n}} - \frac{8}{\sqrt{n}} < 1$ for $n \geq 2$ it follows that

$$(r + 1)(r + 2) \log(4rn) \leq n - \sqrt{(3n \log n)/2} = n - \rho_n.$$

This proves (A). $\square$

For further applications we need an auxiliary result from number theory.

(9.20) Lemma. *Let $k := \mathbb{Q}(\zeta)$ where ζ is a primitive p^{th} root of unity, p prime, and let s be an integer not divisible by p. Furthermore let $p_1, \ldots, p_r$ be different primes. Then $k(p_1^{s/p}, \ldots, p_r^{s/p})$ is a Galois extension of k with Galois group isomorphic to $(\mathbb{Z}/p\mathbb{Z})^r$. In particular, $[k(p_1^{s/p}, \ldots, p_r^{s/p}) : k] = p^r$.*

Proof. Let D be the subgroup of $k^\times$ generated by $p_1^s, \ldots, p_r^s$ and $(k^\times)^p$. The field $k(D^{1/p}) = k(p_1^{s/p}, \ldots, p_r^{s/p})$ is a *Kummer extension* over k, and it is well-known that $k(D^{1/p})/k$ is Galois with Galois group isomorphic to $D/(k^\times)^p$ (cf. Lang [316, Chap. VIII, §8]). It is thus sufficient to show that the surjective group morphism

$$\mathbb{Z}^r \to D/(k^\times)^p, \ (e_1, \ldots, e_r) \mapsto p_1^{e_1} \cdots p_r^{e_r} \bmod (k^\times)^p$$

has the kernel $(p\mathbb{Z})^r$. This, however, follows by considering the prime divisor decomposition of the p_i in the ring $\mathcal{O}$ of integers of $k = \mathbb{Q}(\zeta)$: if p_i is different from p, then p_i decomposes into a product of pairwise different prime divisors $\wp_{ij}$, whereas p decomposes to the $(p-1)$th power of a prime divisor $\wp$ (cf. Hasse [234, Chap. 27]).   □

We turn to the investigation of the polynomials $f_r := \sum_{j=1}^{n} j^r X^j$, $r \in \mathbb{Q}$. We note first that

$$L_{\mathbb{C}(X)}^{\text{tot}}(f_r | X) = O(\log n) \quad \text{for } r \in \mathbb{N}.$$

To see this, observe that $f_r = X(d/dX)f_{r-1}$, which implies inductively $f_r = (X^{n+1}g + h)/(X - 1)^{r+1}$ with polynomials g, h of degree $\leq r + 1$. Hence the claim follows.

However, for $r \in \mathbb{Q} \setminus \mathbb{Z}$ the polynomials f_r are hard to compute, as the next proposition shows. It is an open problem (cf. Problem 9.2) whether the polynomials f_r are also hard for $r \in \mathbb{Z} \setminus \mathbb{N}$. For the Taylor approximations $f_{-1} = \sum_{j=1}^{n} j^{-1}X^j$ of the natural logarithm this would be particularly interesting to know.

(9.21) Corollary. *Let $r \in \mathbb{Q} \setminus \mathbb{Z}$, and p be a prime occurring in r with negative multiplicity. Then we have for $n \to \infty$*

$$L_{\mathbb{C}(X)}\left(\sum_{j=1}^{n} j^r X^j\right) \gtrsim \sqrt{\frac{2 \ln p}{3} \frac{n}{(\ln n)^2}}.$$

Proof. Let $r = s/t$ with $s \in \mathbb{Z}$, $t \in \mathbb{N}$ being relatively prime. By assumption $p \mid t$. Denote by $p_1, \ldots, p_{\pi(n)}$ the prime numbers $\leq n$. We apply Thm. (9.15) with $k := \mathbb{Q}(\zeta)$, where ζ is a primitive p root of unity. Obviously $M \leq t$. We have

$$
\begin{aligned}
N &= [k(1^r, 2^r, \ldots, n^r) : k] \\
&= [k(p_1^r, \ldots, p_{\pi(n)}^r) : k] \\
&\geq [k(p_1^{s/p}, \ldots, p_{\pi(n)}^{s/p}) : k] \\
&= p^{\pi(n)},
\end{aligned}
$$

where the last equality is a consequence of Lemma (9.20). The assertion follows now from Thm. (9.15) and the prime number theorem (9.5). □

For our final application of Thm. (9.15) we need an auxiliary result. A domain D is called a *Dedekind domain* iff every proper ideal of D is a product of prime ideals. In that case the factorization is even unique. (See Jacobson [264, Thm. 10.1]). Two important examples of Dedekind domains are provided by the ring $\mathbb{Z}$ of integers and the ring $k[X]$ of univariate polynomials over a field k. If D is a Dedekind domain with field of fractions K, F is a finite extension of K, and D' is the integral closure of D in F, then D' is also Dedekind (cf. [264, Thm. 10.7]). Hence, integral closures of the ring $\mathbb{Z}$ of integers in finite extensions of $\mathbb{Q}$, or integral closures of the univariate polynomial ring $k[X]$ in finite extensions of $k(X)$ are Dedekind domains as well. We shall need the following simple result.

(9.22) Proposition. *Let D be a Dedekind domain with field of fractions K, and $a \in K^\times$. If $a \notin D^\times$, then $\{n \mid \exists\, b \in K,\ a = b^n\}$ is finite.*

Proof. It suffices to show the assertion for $a \in D \setminus \{0\}$. Suppose that $a \notin D^\times$. Then we have $aD = \wp_1^{e_1} \cdots \wp_t^{e_t}$ for some $t \geq 1$, prime ideals $\wp_1, \ldots, \wp_t$ of D, and positive integers $e_1, \ldots, e_t$. If $a = b^n$, then the unique factorization of proper ideals into prime ideals implies that all $e_1, \ldots, e_t$ are divisible by n. This gives the assertion. □

Let us now study more concrete examples of Dedekind domains.

(9.23) Lemma. *Let a be a nonzero complex number which is not a root of unity and K be a finite extension of $\mathbb{Q}(a)$. Then for sufficiently large p we have $[K(\alpha):K] = p$ for any root α of $X^p - a$. In particular, if $p_1, \ldots, p_t$ are different primes and $\alpha_j \in \mathbb{C}$ is a jth root of a, then $[\mathbb{Q}(\alpha_{p_1}, \ldots, \alpha_{p_t}):\mathbb{Q}(a)] = \prod_{j=1}^t p_j$ for sufficiently large $p_1, \ldots, p_t$.*

Proof. By Lang [316, Chap. VIII, §9] the polynomial $X^p - a$ is irreducible over K if a is not a pth power in $K^\times$, i.e., $a \notin (K^\times)^p$. Thus we only need to show that the latter condition is satisfied for sufficiently large p. Suppose that on the contrary $a \in (K^\times)^p$ for infinitely many primes p.

Assume first that a is transcendental over $\mathbb{Q}$. The integral closure of $\mathbb{Q}[a]$ in K is a Dedekind domain D. Our assumption implies that $a \in D^\times$ by Prop. (9.22). Hence a^{-1} is integral over $\mathbb{Q}[a]$, say

$$a^{-m} + \lambda_{m-1} a^{-(m-1)} + \ldots + \lambda_1 a^{-1} + \lambda_0 = 0$$

with $\lambda_i \in \mathbb{Q}[a]$. By multiplying with a^m and applying the morphism $\mathbb{Q}[a] \to \mathbb{Q}, a \mapsto 0$ we get a contradiction.

We may thus suppose that a is algebraic over $\mathbb{Q}$. Let Γ be the integral closure of $\mathbb{Z}$ in K. Then Γ is a Dedekind domain and hence $a \in \Gamma^\times$ by Prop. (9.22). Dirichlet's unit theorem (cf. Hasse [234, Chap. 28]) implies that $\Gamma^\times$ is a *finitely generated* abelian subgroup of $K^\times$, hence is of the form $\Gamma^\times \simeq T \times \mathbb{Z}^r$ for some finite abelian subgroup T (which necessarily consists of roots of unity) and some $r \in \mathbb{N}$. But $a \in (K^{\times})^p$ for infinitely many p implies that $a \in (\Gamma^{\times})^p$ for infinitely many p, hence $a \in T$, a contradiction.

The second assertion is easily obtained by induction on t and the first part. $\square$

(9.24) Corollary. *Assume that $a \in \mathbb{C} \setminus \{0\}$ is not a root of unity and $\alpha_j \in \mathbb{C}$ is a jth root of a (i.e., $\alpha_j^j = a$) for $1 \leq j \leq n$. Then we have for $n \to \infty$*

$$L_{\mathbb{C}(X)}\Big(\sum_{j=1}^n \alpha_j X^j \Big) \gtrsim \sqrt{\frac{n}{5 \ln n}}.$$

Proof. The assertion follows from Thm. (9.15), once we have shown that

$$\ln [\mathbb{Q}(\alpha_1, \ldots, \alpha_n) : \mathbb{Q}(a)] \gtrsim n/2 \quad (n \to \infty).$$

But we have $[\mathbb{Q}(\alpha_1, \ldots, \alpha_n) : \mathbb{Q}(a)] \geq [\mathbb{Q}(\alpha_{p_1}, \ldots, \alpha_{p_t}) : \mathbb{Q}(a)] = \prod_{i=1}^t p_j$ for sufficiently large n by Lemma (9.23), where $p_1, \ldots, p_t$ are the primes between $n/2$ and n. Hence we have for $n \to \infty$

$$\ln [\mathbb{Q}(\alpha_1, \ldots, \alpha_n) : \mathbb{Q}(a)] \geq \sum_{n/2 \leq p \leq n} \ln p \geq (\pi(n) - \pi(n/2)) \ln(n/2) \gtrsim n/2,$$

and we are done. $\square$

We remark that the assumption that a is not a root of unity is necessary in the previous corollary (see Ex. 9.13).

9.4* Polynomials with Rapidly Growing Integer Coefficients

In this section K denotes an algebraically closed field of characteristic zero.

Our goal is to prove nontrivial lower bounds for the multiplicative complexity of specific polynomials with rapidly growing integer coefficients, such as $\sum_{j=0}^n 2^{2^{jn}} X^j$. We remark that in Ex. 9.12 the reader may find instructions for a different, less technical approach which, however, yields only considerably weaker statements. Let us give an outline of ideas. We recall that $\overline{C}_{r,n}$ denotes the closure of the set of $f \in A_n := K[X]/(X^{n+1})$ satisfying $L_{A_n}(f) \leq r$. Lemma (9.8) tells us that in order to prove that $L_{K(X)}(g) > r$ for a polynomial $g \in K[X]$, it suffices to show that $g \bmod X^{n+1} \notin \overline{C}_{r,n}$. Assume $(r+2)^2 < n$ and let $D \in \mathbb{N}$ be large enough to satisfy

(9.25) $$D^{n-(r+2)^2} \geq (2n^3)^{(r+2)^2} n^n.$$

Under this assumption we will show the existence of a form $H \in \mathbb{Z}[Y_0, \ldots, Y_n]$ which lies in the vanishing ideal of $\overline{C}_{r,n}$ and has degree D and height at most three. The proof will be based on the representation theorem (9.9) and on a lemma dealing with integral solutions of linear homogeneous equations. Consider now the sequence $f_n := \sum_{j=1}^{n} 2^{2^{jn}} X^j \in A_n$ and put $D_n := \lfloor 2^n/n \rfloor$, $r_n := L_{A_n}(f_n)$. We wish to show that $r_n \geq \sqrt{n}(1 + o(1))$ for $n \to \infty$. So we may suppose w.l.o.g. that $(r_n + 2)^2 < n$. If inequality (9.25) holds for $D = D_n$, then there are forms H_n of degree D_n and height ≤ 3 in the vanishing ideal of $\overline{C}_{r_n,n}$. Hence, as $f_n \in \overline{C}_{r_n,n}$, we have

$$H_n(2^{2^n}, 2^{2^{2n}}, \ldots, 2^{2^{n^2}}) = 0.$$

On the other hand, it is not difficult to see that such a relation is impossible. Therefore, we conclude that D_n does not satisfy inequality (9.25), which implies that indeed $r_n \geq \sqrt{n}(1 + o(1))$ as $n \to \infty$.

In fact, this proof bears some similarity with the classical proofs of the transcendence of e and π (cf. Lang [316, Appendix 1], where also the subsequent lemma due to Siegel appears.

(9.26) Lemma. Let $\ell_1, \ldots, \ell_M \in \mathbb{Z}[X_1, \ldots, X_N]$ be linear forms and let $N > M$. Then $\ell_1, \ldots, \ell_M$ have a nontrivial common zero $(\xi_1, \ldots, \xi_N) \in \mathbb{Z}^N$ satisfying

$$\max_{1 \leq i \leq N} |\xi_i| < \left(\max_{1 \leq j \leq M} \text{wt}\,(\ell_j) \right)^{M/(N-M)} + 2.$$

Proof. W.l.o.g. $G := \max_{1 \leq j \leq M} \text{wt}\,(\ell_j) > 0$. Let $s \in \mathbb{N}$ satisfy

$$G^{M/(N-M)} \leq 2s < G^{M/(N-M)} + 2.$$

The image of the cube $[-s, s]^N$ under the map

$$\ell : \mathbb{R}^N \to \mathbb{R}^M, \xi \mapsto (\ell_1(\xi), \ldots, \ell_M(\xi))$$

is contained in the cube $[-Gs, Gs]^M$. Subdivide now each edge of $[-Gs, Gs]^M$ into $t := \lfloor (2s)^{N/M} \rfloor + 1$ segments of equal length, thus subdividing the cube $[-Gs, Gs]^M$ into t^M subcubes. We have

$$|\mathbb{Z}^N \cap [-s, s]^N| = (2s + 1)^N > t^M,$$

the latter inequality holding, since by Rolle's theorem for some $\varepsilon \in (0, 1)$

$$(2s + 1)^{N/M} - (2s)^{N/M} = \frac{N}{M}(2s + \varepsilon)^{N/M-1} > 1.$$

Hence, by the pigeonhole principle there exist different ξ', ξ'' in $\mathbb{Z}^N \cap [-s, s]^N$ such that $\ell(\xi'), \ell(\xi'')$ are contained in the same subcube of $[-Gs, Gs]^M$. In particular, for all j,

$$|\ell_j(\xi') - \ell_j(\xi'')| \leq 2sG/t.$$

But $2sG/t < 1$ and $\ell_j(\xi'), \ell_j(\xi'') \in \mathbb{Z}$, so we have in fact $\ell(\xi') = \ell(\xi'')$. The vector $\xi := \xi' - \xi''$ meets our requirements. $\qquad\square$

(9.27) Remark. We have $\binom{m+n}{n} \leq \min\{m^n, n^m\}$ provided $(m-1)(n-1) > 2$. (Note that the left-hand side equals the number of monomials in m variables of total degree $\leq n$, whereas the right-hand side is the number of monomials in m variables such that the degree in the ith variable is strictly less than n.)   •

(9.28) Lemma. Let $q \geq 5$, $m \geq 2$ and $P_1, \ldots, P_q \in \mathbb{Z}[Z_1, \ldots, Z_m]$ be polynomials such that for all $1 \leq i \leq q$

$$\deg P_i \leq d, \quad \mathrm{wt}\,(P_i) \leq w,$$

where $d \geq 2$, $w \geq 4$ are natural numbers. If D is a natural number satisfying

$$D^{q-m-2} > d^m q^q \log w,$$

then there is a nontrivial form $H \in \mathbb{Z}[Y_1, \ldots, Y_q]$ of degree D and height ≤ 3 such that

$$H(P_1, \ldots, P_q) = 0.$$

Proof. For $i \in \mathbb{N}^q$ with $i_1 + \ldots + i_q = D$ let X_i be indeterminates over $\mathbb{Z}[Z]$ and define

$$Q := \sum_{i_1 + \ldots + i_q = D} X_i P_1^{i_1} \cdots P_q^{i_q} \in \mathbb{Z}[X, Z].$$

Evidently $\deg_Z Q \leq Dd$. Using the subadditivity and submultiplicativity of the weight we see that

$$\mathrm{wt}\,(Q) \leq w^D \binom{D+q-1}{q-1}.$$

Let us write

$$Q = \sum_{j_1 + \ldots + j_m \leq Dd} \ell_j(X) Z_1^{j_1} \cdots Z_m^{j_m}$$

where $\ell_j(X) \in \mathbb{Z}[X]$ are linear forms. By the weight bound on Q we have

$$\mathrm{wt}\,(\ell_j) \leq w^D \binom{D+q-1}{q-1}$$

for all $j \in \mathbb{N}^m$ such that $j_1 + \ldots + j_m \leq Dd$.

If we can find a nontrivial common zero ξ of the $\ell_j(X)$ with $\xi_i \in \mathbb{Z}$, $|\xi_i| \leq 3$, then $H := \sum_i \xi_i Y_1^{i_1} \cdots Y_q^{i_q}$ will work. The existence of such ξ however follows from Lemma (9.26), if the following two conditions are satisfied:

(A) $$N := \binom{D+q-1}{q-1} > \binom{Dd+m}{m} =: M,$$

(B) $$\binom{Dd+m}{m} \log\left(w^D \binom{D+q-1}{q-1}\right) < \binom{D+q-1}{q-1} - \binom{Dd+m}{m}.$$

(Note that (B) is equivalent to $\left(w^D \binom{D+q-1}{q-1}\right)^{M/(N-M)} < 2$.)

Note first that (A) is clearly a consequence of (B). Let us suppose now that (B) is false. Using the estimates

$$\binom{Dd+m}{m} \leq (Dd)^m, \quad \left(\tfrac{D}{q}\right)^{q-1} < \binom{D+q-1}{q-1} < D^q$$

(cf. Rem. (9.27)), we conclude that

$$(Dd)^m D(\log w + q) \geq (Dd)^m \log(w^D D^q) \geq (\frac{D}{q})^{q-1} - (Dd)^m.$$

Since $Dq \log w \geq D \log w + Dq + 1$ (use $D, w \geq 2, q \geq 3$), we obtain

$$d^m q^q \log w \geq D^{q-m-2},$$

which contradicts our assumption.

Therefore (A), (B) are true and the lemma is proved. $\quad\square$

(9.29) Theorem (Strassen). *Let $n, r \in \mathbb{N}, n, r \geq 2$ such that $(r+1)(r+2)+2 < n$. Choose $q \in \mathbb{N}$ with $(r+1)(r+2)+2 < q \leq n$ and let $D \in \mathbb{N}$ be large enough to satisfy*

$$D^{q-(r+1)(r+2)-2} > (2rn)^{(r+1)(r+2)} q^q n^{2r}.$$

Let T_i denote the coordinate function which assigns to f in A_n the coefficient of X^i mod X^{n+1}, where $A_n := K[X]/(X^{n+1})$. Then for any sequence $1 \leq v_1 < v_2 < \ldots < v_q \leq n$ of natural numbers there exists a nontrivial form $H \in \mathbb{Z}[Y_1, \ldots, Y_q]$ of degree D and height at most three such that $H(T_{v_1}, \ldots, T_{v_q})$ vanishes on $\overline{C}_{r,n}$. In particular, we have for all $f = \sum_{i=0}^n \alpha_i X^i \in K[X]$

$$L_{K(X)}(f) \leq r \implies L_{A_n}(f \bmod X^{n+1}) \leq r \implies H(\alpha_{v_1}, \ldots, \alpha_{v_q}) = 0.$$

Proof. We apply Lemma (9.28) to the polynomials

$$F_{rv_1}, \ldots, F_{rv_q} \in \mathbb{Z}[Y_1, \ldots, Y_r, Z_1, \ldots, Z_{(r+1)^2+1}],$$

from representation theorem (9.9). They satisfy the bounds

$$\deg F_{rv_i} \leq (2r-1)v_i + 2 \leq 2rn =: d, \quad \mathrm{wt}(F_{rv_i}) \leq \sum_{v=1}^n \mathrm{wt}(F_{rv}) \leq 2^{n^{2r}} =: w.$$

Hence there exists a form $H \in \mathbb{Z}[Y_1, \ldots, Y_q]$ of degree D and height ≤ 3 such that $H(F_{rv_1}, \ldots, F_{rv_q}) = 0$, provided D is large enough to satisfy the above condition. Therefore $H(T_{v_1}, \ldots, T_{v_q})$ is contained in the vanishing ideal of $\overline{C}_{r,n}$. $\quad\square$

(9.30) Lemma. *Let $\alpha_1, \ldots, \alpha_q \in \mathbb{Z}, D \in \mathbb{N}$ such that $q \geq 10, D \geq 4, |\alpha_1| > 4$ and for $1 < i \leq q$*

$$|\alpha_i| > |q\alpha_{i-1}|^D.$$

Then there is no nontrivial form $H \in \mathbb{Z}[Y_1, \ldots, Y_q]$ of degree D and height ≤ 3 such that

$$H(\alpha_1, \ldots, \alpha_q) = 0.$$

Proof. We show first that any complex root $\alpha \in \mathbb{C}$ of a nonzero univariate polynomial $Q = \sum_{j=0}^{N} \gamma_j Y^j$ with integer coefficients satisfies $|\alpha| \leq 1 + \text{ht}(Q)$. Assume w.l.o.g. $|\alpha| > 1$ and $\gamma_N \neq 0$. Then $\gamma_N \alpha^N = -\sum_{j=0}^{N-1} \gamma_j \alpha^j$ implies

$$|\alpha|^N \leq \text{ht}(Q) \sum_{j=0}^{N-1} |\alpha|^j \leq \text{ht}(Q)|\alpha|^N/(|\alpha|-1),$$

which in turn shows the assertion.

Assume now that there is a nontrivial form $H \in \mathbb{Z}[Y_1, \ldots, Y_q]$ of degree D and height ≤ 3 such that $H(\alpha_1, \ldots, \alpha_q) = 0$. Choose $1 \leq \ell \leq q$ such that

$$H(\alpha_1, \ldots, \alpha_{\ell-1}, Y_\ell, \ldots, Y_q) \neq 0, \quad H(\alpha_1, \ldots, \alpha_\ell, Y_{\ell+1}, \ldots, Y_q) = 0.$$

We write

$$H = \sum_i h_i(Y_1, \ldots, Y_\ell) Y_{\ell+1}^{i_{\ell+1}} \cdots Y_q^{i_q}$$

where $h_i \in \mathbb{Z}[Y_1, \ldots, Y_\ell]$. The univariate polynomial $h_i(\alpha_1, \ldots, \alpha_{\ell-1}, Y_\ell)$ is nonzero for some i, but it vanishes at α_ℓ. Moreover (put $\alpha_0 := 1$)

$$\text{ht}(h_i(\alpha_1, \ldots, \alpha_{\ell-1}, Y_\ell)) \leq 3\binom{D+\ell-1}{D}|\alpha_{\ell-1}|^D.$$

Therefore, by the remark at the beginning of this proof, we obtain

$$|\alpha_\ell| \leq 1 + 3\binom{D+\ell-1}{D}|\alpha_{\ell-1}|^D.$$

If $\ell = 1$, this is a contradiction to our assumption $|\alpha_1| > 4$. So let us assume $\ell > 1$. Using the estimate

$$\binom{D+\ell-1}{D} \leq \binom{D+q-1}{D} = \frac{D+q-1}{D}\binom{D+q-2}{D-1} < \frac{D+q-1}{D}q^{D-1}$$

we obtain

$$\frac{|\alpha_\ell|}{|\alpha_{\ell-1}|^D} \leq \left(\varepsilon + \frac{3(D+q-1)}{D}\right)q^{D-1}$$

where $\varepsilon := q^{-D+1}|\alpha_{\ell-1}|^{-D} < 10^{-3} \cdot 4^{-4}$. But $\varepsilon + 3(D+q-1)D^{-1} \leq q$, for this is equivalent to $6 + 3\varepsilon \leq (D-3)(q-3-\varepsilon)$ which is true, since $D \geq 4$, $q \geq 10$ and $4\varepsilon \leq 1$. We thus obtain $|\alpha_\ell| \leq |q\alpha_{\ell-1}|^D$, in contradiction to our assumption. $\square$

After these preparations, we are finally in a position to prove nontrivial lower bounds on the complexity of polynomials with rapidly growing integer coefficients.

(9.31) Corollary. *The following asymptotic lower bound holds*

$$L_{K(X)}\left(\sum_{j=1}^{n} 2^{2^{j/n}} X^j\right) \gtrsim \sqrt{n} \quad (n \to \infty).$$

Proof. We put $r_n := L_{K(X)}(\sum_{j=1}^n 2^{2^{jn}} X^j)$, $D_n := \lfloor 2^n/n \rfloor$. We may w.l.o.g. assume that $(r_n + 1)(r_n + 2) + 2 < n$. A routine calculation shows that for sufficiently large n and $1 < j \leq n$

$$2^{2^{jn}} > \left(n 2^{2^{(j-1)n}}\right)^{D_n}.$$

Therefore, by Lemma (9.30), there is no form $H \in \mathbb{Z}[Y_1, \ldots, Y_n]$ of degree D_n and height ≤ 3 such that

$$H(2^{2^n}, 2^{2^{2n}}, \ldots, 2^{2^{n^2}}) = 0.$$

This implies by Thm. (9.29)

$$D_n^{n-(r_n+1)(r_n+2)-2} \leq (2r_n n)^{(r_n+1)(r_n+2)} n^n n^{2r_n}.$$

Therefore $D_n^{n-(r_n+2)^2} \leq (2n^3)^{(r_n+2)^2} n^n$ (use the trivial estimate $r_n \leq n$), hence

$$n \frac{\log D_n - \log n}{\log D_n + 3\log n + 1} \leq (r_n + 2)^2.$$

From this we get that $n(1 + o(1)) \leq (r_n + 2)^2$ as $n \to \infty$ (use the fact that $\log D_n = n - \log n + o(1)$), which shows the assertion.   □

(9.32) Corollary. *For sufficiently large n we have*

$$L_{K(X)}\left(\sum_{j=1}^n 2^{2^j} X^j\right) \geq 0.38\sqrt{n/\log n}.$$

Proof. Let $c \in (1, \infty)$. For $n > 1$ we put

$$q_n := \left\lfloor \frac{n}{1 + c\log n} \right\rfloor, \quad r_n := L_{K(X)}\left(\sum_{j=1}^n 2^{2^j} X^j\right), \quad D_n := \lfloor n^{c-n^{-1}} \rfloor,$$

and set $v_j := j\lceil c\log n \rceil$ for $1 \leq j \leq q_n$. (Note that $v_{q_n} \leq n$.) We show that for sufficiently large n and $1 < j \leq n$

(A) $2^{2^{v_j}} > (q_n 2^{2^{v_{j-1}}})^{D_n}.$

Namely, for sufficiently large n, $n^{-1} \ln n > n^{-c} \log n \geq \ln(1 + n^{-c}\log n)$, hence $n^c > n^{c-n^{-1}}(1 + n^{-c}\log n)$. This implies $2^{v_j - v_{j-1}} > D_n(1 + 2^{-v_{j-1}}\log q_n)$, which proves (A).

Therefore, by Lemma (9.30), there is no form $\in \mathbb{Z}[Y_1, \ldots, Y_{q_n}]$ of degree D_n and height ≤ 3 such that

$$H(2^{2^{v_1}}, \ldots, 2^{2^{v_{q_n}}}) = 0.$$

Let us assume for the moment that $(r_n + 1)(r_n + 2) + 2 < q_n$. Then Thm. (9.29) implies

$$D_n^{q_n-(r_n+1)(r_n+2)-2} \leq (2r_n n)^{(r_n+1)(r_n+2)} q_n^{q_n} n^{2r_n}.$$

Using $r_n \le 2\sqrt{n}$ (cf. Prop. (9.1)) we get

$$D_n^{q_n - (r_n+2)^2} \le (4n^{3/2})^{(r_n+2)^2} q_n^{q_n} n^{2r_n},$$

and by taking logarithms we obtain after a short calculation

$$\gamma_n \le (r_n + 2)^2 + (r_n + 2)\beta_n,$$

where

$$\gamma_n := \frac{c-1}{c(3/2+c)} \frac{n}{\log n}(1 + o(1)), \quad \beta_n := \frac{2\log n}{\log(4n^{3/2}D_n)} \to \frac{2}{3/2+c}$$

for $n \to \infty$. By solving the above quadratic equation we conclude $\sqrt{\gamma_n} \lesssim r_n + 2$, hence $h(c)\sqrt{n/\log n} \lesssim r_n$, where $h(c) := \sqrt{(c-1)/c(3/2+c)}$. Clearly, this is also true if $q_n \le (r_n+1)(r_n+2)+2$. The function $h: (1, \infty) \to \mathbb{R}$ attains its absolute maximum at $c_0 := 1 + \sqrt{10}/2$ and $0.38 < h(c_0)$. Therefore $0.38\sqrt{n/\log n} \le r_n$ for sufficiently large n, which was to be shown.   □

9.5* Extension to other Complexity Measures

Throughout this section k denotes a subfield of K. All straight-line programs considered are defined over k.

In Sect. 9.2 we developed a technique to prove lower bounds on the multiplicative complexity $L^*_{K \to K(X)}(f)$ to compute a polynomial $f \in K[X]$ from X. We will modify this technique so that it will give us also lower bounds on the multiplicative complexity $L^*_{k \to K(X)}(f)$ and additive complexity $L^+_{k \to K(X)}(f)$ with respect to a subfield k of K. Thus we view $K(X)$ as an algebra over k. Recall that $L^*_{k \to K(X)}$ counts all the multiplications and divisions except the multiplications with a scalar in the subfield k. (A motivation why to consider this notion can be found in Sect. 5.2.)

Before going into technical details let us sketch the main ideas of the following developments. Let a consistent straight-line program $\Gamma = (\Gamma_1, \ldots, \Gamma_t)$ expecting inputs of length N be given, and denote by ℓ^* its multiplicative length. The subset U consisting of those points $\xi \in K^N$ for which Γ is executable on input $(\xi_{-N+1}, \ldots, \xi_{-1}, X + \xi_0)$ in the ring of formal power series $K[[X]]$ is nonempty and Zariski open (cf. Thm. (4.5)). Γ produces on such an input a result sequence $(b_{-N+1}, \ldots, b_t)$, the $b_i = \sum_{j=0}^{\infty} b_{ij} X^j$ being formal power series whose coefficients $b_{ij} = b_{ij}(\xi)$ depend rationally on ξ. The point is now that the b_{ij} for $j > 0$ can be written in the form

$$b_{ij} = P_{ij}(\zeta_1, \ldots, \zeta_{2\ell^*}),$$

where the P_{ij} are polynomials over k in $2\ell^*$ indeterminates and $\zeta_i : U \to K$ are rational functions ($i \geq 1$). Moreover, the degree of P_{ij} increases at most linearly in j, namely deg $P_{ij} \leq 3\ell^* j$. If we denote the Zariski closure of the image of the morphism

$$K \times K^{2\ell^*} \to \sum_{j=0}^n KX^j, \quad (\rho, \zeta) \mapsto \rho + \sum_{j=1}^n P_{tj}(\zeta)X^j,$$

by $W_n^*(\Gamma)$, then any polynomial of degree $\leq n$ which is computed by Γ on some $(\xi_{-N+1}, \ldots, \xi_{-1}, X + \xi_0)$ in $K[[X]]$ is contained in $W_n^*(\Gamma)$. It is not hard to see that this is even true if the computation takes place in the rational function field $K(X)$. As in the proof of Thm. (9.13) we conclude that

$$\dim W_n^*(\Gamma) \leq 2\ell^* + 1, \quad \deg W_n^*(\Gamma) \leq (3\ell^* n)^{2\ell^* + 1},$$

and that $W_n^*(\Gamma)$ is stable under the action of the group $\mathrm{Aut}(K/k)$. From this information we then obtain an analogue of Thm. (9.15) (see Thm. (9.35)).

The main work to be done consists in proving the following statement resembling Prop. (9.11).

(9.33) Proposition (Schnorr, van de Wiele). *Let* $X, Z_1, Z_2, \ldots$ *be indeterminates over* K *and* $R := K[Z_i \mid i \geq 1]$. *Let* $\Gamma = (\Gamma_1, \ldots, \Gamma_t)$ *be a straight-line program expecting inputs of length* N, *and denote by* ℓ_i^* *the multiplicative length of the initial segment* $(\Gamma_1, \ldots, \Gamma_i)$ *and by* ℓ_i^+ *its additive length.*

(1) *Γ induces a sequence* $(P_{-N+1}, \ldots, P_t)$ *of formal power series in* $R[[X]]$ *without constant term having the following properties:*

    (1a) *Let* $P_i = \sum_{j \geq 1} P_{ij} X^j$ *with* $P_{ij} \in R$. *Then* $P_{ij} \in k[Z_1, \ldots, Z_{2\ell_i^*}]$, *deg* $P_{ij} \leq 3\ell_i^* j$. *Moreover* $P_i = 0$ *for* $-N < i < 0$ *and* $P_0 = X$.

    (1b) *If Γ is executable on*

$$(K[[X]]; \xi_{-N+1}, \ldots, \xi_{-1}, X + \xi_0)$$

    for $(\xi_{-N+1}, \ldots, \xi_{-1}, \xi_0) \in K^N$ *with result sequence* $(b_{-N+1}, \ldots, b_t)$, *then there exists* $\zeta \in K^{2\ell_t^*}$ *such that for all* $i \leq t$

$$b_i = b_i(0) + P_i(\zeta, X).$$

(2) *Γ induces a sequence* $(R_{-N+1}, \ldots, R_t)$ *of formal power series in* $R[[X]]$ *without constant term having the following properties:*

    (2a) *Let* $R_i = \sum_{j \geq 1} R_{ij} X^j$ *with* $R_{ij} \in R$. *Then* $R_{ij} \in k[Z_1, \ldots, Z_{\ell_i^+}]$, *deg* $R_{ij} \leq \ell_i^+ j$. *Moreover* $R_i = 0$ *for* $-N < i < 0$ *and* $R_0 = X$.

    (2b) *If Γ is executable on*

$$(K[[X]]; \xi_{-N+1}, \ldots, \xi_{-1}, \xi_0(1 + X))$$

    for $(\xi_{-N+1}, \ldots, \xi_0) \in K^N$ *such that all elements of the corresponding result sequence* $(b_{-N+1}, \ldots, b_t)$ *are either zero or have nonzero constant term, then there exists* $\zeta \in K^{\ell_t^+}$ *such that for all* $i \leq t$

$$b_i = b_i(0)(1 + R_i(\zeta, X)).$$

Proof. (1) We define the sequence of formal power series $(P_{-N+1}, \ldots, P_t)$ by induction on the length t of Γ. For $t = 0$ we set $P_0 := X$, $P_i := 0$ for $-N < i < 0$. Assume $t > 0$ and let $(P_{-N+1}, \ldots, P_{t-1})$ be the sequence of polynomials assigned to the straight-line program $(\Gamma_1, \ldots, \Gamma_{t-1})$ by the induction hypothesis. We are going to define P_t by distinguishing for each possible operation symbol of the instruction Γ_t a separate case.

Case 1. $\Gamma_t = (\pm; \mu, \nu)$. We set $P_t := P_\mu \pm P_\nu$.

Case 2. $\Gamma_t = (*; \mu, \nu)$. We put

$$P_t := (Z_{2\ell_t^*-1} + P_\mu)(Z_{2\ell_t^*} + P_\nu) - Z_{2\ell_t^*-1}Z_{2\ell_t^*}.$$

(Note that $P_\mu, P_\nu \in k[Z_1, \ldots, Z_{2\ell_{t-1}^*}]$, $\ell_{t-1}^* + 1 = \ell_t^*$.)

Case 3. $\Gamma_t = (/; \mu, \nu)$. We define P_t by

$$
\text{(A)} \qquad
\begin{aligned}
P_t & := & (Z_{2\ell_t^*-1} + P_\mu)(Z_{2\ell_t^*}^{-1} + P_\nu)^{-1} - Z_{2\ell_t^*-1}Z_{2\ell_t^*} \\
& = & (Z_{2\ell_t^*-1} + P_\mu)Z_{2\ell_t^*}(1 + Z_{2\ell_t^*}P_\nu)^{-1} - Z_{2\ell_t^*-1}Z_{2\ell_t^*}.
\end{aligned}
$$

(Observe that $1 + Z_{2\ell_t^*}P_\nu$ is invertible in the ring $k[Z_1, \ldots, Z_{2\ell_t^*}][[X]]$.)

Case 4. $\Gamma_t = (\lambda; \mu)$ where $\lambda \in k$. We put $P_t := \lambda P_\mu$.

Case 5. $\Gamma_t = (\lambda^c)$. We set $P_t := 0$.

It is clear from the definition that the coefficient P_{ij} of X^j in P_i depends only on $Z_1, \ldots, Z_{2\ell_t^*}$ and that $P_{i0} = 0$.

Let us verify the degree estimate in (1a) by induction on t. The start "$t = 0$" being clear assume "$t > 0$." In cases 1, 4, 5 we have $\ell_t^* = \ell_{t-1}^*$ and the statement follows immediately from the induction hypothesis. We use now Lemma (9.10) and the notation introduced there. The induction hypothesis implies in case 2 that $P_t \in B_{3\ell_{t-1}^*,1} \cdot B_{3\ell_{t-1}^*,1} \subseteq B_{3\ell_{t-1}^*,2}$. Hence for $j \geq 1$ we have $\deg P_{tj} \leq 3\ell_{t-1}^* j + 2 \leq 3\ell_t^* j$. Similarly we obtain in case 3

$$P_t \in B_{3\ell_{t-1}^*,1} \cdot B_{0,1} \cdot B_{3\ell_{t-1}^*+1,0} \subseteq B_{3\ell_{t-1}^*+1,2}.$$

Hence we have for $j \geq 1$: $\deg P_{tj} \leq (3\ell_{t-1}^* + 1)j + 2 \leq 3\ell_t^* j$.

We turn to the demonstration of (1b). Again we proceed by induction on t. The start "$t = 0$" being trivial let us suppose $t > 0$. Let $\Gamma = (\Gamma_1, \ldots, \Gamma_t)$ be executable on $(K[[X]]; \xi_{-N+1}, \ldots, \xi_{-1}, \xi_0 + X)$ for some $(\xi_{-N+1}, \ldots, \xi_{-1}, \xi_0) \in K^N$ with result sequence $(b_{-N+1}, \ldots, b_t)$. By the induction hypothesis applied to $(\Gamma_1, \ldots, \Gamma_{t-1})$ there exists $\zeta \in K^{2\ell_{t-1}^*}$ such that for all $i < t$

$$\text{(B)} \qquad\qquad b_i = b_i(0) + P_i(\zeta, x).$$

We just give the argument for case 3 where $\Gamma_t = (/; \mu, \nu)$; the other cases are treated similarly. We define (note that $b_\nu(0) \in K^\times$ since $b_\nu \in K[[X]]^\times$)

$$\zeta_{2\ell_t^*-1} := b_\mu(0), \quad \zeta_{2\ell_t^*} := b_\nu(0)^{-1}, \quad \zeta' := (\zeta, \zeta_{2\ell_t^*-1}, \zeta_{2\ell_t^*}) \in K^{2\ell_t^*}.$$

Applying the k-algebra morphism

$$k[Z_1, \ldots, Z_{2\ell_t^+}][[X]] \to K[[X]], Z_i \mapsto \zeta_i$$

to (A) we obtain

$$b_\mu(0)b_\nu(0)^{-1} + P_t(\zeta', X) = (b_\mu(0) + P_\mu(\zeta, X))(b_\nu(0) + P_\nu(\zeta, X))^{-1}.$$

The right-hand side equals by our assumption (B) $b_\mu b_\nu^{-1} = b_t$, therefore

$$b_t = b_t(0)^{-1} + P_t(\zeta', X).$$

This proves statement (1b).

(2) The proof of the second part of the proposition is quite analogous to the proof of the first part. We define the sequence of formal power series $(R_{-N+1}, \ldots, R_t)$ by induction on the length t of Γ. For $t = 0$ we set $R_0 := X$, $R_i := 0$ for $-N < i < 0$. We assume now $t > 0$ and denote by $(R_{-N+1}, \ldots, R_{t-1})$ the sequence of polynomials assigned to the straight-line program $(\Gamma_1, \ldots, \Gamma_{t-1})$ by the induction hypothesis.

Case 1. $\Gamma_t = (+; \mu, \nu)$ or $\Gamma_t = (-; \mu, \nu)$. We define R_t by

(C) $$1 + R_t := 1 + Z_{\ell_t^+} R_\mu + (1 - Z_{\ell_t^+}) R_\nu.$$

Case 2. $\Gamma_t = (*; \mu, \nu)$. We put

$$1 + R_t := (1 + R_\mu)(1 + R_\nu).$$

Case 3. $\Gamma_t = (/; \mu, \nu)$. We define

$$1 + R_t := (1 + R_\mu)(1 + R_\nu)^{-1}.$$

Case 4. $\Gamma_t = (\lambda; \mu)$ where $\lambda \in k$. We put $R_t := R_\mu$.

Case 5. $\Gamma_t = (\lambda^c)$. We set $R_t := 0$.

From the definition it is clear that the coefficient $R_{ij} \in R$ depends only on $Z_1, \ldots, Z_{\ell_t^+}$ and that $R_{i0} = 0$.

The degree estimate in (2a) is proved by induction on t. As an example we show $R_t \in B_{\ell_t^+, 0}$ in case 1. The start $t = 0$ is trivial. By Lemma (9.10) and the induction hypothesis we obtain

$$1 + R_t \in 1 + B_{\ell_{t-1}^+, 1} + B_{\ell_{t-1}^+, 1} = B_{\ell_{t-1}^+, 1}.$$

We thus have for all $j \geq 1$: $\deg R_{tj} \leq \ell_{t-1}^+ j + 1 \leq \ell_t^+ j$.

We prove now statement (2b) by induction on t. The start "$t = 0$" is trivial, so assume $t > 0$. Let $\Gamma = (\Gamma_1, \ldots, \Gamma_t)$ be executable on

$$(K[[X]]; \xi_{-N+1}, \ldots, \xi_{-1}, \xi_0(1 + X))$$

where $(\xi_{-N+1}, \ldots, \xi_{-1}, \xi_0) \in K^N$, and the elements b_i of the corresponding result sequence $(b_{-N+1}, \ldots, b_t)$ have nonzero constant term unless $b_i = 0$. We only give the argument for case 1 where $\Gamma_t = (+; \mu, \nu)$; the other cases can be checked similarly. The induction hypothesis applied to $(\Gamma_1, \ldots, \Gamma_{t-1})$ implies that there exists $\zeta \in K^{\ell_{t-1}^+}$ such that for all $i < t$

$$b_i = b_i(0)(1 + R_i(\zeta, X)).$$

By our assumption we can assume w.l.o.g that $b_t(0) \neq 0$. We therefore have

$$b_t b_t(0)^{-1} = b_t(0)^{-1}(b_\mu + b_\nu)$$

(D)
$$= b_\mu(0)b_t(0)^{-1}(1 + R_\mu(\zeta, X)) + b_\nu(0)b_t(0)^{-1}(1 + R_\nu(\zeta, X))$$

$$= 1 + b_\mu(0)b_t(0)^{-1}R_\mu(\zeta, X) + (1 - b_\mu(0)b_t(0)^{-1})R_\nu(\zeta, X).$$

Defining $\zeta_{\ell_t^+} := b_\mu(0)b_t(0)^{-1}$, $\zeta' := (\zeta, \zeta_{\ell_t^+ - 1}) \in K^{\ell_t^+}$ and applying the k-algebra morphism

$$k[Z_1, \ldots, Z_{\ell_t^+}][[X]] \to K[[X]], \quad Z_i \mapsto \zeta_i$$

to (C) we see together with (D) that

$$b_t = b_t(0)(1 + R_t(\zeta', X))$$

which was to be shown. $\square$

We denote by $K[X]_{\leq n}$ the subspace of polynomials in $K[X]$ of degree $\leq n$. Note that $K[X]_{\leq n}$ can be considered as an affine space. The next result corresponds to Thm. (9.13).

(9.34) Theorem. *To a straight-line program Γ and a natural number n one can assign irreducible, Zariski closed subsets $W_n^*(\Gamma)$, $W_n^+(\Gamma)$ of $K[X]_{\leq n}$ which are stable under $\mathrm{Aut}(K/k)$ and have the following properties:*

(1) *if ℓ^*, resp. ℓ^+, denote the multiplicative, resp. additive length of Γ, then*

$$\dim W_n^*(\Gamma) \leq 2\ell^* + 1, \qquad \dim W_n^+(\Gamma) \leq \ell^+ + 1,$$
$$\deg W_n^*(\Gamma) \leq (3\ell^* n)^{2\ell^* + 1}, \qquad \deg W_n^+(\Gamma) \leq (\ell^+ n)^{\ell^+ + 1}.$$

(2) *For $r \in \mathbb{N}$ we have*

$$\left\{ f \in K[X]_{\leq n} \,\middle|\, L_{k \to K(X)}^*(f) \leq r \right\} \subseteq \bigcup_\Gamma W_n^*(\Gamma)$$

where the union is taken over all straight-line programs Γ with multiplicative length at most r.

(3) *For $r \in \mathbb{N}$ we have*

$$\left\{ f \in K[X]_{\leq n} \,\middle|\, L_{k \to K(X)}^+(f) \leq r \right\} \subseteq \bigcup_\Gamma W_n^+(\Gamma)$$

where the union is taken over all straight-line programs Γ with additive length at most r.

Proof. Let $n \in \mathbb{N}$ and $\Gamma = (\Gamma_1, \ldots, \Gamma_t)$ be a straight-line program of multiplicative length ℓ^* and additive length ℓ^+. Using the formal power series $P_t = \sum_{j \geq 1} P_{tj} X^j$, $R_t = \sum_{j \geq 1} R_{tj} X^j$ which are assigned to Γ by Prop. (9.33) we define $W_n^*(\Gamma)$ as the closure of the image of the morphism

$$K \times K^{2\ell^*} \to K[X]_{\leq n}, \quad (\rho, \zeta) \mapsto \rho + \sum_{j=1}^n P_{tj}(\zeta) X^j,$$

and we let $W_n^+(\Gamma)$ be the closure of the image of

$$K \times K^{\ell^+} \to K[X]_{\leq n}, \quad (\rho, \zeta) \mapsto \rho\left(1 + \sum_{j=1}^n R_{tj}(\zeta) X^j\right).$$

The bounds on the dimensions of $W_n^*(\Gamma)$, $W_n^+(\Gamma)$ are obvious, whereas the bounds on the degrees follow from $\deg P_{tj} \leq 3\ell^* j$, $\deg R_{tj} \leq \ell^+ j$ by Thm. (8.48). The stability under $\mathrm{Aut}(K/k)$ is a consequence of Lemma (9.12)(3) as the above morphisms are given by polynomials over k.

We are going to show assertion (2). Let $f \in K[X]_{\leq n}$ and let Γ be a straight-line program executable on $(K(X); \xi_{-N+1}, \ldots, \xi_{-1}, X)$ for some $N \in \mathbb{N}$, $(\xi_{-N+1}, \ldots, \xi_{-1}) \in K^{N-1}$, such that

$$r := L_{k \to K(X)}^*(f) = \text{multiplicative length of } \Gamma$$

and $f = b_t$, where $(b_{-N+1}, \ldots, b_t)$ is the corresponding result sequence. Choose $\xi_0 \in K$ such that $b_i(\xi_0) \neq 0$ for all i with $b_i \neq 0$. Let $\mathcal{O}_{\xi_0}$ denote the subring of $K(X)$ consisting of the rational functions defined at ξ_0. Then Γ is even executable on $(\mathcal{O}_{\xi_0}; \xi_{-N+1}, \ldots, \xi_{-1}, X)$ and by virtue of the K-algebra morphism (variable shift)

$$\mathcal{O}_{\xi_0} \to K[[X]], X \mapsto X + \xi_0$$

Γ is also executable on $(K[[X]]; \xi_{-N+1}, \ldots, \xi_{-1}, X+\xi_0)$ having as result sequence

$$(b_{-N+1}(X + \xi_0), \ldots, b_t(X + \xi_0)).$$

Prop. (9.33)(1) implies that for some $\zeta \in K^{2r}$

$$f(X + \xi_0) = b_t(X + \xi_0) = b_t(\xi_0) + \sum_{j=1}^n P_{tj}(\zeta) X^j.$$

Hence $f(X + \xi_0) \in W_n^*(\Gamma)$ for all but finitely many $\xi_0 \in K$, which implies $f \in W_n^*(\Gamma)$. Thus assertion (2) is shown. Assertion (3) can be shown analogously. $\square$

As a consequence we obtain the subsequent analogue of Thm. (9.15). We leave its proof as an exercise to the reader.

(9.35) Theorem (Heintz and Sieveking). *With the assumptions of Thm. (9.15) we have*

$$L_{k \to K(X)}^*(f) \geq \frac{1}{2}\left(\frac{\log N}{\log(3n^2 M)} - 1\right),$$

$$L_{k \to K(X)}^+(f) \geq \frac{\log N}{\log(n^2 M)} - 1.$$

Of course this theorem can be applied to give lower bounds on L^*, L^+ for all examples considered in Sect. 9.3. For instance, we obtain for the polynomial $f := \sum_{j=1}^{n} e^{2\pi i/j} X^j \in \mathbb{C}[X]$ the bounds

$$L^*_{\mathbb{Q} \to \mathbb{C}(X)}(f) \gtrsim \frac{n}{6 \ln n}, \quad L^+_{\mathbb{Q} \to \mathbb{C}(X)}(f) \gtrsim \frac{n}{3 \ln n}.$$

In fact, it is straight-forward to check that we get lower bounds which are – up to a constant factor – the squares of the bounds obtained there.

9.6 Exercises

9.1. Let $\varphi: K \to K^n$ be a polynomial map, $W \subseteq K^n$ Zariski closed, and assume $\varphi(\xi) \in W$ for all but finitely many $\xi \in K$. Show that $\varphi(K) \subseteq W$.

9.2. Starting from the representation theorem (9.9) prove that Thm. (9.3) is also true for the multiplicative complexity $L_{K(X)}$ in the rational function field.

9.3. Show that Thm. (9.3) yields lower bounds of order $\Omega(\sqrt{n})$, respectively $\Omega(\sqrt{n/\log n})$ for the polynomials of Corollaries (9.16)(3) and (9.24).

9.4. Improve the bound of Prop. (9.1) by showing that

$$L_{K[X]}(f) \leq \sqrt{2n} + \log n$$

for $f \in K[X]$ of degree $n \geq 1$. (Hint: Combine the ideas from Thm. (5.1) and Prop. (9.1). Let $X^2, \ldots, X^p$ and $X^p, X^{2p}, \ldots, X^{2^{m-1}p}$ be already computed. By induction on m show that a monic polynomial of degree $p(2^m - 1)$ can be computed from these inputs by $2^{m-1} - 1$ nonscalar multiplications. To compute an arbitrary polynomial of degree n set $p := \lceil \sqrt{n/2} \rceil$ and use the decomposition $= p2^{m_0} + \ldots + p2^{m_s} + r$ with $m_0 > \ldots > m_s$, $0 \leq r < p$.)

9.5. In this exercise we exhibit specific matrices $A \in K^{m \times n}$ of high linear complexity $L_\infty(A)$. (For this notion compare Chap. 13.)

(1) Consider straight-line programs which use only additions, subtractions and multiplications with scalars in K. (We may work as in Def. (13.2) with a notion which is completely symmetric w.r.t. additions and subtractions.) Let $\Gamma = (\Gamma_1, \ldots, \Gamma_t)$ be such a straight-line program expecting inputs of length n with r scalar multiplication instructions. If we execute Γ on the sequence of indeterminates $X_1, \ldots, X_n$, then we get a linear computation sequence $(g_{-n+1}, \ldots, g_t)$. Study the dependence of g_ρ on the scalars $z_1, \ldots, z_r \in K$ used by Γ. Show that $g_\rho = \sum_{j=1}^{n} G_{\rho j}(z_1, \ldots, z_r) X_j$ for polynomials $G_{\rho j} \in \mathbb{Z}[Z_1, \ldots, Z_r]$ of degree $\leq r$.

(2) Let $m, n, r \geq 1$. Show that there is a finite family $F^1, \ldots, F^s$ of matrices $F^\sigma \in \mathbb{Z}[Z_1, \ldots, Z_r]^{m \times n}$ whose entries are of degree $\leq r$ and such that

$$\{A \in K^{m \times n} \mid L_\infty(A) \leq r\} = \bigcup_{\sigma=1}^{s} \{F^\sigma(\zeta) \mid \zeta \in K^r\}.$$

(3) In analogy with Thm. (9.3) prove the following. Let $m, n > 1$, let $A = (a_{ij}) \in K^{m \times n}$, and let d_{ij}, $i, j \in \underline{n}$, be positive natural numbers not all equal to one such that the power products

$$\prod_{(i,j) \in \underline{m} \times \underline{n}} a_{ij}^{\delta_{ij}}, \quad 0 \leq \delta_{ij} < d_{ij} \text{ for } i, j \in \underline{m} \times \underline{n},$$

are linearly independent over a subfield k of K. If we set $N := \prod_{i,j} d_{ij}$ and $M := \sum_{i,j} (d_{ij} - 1)$, then

$$L_\infty(A) \geq \log N / \log(m(2n - 1)M).$$

(4) Show $L_\infty(A) \geq mn/(2 \log mn + 1)$ for $A = (\sqrt{p_{ij}}) \in \mathbb{C}^{m \times n}$, $m, n > 1$, the p_{ij} being different primes.

(5) Prove for a matrix $A \in K^{m \times n}$ with entries which are algebraic over a subfield $k \subseteq K$ that

$$L_\infty(A) \geq \log N / \log(m(2n - 1)M),$$

where M, N are defined similarly as in Thm. (9.15).

(6) Derive the improved bound $L_\infty(A) \geq mn/(\log mn + 2)$ for the matrix from (4).

9.6. In Chap. 14 we will introduce the notion of tensor rank. It is quite easy to see that the rank $R(t)$ of Zariski almost all tensors $t \in K^{n \times n \times n}$ satisfies

$$n^3/(3n - 2) \leq R(t) \leq n^2$$

(cf. Sect. 20.1). The goal of this exercise is to exhibit specific tensors $t \in K^{n \times n \times n}$ of high rank.

(1) In analogy with Thm. (9.3) prove the following result. Let $t = (t_{ij\ell}) \in K^{n \times n \times n}$ and $d_{ij\ell}$, $i, j, \ell \in \underline{n}$, be positive natural numbers not all equal to one such that the power products

$$\prod_{i,j,\ell \in \underline{n}} t_{ij\ell}^{\delta_{ij\ell}}, \quad 0 \leq \delta_{ij\ell} < d_{ij\ell} \text{ for } i, j, \ell \in \underline{n}$$

are linearly independent over a subfield k of K. If we set $N := \prod_{i,j,\ell} d_{ij\ell}$ and $M := \sum_{i,j,\ell} (d_{ij\ell} - 1)$, then

$$R(t) \geq (3n)^{-1} \log N / \log 3M.$$

(2) Conclude that $R(t) \geq n^2/(9 \log n + 3 \log 3)$ for the tensor $t \in \mathbb{C}^{n \times n \times n}$ with $t_{ij\ell} = \sqrt{p_{ij\ell}}$, the $p_{ij\ell}$ being distinct primes.

(3) Prove that for a tensor $t \in K^{n \times n \times n}$ with entries which are algebraic over a subfield $k \subseteq K$ the following holds

$$R(t) \geq (3n-2)^{-1} \log N / \log 3M,$$

where M, N are defined similarly as in Thm. (9.15). In fact, these lower bounds even hold for the border rank $\underline{R}$.

(4) For the tensor t from (2) prove the better bound $R(t) \geq (\log 6)^{-1} n^3/(3n-2)$.

The next two exercises form a block and describe a different (and elementary) way to obtain some of the results of Sect. 9.3.

9.7.* Let $A_n \in K^{(n+1) \times (n+1)}$, $n = 1, 2, \ldots$, be a fixed sequence of matrices. Pick a subset $M \subseteq \{0, 1, \ldots, n\}$ uniformly at random and consider the event that every polynomial $f = \sum_{i=0}^{n} \alpha_i X^i \in K[X]$ satisfying

$$\forall 0 \leq i \leq n : (A_n(\alpha_0, \ldots, \alpha_n)^T)_i \neq 0 \Longleftrightarrow i \in M$$

has complexity $L_{K(X)}(f) \geq \sqrt{n/(3 \log n)}$. Prove that the probability of this event tends to 1 as $n \to \infty$. (So this is an almost sure event in the terminology of probability theory.)

Let additionally to (A_n) a sequence $(d_n)_{n \geq 1}$ of natural numbers with $n/3 < d_n < 2n/3$ be given. Show that $L_{K(X)}(f) \geq \sqrt{n/(8 \log n)}$ is an almost sure event if we pick the subset M with equal probability among all the subsets of $\{0, 1, \ldots, n\}$ of cardinality d_n.

(Hint: Set $r := \lfloor \sqrt{n/(3 \log n)} \rfloor$, $m := (r+1)(r+2)$, $c := 2rn$, and let $F_{r0}, \ldots, F_{rn} \in \mathbb{Z}[Z_1, \ldots, Z_m]$ be the polynomials from the representation theorem (9.9) (we have renamed the Y_i to Z_j). Define $(\tilde{F}_0, \ldots, \tilde{F}_n) := A_n(F_{r0}, \ldots, F_{rn})^T$. Consider the K-space

$$\mathcal{H} := \{ H \in k[Y_0, \ldots, Y_n] \mid H \text{ homogeneous of degree} \leq n,$$
$$\deg_{Y_i} H \leq 1 \text{ for all } i \}$$

of dimension 2^{n+1}. If $2^{n+1} - \binom{m+cn}{m} \geq (1-\varepsilon)2^{n+1}$, then there exists some

$$H = \sum_{M \subseteq \{0, \ldots, n\}} \beta_M \prod_{i \in M} Y_i$$

satisfying

$$H(\tilde{F}_0, \ldots, \tilde{F}_n) = 0 \text{ and } |\{M \subseteq \{0, \ldots, n\} \mid \beta_M \neq 0\}| \geq (1-\varepsilon)2^{n+1}$$

(compare the proof of Thm. (9.3)). Therefore, if $\beta_M \neq 0$ and $f = \sum_{i=0}^{n} \alpha_i X^i$ satisfies $\{i \mid (A_n(\alpha_0, \ldots, \alpha_n)^T)_i \neq 0\} = M$, then $L_{K(X)}(f) > r$.)

9.8. From the previous exercise deduce the following results.

(1) Pick $(\alpha_0, \ldots, \alpha_n) \in \{0, 1\}^{n+1}$ uniformly at random. Then the probability that $L_{K(X)}(\sum_{i=0}^n \alpha_i X^i) \geq \sqrt{n/(3 \log n)}$ tends to 1 as $n \to \infty$. (Compare Cor. (9.19).)

(2) Let $p_1, p_2, \ldots, p_n$ be distinct prime numbers. Then for sufficiently large n we have $L_{C(X)}(\sum_{i=1}^n \sqrt{p_i} X^i) \geq \sqrt{n/(12 \log n)}$. (Hint: for $M \subseteq \underline{n}$ there exists $\sigma \in \mathrm{Aut}(\mathbb{Q}(\sqrt{p_1}, \ldots, \sqrt{p_n})/\mathbb{Q})$ such that $\sigma(\sqrt{p_i}) = \sqrt{p_i}$ if $i \in M$ and $\sigma(\sqrt{p_i}) = -\sqrt{p_i}$ if $i \notin M$.)

(3) Let $\xi_0, \xi_1, \ldots$ be a sequence of mutually distinct elements of $K^\times$. If we pick $M \subseteq \{0, 1, \ldots, n\}$ uniformly at random, then the probability that

$$L_{K(X)}\left(\prod_{i \in M}(X - \xi_i)\right) \geq \sqrt{n/(3 \log n)}$$

goes to 1 as $n \to \infty$. Therefore, "almost all" of the factors of a squarefree polynomial of degree n have high complexity.

(4) Let $k \subseteq K$ be a subfield and assume that the Galois group of $f \in K[X]$ is isomorphic to S_n. Prove that *all* factors h of degree $n/3 < \deg h < 2n/3$ satisfy $L_{K(X)}(h) \geq \sqrt{n/(8 \log n)}$ if n is sufficiently large.

9.9.

(1) In analogy to Ex. 9.7 and Ex. 9.8 show the following: if we pick a tensor $t \in \{0, 1\}^{n \times n \times n}$ uniformly at random, then the probability that $R(t) \geq n^2/\log n$ tends to 1 as $n \to \infty$.

(2) As in the proof of Cor. (9.19) show that

$$\left|\{t \in \{0, 1\}^{n \times n \times n} \mid R(t) \leq n^2/12\}\right| \leq 2^{n^3 - n}.$$

The next two exercises prepare for Ex. 9.12, where an elegant way of proving results in the spirit of Sect. 9.4 is presented.

9.10. Prove that an irreducible projective variety $W \subset \mathbb{P}^n$ which does not lie in any hyperplane satisfies $\deg W \geq n - \dim W + 1$. (Hint: First prove the assertion for curves W. To settle the general case invoke the following result (cf. Harris [228, p. 230]): the intersection $V := W \cap H$ of W with a generic hyperplane H satisfies $\dim V = \dim W - 1$ and $\deg V = \deg W$. Moreover, V is irreducible provided $\dim W \geq 2$.)

9.11.

(1) Let $v_{n.d}: \mathbb{P}^n \to \mathbb{P}^N$ be the Veronese mapping, $N := \binom{n+d}{d} - 1$. Generalize Ex. (8.49) by showing that $\deg v_{n.d}(W) = \deg W \cdot d^{\dim W}$ for an irreducible projective variety $W \subseteq \mathbb{P}^n$.

(2) Let $f: K^n \to K^r$ be a map given by polynomials $f_1, \ldots, f_r$ of degree $\le d$, $d \ge 1$. Prove that for any irreducible affine variety $W \subseteq K^n$ we have

$$\deg \overline{f(W)} \le \deg W \cdot d^{\dim W}.$$

(Hint: use (1) and proceed as in the proof of Thm. (8.48).)

9.12. Show that for all but finitely many $\alpha \in K$ we have

$$L_{K(X)}\left(\sum_{j=0}^{n} \alpha^{2^j} X^j\right) \ge \sqrt{\frac{n-1}{\log 4n^{5/2}}} - 2.$$

(Hint: Let $A := K[X]/(X^{n+1})$ and consider the morphism

$$\varphi: A \to K^{2^n}, \quad \sum_{j=0}^{n} \alpha_j X^j \bmod X^{n+1} \mapsto (\alpha_1^{\varepsilon_1} \cdots \alpha_n^{\varepsilon_n})_{(\varepsilon_1, \ldots, \varepsilon_n) \in \{0,1\}^n}.$$

Let W be an irreducible component of a closed complexity class $\overline{C}_{r,n}$ in A. Using Ex. 9.10, 9.11 show that

$$D - \dim \overline{\varphi(W)} + 1 \le \deg W \cdot n^{\dim W}$$

if D is the dimension of the smallest affine subspace of K^{2^n} containing $\varphi(W)$.)

9.13. In this exercise we show that the assumption that a is not a root of unity is necessary in Cor. (9.24). Let q be prime and $a := \exp(2\pi i/q)$. For $1 \le j \le n$ let $e_j \in \mathbb{N}$ denote the multiplicity of q in j. There is a unique $u_j \in \mathbb{N}$ such that $u_j j q^{-e_j} \equiv 1 \bmod q$, $0 \le u_j < q$. For $1 \le j \le n$ we define $\alpha_j := \exp(2\pi i u_j q^{-e_j-1})$. Show that $f := \sum_{j=1}^{n} \alpha_j X^j$ can be computed from $\mathbb{C} \cup \{X\}$ with $O(\log^2 n)$ arithmetic operations. (Hint: for $0 \le e \le \log n$, $0 \le u < q$ let S_{eu} be the sum of all X^j satisfying $e_j = e$, $u_j = u$. Then f is the sum over all $\exp(2\pi i u q^{-e-1}) S_{eu}$. Write S_{eu} as a geometric series.)

9.14. Prove Thm. (9.35).

9.15.

(1) Let $k \subseteq K$ be a subfield and $V \subseteq K^m$ be an irreducible, Zariski closed subset which is stable under $\mathrm{Aut}(K/k)$. Prove that for any $\alpha := (\alpha_1, \ldots, \alpha_m) \in V$
$$\dim V \ge \mathrm{tr.deg}_k k(\alpha).$$

(2) Deduce from Thm. (9.35) the subsequent lower bounds of Cor. (5.11): for a polynomial $f \in K[X]$ of degree n with coefficients that are algebraically independent over a subfield k we have

$$L^*_{k \to K(X)}(f) \ge n/2, \quad L^+_{k \to K(X)}(f) \ge n.$$

9.16. From Thm. (9.35) deduce lower bounds on

$$\mathcal{L}(f) := \min\left\{2L^*_{\mathbb{Q}\to\mathbb{C}(X)}(f), L^+_{\mathbb{Q}\to\mathbb{C}(X)}(f)\right\}$$

for the polynomials considered in Sect. 9.3. (Caution: for Cor. (9.19) this does not work.)

9.17.* Let char $K = 0$, $n \geq 1$. Our goal is to show the existence of some nonzero polynomials $H \in \mathbb{Z}[Y_0, \ldots, Y_n]$ such that $L^+_{K(X)}(f|X) \geq \sqrt{n} - 2$ for all $f = \sum_{j=0}^n \alpha_j X^j \in K[X]$ satisfying $H(\alpha_0, \ldots, \alpha_n) \neq 0$. In particular, there exist polynomials f of degree n of additive complexity $\geq \sqrt{n} - 2$ which have rational coefficients.

(1) (Cf. Lemma (12.14).) Let $f \in K(X)$ and $r \geq 1$. Then $L^+_{K(X)}(f|X) \leq r$ iff there exists a sequence of nonzero rational functions $g_0 = X, g_1, \ldots, g_r \in K(X)$ such that for all $1 \leq \rho \leq r$

$$g_\rho = \alpha_\rho + \prod_{j=0}^{\rho-1} g_j^{m_{\rho j}}$$

for some $\alpha_\rho \in K$ and integers $m_{\rho j}$, and such that

$$f = \beta \prod_{j=0}^r g_j^{p_j}$$

for some $\gamma \in K$, $p_j \in \mathbb{Z}$.

(2) The idea is to investigate the dependence of the coefficients of f on the parameters introduced in the above computation. Such parameters are $\alpha_1, \ldots, \alpha_r, \beta$. As we do not know any a priori bound on the exponents $m_{\rho j}, p_j$, we will treat them also as parameters. There is a problem, however, since g^m does not make sense for a field element $m \in K$ and $g \in K(X)$. We remedy this by considering computations in $K[[X]]$ and using the following identity holding in $K[[X]]$

$$(\gamma + X)^m = \sum_{\nu \geq 0} \binom{m}{\nu} \frac{\delta}{\gamma^\nu} X^\nu \text{ for } m \in \mathbb{Z}, \gamma \in K^\times, \delta := \gamma^m.$$

Observe that the coefficient of X^ν on the right-hand side is a rational function in γ, δ *and* m!

Prove the existence of a "generic computation" $(G_0, \ldots, G_r, F)$ consisting of formal power series in $\mathbb{Q}(Z_1, \ldots, Z_{(r+2)^2})[[X]]$ having the following property: if $g_0 = \gamma + X, g_1, \ldots, g_r$ is a computation sequence in $K(X)$ as in (1) computing f such that $g_\rho(0)$ is defined and nonzero for all ρ, then there exists some $\zeta \in K^{(r+2)^2}$ such that $g_\rho = G_\rho(\zeta, X)$ for all ρ and $f = F(\zeta, X)$. Conclude that $L^+_{K(X)}(f|X) \geq \sqrt{n} - 2$ for Zariski almost polynomials $f \in K[X]$ of degree n, as asserted at the beginning of this exercise.

9.18. (Generalization of the representation theorem (9.9) to several multivariate polynomials.) For $m, n, r, s \geq 1$ we define the complexity class

$$C_{m,n,r,s} := \{(f_1, \ldots, f_s) \in A^s \mid L_A(f_1, \ldots, f_s \mid X_1, \ldots, X_m) \leq r\},$$

where A denotes the algebra $A := K[X_1, \ldots, X_m]/\mathcal{M}^{n+1}$ and $\mathcal{M}$ is the ideal $(X_1, \ldots, X_m)$ in the polynomial ring.

Let $m, r, s \geq 1$ be given and put $q := r^2 + (2m-1)r + s(r+m+1)$. Prove that for $1 \leq i \leq s$ and $\nu \in \mathbb{N}^m$ there exist polynomials $F_{i\nu} \in \mathbb{Z}[Y_1, \ldots, Y_r, Z_1, \ldots, Z_q]$ of degree $\deg F_{i\nu} \leq (2r-1)|\nu| + 2$ such that for all n

$$C_{m,n,r,s} = \left\{\left(\sum_{|\nu|\leq n} F_{i\nu}(\eta, \zeta)X^\nu \bmod \mathcal{M}^{n+1}\right)_{1 \leq i \leq s} \ \middle| \ \eta \in \{0, 1\}^r, \zeta \in K^q\right\}$$

(we set $|\nu| := \nu_1 + \ldots + \nu_m$). (Hint: Lemma (9.10) remains true when we define

$$B_{s,a} := \left\{\sum_{\nu\in\mathbb{N}^m} f_\nu X^\nu \in R[[X_1, \ldots, X_m]] \ \middle| \ \forall \nu: \deg f_\nu \leq |\nu| \cdot s + a\right\}.$$

Prove an analogue of Prop. (9.11) first.)

The aim of the next three exercises is to prove the subsequent weak version of the prime number theorem, due to Chebyshev. With $\pi(n)$ denoting the number of primes $\leq n$ we have for all $n \geq 2$

$$\frac{1}{6}\frac{n}{\ln n} < \pi(n) < 6\frac{n}{\ln n}.$$

9.19. Let $n \geq 2$ and for a prime p let $\alpha_n(p) := \max\{\ell \mid p^\ell | n!\}$. Show that $\alpha_n(p) = \sum_{m=1}^{c_n(p)}\lfloor\frac{n}{p^m}\rfloor$, where $c_n(p) = \lfloor\frac{\log n}{\log p}\rfloor$.

9.20. Let $n \geq 2$ and $N := \binom{2n}{n}$.

(1) Show that $2^n < N < 4^n$. Deduce that

$$n \ln 2 < \ln(2n)! - 2\ln n! < n \ln 4.$$

(2) Use the previous exercise to show that

$$n \ln 2 < \sum_{p\leq 2n} c_{2n}(p) \ln p \leq \pi(2n) \ln 2n,$$

where the sum in the middle extends over all primes $\leq 2n$.

(3) Deduce that $\pi(2n) \geq \frac{\ln 2}{2}\frac{2n}{\ln 2n} > \frac{1}{4}\frac{2n}{\ln 2n}$. (Hint: use $\lfloor 2x\rfloor - 2\lfloor x\rfloor \leq 1$ for a real number x.)

(4) Prove that $\pi(n) > \frac{1}{6}\frac{n}{\ln n}$. (Hint: $\pi(2n+1) \geq \pi(2n)$.)

9.21. For positive real numbers x define *Chebyshev's ϑ-function* by $\vartheta(x) :=$ $\sum_{p \leq x} \ln p$, where the sum extends over all primes $\leq x$.

(1) Use Ex. 9.19 to show that $\ln(2n)! - 2 \ln n! \geq \vartheta(2n) - \vartheta(n)$. (Hint: use $0 \leq \lfloor 2x \rfloor - 2\lfloor x \rfloor \leq 1$ for a real number x. When does equality hold?)
(2) Show that $\vartheta(2^{\ell+1}) < 2^{\ell+2} \ln 2$ for $\ell \geq 0$. (Hint: use (1) and Ex. 9.20(1).)
(3) Use (2) to deduce that $\vartheta(n) < 4n \ln 2$ for $n \geq 2$.
(4) Let $0 < \alpha < 1$. Show that $(\pi(n) - \pi(n^\alpha)) \ln n^\alpha < 4n \ln 2$.
(5) Use (4) to prove that $\pi(n) < 6\frac{n}{\ln n}$ for $n \geq 2$. (Hint: Prove first that $\pi(n) < \frac{n}{\ln n}(\frac{4 \ln 2}{\alpha} + \frac{\ln n}{n^{1-\alpha}})$, and choose $\alpha = \frac{2}{3}$.)

9.7 Open Problems

Problem 9.1. Let L_n denote the minimal r such that the complexity class $\overline{C}_{r,n}$ equals $A = K[X]/(X^{n+1})$. Thus Zariski almost all $f \in A$ have complexity $L_A(f) = L_n$. The best known estimates on L_n are $\sqrt{n} - 1 \leq L_n \leq \sqrt{2n} + \log n$. Improve them! (Compare Prop. (9.1), (9.2) and Ex. 9.4.)

Problem 9.2. Decide whether the Taylor approximations $\sum_{j=0}^{n} X^j/j!$ and $\sum_{j=1}^{n} X^j/j$ of exp and log, respectively, have a total complexity $O(\log n)$. (Compare Cor. (9.21).)

Problem 9.3. Find a sequence (f_n) of "specific" univariate polynomials with coefficients in $\{0, 1\}$ such that $\deg f_n = n$ and whose complexity does not have order of magnitude $\log n$. (Compare Cor. (9.19).)

Problem 9.4. For a univariate polynomial $f \in \mathbb{Z}[X]$ let $L(f)$ denote the minimal number of additions, subtractions and multiplications needed to compute f in $\mathbb{Z}[X]$ from 1 and X by a straight-line program. Further, let $z(f)$ be the number of distinct integer roots of f. Is there some $\varepsilon > 0$ such that for all $f \in \mathbb{Z}[X]$, $f \neq X$, we have $L(f) \geq z(f)^\varepsilon$?

9.8 Notes

Strassen was the first who exhibited specific polynomials with algebraic or rational coefficients of provably high complexity. In his landmark paper [499] of 1974 he already developed most of the ideas and tools presented in this chapter. His work contains a variant of the representation theorem (9.9) from which he deduced (among other things) lower bounds of order $n^{1/3}$ on the nonscalar complexity of the polynomials $\sum_{j=1}^{n} e^{2\pi i/2^j} X^j$ or $\sum_{j=1}^{n} 2^{2^j} X^j$. Apart from some improvements in estimations we have followed his paper in the section on polynomials with rapidly growing integer coefficients rather closely.

Schnorr [449] improved Strassen's results by showing lower bounds for the nonscalar complexity of order $(n/\log n)^{1/2}$ for the above polynomials. He also gave lower bounds for the total number of multiplications. The representation theorem (9.9) and the first part of Prop. (9.33) are due to him. The second part of this proposition appears in Schnorr and van de Wiele [452], where the proof methods were extended to additive complexity.

A particularly elegant way to prove these results was discovered by Heintz and Sieveking [247]. Their approach uses the concept of the degree of a variety and Bézout's inequality, and culminates in the beautiful Theorems (9.15), (9.35). (We remark that Heintz and Morgenstern [244] have recently published an analogue of this theorem which establishes lower bounds on the complexity of polynomials in terms of their roots.)

The minimum number of nonscalar multiplications required to evaluate a polynomial was investigated before Strassen's paper [499] by Paterson and Stockmeyer [415], who proved Propositions (9.1) and (9.2). Actually, they showed the somewhat sharper upper bound stated in Ex. 9.4. Their result implies the existence of rational polynomials of high nonscalar complexity. A corresponding result for the additive complexity was established later by Borodin and Cook [63] (compare Ex. 9.17).

The existence of polynomials with 0,1-coefficients of high complexity was first pointed out by Lipton [339]. Stronger results with this respect were obtained by Schnorr [449], and Schnorr and van de Wiele [452]. The proof of Cor. (9.19) stating that most polynomials with 0,1-coefficients are hard to compute is due to Heintz and Schnorr [246].

Lipton and Stockmeyer [340] showed the existence of easy to compute polynomials having hard factors. The strengthening of this statement in Cor. (9.17) and Prop. (9.18) is due to Heintz [240].

The examples of polynomials of high complexity in Cor. (9.21) and (9.24) are due to von zur Gathen and Strassen [192]. (For results related to Lemma (9.20) the reader may consult Besicovich [42].)

A different method yielding many of the results in this chapter has been described by Stoß [490, 491]. We have sketched some of his ideas in Ex. 9.7 and 9.8. The short and elementary proof of Thm. (9.3) has been communicated to us by Baur [30].

The elegant way of establishing an "almost specific" integer polynomial of high complexity in Ex. 9.12 is due to Heintz and Matera [242]. It is based on the classical result of algebraic geometry expressed in Ex. 9.10, a proof of which may be found, e.g., in Harris [228, p. 230]. Ex. 9.13 is from von zur Gathen and Strassen [192]. The proof of Chebyshev's inequality given in the Ex. 9.19–9.21 has been taken from Apostol [10].

Problems 9.1, 9.2 and 9.3 appeared in Strassen [510]. Problem 9.4 is stated in Shub and Smale [481] (compare also Strassen [510]). There it is shown that if this problem has an affirmative answer, then $\mathbf{P} \neq \mathbf{NP}$ over the complex numbers in the sense of Blum, Shub, and Smale (see also the forthcoming book by Blum et al. [58]).

Chapter 10. Branching and Degree

We present a degree bound for computation trees with equality testing due to Strassen [504]. More specifically, we prove a general lower bound on the multiplicative complexity of collections for subsets of k^n, where k is algebraically closed, resp. an arbitrary infinite field. Two applications to the problem of computing the Euclidean representation of two polynomials are discussed. First, we give Strassen's optimality proof of the Knuth-Schönhage algorithm which has been presented in Sect. 3.1. Then we discuss Schuster's lower bound [467] for the problem of computing just the degree pattern of the Euclidean representation. The latter relies on an analysis of the Euclidean representation by means of the subresultant theorem.

Unless otherwise specified k denotes in this chapter an algebraically closed field. Only computation trees with equality testing, $P = \{=\}$, are considered. The notion of computation time always refers to the cost function $c_* = 1_{\{*,/\}}$, i.e., only multiplications and divisions are counted.

10.1 Computation Trees and the Degree Bound

In Sect. 4.4 we have formalized the notion of a computation tree. Such a tree T computes a collection (φ, π) for a constructible subset $J \subseteq k^n$ which consists of a partition $\pi = \{J_1, \ldots, J_t\}$ of J and functions $\varphi_i = \varphi_{|J_i} : J_i \to k^{m_i}$ for $i \in \underline{t}$. The intuitive interpretation is that on an input $\xi \in k^n$ the computation tree T decides in which of the subsets J_i the point ξ lies and computes the value $\varphi(\xi)$. By Prop. (4.27) we may assume that π has a refinement $\{D_1, \ldots, D_s\}$ into locally closed subsets such that all $\varphi_{|D_j}$ are restrictions of rational maps, since otherwise (φ, π) cannot be computed by a tree. Thus all the sets J_i are constructible. Now we assume additionally that the J_i are irreducible locally closed sets and that the $\varphi_i : J_i \to k^{m_i}$ are restrictions of rational maps (e.g., J_i might coincide with some D_j). Then we can show that the logarithm of the degree of graph φ_i yields a lower bound on the running time of almost all inputs $a \in J_i$. This is the contents of the following theorem, which combines the degree bound (8.36) with the results of Sect. 4.5 relating the complexity of collections to straight-line complexity.

(10.1) Degree Bound for Computation Trees (Strassen). *Let T be a computation tree over k which computes a collection (φ, π) for a constructible subset $J \subseteq k^n$ in time $\tau : J \to [0, \infty)$.*

(1) *Let $D = D_v$ denote the input set belonging to an output node v, thus D_v is the set of those inputs $\xi \in J$ whose path ends up with v. Then we have for all $\xi \in D$*

$$\tau(\xi) \geq \log \deg \operatorname{graph} \varphi_{|D}.$$

(2) *Assume that $J_1 \in \pi$ is an irreducible, locally closed subset, and that $\varphi_1 := \varphi_{|J_1} : J_1 \to k^{m_1}$ is a restriction of a rational map. Then there exists an open dense subset U of J_1 such that for all $\xi \in U$*

$$\tau(\xi) \geq \log \deg \operatorname{graph} \varphi_1.$$

In particular, for the multiplicative complexity $C(\varphi, \pi)$ we have

$$C(\varphi, \pi) \geq \log \deg \operatorname{graph} \varphi_1.$$

Proof. We first show that it is sufficient to prove the first part of the theorem. For the following reasonings in general topology compare Ex. 4.12 and 4.17 in Chap. 4. The set J_1 is the disjoint union of finitely many of the D_v's. Since J_1 is assumed to be irreducible, there is some output node v such that $D := D_v$ is dense in J_1. D is locally closed, hence it is open in its closure J_1. As $\operatorname{graph} \varphi_{|D}$ is open and dense in $\operatorname{graph} \varphi_1$, we conclude by Lemma (8.26) that the degrees of these locally closed sets coincide. Therefore, the second part of the theorem follows from the first part by choosing $U = D$.

Let v be an output node of T and let $(k[X]_d; X_1, \ldots, X_n)$ be the universal input of the straight-line program Γ obtained by forgetting the test instructions and the output instructions along the path belonging to v in T. By the Facts (4.24) and (4.25) of Sect. 4.5 there are

$$f_1, \ldots, f_m, g_1, \ldots, g_r, h_1, \ldots, h_s \in k[X]_d$$

such that

$$D = \{\xi \in J \mid g_1(\xi) = \ldots = g_r(\xi) = 0, (h_1 \cdots h_s d)(\xi) \neq 0\},$$

and such that for all $\xi \in D$

$$\varphi(\xi) = (f_1(\xi), \ldots, f_m(\xi)),$$

$$\tau(\xi) \geq L(f_1, \ldots, f_m, g_1, \ldots, g_r, h_1, \ldots, h_s),$$

where L denotes the multiplicative complexity in $k(X)$ with respect to the inputs $X_1, \ldots, X_n$. The degree bound (8.36) implies

$$L(f_1, \ldots, f_m, g_1, \ldots, g_r) \geq \log \deg G_1,$$

where G_1 is the irreducible locally closed set

$$G_1 := \{(\xi, \eta, \zeta) \in k^n \times k^m \times k^r \mid d(\xi) \neq 0, \forall i, j : f_i(\xi) = \eta_i, g_j(\xi) = \zeta_j\}.$$

Intersecting with an open subset and then with the affine subspace $k^n \times k^m \times 0$ we conclude by Bézout's inequality (8.28) that $\deg G_1 \geq \deg G_2$, where

$$G_2 := \deg\{(\xi, \eta) \in k^n \times k^m \mid (h_1 \cdots h_s d)(\xi) \neq 0, \forall i, j : f_i(\xi) = \eta_i, g_j(\xi) = 0\}.$$

However, G_2 equals graph $\varphi_{|D}$, so the first part of the theorem follows. □

As an immediate corollary we obtain a lower bound for the complexity of computing a partition of a constructible subset in k^n.

(10.2) Corollary. *Let π be a partition of a constructible subset $J \subseteq k^n$ and let $J_1 \in \pi$ be irreducible and locally closed. Then $C(\pi) \geq \log \deg J_1$.*

(10.3) Example. Consider the problem of verifying whether given elements ξ_i in $k, i \in \underline{n}$, are the roots of a given polynomial $T^n + \sum_{i=1}^{n} \eta_i T^{n-i}$. More formally, we are interested in the complexity of the membership problem $\{W, k^n \setminus W\}$, where

$$W := \{(\xi, \eta) \in k^{2n} \mid \forall i : \eta_i = \sigma_i(\xi)\}$$

is the graph of the elementary symmetric functions $\sigma_1, \ldots, \sigma_n$. By Cor. (8.41) we know that $\deg W \geq n!$. Therefore, by Cor. (10.2) we have

$$C(\{W, k^n \setminus W\}) \geq n(\log n - 2).$$

An extensive discussion of membership problems can be found in Chap. 11 and the Notes given there. ●

We can also say something if we drop the assumption that k is algebraically closed. Let us first call attention to a certain phenomenon. Consider a computation tree which on input (x, y, z) computes $x^3 + y^3 + z^3$ without division and then tests the result for zero. The set $D = \{(x, y, z) \in \mathbb{Q}^3 \mid x^3 + y^3 + z^3 = 0\}$ is finite (cf. Edwards [155]) and thus not dense in the hypersurface $D' = \{(x, y, z) \in \overline{\mathbb{Q}}^3 \mid x^3 + y^3 + z^3 = 0\}$ of $\overline{\mathbb{Q}}^3$. Another example for this phenomenon is, e.g., $D' = \{(x, y) \in \mathbb{C}^2 \mid x^2 + y^2 = 0\}$, $D = D' \cap \mathbb{R}^2 = \{(0, 0)\}$.

The following statement supplements Thm. (10.1) and holds over any infinite field.

(10.4) Theorem (Strassen). *Let k be an infinite field and denote by $\bar{k}$ its algebraic closure. Let T be a computation tree over k which computes a collection (φ, π) for a constructible subset $J \subseteq k^n$ in time $\tau : J \to [0, \infty)$. Assume that $J_1 \in \pi$ is irreducible, locally closed, and that $\varphi_{|J_1} : J_1 \to k^m$ is the restriction of a rational map. Furthermore suppose that the Zariski closure $G \subseteq \bar{k}^{n+m}$ of graph $\varphi_{|J_1}$ is the zeroset of polynomials of degree $\leq d$ where $d \geq 2$. Then there exists an open dense subset U of J_1 such that for all $\xi \in U$*

$$\tau(\xi) \geq \log \deg G - (n - \dim G) \log d.$$

Proof. The reader should be aware that there are nonconstant polynomials $g \in k[X_1, \ldots, X_n]$ without any zero in k^n if k is not algebraically closed. If T is executable on k^n, then it might not be executable on all inputs in $\bar{k}^n$. In order to fix this problem, we modify the computation tree T in the following way. Before each division instruction we insert a test to check whether the divisor is zero. If so, then we stop the computation and end up in an additional output node. Otherwise, we continue as with T. The resulting computation tree T' is executable on $\bar{k}^n$ and computes some collection (φ', π') for $\bar{k}^n$ in time τ', where $\tau'(\xi) = \tau(\xi)$ for all $\xi \in J$. (Note that we do not count test instructions.) Moreover, we have $\varphi'_{|J} = \varphi$.

As in the proof of Thm. (10.1) we see that there is some output node v such that the set D of inputs $\xi \in J$ whose path in T ends up with v is open and dense in J_1. We denote the set of $\xi \in \bar{k}^n$ whose path in T' ends up with v by D'. Note that D is obviously contained in D', but it is not necessarily a dense subset. (Compare the comment right before this theorem.) By applying Thm. (10.1) to the tree T' we get for all $\xi \in D'$

$$\tau'(\xi) \geq \log \deg W,$$

where W is the Zariski closure of graph $\varphi'_{|D'}$. Clearly $\dim W \leq n$. Thm. (8.50) applied to the pair $G \subseteq W$ of closed subsets yields the estimate

$$\log \deg W \geq \log \deg G - (\dim W - \dim G) \log d.$$

Hence the assertion follows.   □

Applications of these results will be given in the next sections, where the problem of computing the continued fraction expansion of two polynomials is discussed.

10.2 Complexity of the Euclidean Representation

The Euclidean representation $(Q_1, \ldots, Q_{t-1}, A_t)$ of a pair of univariate polynomials (A_1, A_2) was defined and discussed in Sect. 3.1. We recall that $Q_1, \ldots, Q_{t-1}$ is the sequence of quotients obtained from A_1, A_2 by the Euclidean algorithm

$$
\begin{aligned}
A_1 &= Q_1 A_2 + A_3, \\
A_2 &= Q_2 A_3 + A_4, \\
&\ \ \vdots \\
A_{t-2} &= Q_{t-2} A_{t-1} + A_t, \\
A_{t-1} &= Q_{t-1} A_t,
\end{aligned}
$$

and A_t is the GCD of A_1 and A_2. The sequence

$$(d_1, \ldots, d_{t-1}, d_t) := (\deg Q_1, \ldots, \deg Q_{t-1}, \deg A_t)$$

is called the corresponding *degree pattern*. The Knuth-Schönhage algorithm – which can be formalized as a computation tree – computes the coefficients of the Euclidean representation of A_1, A_2 with degree pattern $(d_1, \ldots, d_{t-1}, d_t)$ in time $O(n(1 + H))$, where $H := \mathcal{H}(d_1, \ldots, d_t)$ is the entropy of the probability vector $n^{-1}(d_1, \ldots, d_t)$ and $n := \sum_{i=1}^{t} d_i$ (cf. Cor. (3.16)). Our goal is to prove that this is essentially optimal.

Let us first fix some notation. For $n \geq m \geq 0$ let $J(n, m)$ denote the set of pairs of polynomials (A_1, A_2) of degree n, m, respectively. We will identify this set with the open subset

$$J(n, m) := \{(a_n, \ldots, a_0, b_m, \ldots, b_0) \in k^{m+n+2} \mid a_n b_m \neq 0\}$$

of k^{m+n+2}. We recall that a *degree pattern of type* (n, m) is a sequence $(d_1, \ldots, d_t)$ $\in \mathbb{N}^t$ with $t \geq 2$, $d_i > 0$ for $1 < i < t$, and $n = \sum_{i=1}^{t} d_i$, $m = \sum_{i=2}^{t} d_i$. By $D(d_1, \ldots, d_t)$ we denote the set of all pairs of polynomials $(A_1, A_2) \in J(n, m)$ whose Euclidean representation has the degree pattern $(d_1, \ldots, d_t)$. If we want to emphasize the field k we will write $J_k(n, m)$, $D_k(d_1, \ldots, d_t)$, respectively. It is clear that the $D(d_1, \ldots, d_t)$ form a partition of $J(n, m)$ when $(d_1, \ldots, d_t)$ varies over all degree patterns of type (n, m). Moreover, $D(n - m, 1, \ldots, 1, 0)$ is a nonempty open subset of $J(n, m)$, thus $(n - m, 1, \ldots, 1, 0)$ is the "generic degree pattern" (cf. Ex. 5.11). For a degree pattern $(d_1, \ldots, d_t)$ we define the irreducible open subset

$$J(d_1, \ldots, d_t) := \prod_{i=1}^{t} (k^{\times} \times k^{d_i})$$

of k^{t+n}, whose elements are interpreted as sequences $(Q_1, \ldots, Q_{t-1}, A_t)$ of polynomials of degrees $d_1, \ldots, d_t$. The Euclidean algorithm gives us for each degree pattern $d = (d_1, \ldots, d_t)$ a bijective map

$$\varphi_d : D(d_1, \ldots, d_t) \to J(d_1, \ldots, d_t),$$

which assigns to (A_1, A_2) its Euclidean representation. These maps together with the partition of $J(n, m)$ into the $D(d_1, \ldots, d_t)$ define a collection for $J(n, m)$, which we call the *Euclidean representation for* $J(n, m)$. The Euclidean algorithm, translated to a computation tree T, computes this collection in time $O(n^2)$. The output nodes of T can be indexed by the degree patterns for (n, m), and the set of inputs whose path ends up with the output node corresponding to $(d_1, \ldots, d_t)$ is exactly $D(d_1, \ldots, d_t)$. Therefore, the sets $D(d_1, \ldots, d_t)$ are locally closed and the maps φ_d are restrictions of rational maps (see the comments preceding Prop. (4.27)). The inverse ψ_d of φ_d is much easier to handle; it can be described in a concise manner by means of the Euler brackets $[Q_i, \ldots, Q_{t-1}]$ introduced in Sect. 6.3. Recall that they are defined recursively by

$$[Q_{i-1}, \ldots, Q_{t-1}] = Q_{i-1}[Q_i, \ldots, Q_{t-1}] + [Q_{i+1}, \ldots, Q_{t-1}] \quad (i < t),$$

$[Q_{t-1}] = Q_{t-1}$, $[\ ] = 1$. By Lemma (6.14)(2) we know that $[Q_i, \ldots, Q_{t-1}]$ is the sum of the product $Q_i \cdots Q_{t-1}$ and of all subproducts obtained from this by deleting disjoint pairs of adjacent factors. It is straightforward to verify that ψ_d sends $(Q_1, \ldots, Q_{t-1}, A_t)$ to (A_1, A_2), where

(10.5) $$A_1 = [Q_1, \ldots, Q_{t-1}]A_t, \quad A_2 = [Q_2, \ldots, Q_{t-1}]A_t.$$

In particular, ψ_d is a polynomial map. The following observation will be crucial in the proof of Lemma (10.6): the homogeneous part of ψ_d of degree t maps $(Q_1, \ldots, Q_{t-1}, A_t)$ to $(A_t \prod_{i=1}^{t-1} Q_i, 0)$.

Summarizing, we have seen that φ_d is an isomorphism of the irreducible locally closed subsets $D(d_1, \ldots, d_t)$ and $J(d_1, \ldots, d_t)$.

(10.6) Lemma. *For any degree pattern $d = (d_1, \ldots, d_t)$ of type (n, m) we have*

$$\log \deg \operatorname{graph} \varphi_d \geq n(\mathcal{H}(d_1, \ldots, d_t) - 2).$$

Proof. Recall that the morphism $\psi_d \colon J(d_1, \ldots, d_t) \to D(d_1, \ldots, d_t)$ is the inverse of φ_d, hence $\operatorname{graph} \varphi_d$ is the same as $\operatorname{graph} \psi_d$ up to a coordinate permutation, so their degrees coincide. So we can as well consider the map ψ_d. We already noted that the homogeneous part of degree t of ψ_d is essentially the polynomial multiplication,

$$\psi_d^{(t)} \colon (Q_1, \ldots, Q_{t-1}, A_t) \mapsto (Q_1 \cdots Q_{t-1}A_t, 0).$$

By Thm. (8.51) we have $\deg \operatorname{graph} \psi_d \geq \deg \operatorname{graph} \psi_d^{(t)}$. To estimate the latter degree we argue exactly as in the proof of Prop. (8.44).

Let B be a monic polynomial of degree n with zeroset M such that $|M| = n$. We intersect the graph of $\psi_d^{(t)}$ with an affine linear subspace by requiring

$$Q_1 \cdots Q_{t-1}A_t = B,$$
$$Q_1, \ldots, Q_{t-1}, A_t \text{ are monic having degree } d_1, \ldots, d_{t-1}, d_t, \text{ respectively.}$$

This intersection is finite and consists of the $n!/(d_1! \cdots d_t!)$ sequences $(Q_1, \ldots, Q_{t-1}, A_t; B, 0)$, where

$$Q_i = \prod_{\theta \in M_i} (X - \theta) \ (i < t), \quad A_t = \prod_{\theta \in M_t} (X - \theta)$$

and $\{M_1, \ldots, M_t\}$ is a partition of M into subsets of cardinalities $d_1, \ldots, d_t$, respectively. Therefore, by Bézout's inequality (8.28), we have

$$\deg \psi_d^{(t)} \geq \frac{n!}{d_1! \cdots d_t!}.$$

The assertion of the lemma follows now from the estimate in Rem. (8.43).   □

From this proposition and the degree bound for computation trees (10.1) we immediately conclude the main result of this section.

(10.7) Theorem (Strassen). *Let the Euclidean representation for $J(n, m)$ be computable in time τ. Then any $D(d_1, \ldots, d_t)$ contains an open dense subset U such that for all $(A_1, A_2) \in U$ we have*

$$\tau(A_1, A_2) \geq n(\mathcal{H}(d_1, \ldots, d_t) - 2).$$

Therefore, the Knuth-Schönhage algorithm has optimal running time (up to a constant factor) on each of the sets $D(d_1, \ldots, d_t)$.

We emphasize that this result is much stronger than a statement about the worst case, since it bounds the running time for almost all inputs in each of the sets $D(d_1, \ldots, d_t)$.

From Thm. (10.4) we obtain a lower bound holding over any infinite field k.

(10.8) Theorem (Strassen). *Assume k to be infinite and let the Euclidean representation for $J_k(n, m)$ over k be computable in time τ. Then any $D_k(d_1, \ldots, d_t)$ contains an open dense subset U such that for all $(A_1, A_2) \in U$ we have*

$$\tau(A_1, A_2) \geq (2t/m - 1)n\mathcal{H}(d_1, \ldots, d_t) - 5n.$$

Proof. We may assume w.l.o.g. $t > m/2$ since the bound is trivial otherwise. The Zariski closure G of the graph of the map $(\varphi_d)_k \colon D_k(d_1, \ldots, d_t) \to J_k(d_1, \ldots, d_t)$ in $\overline{k}^{(m+n+2)+(n+t)}$ equals the graph of

$$\varphi_d := (\varphi_d)_{\overline{k}} \colon D_{\overline{k}}(d_1, \ldots, d_t) \to J_{\overline{k}}(d_1, \ldots, d_t),$$

since $J_k(d_1, \ldots, d_t)$ is dense in $J_{\overline{k}}(d_1, \ldots, d_t)$ and the isomorphism φ_d is defined over k.

We apply now Thm. (10.4). $G = \text{graph}\,\varphi_d$ has dimension $n + t$ and Lemma (10.6) gives the degree estimate $\log \deg G \geq n(H - 2)$, where $H = \mathcal{H}(d_1, \ldots, d_t)$. Moreover, Eq. (10.5) shows that $\text{graph}\,\psi_d$, and therefore also $\text{graph}\,\varphi_d$, is the zeroset of polynomials of degree $\leq t$. Therefore, it is sufficient to verify the numerical estimate

(A) $n(H - 2) - (n + m + 2 - (n + t)) \log t \geq (2t/m - 1)nH - 5n.$

Let $\varepsilon := t/m - 1/2 > 0$. Since $1/n \leq d_i/n \leq 1/2$ for all $1 < i < t$ with at most one exception, we have $nH \geq (t - 3) \log n$. Hence $nH \geq (m/2 - 3) \log n$, which implies $2nH + 6 \log n \geq m \log n$. Therefore

$$
\begin{aligned}
&n(H - 2) - (m + 2 - t) \log t \\
&\geq \quad n(H - 2) - (1/2 - \varepsilon)m \log n - 2 \log n \\
&\geq \quad n(H - 2) - (1 - 2\varepsilon)nH - 5 \log n \\
&\geq \quad 2\varepsilon nH - 5n,
\end{aligned}
$$

which shows (A). $\square$

The reader should compare our results with the following assertion saying that the Euclidean algorithm is essentially optimal if one insists on computing the remainders A_i in addition to the quotients Q_i.

(10.9) Proposition. *Let the collection (φ, π) consist of the partition of $J(n, n-1)$ into the $D(d_1, \ldots, d_t)$ together with the functions*

$$D(d_1, \ldots, d_t) \ni (A_1, A_2) \mapsto (A_1, A_2, \ldots, A_t, Q_1, \ldots, Q_{t-1}).$$

Then we have $C(\varphi, \pi) \geq n(n+3)/2$.

Proof. As usual, we denote the set of coefficients of a set of polynomials $F \subseteq k[X]$ by $C(F)$. (This is not to be confused with the branching complexity!) Let $F := \{1, A_1, A_2, \ldots, A_{n+1}, Q_1, \ldots, Q_n\}$. We may view the coefficients of the polynomials in F as rational functions in the coefficients of $(A_1, A_2) \in D(1, \ldots, 1, 0)$. (Recall that $D(1, \ldots, 1, 0)$ is open and dense in $J(n, n-1)$.) As in the proof of Prop. (4.26) we see that

$$L\big(C(F) \mid C(A_1, A_2)\big) \leq C(\varphi, \pi).$$

By the dimension bound (4.12) this multiplicative complexity is not smaller than

$$\dim\big(k + \langle C(F)\rangle_k\big) - (2n+2).$$

We want to show that this has the maximal possible value

$$1 + \sum_{i=1}^{n+1}(1 + \deg A_i) + \sum_{j=1}^{n}(1 + \deg Q_j) - (2n+2) = n(n+3)/2,$$

i.e., all the coefficients of the polynomials in F are linearly independent. By factoring all $f \in C(F)$ over the isomorphism $J(1, \ldots, 1, 0) \xrightarrow{\sim} D(1, \ldots, 1, 0)$ we may now view the coefficients of F as rational functions in the coefficients of $Q_1, \ldots, Q_n, A_{n+1}$, and we need to show that they are linearly independent. (Note the following: let $\psi: J \to D$ be a surjective map and $f_1, \ldots, f_r$ be functions $D \to k$. Then $f_1, \ldots, f_r$ are linearly independent over k iff $f_1 \circ \psi, \ldots, f_r \circ \psi$ are linearly independent over k.)

Let $A_{n+1} = \varepsilon$, $Q_i = \alpha_i X + \beta_i$. From the relation

$$
\begin{aligned}
A_n &= (\alpha_n X + \beta_n)\varepsilon, \\
A_{n-1} &= (\alpha_{n-1} X + \beta_{n-1})A_n + \varepsilon, \\
&\vdots \\
A_1 &= (\alpha_1 X + \beta_1)A_2 + A_3
\end{aligned}
$$

one can conclude by reverse induction on i that the coefficients of A_i are polynomials of degree $n + 2 - i$ in $\varepsilon, \alpha_1, \beta_1, \ldots, \alpha_n, \beta_n$, which are linearly independent over k. (The details are left as an exercise to the reader, cf. Ex. 10.2.) Thus the assertion follows. $\qquad\square$

10.3* Degree Pattern of the Euclidean Representation

We discuss here the problem of computing just the degree pattern of the Euclidean representation of given polynomials A_1, A_2. Formally, we study the complexity of the partition of $J(n, m)$ into the sets $D(d_1, \ldots, d_t)$. Cor. (10.2) tells us that the logarithms of the degrees of the locally closed subsets $D(d_1, \ldots, d_t)$ constitute lower bounds on this complexity. In the sequel, we will estimate these degrees from below. For doing this we need to collect some facts concerning subresultants.

In the sequel let $n \geq m \geq 0$ and $A = \sum_{i=0}^{n} a_i X^i$, $B = \sum_{j=0}^{m} b_j X^j \in k[X]$ be univariate polynomials of degree n, m, respectively. Our first goal is to find an explicit condition on the coefficients of A and B expressing that A and B have at least $\ell + 1$ common roots, that is, $\deg \mathrm{GCD}(A, B) \geq \ell + 1$.

(10.10) Lemma. *Let $0 \leq \ell < m$. The GCD of A and B has degree $\geq \ell + 1$ iff there are nonzero polynomials $U, V \in k[X]$ with $\deg U < m - \ell$, $\deg V < n - \ell$ such that $UA + VB = 0$.*

Proof. Let $C := \mathrm{GCD}(A, B)$. If $\deg C \geq \ell + 1$, then $\tilde{U} := B/C$, $\tilde{V} := -A/C$ satisfy the desired condition. To show the reverse direction note that $UA + VB = 0$ implies $V\tilde{U} = U\tilde{V}$. Hence, as $\tilde{U}, \tilde{V}$ are coprime, $\tilde{U}$ must be a divisor of U and therefore $\deg \tilde{U} \leq \deg U$, which implies $\deg C \geq \ell + 1$. $\qquad\square$

For the moment let $\mathcal{P}_i$ denote the k-space of univariate polynomials of degree less than or equal to i. Consider the linear map

$$\Lambda_\ell(A, B) : \mathcal{P}_{m-\ell-1} \times \mathcal{P}_{n-\ell-1} \to \mathcal{P}_{m+n-\ell-1}, \quad (U, V) \mapsto UA + VB.$$

Lemma (10.10) tells us that

$$\deg \mathrm{GCD}(A, B) > \ell \iff \ker \Lambda_\ell(A, B) \neq 0.$$

How does the matrix of $\Lambda_\ell(A, B)$ with respect to the monomial bases of $\mathcal{P}_i$ look like? Let

$$U = \sum_{i < m-\ell} u_i X^i, \quad V = \sum_{i < n-\ell} v_i X^i, \quad W = \sum_{i < m+n-\ell} w_i X^i = UA + VB.$$

It is straightforward to check that

$$(u_{m-\ell-1}, \ldots, u_0; v_{n-\ell-1}, \ldots, v_0) \cdot M_\ell(A, B) = (w_{m+n-\ell-1}, \ldots, w_0),$$

where $M_\ell(A, B)$ is the following matrix:

$$M_\ell(A, B) := \left. \begin{array}{c} \underbrace{\begin{array}{c} \left. \begin{array}{cccccccc} a_n & a_{n-1} & \cdots & & a_0 & & & \\ & a_n & a_{n-1} & \cdots & & a_0 & & \\ & & \ddots & & & & \ddots & \\ & & & a_n & a_{n-1} & \cdots & & a_0 \\ b_m & b_{m-1} & \cdots & & b_0 & & & \\ & b_m & b_{m-1} & \cdots & & b_0 & & \\ & & \ddots & & & & \ddots & \\ & & & b_m & b_{m-1} & \cdots & & b_0 \end{array} \right\} \end{array}}_{m+n-\ell} \end{array} \right. \begin{array}{c} \left. \begin{array}{c} \\ \\ \\ \end{array} \right\} m - \ell \\ \left. \begin{array}{c} \\ \\ \\ \end{array} \right\} n - \ell \end{array}$$

(blank spaces stand for zero entries). For $0 \le i \le \ell$ let $S_{\ell,i}(A, B)$ be the determinant of the $(m + n - 2\ell)$-square matrix obtained from $M_\ell(A, B)$ by deleting the columns $0, 1, \ldots, i - 1, i + 1, \ldots, \ell$ when we enumerate the columns from the right starting with zero.

(10.11) Definition. Let $0 \le \ell \le m = \deg B$. The polynomial $\mathrm{SRes}_\ell(A, B) := \sum_{i=0}^{\ell} S_{\ell,i}(A, B) X^i$ is called the ℓth *subresultant* of the polynomials A and B. Its coefficients $S_{\ell,i}(A, B)$ are called *subresultant coefficients* and $S_{\ell,\ell}(A, B)$ is the ℓth *principal subresultant coefficient*. The *resultant* of A and B is defined as the 0th subresultant: $\mathrm{Res}(A, B) := \mathrm{SRes}_0(A, B)$.   •

(10.12) Remarks.
(1) If the coefficients a_μ, b_ν of A, B are indeterminates, then $S_{\ell,i}(A, B)$ is an integer polynomials which is homogeneous of degree $\deg B - \ell = m - \ell$ in the a_μ's and homogeneous of degree $\deg A - \ell = n - \ell$ in the b_ν's.
(2) We have $\mathrm{SRes}_m(A, B) = b_m^{n-m-1} B$.
(3) If $\deg \mathrm{GCD}(A, B) > \ell$, then we have $\mathrm{SRes}_j(A, B) = 0$ for all $0 \le j < \ell$. In Ex. 10.8 we will show that the converse is also true.   •

(10.13) Example. Let $n = 3, m = 2, \ell = 1$. Then

$$M_1(A, B) = \begin{pmatrix} a_3 & a_2 & a_1 & a_0 \\ b_2 & b_1 & b_0 & 0 \\ 0 & b_2 & b_1 & b_0 \end{pmatrix},$$

$$S_{1,0}(A, B) = \det \begin{pmatrix} a_3 & a_2 & a_0 \\ b_2 & b_1 & 0 \\ 0 & b_2 & b_0 \end{pmatrix} = a_0 b_2^2 - a_2 b_0 b_2 + a_3 b_0 b_1,$$

$$S_{1,1}(A, B) = \det \begin{pmatrix} a_3 & a_2 & a_1 \\ b_2 & b_1 & b_0 \\ 0 & b_2 & b_1 \end{pmatrix} = a_1 b_2^2 - a_2 b_1 b_2 - a_3 b_0 b_2 + a_3 b_1^2.   •$$

The subsequent theorem relates the subresultants of A and B to the subresultants of B and the remainder R of the division of A by B. (Compare also Ex. 10.7.)

(10.14) Subresultant Theorem (Habicht). *Let $A, B \in k[X]$ with $n := \deg A \geq \deg B =: m \geq 0$, and assume that the remainder $R := A \bmod B$ is nonzero of degree p; put $\sigma_\ell := (-1)^{(m+n-1)(m-\ell)}$. Then*

$$\mathrm{SRes}_\ell(A, B) = \begin{cases} \sigma_\ell b_m^{n-p} \, \mathrm{SRes}_\ell(B, R) & \text{if } 0 \leq \ell \leq p, \\ 0 & \text{if } p < \ell < m - 1, \\ (-1)^{n-m+1} b_m^{n-m+1} R & \text{if } \ell = m - 1, \\ b_m^{n-m-1} B & \text{if } \ell = m. \end{cases}$$

Proof. Let $A = QB + R$ where $Q = \sum_{i=0}^{n-m} q_i X^i$ and $R = \sum_{j=0}^n r_j X^j$ with $r_{p+1} = \ldots = r_n = 0$. From $R = A - \sum_{i=0}^{n-m} q_i(X^i B)$ we conclude that for any $0 \leq \ell \leq m$ we may obtain the $(m - \ell + n - \ell) \times (m + n - \ell)$ matrix

from $M_\ell(A, B)$ by elementary row transformations, i.e., by adding multiples of one row to another one. Recall that $S_{\ell,i}(A, B)$ is the determinant of a submatrix of $M_\ell(A, B)$ obtained by deleting certain columns. Hence the set of such determinants is the same for M'_ℓ and $M_\ell(A, B)$.

Case 1. $0 \leq \ell \leq p$. Let $0 \leq i \leq \ell$. By a permutation of the rows of M'_ℓ we see that $S_{\ell,i}(A, B)$ equals up to a sign the determinant of the submatrix of obtained by deleting the columns $0, 1, \ldots, i-1, i+1, \ldots, \ell$ (we enumerate the columns from the right, starting with zero). Therefore

$$S_{\ell,i}(A, B) = \sigma_\ell b_m^{n-p} S_{\ell,i}(B, R)$$

where the sign σ_ℓ equals $(-1)^{(m+n-2\ell-1)(m-\ell)} = (-1)^{(m+n-1)(m-\ell)}$. (The permutation of the rows is the $(m-\ell)$th power of a cyclic permutation of length $m+n-2\ell$.) This shows the assertion in this case.

Case 2. $p < \ell < m - 1$. In this case the matrix M'_ℓ looks like

The leftmost column of the upper right rectangular matrix equals zero. As $\ell < m - 1$ this column is not going to be deleted. Therefore $S_{\ell,i}(A, B) = 0$ for all $0 \le i \le \ell$.

Case 3. $\ell = m - 1$. Here the matrix obtained from M'_ℓ by deleting the columns $0, 1, \ldots, i-1, i+1, \ldots, \ell$ (enumerated from the right, beginning with zero) equals

Therefore $S_{\ell,i}(A, B) = (-1)^{n-m+1} b_m^{n-m+1} r_i$ for $0 \le i \le \ell$, which shows the assertion in this case.

Case 4. $\ell = m$. This is settled by Rem. (10.12)(2).   □

(10.15) Remark. The above proof shows that $S_{\ell,i}(A, B) = 0$ for $0 \le \ell < m$ if $R = 0$.   •

The following consequence of the subresultant theorem (10.14) will be crucial for us. (For other applications of the subresultant theorem compare the Exercises and the Notes.) We denote the leading coefficient of a nonzero polynomial A by $\mathrm{lc}(A)$.

(10.16) Lemma. *Let $(Q_1, \ldots, Q_{t-1}, A_t)$ be the Euclidean representation of the polynomials A, B of degrees n, m, respectively, $n \geq m \geq 0$, and let $(d_1, \ldots, d_t)$ be the corresponding degree pattern. Set $n_i := \sum_{j=i}^{t} d_j$ for $i \in \underline{t}$. Then we have for all $1 < i < t$*

$$\mathrm{lc}(A)^2 \, S_{n_i, n_i}(A, B) = \varepsilon_i \, \mathrm{lc}(Q_i) \prod_{j=1}^{i-1} \mathrm{lc}(Q_j)^2 \, S_{n_i-1, n_{i+1}}(A, B),$$

where the sign ε_i only depends on the degree pattern $(d_1, \ldots, d_t)$.

Proof. Let $A_1 = A, A_2 = B, A_3, \ldots, A_t$ be the sequence of remainders produced by the Euclidean algorithm. Then $n_i := \deg A_i$. By the subresultant theorem (10.14) we have for all $1 < i < t$

$$\begin{aligned}
\mathrm{SRes}_{n_i-1}(A_{i-1}, A_i) &= (-1)^{n_{i-1}-n_i+1} \mathrm{lc}(A_i)^{n_{i-1}-n_i+1} A_{i+1}, \\
\mathrm{SRes}_{n_i}(A_{i-1}, A_i) &= \mathrm{lc}(A_i)^{n_{i-1}-n_i-1} A_i.
\end{aligned}$$

The subresultant theorem further implies

$$\begin{aligned}
\mathrm{SRes}_{n_i}(A_1, A_2) &= \pm \mathrm{lc}(A_2)^{n_1-n_3} \, \mathrm{SRes}_{n_i}(A_2, A_3) \\
&\;\;\vdots \\
&= \varepsilon_i' \delta \, \mathrm{SRes}_{n_i}(A_{i-1}, A_i),
\end{aligned}$$

where $\delta = \prod_{j=2}^{i-1} \mathrm{lc}(A_j)^{n_{j-1}-n_j+1}$ and $\varepsilon_i' \in \{-1, +1\}$ depends only on the degree pattern. In the same way we obtain

$$\mathrm{SRes}_{n_i-1}(A_1, A_2) = \varepsilon_i'' \delta \, \mathrm{SRes}_{n_i-1}(A_{i-1}, A_i)$$

with some other sign $\varepsilon_i'' \in \{-1, +1\}$, but with *the same* $\delta \in k$.

Altogether we see that

$$\begin{aligned}
\mathrm{SRes}_{n_i-1}(A_1, A_2) &= \varepsilon_i''(-1)^{n_{i-1}-n_i+1} \delta \, \mathrm{lc}(A_i)^{n_{i-1}-n_i+1} A_{i+1}, \\
\mathrm{SRes}_{n_i}(A_1, A_2) &= \varepsilon_i' \delta \, \mathrm{lc}(A_i)^{n_{i-1}-n_i-1} A_i.
\end{aligned}$$

By taking the quotients of the leading coefficients of these two polynomials, we arrive at

$$\frac{S_{n_i-1, n_{i+1}}(A_1, A_2)}{S_{n_i, n_i}(A_1, A_2)} = \varepsilon_i \, \mathrm{lc}(A_i) \, \mathrm{lc}(A_{i+1})$$

with the sign $\varepsilon_i = \varepsilon_i' \varepsilon_i'' (-1)^{n_{i-1}-n_i+1}$. On the other hand we clearly have $\mathrm{lc}(A_i) = \mathrm{lc}(Q_i) \, \mathrm{lc}(A_{i+1})$, which easily implies

$$\mathrm{lc}(A_i) \, \mathrm{lc}(A_{i+1}) = \frac{\mathrm{lc}(A_1)^2}{\mathrm{lc}(Q_i) \prod_{j=1}^{i-1} \mathrm{lc}(Q_j)^2},$$

and proves the lemma. □

We remark that Lemma (10.16) allows to explicitly express the leading coefficients $\mathrm{lc}(Q_j)$ of the quotients (as well as the leading coefficients of the remainders) in the Euclidean representation of $(A, B) \in D(d_1, \ldots, d_t)$ as rational functions in the coefficients of A and B.

Now we are in a position to prove a lower bound on the degree of $D(d_1, \ldots, d_t)$.

(10.17) Proposition. *Let $(d_1, \ldots, d_t)$ be a degree pattern of type (n, m). Then we have*

$$\deg D(d_1, \ldots, d_t) \geq \frac{n!}{d_t! \prod_{i=1}^{t-1} d_i!(d_1 + \ldots + d_i)} \geq \frac{n^s}{2^{3n}},$$

where $s := |\{i \in \underline{t} \mid d_i > 1\}|$.

Proof. We prove the left-hand inequality first. For shortness let us write $D := D(d_1, \ldots, d_t)$, $J := J(d_1, \ldots, d_t)$. To $\lambda \in k^\times$ and a polynomial $A \in k[X]$ of degree n we assign the subset $W_{\lambda, A} \subseteq D$ consisting of all pairs $(A, A_2) \in D$ such that all quotients $Q_1, \ldots, Q_{t-1}$ of the Euclidean representation of (A, A_2) have leading coefficient λ^{-1}. By Lemma (10.16) we can characterize $W_{\lambda, A}$ in the following way

$$W_{\lambda, A} := \left\{ (A, A_2) \in D \ \middle| \ \begin{array}{l} \mathrm{lc}(A_2) = \lambda \, \mathrm{lc}(A), \text{ and for } 1 < i < t \\ S_{n_i-1, n_{i+1}}(A, A_2) = \varepsilon_i \lambda^{2i-1} \, \mathrm{lc}(A)^2 S_{n_i, n_i}(A, A_2) \end{array} \right\}.$$

In particular, $W_{\lambda, A}$ is locally closed. According to Rem. (10.12) the degrees of $S_{n_i, n_i}(A, A_2)$, $S_{n_i-1, n_{i+1}}(A, A_2)$ in the coefficients of A_2 are $n - n_i$, $n - n_i + 1$, respectively, and $n - n_i + 1 \leq d_1 + \ldots + d_i$. Bézout's inequality (8.28) implies therefore

$$\deg W_{\lambda, A} \leq \deg D \cdot \prod_{i=2}^{t-1} (d_1 + \ldots + d_i).$$

If we can show that there exist λ and A such that

$$\frac{n!}{d_1! \cdots d_t!} \leq |W_{\lambda, A}| < \infty,$$

then the left-hand inequality in the statement of the proposition follows.

Recall that we have an isomorphism $\psi_d : J \to D$. We denote the composition of ψ_d with the projection $(A_1, A_2) \mapsto A_1$ by ρ, i.e.,

$$\rho : J \to k^{n+1}, \ (Q_1, \ldots, Q_{t-1}, A_t) \mapsto [Q_1, \ldots, Q_{t-1}]A_t$$

(compare Eq. (10.5)). Further, we consider for $\lambda \in k^\times$ the $(n + 1)$-dimensional closed subsets

$$J_\lambda := \{(Q_1, \ldots, Q_{t-1}, A_t) \mid \forall 1 \leq i < t: \ \mathrm{lc}(Q_i) = \lambda^{-1}\}.$$

Now observe that

$$\psi_d(\rho^{-1}(A) \cap J_\lambda) = W_{\lambda, A}.$$

It is therefore sufficient to prove that for some λ, A

(A)
$$\frac{n!}{d_1! \cdots d_t!} \le |\rho^{-1}(A) \cap J_\lambda| < \infty.$$

For this we embed ρ into a family of morphisms $\rho_\lambda: J \to k^{n+1}$, $\lambda \in k$. To be more specific, we define for $\lambda \in k^\times$ and $z \in J$

$$\rho_\lambda(z) := \lambda^t \rho(\lambda^{-1}z) = \rho^{(t)}(z) + \lambda \rho^{(t-1)}(z) + \ldots + \lambda^t \rho^{(0)}(z),$$
$$\rho_0(z) := \rho^{(t)}(z).$$

Then $\rho_1 = \rho$, and ρ_0 maps the sequence of polynomials $(Q_1, \ldots, Q_{t-1}, A_t)$ to their product, see the comments preceding Lemma (10.6). As in the proof of that lemma we can argue that there is some finite fiber of $\rho_{0|J_1}$ containing at least $n!/(d_1! \cdots d_t!)$ elements. By the theorem on fibers (8.38) we conclude that there exist some $\lambda \in k^\times$ and some $A \in k[X]$ of degree n such that $\rho_\lambda^{-1}(\lambda^t A) \cap J_1$ is finite and contains at least that many elements. (Consider the morphism $k \times J_1 \to k \times k^{n+1}$, $(\lambda, z) \mapsto (\lambda, \rho_\lambda(z))$.) But the bijective map $z \mapsto \lambda^{-1}z$ sends $\rho_\lambda^{-1}(\lambda^t A) \cap J_1$ to $\rho^{-1}(A) \cap J_\lambda$. Hence the estimate (A) follows.

It remains to verify the numerical estimate

$$\log \frac{n!}{d_1! \cdots d_t!} - \sum_{i=1}^{t-1} \log(d_1 + \ldots + d_i) \ge s \log n - 3n.$$

By Rem. (8.43) the left-hand side is at least

$$n\mathcal{H}(d_1, \ldots, d_t) - t \log n - 2n = \sum_{d_i > 1} d_i \log \frac{n}{d_i} - s \log n - 2n.$$

Together with the estimate

$$d \log \frac{n}{d} \ge 2 \log n - d$$

holding for all $2 \le d \le n$, this implies the assertion. To verify the latter estimate, note that it is equivalent to $d(d/2)^{2/(d-2)} \le 2n$, which is a consequence of $(d/2)^{2/(d-2)} \le 2$. $\qquad\square$

From the degree bound for computation trees (10.1) and the last proposition we immediately obtain the subsequent lower bound result for the problem of computing the degree pattern of the Euclidean representation of two given polynomials.

(10.18) Theorem (Schuster). *Let us assume that the partition of $J(n, m)$ into the sets $D(d_1, \ldots, d_t)$ is computable in time τ. Then any $D(d_1, \ldots, d_t)$ contains an open dense subset U such that for all $(A_1, A_2) \in U$*

$$\tau(A_1, A_2) \ge s(d_1, \ldots, d_t) \log n - 3n,$$

where $s(d_1, \ldots, d_t) := |\{i \in \underline{t} \mid d_i > 1\}|$ is the number of degree jumps greater than one.

10.4 Exercises

10.1. The inverse ψ_d of φ_d describes the conversion of a continued fraction into a rational function. In Ex. 3.6 we have shown that this conversion can be performed in time $O(n(1 + \mathcal{H}(d_1, \ldots, d_t)))$. Prove a lower bound of this order for this task.

10.2. Complete the proof of Prop. (10.9).

The next two exercises are preparations for Ex. 10.5.

10.3. In the following we denote by $\mathbb{F}_p$ the finite field with p elements. Let $n \in \mathbb{N}'$. We define the disjoint union $\mathcal{A}^n$ of $\mathbb{Q}^n$ and of all $\mathbb{F}_p^n$ as the set consisting of the elements $(k; \xi)$, where $\xi \in k^n$, and $k = \mathbb{Q}$ or $k = \mathbb{F}_p$ for some prime p. $\mathcal{A}^n$ carries a Zariski topology, a basis being given by the sets

$$\{(k; \xi) \in \mathcal{A}^n \mid f(\xi) \neq 0\},$$

where $f \in \mathbb{Z}[X_1, \ldots, X_n]$. It is easy to see that $\mathcal{A}^n$ is irreducible. (For some general facts concerning the Zariski topology compare the Exercises of Chap. 4.)

$J_k(n, m)$ describes the set of pairs of polynomials (A, B) over k of degrees $n \geq m$. Let $\mathcal{J}(n, m)$ denote the disjoint union of $J_{\mathbb{Q}}(n, m)$ and of all $J_{\mathbb{F}_p}(n, m)$ where p ranges over all primes. By identifying a polynomial with its coefficient sequence we may interpret $\mathcal{J}(n, m)$ as an open subset of $\mathcal{A}^{n+m+2}$. Similarly, we may define the topological spaces $\mathcal{J}(d_1, \ldots, d_t)$ and $\mathcal{D}(d_1, \ldots, d_t)$ for a degree pattern for (n, m). The Euclidean representation defines a bijection $\varphi_d \colon \mathcal{D}(d_1, \ldots, d_t) \to \mathcal{J}(d_1, \ldots, d_t)$. This defines a collection for $\mathcal{J}(n, m)$ (in a more general sense than defined in Sect. 4.4), which we will call the Euclidean representation for $\mathcal{J}(n, m)$. Prove the following facts:

(1) φ_d is a homeomorphism.
(2) $\mathcal{D}(d_1, \ldots, d_t)$ is irreducible.
(3) Set $D_k(d_1, \ldots, d_t) := \mathcal{D}(d_1, \ldots, d_t) \cap J_k(n, m)$. Show that $D_{\mathbb{Q}}(d_1, \ldots, d_t)$ and the union of the $D_{\mathbb{F}_p}(d_1, \ldots, d_t)$ taken over all primes p are dense subsets of $\mathcal{D}(d_1, \ldots, d_t)$.

10.4. We define a computation tree T over $\mathbb{Z}$ as in Def. (4.19) just by replacing the field k by the ring of integers $\mathbb{Z}$, and by requiring $P = \{=\}$. An input to such a computation tree T (expecting inputs of length n) is defined as a pair $(A; \xi)$, where A is a ring and $\xi \in A^n$. In particular, we may take $A = \mathbb{Q}$ or $A = \mathbb{F}_p$ for some prime p. Show the following:

(1) If T is executable on all elements of $\sqcup_p \mathbb{F}_p^n$, then T is executable on $\mathbb{Q}^n$.
(2) Assume T computes the Euclidean representation for $\sqcup_p J_{\mathbb{F}_p}(n, m)$. Then T also computes the Euclidean representation for $\mathcal{J}(n, m)$.

(Hint: Transfer the discussion in Sect. 4.4–4.5 to computation trees over $\mathbb{Z}$. Use the ring morphisms mod p: $\mathbb{Z}_N := \{aN^{-s} \mid a \in \mathbb{Z}, s \in \mathbb{N}\} \rightarrow \mathbb{F}_p$ defined for primes p not dividing N.)

10.5. Show that Thm. (10.8) is also true for the Euclidean representation over finite fields $\mathbb{F}_p$ with p varying. More specifically, it holds for computation trees over $\mathbb{Z}$ computing the Euclidean representation for $\sqcup_p J_{\mathbb{F}_p}(n, m)$. (Hint: use Ex. 10.3 and 10.4.)

10.6. In this exercise we prove Thm. (10.4) for $k = \mathbb{R}$ and computation trees where $\leq$-branching is allowed.

(1) Let T be a computation tree over $\mathbb{R}$ which computes a collection (φ, π) for a semi-algebraic subset $J \subseteq \mathbb{R}^n$ in time $\tau: J \rightarrow [0, \infty)$. Assume that $J_1 \in \pi$ is irreducible and that $\varphi_{|J_1}: J_1 \rightarrow k^m$ is the restriction of a rational map. Furthermore suppose that the Zariski closure $G \subseteq \mathbb{C}^{n+m}$ of graph $\varphi_{|J_1}$ is the zeroset of polynomials of degree $\leq d$ for some $d \geq 2$. Then there exists a semi-algebraic subset $Z \subseteq J_1$ with $\dim Z < \dim J_1$ such that for all $\xi \in J_1 \setminus Z$

$$\tau(\xi) \geq \log \deg G - (n - \dim G) \log d.$$

(2) Apply this result to the Euclidean representation as in Thm. (10.8).

10.7. (Structure of the subresultant chain) Let $(A, B) \in D(d_1, \ldots, d_t)$ and $A_1 = A, A_2 = B, A_3, \ldots, A_t$ be the sequence of remainders produced by the Euclidean algorithm, $n_i := \deg A_i = \sum_{j=i}^t d_j$. Two polynomials $P, Q \in k[X]$ are called associated, $P \sim Q$, iff $P = \lambda Q$ for some $\lambda \in k^\times$. Conclude from the subresultant theorem (10.14) that for $1 < i < t$

$$\begin{aligned}
\mathrm{SRes}_{n_i}(A, B) &\sim A_i, & \mathrm{SRes}_{n_i-1}(A, B) &\sim A_{i+1}, \\
\mathrm{SRes}_\ell(A, B) &= 0 \text{ for } n_{i+1} < \ell < n_i - 1, \\
\mathrm{SRes}_{n_t}(A, B) &\sim A_t, \\
\mathrm{SRes}_\ell(A, B) &= 0 \text{ for } 0 \leq \ell < n_t.
\end{aligned}$$

(The occurring factors $\lambda \in k^\times$ may be explicitly expressed as power products of the leading coefficients of the A_i up to a sign. Together with Lemma (10.16) this gives explicit rational expressions for the coefficients of the A_i in terms of the coefficients of A, B.)

10.8. Let A, B be nonzero polynomials of degrees n, m, respectively, assume $n \geq m$, and let $0 \leq \ell \leq m$. Show the equivalence of the following three conditions:

(1) $\deg \mathrm{GCD}(A, B) = \ell$,
(2) $\mathrm{SRes}_j(A, B) = 0$ for $0 \leq j < \ell$ and $\mathrm{SRes}_\ell(A, B) \neq 0$,
(3) $S_{j,j}(A, B) = 0$ for $0 \leq j < \ell$ and $S_{\ell,\ell}(A, B) \neq 0$.

Moreover, $\mathrm{SRes}_\ell(A, B) \sim \mathrm{GCD}(A, B)$ under these conditions. (Hint: look at the structure of the subresultant chain.)

10.9. Let A, B be nonzero polynomials of degrees n, m, respectively, assume $n \geq m$, and let $A_1 = A, A_2 = B, A_3, \ldots, A_t$ be the remainders produced by the Euclidean algorithm, $n_i := \deg A_i$. Show that there are $U_i, V_i \in k[X]$ with $\deg U_i \leq m - n_{i-1}$, $\deg V_i \leq n - n_{i-1}$ such that $U_i A + V_i B = A_i$ for $1 < i \leq t$. Conclude that for all $0 \leq \ell \leq m$ there are $\tilde{U}_\ell, \tilde{V}_\ell \in k[X]$ with $\deg \tilde{U}_\ell < m - \ell$, $\deg \tilde{V}_\ell < n - \ell$ such that $\tilde{U}_\ell A + \tilde{V}_\ell B = \mathrm{SRes}_\ell(A, B)$. (Hint: use (3.9), (3.10) and Ex. 10.7. Compare also Ex. 3.7(4).)

The next two exercises prepare for Ex. 10.12.

10.10. For $m \leq n$ consider the Zariski closed subset

$$W(m, n) := \{A \in k^{m \times n} \mid \mathrm{rk}(A) < m\}.$$

Prove that $\deg W(m, n) \leq 2^n$. For an exact determination of the degree of $W(m, n)$ you may consult Harris [228, p. 243]. (Hint: $W(m, n)$ is the image under a projection of the set of all $((a_{ij}), \lambda) \in k^{m \times n} \times (k^m \setminus 0)$ satisfying $\sum_{i=1}^m \lambda_i a_{ij}$ for all $j \in \underline{n}$.)

10.11.* For nonzero polynomials A, B of degrees n, m, respectively, and $\ell \in \mathbb{N}$ such that $0 \leq \ell \leq m \leq n$ and $0 \leq i \leq m + n - \ell$, we denote by $N_{\ell,i}(A, B)$ the matrix obtained from $M_\ell(A, B)$ by deleting the last i columns. (Hence $\det N_{\ell,\ell}(A, B) = S_{\ell,\ell}(A, B)$.) Let $(d_1, \ldots, d_t)$ be a degree pattern of type (n, m) and put $n_i := \sum_{j=i}^t d_j$ for $1 \leq i \leq t$. Prove:

(1) For $(A, B) \in D(d_1, \ldots, d_t)$, $2 < i \leq t$ we have

$$\mathrm{rk}(N_{n_{i-1}-1,n_i}(A, B)) = m + n - 2(n_{i-1} - 1)$$

(this is the number of rows of this matrix), and

$$\mathrm{rk}(N_{n_{i-1}-1,n_i+1}(A, B)) < m + n - 2(n_{i-1} - 1).$$

(Hint: by Ex. 10.7 we have $S_{n_{i-1}-1,n_i}(A, B) \neq 0$. By deleting some of the columns of $N_{n_{i-1}-1,n_i}(A, B)$ we get the matrix whose determinant is $S_{n_{i-1}-1,n_i}(A, B)$. This shows the equality. To prove the inequality note that by Ex. 10.9 there are polynomials U, V with $\deg U \leq m - n_{i-1}$, $\deg V \leq n - n_{i-1}$ such that $AU + BV = A_i$. The comparison of the coefficients of X^ν for $n_i < \nu \leq m + n - n_{i-1}$ yields a homogeneous system of linear equations for the coefficients of U, V which has a nontrivial solution. The matrix of this system is $N_{n_{i-1}-1,n_i+1}(A, B)$.)

(2) (A, B) lies in $D(d_1, \ldots, d_t)$ iff for all $2 < i \leq t$ we have $S_{n_{i-1}-1,n_i}(A, B) \neq 0$, and

$$\mathrm{rk}(N_{n_{i-1}-1,n_i+1}(A, B)) < m + n - 2(n_{i-1} - 1)$$

provided $d_i > 1$.

10.12.* Prove the following statements.

(1) $\log \deg D(d_1, \ldots, d_t) \leq (m+n)s(d_1, \ldots, d_t)$. (Hint: Use the characterization of $D(d_1, \ldots, d_t)$ given in Ex. 10.11 and apply Ex. 10.10.)

(2) The complement of $D(d_1, \ldots, d_t)$ in $J(n, m)$ is dense in $J(n, m)$ except in the generic case where $(d_1, \ldots, d_t) = (n - m, 1, \ldots, 1, 0)$. In this case the degree of the complement does not exceed $(m - 1)2^{m+n}$. Hence no interesting lower bounds can be obtained from the degree bound in Cor. (10.2). (Hint: Deduce a characterization of the complement of $D(n - m, 1, \ldots, 1, 0)$ similarly as in Ex. 10.11.)

10.5 Open Problems

Problem 10.1. Does Thm. (10.7) carry over to computation trees over $\mathbb{R}$ (with $P = \{=, \leq\}$) in the sense that the computation time is bounded from below by $n(\mathcal{H}(d_1, \ldots, d_t) - 2)$ (or a function of this order of magnitude) on all inputs in $D_\mathbb{R}(d_1, \ldots, d_t)$ outside a semi-algebraic subset of smaller dimension? (Compare Ex. 10.6 where this is shown for degree patterns of length $t \geq (1 + \varepsilon)m/2$, with $\varepsilon > 0$.)

Problem 10.2. The Knuth-Schönhage algorithm computes the GCD of two univariate polynomials of degree at most n with $O(n \log n)$ nonscalar multiplications and divisions. Is this algorithm optimal up to order of magnitude?

Problem 10.3. Let $f_1, \ldots, f_n$ be n quadratic forms in $X_1, \ldots, X_n$ with indeterminate coefficients. Their *resultant* R_n is an absolutely irreducible integer polynomial in the coefficients of $f_1, \ldots, f_n$, which is homogeneous of degree 2^{n-1} in the coefficients of f_i for each i. If the coefficients of $f_1, \ldots, f_n$ are specialized to elements in an algebraically closed field k, then the vanishing of this resultant is a necessary and sufficient condition for the existence of a nontrivial solution of the system of equations $f_1 = 0, \ldots, f_n = 0$ over k. (Compare van der Waerden [540, §82] or Lang [317, Chap. IX].) Thus the zeroset of R_n is an irreducible hypersurface of degree $n2^{n-1}$ in affine space of dimension $n^2(n + 1)/2$ over k which just consists of the coefficient systems of consistent systems of n quadratic equations in n variables. Decide whether the complexity to test membership to the resultant hypersurface is polynomially bounded in n.

Problem 10.4. Is there a prime ideal $\wp \subset k[X] := k[X_1, \ldots, X_n]$ such that

$$\min\{L(F) \mid F \subseteq A \text{ finite}, \sqrt{(F)} = \wp A\}$$
$$< \min\{L(F) \mid F \subseteq A \text{ finite}, (F) = \wp A\},$$

where $A = k[X]$ (or the localization $A = k[X]_\wp$) and $L(F) := L_A^*(F|X)$ denotes the multiplicative complexity of a finite subset F in A.

10.6 Notes

The first two sections of this chapter are completely based on Strassen [504]. The material of the third section is due to Schuster [467]. The subresultant theorem (10.14) goes back to Habicht [225]. It is an important tool of computer algebra for analyzing variants of the Euclidean algorithm. For references on this topic see the Notes of Chap. 2 and the survey by Loos [341]. An extensive treatment of subresultants may also be found in the textbooks by Akritas [4] or Mishra [372].

Ex. 10.1–10.5 are taken from Strassen [504]. Ex. 10.11 and 10.12 are from Schuster [467]. Problem 10.1 appeared in Strassen [510]. With respect to Problem 10.3, we remark that if the complexity to test membership to the resultant hypersurface is not polynomially bounded in n (over the field $\mathbb{C}$, say), then the analogue of Cook's hypothesis $\mathbf{P} \neq \mathbf{NP}$ over the complex numbers in the sense of Blum, Shub, and Smale [59] is true (cf. Shub [479]).

For a discussion of further developments and results on branching complexity we refer to the detailed Notes of Chap. 11.

Chapter 11. Branching and Connectivity

In the first section we derive the Milnor-Thom bound [366, 516] which gives a quantitative estimate on the number of connected components of a semi-algebraic set in $\mathbb{R}^n$ described by polynomial equalities and inequalities. This estimate depends on the number of variables n, the number of inequalities and the maximum of the degrees of the occurring polynomials. Its proof is based on Bézout's inequality and on the Morse-Sard theorem. In the next section we investigate computation trees over $\mathbb{R}$ which solve the membership problem for a semi-algebraic subset W of $\mathbb{R}^n$. Ben-Or's lower bound [37] on the multiplicative branching complexity of such membership problems in terms of the number of connected components of W is deduced from the Milnor-Thom bound. Then we discuss applications to the real knapsack problem and to several problems of computational geometry (such as the computation of the convex hull of a finite set of points in the plane).

In this chapter all computation trees are defined over $\mathbb{R}$ and test for equality and inequality. The notion computation time always refers to the cost function $c = 1_{\{*,/,=,\leq\}}$, that is, multiplications, divisions, and comparisons are counted.

11.1* Estimation of the Number of Connected Components

We recall that a topological space X is called *connected* iff it cannot be decomposed into a disjoint union of nonempty closed subsets. The *connected components* of X are defined as the maximum connected subsets of X. We denote by $b_0(X) \in \mathbb{N} \cup \{\infty\}$ the number of connected components of X.

The goal of this section is to give a quantitative estimate on the number of connected components of a semi-algebraic set of the form

$$\{\xi \in \mathbb{R}^n \mid f_1(\xi) = \ldots = f_p(\xi) = 0, g_1(\xi) > 0, \ldots, g_q(\xi) > 0\},$$

where $p, q \in \mathbb{N}$, $f_1, \ldots, f_p, g_1, \ldots, g_q \in \mathbb{R}[X_1, \ldots, X_n]$: in other words, we look for an estimate in terms of n and the number as well as the maximum degree of the polynomials describing the above set. More precisely, we want to show the following statement.

(11.1) Theorem. *Let $n, d \geq 1$ and let $f_1, \ldots, f_p, g_1, \ldots, g_q$ be real polynomials in n variables such that $\deg f_i \leq d$, $\deg g_j < d$. Let the semi-algebraic set $W \subseteq \mathbb{R}^n$ be defined by the following equalities, non-equalities, and inequalities ($0 \leq q_1 \leq q_2 \leq q$)*

$$f_1 = 0, \ldots, f_p = 0,$$
$$g_1 \neq 0, \ldots, g_{q_1} \neq 0,$$
$$g_{q_1+1} > 0, \ldots, g_{q_2} > 0,$$
$$g_{q_2+1} \geq 0, \ldots, g_q \geq 0.$$

Then W has at most $d(2d - 1)^{n+q-1}$ connected components.

Note that p, the number of inequalities, does not enter in the above bound. We further remark that this theorem implies that the number of connected components of a semi-algebraic set is always finite (compare Ex. 4.20).

A version of Bézout's inequality will enter in the proof of Thm. (11.1). Let $f_1, \ldots, f_n \in \mathbb{R}[X_1, \ldots, X_n]$. A *non-degenerate solution* $\xi \in \mathbb{R}^n$ of the system of equations $f_1 = 0, \ldots, f_n = 0$ is a solution ξ such that the Jacobian matrix $[\partial_j f_i(\xi)]_{1 \leq i, j \leq n}$ is invertible. (Geometrically, this means that the intersection of the tangent spaces of $Z(f_1), \ldots, Z(f_n)$ at ξ is zero.)

(11.2) Proposition. *Let $f_1, \ldots, f_n \in \mathbb{R}[X_1, \ldots, X_n]$, $d_i := \deg f_i \geq 1$. Then the system $f_1 = 0, \ldots, f_n = 0$ has at most $d_1 \cdots d_n$ non-degenerate real solutions.*

Proof. Let $\xi_0 \in \mathbb{R}^n$ be a non-degenerate solution. The inverse function theorem implies that ξ_0 is an isolated solution of the complex zeroset of $f_1, \ldots, f_n$, i.e., for some open ball $B \subseteq \mathbb{C}^n$ around ξ_0 we have

$$B \cap \{\xi \in \mathbb{C}^n \mid f_1(\xi) = \ldots = f_n(\xi) = 0\} = \{\xi_0\}.$$

In particular, $\{\xi_0\}$ is an irreducible component of the complex zeroset of $f_1, \ldots, f_n$. However, by Bézout's inequality (8.28), their number is at most $d_1 \cdots d_n$. $\qquad\square$

(11.3) Remark. The statement of the above proposition becomes wrong, if we replace "non-degenerate" by "isolated!" For instance, consider the system of two equations in the variables X, Y

$$[(X - 1)(X - 2) \cdots (X - d)]^2 + [(Y - 1)(Y - 2) \cdots (Y - d)]^2 = 0, \; 0 = 0$$

having d^2 real solutions, all being isolated in the real domain. The product of the degrees $2d$ is strictly less than d^2 if $d > 2$. $\qquad\bullet$

A map $\varphi : V \to W$ between semi-algebraic sets $V \subseteq \mathbb{R}^n$, $W \subseteq \mathbb{R}^m$ is called *semi-algebraic* iff the graph of φ is a semi-algebraic set in $\mathbb{R}^{m+n}$. Note that morphisms of real algebraic sets are semi-algebraic. For a smooth morphism $\varphi : V \to W$ and $\xi \in V$ the linear map $d_\xi \varphi : T_\xi V \to T_{\varphi(\xi)} W$ between tangent spaces denotes the differential of φ at ξ.

Our proof of Thm. (11.1) will also rely on the following semi-algebraic version of the Morse-Sard theorem.

(11.4) Semi-algebraic Morse-Sard Theorem. *Let* $V \subseteq \mathbb{R}^n$, $W \subseteq \mathbb{R}^m$ *be semi-algebraic subsets and smooth submanifolds, and let* $\varphi\colon V \to W$ *be a smooth, semi-algebraic map. Let*

$$\Sigma := \{\xi \in V \mid \mathrm{rk}\, d_\xi \varphi < \dim W\}$$

denote the set of critical points of φ. *Then the set* $\varphi(\Sigma)$ *of critical values of* φ *is semi-algebraic of dimension strictly less than* $\dim W$.

For a proof of this theorem, which is more elementary than the corresponding result for C^∞-functions, we refer the reader to Benedetti and Risler [39, 2.5.12]. We emphasize the following special case of the above theorem. Let $f \in \mathbb{R}[X_1, \ldots, X_n]$, $\Sigma := \{\xi \in \mathbb{R}^n \mid \mathrm{grad}\, f(\xi) = 0\}$. Then $f(\Sigma)$ is a finite subset of $\mathbb{R}$. By the implicit function theorem we also know that $Z(f - \eta)$ is a smooth hypersurface for all $\eta \in \mathbb{R} \setminus f(\Sigma)$, provided $Z(f - \eta) \neq \emptyset$.

We turn now to the proof of Thm. (11.1). In a first step we study the case of one equation $f = 0$ describing a smooth and compact hypersurface.

(11.5) Proposition. *Let* $n \geq 2$ *and* $f \in \mathbb{R}[X_1, \ldots, X_n]$ *be of degree* $d \geq 2$ *with compact zeroset* $Z(f)$ *and such that* $\mathrm{grad}\, f(\xi) \neq 0$ *for all* $\xi \in Z(f)$. *Then* $b_0(Z(f)) \leq \frac{1}{2}d(d-1)^{n-1}$.

Proof. Let $V_1, \ldots, V_s$ be different connected components of $V = Z(f)$. (The case $b_0(V) = \infty$ is not excluded in advance.) By the implicit function theorem the V_i are compact, smooth hypersurfaces. The projection $\mathbb{R}^n \to \mathbb{R}$ onto the last coordinate restricted to V_i attains there an absolute maximum and absolute minimum, say in $p_i, q_i \in V_i$. Therefore, the tangent space of V_i in p_i equals $\mathbb{R}^{n-1} \times \{0\}$. Since the gradient of f in p_i is orthogonal to $T_{p_i} V$ we conclude that $\partial_1 f, \ldots, \partial_{n-1} f$ vanish at p_i. Analogously, one sees that these functions vanish at q_i. Moreover, $p_1, q_1, \ldots, p_s, q_s$ are pairwise different. (Note that if $p_i = q_i$ for some i, then V_i is contained in an affine hyperplane. Since $\dim V_i = n - 1 > 0$ this contradicts the compactness of V_i.) Therefore, we have shown that

(A) $$b_0(V) \leq \frac{1}{2}|Z(f, \partial_1 f, \ldots, \partial_{n-1} f)|.$$

If all the solutions of the system

(B) $$f = 0, \partial_1 f = 0, \ldots, \partial_{n-1} f = 0$$

happen to be non-degenerate, then we can apply Prop. (11.2) and obtain

$$|Z(f, \partial_1 f, \ldots, \partial_{n-1} f)| \leq d(d-1)^{n-1},$$

which together with (A) yields the assertion.

To finish the proof we show that we may achieve by a suitable linear coordinate transformation that the above system (B) has only non-degenerate solutions.

We consider the unit normal vector field

$$g : V \to S^{n-1}, \; \xi \mapsto \frac{\mathrm{grad}\, f}{\|\,\mathrm{grad}\, f\,\|}(\xi),$$

where $S^{n-1} := \{x \in \mathbb{R}^n |\; \|x\| = 1\}$. g is semi-algebraic (cf. Ex. 11.1). By the semi-algebraic Morse-Sard theorem (11.4) there is some $w \in S^{n-1}$ such that w and $-w$ are not critical values of g. (Consider the maps g and $-g$.) After a linear change of coordinates we may assume that $w = (0, \ldots, 0, 1)$. Then we have

$$g^{-1}(w) \cup g^{-1}(-w) = Z(f, \partial_1 f, \ldots, \partial_{n-1} f).$$

Let $\xi \in \mathbb{R}^n$ be a solution of the system (B). We want to show that ξ is non-degenerate. W.l.o.g. $\partial_n f(\xi) > 0$. For $x = (x_1, \ldots, x_n) \in \mathbb{R}^n$ we write $x' = (x_1, \ldots, x_{n-1})$. By the implicit function theorem there exists an open connected neighborhood $U \subseteq \mathbb{R}^{n-1}$ of ξ' and a C^∞-function $h : U \to \mathbb{R}$ such that the map

$$U \to V, \; x' \mapsto (x', h(x'))$$

is a diffeomorphism of U onto an open neighborhood of ξ in V. We have $f(x', h(x')) = 0$ for all $x' \in U$. By differentiating we get for all $i < n$, $x' \in U$

$$(C) \qquad \partial_i f(x', h(x')) = -\partial_n f(x', h(x')) \, \partial_i h(x').$$

Since $\mathrm{grad}\, f \neq 0$ on V this implies $\partial_n f(x', h(x')) > 0$ for all $x' \in U$. Using this, we conclude from (C)

$$g(x', h(x')) = -\frac{(\partial_1 h, \ldots, \partial_{n-1} h, -1)}{[1 + \sum_{j=1}^{n-1} (\partial_j h)^2]^{1/2}}(x').$$

By taking into account that $\partial_1 h, \ldots, \partial_{n-1} h$ vanish at ξ', we obtain from this for $i, j < n$

$$(D) \qquad \partial_i g_j(\xi) = -\partial^2_{i,j} h(\xi'), \quad \partial_i g_n(\xi) = 0.$$

Our assumptions state that ξ is a regular point of g, i.e., $\mathrm{rk}[\partial_i g_j(\xi)]_{i < n, j \le n} = n - 1$. By (D), this means that the Hessian matrix $[\partial^2_{i,j} h(\xi')]_{i,j < n}$ is invertible.

On the other hand, Equation (C) implies for $i, j < n$

$$\partial^2_{i,j} f(\xi) = -\partial_n f(\xi) \, \partial^2_{i,j} h(\xi').$$

Therefore the Jacobian matrix of $\partial_1 f, \ldots, \partial_{n-1} f, f$ is of the form

$$-\partial_n f(\xi) \cdot
\begin{bmatrix}
 & & & * \\
 [\partial^2_{i,j} h(\xi')]_{i,j<n} & & & \vdots \\
 & & & * \\
 0 & \cdots & 0 & -1
\end{bmatrix}$$

and thus invertible. So ξ is indeed a non-degenerate solution of the system (B). $\square$

In a second step we bound the number of connected components of an arbitrary real algebraic subset $V = Z(f_1, \ldots, f_p)$ of $\mathbb{R}^n$. The idea is roughly as follows. Set $F := f_1^2 + \ldots + f_p^2$. If V is compact, then we approximate V by the zeroset W of $F - \varepsilon^2$, where $\varepsilon^2 > 0$ is a small regular value of F. We may then apply Prop. (11.5) to W. (See Fig. 11.1.)

If V is not compact, then we choose $R > 0$ and approximate the set $V_R := V \cap \{\xi \in \mathbb{R}^n \mid \|\xi\| \leq R\}$ by a semi-algebraic set $K := \{F(x) + \varepsilon^2 \|x\|^2 \leq \varepsilon^2 R^2\}$ ($\| \cdot \|$ denotes the Euclidean norm). We then use that the boundary $W := \{F(x) + \varepsilon^2 \|x\|^2 = \varepsilon^2 R^2\}$ of K has not fewer connected components than K. Moreover, we will achieve that W is a smooth and compact hypersurface. (See Fig. 11.2.)

Fig. 11.1. $f_1 = X^2 + Y^2 - 1$, $f_2 = Z$. $V = Z(f_1, f_2)$ is a circle, $Z(f_1^2 + f_2^2 - \varepsilon^2)$ is a torus.

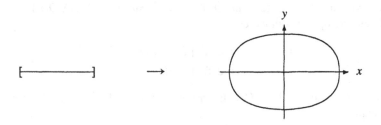

Fig. 11.2. $f_1 = Y$, $R = 1$. V_R is a line segment, $K = \{Y^2 + \varepsilon^2(X^2 + Y^2) \leq \varepsilon^2\}$, $W = \{\varepsilon^2 X^2 + (1 + \varepsilon^2)Y^2 = \varepsilon^2\}$ is an ellipse.

In order to give a formal proof we need the subsequent Lemma (11.7). For the reader's convenience we recall some basic facts from general topology before proving this lemma. (Cf. Kelley [299].)

(11.6) Fact. For every topological space X we have:

(1) The connected components of X are closed subsets.
(2) The set of connected components of X form a partition of X, and this is the only partition of X into closed and connected subsets.
(3) If $\{X_i \mid i \in I\}$ form a finite partition of X into closed subsets, then $b_0(X) \geq |I|$.
(4) If $\{X_i \mid i \in I\}$ is the set of connected components of X and $Y \subseteq X$, then $b_0(Y) \geq |\{i \in I \mid X_i \cap Y \neq \emptyset\}|$.

(5) If $X = \cup_{i \in I} X_i$, the X_i being arbitrary nonempty subsets of X, then $b_0(X) \leq \sum_{i \in I} b_0(X_i)$.

(6) If $\varphi : X \to Y$ is a continuous surjective map of topological spaces, then $b_0(Y) \leq b_0(X)$.   •

(11.7) Lemma. *Let $K_m \subseteq \mathbb{R}^n$ for $m \in \mathbb{N}$.*

(1) *If $K_m \subseteq K_{m+1}$ for $m \in \mathbb{N}$, then*

$$b_0(\cup_{m \in \mathbb{N}} K_m) \leq \sup_{m \in \mathbb{N}} b_0(K_m).$$

(2) *If the K_m are compact and $K_m \supseteq K_{m+1}$ for $m \in \mathbb{N}$, then*

$$b_0(\cap_{m \in \mathbb{N}} K_m) \leq \liminf_{m \to \infty} b_0(K_m).$$

Proof. (1) Let $C_1, \ldots, C_s$ be different connected components of $\cup_m K_m$. By the monotonicity there is some m_0 such that $C_i \cap K_m \neq \emptyset$ for $1 \leq i \leq s$, $m \geq m_0$. Hence by Fact (11.6)(4) $b_0(K_m) \geq s$ for $m \geq m_0$.

(2) Put $D := \cap_m K_m$ and let $s \in \mathbb{N}$, $s \leq b_0(D)$. Then D is a disjoint union of nonempty closed subsets $D_1, \ldots, D_s$. The distances

$$\text{dist}(D_i, D \setminus D_i) := \inf \{ \|x - y\| \mid x \in D_i, y \in D \setminus D_i \}$$

are positive since the D_i are compact. Put $\delta := \frac{1}{2} \min\{\text{dist}(D_i, D \setminus D_i) | 1 \leq i \leq s\}$ and define the open neighborhoods

$$
\begin{aligned}
U_i &:= \{x \in \mathbb{R}^n \mid \text{dist}(x, D_i) < \delta\}, \\
U &:= \{x \in \mathbb{R}^n \mid \text{dist}(x, D) < \delta\}
\end{aligned}
$$

of D_i, D, respectively. Then U is the disjoint union of the $U_1, \ldots, U_s$. If we can show that

(A) $$K_{m_0} \subseteq U \text{ for some } m_0,$$

then Fact (11.6)(3) implies $s \leq b_0(K_m)$ for $m \geq m_0$ and we are done. However, if (A) were false, then there existed a sequence $x_m \in K_m$ with $\text{dist}(x_m, D) \geq \delta$. This sequence has a limit point ξ which on the one hand satisfies $\text{dist}(\xi, D) \geq \delta > 0$, and on the other hand, by the monotonicity of the K_m, is contained in every K_m. Thus $\xi \in D$, a contradiction. $\square$

(11.8) Milnor-Thom Bound. *Let $f_1, \ldots, f_p \in \mathbb{R}[X_1, \ldots, X_n]$ of degree at most d, where $n, d \geq 1$. Then*

$$b_0(Z(f_1, \ldots, f_p)) \leq d(2d - 1)^{n-1}.$$

Proof. W.l.o.g. $n \geq 2$. We have $V := Z(f_1, \ldots, f_p) = Z(F)$ with $F := \sum_{j=1}^p f_j^2$. For $R > 0$ we define $V_R := Z(F) \cap \overline{B}(0, R)$, where $\overline{B}(0, R) := \{\xi \in \mathbb{R}^n \mid \|\xi\| \leq R\}$.

By Lemma (11.7)(1) it suffices to show that $b_0(V_R) \leq d(2d - 1)^{n-1}$ for every $R > 0$.

Let now $R > 0$ be fixed. We define for $\varepsilon, r > 0$ compact subsets

$$K(\varepsilon, r) := \{\xi \in \mathbb{R}^n \mid F(\xi) + \varepsilon^2 \|\xi\|^2 \leq r^2 \varepsilon^2\},$$
$$W(\varepsilon, r) := \{\xi \in \mathbb{R}^n \mid F(\xi) + \varepsilon^2 \|\xi\|^2 = r^2 \varepsilon^2\}.$$

We have $V_R \subseteq K(\varepsilon, r)$ if $R \leq r$, and $K(\varepsilon', r') \subseteq K(\varepsilon, r)$ if $0 < \varepsilon' \leq \varepsilon$, $0 < r' \leq r$. By the semi-algebraic Morse-Sard theorem (11.4) it is possible to choose sequences (ε_m), (r_m) having the following properties:

- $0 < \varepsilon_{m+1} < \varepsilon_m$, $R < r_{m+1} < r_m$,
- $\lim_{m \to \infty} \varepsilon_m = 0$, $\lim_{m \to \infty} r_m = R$,
- $(r_m \varepsilon_m)^2$ is not a critical value of $F(x) + \varepsilon_m^2 \|x\|^2$.

It is clear that $V_R = \bigcap_{m \in \mathbb{N}} K(\varepsilon_m, r_m)$ and Lemma (11.7)(2) tells us that $b_0(V_R) \leq \liminf_{m \to \infty} b_0(K(\varepsilon_m, r_m))$. It is therefore sufficient to bound $b_0(K(\varepsilon_m, r_m))$.

We note that for all $\varepsilon, r > 0$

$$b_0(K(\varepsilon, r)) \leq b_0(W(\varepsilon, r)).$$

We prove this by showing that $W(\varepsilon, r)$ meets every connected component C of $K(\varepsilon, r)$. Suppose the contrary. Then the compact set C is contained in the open set $K(\varepsilon, r) \setminus W(\varepsilon, r)$. Therefore, there is some $\delta > 0$ such that

$$\{x \in \mathbb{R}^n \mid \text{dist}(x, C) < \delta\} \subseteq K(\varepsilon, r) \setminus W(\varepsilon, r).$$

Moreover, the left-hand set is connected, which contradicts the maximality of C.

Prop. (11.5) applied to the polynomials $F(x) + \varepsilon_m^2(\|x\|^2 - r_m^2)$ of degree $\leq 2d$ implies $b_0(W(\varepsilon_m, r_m)) \leq d(2d - 1)^{n-1}$ for all m.

Altogether, we have for all m

$$b_0(K(\varepsilon_m, r_m)) \leq b_0(W(\varepsilon_m, r_m)) \leq d(2d - 1)^{n-1}$$

and therefore $b_0(V_R) \leq d(2d - 1)^{n-1}$. Since $R > 0$ was arbitrary, the assertion follows.   □

Proof of Thm. (11.1). W.l.o.g. $d \geq 2$. Let $W_1, \ldots, W_s$ be different connected components of W. For $1 \leq i \leq s$ choose $\xi_i \in W_i$ and let $\varepsilon > 0$ denote the minimum of $\{g_j(\xi_i) \mid i \leq s, q_1 < j \leq q_2\}$. We denote by $V \subseteq \mathbb{R}^{n+q}$ the zeroset of the following equations of degree $\leq d$

$$f_1 = 0, \ldots, f_p = 0,$$
$$g_1 Y_1 = 1, \ldots, g_{q_1} Y_{q_1} = 1,$$
$$g_{q_1+1} = Y_{q_1+1}^2 + \varepsilon, \ldots, g_{q_2} = Y_{q_2}^2 + \varepsilon,$$
$$g_{q_2+1} = Y_{q_2+1}^2, \ldots, g_q = Y_q^2,$$

where $Y_1, \ldots, Y_q$ are further indeterminates. By Thm. (11.8) we have $b_0(V) \leq d(2d-1)^{n+q-1}$. Let $\pi : \mathbb{R}^{n+q} \to \mathbb{R}^n$ be the projection onto the first n coordinates. Then $\pi(V)$ is the zeroset of

$$f_1 = 0, \ldots, f_p = 0,$$

$$g_1 \neq 0, \ldots, g_{q_1} \neq 0,$$

$$g_{q_1+1} \geq \varepsilon, \ldots, g_{q_2} \geq \varepsilon,$$

$$g_{q_2+1} \geq 0, \ldots, g_q \geq 0.$$

The subset $\pi(V)$ of W meets the components $W_1, \ldots, W_s$, hence, $s \leq b_0(\pi(V))$ by Fact (11.6)(4). Moreover, since π is a continuous map we have $b_0(\pi(V)) \leq b_0(V)$ by Fact (11.6)(6). Therefore $s \leq d(2d-1)^{n+q-1}$.   □

11.2 Lower Bounds by the Number of Connected Components

We study computation trees over $\mathbb{R}$ which solve membership problems $\pi = \{W, \mathbb{R}^n \setminus W\}$. Intuitively, such a computation tree decides for an input in $\mathbb{R}^n$ whether it lies in W or not. By Prop. (4.28), the set W must be semi-algebraic in order for the membership problem to be solvable. We recall that the *multiplicative branching complexity* $C^{*, \leq}(\pi)$ is the minimum number of nonscalar multiplications, divisions and comparisons needed by a computation tree to solve the membership problem π. For convenience we will write $C^{*, \leq}(W) := C^c(\pi)$.

Thm. (11.1) of the previous section allows us to deduce a lower bound on $C^{*, \leq}(W)$ in terms of the sum $b_0(W) + b_0(\mathbb{R}^n \setminus W)$ of the number of connected components of W and its complement. This result may be viewed as a real analogue of the degree bound (10.2) which holds for membership problems over algebraically closed fields. In its proof, Thm. (11.1) plays the same decisive role as Bézout's inequality did in the proof of the degree bound.

(11.9) Theorem (Ben-Or). *For a semi-algebraic set W of $\mathbb{R}^n$ we have*

$$C^{*, \leq}(W) \geq \frac{1}{\log 6} \left(\log(b_0(W) + b_0(\mathbb{R}^n \setminus W)) - n \log 3 \right).$$

Proof. Let $r \in \mathbb{N}$ and T be a computation tree of cost at most r on $\mathbb{R}^n$ which decides membership to W. Let the output nodes of T be partitioned into accepting and rejecting ones. An input $\xi \in \mathbb{R}^n$ lies in W iff its path in T ends up with an accepting output node. Fix an output node v of T and consider the set $D := D_v$ of those inputs $\xi \in \mathbb{R}^n$ whose common path ends up with v. We want to describe this set by an explicit system of equalities and inequalities.

To the input nodes $v_1, \ldots, v_n$ of T we assign the indeterminates $R_{v_1} := X_1$, $\ldots$, $R_{v_n} := X_n$. Starting at the root and going down along the unique path in T leading to v we assign to each computation node w on this path a value R_w as follows: if w carries the operational instruction

$$
\begin{array}{lll}
(\lambda^c) & \text{then} & R_w := \lambda, \\
(\pm; u_1, u_2) & \text{then} & R_w := R_{u_1} \pm R_{u_2}, \\
(\lambda; u) & \text{then} & R_w := \lambda R_u \ (\lambda \in \mathbb{R}), \\
(*; u_1, u_2) & \text{then} & R_w := Y_w, \\
(/; u_1, u_2) & \text{then} & R_w := Y_w,
\end{array}
$$

where the Y_w are additional indeterminates. We see that all R_w are affine linear in the X- and Y-variables. The number of Y-variables introduced equals the number $m(v)$ of multiplication and division instructions along the path considered.

Now we assign to each computation node w with multiplication instruction $(*; u_1, u_2)$ the quadratic equation $Y_w - R_{u_1} R_{u_2} = 0$ and to each node w with division instruction $(/; u_1, u_2)$ the quadratic equation $Y_w R_{u_2} - R_{u_1} = 0$. Furthermore, to each branching node with test instruction $(\leq; u_1, u_2)$ we assign the linear inequality $R_{u_1} - R_{u_2} \leq 0$ (resp. $R_{u_1} - R_{u_2} > 0$) if the path continues with the left (resp. right) son. Finally, we assign to each branching node with the test instruction $(=; u_1, u_2)$ the linear equality $R_{u_1} - R_{u_2} = 0$ (resp. the non-equality $R_{u_1} - R_{u_2} \neq 0$) if the path continues with the left (resp. right) son. We note that the number of inequalities introduced is bounded from above by the number $t(v)$ of test instructions occurring along the path to v.

Let E denote the semi-algebraic set in $\mathbb{R}^{n+m(v)}$ consisting of the solutions of the above equalities and inequalities. Thm. (11.1) implies that

$$
b_0(E) \leq 2 \cdot 3^{n+m(v)+t(v)-1}.
$$

It is easy to verify that the projection of $\mathbb{R}^{n+m(v)}$ onto the first n coordinates maps E to D (compare the description of D in (4.24)). Therefore, by Fact (11.6)(6), $b_0(D) \leq b_0(E)$.

Since our computation tree T has cost $\leq r$ on $\mathbb{R}^n$ and we we count only multiplications, divisions, and comparisons, we have $m(v) + t(v) \leq r$ for every output node v. We conclude that

$$
b_0(D_v) \leq 3^{n+r}
$$

for every output node v. The set W is the union of all sets D_v belonging to accepting output nodes. Therefore, by Fact (11.6)(5), $b_0(W)$ is at most 3^{n+r} times the number of accepting output nodes. Analogously, $b_0(\mathbb{R}^n \setminus W)$ is at most 3^{n+r} times the number of rejecting output nodes. Since there are at most $2^{t(v)} \leq 2^r$ output nodes, we conclude that $b_0(W) + b_0(\mathbb{R}^n \setminus W) \leq 2^r \cdot 3^{n+r}$. This implies the assertion. □

Before proceeding with first applications of this lower bound, we remark that there are computation trees which sort a list of n real numbers with $O(n \log n)$ comparisons (and no arithmetic operations at all). (See Ex. 4.6.) In the following let us call such a tree a *fast sorting tree*. We note that the number of distinct elements of n given real numbers can be determined from a sorted list with $O(n)$ comparisons.

We consider now the problem of deciding whether the components of a given vector $(x_1, \ldots, x_n) \in \mathbb{R}^n$ are pairwise distinct.

(11.10) Corollary. *(Element Distinctness) We have $C^{*,\leq}(W) = \Theta(n \log n)$ for the set $W := \{(x_1, \ldots, x_n) \in \mathbb{R}^n \mid \forall i < j : x_i \neq x_j\}$.*

Proof. We define the open and convex sets $W_\pi := \{x \in \mathbb{R}^n \mid x_{\pi(1)} < x_{\pi(2)} < \ldots < x_{\pi(n)}\}$ for a permutation $\pi \in S_n$. The set W is the disjoint union of the W_π, therefore the W_π are exactly the connected components of W and $b_0(W) = n!$. (The complement of W is connected.) The lower bound is thus a consequence of Thm. (11.9).

The upper bound follows easily from the existence of fast sorting trees. Another possibility for verifying the upper bound is to observe that

$$W = \left\{(x_1, \ldots, x_n) \in \mathbb{R}^n \;\middle|\; \prod_{i<j}(x_i - x_j)^2 \neq 0\right\}.$$

The discriminant $\prod_{i<j}(X_i - X_j)^2$ can be computed by a straight-line program in the polynomial ring from $X_1, \ldots, X_n$ with $O(n \log n)$ nonlinear operations (cf. Ex. 8.23). By adding one test for zero we get a computation tree of the desired cost which decides membership in W.   □

Next we study the problem of deciding whether two given sets of real numbers are disjoint.

(11.11) Corollary. *(Set Disjointness) We have $C^{*,\leq}(W) = \Theta(n \log n)$ for the set $W := \{(x, y) \in \mathbb{R}^{2n} \mid \{x_1, \ldots, x_n\} \cap \{y_1, \ldots, y_n\} = \emptyset\}$.*

Proof. For $I \subseteq \underline{n}^2$ we define the open convex set

$$W_I := \{(x, y) \in \mathbb{R}^{2n} \mid \forall(i, j) \in I : x_i < y_j, \forall(i, j) \notin I : x_i > y_j\}.$$

The set W_I is the union of these sets, hence the nonempty among the W_I are just the connected components of W. Note that $W_{I_0} \neq \emptyset$ for $I_0 := \{(i, j) \in N \mid i \leq j\}$, where $N := \underline{n} \times \underline{n}$. The group of permutations $S_n \times S_n$ acts on N as well as on $\mathbb{R}^n \times \mathbb{R}^n$ and for $(\pi, \sigma) \in S_n \times S_n$, $I \subseteq N$ we have $(\pi, \sigma)(W_I) = W_{(\pi,\sigma)(I)}$. Therefore $W_{(\pi,\sigma)(I_0)} \neq \emptyset$ for all $(\pi, \sigma) \in S_n \times S_n$. It is easy to verify that the stabilizer of I_0 is trivial, i.e.,

$$(\pi, \sigma)(I_0) = I_0 \implies (\pi, \sigma) = (\text{id}, \text{id}).$$

We conclude that there are at least $n!^2$ different index sets I for which W_I is nonempty, thus $b_0(W) \geq n!^2$. The asserted lower bound is a consequence of Thm. (11.9).

To show the upper bound we may use fast sorting trees to determine the cardinalities of the sets $A = \{x_1, \ldots, x_n\}$, $B = \{y_1, \ldots, y_n\}$ and $A \cup B$. Note that A and B are disjoint iff $|A \cup B| = |A| + |B|$.

Alternatively, we may use

$$W = \left\{(x, y) \in \mathbb{R}^{2n} \;\middle|\; \prod_{i,j}(x_i - y_j) \neq 0\right\}$$

and the fact that the resultant $\prod_{i,j}(X_i - Y_j)$ can be computed in the polynomial ring $\mathbb{R}[X_1, \ldots, X_n]$ from $X_1, \ldots, X_n$ with $O(n \log n)$ nonlinear operations (cf. Cor. (8.13)). By adding one test for zero we get a tree deciding membership to W with cost $O(n \log n)$. $\qquad \square$

11.3 Knapsack and Applications to Computational Geometry

We first discuss the basic problem of computing the convex hull of a given finite set of points in the plane. A point x of a convex set $C \subseteq \mathbb{R}^d$ is called an *extreme point* of C iff there are no two points $y, z \in C$ such that x lies in the open line segment determined by y, z. It is clear that all extreme points of the convex hull $C := \text{conv}(M)$ of a finite subset $M \subseteq \mathbb{R}^d$ are contained in M. Moreover, it is a well-known fact that the convex hull of the set of extreme points of C equals C. (Cf. Grünbaum [223].)

We will restrict our discussion to the plane $\mathbb{R}^2$. One can devise a computation tree which for n points in the plane, encoded as an element $(x_1, \ldots, x_n) \in (\mathbb{R}^2)^n$, outputs the extreme points of the convex hull of these points in time $O(n \log n)$. For a description of such an algorithm we refer the reader to Ex. 11.5. Our aim is to show that this algorithm is essentially optimal.

We can express the signed area of the triangle formed by three points $x, y, z \in \mathbb{R}^2$ by the formula

$$\Delta(x, y, z) := \frac{1}{2} \det \begin{pmatrix} 1 & 1 & 1 \\ x_1 & y_1 & z_1 \\ x_2 & y_2 & z_2 \end{pmatrix}.$$

The point z lies to the left of the oriented line from y to z iff $\Delta(x, y, z) > 0$.

We will use the following fact which is intuitively clear. If $x_1, \ldots, x_n$ are the extreme points of a planar convex set, then, by reordering these points, we may achieve that $\Delta(x_i, x_j, x_k) > 0$ for all $1 \le i < j < k \le n$. The converse is also true.

(11.12) Corollary. *(Convex Hull) Let W be the set of all $(x_1, \ldots, x_n) \in (\mathbb{R}^2)^n$ with the property that the x_i are pairwise distinct and are extreme points of the convex hull of $\{x_1, \ldots, x_n\}$. Then $C^{*, \le}(W) = \Omega(n \log n)$.*

Proof. Consider the open set

$$C := \{(x_1, \ldots, x_n) \in (\mathbb{R}^2)^n \mid \forall i < j < k : \Delta(x_i, x_j, x_k) > 0\}.$$

The group $S_{n-1} := \{\pi \in S_n \mid \pi(1) = 1\}$ operates on $(\mathbb{R}^2)^n$ by permuting the coordinates: $\pi(x_1, \ldots, x_n) := (x_{\pi^{-1}(1)}, \ldots, x_{\pi^{-1}(n)})$. The fact just mentioned before implies that W is the union of all permuted sets $\pi(C)$ where $\pi \in S_{n-1}$. Let us show that this union is disjoint. Note that

$$\pi(C) = \{(x_1, \ldots, x_n) \in (\mathbb{R}^2)^n \mid \forall i < j < k : \Delta(x_{\pi(i)}, x_{\pi(j)}, x_{\pi(k)}) > 0\}.$$

If $\pi \neq \text{id}$, then there exist $1 < i < j$ such that $\pi(i) > \pi(j) > 1$. Hence $C \cap \pi(C) = \emptyset$. This implies that the $\pi(C)$, $\pi \in S_{n-1}$, are pairwise disjoint. By Fact (11.6)(3) we obtain $b_0(W) \geq (n-1)!$. (It is not hard to see that C is connected, so we have actually established the decomposition of W into its connected components.) The lower bound follows by Thm. (11.9).   □

Let us turn to another problem of computational geometry. The *largest empty circle problem* is the following. Given n points in the plane $\mathbb{R}^2$, find a largest circle that contains none of these points in its interior and whose center lies in the convex hull of the given points. This problem can be algorithmically solved in time $O(n \log n)$ using the so-called Voronoi polygon. However, the corresponding problem of finding a smallest enclosing circle for n given points in the plane can be solved in linear time $O(n)$. (See Preparata and Shamos [425].)

(11.13) Corollary. *Any computation tree solving the largest empty circle problem with n points has worst case complexity $\Omega(n \log n)$.*

Proof. Let

$$W := \{(x_1, \ldots, x_n) \in \mathbb{R}^n \mid \exists \pi \in S_n \, \forall i > 1 : x_{\pi(i)} = x_{\pi(i-1)} + 1\}.$$

Assume we have a computation tree T solving the largest empty circle problem with n points in (constant) time $\tau(n)$. Then we can compute $\{W, \mathbb{R}^n \setminus W\}$ in the following way. By means of T we compute the radius ρ of a largest empty circle for the points $(x_1, 0), \ldots, (x_n, 0) \in \mathbb{R}^2$. We reject if $\rho > 1$. If $\rho \leq 1$, then we compute $x_{\min} := \min_i x_i$ and $x_{\max} := \max_i x_i$ in time $O(n)$. We accept if $x_{\max} - x_{\min} = n - 1$ and reject otherwise. This shows that $C^{*, \leq}(W) \leq \tau(n) + O(n)$.

On the other hand, the set W clearly has $n!$ connected components. Thus we conclude $\tau(n) = \Omega(n \log n)$ by Thm. (11.9).   □

Our last application of Thm. (11.9) concerns the n-dimensional *real knapsack problem*. It is to decide whether a given point in $\mathbb{R}^n$ lies in the set

$$\text{KS}_n := \left\{ x \in \mathbb{R}^n \;\middle|\; \exists S \subseteq \underline{n} : \sum_{i \in S} x_i = 1 \right\}$$

or not. Cor. (3.27) states the astonishing fact that there is an algorithm solving this problem in time $O(n^5 \log^2 n)$. This algorithm may be formalized as a computation tree, hence the complexity of the n-dimensional real knapsack problem is polynomially bounded in n: we have $C^{*, \leq}(\text{KS}_n) = O(n^5 \log^2 n)$. In the sequel we will derive a lower bound of order n^2 for this problem. As KS_n is connected (check this!), it remains the problem to estimate $b_0(\mathbb{R}^n \setminus \text{KS}_n)$. Let B_n denote the set of all Boolean functions $\{0, 1\}^n \to \{0, 1\}$. For a Boolean function $f \in B_n$ we set

$$E_f := \bigcap \left\{ x \;\middle|\; \sum_{i=1}^n \xi_i x_i < 1 \right\} \cap \bigcap \left\{ x \;\middle|\; \sum_{i=1}^n \xi_i x_i > 1 \right\},$$

where the intersection on the left-hand side is over all $\xi \in \{0, 1\}^n$ satisfying $f(\xi) = 0$, whereas the intersection on the right-hand side is over all $\xi \in \{0, 1\}^n$ satisfying $f(\xi) = 1$. It is easy to verify that

$$
\begin{aligned}
\mathbb{R}^n \setminus \mathrm{KS}_n &= \bigcap_{\xi \in \{0,1\}^n} \left(\left\{ x \mid \sum_{i=1}^n \xi_i x_i < 1 \right\} \cup \left\{ x \mid \sum_{i=1}^n \xi_i x_i > 1 \right\} \right) \\
&= \bigcup_{f \in B_n} E_f.
\end{aligned}
$$

The E_f are open, convex and pairwise disjoint, hence the nonempty among them are exactly the connected components of $\mathbb{R}^n \setminus \mathrm{KS}_n$.

A Boolean function $f \in B_n$ is called a *threshold function* if there exist $x_1, \ldots, x_n, \theta \in \mathbb{R}$ such that

$$
\forall \xi \in \{0, 1\}^n : f(\xi) = 0 \iff \sum_{i=1}^n x_i \xi_i < \theta.
$$

This means that there exists an affine hyperplane such that all zeros of f (considered as elements of $\mathbb{R}^n$) lie on one side of the hyperplane while the ones lie on the other side. $(x_1, \ldots, x_n)$ is called a weight vector and θ a threshold of f. (Of course they are not uniquely determined by f.) We denote the set of threshold functions of n variables by T_n. It is obvious that E_f is nonempty iff $f \in B_n$ is a threshold function with $f(0) = 0$. Therefore

$$
b_0(\mathbb{R}^n \setminus \mathrm{KS}_n) = \frac{1}{2} |T_n|
$$

(consider the involution $T_n \to T_n$, $f \mapsto 1 - f$).

(11.14) Lemma. *The number $|T_n|$ of threshold functions of n variables satisfies $|T_n| \geq 4 \cdot 2^{\binom{n}{2}}$.*

Proof. We first show that every $f \in T_n$ can be described with a weight vector x such that

(A)
$$
\left| \left\{ \sum_{i=1}^n x_i \xi_i \;\middle|\; \xi \in \{0, 1\}^n \right\} \right| = 2^n.
$$

In fact, the set of weights describing f is open. Moreover, the above condition (A) is satisfied if x is not contained in the finite union of hyperplanes

$$
\bigcup \left\{ x \;\middle|\; \sum_i u_i x_i = 0 \right\}.
$$

taken over all nonzero $u \in \{-1, 0, 1\}^n$.

We have a bijection $B_n \times B_n \to B_{n+1}$ which assigns to $(f, g) \in B_n^2$ the Boolean function

$$
F(\xi_1, \ldots, \xi_n, \xi_{n+1}) := f(\xi_1, \ldots, \xi_n) \xi_{n+1} + g(\xi_1, \ldots, \xi_n)(1 - \xi_{n+1}).
$$

Moreover, $F \in T_{n+1}$ iff f, g are threshold functions that can be described by the same weight vector.

From this and the preceding observation (A) we obtain the recursion

$$|T_{n+1}| \geq (2^n + 1) \cdot |T_n|.$$

Therefore $|T_n| \geq 2^{(n-1)+(n-2)+\ldots+1} \cdot |T_1| = 4 \cdot 2^{\binom{n}{2}}$. $\square$

From this lemma and Thm. (11.9) we obtain the following result.

(11.15) Corollary. *(Real Knapsack) The n-dimensional real knapsack problem has worst case complexity $\Omega(n^2)$, that is $C^{*, \leq}(KS_n) = \Omega(n^2)$.*

11.4 Exercises

11.1. Prove that the unit normal vector field

$$g : V \to S^{n-1}, \quad \xi \mapsto \frac{\operatorname{grad} f}{\|\operatorname{grad} f\|}(\xi)$$

considered in the proof of Thm. (11.5) is semi-algebraic.

11.2.* Prove the following variant of Thm. (11.8): we have $b_0(Z(f)) \leq d^n$ for $f \in \mathbb{R}[X_1, \ldots, X_n]$ of degree $d \geq 1$. (Hint: As in the proof of Thm. (11.8) we see that it is sufficient to bound $b_0(W(\varepsilon, r))$, where $W(\varepsilon, r) = Z(g)$, $g := f^2 + \varepsilon^2(\|X\|^2 - r^2)$. Proceed as in the proof of Thm. (11.5), but replace the system $g = 0, \partial_1 g = 0, \ldots, \partial_{n-1} g = 0$ of equations of degree $\leq 2d$ by a system of equations of smaller degrees by introducing a new variable.)

11.3. Prove a lower bound $\Omega(n \log n)$, resp. $\Omega(n^2)$, for the multiplicative branching complexity of the following problems.

(1) (Set equality) There are given $(x_1, \ldots, x_n, y_1, \ldots, y_n) \in \mathbb{R}^{2n}$, decide whether $\{x_1, \ldots, x_n\} = \{y_1, \ldots, y_n\}$.
(2) (Set inclusion) There are given $(x_1, \ldots, x_n, y_1, \ldots, y_n) \in \mathbb{R}^{2n}$, decide whether $\{x_1, \ldots, x_n\} \subseteq \{y_1, \ldots, y_n\}$.
(3) (Measure problem) For given $(x_1, \ldots, x_n, y_1, \ldots, y_n) \in \mathbb{R}^{2n}$ such that $x_i < y_i$ compute the Lebesgue measure of $\cup_{i=1}^n [x_i, y_i]$.
(4) (ε-knapsack) For given $(x_1, \ldots, x_n, \varepsilon) \in \mathbb{R}^{n+1}$ such that $\varepsilon > 0$ decide whether there is some $S \subseteq \underline{n}$ satisfying

$$\left| \sum_{i \in S} x_i - 1 \right| \leq \varepsilon.$$

11.4. (Algebraic decision trees) In Ex. 3.15 we defined the notion of a linear decision tree. In analogy we obtain the concept of an *(algebraic) decision tree* if we allow not only for linear test functions f_v, but for arbitrary polynomials $f_v \in \mathbb{R}[X_1, \ldots, x_n]$. At the node v the test $f_v(\xi) \geq 0$ (or $f_v(\xi) = 0$) on input $\xi \in \mathbb{R}^n$ is performed and one branches according to the outcome of the test. If all occurring polynomials have degree $\leq d$ we speak of a *degree d decision tree*. Let the (degree d) *decision complexity* $C_d(W)$ be the minimum depth of a degree d decision tree accepting the subset $W \subset \mathbb{R}^n$.

(1) Show that $C_\infty(W)$ equals the worst case complexity to test membership to W by computation trees over $\mathbb{R}$ when only comparisons are counted. That is, $C_\infty(W) = C^c(W)$ where c is the cost function $c = 1_{\{=, \leq\}}$.
(2) Let $W \subset \mathbb{R}^2$ be a polygon (i.e., the convex hull of finitely many points in the plane). Find two polynomials $f, g \in \mathbb{R}[X, Y]$ such that W can be described by the inequalities $f \geq 0, g \geq 0$. Conclude that $C_\infty(W) \leq 2$.
(3) Show that Ben-Or's lower bound (11.9) does not hold for C_∞. However, prove that for fixed $d < \infty$ and $W \subset \mathbb{R}^n$ we have

$$C_d(W) \geq c_1 \log(b_0(W) + b_0(\mathbb{R}^n \setminus W)) - c_2 n,$$

where the constants $c_1, c_2 > 0$ may depend on d.

11.5.* The purpose of this exercise is to describe an algorithm which computes the extreme points of the convex hull $\text{conv}(M)$ of a given finite set $M \subset \mathbb{R}^2$ of n points with $O(n \log n)$ arithmetic operations and tests.

(1) Find a point z in the interior of $\text{conv}(M)$ in time $O(n)$; for ease of notation assume $z = 0$.
(2) Express each point in M in polar coordinates r, θ in time $O(n)$.
(3) Sort the elements of M with respect to increasing angle θ in time $O(n \log n)$, say $M = \{r_1 e^{i\theta_1}, \ldots, r_n e^{i\theta_n}\}$, where $0 \leq \theta_1 \leq \ldots \leq \theta_n < 2\pi$ and $r_i \geq 0$. By eliminating some points we may assume $\theta_i < \theta_{i+1}$ and $r_i > 0$. W.l.o.g. $r_1 = \max_i r_i$, hence $r_1 e^{i\theta_1}$ is an extreme point of M.
(4) Consider the following procedure.
   START: Put $\ell := 1$.
   LOOP: If $\alpha_\ell + \beta_\ell < \pi$ (see Fig. 11.3) and $\ell = n$ then STOP. If $\alpha_\ell + \beta_\ell < \pi$ and $\ell < n$ then replace ℓ by $\ell + 1$ and goto LOOP. If $\alpha_\ell + \beta_\ell \geq \pi$ then eliminate the point $r_\ell e^{i\theta_\ell}$ from M, replace ℓ by $\ell - 1$, and goto LOOP (with the smaller set M of $n - 1$ elements).

Prove that the procedure in (4) stops after at most $2n - 3$ iterations of LOOP and that M contains then exactly the desired extreme points! Note that this algorithm can be turned into one using only rational operations by working with $r^2 = x^2 + y^2$, $\tan \theta = y/x$ instead of with r, θ. In this way, it can be formalized as a computation tree over $\mathbb{R}$.

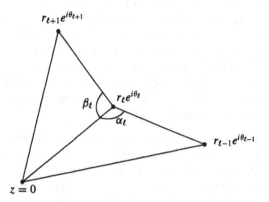

Fig. 11.3. Convex hull algorithm.

11.6. Let us consider computation trees which, in addition to their ability of performing arithmetic operations and comparisons, may take square roots of non-negative real numbers. Moreover let us charge all nonlinear operations by one. Prove that any such computation tree deciding membership in a semi-algebraic subset $W \subset \mathbb{R}^n$ has cost at least

$$\frac{1}{\log 9} \log\big(b_0(W) + b_0(\mathbb{R}^n \setminus W)\big) - \frac{n}{2}.$$

11.7.* (Complexity of constructions by ruler and compass)

(1) Formalize geometric constructions with ruler and compass in the plane by computation trees which are capable of performing the following basic operations:

    (1a) Intersecting two lines or circles. (A line is considered to be given by two of its points, a circle by one of its points and its center.)

    (1b) Deciding whether a point is to the left, to the right, or on a directed line.

    (1c) Deciding whether a point is inside, outside, or on a circle.

(2) Show that we may simulate any such computation tree by a computation tree as considered in Ex. 11.6 within a constant time factor. Conclude that the problem of computing the extreme points of the convex hull of a given set of n points in the plane by geometric constructions with ruler and compass needs at least $\Omega(n \log n)$ operations.

11.8. Prove that $b_0(\mathbb{R}^n \setminus KS_n) \leq 2^{n^2}$. Thus no better bound than $\Omega(n^2)$ can be obtained for the knapsack problem by applying Thm. (11.9). (Hint: use Lemma (3.32) and Ex. 3.19.)

11.9.* Let $W \subset \mathbb{R}^n$ be an n-dimensional polytope (that is a polytope with affine hull $\mathbb{R}^n$). We denote by N the number of facets, i.e., the number of $(n-1)$-

dimensional faces of W. (For some notation and facts concerning polytopes compare Sect. 3.4.) Prove that $C^{*,\leq}(W) \geq \frac{1}{3} \log N$. (Hint: Let T be a tree deciding membership in W and denote by D_v the set of those $\xi \in \mathbb{R}^n$ whose path ends up with the output node v. For each facet F of W there is some v such that $D_v \subseteq W$ and $\dim(F \cap D_v) = n - 1$. Let the affine hull of F be the zeroset of the affine linear function ℓ_F. The straight-line program corresponding to v computes a nonzero rational function $g \in \mathbb{R}(X_1, \ldots, X_n)$ such that g vanishes on the zeroset of ℓ_F and $L(g) \leq C^{*,\leq}(W)$. Hence ℓ_F is a divisor of the nominator of g.)

11.10.* (Generic width) Let $S \subseteq \mathbb{R}^n$ be a closed semi-algebraic set. The finiteness theorem for semi-algebraic sets says that there are natural numbers $t, r \in \mathbb{N}'$ and polynomials $p_{ij} \in \mathbb{R}[X_1, \ldots, X_n]$ for $i \in \underline{t}$ and $j \in \underline{r}$ such that S can be written in the form

$$S = \bigcup_{i=1}^{t} \{p_{i1} \geq 0, \ldots, p_{ir} \geq 0\}.$$

(Cf. Bochnak et al. [61].) The minimum $r \in \mathbb{N}'$ for which such a representation for S is possible is called the *width* of S. The width of $\emptyset$ and $\mathbb{R}^n$ are zero by convention. Now let $W, U \subseteq \mathbb{R}^n$ be semi-algebraic subsets and assume U to be open. The *generic width* $w(W, U)$ of W in U is the minimum $r \in \mathbb{N}$ such that there exists a closed semi-algebraic subset S of width r such that $\dim U \cap ((S \setminus W) \cup (W \setminus S)) < n$. (That is, W coincides with S in U up to a semi-algebraic set of dimension $< n$.)

Prove that $w(\Delta_n, U) = n$ for the positive orthant $\Delta_n := \{X_1 \geq 0, \ldots, X_n \geq 0\}$ in any open semi-algebraic neighbourhood U of the origin $0 \in \mathbb{R}^n$. (Hint: Proceed by induction on n. Assume that

$$\Delta_n \cap \{q \neq 0\} \cap U = S \cap \{q \neq 0\} \cap U$$

where S is as above and q is a nonzero polynomial. We may assume that all p_{ij} and q are squarefree. Let $p_{ij} = X_n^{\mu_{ij}} \tilde{p}_{ij}$, $q = X_n^{\nu} \tilde{q}$ with $\mu_{ij}, \nu \in \{0, 1\}$ such that X_n does not divide $\tilde{p}_{ij}, \tilde{q}$. Set $Q := \tilde{q} \prod_{i,j} \tilde{p}_{ij}$. First show that

$$\Delta_n \cap \{Q \neq 0\} \cap U = \tilde{S} \cap \{X_n \geq 0\} \cap \{Q \neq 0\} \cap U$$

where $\tilde{S}$ is obtained from S by replacing p_{ij} with $\tilde{p}_{ij}$. Now restrict to $\{X_n = 0\}$ and note that Q does not vanish on $\{X_n = 0\}$. Show that each nonempty set

$$\{\tilde{p}_{i1} \geq 0, \ldots, \tilde{p}_{ir} \geq 0\} \cap \{X_n = 0\} \cap \{Q \neq 0\} \cap U$$

may be described by less than r inequalities in $\{X_n = 0\} \cap \{Q \neq 0\} \cap U$.)

11.11.
(1) Show that the generic width constitutes a lower bound for decision complexity: for any subset $W \subseteq \mathbb{R}^n$ we have $C_\infty(W) \geq w(W, \mathbb{R}^n)$.
(2) We have $C_\infty(W) = n - 1$ for $W = \{X_1 \geq X_2, \ldots, X_1 \geq X_n\}$. In particular, any computation tree which finds the maximum of n given real numbers needs $n - 1$ comparisons.

11.12. * (Generic Nash width) This exercise requires some familiarity with the theory of Nash functions (cf. Bochnak et al. [61, Chap. 8]). Let $U \subseteq \mathbb{R}^n$ be an open semi-algebraic set. A *Nash function* on U is a C^∞-differentiable semi-algebraic map $U \to \mathbb{R}$. The Nash functions on U form a ring denoted by $\mathcal{N}(U)$. The *generic Nash width* $w^\mathcal{N}(W, U)$ of a semi-algebraic set $W \subseteq \mathbb{R}^n$ in U is defined as the minimum $r \in \mathbb{N}$ such that there exist $t \in \mathbb{N}$ and Nash functions $p_{ij} \in \mathcal{N}(U)$ for $i \in \underline{t}$, $j \in \underline{r}$, such that $W \cap U$ coincides with the set

$$\bigcup_{i=1}^{t} \{\xi \in U \mid p_{i1}(\xi) \geq 0, \dots, p_{ir} \geq 0\}$$

up to a semi-algebraic set of dimension $< n$. The generic Nash width is obviously invariant under Nash diffeomorphisms (that is, semi-algebraic diffeomorphisms).

(1) A point $\xi \in W$ is called a *Nash m-corner point* iff there is a Nash diffeomorphism from an open ball B of the origin in $\mathbb{R}^n$ to an open semi-algebraic neighbourhood U of ξ in $\mathbb{R}^n$ which maps $\{X_1 \geq 0, \dots, X_n \geq 0\} \cap B$ onto $W \cap U$. Let $f_1, \dots, f_m \in \mathbb{R}[X_1, \dots, X_n]$ and $\xi \in \mathbb{R}^n$ be such that $f_1(\xi) = 0, \dots, f_m(\xi) = 0$ and such that the rank of the Jacobian matrix of $(f_1, \dots, f_m)$ in ξ equals m. Conclude by the implicit function theorem that ξ is a Nash m-corner point of the set $\{f_1 \geq 0, \dots, f_m \geq 0\}$.

(2) Prove that $w^\mathcal{N}(\{X_1 \geq 0, \dots, X_n \geq 0\}, U) = n$ for any open semi-algebraic neighbourhood U of $0 \in \mathbb{R}^n$. (Hint: The germs of Nash functions at the origin of $\mathbb{R}^n$ can be identified with the formal power series in $\mathbb{R}[[X_1, \dots, X_n]]$ which are algebraic over $\mathbb{R}[X_1, \dots, X_n]$. These formal power series form a factorial ring. (Cf. Bochnak et al. [61, Chap. 8]).) Proceed as in Ex. 11.10.)

(3) Conclude that $C_\infty(W) \geq m$ if the set $W \subseteq \mathbb{R}^n$ contains a Nash m-corner point.

(4) (Oriented convex hull problem) We study the following problem: given a sequence $(x_1, \dots, x_n) \in (\mathbb{R}^2)^n$ of n points x_i in the real plane, decide whether they are the clockwise oriented vertices of their convex hull. Prove that the decision complexity C_∞ of this problem satisfies $n - 3 \leq C_\infty \leq n$. (Hint: verify that $((1, 0), (2, 0), \dots, (n-1, 0), (n, 1))$ is a Nash $(n-3)$-corner point of $W := \{\Delta_i \geq 0 \mid i \in \underline{n}\}$, where Δ is defined in Sect. 11.3 and $\Delta_i := \Delta(X_i, X_{i+1}, X_{i+2})$, the addition of indices being mod n.)

11.5 Open Problems

Problem 11.1. For a convex polytope W let $f_i(W)$ denote its number of faces of dimension i and $f(W) := \sum_{i \geq 0} f_i(W)$ its total number of faces. For instance $f_i(W) = \binom{n+1}{i}$, $f(W) = 2^{n+1}$ for a simplex W of dimension n. In Ex. 11.9 we proved that

$$C^{*, \leq}(W) \geq \frac{1}{3} \cdot \log f_{n-1}(W)$$

for a polytope $W \subset \mathbb{R}^n$ of dimension n. Does an analogous lower bound hold for the total number of faces $f(W)$?

Problem 11.2. Let $W \subseteq \mathbb{R}^n$ be obtained from open halfspaces by finitely many union, intersection and complement operations. Membership to W can be tested by a linear decision tree. It is natural to conjecture that computation trees do not help for solving such "linear problems." More specifically, we have the following relation between the linear decision complexity $C_1(W)$ and the multiplicative branching complexity: $C^{*,\leq}(W) \leq C_1(W)$ (compare Ex. 11.4). How are these two quantities related? When does equality hold?

11.6 Notes

The upper bound on the number of connected components of a real algebraic variety in Thm. (11.8) is due to Milnor [366] and Thom [516]. They even showed that not only the number of connected components, but also the sum of all Betti numbers can be bounded in this way. Similar bounds for a hypersurface had been obtained earlier by Oleĭnik and Petrovskii [396], and Oleĭnik [395]. The proof of the bounds on the Betti numbers requires some familiarity with algebraic topology and Morse theory. In order to keep our presentation at an elementary level, we have confined our discussion to the number of connected components. We also remark that the proof of this result is the basis for the asymptotically fastest known algorithms for dealing with semi-algebraic sets, compare the survey by Heintz et al. [245]. Thm. (11.1), which is particularly useful for complexity estimates, is due to Ben-Or [37]. In writing Sect. 11.1, the book by Benedetti and Risler [39] has been very helpful.

Rabin [429] seems to be the first who introduced the model of *decision trees* to study the complexity of membership problems for subsets W of $\mathbb{R}^n$. (Compare Ex. 3.15, 11.4 for a definition of this notion.) For subsets W which are defined by linear constraints it is quite natural to restrict oneself to linear decision trees. Within this restricted class of algorithms the lower bound $\log b_0(W)$ for the linear decision complexity is straightforward. Reingold [433] applied this to various problems related to the element distinctness and the set disjointness problem. An $\Omega(n \log n)$ lower bound for the linear decision complexity of the real knapsack problem was obtained by Dobkin and Lipton [146]. The convex hull problem cannot be solved by linear decision trees but certainly by quadratic ones. Yao [564] was able to prove the optimality of the known algorithms for the convex hull problem in the model of quadratic decision trees. Steele and Yao [487] were the first who recognized that the Milnor-Thom bound can be applied to prove nontrivial lower bounds valid for algebraic decision trees of any order. This idea was taken up by Ben-Or [37] who proved the crucial Thm. (11.9).

Our proof of the lower bound for the convex hull problem in Cor. (11.12) follows Steele and Yao [487]. Cor. (11.13), which is concerned with the largest empty circle problem, was first shown by Montaña et al. [377] by a different method. Our proof of this result is based on a reduction appearing in the book by Preparata and Shamos [425], which is also a good source for more information about computational geometry. The lower bound estimate on the number of connected components for the knapsack problem (Lemma (11.14)), on which Cor. (11.15) is based, is due to Dobkin and Lipton [146].

Let us now proceed with a discussion of further developments and results on decision and branching complexity which could not be covered in this and the previous chapter.

Rabin [429] studied the optimality of decision algorithms for subsets given by linear constraints. One of his results was that $n - 1$ comparisons are necessary to find the maximum of n given real numbers. Rabin's approach was generalized by Jaromczyk [274] to nonlinear problems under strong restrictions. However, the proofs in both papers contain a gap. In Montaña et al. [378] this was corrected, and a general lower bound on the decision complexity of a semi-algebraic set in terms of the so-called generic width was proved. We have sketched their results in Ex. 11.10–11.12. This is related to a result due to Bröcker [77] and Scheiderer [448] which states that any semi-algebraic subset of $\mathbb{R}^n$ defined by a conjunction of strict polynomial inequalities may be defined by at most n such inequalities. (See also Mahé [346].) In particular, the decision complexity of any such semi-algebraic set is at most n. (Cf. Ex. 11.4.) We remark that Fleischer [175] independently gave an alternative proof of Rabin's result just for linear problems.

Several recent papers deal with generalizations of Ben-Or's result (11.9) to higher Betti numbers. The first step in this direction was taken by Björner et al. [55]. To state their result, let us introduce some notation. A *polyhedron P* in $\mathbb{R}^n$ is defined as a set obtained from open halfspaces in $\mathbb{R}^n$ by finitely many union, intersection, and complement operations. Let $\hat{P}$ denote the compactification of P obtained by adding a simple point at infinity, and let $\chi(\hat{P})$ be its Euler characteristic. Björner et al. [55] proved the following result: let T be a linear decision tree deciding membership to a closed polyhedron P in $\mathbb{R}^n$. Then the number ℓ^- of no-leaves of T satisfies $\ell^- \geq |\chi(\hat{P}) - 1 + (-1)^{n-1}|$. In Björner and Lovász [54] this result was generalized and refined as follows. Again let T be a linear decision tree deciding membership to a closed polyhedron P in $\mathbb{R}^n$. A leaf of T is said to have dimension i iff the set of inputs belonging to this leaf has dimension i. Let ℓ_i^- denote the number of no-leaves of dimension i. Then the Betti numbers of the complement of P provide the following lower bound

$$\ell_i^- \geq b_{n-i}(\mathbb{R}^n \setminus P).$$

(By Alexander duality in $S^n \simeq \hat{\mathbb{R}}^n$ this implies the strengthening $\ell^- \geq \sum_{i=0}^n b_i(\hat{P})$ of the previous result.) The applications of these general topological bounds require quite involved computations of the Möbius function of intersection lattices of certain subspace arrangements. Compare also the survey by Björner [53].

Yao [566, 567] succeeded in proving that the higher Betti numbers b_i yield lower bounds for *arbitrary* computation trees. He showed the following: there are positive constants $c_1, c_2 > 0$ such that for all semi-algebraic and compact subsets

$W \subset \mathbb{R}^n$ we have for all $0 \leq i \leq n$

$$C^{*, \leq}(W) \geq c_1 \log b_i(W) - c_2 n.$$

We remark that this and Ben-Or's result have been extended to parallel complexity by Montaña and Pardo [376] and Montaña et al. [375].

Several papers deal with the problem of testing membership to a convex polytope. Yao and Rivest [568] showed that the logarithm of the number of faces of a fixed dimension of a convex polytope W constitute a lower bound on the number of comparisons needed to test membership to W for the restricted class of linear decision trees. This was partially generalized to algebraic decision trees by Grigoriev and Karpinski [212] and Grigoriev et al. [213, 214]. (Compare Ex. 11.9 and Problem 11.1.)

Lickteig [334, 335] deduced various lower bounds on branching complexity by combining the derivative inequality (7.7), the degree bound (8.35) and the concept of approximative complexity. For instance, if the hypersurface $W \subset \mathbb{R}^n$ is the zeroset of the irreducible polynomial f and G is the graph of the gradient of f on the zeroset of f in $\mathbb{C}^n$, then

$$C^*(W) \geq \frac{1}{\log 6} \left(\log \deg G - \log \deg f - \log(n+1) \right).$$

As applications he obtained lower bounds of order $\Omega(n \log q)$ for the problems of deciding whether a power sum $X_1^q + \ldots + X_n^q$, resp. an elementary symmetric function σ_q, have the value one. In Cucker and Lickteig [138] the above lower bound was generalized to wider classes of computation trees where a basic operation may be any Nash function (e.g., taking a square root of a positive number). Another of Lickteig's results is a relative lower bound for the problem of testing whether the determinant of a given matrix equals one. He proved $C^*(\mathrm{SL}(n, \mathbb{R})) \geq c \cdot \underline{R}(\langle n, n, n \rangle)$ where $c > 0$ is a constant and $\underline{R}(\langle n, n, n \rangle)$ is the border rank of $n \times n$-matrix multiplication. Hence the exponent of this sequence of problems (in the sense of Chap. 16) equals the exponent of matrix multiplication.

Bürgisser and Lickteig [102] applied the substitution method to show the following. Let $W \subseteq \mathbb{R}^{s+m}$ be the zeroset of polynomials $f_i = \sum_{j=1}^m g_{ij}(X)Y_j + g_i(X)$, $i \in \underline{p}$, in the variables $X_1, \ldots, X_s, Y_1, \ldots, Y_m$ and assume that the ideal generated by $f_1, \ldots, f_p$ in $\mathbb{R}(X)[Y]$ is nontrivial. Then the number of $\mathbb{R}$-linearly independent columns of the matrix $(g_{ij}(X))$ minus the rank of this matrix is a lower bound for the branching complexity $C^*(W)$ of W. An application is the optimality of Horner's rule for verifying whether a number is a zero of (several) polynomials, if this number and the coefficients of these polynomials are given as input data. In Bürgisser et al. [103] and Bürgisser [99] lower bounds based on the concept of transcendence degree are deduced. The main result is as follows. Let $W \subseteq \mathbb{R}^n$ be an irreducible algebraic subset with minimal field of definition K. Then any computation tree over $\mathbb{Q}$ deciding membership in W uses at least $(1 + \mathrm{tr.deg}_\mathbb{Q} K)/2$ multiplications, divisions or comparisons, and at least $\mathrm{tr.deg}_\mathbb{Q} K$ additions, subtractions or comparisons. This bound allows the exact determination of branching complexities in generic situations.

Ben-Or [38] has recently proved a lower bound for membership testing to algebraic subsets $W \subseteq k^n$, where k is an algebraically closed field of positive characteristic. His lower bound involves the total degree of the zeta function of W as a rational function.

Yao [565] generalized Ben-Or's result (11.9) to the integer constrained form of semi-algebraic membership problems. (For a simplified proof see Hirsch [249].) In Montaña et al. [377] some general remarks about complexities of semi-algebraic sets can be found. E.g., there are only finitely many topological types of semi-algebraic sets $W \subseteq \mathbb{R}^n$ of bounded multiplicative (or total) complexity.

For a study of probabilistic computation trees we refer to Manber and Tompa [348], Meyer auf der Heide [361], Bürgisser et al. [101], and Cucker et al. [137]. Roughly speaking, these papers discuss derandomization techniques which show that randomization does not help much for solving membership problems.

An important line of research is concerned with complexity aspects of the fundamental theorem of algebra, in other words, with the problem of approximating the roots of a complex univariate polynomial. The reader may find more information on this in Schönhage [461, 462], Smale [485], Renegar [435, 436], Shub and Smale [480], and the references given there. We confine ourselves to describe an approach initiated by Smale [486] to obtain lower complexity bounds. For $\varepsilon > 0$ and $d \in \mathbb{N}'$ let $\Pi(\varepsilon, d)$ denote the following problem: for a given monic univariate polynomial f of degree d having complex coefficients of absolute value less than one, find d complex numbers z_i such that the roots ζ_i of f can be ordered in such a way that $|z_i - \zeta_i| \le \varepsilon$ for all i. The model of algorithms Smale uses is essentially a computation tree T over the reals (a complex number is represented by its real and imaginary part). He charges only for branchings, but in a different way we did in this chapter: the topological cost t of T is defined as the number of leaves of T minus one. (Note that $c \le t < 2^c$ if c is the maximum number of branchings along a path of T.) Let the *topological complexity* $\tau(\varepsilon, d)$ of $\Pi(\varepsilon, d)$ be the minimal topological cost of an algorithm solving this problem. Smale first estimated $\tau(\varepsilon, d)$ from below by a purely topological quantity, the so-called Schwarz genus. (The Schwarz genus of a continuous map $\pi: X \to Y$ of topological spaces is the minimal number of open subsets of Y needed to cover Y such that there exists a continuous section of π over any of them.) By an involved algebraic topology computation he showed $\tau(\varepsilon, d) \ge (\log d)^{2/3}$ for $\varepsilon > 0$ small enough. Vassiliev [535, 536] vastly improved this result and generalized it to systems of polynomial equations. E.g., he was able to prove that $\tau(\varepsilon, d) = d - 1$ for d being a prime power and $\varepsilon > 0$ small enough. For details we refer to Vassiliev's book [537]. Results on the topological complexity of other problems may be found in Hirsch [250].

Ex. 11.2 is taken from Benedetti and Risler [39]. Ex. 11.3 is from Ben-Or [37]. The algorithm in Ex. 11.5 is due to Graham [203]. Ex. 11.6 and Ex. 11.7 are adapted from Ben-Or [37]. Ex. 11.9 is due to Grigoriev et al. [213]. Ex. 11.10–11.12 describe results of Montaña et al. [378]. Problem 11.1 is due to Grigoriev et al. [213]. For Problem 11.2 compare Ramanan [432].

Chapter 12. Additive Complexity

We prove Khovanskii's theorem [300] which gives an upper bound on the number of non-degenerate real solutions of a system of n polynomial equations in n variables which depends only on n and the number of distinct terms occurring in the polynomials. This result is in fact a consequence of a more general result dealing with certain systems of transcendental equations. A variant of Rolle's theorem and Bézout's inequality enter in the proof. As a consequence we deduce Grigoriev's and Risler's lower bound [209, 438] on the additive complexity of a univariate real polynomial in terms of the number of its real roots.

12.1 Introduction

For investigating the multiplicative complexity of polynomials the notion of the degree is crucial. We have seen in Sect. 8.1 that the logarithm of the degree of a univariate polynomial is a lower bound for its multiplicative complexity. By extending this to systems of multivariate polynomials we obtained numerous interesting results; for instance, we determined the multiplicative complexity for evaluating a polynomial at several points. The motivation for the subsequent developments is the question whether there exists an appropriate concept (such as degree) which allows us to derive lower bounds on the additive complexity. For example, does the evaluation of an nth degree polynomial at n points require $\Omega(n \log n)$ additions or subtractions? At the present time only very little is known, so we will solely discuss the additive complexity $L^+(f)$ of a univariate polynomial f.

The polynomials $X^n - 1$ have additive complexity one for arbitrarily large degree n, hence the degree cannot be used to bound the additive complexity. However, we remark that this polynomial has at most two real roots. In fact, we will prove that the number of distinct *real* roots of a real polynomial f is bounded from above by a function in $L^+(f)$. The following statement, already discovered by Descartes, can be viewed as a first step towards this result, with $L^+(f)$ replaced by the number of distinct terms of f.

(12.1) Descartes' Rule. *Let $f \in \mathbb{R}[X]$ be a nonzero polynomial with t distinct terms. Then f has at most $t - 1$ positive real roots (counted with multiplicity).*

Proof. We proceed by induction on the number t of terms. The start being clear assume $t > 1$. By dividing through a suitable power of X we may w.l.o.g. assume that f has a nonzero constant term. Thus the derivative f' of f has $t - 1$ distinct terms. The assertion follows from the induction hypothesis applied to f', once we have shown that the number of positive real roots of f exceeds the number of positive real roots of f' at most by one. However, this follows from Rolle's theorem stating that there is always a root of f' between two distinct roots of f, and from the fact that if ξ is a root of f with multiplicity μ, then ξ is a root of f' with multiplicity $\mu - 1$ (cf. Fig. 12.1).   □

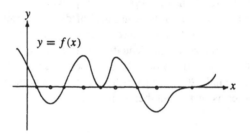

Fig. 12.1. Descartes' rule.

The reader should note that the bound in the above proposition is obviously sharp. We continue with two examples showing that a polynomial of additive complexity at most r may have as many as 3^r real roots.

(12.2) Example.
(1) Let T_n be the nth Chebyshev polynomial, i.e., T_n is defined by $T_n(\cos t) = \cos nt$ for all $t \in \mathbb{R}$. We have $T_1 = X$, $T_3 = 4X^3 - 3X$, hence $T_{3^r} = 4T_{3^{r-1}}^3 - 3T_{3^{r-1}}$. From this we see that $L^+(T_{3^r}) \le r$. Moreover, T_n has exactly n real roots, namely $\cos((2j + 1)\pi/(2n))$, $0 \le j < n$.
(2) Let $(\varepsilon_r)_{r \ge 0}$ be a sequence of positive numbers. Define $f_0 = X$ and for $r \ge 0$

$$f_{r+1} := (f_r^2 - \varepsilon_r^2)f_r.$$

Then f_r is of degree 3^r and $L^+(f_r) \le r$. By induction it is easy to see that f_r has 3^r distinct real roots, provided the ε_i are chosen sufficiently small.   •

It is not known whether the number of real roots that a polynomial of additive complexity r may possess is bounded from above by a function of the form C^r, where $C > 0$ is a constant (cf. Problem 12.3). We are only able to prove that the number of real roots is bounded by C^{r^2}, $C > 0$ being a constant. (See Thm. (12.12), we may take $C = 2^{16}$.)

12.2* Real Roots of Sparse Systems of Equations

In this section we attempt to generalize Prop. (12.1) to systems of polynomial equations in several variables.

Let $f_1, \ldots, f_n$ be differentiable functions $\mathbb{R}^n \to \mathbb{R}$. Recall that a solution $\xi \in \mathbb{R}^n$ of a system $f_1 = 0, \ldots, f_n = 0$ is called *non-degenerate* iff the Jacobian matrix $[\partial_j f_i(\xi)]$ at ξ is regular. We will show that the number of non-degenerate solutions of a system of polynomial equations $f_1 = 0, \ldots, f_n = 0$ is bounded from above by a function in the number n of variables and the number of terms of the system, irrespective of the degrees of the f_i. By a *term* (in the variables $Z_1, \ldots, Z_n$) we will understand a power product $Z_1^{a_1} \cdots Z_n^{a_n}$ with integer exponents a_i.

(12.3) Theorem (Khovanskii). *Let m be the number of distinct terms occurring in $f_1, \ldots, f_n \in \mathbb{R}[Z_1, Z_1^{-1}, \ldots, Z_n, Z_n^{-1}]$. Then the number of non-degenerate solutions of the system $f_1 = 0, \ldots, f_n = 0$ lying in the positive quadrant $(0, \infty)^n$ is less than or equal to $(n + 4)^m 2^{m(m-1)/2}$.*

Of course the same is true for every quadrant. In the case of one equation in one variable the above theorem gives an exponential bound, whereas Descartes' rule (12.1) yields a linear upper bound. This shows that Thm. (12.3) is far from being sharp.

We will obtain Thm. (12.3) as a consequence of a more general result dealing with certain systems of transcendental equations.

(12.4) Theorem (Khovanskii). *Let $F_i \in \mathbb{R}[X_1, \ldots, X_n, Y_1, \ldots, Y_m]$ for $i \in \underline{n}$ be polynomials of degree d_i and consider the exponential functions*

$$y_j(x) := e^{a_{j1}x_1 + \ldots + a_{jn}x_n}, \quad 1 \le j \le m,$$

where $a_{j\ell} \in \mathbb{R}$. Define the functions $f_i : \mathbb{R}^n \to \mathbb{R}$ by

$$f_i(x) := F_i(x_1, \ldots, x_n, y_1(x), \ldots, y_m(x)).$$

Then the number of non-degenerate solutions of the system $f_1 = 0, \ldots, f_n = 0$ is finite and does not exceed

$$\left(\prod_{i=1}^{n} d_i\right)\left(4 + \sum_{i=1}^{n} d_i\right)^m 2^{m(m-1)/2}.$$

Let us convince ourselves that this result implies Thm. (12.3). Let $f_i = \sum_{j=1}^{m} \lambda_{ij} M_j$ for $i \in \underline{n}$, where $M_j = Z_1^{a_{j1}} \cdots Z_n^{a_{jn}}$ are distinct terms, $a_{ji} \in \mathbb{Z}$ and $\lambda_{ij} \in \mathbb{R}$. The diffeomorphism $\mathbb{R}^n \to (0, \infty)^n$, $x \mapsto z$, $z_i = e^{x_i}$, transforms $M_j(z)$ to the exponential function $e^{a_{j1}x_1 + \ldots + a_{jn}x_n}$, and the system $f_i(z) = 0$, $i \in \underline{n}$, to the system $F_i(y_1(x), \ldots, y_n(x)) = 0$, $i \in \underline{n}$, where F_i denotes the linear form $\sum_{j=1}^{m} \lambda_{ij} Y_j$ in the indeterminates $Y_1, \ldots, Y_m$. By applying Thm. (12.4) to the transformed system the assertion follows.

By a smooth curve in $\mathbb{R}^{n+1}$ we will always understand a closed one-dimensional C^∞-submanifold of $\mathbb{R}^{n+1}$. By a smooth function we mean a C^∞-differentiable function.

A major ingredient of the proof of Thm. (12.4) is the following generalization of Rolle's theorem.

(12.5) Lemma. *Let $\Gamma \subseteq \mathbb{R}^{n+1}$ be a smooth curve which intersects the hyperplane $H := \{X_{n+1} = 0\}$ transversally in finitely many points. Let $v := (v_1, \ldots, v_{n+1}): \Gamma \to \mathbb{R}^{n+1}$ be a smooth nonvanishing tangential vector field to Γ and let $q \leq \infty$ denote the number of noncompact connected components of Γ. Moreover, let $c \in \mathbb{R}$ be such that*

$$|c| < \varepsilon := \min_{\xi \in \Gamma \cap H} |v_{n+1}(\xi)|.$$

Then we have $|\Gamma \cap H| \leq N' + q$, where N' is the number of curve points $\xi \in \Gamma$ with $v_{n+1}(\xi) = c$.

Proof. The connected components of Γ, being one-dimensional connected manifolds, are either diffeomorphic to the line $\mathbb{R}$ or to the circle S^1 (cf. Hirsch [251, p. 20]). Let Γ_1 be a noncompact component of Γ and $\phi = (\phi_1, \ldots, \phi_{n+1}): \mathbb{R} \to \Gamma_1$ be a diffeomorphism. By a suitable change of parameters we may achieve that $\phi'(s) = v(\phi(s))$ for all $s \in \mathbb{R}$. Let $s_1 < s_2$ be consecutive zeros of ϕ_{n+1}. Then we have $\phi'_{n+1}(s_1)\phi'_{n+1}(s_2) < 0$. Since $|c|$ is strictly smaller than $|\phi'_{n+1}(s_1)|$ and $|\phi'_{n+1}(s_2)|$, we conclude by the mean value theorem that there is some s between s_1 and s_2 such that $\phi'_{n+1}(s) = v_{n+1}(\phi(s)) = c$. We therefore see that (compare Fig. 12.2)

$$\left|\Gamma_1 \cap H\right| \leq \left|\{\xi \in \Gamma_1 \mid v_{n+1}(\xi) = c\}\right| + 1.$$

If Γ_2 is a component of Γ diffeomorphic to S^1, then a similar argument shows that

$$\left|\Gamma_2 \cap H\right| \leq \left|\{\xi \in \Gamma_2 \mid v_{n+1}(\xi) = c\}\right|. \qquad \square$$

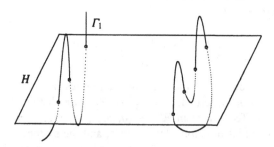

Fig. 12.2. A smooth curve intersecting a hypersurface transversally.

Let the smooth curve $\Gamma \subseteq \mathbb{R}^{n+1}$ be given as the zeroset of a smooth map $g: \mathbb{R}^{n+1} \to \mathbb{R}^n$ with regular value 0 (i.e., the Jacobian matrix of g has rank n in all points of Γ). Then we can explicitly describe a nonvanishing tangential vector field $v = (v_1, \ldots, v_{n+1}): \Gamma \to \mathbb{R}^{n+1}$ to Γ by

$$v_i(\xi) := (-1)^i \det \frac{\partial(g_1, \ldots, g_n)}{\partial(x_1, \ldots, \hat{x}_i, \ldots, x_{n+1})}(\xi).$$

This is a consequence of the following observation: let the matrix $A \in \mathbb{R}^{n \times (n+1)}$ be of rank n and v_i be $(-1)^i$ times the subdeterminant of A arising by deleting the ith column. Then $(v_1, \ldots, v_{n+1})$ is a nonzero element of the kernel of A. (See Ex. 12.5.)

We will also need the following result in order to bound the number of non-compact components of a curve.

(12.6) Lemma. *Let $\Gamma \subseteq \mathbb{R}^{n+1}$ be a smooth curve having q noncompact components. Then there is a hyperplane H which intersects Γ in at least q points.*

Proof. Each noncompact component Γ_j of Γ, $j \in q$, is diffeomorphic to $\mathbb{R}$; let $\gamma_j: \mathbb{R} \to \Gamma_j$ be a diffeomorphism (cf. Hirsch [251, p. 20]). As Γ is assumed to be closed, also Γ_j is closed and we may conclude that $\|\gamma_j(t)\|$ is unbounded for $t \to +\infty$ as well as for $t \to -\infty$. To each Γ_j we assign an accumulation point ξ_j^+ on the sphere S^n of the sequence $\mathbb{N} \to S^n$, $n \mapsto \gamma_j(n)/\|\gamma_j(n)\|$, as well as an accumulation point ξ_j^- of $\mathbb{N} \to S^n$, $n \mapsto \gamma_j(-n)/\|\gamma_j(-n)\|$. If a point among $\xi_1^+, \xi_1^-, \ldots, \xi_q^+, \xi_q^-$ appears p times we say that it has "multiplicity p." Thus, counting these points with their multiplicities, we obtain $2q$ points on S^n. Now let H_0 be a hyperplane through the origin which does not contain any of these points. H_0 determines two half spheres from which one contains at least q points of those $2q$ points. If we move H_0 parallel and far enough in the direction of this half sphere, it meets Γ in at least q points. (Use the mean value theorem to prove this.) $\qquad \square$

Let us now turn to the proof of Thm. (12.4). In order not to conceil the main ideas with technical details, we will make some simplifying assumptions in a first sketch of the proof. Details will be supplied later.

We proceed by induction on the number m of exponential functions. If m equals zero, then the statement is a consequence of Bézout's theorem (compare Prop. (11.2)). So assume $m > 0$. We will bound the number of non-degenerate solutions of

$$(12.7) \qquad f_i(x) = F_i(x, y_1(x), \ldots, y_{m-1}(x), y_m(x)) = 0, \quad 1 \le i \le n,$$

by the number of non-degenerate solutions of a system of the same structure with $m - 1$ exponentials and $n + 1$ variables, and then apply induction. For this we embed the functions f_i in a one-parameter family $g_i: \mathbb{R}^n \times \mathbb{R} \to \mathbb{R}$ by setting

$$g_i(x, t) = F_i(x, y_1(x), \ldots, y_{m-1}(x), t y_m(x)), \quad (x, t) \in \mathbb{R}^n \times \mathbb{R}.$$

We make now the simplifying assumptions that the system (12.7) has only finitely many solutions, say N, all of which are non-degenerate, and that 0 is a regular value for $g := (g_1, \ldots, g_n): \mathbb{R}^{n+1} \to \mathbb{R}^n$. Then $\Gamma := g^{-1}(0)$ is a smooth curve in $\mathbb{R}^{n+1}$ whose intersection with the hyperplane $H := \{t = 1\} = \mathbb{R}^n \times \{1\}$ consists exactly of the solutions of (12.7). We can describe a nonvanishing tangential vector field $v: \Gamma \to \mathbb{R}^{n+1}$ to Γ explicitly by

$$v_i(x, t) := (-1)^i \det \partial(g_1, \ldots, g_n)/\partial(x_1, \ldots, \hat{x}_i, \ldots, x_n, t)(x, t), \; i \in \underline{n},$$
$$v_{n+1}(x, t) := (-1)^{n+1} \det \partial(g_1, \ldots, g_n)/\partial(x_1, \ldots, x_n)(x, t).$$

Since by our assumptions the Jacobian matrix of $(f_1, \ldots, f_n)$ is regular in all points of $\Gamma \cap H$, we see that $v_{n+1} \neq 0$ in these points, i.e., Γ intersects H transversally. We can apply Lemma (12.5) (with $c = 0$) and conclude that $N \leq N' + q$, where q denotes the number of noncompact components of Γ, and where N' is the number of solutions of the following system in the variables $x_1, \ldots, x_n, t$

(12.8) $g_1 = 0, \ldots, g_n = 0, v_{n+1} = 0.$

Let us have a closer look at the derivatives $\partial g_i / \partial x_j$. Since the exponentials $y_\ell(x) = e^{a_{\ell 1} x_1 + \ldots + a_{\ell n} x_n}$ satisfy the differential equations $\partial y_\ell / \partial x_j = a_{\ell j} y_\ell$ we get

$$\partial g_i / \partial x_j = \partial F_i / \partial x_j + \sum_{\ell=1}^{m-1} \partial F_i / \partial y_\ell \cdot a_{\ell j} y_\ell + \partial F_i / \partial y_m \cdot a_{mj} t y_m$$
$$=: G_{ij}(x, y_1(x), \ldots, y_{m-1}(x), t y_m(x)),$$

where $G_{ij} \in \mathbb{R}[X_1, \ldots, X_n, Y_1, \ldots, Y_{m-1}, U]$ is a polynomial of degree $\leq d_i$. Therefore

$$v_{n+1}(x, t) = P(x, y_1(x), \ldots, y_{m-1}(x), t y_m(x)),$$

with some polynomial $P \in \mathbb{R}[X_1, \ldots, X_n, Y_1, \ldots, Y_{m-1}, U]$ of degree at most $D := \sum_{i=1}^n d_i$.

The diffeomorphism $(x_1, \ldots, x_n, t) \overset{\Phi}{\mapsto} (x_1, \ldots, x_n, u)$, $u = t y_m(x)$ transforms the system (12.8) to the system

(12.9) $\begin{aligned} F_i(x, y_1(x), \ldots, y_{m-1}(x), u) &= 0, \; i \in \underline{n}, \\ P(x, y_1(x), \ldots, y_{m-1}(x), u) &= 0, \end{aligned}$

which has the same number of solutions as (12.8). Assuming for simplicity that all the solutions of this system are non-degenerate, we get for their number N' by the induction hypothesis

$$N' \leq d_1 \cdots d_n D (4 + 2D)^{m-1} 2^{(m-1)(m-2)/2}.$$

In order to bound the number q of noncompact components of Γ (= number of noncompact components of $\Phi(\Gamma)$) we choose a hyperplane intersecting $\Phi(\Gamma)$ in q points, which is possible according to Lemma (12.6). It follows that q is smaller than or equal to the number of solutions of a system

$$F_i(x, y_1(x), \ldots, y_{m-1}(x), u) = 0, \quad 1 \le i \le n,$$
$$\lambda_1 x_1 + \ldots + \lambda_n x_n + \lambda_{n+1} u + \mu = 0$$

for suitable $\lambda_i \in \mathbb{R}$. Assuming that all solutions of this system are non-degenerate, we conclude by the induction hypothesis that

$$q \le d_1 \cdots d_n \cdot 1 \cdot (5+D)^{m-1} 2^{(m-1)(m-2)/2}.$$

Altogether, we obtain $N \le N' + q \le d_1 \cdots d_n 2^{(m-1)(m-2)/2} R$, where

$$
\begin{aligned}
R &:= D(4+2D)^{m-1} + (5+D)^{m-1} \\
&\le (D+3)(8+2D)^{m-1} + (8+2D)^{m-1} = (4+D)^m 2^{m-1}.
\end{aligned}
$$

Therefore, the desired estimate $N \le d_1 \cdots d_n (4+D)^m 2^{m(m-1)/2}$ follows. This finishes our first sketch of the proof.

To complete our argument we rely on the Morse-Sard theorem for C^∞-functions stated below, a proof of which may be found in Hirsch [251, p. 68]. (Compare also with the semi-algebraic version (11.4) of this theorem formulated in Chap. 11.) Before stating it we collect the necessary definitions and facts. Let $\phi \colon \mathbb{R}^m \to \mathbb{R}^n$ be smooth. A *regular point* ξ of ϕ is a point $\xi \in \mathbb{R}^m$ with the property that the Jacobian matrix of ϕ at ξ has the maximal possible rank $\min\{m, n\}$. The regular points of ϕ form an open subset of $\mathbb{R}^m$. An element $\eta \in \mathbb{R}^n$ is called a *regular value* of ϕ iff all $\xi \in \phi^{-1}(\eta)$ are regular points of ϕ. We denote the set of regular values of ϕ by $\mathcal{R}(\phi)$. A continuous map $\mathbb{R}^m \to \mathbb{R}^n$ is called *proper* iff inverse images of compact subsets under this map are again compact. This is equivalent to saying that inverse images of bounded subsets are again bounded. We leave it as an easy exercise to the reader to show that the image of a closed set under a proper map is again closed (cf. Bourbaki [70, I §10]).

A subset of $\mathbb{R}^n$ is called *residual* iff it contains the intersection of a countable family of dense open subsets. It is clear that a countable intersection of residual sets is again residual. Baire's theorem (cf. Lang [315, p. 207]) states that a residual subset of $\mathbb{R}^n$ (or more generally of a complete metric space) is dense.

(12.10) Morse-Sard Theorem. *The set $\mathcal{R}(\phi)$ of regular values of a smooth map $\phi \colon \mathbb{R}^m \to \mathbb{R}^n$ is residual and therefore dense. If ϕ is proper, then $\mathcal{R}(\phi)$ is even open and dense.*

We will also need the subsequent auxiliary result.

(12.11) Lemma. *Let $\phi \colon \mathbb{R}^m \to \mathbb{R}^n$ be smooth and proper. For $b \in \mathbb{R}^n$ we denote the number of regular points in $\phi^{-1}(b)$ by $N(b)$ (i.e., $N(b)$ is the number of non-degenerate solutions of the system $\phi(x) = b$). Then:*

(1) *$N(b)$ is finite for all $b \in \mathbb{R}^n$.*
(2) *Every $b_0 \in \mathbb{R}^n$ has a neighborhood V such that $N(b) \ge N(b_0)$ for all $b \in V$.*

(3) *The restriction of ϕ to $\phi^{-1}(\mathcal{R}(\phi))$ is a covering projection with base space $\mathcal{R}(\phi)$ and finitely many sheets. This means: for all $b_0 \in \mathcal{R}(\phi)$ there is an open neighborhood $V \subseteq \mathcal{R}(\phi)$ of b_0 such that $\phi^{-1}(V)$ is a disjoint union of $N := N(b_0)$ open sets $U_1, \ldots, U_N$ with the property that $\phi_{|U_i} : U_i \to V$ is a diffeomorphism for all i.*

Proof. (1) The subset consisting of the regular points in $\phi^{-1}(b)$ is discrete by the inverse function theorem, and compact since ϕ is proper. Hence it must be finite.

(2) Let $\xi_1, \ldots, \xi_N$ be the regular points in $\phi^{-1}(b_0)$, $N := N(b_0)$. By the inverse function theorem there are open neighborhoods U_i of ξ_i, and V_i of b_0 such that ϕ induces a diffeomorphism of U_i onto V_i. By shrinking the neighborhoods we may assume that the U_i are pairwise disjoint. Then for all $b \in V := V_1 \cap \ldots \cap V_N$ we have $N(b) \geq N$.

(3) We use the same notation as in the proof of (2). Since ϕ is proper, $\mathcal{R}(\phi)$ is open. By shrinking we may therefore assume that $V \subseteq \mathcal{R}(\phi)$ is an open connected neighborhood of b_0 such that ϕ induces a diffeomorphism of U_i onto V for all i. It remains to show that $\phi^{-1}(V) = U_1 \cup \ldots \cup U_N$. This can be seen as follows: let $b' \in V$, $\xi' \in \mathbb{R}^n$ be such that $\phi(\xi') = b'$. Choose a path γ connecting b' to b_0. Using that ϕ restricted to $\phi^{-1}(\mathcal{R}(\phi))$ is a local diffeomorphism, we see that γ can be lifted to a path $\tilde{\gamma}$ connecting ξ' to some ξ_i. On the other hand, $(\phi_{|U_i})^{-1} \circ \gamma$ is a path in U_i also ending with ξ_i and mapping to γ under ϕ. Therefore this path must coincide with $\tilde{\gamma}$ and hence $\xi' \in U_i$. $\square$

After these preparations we can now give a complete proof of Thm. (12.4).

Proof of Thm. (12.4). We use induction on m. The start "$m = 0$" being a consequence of Bézout's theorem, let us assume $m > 0$. We want to bound the number of non-degenerate solutions of the following extended system

$$f_i(x_1, \ldots, x_n) = 0, \quad 1 \leq i \leq n,$$

(A)

$$f_{n+1}(x_0, x_1, \ldots, x_n) := \sum_{i=0}^{n} x_i^2 - R^2 = 0,$$

where x_0 is a new variable and $R > 0$. This number is exactly twice the number of non-degenerate solutions of the original system

$$f_i(x_1, \ldots, x_n) = 0 \ (1 \leq i \leq n), \quad x_1^2 + \ldots + x_n^2 < R^2$$

which lie in the open ball of radius R centered at the origin. The additional component f_{n+1} guarantees that the map $f := (f_1, \ldots, f_{n+1}) : \mathbb{R}^{n+1} \to \mathbb{R}^{n+1}$ is proper. Hence, by Thm. (12.10), its set $\mathcal{R}(f)$ of regular values is open and dense. For $b \in \mathbb{R}^{n+1}$ we denote by $N(b)$ the number of non-degenerate solutions of the system $f(x) = b$. Lemma (12.11) tells us that $N(0) \leq N(b)$ for all b lying in some open neighbourhood V_0 of 0. We follow now our preliminary proof sketch rather closely. As earlier, we introduce a one-parameter family of functions $g_i : \mathbb{R}^{n+1} \times \mathbb{R} \to \mathbb{R}$ by defining (set $F_{n+1} := f_{n+1}$)

$$g_i(x, t) := F_i(x, y_1(x), \ldots, y_{m-1}(x), ty_m(x)),$$

and consider the map $g := (g_1, \ldots, g_{n+1}) \colon \mathbb{R}^{n+2} \to \mathbb{R}^{n+1}$. If $b \in \mathcal{R}(g)$, then $\Gamma_b := g^{-1}(b)$ is a smooth curve in $\mathbb{R}^{n+2}$. If additionally $b \in \mathcal{R}(f)$, then Γ_b intersects the hyperplane $\{t = 1\}$ transversally in the $N(b)$ points $\xi \in f^{-1}(b)$. Again, we have a nonvanishing tangential vector field $v = (v_0, \ldots, v_{n+1}) \colon \Gamma_b \to \mathbb{R}^{n+2}$ to Γ_b whose last component is given by

$$v_{n+1}(x, t) = (-1)^{n+2} \det \partial(g_1, \ldots, g_{n+1})/\partial(x_0, \ldots, x_n)(x, t).$$

We define for $b \in \mathcal{R}(f)$

$$\varepsilon(b) := \min_{\xi \in f^{-1}(b)} |v_{n+1}(\xi)| > 0.$$

Lemma (12.11)(3) states that f restricted to $f^{-1}(\mathcal{R}(f))$ is a covering projection. This implies that $\varepsilon \colon \mathcal{R}(f) \to \mathbb{R}$ is continuous. By Thm. (12.10) there exists $(b, c) \in \mathbb{R}^{n+1} \times \mathbb{R}$ satisfying

(1) b is a regular value of f and g,
(2) (b, c) is a regular value of $(g_1, \ldots, g_{n+1}, v_{n+1})$,
(3) $|c| < \varepsilon(b)$, $b \in V_0$.

Let (b, c) be chosen in this way. Lemma (12.5) implies that $N(b) \le N' + q$, where N' is the number of non-degenerate solutions of

(B) $$g(x, t) = b, \quad v_{n+1}(x, t) = c,$$

and q denotes the number of noncompact connected components of $\Gamma_b = g^{-1}(b)$. As before, we may express $v_{n+1}(x, t)$ in the form

$$v_{n+1}(x, t) = P(x, y_1(x), \ldots, y_{m-1}(x), ty_m(x))$$

with a polynomial $P \in \mathbb{R}[X_0, \ldots, X_n, Y_1, \ldots, Y_{m-1}, U]$ of degree $\le 2 + D$, where $D := \sum_{i=1}^n d_i$ (note that $F_{n+1} = f_{n+1}$ is quadratic). The system (B) transforms to

(C) $$\begin{aligned} F_i(x, y_1(x), \ldots, y_{m-1}(x), u) &= b, \quad 1 \le i \le n+1, \\ P(x, y_1(x), \ldots, y_{m-1}(x), u) &= c \end{aligned}$$

by the change of variables $u = ty_m(x)$. The induction hypothesis applied to this system gives us

$$N' \le d_1 \cdots d_n \cdot 2 \cdot (2 + D)(8 + 2D)^{m-1} 2^{(m-2)(m-1)/2}.$$

The number q can be bounded similarly as in the sketch, yielding

$$q \le d_1 \cdots d_n \cdot 2 \cdot 1 \cdot (7 + D)^{m-1} 2^{(m-2)(m-1)/2}.$$

Summarizing, we obtain

$$N(0) \le N(b) \le N' + q \le 2 \cdot d_1 \cdots d_n (4 + D)^m 2^{m(m-1)/2}.$$

We already know that the number of non-degenerate solutions of the original system $f_i(x_1, \ldots, x_n) = 0, i \in \underline{n}$, satisfying $\sum_{i=1}^n x_i^2 < R$ equals $N(0)/2$. Since the above bound on $N(0)$ does not depend on R, our claim follows. $\qquad\square$

12.3 A Bound on the Additive Complexity

Our goal is the following lower bound result for the additive complexity of a real univariate polynomial.

(12.12) Theorem (Grigoriev, Risler). *Let* $f \in \mathbb{R}[X]$ *be a univariate nonzero polynomial with N real roots. Then*

$$\frac{1}{4}\sqrt{\log N} \le L^+(f).$$

(12.13) Example. For the Chebyshev polynomials T_{3^r} considered in Example (12.2)(1) the above theorem implies $\sqrt{\log 3}/4 \cdot \sqrt{r} \le L^+(f) \le r$.   •

Before proving Thm. (12.12) we state two auxiliary results. The first one characterizes the additive complexity $L^+(f) := L^+_{k(X)}(f|X)$ of univariate polynomials and follows easily by grouping together the multiplicative steps in a computation of f. (The details are left to the reader, cf. Ex. 12.7.)

(12.14) Lemma. *Let k be a field, $f \in k(X)$. The additive complexity $L^+(f)$ of f is the smallest natural number r such that there exists a sequence of nonzero rational functions $g_0 = X, g_1, \ldots, g_r \in k(X)$ built in the following way: for all $\rho \in \underline{r}$ there are $\alpha, \beta \in k$ and integers $m_0, \ldots, m_{\rho-1}, n_0, \ldots, n_{\rho-1}$ satisfying*

$$g_\rho = \alpha \prod_{i=0}^{\rho-1} g_i^{m_i} + \beta \prod_{i=0}^{\rho-1} g_i^{n_i},$$

and such that f can be obtained from the g_i in the form

$$f = \gamma \prod_{i=0}^{r} g_i^{p_i},$$

with $\gamma \in k$, $p_0, \ldots, p_r \in \mathbb{Z}$.

(12.15) Lemma. *Let $f \in \mathbb{R}[X]$, $f \ne 0$. Then there exists a sign $\nu \in \{-1, +1\}$ such that for all sufficiently small $\varepsilon > 0$ we have*

(1) *all the (real) roots of $f - \nu\varepsilon$ are simple,*
(2) *the number of real roots of $f - \nu\varepsilon$ is greater than or equal to the number of real roots of f.*

Proof. (1) It is sufficient to choose ε such that $\nu\varepsilon \ne f(\xi)$ for all roots ξ of the derivative of f.

(2) Let us assume that among the n real roots of f there are n_1 local minima, n_2 local maxima and n_3 points where the sign of f changes in a neighbourhood of these points. Then, for $\varepsilon > 0$ sufficiently small, the number of real roots of $f - \varepsilon$, $f + \varepsilon$ equals $2n_1 + n_3$, $2n_2 + n_3$, respectively. (Think of the graph of f as being intersected by the line $y = \nu\varepsilon$, see Fig. 12.3.) However, $2n_1 + n_3 \ge n = n_1 + n_2 + n_3$ if $n_1 \ge n_2$ and $2n_2 + n_3 \ge n$ if $n_2 \ge n_1$.   □

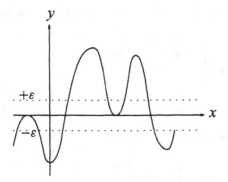

Fig. 12.3. $n_1 = 1, n_2 = 1, n_3 = 2$.

Proof of Thm. (12.12). Let $f \in \mathbb{R}[X]$ be a nonzero polynomial with additive complexity r. By Lemma (12.14) there exist $g_0 = X, g_1, \ldots, g_r \in \mathbb{R}(X)$ such that for all ρ

$$g_\rho = \alpha_\rho \prod_{i=0}^{\rho-1} g_i^{m_{\rho i}} + \beta_\rho \prod_{i=0}^{\rho-1} g_i^{n_{\rho i}}, \quad f = \gamma \prod_{i=0}^{r} g_i^{p_i},$$

for suitable $\alpha_\rho, \beta_\rho, \gamma \in \mathbb{R}$ and $m_{\rho i}, n_{\rho i}, p_i \in \mathbb{Z}$. Let $\varepsilon > 0$ and $\nu \in \{-1, +1\}$ be chosen as in Lemma (12.15). Moreover, we may assume that $(g_0 \cdots g_r)(\xi) \neq 0$ for all roots of $f - \nu\varepsilon$. We are now going to apply Thm. (12.3) to the system of $r + 1$ equations

(A)
$$
\begin{aligned}
-Y_\rho + \alpha_\rho \prod_{i=0}^{\rho-1} Y_i^{m_{\rho i}} + \beta_\rho \prod_{i=0}^{\rho-1} Y_i^{n_{\rho i}} &= 0, \quad 1 \leq \rho \leq r, \\
\gamma \prod_{i=0}^{r} Y_i^{p_i} - \nu\varepsilon &= 0
\end{aligned}
$$

in the unknowns $Y_0 = X, Y_1, \ldots, Y_r$. The projection $(Y_0, Y_1, \ldots, Y_r) \mapsto Y_0$ clearly gives a bijection between the set of solutions of (A) and the set of roots of $f - \nu\varepsilon$. Note that by our assumption the solutions of (A) have nonzero components.

The next lemma guarantees that all the solutions of (A) are non-degenerate.

(12.16) Lemma. *The Jacobian determinant $J(Y_0, \ldots, Y_r)$ of the system (A) satisfies*

$$f' = J(g_0, \ldots, g_r).$$

Proof. We consider a more general situation. Let $G_1, G_2, \ldots$ be any sequence of rational functions in indeterminates $Y_0, Y_1, \ldots$ such that G_ρ only depends on $Y_0, \ldots, Y_{\rho-1}$. For $r \geq 1$ consider the system

$$
\begin{aligned}
-Y_\rho + G_\rho &= 0, \quad 1 \leq \rho < r, \\
G_r &= 0
\end{aligned}
$$

in the variables $Y_0, \ldots, Y_{r-1}$ and denote by J_r its Jacobian determinant. We have

$$J_r = \det \begin{pmatrix} \partial_0 G_1 & -1 & 0 & \cdots & 0 \\ \partial_0 G_2 & \partial_1 G_2 & -1 & \cdots & 0 \\ \vdots & \vdots & \vdots & \ddots & \vdots \\ \partial_0 G_{r-1} & \partial_1 G_{r-1} & \cdots & \cdots & -1 \\ \partial_0 G_r & \partial_1 G_r & \cdots & \cdots & \partial_{r-1} G_r \end{pmatrix},$$

and by developing with respect to the last row we obtain

(B)
$$\begin{aligned} J_r &= \sum_{j=0}^{r-1} (-1)^j \partial_{r-1-j} G_r \cdot (-1)^j J_{r-1-j} \\ &= \sum_{j=0}^{r-1} \partial_i G_r \cdot J_i. \end{aligned}$$

Let now $g_0 = X, g_1, g_2, \ldots \in \mathbb{R}(X)$ be defined recursively by

$$g_\rho := G_\rho(g_0, \ldots, g_{\rho-1}).$$

The chain rule implies $g_r' = \sum_{i=0}^{r-1} \partial_i G_r(g_0, \ldots, g_{r-1}) \cdot g_i'$, hence, using (B) and induction, we get

$$g_r' = J_r(g_0, \ldots, g_{r-1}).$$

From this the statement of the lemma follows.   □

The number of distinct terms in the system (A) is at most $3r + 2$. Thm. (12.3) implies therefore that the number of non-degenerate solutions of (A) in one quadrant of $\mathbb{R}^{r+1}$ is less than or equal to $B := (r+5)^{3r+2} 2^{(3r+2)(3r+1)/2}$. Since there are 2^{r+1} quadrants, the number of real roots of (A) does not exceed $2^{r+1} B$. A calculation shows that this is smaller than 2^{16r^2} for all $r \geq 2$. Therefore, the number N of real roots of f is smaller than 2^{16r^2}. (For $r = 1$ one sees directly that $N \leq 3$.) This finishes the proof of Thm. (12.12).   □

12.4 Exercises

12.1. Let $f: \mathbb{R} \to \mathbb{R}, x \mapsto \sum_{i=1}^t a_i e^{\alpha_i x}$, $a_i, \alpha_i \in \mathbb{R}$, $a_i \neq 0$. Show that the number of real roots of f (counted with multiplicity) does not exceed $t - 1$.

12.2. (This exercise generalizes Prop. (12.1).) The *variation* of a sequence $(a_0, \ldots, a_n)$ of real numbers is the number of pairs $0 \leq i < j \leq n$ such that $a_i a_j < 0$ and $a_\ell = 0$ for all $i < \ell < j$. E.g., the sequence $(-2, 0, 3, 1, 0, -1)$ has the variation 2. Let $f \in \mathbb{R}[X] \setminus 0$ and $v(f)$ be the variation of its coefficient sequence. Prove that f has at most $v(f)$ positive real roots (counted with multiplicity).

12.3. Show that $\sum_{i=0}^{2^t-1} X^i$ may be computed from X with t additions without divisions.

12.4. This example shows that the additive complexity may drastically change with respect to a scalar extension $\mathbb{R} \to \mathbb{C}$. Consider the *real* polynomial

$$f_n = i[(1 - iX)^{n+1} - (1 + iX)^{n+1}].$$

f_n can be computed in $\mathbb{C}[X]$ with only 3 additions/subtractions. Prove that f_n has n distinct real roots. Hence, by Thm. (12.12), we have $\frac{1}{4}\sqrt{\log n} \leq L^+_{\mathbb{R}(X)}(f_n)$. (Hint: The Möbius transformation $z \mapsto z' := 2/(1 + iz) - 1$ maps the real line to the unit circle. Moreover,

$$(z')^{n+1} - 1 = [(1 - iz)^{n+1} - (1 + iz)^{n+1}](1 + iz)^{-n-1}.)$$

12.5. Let the matrix $A \in \mathbb{R}^{n \times (n+1)}$ be of rank n and v_i be $(-1)^i$ times the subdeterminant of A arising by deleting the ith column. Then $(v_1, \ldots, v_{n+1})$ is a nonzero element of the kernel of A. (Hint: use Cramer's rule.)

12.6. Show that in the characterization of L^+ in Lemma (12.14) we may assume that all m_i are zero. Conclude that the number N of a real roots of some nonzero $f \in \mathbb{R}[X]$ of additive complexity r satisfies $N \leq (r + 5)^{2r+2}4^{(r+1)^2}$.

12.7. Prove Lemma (12.14) (characterization of L^+). Generalize the statement to several rational functions in several variables.

12.8. Let $f \in k[X_1, \ldots, X_n]$ and $a \in k^{n \times n}$, $v \in k^n$. Show that

$$L^+(f(a(X_1, \ldots, X_n)^T + v)) \leq L^+(f) + n^2.$$

Hence additive complexity behaves well under a linear change of variables, while the number of terms does not.

12.9. Let $f \in k(X_1, \ldots, X_n)$, $L^+(f) \leq r$, $1 \leq i \leq n$. Show that $L^+(f, \partial_i f) \leq r(r + 3)/2$. (Hint: use Ex. 12.7.)

12.10.* (Generalization of Thm. (12.12) to several variables) Prove that there is a function $\varphi: \mathbb{N}^2 \to \mathbb{N}$ having the following property. For any $n, r \geq 1$, $f_1, \ldots, f_n \in \mathbb{R}(X_1, \ldots, X_n)$ such that $L^+(f_1, \ldots, f_n) \leq r$ the number of non-degenerate real solutions of the system $f_1 = 0, \ldots, f_n = 0$ is at most $\varphi(n, r)$. (You might explicitly describe such a function if you wish.) (Hint: use Thm. (12.3) and proceed as in Sect. 12.3.)

12.11.* Imitate the proofs given in Sect. 11.1 (Milnor-Thom) to show the following.

(1) Let $f_1, \ldots, f_p \in \mathbb{R}[X_1, \ldots, X_n]$ with $L^+(f_1, \ldots, f_p) \leq r$. Then the number of connected components of the zeroset of $f_1, \ldots, f_p$ is bounded from above by a function in n and r only.

(2) (Compare Thm. (11.1)).) Let $f_1, \ldots, f_p$, and $g_1, \ldots, g_q$ be polynomials in $\mathbb{R}[X_1, \ldots, X_n]$ of additive complexity $\leq r$. Let the semi-algebraic set $W \subseteq \mathbb{R}^n$ be defined by the following equalities, non-equalities, and inequalities $(1 \leq q_1 \leq q_2 \leq q)$

$$f_1 = 0, \ldots, f_p = 0,$$

$$g_1 \neq 0, \ldots, g_{q_1} \neq 0,$$

$$g_{q_1+1} > 0, \ldots, g_{q_2} > 0,$$

$$g_{q_2+1} \geq 0, \ldots, g_q \geq 0.$$

Then the number of connected components of W is bounded from above by a function in n, p, q and r only.

(Hint: use Ex. 12.9 and Ex. 12.10 and proceed as in Sect. 11.1.)

12.5 Open Problems

Problem 12.1. Does the computation of $\sum_{i=0}^{n} X^i$ require $\Omega(\log n)$ additions or subtractions if the computation has to take place in the polynomial ring $k[X]$?

Problem 12.2. Let $f_i \in \mathbb{R}[Z_1, Z_1^{-1}, \ldots, Z_n, Z_n^{-1}]$ have m_i distinct terms. Is the number of non-degenerate solutions of the system $f_1 = 0, \ldots, f_n = 0$ lying in the positive quadrant $(0, \infty)^n$ at most $\prod_{i=1}^{n} (m_i - 1)$?

Problem 12.3. Decide whether there is a universal constant $c > 0$ such that for all univariate polynomials $f \in \mathbb{R}[X]$ with N real roots

$$L^+(f) \geq c \log N.$$

Note that this statement follows if Problem 12.2 has an affirmative answer. The Chebyshev polynomials T_{3^r} considered in Example (12.2)(1) show that this is the best result of its kind which can be expected.

Problem 12.4. Is there an an algorithm which computes the additive complexity of a given polynomial $f \in \mathbb{R}[X]$? The difficulty here stems from the fact that we do not know any a priori bound on the size of the exponents m_i, n_i which might occur in the characterization of the additive complexity in Lemma (12.14). If we could compute such a priori bounds, then the existence of a desired algorithm would follow immediately from the the fact that the theory of real closed fields allows quantifier elimination (cf. [118]).

Problem 12.5. Is the number $T(r, s, n)$ of different topological types of semi-algebraic sets defined by r equalities or inequalities of polynomials of additive complexity at most r in n variables finite?

12.6 Notes

The central result of this chapter is Thm. (12.3), which is due to Khovanskii [300]. It can be viewed as a refinement of Bézout's inequality for the real setting (cf. Prop. (11.2)) with the quantity degree replaced by the number of terms. Unfortunately, Khovanskii's bound is far from being optimal, as is easily seen by specializing it to the case of one variable and comparing it with Descartes rule. A straightforward generalization of Descartes rule has been conjectured by Kusnirenko (cf. Problem 12.2). In fact, Khovanskii's theorem resulted from unsuccessful attempts to prove this conjecture. Our presentation of Sect. 12.2 follows closely the book by Benedetti and Risler [39]. (See also Khovanskii's book [301].)

Borodin and Cook [63] were the first who related additive complexity to the number of real zeros. They showed that the number of real zeros of a nonvanishing real univariate polynomial of additive complexity at most r is bounded from above by a function in r only. In this paper the authors also expressed the hope that Problem 12.3 has an affirmative answer. We have taken Example (12.2)(2) from that paper. Grigoriev [209] and Risler [438] independently applied Khovanskii's theorem to obtain Thm. (12.12), which is a substantial improvement upon the results of Borodin and Cook.

Ex. 12.10 and 12.11 generalize Grigoriev and Risler's theorem and constitute important results on their own. There, we claim the existence of an upper bound on the number of connected components of a basic semi-algebraic set in terms of the number of variables, number of the defining polynomials and their additive complexity. We remark that the sum of all Betti numbers of a semi-algebraic set may be bounded in the same way. Benedetti and Risler conjectured that even the number of topological types may be thus bounded (cf. Problem 12.5). In fact, this is true when we replace additive complexity by degree. Solutions to Ex. 12.10, 12.11 and more information about this topic may be found in Benedetti and Risler [39]. We note that Ngọc-Minh Lê [391] has applied some of the results on additive complexity to bound the number of vertices of the Voronoi diagram of a set of points in general position with respect to the L_p-metric.

There has been a recent progress with respect to Problem 12.4. Grigoriev and Karpinski [211] defined a generalized additive complexity by allowing also for root extraction and were able to prove that this quantity is computable. Their proof relies on quantifier elimination in differentially closed fields.

Problem 12.1 was raised by Borodin [68]. For a partial answer see Hyafil and van de Wiele [261]. Ex. 12.4 is due to van de Wiele [531].

Part IV

Low Degree

Chapter 13. Linear Complexity

In this chapter we study the linear complexity of a matrix A, which is the minimum number of additions, subtractions, and scalar multiplications to compute the product AX, where X is a generic input vector. In Sect. 13.2 we first follow Winograd to determine the exact linear complexity of a generic matrix, and then proceed with deriving reasonable upper complexity bounds for some classes of structured matrices. There, we also prove Morgenstern's theorem [382], which yields a lower bound of order $n \log n$ for the number of additions and multiplications by scalars of bounded absolute value to compute the discrete Fourier transform of length n over the complex field. A reformulation of the linear computational model in terms of graph theory is presented in Sect. 13.3. As a first application of this approach, we give a short proof of Tellegen's theorem, which (in a special case) states that an invertible matrix and its transpose have the same linear complexity. Sect. 13.4 discusses the linear complexity of superregular and totally regular matrices in terms of graph theory. We prove Shoup and Smolensky's theorem [477] that sometimes gives nonlinear lower bounds for superregular matrices, as well as Valiant's theorem [525] that yields a lower bound for the linear complexity of totally regular $n \times n$-matrices as a linear function of the minimal number of edges in an n-superconcentrator. We also prove Pinsker and Valiant's theorem [418, 525] stating that n-superconcentrators with $O(n)$ edges exist. The last section is concerned with discrete Fourier transforms for arbitrary finite groups. We present results of Baum, Beth, and Clausen that generalize the classical Cooley-Tukey FFT to wider classes of finite groups. In particular, Baum's theorem [24] shows that every supersolvable group of order n has a DFT of linear complexity $O(n \log n)$.

Throughout this chapter, K denotes an infinite field with prime subfield k.

13.1 The Linear Computational Model

Many problems in computational linear algebra involve evaluating the product of a given matrix and a vector of indeterminates. Often, these matrices are structured and reflect symmetries of the problem under investigation. Among these structured matrices are DFT, circulant, Vandermonde, Toeplitz, and Hankel matrices. Obviously, the number of arithmetic operations used in the evaluation of such a product depends on the organization of the computation. There are efficient algorithms solving this task for every $n \times n$-Vandermonde matrix with $O(n \log^2 n)$ arithmetic operations and each of the other tasks with $O(n \log n)$ arithmetic operations, see the next section. As these problems occur frequently in many applications, it is of practical significance to ask for the optimality of such algorithms. For investigating this question, we have to fix an adequate computational model. This will be our first goal.

Let $A = (a_{ij}) \in K^{m \times n}$ be a fixed matrix and let $X = (X_1, \ldots, X_n)^\top$ be a vector of n indeterminates over K. Then the ith entry $f_i := \sum_{j=1}^n a_{ij} X_j$ of the resulting vector AX is a linear form in these indeterminates. Thus our task is to compute from the inputs $X_1, \ldots, X_n$ a finite set $F = \{f_1, \ldots, f_m\}$ of linear forms. Obviously, such an F can be computed by a *linear* straight-line program that uses only K-linear operations (additions, subtractions and scalar multiplications). In other words, using the language of Chap. 4, the set of instructions of such a program is $K \cup \{+, -\}$. For the moment, let us denote by $\Lambda(F)$ the length of the shortest linear straight-line program that on input X computes F. In analogy to Thm. (7.1) we will now show that non-linear operations (multiplications or divisions outside K) do not help very much when computing linear forms.

(13.1) Theorem. *Let $F \subseteq K[X] := K[X_1, \ldots, X_n]$ be a finite set of linear forms. Then*

$$\tfrac{1}{3}\Lambda(F) \le L_{K(X)}^{\text{tot}}(F \mid X_1, \ldots, X_n) \le \Lambda(F).$$

Proof. W.l.o.g. we may assume that F consists of m nonzero linear forms, $F = \{f_1, \ldots, f_m\}$. Obviously, $\Lambda(F) \ge L_{K(X)}^{\text{tot}}(F)$. To prove the other inequality, let $\Gamma = (\Gamma_1, \ldots, \Gamma_r)$ be a straight-line program over K which computes F on input $X_1, \ldots, X_n$. Let ℓ denote the total length of Γ. (Recall from Sect. 4.1 that Γ has a set of instructions $\Omega = K^c \cup K \cup \{+, -, *, /\}$ and that the total length is measured w.r.t. the cost function tot $= 1_{K \cup \{+, -, *, /\}}$.) It suffices to show that Γ can be transformed into a linear straight-line program Γ' of length $\le 3\ell$ that also computes F. To this end let $b = (b_{-n+1} = X_1, \ldots, b_0 = X_n, b_1, \ldots, b_r)$ be the result sequence corresponding to Γ on input $X_1, \ldots, X_n$. W.l.o.g. we may assume that all b_i are nonzero. As K is infinite, there exists $\lambda \in K^n$ such that all b_i are defined at λ and $b_i(\lambda) \ne 0$. Then the b_i are units of the local subring R of $K(X)$ consisting of those elements of $K(X)$ which are defined at λ. Hence Γ is executable on $(R; X_1, \ldots, X_n)$. The embedding of the polynomial ring $K[X]$ into the ring $K[[X]]$ of formal power series in $X_1, \ldots, X_n$ mapping X_i to $X_i + \lambda$ extends to an embedding $\Phi: R \to K[[X]]$. Let $(B_{-n+1}, \ldots, B_0, B_1, \ldots, B_r)$ be the result sequence of Γ on input $(K[[X]]; X_1 + \lambda_1, \ldots, X_n + \lambda_n)$. Observe that

$$\Phi(f_i)^{(1)} = f_i(X + \lambda)^{(1)} = f_i \quad \text{and} \quad \Phi(X_j)^{(1)} = (X_j + \lambda_j)^{(1)} = X_j,$$

for all $1 \le i \le m$ and $1 \le j \le n$. (Recall that for $g \in K[[X]]$ and $d \in \mathbb{N}$, $g^{(d)}$ denotes the homogeneous part of g of degree d.) As each $\Phi(f_j)$ is equal to a suitable B_ρ, we are only interested in the linear parts of the B_ρ, $1 \le \rho \le r$. If

$B_\rho = B_i \omega_\rho B_j$ with $\omega_\rho \in \{*, /\}$ and suitable $-n < i, j < \rho$, then (see the proof of Lemma (7.2))

$$B_\rho^{(1)} \in K \cdot B_i^{(1)} + K \cdot B_j^{(1)}.$$

Hence in both cases we can compute the linear part of B_ρ from the linear parts of B_i and B_j with at most two scalar multiplications and one addition. So every non-linear step in Γ can be simulated by at most three linear steps. This proves the theorem. $\square$

According to the last result we can restrict ourselves to linear operations when computing linear forms. For technical reasons we will work (in terms of computation sequences) with a slightly modified definition of linear complexity, which is completely symmetric w.r.t. additions and subtractions.

(13.2) Definition. A sequence $(g_{-n+1}, \ldots, g_0, g_1, \ldots, g_r)$ of linear forms in indeterminates $X_1, \ldots, X_n$ over K is called a *linear computation sequence* (over k with n inputs) of length r iff $g_{-n+1} = X_1, \ldots, g_0 = X_n$ and, for every $1 \le \rho \le r$, either

$$g_\rho = z_\rho \cdot g_i \quad \text{or} \quad g_\rho = \varepsilon_\rho g_i + \delta_\rho g_j,$$

for suitable $0 \ne z_\rho \in K$, $\varepsilon_\rho, \delta_\rho \in \{+1, -1\}$, and $-n < i, j < \rho$. Such a sequence *computes* a set F of linear forms iff F is a subset of $\{0, \pm g_\rho \mid -n < \rho \le r\}$. In this case, F is said to be *computable* with r *linear operations*.

The *linear complexity* $L_\infty(f_1, \ldots, f_m)$ of the linear forms $f_1, \ldots, f_m$ is the minimum $r \in \mathbb{N}$ such that there is a linear computation sequence of length r computing $\{f_1, \ldots, f_m\}$. The *linear complexity* of a matrix $A = (a_{ij}) \in K^{m \times n}$ is defined by $L_\infty(A) := L_\infty(f_1, \ldots, f_m)$, where $f_i := \sum_{j=1}^n a_{ij} X_j$. $\bullet$

(13.3) Remarks.
(1) By abuse of notation we will call an operation of type $g_\rho = \varepsilon_\rho g_i + \delta_\rho g_j$ an addition.
(2) Let $\tilde\Omega := K^c \cup K \cup \{+, -\}$ and $c: \tilde\Omega \to \mathbb{N}$ be the cost function which is zero on $K^c \cup \{\pm 1, 0\}$ and equal to 1 elsewhere. The linear complexity of a finite set F of linear forms is then the c-length of the shortest linear straight-line program that on input X computes F. In particular $\Lambda(F) \ge L_\infty(F)$ for any finite set of linear forms. On the other side, one can prove that $\Lambda(F) \le L_\infty(F) + |F|$, see Ex. 13.1.
(3) $L_\infty(f_1, \ldots, f_m) = 0$ iff $\{f_1, \ldots, f_m\} \subseteq \{0, \pm X_1, \ldots, \pm X_n\}$.
(4) If $\{f_1, \ldots, f_m\} \cap \{0, \pm X_1, \ldots, \pm X_n\} = \emptyset$ and $f_i \ne \pm f_j$ for all $i \ne j$, then $L_\infty(f_1, \ldots, f_m) \ge m$.
(5) For $c \in K$ and $A \in K^{m \times n}$ we have $L_\infty(c \cdot A) \le m L_\infty(A)$. $\bullet$

In the sequel we will frequently discuss the linear complexity of DFT-matrices. Recall from Chap. 2 that if ω is a primitive nth root of unity in K, then the corresponding *DFT-matrix* is defined by

$$DFT(n, \omega) := (\omega^{ij})_{0 \le i, j < n}.$$

If there is no need to specify such an ω, we let DFT_n denote any of the $n \times n$-DFT-matrices.

(13.4) Lemma. *Let ω be a primitive n^{th} root of unity in K, and let P be the permutation matrix corresponding to the permutation π defined by $\pi(0) := 0$ and $\pi(i) := n - i$, for $1 \le i < n$. Then the following holds.*

(1) ω^{-1} *is a primitive n^{th} root of unity and $DFT(n, \omega)^{-1} = n^{-1}DFT(n, \omega^{-1})$.*
(2) $DFT(n, \omega^{-1}) = DFT(n, \omega) \cdot P = P \cdot DFT(n, \omega)$.
(3) $DFT(n, \omega)^2 = n \cdot P$.

Proof. Let $(p_{i\ell}) := DFT(n, \omega) \cdot \frac{1}{n}DFT(n, \omega^{-1})$. For $0 \le i, \ell < n$ we have $p_{i\ell} = \sum_{j=0}^{n-1} \omega^{ij} \frac{1}{n} \omega^{-j\ell}$. If $i = \ell$, then obviously $p_{i\ell} = 1$. For $i \ne \ell$, we have $q := \omega^{i-\ell} \ne 1$ and since $q^n = 1$, we obtain

$$p_{i\ell} = \frac{1}{n} \sum_{j=0}^{n-1} q^j = \frac{q^n - 1}{q - 1} = 0.$$

This proves (1) and the other claims follow easily. □

(13.5) Example. (char $K \ne 2, 3$) Let us determine the linear complexity of $DFT_2 = \begin{pmatrix} 1 & 1 \\ 1 & -1 \end{pmatrix}$ and its inverse $DFT_2^{-1} = \frac{1}{2} \begin{pmatrix} 1 & 1 \\ 1 & -1 \end{pmatrix}$. We claim that $L_\infty(DFT_2) = 2$. In fact, the linear computation sequence

$$(g_{-1}, g_0, g_1, g_2) := (X_1, X_2, X_1 + X_2, X_1 - X_2)$$

shows that $L_\infty(DFT_2) \le 2$. As char $K \ne 2$, the lower bound results from Rem. (13.3)(2).

The inverse transform is actually harder to evaluate: $L_\infty(DFT_2^{-1}) = 3$. We have to compute the linear forms $f_1 := \frac{1}{2}(X_1 + X_2)$ and $f_2 := \frac{1}{2}(X_1 - X_2)$. The upper bound $L_\infty(DFT_2^{-1}) \le 3$ is obtained by the sequence $g_{-1} = X_1$, $g_0 = X_2$, $g_1 := g_{-1} + g_0 = X_1 + X_2$, $g_2 := \frac{1}{2}g_1 = f_1$, $g_3 := g_2 - g_0 = f_2$. By Rem. (13.3)(3), $L_\infty(f_1, f_2) \ge 2$. Suppose that equality holds. Then $L_\infty(f_1) = 1$ or $L_\infty(f_2) = 1$. But this is obviously wrong, for char $K \ne 3$ implies $\frac{1}{2} \ne -1$; hence $L_\infty(DFT_2^{-1}) \ge 3$. ●

More generally, if the field K contains a primitive nth root of unity ω, then by Rem. (13.3)(3) we have for $n \ge 2$

(13.6) $$L_\infty(DFT(n, \omega)) \ge n.$$

We summarize some fundamental properties of the linear complexity. (Recall that $A \otimes B$ denotes the Kronecker product of A and B, see Chap. 1, $A \oplus B = \begin{pmatrix} A & 0 \\ 0 & B \end{pmatrix}$ denotes the *direct sum* of matrices.)

(13.7) Lemma. (1) $L_\infty(P) = 0$, *if P is a permutation matrix.*
(2) *If B is a submatrix of A, i.e., $B = A$ or B is obtained from A by deleting some rows and/or columns, then $L_\infty(B) \le L_\infty(A)$.*
(3) $L_\infty(A \oplus B) \le L_\infty(A) + L_\infty(B)$ *for any two matrices A and B.*
(4) $L_\infty(AB) \le L_\infty(A) + L_\infty(B)$, *if the matrices A and B can be multiplied.*

(5) $L_\infty(A \otimes B) \leq aL_\infty(B) + bL_\infty(A)$, if A has exactly a rows and B exactly b columns.

Proof. The statements (1) to (4) are easy exercises. For proving (5) we observe that $A \otimes B = (I_a \otimes B)(A \otimes I_b)$ and $I_a \otimes B = B \oplus \ldots \oplus B$ (a times). On the other hand, we have $A \otimes I_b = P(I_b \otimes A)Q$ for suitable permutation matrices P and Q. (Check this!) Now (5) follows from (1), (3), and (4). $\square$

In general, $L_\infty(A \oplus B) = L_\infty(A) + L_\infty(B)$ does not hold. To see this, let A be a generic $n \times n$ matrix. Then the evaluation of $A^{\oplus n} := A \oplus \cdots \oplus A$ (n times) at a generic input vector of length n^2 just means the multiplication of two generic $n \times n$ matrices. In the next section we will see that $L_\infty(A) = 2n^2 - n$. So if equality in (13.7)(3) would hold, then $L_\infty(A^{\oplus n}) = 2n^3 - n^2$. But this would contradict the fact that multiplication of $n \times n$ matrices can be performed with $O(n^\alpha)$ arithmetic operations, for some $\alpha < 2.38$, see Sect. 15.8.

Let us now proceed with two simple applications of the last lemma.

(13.8) Corollary. *If K has a primitive nth root of unity ω, then*

$$|L_\infty(DFT(n, \omega)^{-1}) - L_\infty(DFT(n, \omega))| \leq n.$$

Proof. By Lemma (13.4) we have $DFT(n, \omega)^{-1} = n^{-1}DFT(n, \omega) \cdot P$ for a suitable permutation matrix P. Hence our claim follows by the last lemma and Rem. (13.3)(5). $\square$

(13.9) Example. We apply Lemma (13.7)(5) to the matrix

$$H_n := DFT_2 \otimes \ldots \otimes DFT_2$$

(n factors), which is the matrix corresponding to the Walsh-Hadamard transform. An easy induction combined with $L_\infty(DFT_2) = 2$ gives $L_\infty(H_n) \leq N \log N$, where $N := 2^n$ is the size of H_n.   •

13.2 First Upper and Lower Bounds

The naive algorithm for the multiplication of $A \in K^{m \times n}$ with a generic input vector X uses mn multiplications and $m(n - 1)$ additions, hence $L_\infty(A) \leq m(2n - 1)$. The following theorem shows that this algorithm is optimal for *generic* matrices. (Recall that A is generic iff all entries of A are algebraically independent over the prime subfield k of K.) For a matrix A over K, we denote by $k(A)$ the smallest subfield of K containing all entries of A.

(13.10) Theorem (Winograd). $L_\infty(A) \geq 2 \operatorname{tr.deg}_k k(A) - m$ *for a matrix $A \in K^{m \times n}$. In particular, $L_\infty(A) = m(2n - 1)$ for generic A.*

Proof. Let $g = (g_{-n+1}, \ldots, g_0, g_1, \ldots, g_r)$ be an optimal linear computation sequence for the linear forms $f_i := \sum_{j=1}^n a_{ij}X_j$, $1 \leq i \leq m$, using α additions and μ scalar multiplications, so $r = \alpha + \mu$. With the notation of Chap. 5 we have

$$\begin{aligned}
\text{tr.deg}_k \, k(A) \quad &= \quad \text{tr.deg}_k \, \text{Coeff}(f_1, \dots, f_m) \\
&\le \quad \text{tr.deg}_k \, \text{Coeff}(g_{-n+1}, \dots, g_0, g_1, \dots, g_r) \le \mu.
\end{aligned}$$

Furthermore, as in the proof of Thm. (5.9), an easy induction on r shows that there exist $0 \ne c_i \in K$ and $h_i \in \langle X_1, \dots, X_n \rangle_K$ such that $g_i = c_i h_i$, for $-n < i \le r$ and $\text{tr.deg}_k \, \text{Coeff}(h_{-n+1}, \dots, h_r) \le \alpha$. As each $f_j \ne 0$ is equal to some $\pm g_i$ and $g_i = c_i h_i$, we have

$$\text{tr.deg}_k \, \text{Coeff}(f_1, \dots, f_m) - m \le \text{tr.deg}_k \, \text{Coeff}(h_{-n+1}, \dots, h_r).$$

Altogether, we get $2 \, \text{tr.deg}_k \, \text{Coeff}(f_1, \dots, f_m) - m \le \alpha + \mu = r$, which proves our claims. $\qquad \square$

The following result gives an improved general upper bound for the linear complexity of $m \times n$ matrices having entries belonging to a finite subset R of K, where $|R|$ is small compared to m.

(13.11) Theorem (Savage). *Let R be a finite subset of K with $r \ge 2$ elements. Then there is a constant γ such that for all matrices $A \in R^{m \times n}$ with $m \ge r^2$ the following holds:*

$$L_\infty(A) \le \gamma \frac{mn}{\log_r m}.$$

Proof. Below we will need the inequalities $2 \log_r \log_r m \le \log_r m$ and $r \le m / \log_r m$, which follow from our assumption $m \ge r^2$ by a straightforward computation. (Use $\sqrt{m} \ge \log m$ for $m \ge 4$.) Ignoring diophantine constraints, we let $\ell := \log_r(m / \log_r m)$, put $p := n / \ell$, and partition A as well as the generic input vector $X = (X_1, \dots, X_n)^\top$ into blocks

$$A = [B_1 | B_2 | \dots | B_p], \quad X = (Y_i)_{1 \le i \le p}.$$

Here $B_1, \dots, B_{p-1}$ are $m \times \ell$, $Y_1, \dots, Y_{p-1}$ are $\ell \times 1$, and the tails B_p and Y_p of A and X are $m \times \ell'$ and $\ell' \times 1$, respectively, for some $\ell' \le \ell$.

In a first step compute αY_i, for all $1 \le i < p$ and *all* $\alpha \in R^{1 \times \ell}$, as well as βY_p, for *all* $\beta \in R^{1 \times \ell'}$. This can be done recursively as follows (we illustrate this only for the case $i = 1$, where $Y_1 = (X_1, \dots, X_\ell)^\top$): compute for all $\lambda \in R$ and $1 \le j \le \ell$ the $r\ell$ products λX_j. Suppose we know for some $t < \ell$ already all $\gamma(X_1, \dots, X_t)^\top$, $\gamma \in R^{1 \times t}$, then we can compute all $\delta(X_1, \dots, X_{t+1})^\top$, $\delta \in R^{1 \times (t+1)}$, with additional $r^t \cdot r$ additions. The number of linear operations to perform the first step is thus at most

$$(p-1)r\ell + r\ell' + (p-1)\sum_{t=1}^{\ell-1} r^{t+1} + \sum_{t=1}^{\ell'-1} r^{t+1}$$

$$\le \quad rn + p\left(\frac{r^{\ell+1} - 1}{r - 1} - (r+1)\right)$$

$$= \quad O\left(\frac{m \cdot n}{\log_r m}\right).$$

(For the last equality use the fact that $r^\ell = m / \log_r m$.)

In a second step we compute AX according to the formula $AX = \sum_{i=1}^{p} B_i Y_i$. As we already know all $B_i Y_i$, this can be done with at most $(p-1)m \leq 2mn/\log_r m$ additions (use $2\log_r \log_r m \leq \log_r m$), which completes the proof of the theorem. $\qquad\square$

It can be shown, see Ex. 13.5, that for sufficiently large n there exist matrices $A \in \{0,1\}^{n \times n}$ such that $L_\infty(A) \geq n^2/\log n$. So the last theorem cannot be improved in general.

Thm. (13.11) does not apply to, e.g., DFT and Vandermonde matrices. Nevertheless, these types of matrices are so well-structured that still better upper bounds can be achieved. For example, for an $n \times n$-Vandermonde matrix A we have $L_\infty(A) = O(n \log^2 n)$, see Sect. 3.3. The Cooley-Tukey FFT algorithm gives the following upper bound on the linear complexity of $n \times n$-DFT-matrices, n a power of 2 (see Chap. 1):

$$L_\infty(DFT(n,\omega)) \leq \tfrac{3}{2}n \log n - n + 1,$$

where ω is a primitive nth root of unity. (For an extension of this result to arbitrary n, see Ex. 13.9.) This result is the basis for efficient algorithms involving other types of matrices. Recall that a matrix of the form

$$\begin{pmatrix} a_0 & a_1 & a_2 & \cdots & a_{n-1} \\ a_{n-1} & a_0 & a_1 & \cdots & a_{n-2} \\ a_{n-2} & a_{n-1} & a_0 & \cdots & a_{n-3} \\ \cdots & \cdots & \cdots & \cdots & \cdots \\ a_1 & a_2 & a_3 & \cdots & a_0 \end{pmatrix}$$

is called *circulant*. Furthermore, matrices of the form

$$\begin{pmatrix} a_0 & a_1 & \cdots & a_{n-1} \\ a_{-1} & a_0 & \cdots & a_{n-2} \\ a_{-2} & a_{-1} & \cdots & a_{n-3} \\ \cdots & \cdots & \cdots & \cdots \\ a_{-n+1} & a_{-n+2} & \cdots & a_0 \end{pmatrix}, \qquad \begin{pmatrix} a_0 & a_1 & \cdots & a_{n-1} \\ a_1 & a_2 & \cdots & a_n \\ a_2 & a_3 & \cdots & a_{n+1} \\ \cdots & \cdots & \cdots & \cdots \\ a_{n-1} & a_n & \cdots & a_{2n-2} \end{pmatrix}$$

are called *Toeplitz* and *Hankel* matrices, respectively. There are close connections between such matrices and DFT matrices.

(13.12) Remarks.
(1) Let $\omega \in K$ be a primitive nth root of unity with corresponding DFT D_n that maps the algebra $\mathcal{A} := K[Y]/(Y^n - 1)$ isomorphically onto the algebra K^n. For a fixed $a = \sum_{i<n} a_i Y^i \in \mathcal{A}$ let $\mu_a : \mathcal{A} \to \mathcal{A}$ denote the multiplication by a. Analogously, let μ_A denote the (componentwise) multiplication by $A := D_n(a) = (A_0, \ldots, A_{n-1})^\top$ in the algebra K^n. Then

$$\mu_a = D_n^{-1} \circ \mu_A \circ D_n.$$

The matrix of μ_a with respect to the basis $1, Y, \ldots, Y^{n-1}$ is the circulant $C(a) = C(a_0, \ldots, a_{n-1})$; the representation matrix of μ_A with respect to the standard basis is the diagonal matrix $\text{diag}(A) = \text{diag}(A_0, \ldots, A_{n-1})$; the representation matrix of D_n with respect to these bases is $DFT(n, \omega)$; so we obtain by Lemma (13.4)

$$
\begin{aligned}
C(a) &= DFT(n, \omega)^{-1} \cdot \text{diag}(A) \cdot DFT(n, \omega) \\
&= P \cdot DFT(n, \omega) \cdot \text{diag}(n^{-1}A) \cdot DFT(n, \omega).
\end{aligned}
$$

Together with (13.6) this yields

$$
L_\infty(C) \le 2L_\infty(DFT(n, \omega)) + n \le 3L_\infty(DFT(n, \omega)),
$$

for every $n \times n$-circulant matrix C.

(2) If $T = T(a_i \mid -n < i < n)$ is a Toeplitz matrix and C is the circulant corresponding to $(a_0, a_1, \ldots, a_{n-1}, 0, a_{-n+1}, \ldots, a_{-2}, a_{-1})$, then

$$
C \cdot \begin{pmatrix} X \\ 0 \end{pmatrix} = \begin{pmatrix} T \cdot X \\ Y \end{pmatrix},
$$

for some vector Y.

(3) If $H = H(a_i \mid 0 \le i \le 2n - 2)$ is a Hankel matrix, and P the permutation matrix corresponding to the permutation $(i \mapsto n - 1 - i)$, $0 \le i < n$, then PH is a Toeplitz matrix.   ●

(13.13) Corollary. *If the field K supports FFTs, then circulant, Toeplitz, and Hankel matrices of format $n \times n$ can be multiplied by a vector in $K^{n \times 1}$ with $O(n \log n)$ arithmetic operations.*

In the remaining of this section we will study the evaluation of matrices over the field $\mathbb{C}$ of complex numbers under a restricted linear complexity model. To this end we fix a real constant $c \ge 2$ and instead of the unrestricted scalar multiplication $g_\ell = z_\ell \cdot g_i$ for some nonzero $z_\ell \in \mathbb{C}$ and $1 \le i < \ell$ (compare Def. (13.2)), we allow only multiplications by scalars z_ℓ which are of absolute value $\le c$. The corresponding computation sequences are called *c-restricted linear computation sequences*, and the *c-restricted linear complexity* $L_c(A)$ is the minimum r such that there is a c-restricted linear computation sequence of length r computing A. It is an easy exercise to check that Lemma (13.7) remains valid when using L_c instead of L_∞. As already indicated in Chap. 1, one can prove by a volume argument a general lower bound on the c-restricted linear complexity of an invertible matrix.

(13.14) Theorem (Morgenstern). $L_c(A) \ge \log_c |\det(A)|$ *for any invertible complex matrix A and $2 \le c < \infty$.*

Proof. Let $f_i = \sum_{j=1}^n a_{ij} X_j$ denote the linear form corresponding to the ith row of A. Suppose that $L_c(A) = r$ and let $(g_{-n+1}, \ldots, g_r)$ be an optimal c-restricted linear computation sequence for $f_1, \ldots, f_n$ with corresponding coefficient matrix $B = (b_{\ell j}) \in \mathbb{C}^{(n+r) \times n}$ defined by $g_\ell = \sum_{j=1}^n b_{\ell j} X_j$, $-n < \ell \le r$. For simplicity, let $B(j_1, \ldots, j_n)$ denote the $n \times n$-matrix consisting of rows $j_1, \ldots, j_n$ of B:

$$B(j_1, \ldots, j_n) := \begin{pmatrix} b_{j_1 1} & \cdots & b_{j_1 n} \\ \vdots & & \vdots \\ b_{j_n 1} & \cdots & b_{j_n n} \end{pmatrix}.$$

By definition, $\{f_1, \ldots, f_n\} \subseteq \{0, \pm g_{-n+1}, \ldots, \pm g_r\}$. As A is invertible, we can assume that both $\{f_1, \ldots, f_n\} \subseteq \{g_{-n+1}, \ldots, g_r\}$ and $A = B(j_1, \ldots, j_n)$ for suitable $1 \le j_1 < \ldots < j_n = r$. ($j_n = r$ as the sequence is optimal.)

Next, we are going to show that for all ℓ with $0 \le \ell \le r$, we have

(A) $\forall \ -n < i_1 < \ldots < i_n \le \ell: \quad |\det(B(i_1, \ldots, i_n))| \le c^\ell.$

We proceed by induction on ℓ. For $\ell = 0$, we necessarily have $(i_1, \ldots, i_n) = (-n + 1, \ldots, 0)$. By the definition of a linear computation sequence, $g_{-n+1} = X_1, \ldots, g_0 = X_n$. Hence $B(-n + 1, \ldots, 0) = I_n$, and our claim follows.

Now let $\ell > 0$. If $i_n < \ell$, the induction hypothesis directly applies, so we may assume that $i_n = \ell$. We have to consider two cases. If $g_\ell = \lambda g_i$ for some $i < \ell$ with $|\lambda| \le c$, then $(b_{\ell 1}, \ldots, b_{\ell n}) = \lambda(b_{i1}, \ldots, b_{in})$. Hence

$$\begin{aligned} |\det(B(i_1, \ldots, i_n))| &= |\lambda| \cdot |\det(B(i_1, \ldots, i_{n-1}, i))| \\ &\le |\lambda| \cdot c^{\ell-1} \quad \text{(induction hypothesis)} \\ &\le c^\ell \quad (|\lambda| \le c). \end{aligned}$$

If $g_\ell = \pm g_i \pm g_j$ for suitable $i, j < \ell$, then by the multilinearity of the determinant,

$$\begin{aligned} |\det(B(i_1, \ldots, i_n))| &= \\ |\pm \det(B(i_1, \ldots, i_{n-1}, i)) &\pm \det(B(i_1, \ldots, i_{n-1}, j))| \\ \le |\det(B(i_1, \ldots, i_{n-1}, i))| &+ |\det(B(i_1, \ldots, i_{n-1}, j))| \\ \le 2 \cdot c^{\ell-1} \quad \text{(induction hypothesis)} \\ \le c^\ell \quad (c \ge 2). \end{aligned}$$

This shows (A). To prove the theorem, we apply (A) to $A = B(j_1, \ldots, j_n)$ and $\ell = r$ and obtain $|\det(A)| \le c^r$.   □

It is a nice exercise to rewrite this proof in terms of the volume of a linear computation sequence, compare Chap. 1. How realistic are lower bounds in the c-restricted model? If the matrix to be evaluated has entries which are very large in absolute value, then it is not realistic to study lower bounds in the c-restricted model for small c: consider for example the matrix $A = nI_n$, I_n being the $n \times n$-identity matrix. Then $L_2(A) \ge n \log n$ by the above theorem, but $L_\infty(A) = n$. Whether the c-restricted model of computation is adequate for matrices which only have entries of small absolute value, is an open problem. Prominent examples of such matrices are the DFT-matrices. By Lemma (13.4) we know that $DFT_n^2 = n \cdot P$ for a permutation matrix P. Hence, application of the previous theorem yields the following.

(13.15) Corollary (Morgenstern). $L_c(DFT_n) \ge \frac{1}{2} n \log_c n$ for $2 \le c < \infty$.

As the Cooley-Tukey FFT only uses scalar multiplications by roots of unity (besides additions and subtractions), we see that for n a power of 2

$$\tfrac{1}{2}n \log n \leq L_2(DFT_n) \leq \tfrac{3}{2}n \log n - n + 1.$$

This shows that in the L_2-model the Cooley-Tukey FFT has optimal order of magnitude. Obviously, Thm. (13.14) is valid for any valuated field $(K, |\cdot|)$. The proof shows that if the valuation is non-archimedean and there is a c-restricted linear computation sequence for A with μ scalar multiplications, then even $\mu \geq \log_c |\det(A)|$.

13.3* A Graph Theoretical Approach

There is an alternative characterization of linear straight-line programs Γ in terms of certain digraphs, which will be of some use in the sequel. It is a slight modification of the directed acyclic multigraph corresponding to a straight-line program Γ, which was introduced in Chap. 4. As a first application we shall deduce a close connection between the linear complexities of a matrix A and its transpose $A^\top$.

(13.16) Definitions.
(1) (Syntax) A K-*linear computation graph* (for short: K-*DAG*) over (n, m) is a quintuple $G = (V, E, I, O, \lambda)$ consisting of a finite directed acyclic graph (V, E), a weight function $\lambda: E \to K^\times$, an ordering $I = (x_1, \ldots, x_n)$ of the nodes x_i of in-degree zero (these are the *input nodes*), and an ordering $O = (F_1, \ldots, F_m)$ of the nodes F_j of out-degree zero (these are the *output nodes*).
(2) (Semantics) Let $G = (V, E, I, O, \lambda)$ be a K-DAG over (n, m). By induction on the depth of the nodes, every node $v \in V$ is assigned to a linear form h_v in indeterminates $X_1, \ldots, X_n$ over K as follows: to the input node x_j assign the linear form X_j. Inductively, suppose that $v \in V$ has in-degree $q > 0$. Let $h_1, \ldots, h_q$ be the linear forms assigned to the q nodes $v_1, \ldots, v_q$ with $(v_i, v) \in E$. Then the linear form $h_v := \sum_{i=1}^q \lambda(v_i, v) h_i$ is assigned to v. (We also say that G *computes* h_v at v.) A set F of linear forms is said to be *computable* by G iff it is contained in $\{0, \pm h_v \mid v \in V\}$. G *outputs* the linear forms attached to the output nodes.
(3) (Cost) The number $c(G) := |\{e \in E \mid \lambda(e) \neq \pm 1\}| + |E| - |V| + n$ is called the *cost* of the K-DAG $G = (V, E, I, O, \lambda)$ over (n, m).   ●

To motivate the last definition consider a vertex v of in-degree $d(v) > 0$ with $w(v)$ incoming edges with weights $\neq \pm 1$. This vertex v gives rise to $w(v)$ scalar multiplications and $d(v) - 1$ additions. Hence the number of linear operations caused by G is

$$\sum_{d(v) \geq 1} (d(v) - 1 + w(v)) = |E| - (|V| - n) + |\{e \in E \mid \lambda(e) \neq \pm 1\}|,$$

which equals $c(G)$.

In the sequel we will (without further mention) only consider DAGs, in which every input node x_j (resp. output node F_i) is connected to at least one output node F_i (resp. input node x_j) by a directed path $x_j \to \ldots \to F_i$.

(13.17) Lemma. *Let F be a finite set of linear forms over K. Then*

$$L_\infty(F) = \min\{c(G) \mid G \text{ is a } K\text{-DAG computing } F\}.$$

Proof. Let $F = \{f_1, \ldots, f_m\}$, where $f_i := \sum_{j=1}^n a_{ij} X_j$. W.l.o.g. we can assume that all $f_i \neq 0$ and every X_j occurs in some f_i. Let $g = (g_{-n+1}, \ldots, g_0, g_1, \ldots, g_r)$ be an optimal linear computation sequence for F using α additions and μ scalar multiplications, thus $L_\infty(F) = r = \alpha + \mu$. We will use g to define a K-DAG $G = (V, E, I, O, \lambda)$ over (n, m) computing F with cost r. To this end let $V := \{v \in \mathbb{Z} \mid -n < v \leq r+m\}$, where $I = (-n+1, \ldots, 0)$ and $O = (r+1, \ldots, r+m)$ are the total orderings of the input and output nodes, respectively. The set E consists of all the edges (i, v) of weight z_v such that $g_v = z_v g_i$ ($z_v \neq \pm 1$ by the optimality of g), and all edges (i, v) and (j, v) of respective weights ε_v and δ_v, such that $g_v = \varepsilon_v g_i + \delta_v g_j$. (Although the description of G is not complete, we already mention that so far g_v is computed at the node v, for all $-n < v \leq r$.) The last m nodes of V correspond to the outputs $f_1, \ldots, f_m$: if $f_j = \pm g_i$, we add an edge $(i, r + j)$ of weight ± 1. By this construction, the nodes $-n + 1, \ldots, 0$ are exactly the nodes of in-degree 0 in G corresponding to the standard input $(X_1, \ldots, X_n)$, whereas $r + 1, \ldots, r + m$ are exactly the nodes of out-degree 0 in G corresponding to the outputs $f_1, \ldots, f_m$. A K-DAG constructed in this way is said to be a *standard K-DAG for F*. As G has μ edges of weight $\neq \pm 1$ and $2\alpha + m$ edges of weight ± 1, we obtain $c(G) = \mu + (\mu + 2\alpha + m) - (n + r + m) + n = r$. Hence $L_\infty(F) \geq \min\{c(G) \mid G \text{ computes } F\}$.

To prove the other direction, let $G = (V, E, I, O, \lambda)$ be a K-DAG over (n, m) with $I = (x_1, \ldots, x_n)$ and $O = (F_1, \ldots, F_m)$ that computes F with minimum cost $c(G)$. We associate to G a linear computation sequence g computing F in $c(G)$ steps. Let $\sqsubset$ be any total ordering of V extending the partial ordering on V induced by G in such a way that the input nodes are the n smallest elements of $(V, \sqsubset)$. Put $(g_{-n+1}, \ldots, g_0) := (X_1, \ldots, X_n)$. We then traverse the remaining nodes v in $\sqsubset$-order, breaking up each computation $h_v := \sum_{i=1}^q \lambda(v_i, v) h_i$ into single scalar multiplications $\lambda(v_i, v) h_i$ (for the weights $\neq \pm 1$) and $q - 1$ binary additions. It is obvious that the number r of linear operations in the resulting linear computation sequence only depends on G and not on the choice of $\sqsubset$. This number r can be expressed purely in terms of G: clearly, the number of nontrivial scalar multiplications equals the number of edges with weight $\neq \pm 1$. At each node of in-degree $d > 0$, we need $d - 1$ additions. With a computation similar to that preceding this lemma one sees that $r = c(G)$, which completes the proof of the lemma. $\square$

Note that every finite set F of m linear forms in n indeterminates can be computed optimally by a standard K-DAG G over (n, m). (If F is the set of linear forms corresponding to the matrix A, then G is also called a *standard K-DAG for the matrix A*.) A standard K-DAG for F computes F optimally by definition, and has thus the property that all edges pointing to an inner node of in-degree 1 have weight $\neq \pm 1$. The next result tells us that the number of edges in such a G is a good approximation for the linear complexity of F.

(13.18) Lemma. *Let $F \subseteq K[X_1, \ldots, X_n]$ be a finite set of linear forms computed by the standard K-DAG $G = (V, E, I, O, \lambda)$ over (n, m). Then the following holds.*

(1) $\frac{1}{2}(|E| - |F|) \leq L_\infty(F) \leq |E| - |F|$.

(2) *If $L_\infty(F) \geq |F|$, then $\frac{1}{3}|E| \leq L_\infty(F) < |E|$. Hence $L_\infty(F) = \Theta(|E|)$.*

Proof. According to our assumptions and with the notation of the last proof we have $r := L_\infty(F) = \alpha + \mu$, $|F| = m$, $n + m \leq |V| = n + m + r$, $|E| = \mu + 2\alpha + m$, and $|\{e \in E \mid \lambda(e) \neq \pm 1\}| = \mu \leq |E| - m$. Now $L_\infty(F) = c(G)$ gives $\frac{1}{2}(|E| - m) \leq \frac{1}{2}(\mu + |E| - m) = r \leq \frac{1}{2}(|E| - m + |E| - m) = |E| - m$. This proves (1). If $r \geq m$, then $\frac{1}{2}(|E| - r) \leq \frac{1}{2}(|E| - m) \leq r$; hence $\frac{1}{3}|E| \leq r \leq |E| - m \leq |E|$. This proves (2). □

The following remark, easily proved by induction, gives a useful formula for the linear forms computed by a K-DAG.

(13.19) Remark. Let $G = (V, E, I, O, \lambda)$ be a K-DAG over (n, m) with $I = (x_1, \ldots, x_n)$. For nodes $u, v \in V$, let $G(u, v)$ denote the set of all paths $p = (u \to \cdots \to v)$ in (V, E). If h_v is the linear form assigned to $v \in V$ by G, then for every $v \in V$, which is not an input node, we have

$$h_v = \sum_{j=1}^{n} \left(\sum_{p \in G(x_j, v)} \lambda(p) \right) X_j,$$

where $\lambda(p)$ is the product of the weights of all edges involved in p. (By definition, the empty sum is zero and the empty product has the value 1.)   •

Now we are going to apply this graph theoretic interpretation to prove a property of linear complexity which is less obvious. In Thm. (13.10) we have seen that $L_\infty(A) = m(2n - 1)$, for a generic matrix $A \in K^{m \times n}$. But the transpose of A is also generic, hence $L_\infty(A^\top) = n(2m - 1)$. This shows that in the generic case $L_\infty(A^\top) = L_\infty(A) - n + m$. The following result, one of the basic theorems in network theory, is a generalization of this fact. Roughly speaking, it says that reversing all arrows in a standard K-DAG for a matrix A without changing the weights gives an optimal K-DAG for $A^\top$.

(13.20) Tellegen's Theorem. *If $z(A)$ denotes the number of zero rows of $A \in K^{m \times n}$, then $L_\infty(A^\top) = L_\infty(A) - n + m - z(A) + z(A^\top)$. In particular, if A is invertible, then*

$$L_\infty(A) = L_\infty(A^\top).$$

Proof. We first prove the above claim for the non-degenerate case when $z(A) = z(A^\top) = 0$. Let $f_i := \sum_{j=1}^{n} a_{ij} X_j$ be the linear forms corresponding to $A = (a_{ij})$. Let $G = (V, E, I, O, \lambda)$ be a standard K-DAG for $f_1, \ldots, f_m$ with $I = (x_1, \ldots, x_n)$ and $O = (F_1, \ldots, F_m)$. Define the transposed K-DAG

$$G^\top = (V, E^\top, I^\top, O^\top, \lambda^\top)$$

of G by $E^T := \{(w, v) \mid (v, w) \in E\}$, $I^T = O$, $O^T = I$, and $\lambda^T(w, v) := \lambda(v, w)$ for every $(v, w) \in E$. In other words, G^T results from G by reversing the arrows without changing the weights. Thus G^T computes n linear forms $X_1^T, \ldots, X_n^T$ in m indeterminates $f_1^T, \ldots, f_m^T$.

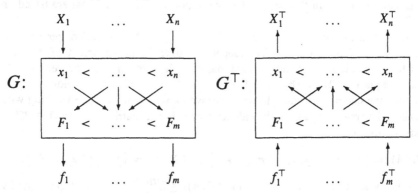

Reversing the arrows in a path $p \in G(v, w)$ yields a path $p^T \in G^T(w, v)$. This defines a natural bijection $G(v, w) \to G^T(w, v)$ satisfying $\lambda(p) = \lambda^T(p^T)$. According to Rem. (13.19), $f_i = \sum_{j=1}^n (\sum_{p \in G(x_j, F_i)} \lambda(p)) X_j$, thus $a_{ij} = \sum_{p \in G(x_j, F_i)} \lambda(p)$. On the other hand,

$$X_j^T = \sum_{i=1}^m \left(\sum_{p^T \in G^T(F_i, x_j)} \lambda^T(p^T) \right) f_i^T$$

$$= \sum_{i=1}^m \left(\sum_{p \in G(x_j, F_i)} \lambda(p) \right) f_i^T = \sum_{i=1}^m a_{ij} f_i^T.$$

Therefore, G^T computes the linear forms which correspond to A^T. By Lemma (13.17),

$$
\begin{aligned}
L_\infty(A^T) \le c(G^T) &= |\{e^T \in E^T \mid \lambda^T(e^T) \ne \pm 1\}| + |E^T| - |V| + m \\
&= |\{e \in E \mid \lambda(e) \ne \pm 1\}| + |E| - |V| + n - n + m \\
&= L_\infty(A) - n + m.
\end{aligned}
$$

Applying the same argument to A^T, we obtain $L_\infty(A) \le L_\infty(A^T) - m + n$. Altogether, this proves the non-degenerate case. In the general case, deleting the zero rows and columns of A results in a non-degenerate $(m - z(A)) \times (n - z(A^T))$-matrix B satisfying $L_\infty(B) = L_\infty(A)$ and $L_\infty(B^T) = L_\infty(A^T)$ (check this!). Combining this with the fact that $L_\infty(B^T) = L_\infty(B) - (n - z(A^T)) + (m - z(A))$, the general claim follows. $\qquad\square$

The last theorem remains valid, if K is the field of complex numbers and L_∞ is replaced by L_c, $2 \le c < \infty$. An interesting corollary of the last theorem is the determination of the linear complexity of a single linear form:

(13.21) Corollary. $L_\infty(\sum_{i=1}^n a_i X_i) = n - 1 + |\{|a_1|, \ldots, |a_n|\} \setminus \{1\}|$, *if all a_i are nonzero.*

13.4* Lower Bounds via Graph Theoretical Methods

Various attempts have been made to prove superlinear lower bounds for the unrestricted linear complexity of structured matrices like DFT, circulant and Vandermonde matrices. In this section we will report on techniques that are based on K-DAGs.

A Vandermonde matrix $A = (a_i^{j-1})_{1 \le i, j \le n} \in K^{n \times n}$ is called *generic*, if $a_1, \ldots, a_n \in K$ are algebraically independent over the prime field k of K. Note that for a generic $n \times n$-Vandermonde matrix A the transcendence degree of $k(A)$ over k equals n. Thus Thm. (13.10) gives only a trivial result. Below we will see that one can occasionally deduce improved lower bounds, when working with weakened versions of algebraic independence. For a matrix $A = (a_{ij}) \in K^{n \times n}$ consider the sets

$$\Pi(A) := \left\{ a_\pi := \prod_{i=1}^n a_{i\pi(i)} \mid \pi : \underline{n} \to \underline{n} \right\} \text{ and } \Sigma(A) := \left\{ \sum_{t \in T} t \mid T \subseteq \Pi(A) \right\}.$$

It is easily seen that $|\Pi(A)| \le n^n$ and $|\Sigma(A)| \le 2^{|\Pi(A)|} \le 2^{n^n}$, and that equality holds in the generic case. Below we shall develop lower bounds for the complexity of matrices which share this property with the generic ones.

(13.22) Definition. Let $A \in K^{n \times n}$. A is called *totally regular* iff every square submatrix of A is invertible; A is *superregular* iff $|\Sigma(A)| = 2^{n^n}$. •

Note that the entries of a totally regular matrix are all nonzero. Furthermore, $A \in K^{1 \times 1}$ is superregular iff $A \ne 0$. Generic Vandermonde matrices are superregular. It can also be verified quickly that a Vandermonde matrix corresponding to $a_i = 2^{n^{i-1}}$, $1 \le i \le n$, is superregular as well, see Ex. 13.6. The 2×2-DFT-matrix is an example of a totally regular matrix which is not superregular.

(13.23) Lemma. *Every superregular matrix is totally regular.*

Proof. Let $A = (a_{ij}) \in K^{n \times n}$ be superregular. We first show that all principal submatrices $B_m = (a_{ij})_{1 \le i, j \le m}$, $1 \le m \le n$, are invertible. To this end we embed the symmetric group S_m in an obvious way into S_n and let A_m denote the alternating subgroup of S_m. Recall that $a_\pi := \prod_{i=1}^n a_{i\pi(i)} \in \Pi(A)$ for every mapping $\pi : \underline{n} \to \underline{n}$. This applies in particular to permutations $\pi \in S_m$. As A is superregular, $\sum_{\pi \in A_m} a_\pi$ and $\sum_{\pi \in S_m \setminus A_m} a_\pi$ are different elements in $\Sigma(A)$ and $0 \ne \sum_{\pi \in A_m} a_\pi - \sum_{\pi \in S_m \setminus A_m} a_\pi = \det(B_m) \cdot \prod_{j=m+1}^n a_{jj}$, hence $\det(B_m) \ne 0$. As the set of superregular matrices in $K^{n \times n}$ is invariant under row and column permutations, the general case reduces to the case of principal submatrices. □

Combination of Rem. (13.3)(3) and the previous lemma yields the following.

(13.24) Corollary. *If $A \in K^{n \times n}$ is superregular, $n > 1$, then $L_\infty(A) \ge n$.*

The *depth* of a K-DAG is the length of the longest directed path in G. For a superregular matrix computable by a standard K-DAG of small depth, the above lower bound can be improved as follows.

(13.25) Theorem (Shoup and Smolensky). *Let G be a standard K-DAG over (n, n) for the superregular matrix $A \in K^{n \times n}$. If G has depth $d > 1$, then $c(G) = \Omega(n \log n / \log d)$. In particular, $L_\infty(A) = \Omega(n \log n / \log d)$.*

Proof. We may assume that $n \geq 2$. Let $G = (V, E, I, O, \lambda)$ be a standard K-DAG for $A = (a_{ij})$ with $I = (x_1, \ldots, x_n)$ and $O = (F_1, \ldots, F_n)$. Thus $f_i := \sum_{j=1}^{n} a_{ij} X_j$ is the linear form computed at the output node F_i. Let d be the depth of (V, E). By Cor. (13.24), $L_\infty(A) =: r \geq n$. Hence the number $\ell := |E|$ of edges in G is up to a factor c, $1/3 \leq c < 1$, equal to r, see Lemma (13.18). Thus our claim will follow once we have shown that

(A) $$2^{n^n} = |\Sigma(A)| \leq \ell^{n(d+1)|S|},$$

where S is a set satisfying

(B) $$|S| \leq (12d)^\ell (2n)^{n/2}.$$

The proof proceeds in several steps.

Step 1. In this step we introduce S as well as a first upper bound for $|S|$.

For $1 \leq i \leq d$ let $e_{i1}, \ldots, e_{i\ell_i}$ be those edges in G having a target node of depth i. These edges form the *level i* in G. Let $T_i := \{\lambda_{ij} \mid 1 \leq j \leq \ell_i\}$ be the set of edge weights $\lambda_{ij} := \lambda(e_{ij})$ that appear at level i. If a path p in G contributes the edge e to level i, then we set $p^i := e$. In particular, $\lambda(p^i) = \lambda(e)$. If no edge of level i is involved in p, we put $\lambda(p^i) := 1$. The following example illustrates this.

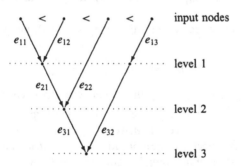

Here, $\ell_1 = 3$, $\ell_2 = 2 = \ell_3$; $T_1 = \{\lambda(e_{11}), \lambda(e_{12}), \lambda(e_{13})\}$, $T_2 = \{\lambda(e_{21}), \lambda(e_{22})\}$, and $T_3 = \{\lambda(e_{31}), \lambda(e_{32})\}$.

We continue with the proof. With the above notation we can write, see Rem. (13.19),

$$a_{ij} = \sum_{p \in G(x_j, F_i)} \lambda(p^1) \cdots \lambda(p^d).$$

Thus each a_{ij} can be expressed as a sum of products of the form $\alpha_1 \cdots \alpha_d$ where $\alpha_i \in T_i \cup \{1\}$. For $\pi : \underline{n} \to \underline{n}$ let $\mathcal{P}(\pi) := \{p := (p_1, \ldots, p_n) \mid p_i \in G(x_{\pi(i)}, F_i)\}$. Then the element $a_\pi = \prod_i a_{i\pi(i)}$ of $\Pi(A)$ can be rewritten as

$$a_\pi = \sum_{\underline{p} \in \mathcal{P}(\pi)} \lambda(p_1^1) \cdots \lambda(p_1^d) \cdots \lambda(p_n^1) \cdots \lambda(p_n^d)$$

$$= \sum_{\underline{p} \in \mathcal{P}(\pi)} \left(\lambda(p_1^1) \cdots \lambda(p_n^1) \right) \cdots \left(\lambda(p_1^d) \cdots \lambda(p_n^d) \right).$$

Thus each element in $\Pi(A)$ can be expressed as the sum of products of the form

$$(\alpha_{11} \cdots \alpha_{1n}) \cdots (\alpha_{d1} \cdots \alpha_{dn}),$$

where $\alpha_{ij} \in T_i \cup \{1\}$. We let S denote the collection of all such products

$$S := (T_1 \cup \{1\})^n \cdots (T_d \cup \{1\})^n.$$

It is obvious that the set $(T_i \cup \{1\})^n$ consists of the evaluations at the λ_{ij} of all the monomials in ℓ_i indeterminates with degree at most n. There are $\binom{n+\ell_i}{\ell_i}$ such monomials; hence we obtain

(C)
$$|S| \le \prod_{i=1}^d \binom{n+\ell_i}{\ell_i},$$

with equality holding if the λ_{ij} are algebraically independent.

Step 2. We show that every $a_\pi \in \Pi(A)$ can be written as an integer linear combination of elements of S, $a_\pi = \sum_{s \in S} c_{\pi,s} s$, with $0 \le c_{\pi,s} < \ell^{nd}$.

The reader may easily verify that it is sufficient to show this in the case where all the occurring weights λ_{ij} are algebraically independent. Thus we shall assume this from now on.

For $\underline{p} \in \mathcal{P}(\pi)$ there are non-negative integers $\varepsilon_{ij}(\underline{p})$, uniquely determined by $\underline{p}$, such that

$$\lambda(p_1^i) \cdots \lambda(p_n^i) = \lambda_{i1}^{\varepsilon_{i1}(\underline{p})} \cdots \lambda_{i\ell_i}^{\varepsilon_{i\ell_i}(\underline{p})}.$$

Furthermore, $\sum_{j=1}^{\ell_i} \varepsilon_{ij}(\underline{p}) \le n$. By the algebraic independence of the λ_{ij} we see that each $\underline{p} \in \mathcal{P}(\pi)$ is uniquely determined by the sequence $(\lambda(p_i^j))_{(i,j)}$. Thus for fixed $\pi : \underline{n} \to \underline{n}$ and a given multi-vector $\delta = (\delta_1, \ldots, \delta_d)$, where each $\delta_i = (\delta_{i0}, \delta_{i1}, \ldots, \delta_{i\ell_i})$ is a vector of non-negative integers that sum up to n, there are (by the definition of the multinomial coefficients) at most

$$\prod_{i=1}^d \binom{n}{\delta_{i0}, \delta_{i1}, \ldots, \delta_{i\ell_i}} =: \binom{n}{\delta}$$

many $\underline{p} \in \mathcal{P}(\pi)$ with

$$\prod_{i=1}^d (\lambda(p_1^i) \cdots \lambda(p_n^i)) = \prod_{i=1}^d \prod_{j=1}^{\ell_i} \lambda_{ij}^{\delta_{ij}} =: \lambda^\delta.$$

Thus $a_\pi = \sum_\delta c_{\pi,\delta} \lambda^\delta$, for suitable integers $0 \le c_{\pi,\delta} \le \binom{n}{\delta}$. To complete the proof of Step 2 it remains to show that $\binom{n}{\delta} < \ell^{nd}$, for every such multi-vector δ. By the multinomial theorem, $\binom{n}{\delta_{i0}, \delta_{i1}, \ldots, \delta_{i\ell_i}} \le (\ell_i + 1)^n$. Furthermore, as $d > 1$ and all ℓ_i are positive integers, we have $\ell_i + 1 \le \ell = \sum_i \ell_i$. As $n \ge 2$ by assumption and G is a K-DAG for a superregular matrix, we obtain $\ell_i + 1 < \ell$ for all i. Altogether we have shown that $\binom{n}{\delta} \le \prod_{i=1}^d (\ell_i + 1)^n < \ell^{nd}$.

Step 3. Proof of (A).

Let T be a set of functions $\underline{n} \to \underline{n}$. Then the corresponding element in $\Sigma(A)$ looks as follows:

$$\sum_{\pi \in T} a_\pi = \sum_{\pi \in T} \sum_\delta c_{\pi,\delta} \lambda^\delta = \sum_\delta \left(\sum_{\pi \in T} c_{\pi,\delta} \right) \lambda^\delta.$$

Next we estimate the inner sum. For $1 \le i \le \ell_1$ there is a unique $j(i) \in \underline{n}$ and a node v_i at level 1 in G such that the edge $X_{j(i)} \to v_i$ has weight λ_{1i}. Thus a mapping $\pi : \underline{n} \to \underline{n}$ with $c_{\pi,\delta} \ne 0$ has the property that for every $i \in \underline{n}$ there are exactly δ_{1i} elements in $\underline{n}$ which are mapped by π onto $j(i)$. Thus for fixed $\delta = (\delta_{10}, \delta_{11}, \ldots, \delta_{1\ell_1})$ the number of π with $c_{\pi,\delta} \ne 0$ is at most

$$\binom{n}{\delta_{10}, \delta_{11}, \ldots, \delta_{1\ell_1}} \le (\ell_1 + 1)^n < \ell^n.$$

Hence $\sum_{\pi \in T} c_{\pi,\delta} < \ell^n \cdot \ell^{nd} = \ell^{n(d+1)}$. Thus each element of $\Sigma(A)$ is a linear combination of elements of S with non-negative integer coefficients smaller than $\ell^{n(d+1)}$. This proves (A).

Step 4. Proof of (B).

We use Stirling's approximation,

$$\sqrt{2\pi n}(n/e)^n \le n! \le \sqrt{2\pi n}(n/e)^n e^{1/(12n)},$$

which gives $\binom{n+\ell_i}{\ell_i} \le (1 + \ell_i/n)^n \cdot (1 + n/\ell_i)^{\ell_i}$. Therefore, by (C),

$$|S| \le \left(\prod_{i=1}^d (1 + n/\ell_i)^{\ell_i} \right) \cdot \left(\prod_{i=1}^d (1 + \ell_i/n)^n \right).$$

We have $\Pi_2 := \prod_{i=1}^d (1 + \ell_i/n)^n \le e^\ell < 3^\ell$. To bound $\Pi_1 := \prod_{i=1}^d (1 + n/\ell_i)^{\ell_i}$, let $s \in \mathbb{R}$, $0 < s < n$. The contribution to Π_1 from those factors with $\ell_i \ge s$ is at most $(1 + n/s)^\ell \le 2^\ell (n/s)^\ell$. As all $\ell_i \ge 1$, the contribution to Π_1 from those factors with $\ell_i < s$ is at most $(1 + n)^{sd} \le (2n)^{sd}$. Now set $s = n/(2d)$. Then $\Pi_1 \le (4d)^\ell (2n)^{n/2}$. Combining the estimates for Π_1 and Π_2 we obtain (B). This completes the proof of the theorem. $\square$

The attempt to prove non-linear lower bounds for $L_\infty(DFT(n, \omega))$ has directed attention to a class of interesting graphs, the n-superconcentrators.

(13.26) Definition. An *n-superconcentrator* is a directed acyclic graph (V, E) having both n nodes of indegree zero (the input nodes) and n nodes of outdegree zero (the output nodes) such that any set X of $q \leq n$ input nodes can be connected to any set Y of q output nodes by q node-disjoint directed paths $p_1, \ldots, p_q$. (These q paths form a *q-flow* from X to Y.)   •

With the above notation, a subset C of V is called an (X, Y)-*cut* in G iff every path from X to Y meets C. Below we will need the following version of Menger's theorem. (For a proof consult, e.g., Bollobás [62].)

(13.27) Menger's Theorem. *Let $G = (V, E)$ be a digraph, X and Y disjoint subsets of V. Then*

$$\max\{q \mid \exists \ q\text{-flow from } X \text{ to } Y \text{ in } G\} = \min\{|C| \mid C \text{ is an } (X, Y)\text{-cut in } G\}.$$

(13.28) Theorem (Valiant). *Let $G = (V, E, I, O, \lambda)$ be any standard K-DAG over (n, n) computing a totally regular matrix in $K^{n \times n}$. Then (V, E) is an n-superconcentrator.*

Proof. Let $A \in K^{n \times n}$ be a totally regular matrix computed by G. The proof of our claim proceeds in two steps.

Step 1. Let $v \in V$ and $U \subseteq V$ be such that every path from an input node to v meets U. We claim that the linear form h_v assigned to v by G is a linear combination of the linear forms h_u assigned by G to the nodes $u \in U$. In fact, if we denote by $G_u(x_j, v)$ the set of all paths from the jth input node to v meeting U for the last time at $u \in U$ and, for $u \in U$, by $G^U(u, v)$ the set of all paths $p \in G(u, v)$ meeting U only at u, then, by our assumptions, $G(x_j, v) = \sqcup_{u \in U} G_u(x_j, v)$, and according to Rem. (13.19) we get

$$\begin{aligned}
h_v &= \sum_{j=1}^{n} \sum_{p \in G(x_j, v)} \lambda(p) X_j = \sum_{j=1}^{n} \sum_{u \in U} \sum_{p \in G_u(x_j, v)} \lambda(p) X_j \\
&= \sum_{j=1}^{n} \sum_{u \in U} \sum_{p_2 \in G^U(u, v)} \lambda(p_2) \sum_{p_1 \in G(x_j, u)} \lambda(p_1) X_j \\
&= \sum_{u \in U} \sum_{p_2 \in G^U(u, v)} \lambda(p_2) \sum_{j=1}^{n} \sum_{p_1 \in G(x_j, u)} \lambda(p_1) X_j \\
&= \sum_{u \in U} \left(\sum_{p_2 \in G^U(u, v)} \lambda(p_2) \right) h_u.
\end{aligned}$$

Step 2. Now suppose that (V, E) were not an n-superconcentrator. W.l.o.g. we can assume that for some $r \leq n$ there are no r node-disjoint paths from the input nodes $x_1, \ldots, x_r$ to the output nodes $F_1, \ldots, F_r$. By Menger's theorem (13.27), there is an $(r - 1)$-subset U of V such that every path from $x_1, \ldots, x_r$ to $F_1, \ldots, F_r$ meets

U. Hence, by Step 1, $f_1, \ldots, f_r \in \langle \{h_u \mid u \in U\} \cup \{X_{r+1}, \ldots, X_n\} \rangle_K$, and with $f_i^* := f_i(X_1, \ldots, X_r, 0, \ldots, 0)$ and $h_u^* := h_u(X_1, \ldots, X_r, 0, \ldots, 0)$ we get

$$f_1^*, \ldots, f_r^* \in \langle h_u^* \mid u \in U \rangle_K,$$

i.e., $f_1^*, \ldots, f_r^*$ are K-linearly dependent, since $|U| = r - 1$. In other words, $\det(B) = 0$, where B is the principal $r \times r$-submatrix of A. This contradiction completes the proof of the theorem. $\qquad\square$

By Lemma (13.23), the last theorem applies in particular to all superregular matrices. Our next goal is to prove that, after a slight modification, the last theorem also applies to the $n \times n$ DFT-matrices. By Rem. (13.12)(1) we know that $L_\infty(A) \leq 2L_\infty(DFT_n) + n \leq 3L_\infty(DFT_n)$, for every $n \times n$-circulant matrix A. Thus the linear complexity of any circulant $A \in K^{n \times n}$ is up to a constant factor a lower bound for $L_\infty(DFT_n)$. In particular, this applies to a *generic* circulant. (A circulant matrix $A = C(a) \in K^{n \times n}$, $a = \sum_{i < n} a_i Y^i \in K[Y]/(Y^n - 1)$, is called *generic* if $a_0, \ldots, a_{n-1} \in K$ are algebraically independent over k.)

(13.29) Lemma. *Every generic circulant matrix $A \in K^{n \times n}$ is totally regular.*

Proof. By induction on m we show that every $m \times m$ submatrix B of A is invertible. The start $m = 1$ being clear, assume $m > 1$. Since every a_i occurs in every row and every column of B at most once, we can assume (after suitable permutations of rows and columns and renaming of the a_i) that a_0 occurs in B exactly at the first $d \geq 1$ diagonal positions. Then $\det(B)$, viewed as a polynomial in $k(a_1, \ldots, a_{n-1})[a_0]$, has degree d and its leading coefficient is up to a sign the determinant of an $(m - d) \times (m - d)$-submatrix of A, which is nonzero by induction. (For $d = m$ the leading coefficient is 1.) $\qquad\square$

For a generic circulant matrix $A \in K^{n \times n}$, $\Pi(A)$ can be viewed as the set of all monomials of degree n in n indeterminates; hence $|\Pi(A)| = \binom{2n-1}{n} \leq 4^{n-1}$. This shows that an $n \times n$ generic circulant matrix is superregular iff $n = 1$.

(13.30) Corollary (Valiant). *If $s(n)$ denotes the minimum possible number of edges in an n-superconcentrator, then $L_\infty(DFT_n) \geq s(n)/6 - n/2$.*

Proof. Combine Rem. (13.12)(1) and Lemma (13.18). $\qquad\square$

The last result shows that one would get a nonlinear lower bound for the linear complexity of the DFT-matrices if one were able to prove a nonlinear lower bound on the minimal number of edges in an n-superconcentrator. For some while, it had been conjectured that $s(n) = \Omega(n \log n)$. As a surprise it turned out that linear-sized superconcentrators exist, see the next theorem. Though a failure for their original purpose, the existence proof for (and beyond that the construction of) linear-sized superconcentrators is a beautiful result in graph theory with many applications, for which the reader is referred to the notes of this chapter.

(13.31) Theorem (Pinsker, Valiant). $s(n) \leq 39n + O(\log n)$.

Below we shall give a non-constructive proof. A major ingredient is the existence of a certain type of bipartite graphs. Let us begin by defining them. For every m and every permutation π of $\{0, \ldots, 36m - 1\}$ we define a bipartite graph $G_\pi = (A, B, R_\pi)$ with $A = \{0, \ldots, 6m - 1\}$, $B = \{0, \ldots, 4m - 1\}$ and

$$R_\pi := \{(x \bmod 6m, \pi x \bmod 4m) \mid 0 \le x < 36m\}.$$

In G_π, every input $a \in A$ has out-degree ≤ 6 and each output $b \in B$ has in-degree ≤ 9. We will say that G_π is *good* iff for every $j \le 3m$ and every j-set X of inputs, there exists a j-set Y of outputs and a j-flow from X to Y in G_π; otherwise, G_π is called *bad*.

(13.32) Lemma. *For every m, there exists a permutation π of $\{0, \ldots, 36m - 1\}$ such that G_π is good.*

Proof. It suffices to show that for all m the fraction β_m of all permutations π of $\mathcal{M} := \{0, \ldots, 36m - 1\}$ for which G_π is bad is < 1.

If G_π is bad, then P. Hall's marriage theorem (cf. [2]) ensures that there exist $Y \subseteq A$ and $Z \subseteq B$ such that $|Z| = |Y| \le 3m$ and $R_\pi Y \subset Z$, where for any $Y \subseteq A$

$$R_\pi Y := \{b \in B \mid \exists a \in Y : (a, b) \in R_\pi\}.$$

Now consider the maps $\phi_6 := (\mathcal{M} \ni x \mapsto x \bmod 6m \in A)$ and $\pi_4 := (\mathcal{M} \ni x \mapsto \pi x \bmod 4m \in B)$, and observe that for any j-set Y of inputs the preimage $\phi_6^{-1}(Y)$ is a $6j$-subset $\mathcal{Y}$ of $\mathcal{M}$, and for any j-set Z of outputs the preimage $\pi_4^{-1}(Z)$ is a $9j$-subset $\mathcal{Z}$ of $\mathcal{M}$. Every edge of G_π directed out of Y will be directed into Z if and only if π sends every element of $\mathcal{Y}$ into $\mathcal{Z}$. Out of the $(36m)!$ permutations π of $\mathcal{M}$, there are $[9j]_{6j}(36m - 6j)!$ that satisfy the condition $\pi\mathcal{Y} \subseteq \mathcal{Z}$, where $[n]_r := n(n - 1) \cdots (n + 1 - r)$. For a given value of j, there are $\binom{6m}{j}$ possible choices for Y and $\binom{4m}{j}$ possible choices for Z. Thus

$$\beta_m \le \frac{1}{(36m)!} \sum_{j=1}^{3m} \binom{6m}{j}\binom{4m}{j}[9j]_{6j}(36m - 6j)! = \sum_{j=1}^{3m} \frac{\binom{6m}{j}\binom{4m}{j}\binom{9j}{6j}}{\binom{36m}{6j}}.$$

As $\binom{6m}{j}\binom{4m}{j}\binom{26m}{4j} \le \binom{36m}{6j}$, we get with $L_j := \binom{9j}{6j}/\binom{26m}{4j}$

$$\beta_m \le \sum_{j=1}^{3m} L_j \le 3m \max_{j \le 3m} L_j.$$

It remains to prove that $\beta_m < 1$ for all m. A direct calculation shows that $\beta_1 < 1$ and $\beta_2 < 1$. So we can assume that $m \ge 3$. As

$$\frac{L_{j+1}}{L_j} = \frac{(9j + 9) \cdots (9j + 7)(9j + 6) \cdots (9j + 1)}{(6j + 6) \cdots (6j + 1)}.$$

$$\frac{(4j + 4)(4j + 3) \cdots (4j + 1)}{(3j + 3) \cdots (3j + 1)(26m - 4j) \cdots (26m - 4j - 3)}.$$

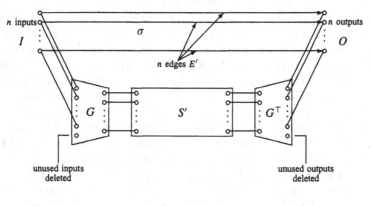

Fig. 13.1. Superconcentrator S.

The function $(\{1, \ldots, 3m\} \ni j \mapsto L_{j+1}/L_j)$ is increasing, hence we obtain $\max_{1 \le j \le 3m} L_j = \max\{L_1, L_{3m}\}$. If L_1 is the maximum, then $\beta_m \le 3mL_1 < 1$ for all those m. If L_{3m} is the maximum, then

$$\beta_m \le 3mL_{3m} = 3m\binom{27m}{18m}\binom{26m}{12m}^{-1} = 3m\frac{(27m)!(12m)!(14m)!}{(18m)!(9m)!(26m)!}.$$

Now Stirling's formula combined with $e^x \le \frac{1}{1-x}$ for $x < 1$ implies

$$\sqrt{2\pi n}\left(\frac{n}{e}\right)^n \le n! \le \frac{12n}{12n-1}\sqrt{2\pi n}\left(\frac{n}{e}\right)^n.$$

These inequalities give

$$\beta_m \le 3m \cdot \frac{324m}{324m-1} \cdot \frac{144m}{144m-1} \cdot \frac{168m}{168m-1}$$
$$\cdot \left(\frac{27 \cdot 12 \cdot 14}{18 \cdot 9 \cdot 26}\right)^{1/2} \cdot \left(\frac{27^{27} \cdot 12^{12} \cdot 14^{14}}{18^{18} \cdot 9^9 \cdot 26^{26}}\right)^m.$$

By a direct calculation one shows that $\beta_3 < 1$. Furthermore, as the right-hand side is a decreasing function of m, we obtain $\beta_m < 1$ for all m.   □

Proof of Thm. (13.31). Let $m = \lceil n/6 \rceil$. An easy induction shows that it suffices to prove for every n

$$s(n) \le 13n + s(4m).$$

Let S' be a $4m$-superconcentrator with $s(4m)$ edges and let $G = G_\pi = (A, B, R)$ be a good bipartite graph corresponding to a permutation π of $\{0, 1, \ldots, 36m - 1\}$. Then the transposed bipartite graph $G^\top = (B, A, R^\top)$, $R^\top := \{(b, a)|(a, b) \in R\}$, has the "dual" properties of G. The graph S with input set I and output set O (both of cardinality n) is obtained by deleting $6m - n(\leq 6)$ inputs (and the edges directed out of them) from G, identifying the $4m$ outputs of G with the $4m$ inputs of S', identifying the $4m$ outputs of S' with the $4m$ inputs of $G^\top$, deleting $6m - n$ outputs (and the edges directed into them) from $G^\top$, and adding a set E' of n edges from the surviving inputs of G to the surviving outputs of $G^\top$ such that $E' = \{(i, \sigma i)|i \in I\}$ for a bijection $\sigma : I \to O$. This is illustrated in Fig. 13.1. As each input node in G has out-degree ≤ 6, the modified G contributes to S at most $6n$ edges; the same is true for the modified $G^\top$. Thus S has at most $13n + s(4m)$ edges. All that remains is to verify that S is an n-superconcentrator. For some $j \leq n$, let $X \subseteq I$ be a set of j inputs and $Y \subseteq O$ be a set of j outputs in S. We partition both X and Y into two parts $X = X_1 \sqcup X_2$ and $Y = Y_1 \sqcup Y_2$ such that $\sigma X_1 = Y_1$ and $Y_2 \cap \sigma X_2 = \emptyset$. Now σ constitutes a $|X_1|$-flow from X_1 through E' to Y_1. Secondly, as σ is a bijection, $Y_2 \cap \sigma X_2 = \emptyset$ and $|X_2| = |Y_2|$ imply

$$|X_2| = |Y_2| \leq n/2 \leq 3m.$$

Hence, as G is good, we can find a $|X_2|$-subset X_2' of B and a $|X_2|$-flow in the modified G from X_2 to X_2'. Analogously, there exists a $|Y_2|$-subset Y_2' of A and a $|Y_2|$-flow in the modified $G^\top$ from Y_2' to Y_2. Finally, as S' is a superconcentrator, there is a $|X_2'|$-flow in S' from X_2' to Y_2'. These four flows constitute an r-flow from X to Y. Thus S is an n-superconcentrator. $\square$

13.5* Generalized Fourier Transforms

Throughout this section we will work over the field $\mathbb{C}$ of complex numbers.

So far we have studied lower and upper bounds for the linear complexities of DFT-like matrices. Now we aim at generalizing these results. Every finite group G (abelian or non-abelian) of order n defines a set of $n \times n$-DFT-matrices and we are going to derive $\Omega(n \log n)$ lower bounds in the restricted linear computation models as well as similar upper bounds for the linear complexities of suitable DFT-matrices of certain types of (non-abelian) groups.

Let G be a finite group of order n. Then the set $\mathbb{C}G$ of all $\mathbb{C}$-valued functions on G is a $\mathbb{C}$-space under pointwise addition and scalar multiplication. The indicator functions $\chi_g := (G \ni h \mapsto \delta_{g,h})$ of the group elements $g \in G$ form a $\mathbb{C}$-basis of $\mathbb{C}G$ and every function $a \in \mathbb{C}G$ can be uniquely written as $a = \sum_{g \in G} a(g)\chi_g$. By abuse of notation we identify χ_g with g and write $a \in \mathbb{C}G$ as a formal sum

$$a = \sum_{g \in G} a(g)\chi_g =: \sum_{g \in G} a_g g.$$

$\mathbb{C}G$ becomes an associative $\mathbb{C}$-algebra (called the complex *group algebra* of G) via

$$\left(\sum_{g\in G} a_g g\right) \cdot \left(\sum_{h\in G} b_h h\right) := \sum_{\ell\in G}\left(\sum_{gh=\ell} a_g b_h\right)\ell.$$

This multiplication, called the *convolution* with respect to G, is commutative iff G is an abelian group. The following result, due to Wedderburn, describes the structure of complex group algebras. (For a proof see, e.g., Curtis and Reiner [139], or Serre [470].)

(13.33) Theorem. *Let G be a finite group with h conjugacy classes. Then there are positive integers $1 = d_1 \le \ldots \le d_h$, uniquely determined by G, such that $\mathbb{C}G$ is isomorphic to the algebra $\oplus_{i=1}^{h}\mathbb{C}^{d_i\times d_i}$ of block-diagonal matrices.*

Our main object of study in this section are the isomorphisms whose existence is guaranteed by the above theorem. They deserve a special name.

(13.34) Definition. Every algebra isomorphism $D\colon \mathbb{C}G \to \oplus_{i=1}^{h}\mathbb{C}^{d_i\times d_i}$ is called a *(generalized) discrete Fourier transform* (DFT) for $\mathbb{C}G$. The representation matrix of a DFT D for $\mathbb{C}G$ w.r.t. the canonical bases in $\mathbb{C}G$ and in $\oplus_{i=1}^{h}\mathbb{C}^{d_i\times d_i}$ is called a *DFT-matrix for G*. $\mathrm{DFT}(G) \subseteq \mathbb{C}^{|G|\times|G|}$ denotes the set of all DFT-matrices for G. •

A DFT D for $\mathbb{C}G$ links the convolution in $\mathbb{C}G$ with the multiplication of block diagonal matrices:

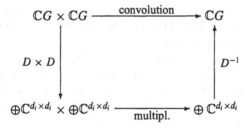

Thus a fast algorithm to evaluate D and its inverse, together with a fast procedure to multiply block-diagonal matrices, yields a fast convolution in $\mathbb{C}G$. The other way round, a fast algorithm to evaluate D and its inverse together with a fast procedure for the convolution in $\mathbb{C}G$ yields a fast algorithm for the multiplication of block-diagonal matrices. We have already used this strategy in the case of the cyclic group $G = C_n = \langle X \mid X^n = 1\rangle = \{1, X, X^2, \ldots, X^{n-1}\}$ of order n. Here, the group algebra $\mathbb{C}C_n$ is isomorphic to the algebra $\mathbb{C}[X]/(X^n - 1)$ of all polynomials of degree smaller than n with multiplication modulo $X^n - 1$. This algebra is isomorphic to the algebra $\mathbb{C}^n$ of diagonal matrices. For every primitive nth root of unity ω there is a unique algebra isomorphism $\mathbb{C}C_n \to \mathbb{C}^n$ defined by $X \mapsto \omega$, and this isomorphism is described by the DFT-matrix $DFT(n, \omega)$ (w.r.t. canonical bases). In turn, every algebra isomorphism $\mathbb{C}C_n \to \mathbb{C}^n$ is of this form. As we know, these DFT-matrices differ only by permutation matrices, thus their linear complexities are equal. Our discussion shows that in the cyclic case (and more generally in the abelian case) the DFT is unique up to these trivial modifications.

For a non-abelian group G the situation looks quite different. Here $d_h \geq 2$, i.e., at least the block $\mathbb{C}^{d_h \times d_h}$ is a non-commutative algebra. Every DFT $D: \mathbb{C}G \to \oplus_{i=1}^{h} \mathbb{C}^{d_i \times d_i}$ can be modified in a non-trivial way by composing D with a non-trivial automorphism α of the target algebra. By the Skolem-Noether theorem, cf. [264, p. 222], every such α is of the form $(\oplus_i \mathbb{C}^{d_i \times d_i} \ni \oplus_i M_i \mapsto \oplus_i T_i M_i T_i^{-1})$, where $T_i \in \mathrm{GL}(d_i, \mathbb{C})$. So in the non-abelian case there are infinitely many DFTs for $\mathbb{C}G$ and thus there are also infinitely many DFT-matrices corresponding to G (w.r.t. the canonical bases). These DFT-matrices might differ substantially in their linear complexities. As we are interested in a fast transition from the group algebra $\mathbb{C}G$ to the target algebra $\oplus_{i=1}^{h} \mathbb{C}^{d_i \times d_i}$, it makes sense to choose a DFT with a minimum linear complexity. This motivates the following.

(13.35) Definition. Let G be a finite group of order n, $2 \leq c \leq \infty$. $L_c(G) := \min\{L_c(A) \mid A \in \mathrm{DFT}(G)\}$ is called the *c-restricted linear complexity of the group* G. ($L_\infty(G)$ will also be called the *linear complexity* of G.)   •

Example (13.5) and the discussions preceding the above definition show that for all $2 \leq c \leq \infty$, $L_c(C_2) = L_2(DFT_2) = 2$.

The fact that for a non-abelian group G the target algebra of $\mathbb{C}G$ has at least one non-commutative block has another interesting aspect. In Chap. 15 we shall see that every fast algorithm for the multiplication of block-diagonal matrices with at least one block of size > 1 can automatically be transformed into an algorithm to multiply *general* square matrices of large size. (This is the contents of Schönhage's asymptotic sum inequality, Thm. (15.11).) Occasionally the resulting algorithm is more efficient than the naive algorithm.

The rest of this section is devoted to the derivation of lower and upper bounds for $L_c(G)$. We begin by recalling some basic facts from the ordinary representation theory of finite groups. A *representation* of $\mathbb{C}G$ with *representation space* $M \simeq \mathbb{C}^d$ is an algebra morphism $\mathrm{R}: \mathbb{C}G \to \mathrm{End}_{\mathbb{C}}(M)$; $\deg(\mathrm{R}) = d$ is called the *degree* or *dimension* of R. Choosing a $\mathbb{C}$-basis in M, each $\mathrm{R}(a)$, $a \in \mathbb{C}G$, is described by a $d \times d$ matrix $R(a)$, and $a \mapsto R(a)$ is an algebra morphism $R: \mathbb{C}G \to \mathbb{C}^{d \times d}$, a so-called (matrix) representation of $\mathbb{C}G$ of degree d. A group morphism $D: G \to \mathrm{GL}(d, \mathbb{C})$ is called a (linear) *representation* of G of *degree* d. (Restricting a representation of $\mathbb{C}G$ to G yields a group representation of G. In turn, every representation of G defines via linear extension a representation of $\mathbb{C}G$. According to this one-to-one correspondence we sometimes make no difference between a representation of $\mathbb{C}G$ and its restriction to G.) Matrix representations are fine for doing computations. For structural purposes a more abstract language is sometimes more convenient. A *left* $\mathbb{C}G$-*module* of dimension d is a $\mathbb{C}$-space M equipped with a mapping $\mathbb{C}G \times M \to M$, $(a, m) \mapsto am$, satisfying $a(m_1 + m_2) = am_1 + am_2$, $(a_1 + a_2)m = a_1 m + a_2 m$, $(a_1 a_2)m = a_1(a_2 m)$, $(\lambda a)m = a(\lambda m) = \lambda(am)$, and $1m = m$, for all $a, a_i \in \mathbb{C}G$, $m, m_i \in M$, $\lambda \in \mathbb{C}$. A left $\mathbb{C}G$-module M defines a representation R of $\mathbb{C}G$ with representation space M via $\mathrm{R}(a)(m) := am$, for $a \in \mathbb{C}G$ and $m \in M$, and vice versa. We sometimes say that M *affords* R. A subspace U of the left $\mathbb{C}G$-module M is called a *submodule* of M iff $au \in U$ for all $a \in \mathbb{C}G$ and $u \in U$. Trivially, 0 and M are always submodules of M. M is called *simple* iff $0 \neq M$ and M has no submodules other than 0 and M. A representation corresponding to a simple module is called *irreducible*. In particular, a matrix representation R of $\mathbb{C}G$ of degree d is irreducible iff no nonzero proper linear subspace of $\mathbb{C}^{d \times 1}$ is invariant under all matrices in the image $R(\mathbb{C}G)$ of R. It is easy to see that surjective representations of $\mathbb{C}G$ are irreducible. We mention in passing that the converse is also true, though not as easy to see. Every representation of degree 1 is irreducible; this applies in particular to the trivial representation $\sum_{g \in G} a_g g \mapsto \sum_{g \in G} a_g$. Two matrix representations R and R', both of degree d, are *equivalent* ($R \sim R'$) iff there is a $T \in \mathrm{GL}(d, \mathbb{C})$ such that $R'(a) = T R(a) T^{-1}$, for all $a \in \mathbb{C}G$. (Note that two representations of degree 1 are equivalent iff they are equal.) With these notions we can make Wedderburn's structure theorem for complex group algebras more precise. (For a proof see, e.g., Curtis and Reiner [139], or Serre [470].)

(13.36) Wedderburn's Structure Theorem. *Let G be a finite group, h the number of conjugacy classes of G. Then there are exactly h equivalence classes of irreducible matrix representations of $\mathbb{C}G$. If $D_1, \ldots, D_h$ is a complete set of representatives of these classes such that $d_1 := \deg(D_1) \leq \ldots \leq d_h := \deg(D_h)$, then*

$$D = \oplus_{i=1}^{h} D_i := \left(\mathbb{C}G \ni a \mapsto \oplus_i D_i(a) \in \oplus_i \mathbb{C}^{d_i \times d_i} \right)$$

is a DFT for $\mathbb{C}G$ and every DFT for $\mathbb{C}G$ is obtained in this way.

We note that the above theorem in particular implies that $\sum_{i=1}^{h} d_i^2 = |G|$. In the sequel we shall make tacitly use of this observation. Let us illustrate the above theorem by an example.

(13.37) Example. The symmetric group S_n, which is the group of all $n!$ permutations π of $\{1, 2, \ldots, n\}$, is non-abelian iff $n \geq 3$. For constructing a DFT for $\mathbb{C}S_3$, we first determine the degrees $d_1 \leq d_2 \leq \ldots \leq d_h$ of the irreducible representations of $\mathbb{C}S_3$. As S_3 is non-abelian, we already know that $d_h > 1$. On the other hand, $|S_3| = 6 = \sum_j d_j^2$, thus $d_h = 2$ and consequently $6 - d_h^2 = 2 = \sum_{j<h} d_j^2$. This forces $h = 3$ and $d_1 = d_2 = 1$. Hence, by Wedderburn's structure theorem (13.36),

$$\mathbb{C}S_3 \simeq \mathbb{C}^{1 \times 1} \oplus \mathbb{C}^{1 \times 1} \oplus \mathbb{C}^{2 \times 2}.$$

The two irreducible representations of degree 1 are the trivial representation $\iota : \pi \mapsto 1$ and the alternating representation $\varepsilon : \pi \mapsto \mathrm{sgn}(\pi)$; here we have identified $\mathbb{C}$ and $\mathbb{C}^{1 \times 1}$. Finally, there is an irreducible representation Δ of degree 2 realizing S_3 as the symmetry group of a regular triangle T. If we take the center of gravity of T as the origin in the Euclidean 2-space and denote its vertices by e_1, e_2, e_3, then (e_1, e_2) is a basis and $e_3 = -e_1 - e_2$. The natural S_3-action $\pi e_i := e_{\pi(i)}$ yields (we write permutations in their cycle representations)

$$\Delta((1)) = \begin{pmatrix} 1 & 0 \\ 0 & 1 \end{pmatrix}, \quad \Delta((123)) = \begin{pmatrix} 0 & -1 \\ 1 & -1 \end{pmatrix}, \quad \Delta((132)) = \begin{pmatrix} -1 & 1 \\ -1 & 0 \end{pmatrix},$$

$$\Delta((12)) = \begin{pmatrix} 0 & 1 \\ 1 & 0 \end{pmatrix}, \quad \Delta((23)) = \begin{pmatrix} 1 & -1 \\ 0 & -1 \end{pmatrix}, \quad \Delta((13)) = \begin{pmatrix} -1 & 0 \\ -1 & 1 \end{pmatrix}.$$

Δ is surjective, hence irreducible. Thus

$$
\begin{array}{c}
 \\
\iota \\
\varepsilon \\
\Delta_{11} \\
\Delta_{12} \\
\Delta_{21} \\
\Delta_{22}
\end{array}
\begin{array}{cccccc}
(1) & (123) & (132) & (12) & (23) & (13) \\
\left(\begin{array}{cccccc}
1 & 1 & 1 & 1 & 1 & 1 \\
1 & 1 & 1 & -1 & -1 & -1 \\
1 & 0 & -1 & 0 & 1 & -1 \\
0 & -1 & 1 & 1 & -1 & 0 \\
0 & 1 & -1 & 1 & 0 & -1 \\
1 & -1 & 0 & 0 & -1 & 1
\end{array}\right)
\end{array}
$$

is a DFT-matrix for $\mathbb{C}S_3$.   ●

After these preliminaries we can prove first upper and lower bounds.

(13.38) Lemma. *We have* $n - 1 \leq L_\infty(G) \leq 2n(n - 1)$, *for every finite group* G *of order* n.

Proof. As the trivial representation that maps every $g \in G$ to $(1) \in \mathbb{C}^{1 \times 1}$ is always involved in a DFT for $\mathbb{C}G$, one row of *every* $A \in DFT(G)$ is the vector $B := (1, \ldots, 1) \in \mathbb{C}^{1 \times n}$. Since $L_\infty(B) = n - 1$, Lemma (13.7)(2) gives the lower bound for $L_\infty(G)$. To prove the upper bound, observe that $A \in DFT(G)$ is an $n \times n$-matrix with one row equal to B. To multiply such a matrix by a generic vector with n components, we need at most $(n - 1)n$ scalar multiplications and $n(n - 1)$ additions.   □

As $L_\infty(G) \leq L_2(G)$, we also have the lower bound $|G| - 1 \leq L_2(G)$. Our next goal is to improve this result. For deriving a lower bound for $L_2(G)$ with the aid of Thm. (13.14) we need to know the infimum of $|\det(A)|$ for all $A \in DFT(G)$. The following result shows that $|\det(A)|$ only depends on G.

(13.39) Schur Relations. *Let* G *be a finite group of order* n, $\mathbb{C}G \simeq \oplus_{i=1}^{h} \mathbb{C}^{d_i \times d_i}$. *Then for every* $A \in DFT(G)$ *there exist permutation matrices* P *and* Q *such that*

$$APA^\top Q = \bigoplus_{i=1}^{h} \frac{n}{d_i} I_{d_i^2}.$$

In particular, $|\det(A)| = \prod_{i=1}^{h}(n/d_i)^{d_i^2/2}$.

Proof. Let $D_1, \ldots, D_h$ be a complete list of inequivalent irreducible representations of $\mathbb{C}G$, $d_i := \deg(D_i)$. Then the Schur relations in its classical form, cf. [470], state that

$$\sum_{g \in G} D_a(g)_{ij} D_b(g^{-1})_{\ell m} = \delta_{a,b}\delta_{i,m}\delta_{j,\ell}\frac{n}{d_a},$$

for all $1 \leq a, b \leq h, 1 \leq i, j \leq d_a$ and $1 \leq \ell, m \leq d_b$. Now let $A \in \mathbb{C}^{n \times n}$ be the DFT- matrix of G corresponding to $D_1, \ldots, D_h$. Since the columns of A

are parameterized by the elements of G whereas the rows of A correspond to $\{(i, j, \ell) \mid 1 \leq i \leq h, 1 \leq j, \ell \leq d_i\}$, (i.e., (i, j, ℓ) describes the position (j, ℓ) in D_i), we have

$$A_{((i,j,\ell),g)} = D_i(g)_{j\ell}.$$

Now let $B := PA^TQ \in \mathbb{C}^{n \times n}$ be the matrix obtained by first transposing A and then performing in A^T the permutation P of the rows corresponding to the inversion $(G \ni g \mapsto g^{-1})$ and the permutation Q of the columns corresponding to $(i, j, \ell) \mapsto (i, \ell, j)$. Then

$$B_{(g,(i,j,\ell))} = D_i(g^{-1})_{\ell j}.$$

According to the Schur relations, AB is a diagonal matrix with d_i^2 occurrences of n/d_i, for $1 \leq i \leq h$, which proves the theorem. $\square$

This result has an interesting corollary. According to our general strategy to design fast algorithms for the multiplication in $\mathbb{C}G$ via a DFT D by $ab = D^{-1}(D(a)D(b))$, we are only interested in those $A \in \mathrm{DFT}(G)$, where both A and A^{-1} can be multiplied with a generic vector X at minimum arithmetic costs. At first sight our definition of $L_\infty(G)$ looks asymmetric for it only minimizes $L_\infty(A)$, $A \in \mathrm{DFT}(G)$. However the following result justifies our definition.

(13.40) Corollary (Baum and Clausen). *Let G be a finite group of order n, and $A \in \mathrm{DFT}(G)$. Then*

$$|L_\infty(A) - L_\infty(A^{-1})| \leq n.$$

Proof. Let $\mathbb{C}G \simeq \oplus_i \mathbb{C}^{d_i \times d_i}$. By the last theorem $A^{-1} = PA^TQ(\oplus_i \frac{d_i}{n} I_{d_i^2})$. As A is invertible, we have $L_c(A^T) = L_c(A)$ by Thm. (13.20). Applying Lemma (13.7) our claim follows. $\square$

Now we can prove general lower bounds in the restricted linear computation model.

(13.41) Theorem (Baum and Clausen). *Let G be a finite group of order n, and $2 \leq c < \infty$. Then the following holds.*

(1) $L_c(G) > 0.25n \log_c n$.
(2) $L_c(G) \geq 0.5n \log_c n$, *if G is abelian.*

Proof. Let $\mathbb{C}G \simeq \oplus_{i=1}^h \mathbb{C}^{d_i \times d_i}$. Combining Thm. (13.14) and the Schur relations (13.20) with the fact that $n = \sum_{i=1}^h d_i^2$, we get

$$L_c(G) \geq \log_c \left(\prod_{i=1}^h \left(\frac{n}{d_i} \right)^{d_i^2/2} \right)$$

$$= \frac{n}{2} \log_c n - \sum_{i=1}^h \frac{d_i^2}{4} \log_c d_i^2$$

$$\geq \frac{n}{2} \log_c n - \sum_{d_i > 1} \frac{d_i^2}{4} \log_c n.$$

Using the fact that $GL(1, \mathbb{C})$ is abelian and that every group morphism of G into an abelian group factors through G/G', the commutator factor group of G, one easily sees that the number of irreducible representations of $\mathbb{C}G$ of degree 1 equals the order $[G : G']$ of the commutator factor group. Thus we obtain

$$
\begin{aligned}
L_c(G) & \geq \frac{n}{2} \log_c n - \sum_{d_i > 1} \frac{d_i^2}{4} \log_c n \\
& = \frac{n}{2} \log_c n - \frac{1}{4}(|G| - [G : G']) \log_c n \\
& = \frac{n}{4}(1 + |G'|^{-1}) \log_c n.
\end{aligned}
$$

As $|G'| = 1$, if G is abelian, our claims follow. $\square$

Now we are going to derive improved upper bounds for $L_c(G)$. It will be of some advantage to look at our computational problem from a slightly different point of view that better reflects the structure of the linear forms to be computed. More precisely, let $D_1, \ldots, D_h$ be a complete list of pairwise inequivalent irreducible representations of $\mathbb{C}G$. Then the linear forms corresponding to this list are

$$
f_{(i,a,b)} := \sum_{g \in G} D_i(g)_{ab} X_g,
$$

for $1 \leq i \leq h$ and $1 \leq a, b \leq d_i := \deg(D_i)$. Instead of computing the mere *set* of all such $f_{(i,a,b)}$ we reformulate our problem as follows: given for each $g \in G$ the block-diagonal matrix $\bigoplus_{i=1}^h D_i(g)$ (this corresponds to the given matrix A to be evaluated) and given a generic input vector $(X_g)_{g \in G}$ of indeterminates over $\mathbb{C}$, we have to compute with the minimum number of linear operations the entries of the block-diagonal matrix

$$
\bigoplus_{i=1}^h \left(f_{(i,a,b)} \right)_{1 \leq a, b \leq d_i} = \sum_{g \in G} X_g \left(\bigoplus_{i=1}^h D_i(g) \right) = \bigoplus_{i=1}^h \left(\sum_{g \in G} X_g D_i(g) \right).
$$

As for a non-abelian group G there are infinitely many DFT-matrices which might differ dramatically in their linear complexities, our first task is to choose a promising candidate. Fortunately, there is both a systematic way to find such DFT-matrices A, as well as a uniform approach to the design of efficient algorithms for their evaluation. In fact, these topics are closely related. The results obtained in this way range from $L_\infty(G) = O(|G|^{3/2})$ for arbitrary finite groups G to $L_2(G) = O(|G| \log |G|)$, if G is, e.g., supersolvable. This uniform approach is a development of the following trivial remark.

(13.42) Remark. Let A_i and B_i denote the ith row of $A \in \mathbb{C}^{m \times n}$ and $B \in \mathbb{C}^{M \times n}$, respectively. If $\{A_1, \ldots, A_m\} = \{B_1, \ldots, B_M\}$, then the set of linear forms corresponding to A and B are equal. In particular $L_c(A) = L_c(B)$, for $2 \leq c \leq \infty$.   •

DFT-matrices are invertible. In particular their rows are pairwise different. This shows that the last remark cannot directly be applied to these matrices. Before we can present suitable modifications that make things work, we have to recall some further facts from representation theory. If R and R' are representations of $\mathbb{C}G$ then $(R \oplus R')(a) := R(a) \oplus R'(a)$, for $a \in \mathbb{C}G$, defines a new representation of $\mathbb{C}G$, the *direct sum* of R and R'.

(13.43) Maschke's Theorem. *Every representation R of $\mathbb{C}G$ is equivalent to a direct sum of irreducible representations. More precisely, if $D_1, \ldots, D_h$ is a complete list of pairwise inequivalent irreducible representations of $\mathbb{C}G$, then there are non-negative integers $m_1, \ldots, m_h$, uniquely determined by R, and an invertible matrix T, such that*

$$TR(a)T^{-1} = \bigoplus_{i=1}^{h} \underbrace{(D_i(a) \oplus \ldots \oplus D_i(a))}_{m_i \ times} = \bigoplus_{i=1}^{h}(I_{m_i} \otimes D_i(a)) =: \left(\bigoplus_i m_i D_i\right)(a),$$

for all $a \in \mathbb{C}G$.

(For a proof of this theorem the reader is referred to Serre [470].) The number $\langle D_i \mid R \rangle := m_i$ is called the *multiplicity* of D_i in R. D_i is said to *occur* in R iff $\langle D_i \mid R \rangle \geq 1$, R is called *multiplicity-free* iff $\langle D_i \mid R \rangle \leq 1$, for all i. Note that DFTs for $\mathbb{C}G$ are exactly those multiplicity-free representations of $\mathbb{C}G$ of highest degree that are written as a direct sum of irreducible representations. Now suppose H is a subgroup of G and let $F_1, \ldots, F_r$ be a full set of inequivalent irreducible representations of $\mathbb{C}H$. Then $\mathbb{C}H$ can be viewed as a subalgebra of $\mathbb{C}G$ and the restriction $D_i \downarrow \mathbb{C}H$ of D_i to this subalgebra defines a representation of $\mathbb{C}H$. By Maschke's Thm. (13.43), $D_i \downarrow \mathbb{C}H$ is equivalent to $\oplus_{j=1}^r \langle F_j \mid D_i \downarrow \mathbb{C}H \rangle F_j$. If we require that this equivalence is an equality, $D_i \downarrow \mathbb{C}H = \oplus_{j=1}^r \langle F_j \mid D_i \downarrow \mathbb{C}H \rangle F_j$, then the minimum number of linear operations to evaluate D_i at an element b in $\mathbb{C}H$ is exactly the same as that of the evaluation of $\oplus_j F_j$, where the direct sum is over all j such that F_j occurs in $D_i \downarrow \mathbb{C}H$. In terms of the DFT-matrix A corresponding to $D_1, \ldots, D_h$ this requirement says that the $|G| \times |H|$-submatrix B of A corresponding to those columns of A that are indexed by the elements of H, has typically a number of multiple rows. So if one gets rid of the multiplicities one comes typically to a much smaller matrix-vector multiplication problem. This basic idea can be extended in two directions. Instead of *one* D_i we consider the restrictions of *all* D_i simultaneously and instead of *one* subgroup we consider *chains* of subgroups.

(13.44) Definition. Let $\mathcal{T} = (G = G_\ell \supset \ldots \supset G_0 = \{1\})$ be a chain of subgroups of G. A representation D of $\mathbb{C}G$ is called $\mathcal{T}$-*adapted* iff for all j, $0 \leq j \leq \ell$, the following conditions hold:

(1) The restriction $D \downarrow \mathbb{C}G_j$ of D to $\mathbb{C}G_j$ is *equal* to a direct sum of irreducible representations of $\mathbb{C}G_j$, i.e., $D \downarrow \mathbb{C}G_j = \oplus_{p=1}^{r_j} F_{jp}$, with irreducible representations F_{jp}.

(2) Equivalent irreducible constituents of $D \downarrow CG_j$ are *equal*, i.e., if $F_{jp} \sim F_{jq}$, then $F_{jp} = F_{jq}$. (Warning: in general, this does not imply $p = q$.)   •

(13.45) Theorem. *Let* $T = (G = G_\ell \supset \ldots \supset G_0 = \{1\})$ *be a chain of subgroups of the finite group* G. *Then every representation* D *of* CG *is equivalent to a* T-*adapted representation.*

Proof. The proof is by induction on the length ℓ of T. Let $D_1, \ldots, D_h$ be a full set of pairwise inequivalent irreducible representations of CG_ℓ. By Maschke's theorem, $T D(a) T^{-1} = (\oplus_i \langle D_i \mid D \rangle D_i)(a)$, for a suitable invertible matrix T and for all $a \in CG$. Hence it suffices to show that the representation $\oplus_{i=1}^h D_i$ is equivalent to a T-adapted representation. Let $F_1, \ldots, F_r$ be a full set of inequivalent irreducible representations of $CG_{\ell-1}$. By induction, we can assume that $F_1 \oplus \ldots \oplus F_r$ is adapted to the chain $(G_{\ell-1} \supset \ldots \supset G_0)$. For every i, we have $D_i \downarrow CG_{\ell-1} \sim \oplus_{j=1}^r \langle F_j \mid D_i \rangle F_j$. (For the sake of simplicity we have set $\langle F_j \mid D_i \rangle := \langle F_j \mid D_i \downarrow CG_{\ell-1} \rangle$.) By a suitable basis transformation T_i we can obtain equality: $T_i D_i(a) T_i^{-1} = (\oplus_{j=1}^r \langle F_j \mid D_i \rangle F_j)(a)$, for all $a \in CG_{\ell-1}$. If we set $\Delta_i(a) := T_i D_i(a) T_i^{-1}$ for all $a \in CG_\ell$, then $\Delta_1 \oplus \ldots \oplus \Delta_h$ is a T-adapted representation equivalent to $D_1 \oplus \ldots \oplus D_h$.   □

Trivially, one always has $L_\infty(G) = O(|G|^2)$ for any finite group G. With the concept of T-adapted representations one can however reduce the exponent 2 to roughly 3/2, as will be shown in Thm. (13.48). Its proof makes use of a special case of Jensen's inequality and of Frobenius reciprocity. For convenience, we first recall these topics. Let $d := \max(d_1, \ldots, d_h)$ denote the maximum degree of an irreducible representation of CG. Then $|G| = d_1^2 + \ldots + d_h^2$ implies $d \leq |G|^{1/2}$. Hence for all real $r \geq 2$, we have

$$d^r(G) := \sum_{i=1}^h d_i^r \leq d^{r-2} \sum_i d_i^2 \leq |G|^{r/2}.$$

Now let us turn to Frobenius reciprocity. One important tool in constructing representations is the process of *induction*, where a representation of a group G is constructed from a given representation of a subgroup H. The basic idea of this construction is straightforward: Let L be a left ideal in CH. Then $CGL := \{\sum_{g \in G} g\ell_g \mid \ell_g \in L\}$ is a left ideal in CG and with $G = \cup_{i=1}^r g_i H$ one obtains the decomposition $CGL = \oplus_{i=1}^r g_i L$ of CGL into C-spaces. In particular, CGL has dimension $r \dim L$. The left CG-module CGL is said to be *induced* by L. In terms of matrix representations this construction can be generalized as follows.

(13.46) Remark. Let H be a subgroup of the group G, $T := (g_1, \ldots, g_r)$ be a transversal of the left cosets of H in G, and let F be a matrix representation of H of degree f. If $\dot{F}(y) := F(y)$, for $y \in H$, and $\dot{F}(y)$ is the $f \times f$ zero-matrix for $y \in G \setminus H$, then for $x \in G$

$$(F \uparrow_T G)(x) := (\dot{F}(g_i^{-1} x g_j))_{1 \leq i, j \leq r} \in (C^{f \times f})^{r \times r} = C^{fr \times fr}$$

defines a representation of $\mathbb{C}G$, the *induced representation* $F \uparrow_T G$ of degree fr. •

In particular, if F has degree 1, then $F \uparrow_T G$ is a *monomial* representation, i.e., $(F \uparrow_T G)(g)$ is a monomial matrix, for every $g \in G$. If S is another transversal of H in G, one easily shows that $F \uparrow_T G$ and $F \uparrow_S G$ are equivalent. We therefore write sometimes $F \uparrow G$. There is a fundamental result of Frobenius which relates the "reciprocal" concepts of induction and subduction (i.e. restriction) of representations.

(13.47) Frobenius Reciprocity Theorem. *Let H be a subgroup of G and let F and D be irreducible representations of H and G, respectively. Then the multiplicity of F in the restriction $D \downarrow H$ equals the multiplicity of D in the induced representation $F \uparrow G$: $\langle D \mid F \uparrow G \rangle = \langle F \mid D \downarrow H \rangle$.*

For a proof of this theorem see, e.g., Serre [470]. Now we are well-prepared to show the following improved upper bound.

(13.48) Theorem (Beth, Clausen). *Let $T = (G_\ell \supset \ldots \supset G_0 = \{1\})$ be a chain of subgroups of the finite group $G = G_\ell$ and define $q_j := [G_j : G_{j-1}]$ and $q := \max_j q_j$. Then*

$$L_\infty(G) < 2 \sum_{j=1}^\ell \left(q_\ell \cdots q_{j+1} \cdot q_j^2 \cdot d^3(G_{j-1}) \right) < 7\sqrt{q} \cdot |G|^{3/2}.$$

Proof. Let $W_\ell = \oplus_{i \leq h} D_i$ be a T-adapted Fourier transform for $\mathbb{C}G_\ell$. If $F_1, \ldots, F_r$ are the distinct irreducible constituents of $W_\ell \downarrow \mathbb{C}G_{\ell-1}$, then, by Frobenius reciprocity, $W_{\ell-1} := \oplus_{\rho \leq r} F_\rho$ is a DFT for $\mathbb{C}G_{\ell-1}$, which in addition is adapted to $(G_{\ell-1} \supset G_{\ell-2} \supset \ldots \supset G_0)$. (For a different proof of this fact see Ex. 13.17.) Since W_ℓ is T-adapted and copying is free in our computational model, we have

$$L_\infty(W_\ell \downarrow \mathbb{C}G_{\ell-1}) = L_\infty(W_{\ell-1}).$$

(In terms of matrices, $W_\ell \downarrow \mathbb{C}G_{\ell-1}$ is the submatrix of W_ℓ obtained by selecting the columns indexed by the elements of $G_{\ell-1}$.) If $1 = g_1, \ldots, g_{q_\ell}$ is a transversal of the left cosets of $G_{\ell-1}$ in G_ℓ, every $a \in \mathbb{C}G_\ell$ can be written as $a = \sum_{j \leq q_\ell} g_j a_j$ with $a_j \in \mathbb{C}G_{\ell-1}$. Hence

$$W_\ell(a) = \sum_{j \leq q_\ell} \bigoplus_{i \leq h} D_i(g_j) \cdot (D_i \downarrow \mathbb{C}G_{\ell-1})(a_j).$$

Denoting the degrees of D_i and F_p by d_i and f_p, respectively, we get (see below)

$$
\begin{aligned}
L_\infty(W_\ell) \leq\ & q_\ell L_\infty(W_{\ell-1}) + (q_\ell - 1)|G_\ell| \\
& + (q_\ell - 1) \sum_{i,p} (2f_p - 1) f_p d_i \langle F_p \mid D_i \downarrow \mathbb{C}G_{\ell-1} \rangle.
\end{aligned}
$$

In this formula, $q_\ell \cdot L_\infty(W_{\ell-1})$ estimates the complexity of computing all $W_\ell(a_j)$, whereas $(q_\ell - 1) \cdot |G_\ell|$ corresponds to the summation of q_ℓ block matrices in

$\oplus_i \mathbb{C}^{d_i \times d_i}$ each having at most $\sum_i d_i^2 = |G_\ell|$ nonzero entries. The last term estimates the number of arithmetic operations sufficient to multiply for each $2 \leq j \leq q_\ell$ the $d_i \times d_i$-matrices $D_i(g_j)$ with the block diagonal matrices $(D_i \downarrow \mathbb{C}G_{\ell-1})(a_j) = \oplus_p \langle F_p \mid D_i \downarrow \mathbb{C}G_{\ell-1} \rangle F_p(a_j)$. By the Frobenius reciprocity theorem (13.47)

$$\sum_{i,p} (2f_p - 1)f_p \cdot d_i \langle F_p \mid D_i \downarrow \mathbb{C}G_{\ell-1} \rangle =$$

$$= \sum_p (2f_p - 1)f_p \sum_i d_i \langle D_i \mid F_p \uparrow \mathbb{C}G_\ell \rangle$$

$$= \sum_p (2f_p - 1)f_p \deg(F_p \uparrow \mathbb{C}G_\ell)$$

$$= \sum_p (2f_p - 1)f_p \cdot f_p q_\ell$$

$$= 2q_\ell \cdot d^3(G_{\ell-1}) - |G_\ell|.$$

Hence

$$L_\infty(W_\ell) \leq q_\ell L_\infty(W_{\ell-1}) + (q_\ell - 1)\left(|G_\ell| + 2q_\ell d^3(G_{\ell-1}) - |G_\ell|\right)$$
$$< q_\ell L_\infty(W_{\ell-1}) + 2q_\ell^2 d^3(G_{\ell-1}),$$

and an induction proves that

$$L_\infty(W_\ell) < 2\sum_{j=1}^{\ell} \left(q_\ell \cdots q_{j+1} \cdot q_j^2 \cdot d^3(G_{j-1})\right).$$

Using the fact that $d^3(G_{j-1}) \leq |G_{j-1}|^{3/2}$, the last formula can further be estimated:

$$L_\infty(W_\ell) \leq 2|G_\ell|^{3/2} \sum_{j=1}^{\ell} \frac{\sqrt{q_j}}{\sqrt{q_\ell \cdot \ldots \cdot q_{j+1}}} \leq 2|G_\ell|^{3/2} \sum_{j=1}^{\ell} \frac{\sqrt{q}}{\sqrt{2^{\ell-j}}}$$

$$< 2|G_\ell|^{3/2}\sqrt{q} \sum_{j=0}^{\infty} 2^{-j/2} < 7\sqrt{q}|G_\ell|^{3/2}. \quad \square$$

Using asymptotically fast matrix multiplication methods and the classification of finite simple groups, one can prove that $L_\infty(G) = O(|G|^{1.44})$ for arbitrary finite groups G, see Ex. 13.16.

The goal of the remaining part of this section is to improve the above result for special classes of groups. To this end, we need further facts from representation theory. The *character* $\chi : G \to \mathbb{C}$ of a representation D of $\mathbb{C}G$ is defined by $\chi(g) := \mathrm{Tr}(D(g))$, $g \in G$. The character of an irreducible representation of $\mathbb{C}G$ is called an *irreducible character* of G, and $\mathrm{Irr}(G)$ denotes the set of all irreducible characters of G. A character corresponding to a one-dimensional representation of G is called a *linear character*. As $\mathrm{Tr}(AB) = \mathrm{Tr}(BA)$, one easily sees that equivalent representations have the same character and that each character χ of G

is a *class function* on G, i.e., $\chi: G \to \mathbb{C}$ is constant on the conjugacy classes of G. Moreover, as $D(g)$ has finite order for every $g \in G$ we have $\chi(g^{-1}) = \overline{\chi(g)}$. The space $CF(G, \mathbb{C})$ of all complex-valued class functions on G becomes an inner product space by $\langle \chi \mid \psi \rangle := |G|^{-1} \sum_{g \in G} \chi(g)\overline{\psi(g)}$. For a proof of the following facts we refer to Serre [470].

(13.49) Theorem. *Let G be a finite group, h the number of conjugacy classes of G. Then the following is true.*

(1) *G has h irreducible characters $\chi_1, \ldots, \chi_h$. These characters form an orthonormal basis of the space $CF(G, \mathbb{C})$.*

(2) *If D_i and F are representations of $\mathbb{C}G$ with characters χ_i and χ, respectively, D_i irreducible, then $\langle D_i \mid F \rangle = \langle \chi_i \mid \chi \rangle$.*

(3) *If $e_i = e_{\chi_i} := \chi_i(1)|G|^{-1} \sum_{g \in G} \chi_i(g)g$, then $e_1, \ldots, e_h$ are a basis of the center $\{c \in \mathbb{C}G \mid \forall a \in \mathbb{C}G: ac = ca\}$ of $\mathbb{C}G$. Moreover, $1 = e_1 + \ldots + e_h$ and $e_i e_j = \delta_{ij} e_i$. (The e_i are called the* central primitive idempotents *in $\mathbb{C}G$.)*

(4) *If M is a left $\mathbb{C}G$-module affording the representation F with character χ, then $M = \oplus_{i=1}^h e_i M$ (isotypic decomposition). If M_i is a simple module affording the character χ_i, then $e_i M$, the isotypic component of type χ_i, is a submodule of M which is isomorphic to the direct sum of $\langle \chi_i \mid \chi \rangle$ copies of M_i.*

Let $T = (G = G_\ell \supset \ldots \supset G_0 = \{1\})$ be a chain of subgroups of G, and let M be a left $\mathbb{C}G$-module. A $\mathbb{C}$-basis B of M is called T-*adapted* iff the matrix representation of $\mathbb{C}G$ corresponding to M and B is T-adapted. According to Thm. (13.45) T-adapted bases always exist. Of particular importance will be those T for which T-adapted bases are unique up to ordering and scaling. This will be our next topic. We will use the following terminology: A basis B of a $\mathbb{C}$-space M *respects* the direct sum decomposition $M = \oplus M_i$ iff for every i, there is a subsequence B_i of B forming a basis of M_i. For $\chi \in \mathrm{Irr}(G)$, we let

$$T(\chi) := \{(\chi_0, \ldots, \chi_\ell = \chi) \mid \forall i : \chi_i \in \mathrm{Irr}(G_i) \text{ and } \langle \chi_i \mid \chi_{i+1} \downarrow G_i \rangle > 0\},$$

and $e(w) := e_{\chi_0} \cdot \ldots \cdot e_{\chi_\ell}$ denotes the product of all central primitive idempotents corresponding to $w := (\chi_0, \ldots, \chi_\ell) \in T(\chi)$. Obviously, the $e(w)$, $w \in \bigcup_{\chi \in \mathrm{Irr}(G)} T(\chi)$, are pairwise orthogonal idempotents in $\mathbb{C}G$. Note that if M is a left $\mathbb{C}G$-module and B is a T-adapted basis of M, then B respects the direct sum decomposition $M = \oplus'_\chi e_\chi M$, where the sum is over all those irreducible characters of G such that $e_\chi M \neq 0$. (In this case we also say that χ *occurs* in M.) As a generalization of this we have the following result.

(13.50) Lemma. *Every T-adapted basis of a left $\mathbb{C}G$-module M respects the direct sum decomposition*

$$M = \bigoplus_{\chi \in X(M)} \bigoplus_{w \in T(\chi)} e(w)M,$$

where $X(M)$ denotes the set of all $\chi \in \mathrm{Irr}(G)$ that occur in M.

Proof. Let B be a $\mathcal{T}$-adapted basis of M. Recall that every decomposition of M into the direct sum of simple submodules is a refinement of the isotypic decomposition $M = \oplus'_\chi e_\chi M$, where the direct sum is over all $\chi \in \mathrm{Irr}(G)$ that occur in M. Since B respects this direct sum decomposition, there is for all χ occuring in M a subset B_χ of B which forms a basis of $e_\chi M$. B_χ needs to be adapted to $(G_{\ell-1} \supset \ldots \supset G_0)$. Now use induction on the length of $\mathcal{T}$ to complete the proof. $\qquad\square$

As a consequence we get the following crucial quasi-uniqueness result for $\mathcal{T}$-adapted representations.

(13.51) Theorem. *Let $\mathcal{T} = (G = G_\ell \supset \ldots \supset G_0 = \{1\})$ be a chain of subgroups of G, and let M be a multiplicity-free left $\mathbb{C}G$-module such that for every $j \in \underline{\ell}$ and every simple $\mathbb{C}G_j$-submodule U of M, U viewed as a $\mathbb{C}G_{j-1}$-module is multiplicity-free. Then $M = \oplus_{\chi \in X(M)} \oplus_{w \in \mathcal{T}(\chi)} e(w)M$ is a direct decomposition of M into one-dimensional subspaces $e(w)M$. In particular, a $\mathcal{T}$-adapted basis for M is unique up to ordering and scaling, i.e., if D and Δ are two $\mathcal{T}$-adapted representations affording M, then there exists a monomial matrix X such that $D(a) = X^{-1}\Delta(a)X$ for all $a \in \mathbb{C}G$. (In this case, we call D and Δ monomially equivalent.)*

Proof. According to Lemma (13.50), every $\mathcal{T}$-adapted basis has to respect the direct decomposition $M = \oplus_{\chi \in X(M)} \oplus_{w \in \mathcal{T}(\chi)} e(w)M$. Thus all we have to prove is that for $\chi \in X(M)$ and $w = (\chi_0, \ldots, \chi_\ell = \chi) \in \mathcal{T}(\chi)$ the space $e(w)M$ is one-dimensional. As M is multiplicity-free and χ occurs in M we know that $e_\chi M$ is a simple left $\mathbb{C}G$-module, and induction shows that $e_{\chi_j} \cdot \ldots \cdot e_{\chi_\ell} M$ is a simple left $\mathbb{C}G_j$-module. (Use the properties of isotypic decompositions and the fact that each χ_i restricted to G_{i-1} is multiplicity-free.) In particular, $e(w)M$ is a simple $\mathbb{C}G_0$-module, hence one-dimensional. $\qquad\square$

We are going to apply this theorem to solvable groups. To this end we need to recall some rudimentary facts from Clifford theory. If N is a normal subgroup of G, then there is a close connection between the irreducible representations of $\mathbb{C}N$ and those of $\mathbb{C}G$. Let F be a representation of $\mathbb{C}N$ and $g \in G$. Then

$$F^g(n) := F(g^{-1}ng) \quad \text{for all } n \in N$$

defines a new representation F^g of $\mathbb{C}N$. As conjugation by g is an automorphism of N, the sets $\{F^g(n) \mid n \in N\}$ and $\{F(n) \mid n \in N\}$ are equal, which proves that F is irreducible iff F^g is irreducible. Let $T = (g_1, \ldots, g_r)$ be a transversal of the cosets of N in G. From the definition of $F \uparrow_T G$ and the fact that N is normal in G, it follows immediately that

$$(F \uparrow_T G) \downarrow N = \bigoplus_{i=1}^{r} F^{g_i}.$$

In the sequel we only need the following special case of a much more general result due to Clifford. (For a proof we refer to Isaacs [263], and Clausen and Baum [115].)

(13.52) Theorem (Clifford). *Let N be normal in G of prime index p and F an irreducible representation of $\mathbb{C}N$. For any fixed $g \in G \setminus N$, let T denote the transversal $(1, g, g^2, \ldots, g^{p-1})$ of the cosets of N in G. Then exactly one of the following two cases applies.*

(1) *The representations F^{g^i} are pairwise inequivalent. In this case, the induced representation $F \uparrow_T G$ is irreducible.*
(2) *All F^{g^i} are equivalent. Then there are exactly p irreducible representations $D_0, \ldots, D_{p-1}$ of G extending F. The D_k are pairwise inequivalent and satisfy $F \uparrow_T G \sim D_0 \oplus \ldots \oplus D_{p-1}$. Moreover, if $\chi^0, \ldots, \chi^{p-1}$ are the linear characters of the cyclic group G/N in a suitable order, we have $D_j(x) := \chi^j(xN)D_0(x) = (\chi^i \otimes D_0)(x)$ for all $x \in G$ and all $1 \le j < p$.*

Recall that a finite group G is *solvable* iff there is a chain $\mathcal{T} = (G = G_\ell \supset \ldots \supset G_0 = \{1\})$ of subgroups of G such that for all $i \in \underline{\ell}$, G_{i-1} is normal in G_i of prime index. Such chains constitute a composition series of G.

(13.53) Corollary. *If G is solvable and $\mathcal{T}$ a composition series of G, then two equivalent multiplicity-free $\mathcal{T}$-adapted representations of $\mathbb{C}G$ are monomially equivalent.*

Proof. The composition factors G_i/G_{i-1} of G have prime order. By Clifford's theorem, the restriction of an irreducible representation of $\mathbb{C}G_i$ to $\mathbb{C}G_{i-1}$ is either irreducible or decomposes into $[G_i : G_{i-1}]$ pairwise inequivalent irreducible constituents. In either case, it is multiplicity-free. Our claim now follows from Thm. (13.51). □

A special class of solvable groups are the so-called supersolvable groups. A group G is *supersolvable* iff G has a chain $\mathcal{T}$ such that for all i, G_i is normal in G and G_i/G_{i-1} has prime order. Such a chain is called a *chief series* of G. (For example, every nilpotent group, in particular every p-group, is supersolvable.) The classical FFT-algorithms generalize to the class of supersolvable groups. The basis of this generalization is the following.

(13.54) Theorem. *Let $\mathcal{T}$ be a chief series of the supersolvable group G. Then every $\mathcal{T}$-adapted DFT $D = \oplus_{i=1}^h D_i : \mathbb{C}G \to \oplus_{i=1}^h \mathbb{C}^{d_i \times d_i}$ is monomial, i.e., for every $i \le h$ and every $g \in G$ the matrix $D_i(g)$ is monomial.*

Proof. Let M be a left $\mathbb{C}G$-module affording the representation D. Then M is multiplicity-free and $e_\chi M \ne 0$ for every $\chi \in \mathrm{Irr}(G)$. As $\mathcal{T}$ is in particular a composition series of the solvable group G, we get by the last corollary a decomposition of M into one-dimensional subspaces:

$$M = \bigoplus_{\chi \in \mathrm{Irr}(G)} \bigoplus_{w \in T(\chi)} e(w)M,$$

and every $\mathcal{T}$-adapted basis has to respect this decomposition. In order to prove monomiality, it suffices to show that for fixed $\chi \in \mathrm{Irr}(G)$ the given linear action of G on M induces an action of G on the set $\{e(w)M \mid w \in T(\chi)\}$ of lines: First of all, as G_i is normal in G, we have a G-action on $\text{Irr}(G_i)$ via

$$g\chi_i := (G_i \ni h \mapsto \chi_i(g^{-1}hg)).$$

As $g\chi = \chi$ for all $g \in G$ and all $\chi \in \text{Irr}(G)$ this induces a G-action on $T(\chi)$:

$$g(\chi_0, \ldots, \chi_\ell) := (g\chi_0, \ldots, g\chi_\ell).$$

Now for $g \in G$ and $\chi_i \in \text{Irr}(G_i)$ we have $ge_{\chi_i}g^{-1} = e_{g\chi_i}$. Hence for $w \in T(\chi)$ we get $ge(w)g^{-1} = e(gw)$, thus G acts on $\{e(w)M \mid w \in T(\chi)\}$ via

$$ge(w)M = e(gw)gM = e(gw)M.$$

(As $e_\chi M$ is a simple $\mathbb{C}G$-module this G-action has to be transitive.) $\square$

(13.55) Theorem (Baum). *If G is a supersolvable group of order n then*

$$L_\infty(G) \le \gamma \cdot n \log n,$$

for some constant $1.5 \le \gamma \le 8.5$ depending on the prime divisors of n.

Proof. Let $T = (G = G_\ell \supset \ldots \supset G_0 = \{1\})$ be a chief series of G. For a prime p let $c_p := L_\infty(C_p)/(p \log p)$. If the chief factor G_i/G_{i-1} is cyclic of prime order p_i, then we prove by induction on the length ℓ of T that

(A) $$L_\infty(G) \le (c_{p_1} \log p_1 + \ldots + c_{p_\ell} \log p_\ell + \ell)|G|.$$

By Thm. (13.54), there is a monomial and T-adapted Fourier transformation $W_\ell = \oplus_{i \le h} D_i$ for $\mathbb{C}G$. As $L_\infty(\{1\}) = 0$, the claim certainly holds for $\ell = 0$.

So let $\ell \ge 1$, $N := G_{\ell-1}$ with $[G : N] = p$ and $g \in G \setminus N$. If $F_1, \ldots, F_r$ are the distinct irreducible constituents of $W_\ell \downarrow \mathbb{C}N$, then $W_{\ell-1} := \oplus_{k \le r} F_k$ is a Fourier transform for $\mathbb{C}N$ adapted to $(G_{\ell-1} \supset \ldots \supset G_0)$, and

$$L_\infty(W_\ell \downarrow \mathbb{C}N) = L_\infty(W_{\ell-1}),$$

see the proof of Thm. (13.48).

We want to evaluate W_ℓ at $a \in \mathbb{C}G$. To this end, we rewrite a according to the coset partition $G = \cup_{j=0}^{p-1} g^j N$; we have $a = \sum_{j=0}^{p-1} g^j a_j$ with suitable $a_j \in \mathbb{C}N$. Then $W_\ell(a) = \oplus_{i=1}^{h} \sum_{j=0}^{p-1} D_i(g^j)(D_i \downarrow \mathbb{C}N)(a_j)$. To compute $W_\ell(a)$, we first evaluate p "smaller" Fourier transforms $W_{\ell-1}(a_j)$, $0 \le j < p$. Then we obtain $W_\ell(a)$ according to the above formula. Now Clifford's theorem suggests to distinguish two cases.

Case 1. The restriction $D_i \downarrow \mathbb{C}N$ decomposes into p inequivalent irreducible components $D_{i1}, \ldots, D_{ip}$ of degree d_i/p each. Here and in the sequel d_i denotes the degree of D_i. The product of the monomial matrix $D_i(g^j)$ and the block diagonal matrix $(D_i \downarrow \mathbb{C}N)(a_j) = \oplus_k(D_{ik}(a_j))$ can be computed using at most d_i^2/p scalar multiplications. As $D_i(g^0) = 1$, we need at most $\frac{p-1}{p}d_i^2$ scalar multiplications to evaluate D_i at a. The concluding summation is free of cost in our model: As D_i is irreducible, we have $\dim_{\mathbb{C}} D_i(\mathbb{C}G) = d_i^2$. On the other hand, each of the p summands has at most d_i^2/p nonzero entries whose input-independent positions are given by the support of the monomial matrix $D_i(g^j)$.

Case 2. $D_i \downarrow \mathbb{C}N =: F$ is irreducible. Then there are exactly p inequivalent irreducible extensions $D_i = D_{i1}, \ldots, D_{ip}$ of F to $\mathbb{C}G$. Because of the T-adaptation, all the D_{ik} have the form $D_{ik} = \chi_k \otimes D_{i1}$, where χ_k runs through all irreducible (=linear) characters $1 = \chi_1, \ldots, \chi_p$ of the cyclic group G/N. This gives an efficient method for the *simultaneous* evaluation of $D_{i1}, \ldots, D_{ip}$ at a: We have

$$\bigoplus_{k=1}^{p} \sum_{j=0}^{p-1} D_{ik}(g^j) F(a_j) = \bigoplus_{k=1}^{p} \sum_{j=0}^{p-1} \chi_k(g^j N) D_{i1}(g^j) F(a_j).$$

The matrices $B_j := D_{i1}(g^j) F(a_j)$ are independent of k and can altogether be computed with at most $(p-1)d_i^2$ scalar multiplications. Then the $D_{ik}(a)$ can be obtained from the B_j by performing a local DFT_p for each of the d_i^2 positions. This can be done with at most $d_i^2 c_p p \log p$ operations. Thus $(p-1+c_p p \log p)d_i^2$ linear operations are sufficient to compute $D_{i1}(a), \ldots, D_{ip}(a)$, which gives an average cost of at most $(\frac{p-1}{p} + c_p \log p)d_i^2$ operations for each d_i-dimensional representation D_{ik}.

Altogether, we can compute $W_\ell(a)$ from the $W_{\ell-1}(a_j)$, $0 \le j < p$, using at most

$$\left(\frac{p-1}{p} + c_p \log p \right) \sum_i d_i^2 = \left(\frac{p-1}{p} + c_p \log p \right) |G|$$

operations. This yields the recurrence relation

(B) $\qquad L_\infty(G) \le L_\infty(W_\ell) \le p L_\infty(W_{\ell-1}) + \left(\frac{p-1}{p} + c_p \log p \right) |G|,$

and an induction proves (A).

With $c := \max(c_{p_1}, \ldots, c_{p_\ell})$ we claim that

$$L_\infty(G) \le (c + 0.5)|G| \log |G|.$$

(In particular, we have $L_\infty(G) \le 1.5|G| \log |G|$ for all 2-groups G, $L_\infty(G) \le 2.2|G| \log |G|$ for all 3-groups G, and $L_\infty(G) \le 8.5|G| \log |G|$ for all supersolvable groups G.) We prove this by induction on ℓ. The claim is trivial for $\ell = 1$. So let $\ell > 1$ and G, N, p as above. According to (B), we have

$$\begin{aligned} L_\infty(G) &\le L_\infty(W_\ell) \le p L_\infty(W_{\ell-1}) + \left(\frac{p-1}{p} + c_p \log p \right) |G| \\ &\le (c + 0.5)p|N| \log |N| + \left(\frac{p-1}{p} + c_p \log p \right) |G| \end{aligned}$$

$$= (c+0.5)|G|\log|G| - (c+0.5)|G|\log p + \left(\frac{p-1}{p} + c_p\log p\right)|G|$$

$$= (c+0.5)|G|\log|G| + \left(\frac{p-1}{p\log p} + c_p - c - 0.5\right)|G|\log p$$

$$\le (c+0.5)|G|\log|G|,$$

where the last step is correct because $\frac{p-1}{p\log p} \le 0.5$. As $c_2 \le 1$, $c_3 \le 1.7$ and $c \le 8$ (see Baum et al. [27]), our claims follow.   □

Finally, we apply a variant of this technique to the symmetric groups. As we shall see below, Thm. (13.51) applies to the symmetric group S_ℓ and the chain

$$\mathcal{T}_\ell := (S_\ell \supset S_{\ell-1} \supset \ldots \supset S_1),$$

where S_{i-1} is the stabilizer of i in S_i for each i. We shall need some facts from the representation theory of symmetric groups. For proofs the reader is referred to James and Kerber [272].

The conjugacy classes as well as the equivalence classes of irreducible representations of $\mathbb{C}S_\ell$ are usually parameterized by the partitions of ℓ: a *partition* $\alpha = (\alpha_1, \alpha_2, \ldots)$ of ℓ, abbreviated $\alpha \vdash \ell$, is a non-increasing sequence of positive integers summing up to ℓ. Partitions can be illustrated by their corresponding diagrams: the diagram of the partition α is the set $\bigcup_i\{(i, j) \mid 1 \le j \le \alpha_i\}$, which can be visualized as a left-justified arrangement of α_i boxes in the ith row. By abuse of notation we will make no difference between partitions and their diagrams. The following figure shows all diagrams for $\ell = 4$:

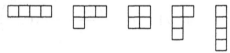

Hence there are 5 inequivalent irreducible representations of S_4. For every partition α of ℓ, we denote by $[\alpha]$ or $[\alpha_1, \alpha_2, \ldots]$ an irreducible representation of S_ℓ of "type" $\alpha = (\alpha_1, \alpha_2, \ldots)$. Inclusion defines a partial ordering on the set of all diagrams, the so-called *Young lattice*:

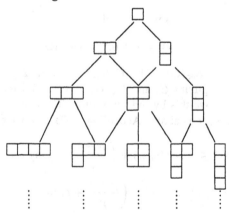

The Young lattice has a representation theoretical meaning according to the following celebrated branching theorem. (See James and Kerber [272] or Clausen [114] for a proof.)

(13.56) Young's Branching Theorem. *Let α be a diagram of ℓ. Then*

$$[\alpha] \downarrow \mathbb{C}S_{\ell-1} \sim \bigoplus_{\beta} [\beta],$$

where the sum is over all diagrams β of $\ell - 1$ contained in α. In particular, $[\alpha] \downarrow \mathbb{C}S_{\ell-1}$ is multiplicity-free, and the degree f_α of $[\alpha]$ equals the number of paths from (1) to α in the Young lattice.

For example, this branching rule says that

$$[5, 2, 2, 1] \downarrow \mathbb{C}S_9 \sim [4, 2, 2, 1] \oplus [5, 2, 1, 1] \oplus [5, 2, 2].$$

As every $[\alpha] \downarrow \mathbb{C}S_{\ell-1}$ is multiplicity-free, there is up to monomial equivalence exactly one $\mathcal{T}_\ell$-adapted DFT σ_ℓ for $\mathbb{C}S_\ell$, see Thm. (13.51). The next theorem summarizes some facts of this adapted DFT.

(13.57) Theorem. *For every diagram α of j, $1 \le j \le \ell$, there exists an irreducible representation σ^α of $\mathbb{C}S_j$, $\sigma^\alpha \sim [\alpha]$, such that*

$$\sigma_j := \bigoplus_{\alpha \vdash j} \sigma^\alpha$$

is an $(S_j \supset S_{j-1} \supset \ldots \supset S_1)$-adapted DFT for $\mathbb{C}S_j$. Furthermore, for all partitions α of j, $2 \le j \le \ell$, the following holds:

(1) $\sigma^\alpha \downarrow \mathbb{C}S_{j-1} = \oplus_\beta \sigma^\beta$, *where β runs through all diagrams of $j - 1$ that are contained in α.*

(2) *For every i, $1 \le i < j$, and every diagram α of j the matrix $\sigma^\alpha((i, i+1))$ is 2-sparse, i.e., this matrix has at most two nonzero entries in each row and in each column. Moreover, these entries are of absolute value ≤ 1.*

Proof. (Sketch) To prove (1) combine Thm. (13.45) and Young's branching theorem (13.56). For proving (2), note that the multiplicity of an irreducible representation of $\mathbb{C}S_{\ell-2}$ in $[\alpha] \downarrow \mathbb{C}S_{\ell-2}$ is at most 2. Combine this with the fact that S_ℓ is generated by the transpositions of consecutive numbers,

$$S_\ell = \langle (1, 2), (2, 3), \ldots, (\ell - 1, \ell) \rangle,$$

and with the fact that the transposition $(\ell - 1, \ell)$ commutes with all elements in $S_{\ell-2}$. Now apply Thm. (13.51) and Schur's lemma. (For more details see Ex. 13.12 and 13.20.) □

Now we are prepared to prove the following result.

(13.58) Theorem (Clausen). *If $G = S_\ell$, then $L_2(G) = o(|G| \log^3 |G|)$.*

Proof. We decompose S_ℓ into left cosets of the subgroup $S_{\ell-1}$:

$$S_\ell = \bigsqcup_{j \le \ell} g_j S_{\ell-1} \quad \text{where} \quad g_j := (j, j+1, \dots, \ell).$$

Hence $a \in \mathbb{C}S_\ell$ can be written as $a = \sum_{j=1}^{\ell} g_j a_j$ with $a_j = \sum_{h \in S_{\ell-1}} a_{g_j h} h \in \mathbb{C}S_{\ell-1}$. As σ_ℓ is a $\mathcal{T}_\ell$-adapted DFT, we obtain

$$
\begin{aligned}
\sigma_\ell(a) &= \sum_{j=1}^{\ell} \bigoplus_{\alpha \vdash \ell} \sigma^\alpha(g_j)(\sigma^\alpha \downarrow \mathbb{C}S_{\ell-1})(a_j) \\
&= \sum_{j=1}^{\ell} \bigoplus_{\alpha \vdash \ell} \sigma^\alpha(g_j) \bigoplus_{\alpha \supset \beta \vdash \ell-1} \sigma^\beta(a_j).
\end{aligned}
$$

Thus every $\sigma^\beta(a_j)$, once computed, can be used in the evaluation of $\sigma^\alpha(a_j)$ for all partitions α of ℓ containing β. In other words, to compute $\sigma_\ell(a)$, we first compute $\sigma_{\ell-1}(a_j) = \bigoplus_{\beta \vdash \ell-1} \sigma^\beta(a_j)$, for $1 \le j \le \ell$. With no further arithmetic cost we know then (for all partitions α of ℓ and all j) every block-diagonal matrix $\bigoplus_{\alpha \supset \beta \vdash \ell-1} \sigma^\beta(a_j)$. This block-diagonal matrix and the matrix $\sigma^\alpha(g_j)$ have then to be multiplied. This can be done efficiently: as $g_j = (j, j+1) \cdot (j+1, j+2) \cdots (\ell-1, \ell)$, the matrix $\sigma^\alpha(g_j)$ is by Thm. (13.57)(2) the product of $(\ell - j)$ many 2-sparse matrices:

$$\sigma^\alpha(g_j) = \sigma^\alpha((j, j+1))\sigma^\alpha((j+1, j+2))\dots\sigma^\alpha((\ell-1, \ell)).$$

Instead of directly multiplying $\sigma^\alpha(g_j)$ (which is typically not a sparse matrix) and $\sigma^\alpha(a_j)$, we multiply step by step $\sigma^\alpha(a_j)$ from the left by the $\ell - j$ many 2-sparse matrices. Thus the multiplication $\sigma^\alpha(g_j)\sigma^\alpha(a_j)$ can be performed with at most $3(\ell-j)f_\alpha^2$ arithmetic operations, where f_α denotes the degree of the representation σ^α. Finally, we have to sum up the ℓ block diagonal matrices $\sigma_\ell(g_j)\sigma_\ell(a_j)$. This takes $(\ell - 1)\sum_{\alpha \vdash \ell} f_\alpha^2 = (\ell - 1) \cdot \ell!$ operations. Let L_ℓ denote the arithmetic cost of our algorithm to evaluate σ_ℓ in the L_2-model. Then our above discussion yields the recursion

$$L_\ell \le \ell \cdot L_{\ell-1} + \sum_{j=1}^{\ell} \sum_{\alpha \vdash \ell} 3(\ell - j)f_\alpha^2 + (\ell - 1) \cdot |S_\ell|.$$

Obviously, $L_1 = 0$. Combining this with $\sum_{\alpha \vdash \ell} f_\alpha^2 = |S_\ell|$ and the well-known formula $\sum_{i=2}^{\ell} \binom{i}{2} = \binom{\ell+1}{3}$, we obtain

$$\frac{L_\ell}{\ell!} \le \frac{L_{\ell-1}}{(\ell-1)!} + 3\binom{\ell}{2} + \ell - 1 \le 3\sum_{j=0}^{\ell-2}\binom{\ell-j}{2} + \sum_{j=1}^{\ell-1}(\ell-j) = 3\binom{\ell+1}{3} + \binom{\ell}{2},$$

and our claim follows. $\square$

13.6 Exercises

13.1. Let F be a finite set of linear forms. Show that $\Lambda(F) \le L_\infty(F) + |F|$.

13.2. Let k be an infinite field, $K := k(X_1, \ldots, X_n)$. If Ω denotes the free K-module with basis $dX_1, \ldots, dX_n$, then the corresponding universal differential is the k-linear map $d \colon K \to \Omega$ satisfying $d(X_i) = dX_i$, for all i, $d(pq) = p\,d(q) + q\,d(p)$, and $d(p/q) = (q\,d(p) - p\,d(q))/q^2$, for all $p, q \in K$, $q \ne 0$. For $\xi \in k^n$ let $\mathcal{O}_\xi$ denote the local subring of K at ξ consisting of all $f \in K$ that are defined at ξ. The pair (d, ξ) induces a k-linear map $d_\xi \colon \mathcal{O}_\xi \to \Omega$ by $d_\xi(X_i) := dX_i$, $d_\xi(pq) = p(\xi)d(q) + q(\xi)d(p)$, and $d_\xi(p/q) = (q(\xi)d(p) - p(\xi)d(q))/q(\xi)^2$, for all $i \in \underline{n}$ and all $p, q \in \mathcal{O}_\xi$; in addition $q(\xi) \ne 0$ in the quotient rule.

(1) Show that the image of d_ξ consists of k-linear combinations of the basis elements $dX_1, \ldots, dX_n$. (So this image might be viewed as a space of linear forms over k.)

(2) Let F be a finite subset of K. Then for Zariski almost all $\xi \in k^n$, $d_\xi(F) := \{d_\xi(f) \mid f \in F\}$ is well-defined and

$$L^{\mathrm{tot}}_{k \to K}(F \mid X_1, \ldots, X_n) \ge \frac{1}{3} L_\infty(d_\xi(F)).$$

(Hint: Let $(k[X]_d, u)$ be a universal input for an optimal straight-line program Γ computing F. If $\xi \in k^n$ is not a zero of the polynomial d, then the result sequence of Γ on this input can be transformed via d_ξ into a linear computation sequence of at most threefold length that computes $d_\xi(F)$.)

13.3. Prove statements (2) and (3) of Rem. (13.12).

13.4. Prove statements (1)-(4) of Lemma (13.7).

13.5. Let K be the field with 2 elements. Show that for sufficiently large n there exist matrices $A \in K^{n \times n}$ with $L_\infty(A) \ge n^2/\log n$. (Hint: Compare the number of possible linear computation sequences over K of a given length with the number of all matrices in $K^{n \times n}$; see also Ex. 9.5.)

13.6. Show that the Vandermonde matrix corresponding to $a_i = 2^{n^{i-1}}$, $1 \le i \le n$, is superregular.

The following three exercises discuss efficient evaluation techniques for complex DFT-matrices of size n, where n is not necessarily a power of 2. We put $L(n) := L_2(DFT_n)$.

13.7. Let n be composite, $n = pq$. Show that

$$L(pq) \leq p \cdot L(q) + q \cdot L(p) + pq - p + q + 1.$$

(Hint: Let $\omega \in \mathbb{C}$ be a primitive nth root of unity, $A_\mu = \sum_{v=0}^{n-1} \omega^{\mu v} a_v$. Write the indices $0 \leq \mu, v < n$ as $\mu = hq + \ell$ and $v = ip + j$ where $0 \leq h, j < p$ and $0 \leq \ell, i < q$.)

13.8.* Let p be prime. Show that

$$L(p) \leq 2 \cdot L(p-1) + 3p - 3.$$

(Hint: Reduce the evaluation of DFT_p to a cyclic convolution of length $p-1$. Use the fact that the multiplicative group of the prime field with p elements is cyclic.)

13.9.* Show that there is a universal constant γ such that for all n

$$L(n) \leq \gamma n \log n.$$

(Hint: Use a cyclic convolution of small but suitable length $N \leq 2n$, and use $2\mu v = \mu^2 + v^2 - (\mu - v)^2$ where μ, v are as in Ex. 13.7.)

13.10.* Show that $L_\infty(DFT_3) = 8$.

13.11.* Let G_n be a standard K-DAG over (n, n) computing the superregular matrix $A_n \in K^{n \times n}$ in depth $d_n > 1$. Prove the following statements:

(1) If $d_n = O(\log n)$, then $L_\infty(A_n) = \Omega(n \log n)$.
(2) If $d_n = O(1)$, then $L_\infty(A_n) = \Omega(n^{1+1/d_n})$.

13.12. Prove Schur's lemma: If $D: \mathbb{C}G \to \mathbb{C}^{d \times d}$ is an irreducible representation, then

$$\left\{ X \in \mathbb{C}^{d \times d} \mid \forall a \in \mathbb{C}G : D(a)X = XD(a) \right\} = \mathbb{C} \cdot I_d.$$

13.13. Let $C := \{x \mid \forall g \in G : xg = gx\}$ be the *center* of the finite non-abelian group G. Show that every DFT for $\mathbb{C}G$ is adapted to $(G \supset C \supset \{1\})$ and

$$L_\infty(G) \leq [G : C]L_\infty(C) + 2([G : C] - 1)|G|.$$

13.14. Prove Maschke's theorem (13.43) for $\mathbb{C}G$. (Hints: Let D be a d-dimensional representation of $\mathbb{C}G$. A suitable modification of the standard inner product in $\mathbb{C}^d$ (with the help of D) yields an inner product $\langle \mid \rangle$ which is G-invariant, i.e. $\langle x \mid y \rangle = \langle D(g)x \mid D(g)y \rangle$, for all $x, y \in \mathbb{C}^d$ and all $g \in G$. Now the orthogonal complement of any G-invariant subspace of $\mathbb{C}^d$ is G-invariant as well.)

13.15. Let $\mathbb{C}G \simeq \oplus_{i=1}^{h} \mathbb{C}^{d_i \times d_i}$ and $r \geq 2$. Show that

$$d^r(G) := \sum_{i=1}^{h} d_i^r \leq [G : G'] + (|G| - [G : G'])^{r/2},$$

and improve the analysis of Thm. (13.48).

13.16. Let G be a finite group. Show that $L_\infty(G) = O(|G|^{1.44})$. (Hint: use the fast matrix multiplication algorithms of Chap. 15 and (without proof) Lev's theorem [327], which states that every finite group G of non-prime order has a proper subgroup H with $|H| \geq \sqrt{|G|}$.)

13.17. Let $\mathcal{T}_\ell = (G = G_\ell \supset G_{\ell-1} \supset \ldots \supset G_0 = \{1\})$ be a chain of subgroups of the finite group G. If $D = \oplus_i D_i$ is a $\mathcal{T}_\ell$-adapted DFT for $\mathbb{C}G$ and if $F = \oplus_j F_j$ is the direct sum of all pairwise inequivalent irreducible constituents of $D \downarrow \mathbb{C}G_{\ell-1}$, then F is a $(G_{\ell-1} \supset \ldots \supset G_0 = \{1\})$-adapted DFT for $\mathbb{C}G_{\ell-1}$. (Hints: Left multiplication in $\mathbb{C}G$ defines a representation of degree $|G|$, the (left) *regular representation* of $\mathbb{C}G$. Consider the left regular representations of $\mathbb{C}G$ and $\mathbb{C}H$ and apply Maschke's theorem.)

13.18. Let G be a finite group of order n, $G = C_1 \cup \ldots \cup C_h$ the partition of G into its conjugacy classes. The *center* of $\mathbb{C}G$ is defined by $C(\mathbb{C}G) := \{x \in \mathbb{C}G \mid \forall a \in \mathbb{C}G : ax = xa\}$. Prove the following statements:

(1) $C(\mathbb{C}G)$ is a commutative subalgebra of dimension h and the class sums $c_i := \sum_{g \in C_i} g$, $1 \leq i \leq h$, form a basis of $C(\mathbb{C}G)$.

(2) Let $D = \oplus_{i=1}^{h} D_i$ be a DFT for $\mathbb{C}G$, and let d_i be the degree of the irreducible representation D_i. If, for $j \in \underline{h}$, $E_j \in \oplus_{i=1}^{h} \mathbb{C}^{d_i \times d_i}$ denotes the matrix all of whose blocks are zero except for the jth block which is the unit matrix of size d_j, then $D^{-1}(E_j)$ is the centrally primitive idempotent e_j corresponding to the irreducible character χ_j of D_j.

13.19. Let D be a d-dimensional irreducible representation of $\mathbb{C}G$ and let $m \in \mathbb{N}'$. Describe the set of all matrices in $\mathbb{C}^{dm \times dm}$ that commute with all $\oplus_{i=1}^{m} D(a)$.

13.20.* Use Thm. (13.56) to prove Thm. (13.57). (Hints: Combine Schur's lemma, Maschke's theorem, and the last exercise.)

13.21.* Let $\mathcal{T} = (S_4 \supset A_4 \supset V_4 \supset \{1\})$, where S_4 denotes the symmetric group on 4 letters, A_4 is the alternating subgroup, V_4 is the Klein 4-group consisting of the permutations (1), $(12)(34)$, $(13)(24)$, $(14)(23)$.

(1) Construct a $\mathcal{T}$-adapted DFT for $\mathbb{C}S_4$ and show that this DFT is necessarily monomial.

(2) On the basis of (1) design an FFT for S_4 and show that $L_2(A_4) \leq 32 < 0.75|A_4| \log |A_4|$ and $L_2(S_4) \leq 84 < 0.77|S_4| \log |S_4|$.

13.22.* The present exercise discusses FFTs for a series of nonabelian groups G_n, where G_n is a subgroup of $GL(2, 2^n)$ defined by

$$G_n := \left\{ \begin{pmatrix} a & b \\ 0 & a^{-1} \end{pmatrix} \Bigg| \ a, b \in \mathbb{F}_{2^n}, a \neq 0 \right\}.$$

(1) $N_n := \left\{ \begin{pmatrix} 1 & b \\ 0 & 1 \end{pmatrix} \mid b \in \mathbb{F}_{2^n} \right\}$ is a normal subgroup of G_n, isomorphic to the additive group of $\mathbb{F}_{2^n}$. $U_n := \left\{ \begin{pmatrix} a & 0 \\ 0 & a^{-1} \end{pmatrix} \mid a \neq 0 \right\}$ is a cyclic subgroup of order $2^n - 1$, and $N_n \cap U_n = \{1\}$. (Thus G_n is the semidirect product of N_n and U_n.)

(2) Show that G_n has exactly 2^n conjugacy classes and exactly $2^n - 1$ one-dimensional representations. G_n has a unique higher-dimensional irreducible character which is of degree $2^n - 1$.

(3) Prove that if F is a non-trivial irreducible representation of N_n, then $F \uparrow G_n$ is an irreducible representation of G_n.

(4) Show that $L_2(G_n) < n(1 + 2^{3-n})|G_n| \leq (2^{-1} + 5 \cdot 2^{-n})|G_n| \log |G_n|$.

13.23. Generalize Baum's theorem (13.55) to the class of finite groups G with an abelian normal subgroup A such that G/A is supersolvable.

13.7 Open Problems

Problem 13.1. What is the order of magnitude of $L_\infty(DFT_n)$?

Problem 13.2. Find a finite group G such that $L_\infty(G) \leq \frac{1}{4}|G| \log |G|$, or show that such a group does not exist.

Problem 13.3. Is there an infinite class **G** of pairwise non-isomorphic finite groups such that for some $\alpha > 1$, $L_\infty(G) \geq n^\alpha$, for all G in **G**?

Problem 13.4. Can a generic nth degree polynomial and all its derivatives be computed with a total number of $O(n)$ arithmetic operations? (Cf. Ex. 3.12.)

13.8 Notes

The study of linear complexity was stimulated by Cooley and Tukey's celebrated paper [127], in which they (re-)discovered the FFT of length $n = 2^\nu$, see also the notes in Chap. 2. By work of Winograd [555, 556, 558, 560, 561], Rader [431], and Bluestein [57] it became clear that the linear complexity of DFT-matrices of arbitrary size n is at most $8n \log n$, see also Baum et al. [27].

    Thm. (13.1) is a special case of Thm. (7.1). Thm. (13.10) is due to Winograd [550, 552]. Thm. (13.11) is from Savage [447]. Thm. (13.14) is due to

J. Morgenstern [382]. Tellegen's theorem (13.20), a basic result in network theory, see, e.g., [142, 416, 9], says that transposing the K-DAG for a matrix A gives a K-DAG for its transpose. Kaminski et al. [295] independently rediscovered this result in the context of complexity theory, thereby generalizing a result on the duality of addition chains in a paper by Papadimitriou and Knuth [414], see the notes in Chap. 1. Thm. (13.25) and Ex. 13.11 are due to Shoup and Smolensky [477]. Thm. (13.28) and Cor. (13.30) are due to Valiant [525]. In the same paper Valiant proved on the basis of Pinsker's work [418] the existence of n-superconcentrators with at most $238n$ edges. The improved version of Pinsker and Valiant's Thm. (13.31) is due to Pippenger [419]. Meanwhile explicit constructions of n-superconcentrators are known, see Margulis [351, 352], Gabber and Galil [178], Lubotzky et al. [343], and M. Morgenstern [384], among others. For further information on n-superconcentrators consult Pippenger's survey [422].

In a pioneering work, Beth [43] started to investigate discrete Fourier transforms for non-abelian finite groups under a complexity point of view. His main result was an upper bound of order $n^{3/2}$ for the linear complexity of solvable groups of order n. Clausen [113] used symmetry adapted DFTs to generalize Beth's result to arbitrary finite groups, see Thm. (13.48). In the same paper, Clausen applied this technique to the class of symmetric groups and obtains the improved upper bound of Thm. (13.58). With the same technique, Baum [24] has shown that all supersolvable groups of order n have linear complexity $O(n \log n)$, see Thm. (13.55). Baum and Clausen [26] have designed an algorithm that on input a power commutator presentation of a supersolvable group of order n computes in time $O(n \log n)$ a complete list of irreducible representations for G, which is well-suited for an FFT. Cor. (13.40) and Thm. (13.41) are from Baum and Clausen [25]. Thm. (13.51) is implicitely in Janusz [273] and explicitly stated in Clausen [112]. Thm. (13.54) is due to Baum [24]. For more details on fast Fourier transforms the reader is referred to Clausen and Baum [115], Diaconis and Rockmore [143], Rockmore [441], and to the recent work by Maslen and Rockmore [353, 354].

Ex. 13.2 follows J. Morgenstern [380], Ex. 13.5 and several other interesting results can be found in Savage [447]. Ex. 13.8 has been taken from Rader [431], Ex. 13.9 is from Bluestein [57], Ex. 13.22 follows Baum and Clausen [25], and Ex. 13.23 is due to Baum [24].

The problem of computing the matrix-vector product for a fixed matrix A and a generic input vector X has also been studied under a different point of view by several authors. To be more specific, let $k \subset K$ be a field extension and $A \in K^{m \times n}$. Consider the problem to compute the components of AX by K-linear straight-line programs, where k-linear operations and taking constants from K are for free. The resulting complexity has been computed exactly for special A. For instance, if $k = \mathbb{Q}$, $K = \mathbb{C}$, and $A = DFT_p$ for a prime p, this complexity turns out to be $2p - 3 - \delta$, where δ is the number of different divisors of $p - 1$, see Winograd [560]. For generalizations of this result consult Auslander et al. [19, 18], Feig et al. [162, 164], Heideman [237], and Oberst [397, 398, 399].

Chapter 14. Multiplicative and Bilinear Complexity

This chapter introduces the framework of the theory of bilinear complexity and is the basis for the study of Chap. 15–20. The language and concepts introduced here will, e.g., allow for a concise treatment of the fast matrix multiplication algorithms which we will discuss in the next chapter. We shall first turn our attention to the computation of a set of quadratic polynomials. According to Cor. (7.5) divisions do not help for the computation of such polynomials (at least when the field is infinite). The proof of that theorem in the case of quadratic polynomials implies a different characterization of the multiplicative complexity of these polynomials. We then associate to a set of quadratic polynomials quadratic maps between finite dimensional k-spaces and study the maps rather than the polynomials; this gives rise to the notion of computation for such maps. Specializing the quadratic maps further to bilinear maps and modifying the computations will lead to the important notion of rank or bilinear complexity of a bilinear map.

Unless otherwise specified, k denotes in this chapter any field and a k-space will always mean a finite-dimensional vector space over k.

14.1 Multiplicative Complexity of Quadratic Maps

In this section L will always denote the multiplicative complexity.

We have seen in Chap. 7 as a consequence of Thm. (7.1) that if Q is a finite set of quadratic forms in the polynomial ring $k[X] = k[X_1, \ldots, X_n]$ and k is an infinite field, then $L_{k[X]}(Q) = L_{k(X)}(Q)$. Our first aim in this section will be a characterization of optimal computation sequences for the set Q.

(14.1) Proposition (Strassen). *Let $Q \subset k[X_1, \ldots, X_n]$ be a finite set of quadratic forms over k. Then*

$$L_{k[X]}(Q) = \min\left\{\rho \;\middle|\; \exists\, f_1, \ldots, f_\rho, g_1, \ldots, g_\rho \in k[X]^{(1)}: \; Q \subseteq \sum_{i=1}^{\rho} k f_i g_i\right\},$$

and $L_{k(X)}(Q) = L_{k[X]}(Q)$ if k is an infinite field. Here $k[X]^{(1)}$ denotes the set of homogeneous linear polynomials in $k[X]$.

Proof. In view of Cor. (7.5) we are left with the proof of the first equality. To ease the notation, we denote by A the polynomial ring $k[X]$. If $a \in A$ and j is an integer, then $a^{(j)}$ denotes the homogeneous part of degree j of a.

Let $\ell := L_A(Q)$ and r be the right-hand side of the above equality. Obviously $\ell \leq r$. For proving $r \leq \ell$ let $(s_1, \ldots, s_\ell)$ be a division free computation sequence for Q. Then for all $i \in \ell$ there exist $a_i, b_i \in k + \sum_{\mu=1}^{n} kX_\mu + \sum_{j=1}^{i-1} ks_j$ such that $s_i = a_i b_i$. For

$$s_i^{(2)} = a_i^{(0)} b_i^{(2)} + a_i^{(1)} b_i^{(1)} + a_i^{(2)} b_i^{(0)}$$

we prove by induction on i that

(A) $$s_i^{(2)} \in \sum_{j=1}^{i} k a_j^{(1)} b_j^{(1)}.$$

The case $i = 1$ is trivial ($a_1^{(2)} = b_1^{(2)} = 0$). For the induction step we use $s_i = a_i b_i$ to obtain $s_i^{(2)} \in \sum_{j=1}^{i-1} ks_j^{(2)} + a_i^{(1)} b_i^{(1)}$. This together with the induction hypothesis gives (A). Now, since Q consists of quadratic forms only, we have

$$Q \subseteq \sum_{j=1}^{\ell} ks_j^{(2)} \subseteq \sum_{j=1}^{\ell} k a_j^{(1)} b_j^{(1)}.$$

Putting $f_j := a_j^{(1)}$ and $g_j := b_j^{(1)}$ we obtain the assertion. $\square$

Sets of quadratic forms induce a special type of maps between k-spaces. We call a map $\Phi : V \to W$ between k-spaces *quadratic* if there exist bases $(v_i)_{i \leq n}$ and $(w_j)_{j \leq p}$ of V and W, respectively, and quadratic forms $\varphi_1, \ldots, \varphi_p \in k[X_1, \ldots, X_n]$ such that for all $\xi = (\xi_1, \ldots, \xi_n) \in k^n$

$$\Phi\left(\sum_{i=1}^{n} \xi_i v_i\right) = \sum_{j=1}^{p} \varphi_j(\xi) w_j.$$

If $W = k$, then Φ is called a *quadratic form on* V.

The reader will have no difficulties in proving that the property of being quadratic does not depend on the specific bases in V and W. Also, if $\alpha : V' \to V$ and $\beta : W \to W'$ are homomorphisms of k-spaces and $\Phi : V \to W$ is quadratic, then $\beta \circ \Phi \circ \alpha$ is quadratic. We remark further that over fields of characteristic other than 2 one may characterize quadratic maps differently (see Ex. 14.1).

In the sequel, we will call the quadratic forms $\varphi_1, \ldots, \varphi_p$ in the above definition the *coordinate functions* of Φ with respect to the bases (v_i) and (w_j). The coordinate functions of Φ are uniquely determined by Φ and the bases in V and W (see Ex. 14.4).

(14.2) Definition. Let $\Phi : V \to W$ be a quadratic map between k-spaces.

(1) Let $\varphi_1, \ldots, \varphi_p \in k[X_1, \ldots, X_n]$ be the coordinate functions of Φ with respect to some bases in V and W. Then

$$L(\Phi/k) := L(\Phi) := L_{k[X]}(\{\varphi_1, \ldots, \varphi_p\})$$

is called the *multiplicative complexity* of Φ.

(2) Let $f_1, \ldots, f_\ell, g_1, \ldots, g_\ell \in V^*$ and $w_1 \ldots, w_\ell \in W$ be such that for all $v \in V$ we have

$$\Phi(v) = \sum_{i=1}^{\ell} f_i(v) g_i(v) w_i.$$

Then $(f_1, g_1, w_1; \ldots; f_\ell, g_\ell, w_\ell)$ is called *a quadratic computation (algorithm) for Φ of length ℓ.*   •

(14.3) Remarks.
(1) $L(\Phi)$ is well-defined since it is independent of the bases chosen in V and W. (Cf. Ex. 14.7.)
(2) If k is an infinite field, we can, using Prop. (14.1), replace $L_{k[X]}$ in the above definition by $L_{k(X)}$.   •

In the sequel we shall work with the following characterization of the multiplicative complexity of a quadratic map.

(14.4) Proposition. *Let V and W be k-spaces and $\Phi: V \to W$ be a quadratic map. Then*

$$L(\Phi) = Length\ of\ a\ shortest\ quadratic\ computation\ for\ \Phi.$$

The proof is easy and follows from Prop. (14.1); see Ex. 14.8.

Relative upper and lower bounds for the multiplicative complexity are often obtained by comparing the complexities of different quadratic maps. The following obvious remark is a first result in this direction.

(14.5) Remark. Let $\alpha: V' \to V$ and $\beta: W \to W'$ be linear maps of k-spaces and $\Phi: V \to W$ be quadratic. Then $L(\beta \circ \Phi \circ \alpha) \leq L(\Phi)$.   •

Let us compute the multiplicative complexity in a concrete case.

(14.6) Example. (char $k \neq 2$.) Let $V \simeq k^n$ and $\varphi: V \to k$ be a quadratic form on V. We want to compute $L(\varphi)$. It is a well-known fact (cf. Lang [316, Chap. XIV]) that there exists a natural number p and $\alpha_1, \ldots, \alpha_{n-2p} \in k$ such that with respect to a suitable basis $(e_1, \ldots, e_n)$ of V we have for all $v := \sum_{i=1}^n v_i e_i \in V$

$$\varphi(v) = v_1 v_2 + \cdots + v_{2p-1} v_{2p} + \alpha_1 v_{2p+1}^2 + \cdots + \alpha_{n-2p} v_n^2.$$

The largest such number p, called the *Witt-index* of (V, φ), is the dimension of a maximal nullspace of φ. The above representation of φ implies $L(\varphi) \leq n - p$.

On the other side, let $\ell := L(\varphi)$ and suppose that $f_1, \ldots, f_\ell, g_1, \ldots, g_\ell \in V^*$, and $w_1, \ldots, w_\ell \in k$ be such that for all $v \in V$ we have $\varphi(v) = \sum_{i=1}^\ell f_i(v) g_i(v) w_i$. The dimension of the space of common zeros of $f_1, \ldots, f_\ell$ is at least $n - \ell$ since the f_i are linear forms. Further, this space lies in the set of all zeros of φ, hence is of dimension less than or equal to the Witt-index p of φ: $n - \ell \leq p$. We thus obtain: $L(\varphi) = n - p$.   •

We proceed with our discussions by specializing further the class of the quadratic maps. To this end let U, V and W be k-spaces. Recall that a map $\phi: U \times V \to W$ satisfying $\phi(\lambda_{11}u_1 + \lambda_{12}u_2, \lambda_{21}v_1 + \lambda_{22}v_2) = \sum_{i, j \leq 2} \lambda_{ij}\phi(u_i, v_j)$ for all $\lambda_{ij} \in k$, $u_i \in U$, and $v_j \in V$ is called *bilinear*. If $W = k$, ϕ is called a *bilinear form* on $U \times V$. The k-space of all bilinear maps from $U \times V$ to W is denoted by $\mathrm{Bil}(U, V; W)$. Its elements are said to have *format* (dim U, dim V, dim W). In the following we will understand by a k-bilinear map always a k-bilinear map between k-spaces. Often we omit the field k in the notation if it is clear from the context.

Note that a bilinear map in $\mathrm{Bil}(U, V; W)$ can be viewed as a quadratic map from $U \times V$ to W. Let $\ell := L(\phi)$ and $f_i, g_i \in (U \times V)^*$, $w_i \in W$, be such that for all $(u, v) \in U \times V$ we have $\phi(u, v) = \sum_{i=1}^{\ell} f_i(u, v)g_i(u, v)w_i$. This implies

$$
\begin{aligned}
\phi(u, v) &= \sum_{i=1}^{\ell} \Big(f_i(u, 0) + f_i(0, v)\Big)\Big(g_i(u, 0) + g_i(0, v)\Big)w_i \\
&= \sum_{i=1}^{\ell} f_i(u, 0)g_i(0, v)w_i + \sum_{i=1}^{\ell} g_i(u, 0)f_i(0, v)w_i,
\end{aligned}
$$

since the terms $\sum_i f_i(u, 0)g_i(u, 0)w_i$ and $\sum_i f_i(0, v)g_i(0, v)w_i$ vanish. (ϕ is bilinear.)

(14.7) Definition. Let $\phi: U \times V \to W$ be a k-bilinear map.

(1) For $i \in \underline{r}$ let $f_i \in U^*$, $g_i \in V^*$, $w_i \in W$ be such that

$$
\phi(u, v) = \sum_{i=1}^{r} f_i(u)g_i(v)w_i
$$

for all $u \in U$, $v \in V$. Then $(f_1, g_1, w_1; \ldots; f_r, g_r, w_r)$ is called a *bilinear computation (algorithm) of length r* for ϕ.

(2) The length of a shortest bilinear computation for ϕ is called the *bilinear complexity* or the *rank* of ϕ and is denoted by $R(\phi/k)$ (or $R(\phi)$ if there is no danger of confusion.) ●

The discussion preceding the above definition shows that

$$(14.8) \qquad\qquad L(\phi) \leq R(\phi) \leq 2L(\phi)$$

for arbitrary bilinear maps ϕ. If ϕ is non-zero, then we even have $R(\phi) < 2L(\phi)$, see Ex. 14.15. If $\alpha: U' \to U$, $\beta: V' \to V$, and $\gamma: W \to W'$ are k-space morphisms and $\phi \in \mathrm{Bil}(U, V; W)$, then $\psi := \gamma \circ \phi \circ (\alpha \times \beta)$ is a bilinear map in $\mathrm{Bil}(U', V'; W')$ and $R(\phi) \geq R(\psi)$. (Note however that if $\pi: U' \times V' \to U \times V$ is an arbitrary morphism of k-spaces, then $\phi \circ \pi$ is a quadratic map, but *need not be bilinear*.)

As an example, let us justify the notion of *rank* for a bilinear map.

(14.9) Example. Let φ be a bilinear form on $U \times V$ and $\tilde{\varphi} \colon U \to V^*$ be the linear map associated with φ, i.e., $\tilde{\varphi}(u)(v) := \varphi(u, v)$ for all $(u, v) \in U \times V$. We are going to show that $R(\varphi) = \mathrm{rk}\,\tilde{\varphi}$.

Let $r = R(\varphi)$. The representation $\varphi(u, v) = \sum_{i=1}^{r} f_i(u)g_i(v)$ (observe that in this case $W = k$ and we can assume $w_i = 1$ for all i) for all $(u, v) \in U \times V$ yields $\tilde{\varphi}(u) = \sum_{i=1}^{r} f_i(u)g_i$ and hence $\mathrm{im}\,\tilde{\varphi} \subseteq \sum_{i=1}^{r} kg_i$. This in turn shows that $\mathrm{rk}\,\tilde{\varphi} \leq R(\varphi)$.

To show the converse, let $(g_1, \ldots, g_m)$ be a basis of the image of $\tilde{\varphi}$. For every $u \in U$ we can write $\tilde{\varphi}(u) = \sum_{i=1}^{m} f_{i,u}g_i$ with uniquely determined $f_{i,u} \in k$. It follows that f_i defined by $f_i(u) := f_{i,u}$ is a linear form on U and hence $\varphi(u, v) = \tilde{\varphi}(u)(v) = \sum_{i=1}^{m} f_i(u)g_i(v)$ which shows that $R(\varphi) \leq \mathrm{rk}\,\tilde{\varphi}$.   ●

Computing the rank of a given bilinear map is a challenging task (see the Notes). While the determination of the rank of one bilinear form is relatively easy as the last example shows, it is not clear how to proceed in case of several bilinear forms. (The case of a pair of bilinear forms can also be handled completely, see Chap. 19). It is therefore of interest to be able to compare the ranks of different bilinear maps.

(14.10) Lemma. *Let $\phi \in \mathrm{Bil}(U, V; W)$ and $\phi' \in \mathrm{Bil}(U', V'; W')$ be bilinear maps and suppose that the following diagram commutes:*

$$
\begin{array}{ccc}
U \times V & \xrightarrow{\ \phi\ } & W \\
{\scriptstyle \alpha \times \beta}\Big\downarrow & & \Big\downarrow{\scriptstyle \gamma} \\
U' \times V' & \xrightarrow[\ \phi'\ ]{} & W'
\end{array}
$$

(1) *If α and β are surjective, then $R(\phi') \leq R(\phi)$.*
(2) *If γ is injective, then $R(\phi) \leq R(\phi')$.*

Proof. (1) Since α and β are epimorphisms, there exist $\alpha' \colon U' \to U$ and $\beta' \colon V' \to V$ such that $\alpha \circ \alpha' = \mathrm{id}_{U'}$ and $\beta \circ \beta' = \mathrm{id}_{V'}$. This shows that $\gamma \circ \phi \circ (\alpha' \times \beta') = \phi'$, hence $R(\phi') \leq R(\phi)$.

(2) Since γ is a monomorphism there exists $\gamma' \colon W' \to W$ such that $\gamma' \circ \gamma = \mathrm{id}_W$. This shows that $\gamma' \circ \phi' \circ (\alpha \times \beta) = \phi$, hence $R(\phi) \leq R(\phi')$.   □

(14.11) Definition. Let U, U', V, V', W, W' be k-spaces. Two bilinear maps $\phi \in \mathrm{Bil}(U, V; W)$ and $\phi' \in \mathrm{Bil}(U', V'; W')$ are called *isomorphic* if there are isomorphisms $\alpha \colon U \to U'$, $\beta \colon V \to V'$ and $\gamma \colon W \to W'$ such that

$$\gamma \circ \phi = \phi' \circ (\alpha \times \beta). \qquad ●$$

As is clear from the preceding lemma, isomorphic bilinear maps have the same rank. If U, V, and W are k-spaces, then

$$G := \mathrm{GL}(U) \times \mathrm{GL}(V) \times \mathrm{GL}(W)$$

acts on $\mathrm{Bil}(U, V; W)$ via

$$(\alpha, \beta, \gamma) \cdot \phi := \gamma \circ \phi \circ (\alpha^{-1} \times \beta^{-1}),$$

and its orbits are the isomorphism classes of bilinear maps $U \times V \to W$.

In the following a *k-algebra* will always denote a *finite-dimensional k-algebra with unity*. To a large extent our interest in bilinear maps focuses on the structural maps (multiplication) of algebras. If there is no danger of confusion we will denote the multiplication in a k-algebra A by the same letter A, hence $L(A)$ $(R(A))$ is the multiplicative complexity (rank) of the multiplication map in A. An exception is the associative k-algebra k^r, r being a positive integer, consisting of r-tuples over k with addition and multiplication defined componentwise: the multiplication in this algebra will be denoted by $\langle r \rangle$. Its rank is easy to determine: if $(e_i)_{i \leq r}$ is the standard basis of k^r with corresponding dual basis $(e_i^*)_{i \leq r}$, then $ab = \sum_{i \leq r} e_i^*(a) e_i^*(b) e_i$ for all $a, b \in k^r$. It follows that $R(\langle r \rangle) \leq r$. It is an easy exercise to prove $R(\langle r \rangle) \geq r$ as well. (This will be a by-product of our considerations in Sect. 14.3.)

A *k*-space homomorphism $\alpha: A \to B$ between k-algebras is called a *k*-algebra homomorphism if $\alpha(1_A) = 1_B$ and for all $a, a' \in A$ we have $\alpha(aa') = \alpha(a)\alpha(a')$. (Here 1_A denotes the unity of A and 1_B the unity of B.) α is called an *isomorphism* if it is bijective. In terms of the structural maps of algebras Lemma (14.10) reads as follows.

(14.12) Proposition. *Let $\alpha: A \to B$ be a morphism between k-algebras. If α is injective, then $R(A) \leq R(B)$. If α is surjective, then $R(A) \geq R(B)$.*

Isomorphisms of k-algebras and of their structural maps are related in the following manner.

(14.13) Proposition. *Two k-algebras are isomorphic as algebras if and only if they are isomorphic as bilinear maps.*

Proof. We only need to show that A and A' are isomorphic as algebras if they are isomorphic as bilinear maps. Let $\alpha, \beta, \gamma: A \to A'$ be k-space homomorphisms satisfying $\gamma(ab) = \alpha(a)\beta(b)$ for all $a, b \in A$. We deduce $\gamma(b) = \alpha(1)\beta(b)$. The surjectivity of γ implies that $\alpha(1)$ is a unit of A'. Analogously $\beta(1)$ is a unit of A'. We define now the k-space isomorphism $\phi: A \to A'$ by setting $\phi(a) := \alpha(1)^{-1}\alpha(a)$ for $a \in A$. Then $\beta(a) = \phi(a)\beta(1)$, $\gamma(a) = \alpha(1)\phi(a)\beta(1)$ and therefore $\phi(ab) = \phi(a)\phi(b)$, $\phi(1) = 1$. $\square$

(14.14) Example. The $\mathbb{C}$-algebra $\mathbb{C}[X]/(X^n - 1)$ is isomorphic to $\mathbb{C}^n$ as an algebra. Hence $R(\mathbb{C}[X]/(X^n - 1)) = R(\langle n \rangle) = n$.   •

A small, but important class of examples of bilinear maps is furnished by matrix multiplication (not just of square matrices). We write $\langle e, h, \ell \rangle$ for the matrix multiplication $k^{e \times h} \times k^{h \times \ell} \to k^{e \times \ell}$. For example, it is easy to see that Strassen's algorithm for multiplying 2×2-matrices is a bilinear computation and hence $R(\langle 2, 2, 2 \rangle) \leq 7$. (See Chap. 1; we will prove equality later in Chap. 17.) Note that $\langle e, h, \ell \rangle$ has format $(eh, h\ell, e\ell)$. Also, $\langle e, e, e \rangle$ equals $k^{e \times e}$ (as bilinear maps).

(14.15) Proposition. *Let* $e \leq e'$, $h \leq h'$, *and* $\ell \leq \ell'$ *be positive integers. Then* $R(\langle e, h, \ell \rangle) \leq R(\langle e', h', \ell' \rangle)$.

Proof. We embed $k^{e \times h}$ into $k^{e' \times h'}$ by assigning to $a \in k^{e \times h}$ the matrix b having a in its $e \times h$ upper left corner and zeros elsewhere. (Note that if $e = h$, $e' = h'$, and $e < e'$, then this is *not* an embedding of algebras.) Similarly, we embed $k^{h \times \ell}$ into $k^{h' \times \ell'}$ and $k^{e \times \ell}$ into $k^{e' \times \ell'}$. Now we obtain the assertion from Lemma (14.10). □

14.2 The Tensorial Notation

In this section we derive an equivalent characterization of the rank of a bilinear map using tensors.

Recall that for k-spaces U and V their *tensor product* is defined as a k-space $U \otimes V$ together with a k-bilinear map $\tau \in \mathrm{Bil}(U, V; U \otimes V)$ satisfying the following universal mapping property: for every k-space W and every k-bilinear map $\phi \in \mathrm{Bil}(U, V; W)$ there exists exactly one k-space homomorphism $\phi': U \otimes V \to W$ such that $\phi' \circ \tau = \phi$, i.e., the diagram

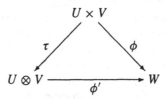

commutes. For $u \in U$ and $v \in V$ we denote by $u \otimes v$ the image $\tau(u, v)$ of (u, v) and call it a *dyad*. $U \otimes V$ is generated by the set of all dyads. If $(u_i)_{i \leq m}$ and $(v_j)_{j \leq n}$ are bases of U and V, then $(u_i \otimes v_j)_{i \leq m, j \leq n}$ is a basis of $U \otimes V$.

The tensor product of a finite number of k-spaces is defined similarly. In particular, if U, V, and W are k-spaces, then $U \otimes V \otimes W$ is generated by the *triads* $u \otimes v \otimes w$, where (u, v, w) runs over the elements of $U \times V \times W$. It is easy to see that there exist canonical isomorphisms

$$U \otimes V \otimes W \simeq (U \otimes V) \otimes W \simeq U \otimes (V \otimes W).$$

If $\alpha: U \to U'$ and $\beta: V \to V'$ are homomorphisms of k-spaces, then by the universal mapping property there is a unique homomorphism $\alpha \otimes \beta: U \otimes V \to U' \otimes V'$ such that $(\alpha \otimes \beta)(u \otimes v) = \alpha(u) \otimes \beta(v)$ for all $u \in U$ and $v \in V$.

Tensor products of k-spaces enjoy various properties. For example, besides the above mentioned associativity, there exist canonical isomorphisms

$$U \otimes V \simeq V \otimes U, \quad U \otimes k \simeq U$$

for k-spaces U and V. For our purposes, the following isomorphism is of particular interest.

(14.16) Proposition. *Let U, V, and W be k-spaces. There exists a unique isomorphism $U^* \otimes V^* \otimes W \to \mathrm{Bil}(U, V; W)$ which sends $f \otimes g \otimes w$ to the bilinear map $(u, v) \mapsto f(u)g(v)w$.*

Proof. Consider the trilinear map $U^* \times V^* \times W \to \mathrm{Bil}(U, V; W)$ sending (f, g, w) to the bilinear map $(u, v) \mapsto f(u)g(v)w$. According to the universal mapping property of the tensor product there exists a unique homomorphism $\sigma: U^* \otimes V^* \otimes W \to \mathrm{Bil}(U, V; W)$ having the stated property. Since σ is obviously injective, it remains to show that it is also surjective. If $(f_i)_{i \leq m}$, $(g_j)_{j \leq n}$, and $(w_\ell)_{\ell \leq p}$ are bases of U^*, V^*, and W, respectively, then the bilinear maps $\phi_{ij\ell}: (u, v) \mapsto f_i(u)g_j(v)w_\ell$ form a basis of $\mathrm{Bil}(U, V; W)$. This completes the proof.   □

(14.17) Definition. The unique tensor $t \in U^* \otimes V^* \otimes W$ associated to the bilinear map $\phi \in \mathrm{Bil}(U, V; W)$ is called the *structural tensor* of ϕ.   •

In the sequel we shall often and implicitly change our point of view and regard tensors as bilinear maps or vice-versa. Also, we agree upon the following convention: a tensor is a quadruple (U, V, W, t), where U, V, and W are k-spaces and $t \in U \otimes V \otimes W$. By abuse of notation, we will often denote such a tensor by t if the spaces U, V, W are clear from the context.

The canonical isomorphism in Prop. (14.16) allows to transfer the concepts developed so far for bilinear maps to tensors. For example, we call $(\dim U, \dim V, \dim W)$ the *format* of the tensor (U, V, W, t). We call two tensors (U, V, W, t) and (U', V', W', t') *isomorphic* if there exist isomorphisms $\alpha: U \to U'$, $\beta: V \to V'$, and $\gamma: W \to W'$ such that $(\alpha \otimes \beta \otimes \gamma)t = t'$. Obviously, two tensors are isomorphic iff their corresponding bilinear maps are isomorphic.

(14.18) Definition. The *rank* $R(t)$ of a tensor (U, V, W, t) is defined as the minimum number r of triads $u_i \otimes v_i \otimes w_i$ such that t can be represented as
$$t = \sum_{i=1}^r u_i \otimes v_i \otimes w_i.$$
•

The isomorphism given in Prop. (14.16) is rank preserving.

One reason why to study tensors besides bilinear maps is the symmetry in the definition of the rank of a tensor. A permutation $\pi \in S_3$, the symmetric group on $\{1, 2, 3\}$, induces an isomorphism $U_1 \otimes U_2 \otimes U_3 \to U_{\pi^{-1}(1)} \otimes U_{\pi^{-1}(2)} \otimes U_{\pi^{-1}(3)}$. (We take π^{-1} instead of π to get an *action* of S_3 if $U_1 = U_2 = U_3$.) The image of $t \in U_1 \otimes U_2 \otimes U_3$ will be denoted by πt. It is clear that $R(t) = R(\pi t)$.

Let us now proceed with an important example.

(14.19) Example. Let e, e', h, h' be positive integers. Comparing dimensions it is clear that $k^{e \times h} \otimes k^{e' \times h'}$ and $k^{ee' \times hh'}$ are isomorphic as k-spaces. Under the various k-space isomorphisms we consider a distinguished one as follows:

Let $(u_{\mu\nu} \mid \mu \in \underline{e}, \nu \in \underline{h})$, $(v_{\rho\sigma} \mid \rho \in \underline{e'}, \sigma \in \underline{h'})$, and $(w_{ij} \mid i \in \underline{ee'}, j \in \underline{hh'})$ be the standard bases of $k^{e \times h}$, $k^{e' \times h'}$, and $k^{ee' \times hh'}$, respectively. Let the isomorphism ι be the k-linear extension of the map defined by

$$u_{\mu\nu} \otimes v_{\rho\sigma} \mapsto w_{(\mu-1)e'+\rho, (\nu-1)h'+\sigma} \in k^{ee' \times hh'}$$

for all $\mu \in \underline{e}, \nu \in \underline{h}, \rho \in \underline{e}'$, and $\sigma \in \underline{h}'$. In other words, ι sends the $u_{\mu\nu} \otimes v_{\rho\sigma}$ to the *Kronecker product* of $u_{\mu\nu}$ and $v_{\rho\sigma}$. In this way, ι can be interpreted as a "block decomposition" of the $k^{ee' \times hh'}$. Note that ι "respects matrix multiplication" in the following sense: if $m, m' \in \mathbb{N}'$, $A \in k^{e \times h}$, $B \in k^{h \times m}$, $A' \in k^{e' \times h'}$, $B' \in k^{h' \times m'}$, and $A \otimes B$ denotes the Kronecker product of A and B, then

$$(A \otimes A')(B \otimes B') = AB \otimes A'B'.$$

If $e = h$ and $e' = h'$, then ι induces an isomorphism of $k^{h \times h} \otimes k^{h' \times h'}$ and $k^{hh' \times hh'}$ as *k-algebras*. We shall use this property later in this section.

Let us now determine a representation of $\langle e, h, \ell \rangle$ as a sum of triads. We identify the *k*-spaces $k^{e \times h}$ and $k^{h \times \ell}$ with their duals. Then, using the above notation, the definition of the matrix multiplication yields the following (we transpose the indices in the third space for ease of notation)

$$(14.20) \qquad \langle e, h, \ell \rangle = \sum_{i,j,m} u_{ij} \otimes v_{jm} \otimes w_{mi} \in k^{e \times h} \otimes k^{h \times \ell} \otimes k^{\ell \times e}.$$

If $\pi \in S_3$ is any permutation, then the above representation of the tensor $\langle h_1, h_2, h_3 \rangle$ shows that

$$(14.21) \qquad \pi \langle h_1, h_2, h_3 \rangle \simeq \langle h_{\sigma^{-1}(1)}, h_{\sigma^{-1}(2)}, h_{\sigma^{-1}(3)} \rangle,$$

where $\sigma = (123)^{-1} \pi (123)$. (See also Ex. 14.10.) $\qquad\qquad\qquad\bullet$

(14.22) Remark. One of the advantages of working with the rank R rather than with the multiplicative complexity L is that $R(\pi t) = R(t)$ for a tensor t, while $L(\pi t)$ may not be equal to $L(t)$. An evidence for this is given by the following observation: let $t_n := \langle n, 2, 2 \rangle$ and $t'_n := \langle 2, n, 2 \rangle$. Then there exists a quadratic computation over $\mathbb{Z}$ of length $3n + 1$ for t_n, while the length of any quadratic algorithm over $\mathbb{Z}$ for t'_n is at least $\lceil 27n/8 \rceil$. (Cf. the Notes.) $\qquad\bullet$

Given bilinear maps $\phi: U \times V \to W$ and $\phi': U' \times V' \to W'$ we can define their *direct sum*

$$\phi \oplus \phi' \quad : \quad (U \oplus U') \times (V \oplus V') \to W \oplus W'$$
$$(u \oplus u', v \oplus v') \mapsto \phi(u, v) \oplus \phi'(u', v'),$$

and their *tensor product*

$$\phi \otimes \phi' \quad : \quad (U \otimes U') \times (V \otimes V') \to W \otimes W'$$
$$(u \otimes u', v \otimes v') \mapsto \phi(u, v) \otimes \phi'(u', v').$$

Equivalently, given tensors (U, V, W, t) and (U', V', W', t') we can define their *direct sum*

$$t \oplus t' := (U \oplus U', V \oplus V', W \oplus W', t + t'),$$

where t, t' are to be interpreted as elements of $(U \oplus U') \otimes (V \oplus V') \otimes (W \oplus W')$, and their *tensor product*

$$t \otimes t' := (U \otimes U', V \otimes V', W \otimes W', s)$$

where $s = t \otimes t'$ under the identification

$$(U \otimes U') \otimes (V \otimes V') \otimes (W \otimes W') \simeq (U \otimes V \otimes W) \otimes (U' \otimes V' \otimes W').$$

It is easy to see that direct sum and tensor product induce operations "+" (addition) and "·" (multiplication) on isomorphism classes of bilinear maps which are associative and commutative, $\otimes$ distributing over $\oplus$. If $\langle 0 \rangle$ denotes the trivial bilinear map (where $U = V = W = 0$) and $\langle 1 \rangle$ denotes the bilinear map "multiplication in k," then the classes of $\langle 0 \rangle$, $\langle 1 \rangle$ are neutral elements for $+, \cdot$, respectively. Hence, an elegant formulation of the concepts developed so far is that the isomorphism classes of tensors (or bilinear maps) form a semiring. Later in Sect. 14.6 we shall study a certain sub-semiring of this semiring.

It is easily shown that the rank is subadditive and submultiplicative on the semiring of isomorphism classes of bilinear maps.

(14.23) Proposition. *Let ϕ_1 and ϕ_2 be bilinear maps.*

(1) $R(\phi_1 \oplus \phi_2) \le R(\phi_1) + R(\phi_2)$.
(2) $R(\phi_1 \otimes \phi_2) \le R(\phi_1)R(\phi_2)$.

The proof is left to the reader, cf. Ex. 14.11.

(14.24) Remarks.
(1) The *additivity conjecture* or *direct sum conjecture* states that we have always equality in Prop. (14.23)(1). Although no counterexamples are known, we will see in Chap. 15 why this conjecture is probably false. However, it can be shown that the second inequality in the above proposition may be strict. (See Ex. 14.12.)
(2) One further advantage of using rank instead of multiplicative complexity is the submultiplicativity given in Prop. (14.23)(2). If ϕ_1 and ϕ_2 are bilinear maps, it is *not known* whether $L(\phi_1 \otimes \phi_2) \le L(\phi_1)L(\phi_2)$.   •

Let us now study our basic examples of k-algebras and matrix multiplication. For k-algebras A and A' the direct sum $A \oplus A'$ as bilinear maps is just the direct product of A and A' as algebras. In particular, it is again an algebra. Conversely, we have the following result.

(14.25) Proposition. *If ϕ, ϕ' are bilinear maps and $\phi \oplus \phi'$ is isomorphic to an associative algebra B, then so are ϕ, ϕ'.*

Proof. We write ϕ and ϕ' as multiplication. Consider an inner direct decomposition of B as a bilinear map, thus $B = U \oplus U' = V \oplus V' = W \oplus W'$ such that $U \cdot V \subseteq W, U' \cdot V' \subseteq W', U \cdot V' = U' \cdot V = 0$. Let $1 = e \oplus e'$ with $e \in U, e' \in U'$. Then for $v \in V$ we have $v = (e \oplus e')v = ev \in W$. Hence $V \subseteq W$ and similarly $V' \subseteq W'$. Therefore $V = W, V' = W'$. Analogously $U = V, U' = W'$. Thus U, resp. U' is an associative k-algebra with unit element e, resp. e'.   □

The tensor product $A \otimes A'$ of k-algebras A and A' (regarded as bilinear maps) is simply their tensor product as algebras. On the other hand $\phi \otimes \phi'$ may be isomorphic to an associative algebra while neither ϕ nor ϕ' are. (This follows, e.g., from the next proposition.)

The direct sum of the matrix tensors $\langle e, h, \ell \rangle$ and $\langle e', h', \ell' \rangle$ corresponds to the multiplication of block diagonal matrices, as is easily verified. The behavior of matrix multiplication under tensor products is as follows.

(14.26) Proposition. *Let $e, e', h, h', \ell, \ell'$ be positive integers. Then we have*

$$\langle e, h, \ell \rangle \otimes \langle e', h', \ell' \rangle \simeq \langle ee', hh', \ell\ell' \rangle.$$

Proof. We use the tensorial representation of $\langle e, h, \ell \rangle$ as derived in (14.20). We have

$$\langle e, h, \ell \rangle \otimes \langle e', h', \ell' \rangle$$

$$= \left(\sum_{i,j,m} u_{ij} \otimes v_{jm} \otimes w_{mi} \right) \otimes \left(\sum_{i',j',m'} u'_{i'j'} \otimes v'_{j'm'} \otimes w'_{m'i'} \right)$$

$$\simeq \sum_{i,j,m,i',j',m'} (u_{ij} \otimes u'_{i'j'}) \otimes (v_{jm} \otimes v'_{j'm'}) \otimes (w_{mi} \otimes w'_{m'i'}).$$

By Example (14.19) we can give an isomorphism of $\underline{e} \times \underline{h} \times \underline{e'} \times \underline{h'}$ and $\underline{ee'} \times \underline{hh'}$ under which we can identify $u_{ij} \otimes u'_{i'j'}$ with $U_{\mu\nu}$ where $(U_{\mu\nu} \mid \mu \in \underline{ee'}, \nu \in \underline{hh'})$ is the canonical basis of $k^{ee' \times hh'}$. Similarly we can process the other terms in the above representation. Denoting by $(V_{\nu\rho})$ and $(W_{\rho\mu})$ the standard bases of $k^{hh' \times \ell\ell'}$ and $k^{\ell\ell' \times ee'}$ respectively, we obtain

$$\langle e, h, \ell \rangle \otimes \langle e', h', \ell' \rangle \simeq \sum_{\mu,\nu,\rho} U_{\mu\nu} \otimes V_{\nu\rho} \otimes W_{\rho\mu}$$

$$= \langle ee', hh', \ell\ell' \rangle. \quad \square$$

14.3 Restriction and Conciseness

We begin with a definition.

(14.27) Definition. Let (U, V, W, ϕ) and (U', V', W', ϕ') be bilinear maps. A *restriction* of ϕ' to ϕ is a triple (α, β, γ') of linear maps $\alpha: U \to U'$, $\beta: V \to V'$, and $\gamma': W' \to W$ such that $\phi = \gamma' \circ \phi' \circ (\alpha \times \beta)$:

$$
\begin{array}{ccc}
U \times V & \xrightarrow{\ \phi\ } & W \\
{\scriptstyle \alpha \times \beta} \downarrow & & \uparrow {\scriptstyle \gamma'} \\
U' \times V' & \xrightarrow{\ \phi'\ } & W'
\end{array}
$$

$\phi \leq \phi'$ (ϕ is a *restriction* of ϕ') means that there exists a restriction of ϕ' to ϕ. Equivalently, for tensors (U, V, W, t) and (U', V', W', t') a *restriction* of t' to t is a triple of linear maps $\alpha \colon U' \to U$, $\beta \colon V' \to V$, and $\gamma \colon W' \to W$ such that $t = (\alpha \otimes \beta \otimes \gamma)t'$; $t \leq t'$ (t is a *restriction* of t') means that there exists a restriction of t' to t.   •

(14.28) Remark. Restriction is a preorder on the set of isomorphism classes of tensors (or bilinear maps) which induces the natural order on $\mathbb{N}$ via $r \mapsto \langle r \rangle$. If $t \leq t'$ and $\pi \in S_3$, then $\pi t \leq \pi t'$. Moreover, $s \leq s'$ and $t \leq t'$ together imply $s \oplus t \leq s' \oplus t'$ and $s \otimes t \leq s' \otimes t'$. (Use the direct sum or the tensor product for each of the three components of the given restrictions, cf. Ex. 14.13.)   •

Our principal motivation for the notion of restriction of tensors is the following fact which shows that tensor rank can be completely characterized by the restriction.

(14.29) Proposition. *If (U, V, W, t) is a tensor and $r \in \mathbb{N}$, then*

$$R(t) \leq r \Longleftrightarrow t \leq \langle r \rangle.$$

In particular, $t \leq t'$ implies $R(t) \leq R(t')$ for tensors t and t'.

Proof. $R(t) \leq r$ means that there are vectors $u_1, \ldots, u_r \in U$, $v_1, \ldots, v_r \in V$, $w_1, \ldots, w_r \in W$ such that $t = \sum_{\rho=1}^r u_\rho \otimes v_\rho \otimes w_\rho$. Let $\alpha \colon k^r \to U$, $\beta \colon k^r \to V$, $\gamma \colon k^r \to W$ be k-linear such that $\alpha(e_\rho) = u_\rho$, $\beta(e_\rho) = v_\rho$ and $\gamma(e_\rho) = w_\rho$ for all ρ, where $(e_1, \ldots, e_r)$ denotes the canonical basis of k^r. Then

$$(\alpha \otimes \beta \otimes \gamma) \left(\sum_{\rho=1}^r e_\rho \otimes e_\rho \otimes e_\rho \right) = t,$$

hence $t \leq \langle r \rangle$. The argument is reversible.   □

As a first example we give a reformulation of Lemma (14.10).

(14.30) Remark. Let (U, V, W, ϕ) and (U', V', W', ϕ') be bilinear maps. Suppose that $\gamma \circ \phi = \phi' \circ (\alpha \times \beta)$. Then $\phi' \leq \phi$ if α and β are surjective. If γ is injective, then $\phi \leq \phi'$.   •

(14.31) Example. If A and B are k-algebras, then $A \subseteq B$ as k-algebras implies $A \leq B$ as tensors, as is easily verified. The converse, however, is not true: For any k-algebra A there exists $r \in \mathbb{N}'$ such that $A \leq \langle r \rangle$, but $A \not\subseteq k^r$ in general.   •

For matrix tensors we have the following result:

(14.32) Proposition. *Let $e \leq e'$, $h \leq h'$, and $\ell \leq \ell'$ be positive integers. Then $\langle e, h, \ell \rangle \leq \langle e', h', \ell' \rangle$.*

The proof is the same as for Prop. (14.15); it uses the variant of Lemma (14.10) given in Rem. (14.30). In the course of studying the rank of bilinear maps a certain class of them will be of special interest.

(14.33) Definition. Let (U, V, W, ϕ) be a bilinear map.

(1) ϕ is called 1-*concise* if its *left-kernel* $\{u \in U \mid \phi(u, V) = 0\}$ equals zero, it is called 2-*concise* if its *right-kernel* $\{v \in V \mid \phi(U, v) = 0\}$ equals zero, and it is called 3-*concise* if the k-span of $\phi(U, V)$ equals W. ϕ is called *concise* if it is i-concise for $i = 1, 2, 3$.

(2) Let U' and V' be the left- and right-kernel of ϕ, respectively and $W' := \langle \phi(U, V) \rangle$. The concise bilinear map $\tilde{\phi} \in \mathrm{Bil}(U/U', V/V'; W')$ defined by $\tilde{\phi}(u + U', v + V') := \phi(u, v)$ for all $(u, v) \in U \times V$ is called the *concise version* of ϕ. $\qquad \bullet$

We remark that a bilinear map is concise iff it coincides with its concise version.

(14.34) Example. Any k-algebra is concise since it contains the unity. $\qquad \bullet$

The following shows that the restriction to concise bilinear maps is not very serious as far as the rank is concerned.

(14.35) Proposition. *Let $\phi \in \mathrm{Bil}(U, V; W)$ and $\tilde{\phi}$ be the concise version of ϕ. Then $R(\tilde{\phi}) = R(\phi)$.*

(See Ex. 14.14.) Conciseness can be interpreted in terms of the tensor associated with a bilinear map in the following way. For a tensor (U, V, W, t) define U_t as the unique minimal subspace of U such that $t \in U_t \otimes V \otimes W$. Analogously, we define V_t and W_t. Then we have the following.

(14.36) Lemma. (1) *Let (U, V, W, t) be a tensor. If (v_i) and (w_j) are bases of V and W respectively and if $t = \sum_{ij} u_{ij} \otimes v_i \otimes w_j$, then*

$$U_t = \langle (\mathrm{id}_U \otimes g \otimes h)t \mid g \in V^*, h \in W^* \rangle = \sum_{i,j} k u_{ij},$$

(where we have identified $U \otimes k \otimes k$ with U).

(2) *Let $\phi \in \mathrm{Bil}(U, V; W)$ be a bilinear map and $t \in U^* \otimes V^* \otimes W$ be the corresponding tensor. Let U' denote the left-kernel of ϕ. Then $(U^*)_t = (U')^\perp \cong (U/U')^*$ and ϕ is 1-concise iff $(U^*)_t = U^*$.*

Proof. (1) Let $\tilde{U} = \sum_{ij} k u_{ij}$. Then $t \in \tilde{U} \otimes V \otimes W$ and therefore $U_t \subseteq \tilde{U}$. On the other hand $t \in U_t \otimes V \otimes W$ implies $u_{ij} = (\mathrm{id}_U \otimes v_i^* \otimes w_j^*)t \in U_t$ ((v_i^*) and (w_j^*) are the dual bases of (v_i) and (w_j) respectively). Therefore $\tilde{U} \subseteq U_t$.

(2) Follows from (1). $\qquad \square$

(14.37) Definition. Let (U, V, W, t) be a tensor and U_t, V_t, and W_t be defined as above.

(1) t is called 1-*concise* if $U_t = U$, 2-*concise* if $V_t = V$, 3-*concise* if $W_t = W$, and *concise* if it is i-concise for $i = 1, 2, 3$.

(2) $\tilde{t} := (U_t, V_t, W_t, t)$ is called the *concise version* of t. $\qquad \bullet$

Note that a tensor is concise iff it coincides with its concise version. In particular, the concise version of a tensor is concise. The following remark, which is an easy consequence of the definition, provides a first tool for proving lower bounds for the rank of bilinear maps.

(14.38) Remark. Let $t = \sum_{i=1}^{r} u_i \otimes v_i \otimes w_i \in U \otimes V \otimes W$. If t is 1-concise, then $\langle u_1, \ldots, u_r \rangle = U$, and $R(t) \geq \dim U$. Similar assertions hold if t is 2- or 3-concise. •

We can completely characterize n-dimensional k-algebras of rank n.

(14.39) Proposition. (1) *Let U, V, and W be k-spaces and $s, t \in U \otimes V \otimes W$. Suppose that $s \leq t$ and s is concise. Then $s \simeq t$.*

(2) *Let A be an n-dimensional k-algebra. Then $R(A) = n$ if and only if $A \simeq k^n$ as k-algebras.*

Proof. (1) We have k-endomorphisms $\alpha \in \text{End}(U)$, $\beta \in \text{End}(V)$, and $\gamma \in \text{End}(W)$ such that $(\alpha \otimes \beta \otimes \gamma)t = s$. The 1-conciseness of s implies that α is surjective. Similarly, β and γ are surjective. Hence they are isomorphisms.

(2) We identify A with a tensor in $k^n \otimes k^n \otimes k^n$. From $R(A) = n$ we deduce from Prop. (14.29) that $A \leq \langle n \rangle$. We obtain from Part (1) (and Example (14.34)) that $A \simeq \langle n \rangle$ as bilinear maps. Now we apply Prop. (14.13) to obtain the assertion. For the other direction note that we have previously shown $R(\langle n \rangle) \leq n$. Since $\langle n \rangle$ is concise, we obtain $R(\langle n \rangle) \geq n$ by Rem. (14.38).   □

Concise versions of tensors have the expected behavior under direct sums and tensor products.

(14.40) Lemma. *Let s, t be tensors. Then*

(1) $\widetilde{s \oplus t} = \tilde{s} \oplus \tilde{t}$,

(2) $\widetilde{s \otimes t} = \tilde{s} \otimes \tilde{t}$,

(3) $s \oplus t$ *is concise* $\Leftrightarrow$ *s and t are concise* $\Leftrightarrow$ $s \otimes t$ *is concise.*

Proof. Statements (1) and (2) follow immediately from the characterization of U_t, V_t, W_t given in Lemma (14.36)(1): if $t = \sum_{i,j} u_{ij} \otimes v_i \otimes w_h$, where $u_{ij} \in U$ and (v_i), (w_j) are bases of V and W, respectively, then

$$U_t = \sum_{i,j} k u_{ij}.$$

(Analogously for V_t and W_t.) The third assertion follows from (1) and (2).   □

An elegant application of the foregoing lemma is the following.

(14.41) Proposition. *Let e, h, and ℓ be positive integers. Then $\langle e, h, \ell \rangle$ is concise. In particular, $R(\langle e, h, 1 \rangle) = eh$.*

Proof. One way of proving the assertion is by verifying it directly. A more elegant way is by applying Prop. (14.26): since

$$\langle e, h, \ell \rangle \otimes \langle h\ell, e\ell, eh \rangle \simeq \langle eh\ell, eh\ell, eh\ell \rangle,$$

and the right-hand side is concise (see Example (14.34)), we obtain the assertion by Lemma (14.40)(3).

As for the second claim, note that the trivial algorithm for matrix by vector multiplication induces a bilinear computation for $\langle e, h, 1 \rangle$ of length eh.   □

14.4 Other Characterizations of Rank

In this section we will derive equivalent coordinate-based characterizations of the rank. Let us agree upon the following notation during this section: $(u_i)_{i \le m}$ is the standard basis of k^m with dual basis $(u_i^*)_{i \le m}$, $(v_j)_{j \le n}$ is the standard basis of k^n with dual basis $(v_j^*)_{j \le n}$, and $(e_\ell)_{\ell \le p}$ is the standard basis of k^p.

Let $\phi \in \text{Bil}(k^m, k^n; k^p)$. There exist $t_{ij\ell} \in k$ such that for all i and j we have

$$\phi(u_i, v_j) = \sum_{\ell=1}^{p} t_{ij\ell} e_\ell.$$

The "3-dimensional matrix" $t = (t_{ij\ell})_{i,j,\ell} \in k^{m \times n \times p}$ is called the *coordinate tensor* of ϕ. One might regard t as a cube containing elements of k. We define the *rank* $R(t)$ of t by $R(t) := R(\phi)$.

As an example, the coordinate tensor of $\langle n \rangle$ is the tensor $t = (t_{ij\ell})$ given by $t_{ij\ell} = 1$ if $i = j = \ell$ and $t_{ij\ell} = 0$ else. Hence, t can be regarded as a cube in which only a space diagonal is occupied.

Let $t = (t_{ij\ell})_{i,j,\ell}$ be the coordinate tensor of a bilinear map $\phi \in \text{Bil}(k^m, k^n; k^p)$. Then $(T_i := (t_{ij\ell})_{j,\ell} \in k^{n \times p} \mid i \in \underline{m})$ is called the sequence of 1-*slices* of ϕ (or of t). Analogously, $((t_{ij\ell})_{i,\ell} \mid j \in \underline{n})$ and $((t_{ij\ell})_{i,j} \mid \ell \in \underline{p})$ are called the 2-*slices*, respectively 3-*slices* of ϕ (or t).

(14.42) Example. The 1-slices (T_1, T_2) of the $\mathbb{R}$-algebra $\mathbb{C}$ with respect to the basis $(1, i)$ are given by

$$T_1 = \begin{pmatrix} 1 & 0 \\ 0 & 1 \end{pmatrix}, \qquad T_2 = \begin{pmatrix} 0 & 1 \\ -1 & 0 \end{pmatrix}. \qquad \bullet$$

A sequence $(A_1, \dots, A_m)$ of $n \times p$-matrices over k defines a coordinate tensor $t = (t_{ij\ell})_{i,j,\ell}$ by requiring that this sequence be the sequence of 1-slices (or 2- or 3-slices) of t. To be more specific, let $A_i = (a_{j\ell}^{(i)})_{j,\ell}$. Then we define $t_{ij\ell} := a_{j\ell}^{(i)}$. In this way we can talk about the *rank* $R(A_1, \dots, A_m)$ of the sequence $(A_1, \dots, A_m)$ which is defined as the rank $R(t)$ of t. A base change in any of the spaces k^n, k^p, or k^m does not effect the rank of the bilinear map corresponding to t. Hence

$$R(A_1, \dots, A_m) = R(PA_1Q, \dots, PA_mQ) = R\left(\sum_{\ell=1}^{p} \alpha_{\pi\ell} A_\ell \,\middle|\, \pi \in \underline{p} \right),$$

if $P \in k^{n \times n}$, $Q \in k^{p \times p}$, and $(\alpha_{\pi\ell}) \in k^{m \times m}$ are invertible matrices. In particular, the rank of a sequence $(A_1, \dots, A_m)$ only depends on $\{A_1, \dots, A_m\}$.

Conciseness of a bilinear map can be formulated in terms of its coordinate tensor.

(14.43) Proposition. *Let $\phi \in \text{Bil}(k^m, k^n; k^p)$ have the 1-slices $(T_1, \dots, T_m)$. Then the following assertions are equivalent:*

(1) ϕ is 1-*concise.*

(2) $T_1, \ldots, T_m$ are linearly independent.

(3) $\{(t_{1j\ell}, \ldots, t_{mj\ell}) \mid j \in \underline{n}, \ell \in \underline{p}\}$ generate k^m.

Analogous assertions hold for 2- and 3-slices of ϕ.

Proof. (1) $\Leftrightarrow$ (2): If ϕ is 1-concise and $\sum_i \alpha_i T_i = 0$ for some $\alpha_1, \ldots, \alpha_m \in k$, then $\phi(u, k^n) = 0$, where $u = \sum_i \alpha_i u_i$. It follows that $u = 0$, hence $\alpha_1, \ldots, \alpha_m = 0$. The argument is reversible.

(2) $\Leftrightarrow$ (3): Obvious.   □

The rank of a bilinear map can be characterized in terms of its associated coordinate tensor.

(14.44) Proposition. *Let $\phi \in \mathrm{Bil}(k^m, k^n; k^p)$ have coordinate tensor $t = (t_{ij\ell})_{i,j,\ell}$. Then $R(\phi) \leq r$ if and only if there exist vectors $(\alpha_{\rho 1}, \ldots, \alpha_{\rho m}) \in k^m$, $(\beta_{\rho 1}, \ldots, \beta_{\rho n}) \in k^n$, $(\gamma_{\rho 1}, \ldots, \gamma_{\rho p}) \in k^p$, $\rho \in \underline{r}$, such that for all $i \in \underline{m}$, $j \in \underline{n}$, and $\ell \in \underline{p}$ we have*

$$t_{ij\ell} = \sum_{\rho=1}^r \alpha_{\rho i} \beta_{\rho j} \gamma_{\rho \ell}.$$

Proof. Suppose that $t_{ij\ell}$ is representable as above. For $\rho \in \underline{r}$ set $f_\rho := \sum_{\mu=1}^m \alpha_{\rho\mu} u_\mu^*$, $g_\rho := \sum_{\nu=1}^n \beta_{\rho\nu} v_\nu^*$, and $w_\rho := \sum_{\pi=1}^p \gamma_{\rho\pi} e_\pi$. Then it is easily verified that $(f_1, g_1, w_1; \ldots; f_r, g_r, w_r)$ is a bilinear computation for ϕ. Reversing the argument, we can construct the vectors given above from a bilinear computation for ϕ.   □

It is sometimes advantageous to work with trilinear forms instead of bilinear maps and their coordinate tensors. Let $X_1, \ldots, X_m, Y_1, \ldots, Y_n$, and $Z_1, \ldots, Z_p$ be indeterminates over k. To the coordinate tensor $t = (t_{ij\ell})_{i,j,\ell}$ of $\phi \in \mathrm{Bil}(k^m, k^n; k^p)$ we associate the trilinear form

$$\tau(X, Y, Z) = \sum_{i,j,\ell} t_{ij\ell} X_i Y_j Z_\ell \in k[X, Y, Z].$$

(14.45) Proposition. *Let $\phi \in \mathrm{Bil}(k^m, k^n; k^p)$ have the coordinate tensor $t = (t_{ij\ell})_{i,j,\ell}$ and the 1-slices $(T_1, \ldots, T_m)$. Then $R(\phi) \leq r$ is equivalent to each of the following assertions:*

(1) *There exist matrices $\Delta_1, \ldots, \Delta_r \in k^{n \times p}$ of rank one such that $\{T_1, \ldots, T_m\} \subseteq \langle \Delta_1, \ldots, \Delta_r \rangle$.*

(2) *There exist diagonal matrices $D_1, \ldots, D_m \in k^{r \times r}$ and matrices $B \in k^{n \times r}$ and $C \in k^{r \times p}$ such that $T_i = B D_i C$ for all $i \in \underline{m}$.*

(3) *There exist linear forms f_ρ in X, g_ρ in Y, and h_ρ in Z, $\rho \in \underline{r}$, such that*

$$\sum_{i,j,\ell} t_{ij\ell} X_i Y_j Z_\ell = \sum_{\rho=1}^r f_\rho(X) g_\rho(Y) h_\rho(Z).$$

Analogous statements hold for the 2- and the 3-slices of ϕ.

Proof. (1) Applying Prop. (14.44) we obtain

$$T_i = (t_{ij\ell})_{j,\ell} = \sum_{\rho=1}^{r} \alpha_{\rho i}(\beta_{\rho j}\gamma_{\rho\ell})_{j,\ell} =: \sum_{\rho=1}^{r} \alpha_{\rho i}\Delta_\rho.$$

Conversely, any rank-one matrix $\Delta \in k^{n\times p}$ can be written as $(\beta_j\gamma_\ell)_{j,\ell}$.

(2) Apply Prop. (14.44) and set $D_i := \mathrm{diag}(\alpha_{1i},\ldots,\alpha_{ri})$, $B := (\beta_{\rho j})_{j,\rho}$, and $C := (\gamma_{\rho\ell})_{\rho,\ell}$. Conversely, the existence of the matrices D_i, B, and C implies the existence of a bilinear computation of length r for ϕ by reversing the above arguments and applying Prop. (14.44).

(3) Apply Prop. (14.44) and set

$$f_\rho(X) := \sum_{i=1}^{m} \alpha_{\rho i}X_i, \quad g_\rho(Y) := \sum_{j=1}^{n} \beta_{\rho j}Y_j, \quad g_\rho(Z) := \sum_{\ell=1}^{p} \gamma_{\rho\ell}Z_\ell.$$

Again, the argument is reversible. □

(14.46) Example. Prop. (14.45)(2) shows that for any matrix A we have $R(A) = \mathrm{rk}\, A$. ●

14.5 Rank of the Polynomial Multiplication

This section serves as a first example for the application of the tools from the previous sections to obtain estimates for the rank of polynomial multiplication. In Chap. 17 we will derive corresponding lower bounds for the problems discussed here.

Let X be an indeterminate over k; for $n \in \mathbb{N}'$ denote by $k[X]_{<n}$ the k-space of polynomials of degree less than n. Let m and n be two positive integers. The restriction of the polynomial multiplication to $k[X]_{<m} \times k[X]_{<n}$ gives a concise bilinear map $\Phi_k^{m,n}: k[X]_{<m} \times k[X]_{<n} \to k[X]_{<m+n-1}$.

(14.47) Proposition. *We have $R(\Phi_k^{m,n}) \geq m+n-1$ with equality holding if k has at least $m+n-2$ elements.*

Proof. We first multiply the highest coefficients of the polynomials involved, and then apply Ex. 2.4 (Interpolation!). This gives a bilinear computation of length $m+n-1$ if $|k| \geq m+n-2$ and proves the upper bound. The lower bound follows from the 3-conciseness of $\Phi_k^{m,n}$. (See Rem. (14.38)). □

(14.48) Proposition (Fiduccia and Zalcstein). *Let $f \in k[X]$ be a univariate polynomial of degree m.*

(1) *The k-algebra $k[X]/(f)$ is a restriction of $\Phi_k^{m,m}$.*

(2) *Let $f = p_1^{\nu_1} \cdots p_r^{\nu_r}$, where the p_i are pairwise non-associated irreducible polynomials in $k[X]$ and suppose that $|k| \geq 2m - 2$. Then $R(k[x]/(f)) \leq 2m - r$.*

(3) *Let $K \supset k$ be a simple field extension of k of degree m and $|k| \geq 2m - 2$. Then $R(K) \leq 2m - 1$.*

Proof. (1) Let $\kappa: k[X] \to k[X]/(f)$ be the canonical map, κ_1 be the restriction of κ to $k[X]_{<m}$ and κ_2 be the restriction of κ to $k[X]_{<2m-1}$. Then κ_i, $i = 1, 2$, are surjective k-space morphisms and we have the following commutative diagram:

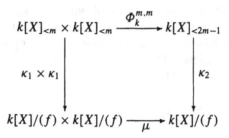

where μ is the multiplication in $k[X]/(f)$. Now Rem. (14.30) implies the assertion.

(2) As $k[X]/(f) \simeq k[X]/(p_1^{\nu_1}) \oplus \cdots \oplus k[X]/(p_r^{\nu_r})$ by the Chinese remainder theorem, we obtain the assertion from (1) and the subadditivity of rank (Prop. (14.23)(1)).

(3) There exists an irreducible polynomial $p \in k[X]$ of degree m such that $K \simeq k[X]/(p)$. We obtain the result by (2).   □

(14.49) Remark. In part (2) of the above proposition it suffices to require that k has at least $2\ell - 2$ elements, where $\ell := \max\{\nu_i \deg p_i \mid i \in \underline{r}\}$.   •

14.6* The Semiring $\mathcal{T}$

In this section we give an elegant and economic description of the concepts developed so far by introducing a language in which assertions on the rank of bilinear maps and tensors can be formulated in a concise manner. We will use the language developed in this section later in Chap. 15 where we investigate the asymptotic complexity of matrix multiplication.

We have already seen that the isomorphism classes of tensors form a commutative semiring. We would like to restrict ourselves to a sub-semiring by considering concise tensors.

(14.50) Definition. A tensor (U, V, W, t) is called a *null tensor* iff $t = 0$. Two tensors s, t are called *equivalent*, in symbols $s \sim t$, iff there are null tensors s_0, t_0 such that $s \oplus s_0 \simeq t \oplus t_0$. Two bilinear maps are called *equivalent* iff their structural tensors are equivalent.   •

Note that $\sim$ is an equivalence relation on the set of isomorphism classes of tensors and that we have $s \sim t$ iff $\tilde{s} \simeq \tilde{t}$. Moreover $t \sim \langle 0 \rangle$ iff t is a null tensor. Thus the equivalence classes of tensors correspond bijectively to the isomorphism classes of concise tensors. We define the semiring $\mathcal{T}$ by

$$\mathcal{T} := \mathcal{T}(k) \quad := \quad \{\text{isomorphism classes of concise tensors}\}$$
$$= \quad \{\text{equivalence classes of tensors}\}.$$

From the preceding reasonings (Lemma (14.40)) it is clear that $\mathcal{T}$ is a commutative semiring, where the addition is induced by the direct sum, the multiplication is induced by the tensor product, 0 is the class of null tensors, and 1 is the class of the multiplication in k. We can embed $\mathbb{N}$ into $\mathcal{T}$ by mapping r to the equivalence class of $\langle r \rangle$. $\mathcal{T}$ is called the *semiring of tensor classes*. It is clear that the action of S_3 on the set of isomorphism classes of tensors induces an action of S_3 on $\mathcal{T}$ by semiring automorphisms. We remark that it is possible to embed $\mathcal{T}$ in a commutative ring (compare Ex. 14.24).

In the following we will not distinguish notationally between concise tensors and their equivalence classes. For example, $\langle e, h, \ell \rangle$ will denote both the matrix multiplication map as well as its equivalence class in $\mathcal{T}$. By considering formats it is easy to check that distinct triples (e, h, ℓ) with $e, h, \ell \geq 1$ lead to distinct classes $\langle e, h, \ell \rangle$.

The following proposition shows that $\leq$ can be interpreted as a partial order on $\mathcal{T}$.

(14.51) Proposition. *Let* (U, V, W, t) *and* (U', V', W', t') *be tensors. Then*

$$t \leq t', \ t' \leq t \iff t \sim t'.$$

Proof. Observe first that $t \leq \tilde{t}$ and $\tilde{t} \leq t$. So we may assume that both t and t' are concise. Then "$\Leftarrow$" is trivial. Conversely, let $(\alpha \otimes \beta \otimes \gamma)t' = t$. Then $t \in \text{im}\,\alpha \otimes \text{im}\,\beta \otimes \text{im}\,\gamma$, hence α, β, γ are surjective because t is concise. Therefore $\dim U \leq \dim U'$, $\dim V \leq \dim V'$, $\dim W \leq \dim W'$. By symmetry we have in fact equalities, hence α, β, γ are isomorphisms. $\qquad \square$

Since the rank function R is characterized by $\leq$ (Prop. (14.29)), it also factors through $\mathcal{T}$. Finally, we note that $s \leq s', t \leq t'$ imply $s \oplus t \leq s' \oplus t', s \otimes t \leq s' \otimes t'$. (See Rem. (14.28).) Let us summarize these facts as follows:

(14.52) Proposition. $\mathcal{T}$ *is a commutative semiring containing* $\mathbb{N}$. *The restriction* $\leq$ *is a partial order on* $\mathcal{T}$ *inducing the natural order on* $\mathbb{N}$, *which satisfies*

$$\forall a, a', b, b' \in \mathcal{T}: \quad a \leq b, \ a' \leq b' \implies a + a' \leq b + b', \ aa' \leq bb'.$$

Moreover

$$\forall a, b \in \mathcal{T}, \ \pi \in S_3: \quad a \leq b \implies \pi a \leq \pi b,$$

and for $a \in \mathcal{T}$

$$R(a) = \min\{r \in \mathbb{N} \mid a \leq r\}.$$

We remark that this proposition implies the well-known properties of the rank function R, such as subadditivity, submultiplicativity, and S_3-invariance. Moreover, R is monotone with respect to $\leq$.

14.7 Exercises

14.1. Let k be a field of characteristic different from two and V, W be k-spaces. Show that a mapping $\Phi: V \to W$ is quadratic iff the following hold:

- For all $\alpha \in k$ and all $v \in V$ we have $\Phi(\alpha v) = \alpha^2 \Phi(v)$, and
- The mapping $\Delta\Phi: V \times V \to W$ defined by $\Delta\Phi(v_1, v_2) := \frac{1}{2}(\Phi(v_1 + v_2) - \Phi(v_1) - \Phi(v_2))$ is bilinear.

14.2. A bilinear map $\phi: U \times V \to W$ is a quadratic map from $U \times V$ to W.

14.3. Let V, W, V', W' be k-spaces, $\alpha: V' \to V$, $\beta: W \to W'$ be homomorphisms, and $\Phi: V \to W$ be quadratic. Show that $\beta \circ \Phi \circ \alpha$ is also quadratic. Deduce that the property of being quadratic does not depend on specific bases.

14.4. Let V and W be k-spaces and $\Phi: V \to W$ be quadratic. Show that Φ and bases in V and W uniquely determine the coordinate functions of Φ. (This is obvious for infinite fields.)

14.5. Verify that Strassen's algorithm is a bilinear computation of length 7 for the algebra $k^{2 \times 2}$. (See Chap. 1.)

14.6. Regarding $\mathbb{C}$ as an $\mathbb{R}$-algebra, give a bilinear algorithm of length 3 for $\mathbb{C}$. (See also Prop. (14.48).)

14.7. Prove Rem. (14.3)(1).

14.8. Prove Prop. (14.4).

14.9. Let U, V, and W be k-spaces. Prove that $U \otimes V \simeq V \otimes U$ and $U \otimes k \simeq U$, $U \otimes V \otimes W \simeq (U \otimes V) \otimes W$ canonically.

14.10. Prove the assertion in (14.21).

14.11. Prove Prop. (14.23).

14.12. Give an example of bilinear maps ϕ_1, ϕ_2 over k such that $R(\phi_1 \otimes \phi_2) < R(\phi_1)R(\phi_2)$.

14.13. Prove Rem. (14.28).

14.14. Prove Prop. (14.35). (Hint: use Lemma (14.10).)

14.15.* Let U, V, and W be k-spaces and let $\Phi: U \times V \to W$ be a concise bilinear map. Define the bilinear map $\phi' \in \mathrm{Bil}(U \oplus V, U \oplus V; W)$ by $\phi'((u_1, v_1), (u_2, v_2)) := \phi(u_1, v_2) + \phi(u_2, v_1)$.

of G by $E^{\mathsf{T}} := \{(w, v) \mid (v, w) \in E\}$, $I^{\mathsf{T}} = O$, $O^{\mathsf{T}} = I$, and $\lambda^{\mathsf{T}}(w, v) := \lambda(v, w)$ for every $(v, w) \in E$. In other words, G^{T} results from G by reversing the arrows without changing the weights. Thus G^{T} computes n linear forms $X_1^{\mathsf{T}}, \ldots, X_n^{\mathsf{T}}$ in m indeterminates $f_1^{\mathsf{T}}, \ldots, f_m^{\mathsf{T}}$.

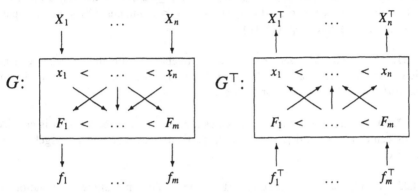

Reversing the arrows in a path $p \in G(v, w)$ yields a path $p^{\mathsf{T}} \in G^{\mathsf{T}}(w, v)$. This defines a natural bijection $G(v, w) \to G^{\mathsf{T}}(w, v)$ satisfying $\lambda(p) = \lambda^{\mathsf{T}}(p^{\mathsf{T}})$. According to Rem. (13.19), $f_i = \sum_{j=1}^{n} (\sum_{p \in G(x_j, F_i)} \lambda(p)) X_j$, thus $a_{ij} = \sum_{p \in G(x_j, F_i)} \lambda(p)$. On the other hand,

$$X_j^{\mathsf{T}} = \sum_{i=1}^{m} \left(\sum_{p^{\mathsf{T}} \in G^{\mathsf{T}}(F_i, x_j)} \lambda^{\mathsf{T}}(p^{\mathsf{T}}) \right) f_i^{\mathsf{T}}$$

$$= \sum_{i=1}^{m} \left(\sum_{p \in G(x_j, F_i)} \lambda(p) \right) f_i^{\mathsf{T}} = \sum_{i=1}^{m} a_{ij} f_i^{\mathsf{T}}.$$

Therefore, G^{T} computes the linear forms which correspond to A^{T}. By Lemma (13.17),

$$\begin{aligned} L_\infty(A^{\mathsf{T}}) \le c(G^{\mathsf{T}}) &= |\{e^{\mathsf{T}} \in E^{\mathsf{T}} \mid \lambda^{\mathsf{T}}(e^{\mathsf{T}}) \ne \pm 1\}| + |E^{\mathsf{T}}| - |V| + m \\ &= |\{e \in E \mid \lambda(e) \ne \pm 1\}| + |E| - |V| + n - n + m \\ &= L_\infty(A) - n + m. \end{aligned}$$

Applying the same argument to A^{T}, we obtain $L_\infty(A) \le L_\infty(A^{\mathsf{T}}) - m + n$. Altogether, this proves the non-degenerate case. In the general case, deleting the zero rows and columns of A results in a non-degenerate $(m - z(A)) \times (n - z(A^{\mathsf{T}}))$-matrix B satisfying $L_\infty(B) = L_\infty(A)$ and $L_\infty(B^{\mathsf{T}}) = L_\infty(A^{\mathsf{T}})$ (check this!). Combining this with the fact that $L_\infty(B^{\mathsf{T}}) = L_\infty(B) - (n - z(A^{\mathsf{T}})) + (m - z(A))$, the general claim follows.   □

The last theorem remains valid, if K is the field of complex numbers and L_∞ is replaced by L_c, $2 \le c < \infty$. An interesting corollary of the last theorem is the determination of the linear complexity of a single linear form:

(13.21) Corollary. $L_\infty(\sum_{i=1}^{n} a_i X_i) = n - 1 + |\{|a_1|, \ldots, |a_n|\} \setminus \{1\}|$, if all a_i are nonzero.

(2) $L(\langle e, 2t, e \rangle) \leq t(e^2 + 2e - 1)$, $L(\langle e, 2t + 1, e \rangle) \leq t(e^2 + 2e - 1) + e^2$.
(3) $L(\langle 3, 3, 3 \rangle) \leq 23$. (This is the best known upper bound on the complexity of 3×3-matrix multiplication.)

Note that (3) does not imply that $R(\langle 3, 3, 3 \rangle) \leq 23$, see also Ex. 15.3. Note further that it it is not clear that $L(\langle h, h, h \rangle) \leq r$ implies $h^\omega \leq r$. (Compare Rem. (14.24)(2) and Prop. (15.3).)

14.21. A module M over an associative algebra A with unit element can be viewed as a bilinear map $A \times M \to M$. Show that it is concise iff M is faithful, i.e., $aM = 0$ implies $a = 0$ for all $a \in A$.

14.22. A Lie algebra $\mathcal{L}$ can be viewed as a bilinear map. Show that it is concise iff its center is zero and $[\mathcal{L}, \mathcal{L}] = \mathcal{L}$. In particular, semi-simple Lie algebras are concise.

14.23. Show that the restriction order $\leq$ in $\mathcal{T}$ does not allow cancellation of nonzero elements, i.e., find $a, b, c \in \mathcal{T}$, $c \neq 0$, such that $ac \leq bc$, but $a \not\leq b$.

14.24.* The aim of this exercise is to show that the semiring $\mathcal{T}$ can be embedded into a ring.

(1) A tensor is called *indecomposable* iff it is not null but $t \sim t_1 \oplus t_2$ implies that either t_1 or t_2 is null. Show that t is indecomposable iff its concise version $\tilde{t}$ is.
(2) Let $s_1, \ldots, s_q, t_1, \ldots, t_r$ be indecomposable tensors such that

$$s_1 \oplus \cdots \oplus s_q \sim t_1 \oplus \cdots \oplus t_r.$$

Show that necessarily $q = r$ and there is a permutation π such that $s_j \sim t_{\pi j}$ for all j.
(3) The semiring $\mathcal{T}$ can be embedded into a ring $\mathcal{B}$.

14.25. Show that the restriction order cannot be extended to the ring $\mathcal{B}$ in a way compatible with the ring operations. (Hint: consider matrix tensors $\langle e, h, 1 \rangle$ and their permutations.)

14.26. Let A be a semisimple associative algebra. Show that A is indecomposable as a bilinear map iff A is simple. (Hint: use Prop. (14.25) and Wedderburn's theorem [316, XVII].)

14.8 Open Problems

Problem 14.1. Decide whether for a finite set $Q \subset k[X_1, \ldots, X_n]$ of quadratic forms over a finite field k one has $L_{k(X)}(Q) = L_{k[X]}(Q)$.

Problem 14.2. Is the additivity conjecture true? (Cf. Rem. (14.24)(1).)

Problem 14.3. Let $(f_1, g_1, w_1; \ldots; f_R, g_R, w_R)$ be an optimal bilinear algorithm for $\phi := \phi_1 \oplus \cdots \oplus \phi_s$, where $\phi_i \in \mathrm{Bil}(U_i, V_i; W_i)$ for $i \in \underline{s}$. The *extended direct sum conjecture* states that then for all $i = 1, \ldots, s$ we have

$$(f_i, g_i, w_i) \in \bigcup_{j \leq s} (U_j^* \times V_j^* \times W_j).$$

Is this conjecture true?

14.9 Notes

The language we have introduced in this chapter is due to Strassen. Sect. 14.1–14.3 have been taken more or less completely from his papers [498, 508]. To some extent, we have also followed de Groote's monograph [219].

The study of the complexity of bilinear problems started with Strassen's discovery [493] of his 2×2-matrix multiplication algorithm. Ever since, matrix multiplication has retained its role as a leading problem in this field. Prior to Strassen's discovery, Winograd [551] had found that $L(\langle n, n, n \rangle) \leq n^3/2 + n^2 - n/2$ for even n (cf. Ex. 14.20). If this algorithm could be used recursively, then choosing $n = 5$ would yield $L(\langle n, n, n \rangle) = O(n^{2.76})$ which would be asymptotically better than Strassen's algorithm. However, Winograd's scheme requires the commutativity of the entries of the matrices to be multiplied, so straightforward recursion is not possible. Due to this fact, quadratic algorithms are called by many authors algorithms which use commutativity of the indeterminates (*commutative algorithms*), while bilinear algorithms are referred to as algorithms which do not use the commutativity (*noncommutative algorithms*).

Pan [406] and independently Hopcroft and Musinski [253] noticed the invariance of the rank of matrix multiplication under the action of S_3; they used this property (also called duality) to transform fast matrix multiplication algorithms from one format to another. It was noticed by Ja'Ja [267] that this property may not be shared by the multiplicative complexity, as is reported in Rem. (14.22). The assertion $L(\langle n, 2, 2 \rangle) \leq 3n + 1$ in that remark is due to Feig [161]. A formulation of the computational complexity of matrix multiplication similar to Strassen's definition of the tensorial rank can also be found in Gastinel [182]. The elegant notation $\langle e, h, \ell \rangle$ for matrix multiplication is due to Schönhage [460].

The results of Sect. 14.4 have been noticed by several authors and those of Sect. 14.5 have appeared for the first time in the paper of Fiduccia and Zalcstein [171]. In this paper the authors also introduce the concept of restriction (though they don't use this name) and prove some of the results in Sect. 14.3. The reader may also wish to consult the book by Heideman [237] for results in this direction, as well as for lower and upper bounds for the complexity of univariate and multivariate polynomial multiplication. (For the latter see also Schönhage [464].)

It is an open problem, whether one has equality in Prop. (14.1) for a finite field k, see Problem 14.1. Note that this proposition does *not* state that all optimal computations for a set of quadratic forms are quadratic. However, Feig [160] has shown that this is true for a single bilinear form. For examples and further investigations we refer the reader to Feig [160, 161] and Tolmieri [519]. Conciseness arguments for proving simple lower bounds on the rank are also known as column-rank theorems and row-rank theorems; see, e.g., Winograd [552] and Ex. 6.7. More sophisticated lower bounds based on similar arguments will be studied in Chap. 17. For other lower bound proofs in the context of bilinear complexity the reader may wish to consult van Leeuwen and van Emde Boas [320].

The additivity conjecture (see Rem. (14.24)(1)) is due to Strassen [498], the extended direct sum conjecture (see Problem 14.3) has been formulated by Winograd [557]. Several papers discuss the additivity conjecture as well as the extended direct sum conjecture, see for instance Auslander et al. [17], Bshouty [87, 90, 92], Feig and Winograd [163, 165], and Ja'Ja and Takche [271]. Strassen [508, p. 427] has proved that if the additivity conjecture is true, then a scalar extension can reduce the rank of a bilinear map by at most a factor of $1/4$. (Cf. Ex. 15.7.) This may be helpful in disproving the conjecture over nonclosed groundfields.

The multiplicative complexity of a set of Boolean quadratic forms has been investigated by Mirwald and Schnorr [369, 370, 371]. For Boolean quadratic forms $f_1, \ldots, f_s$ they define the multiplicative complexity $L(f_1, \ldots, f_s)$ as the minimum number of AND-gates that are sufficient to evaluate $f_1, \ldots, f_s$ over the basis $\{\text{AND}, \text{XOR}, 1\}$. Similarly, the level-one multiplicative complexity $L_1(f_1, \ldots, f_s)$ is defined as the minimum number of AND-gates that suffice to evaluate $f_1, \ldots, f_s$ over the basis $\{\text{AND}, \text{XOR}, 1\}$ by a circuit that has only one level of AND-gates. (L_1 plays the role of R.) For instance, using a counting argument, Mirwald and Schnorr [370] prove that for almost all sets of r n-ary Boolean quadratic forms $f_1, \ldots, f_r$ we have $L(f_1, \ldots, f_r) \geq n\sqrt{r}/3$. The authors also show that the level-one complexity of a set of r n-ary Boolean quadratic forms is bounded from above and from below by the rank of certain $n \times n \times r$-tensors over $\mathbb{F}_2$ associated to this set. Computing the rank of the latter can be a difficult task: Håstad [224] has shown that deciding whether the rank of a coordinate tensor is less than or equal to r is NP-complete over finite fields and NP-hard over the field of rational numbers. This result had been conjectured by Ja'Ja and Gonzalez [269].

Ex. 14.15 is from Ja'Ja [267], Ex. 14.20 from Winograd [551] and Waksman [541]. Laderman [310] has shown that even $R(\langle 3, 3, 3 \rangle) \leq 23$, cf. Ex. 15.3 in Chap. 15. Ex. 14.21–14.26 are from Strassen [508].

Chapter 15. Asymptotic Complexity
of Matrix Multiplication

This chapter is devoted to the discussion of fast matrix multiplication algorithms. We define the exponent ω of matrix multiplication, a quantity measuring the asymptotic complexity of this problem. Strassen's original algorithm gives $\omega \leq 2.81$ (see Chap. 1). The robustness of the exponent with respect to various cost functions, and its invariance with respect to field extensions is shown. (Further motivation for the investigation of ω is given in Chap. 16.) We introduce the concept of border rank of tensors due to Bini, Capovani, Lotti, and Romani [49], and deduce Schönhage's asymptotic sum inequality [460], which has become one of the main tools for gaining upper estimates of the exponent. Then we present the laser method [508] and its generalization by Coppersmith and Winograd [134], and prove their estimate $\omega < 2.39$. Finally, an extension of the asymptotic sum inequality to the multiplication of partially filled matrices is considered. As an application, we describe Coppersmith's construction of astonishingly rapid algorithms for the multiplication of rectangular matrices.

15.1 The Exponent of Matrix Multiplication

Let k be a field. We denote by

$$M_k(h) := L_{k[X,Y]}^{\text{tot}}\left(\left\{\sum_{j=1}^{h} X_{ij}Y_{j\ell} \;\middle|\; i, \ell \in \underline{h}\right\}\right)$$

the total number of arithmetic operations sufficient for multiplying two $h \times h$-matrices of indeterminates $[X_{ij}]$, $[Y_{j\ell}]$ over k without divisions. An exact determination of $M_k(h)$ seems to be outside the range of methods available at the present time. Instead, we focus on the determination of the asymptotic growth of the function $h \mapsto M_k(h)$; we define the *exponent of matrix multiplication* as

$$\omega(k) := \inf\{\tau \in \mathbb{R} \mid M_k(h) = O(h^\tau)\}.$$

The notation $\omega(k)$ indicates a possible dependency on the underlying field k. However, we will see in Sect. 15.3 that $\omega(k)$ can depend only (if at all) on the characteristic of k. By the trivial estimate $M_k(h) \leq h^2(2h-1)$, and the lower bound $h^2 \leq M_k(h)$ resulting from the dimension bound (4.12), we have $\omega(k) \in [2, 3]$.

Matrix multiplication plays a key role in numerical linear algebra. In Chap. 16 we will show that for instance the problems of matrix inversion, LUP–decomposition, computing the determinant or all coefficients of the characteristic polynomial, finding a basis of the kernel of a matrix or transforming a symmetric matrix to diagonal form all have the "exponent" $\omega(k)$. The exponent is thus a measure for the asymptotic complexity of many problems in computational linear algebra.

Let us show now that the exponent is quite robust with respect to the cost function used in its definition. By (14.8) we know that

$$\frac{1}{2} R(\langle h, h, h \rangle) \leq L(\langle h, h, h \rangle) \leq M_k(h),$$

where $L(\langle h, h, h \rangle)$, resp. $R(\langle h, h, h \rangle)$ are the multiplicative complexity, resp. rank of the bilinear map $\langle h, h, h \rangle \colon k^{h \times h} \times k^{h \times h} \to k^{h \times h}$ of matrix multiplication.

(15.1) Proposition. *For every field k we have*

$$\omega(k) = \inf\{\tau \in \mathbb{R} \mid R(\langle h, h, h \rangle) = O(h^{\tau})\}.$$

Proof. Let θ denote the right-hand side of the above equation and write $M(h) := M_k(h)$ for shortness. By the above observation we have $\theta \leq \omega(k)$.

Let us prove that $\omega(k) \leq \theta$. By the definition of θ there exists for every $\varepsilon > 0$ some $m > 1$ such that

$$r := R(\langle m, m, m \rangle) \leq m^{\theta + \varepsilon}.$$

Since $\langle m, m, m \rangle$ is concise we have $r \geq m^2$. In the sequel we will assume that this inequality is strict. (In fact, the forthcoming Prop. (17.9) implies that this is indeed the case. However, the reader may easily check that after obvious modifications our proof will work also under the assumption $r = m^2$.)

There are linear forms $f_\rho, g_\rho \in (k^{m \times m})^*$ and matrices $w_\rho \in k^{m \times m}$ for $\rho \in \underline{r}$, such that for all $a, b \in k^{m \times m}$

$$ab = \sum_{\rho=1}^{r} f_\rho(a) g_\rho(b) w_\rho.$$

Let A be any (not necessarily commutative) k-algebra. The k-linear forms f_ρ, g_ρ extend in the natural way to (left) A-linear forms on $A^{m \times m}$, and we have for all $a, b \in R^{m \times m}$

$$ab = \sum_{\rho=1}^{r} f_\rho(a) g_\rho(b) w_\rho.$$

This equation shows how to compute the product of two $m \times m$-matrices over A using r multiplications of two elements of A, some additions of elements of A, and some multiplications of elements of k with elements of A. We put now $A = k^{m^i \times m^i}$ for $i \in \mathbb{N}$ and observe that $A^{m \times m} \simeq k^{m^{i+1} \times m^{i+1}}$ as k-algebras (block decomposition). From the preceding observation we obtain the recursion

$$M(m^{i+1}) \leq r M(m^i) + c m^{2i},$$

where $c = c(m, r)$ depends on m and r. Now we think of m (and thus r) as being fixed and solve this recursion in dependence on i using Ex. 2.14. Taking into account that $r > m^2$ we get

$$M(m^i) \leq \alpha r^i + \beta m^{2i},$$

where $\alpha = M(1) + m^2 c/(r - m^2)$, $\beta = -m^2 c/(r - m^2)$. Therefore

$$M(m^i) = O(r^i).$$

Using the monotonicity of the sequence M we obtain for all $h \in \mathbb{N}'$

$$M(h) = O(r^{\log_m h}) = O(h^{\log_m r}) = O(h^{\theta + \varepsilon}).$$

Thus $\omega(k) \leq \theta$ and the proof is complete. $\qquad\qquad\square$

(15.2) Remark. Let $F := \{\sum_{j=1}^h X_{ij} Y_{j\ell} \mid 1 \leq i, \ell \leq h\}$ and assume that k is infinite. Then we have by Prop. (14.1) and (14.8)

$$\frac{1}{2} R(\langle h, h, h \rangle) \leq L_{k(X,Y)}(F) \leq L_{k(X,Y)}^{\text{tot}}(F) \leq L_{k[X,Y]}^{\text{tot}}(F) = M_k(h).$$

Therefore, if we allow also for divisions and replace $M_k(h)$ by $L_{k(X,Y)}^{\text{tot}}(F)$ in the definition of $\omega(k)$, we get the same exponent. $\qquad\qquad\bullet$

15.2 First Estimates of the Exponent

We present here some of the main ideas leading to upper estimates on the exponent $\omega(k)$ of matrix multiplication. Since they work over any field, we omit the dependency on k and write ω instead of $\omega(k)$.

We first take a closer look at Strassen's algorithm described in Chap. 1 (see also Ex. 15.1). This algorithm achieves the multiplication of two 2×2-matrices over any ring with only seven multiplications of ring elements. However, we may view the multiplication of $2^N \times 2^N$-matrices over k as a 2×2-matrix multiplication over the ring of $2^{N-1} \times 2^{N-1}$-matrices by taking block decompositions. So we may *recursively* apply Strassen's algorithm, which is a priori designed for 2×2-matrices, to achieve the multiplication of $2^N \times 2^N$-matrices over k with only 7^N multiplications of field elements. Given any $h \times h$-matrix, we can enlarge it artificially to one of size $2^{\lceil \log h \rceil}$ by filling up with zeros, and apply the above algorithm.

In our formal framework this recursion method can be described as follows. Strassen's algorithm implies

$$R(\langle 2, 2, 2 \rangle) \leq 7.$$

Taking tensor powers we obtain (cf. Prop. (14.23)(2) and Prop. (14.26))

$$R(\langle 2^N, 2^N, 2^N \rangle) = R(\langle 2, 2, 2 \rangle^{\otimes N}) \leq R(\langle 2, 2, 2 \rangle)^N \leq 7^N.$$

Furthermore, we know by Prop. (14.15) that $h \leq h'$ implies $R(\langle h, h, h \rangle) \leq R(\langle h', h', h' \rangle)$. Hence we have for all positive h

$$
\begin{aligned}
R(\langle h, h, h \rangle) &\leq R(\langle 2^{\lceil \log h \rceil}, 2^{\lceil \log h \rceil}, 2^{\lceil \log h \rceil} \rangle) \\
&\leq R(\langle 2, 2, 2 \rangle)^{\lceil \log h \rceil} \\
&\leq 7^{\lceil \log h \rceil},
\end{aligned}
$$

and we obtain $R(\langle h, h, h \rangle) \leq 7 \cdot h^{\log 7}$. Hence $\omega \leq \log 7 < 2.81$. In the same manner one proves the following result.

(15.3) Proposition. *If $R(\langle h, h, h \rangle) \leq r$ for positive integers h, r, then $h^\omega \leq r$.*

(15.4) Example. In Ex. 15.3 we describe a bilinear algorithm for multiplying 3×3-matrices using 23 multiplications. The existence of such an algorithm implies $\omega \leq \log_3 23 < 2.86$, which is worse than the bound 2.81.   •

We can generalize the above idea: also upper bounds on the rank of the multiplication of *rectangular* matrices yield upper bounds for ω. To see this note that $R(\langle e, h, \ell \rangle) \leq r$ implies $R(\langle \ell, e, h \rangle) \leq r$ and $R(\langle h, \ell, e \rangle) \leq r$ (cf. equation (14.21)), hence by Prop. (14.26) we have

$$
R(\langle eh\ell, eh\ell, eh\ell \rangle) = R(\langle e, h, \ell \rangle \otimes \langle \ell, e, h \rangle \otimes \langle h, \ell, e \rangle) \leq r^3.
$$

We thus obtain the following generalization of Prop. (15.3)

(15.5) Proposition. *If $R(\langle e, h, \ell \rangle) \leq r$ for positive integers e, h, ℓ, and r, then $(eh\ell)^{\omega/3} \leq r$.*

A significant step in the endeavor of obtaining good upper bounds on ω is the introduction of the concept of "approximative algorithms." We proceed with an example due to Bini, Capovani, Lotti, and Romani by discussing the bilinear map describing the multiplication of a 2×2-matrix having a zero in the lower right corner with a full 2×2-matrix. Let us denote this bilinear map (or the corresponding coordinate tensor) by the symbol $\boxed{} \times \square$. The trilinear form corresponding to $\boxed{} \times \square$ is

$$
\begin{aligned}
F = \sum_{\substack{1 \leq i, j, \ell \leq 2 \\ (i,j) \neq (2,2)}} X_{ij} Y_{j\ell} Z_{\ell i} &= X_{11} Y_{11} Z_{11} + X_{11} Y_{12} Z_{21} + X_{12} Y_{21} Z_{11} \\
&\quad + X_{12} Y_{22} Z_{21} + X_{21} Y_{11} Z_{12} + X_{21} Y_{12} Z_{22},
\end{aligned}
$$

hence $R(\boxed{} \times \square) \leq 6$. (In fact equality holds, see Ex. 17.10.) Now consider the trilinear form

$$
\begin{aligned}
F_1(\varepsilon) &= (X_{12} + \varepsilon X_{11})(Y_{12} + \varepsilon Y_{22})Z_{21} + (X_{21} + \varepsilon X_{11})Y_{11}(Z_{11} + \varepsilon Z_{12}) \\
&\quad - X_{12} Y_{12}(Z_{11} + Z_{21} + \varepsilon Z_{22}) - X_{21}(Y_{11} + Y_{12} + \varepsilon Y_{21})Z_{11} \\
&\quad + (X_{12} + X_{21})(Y_{12} + \varepsilon Y_{21})(Z_{11} + \varepsilon Z_{22})
\end{aligned}
$$

over a rational function field $k(\varepsilon)$. The rank of the corresponding coordinate tensor is obviously at most five, on the other hand we have

$$\varepsilon^{-1}F_1(\varepsilon) = F + \varepsilon G(\varepsilon)$$

with some trilinear form $G(\varepsilon) \in k[\varepsilon][X, Y, Z]$. For a moment assume $k = \mathbb{C}$ and replace ε by a nonzero complex number of small absolute value. Then we see that ⊡ × ⊡ (having rank six) can be approximated with arbitrary precision by tensors having rank at most five!

This leads to the following definition. We call a tensor $t \in k^{m \times n \times p}$ a *degeneration of order* q of $\langle r \rangle$, in symbols $t \trianglelefteq_q \langle r \rangle$, iff there exist vectors $u_\rho(\varepsilon) \in k[\varepsilon]^m$, $v_\rho(\varepsilon) \in k[\varepsilon]^n$, $w_\rho(\varepsilon) \in k[\varepsilon]^p$ for $1 \leq \rho \leq r$ such that

(15.6)
$$\varepsilon^{q-1}t + \varepsilon^q t'(\varepsilon) = \sum_{\rho=1}^{r} u_\rho(\varepsilon) \otimes v_\rho(\varepsilon) \otimes w_\rho(\varepsilon)$$

for some $t'(\varepsilon) \in k[\varepsilon]^{m \times n \times p}$. In this terminology, the above observation can be written as ⊡ × ⊡ $\trianglelefteq_2$ $\langle 5 \rangle$. Before proceeding further with our example, let us explore some of the properties of $\trianglelefteq_q$.

We first show that it is possible to obtain exact algorithms from approximate ones, as discovered by Bini. Let $t \trianglelefteq_q \langle r \rangle$, assume a representation as in (15.6) and write

$$u_\rho(\varepsilon) = \sum_\mu u_\rho^{(\mu)} \varepsilon^\mu, \quad v_\rho(\varepsilon) = \sum_\nu v_\rho^{(\nu)} \varepsilon^\nu, \quad w_\rho(\varepsilon) = \sum_\lambda w_\rho^{(\lambda)} \varepsilon^\lambda,$$

where $u_\rho^{(\mu)} \in k^m$, $v_\rho^{(\nu)} \in k^n$, $w_\rho^{(\lambda)} \in k^p$. By multiplying out we obtain from (15.6) for the coefficient t of ε^{q-1}

$$t = \sum_{\rho=1}^{r} \sum_{\mu, \nu, \lambda} u_\rho^{(\mu)} \otimes v_\rho^{(\nu)} \otimes w_\rho^{(\lambda)},$$

where the second sum is over all (μ, ν, λ) with $\mu + \nu + \lambda = q - 1$. We thus see that

(15.7)
$$t \trianglelefteq_q \langle r \rangle \implies R(t) \leq (q(q+1)/2)r \leq q^2 r.$$

By performing the polynomial multiplication appearing above with the help of Lagrange interpolation, one can even show that $R(t) \leq (2q-1)r$ follows, provided k is infinite (cf. Thm. (2.9) and Prop. (15.26)).

The following properties of $\trianglelefteq_q$ are easy to verify (a proof will be given in Lemma (15.24)).

(15.8)
$$t_1 \trianglelefteq_{q_1} \langle r_1 \rangle, \ t_2 \trianglelefteq_{q_2} \langle r_2 \rangle \implies t_1 \otimes t_2 \trianglelefteq_{q_1+q_2-1} \langle r_1 r_2 \rangle.$$

The smallest natural number r such that $t \trianglelefteq_q \langle r \rangle$ for some $q \in \mathbb{N}'$ is called the *border rank* of t and is denoted by $\underline{R}(t)$. The border rank shares some of the properties of the rank: it is subadditive, submultiplicative, and if π is a permutation in S_3, then $\underline{R}(t) = \underline{R}(\pi t)$. These facts immediately follow from the definitions.

Now let us come back to our example. Similarly as $\boxed{} \times \square \leq_2 \langle 5 \rangle$ one shows that $\boxed{} \times \square \leq_2 \langle 5 \rangle$. Putting these "approximative algorithms" together, we obtain that $\boxed{} \times \square \leq_2 \langle 10 \rangle$. Hence

$$\langle 3, 2, 2 \rangle \leq_2 \langle 10 \rangle.$$

Using symmetrization and observing (15.8) we see that

$$\langle 12, 12, 12 \rangle \simeq \langle 3, 2, 2 \rangle \otimes \langle 2, 3, 2 \rangle \otimes \langle 2, 2, 3 \rangle \leq_4 \langle 1000 \rangle.$$

By switching to exact algorithms with help of (15.7) we get $R(\langle 12, 12, 12 \rangle) \leq 10000$ (or $R(\langle 12, 12, 12 \rangle) \leq 7000$ when using interpolation), which is even worse than the estimate $R(\langle 12, 12, 12 \rangle) \leq 1728$ given by the trivial algorithm. But now we can exploit the idea of recursion: using (15.8) we obtain for all $N \in \mathbb{N}'$

$$\langle 12^N, 12^N, 12^N \rangle \leq_{3N+1} \langle 10^{3N} \rangle,$$

which implies $R(\langle 12^N, 12^N, 12^N \rangle) \leq (3N + 1)^2 \cdot 10^{3N}$ by (15.7). Therefore, by Prop. (15.3), $12^{N\omega} \leq (3N + 1)^2 \cdot 10^{3N}$, and by taking Nth roots and letting N tend to infinity we conclude $12^\omega \leq 1000$. Thus we obtain the following result.

(15.9) Proposition (Bini et al., 1979). $\omega \leq \log_{12} 1000 < 2.78$.

Moreover the above argument shows that in Prop. (15.5) we may replace the rank by the border rank.

(15.10) Proposition (Bini). *If* $\underline{R}(\langle e, h, \ell \rangle) \leq r$ *for positive integers* e, h, ℓ, r*, then we have* $(eh\ell)^{\omega/3} \leq r$.

Another major tool for estimating ω is the subsequent theorem, called asymptotic sum inequality, which is a generalization of Prop. (15.10) to finitely many summands.

(15.11) Asymptotic Sum Inequality (Schönhage). *For positive integers* $r, s, e_1, h_1, \ell_1, \ldots, e_s, h_s, \ell_s$ *we have*

$$\underline{R}\left(\bigoplus_{i=1}^{s} \langle e_i, h_i, \ell_i \rangle \right) \leq r \implies \sum_{i=1}^{s} (e_i h_i \ell_i)^{\omega/3} \leq r.$$

If border rank were additive, then the above theorem would follow immediately from Prop. (15.10). However, we have the following result.

(15.12) Schönhage's Example. *Let* $e, \ell \in \mathbb{N}'$*. Then*

$$\underline{R}(\langle e, 1, \ell \rangle \oplus \langle 1, (e-1)(\ell-1), 1 \rangle) = e\ell + 1.$$

We shall see that for a concise tensor of format (m, n, p) the maximum of the numbers m, n, p is a lower bound for the border rank. (See Lemma (15.23).) This shows that

$$\underline{R}(\langle e, 1, \ell\rangle) = e\ell, \quad \underline{R}(\langle 1, h, 1\rangle) = h.$$

Hence the border rank is not additive! This discovery moreover indicates that the rank might not be additive either.

Before proving the asymptotic sum inequality (15.11) and Schönhage's example (15.12) we are going to discuss the concept of degeneration more thoroughly, which will be done in the subsequent two sections. Let us finish our presentation here by the following estimate of ω.

(15.13) Proposition (Schönhage, 1981). $\omega < 2.55$.

Proof. We apply the asymptotic sum inequality (15.11) to Schönhage's example (15.12) and obtain

$$(e\ell)^{\omega/3} + ((e - 1)(\ell - 1))^{\omega/3} \leq e\ell + 1.$$

A numerical calculation (cf. Ex. 15.16) shows that the best estimate is obtained for $e = \ell = 4$, namely $\omega < 2.55$.   □

15.3 Scalar Restriction and Extension

In order to be able to define the concept of degeneration in a coordinate free way, we have to generalize the discussion of k-bilinear maps and tensors in Chap. 14 by allowing an arbitrary commutative ring $\mathcal{R}$ in place of the ground field k. We will give a detailed development only for bilinear maps, but everything carries over to tensors after obvious modifications.

In the following, we will understand by an *$\mathcal{R}$-bilinear map* always an $\mathcal{R}$-bilinear map between *finitely generated free* $\mathcal{R}$-modules. A *tensor over $\mathcal{R}$* denotes a quadruple (U, V, W, t) where U, V, W are finitely generated free $\mathcal{R}$-modules and $t \in U \otimes V \otimes W$. Direct sum, tensor product, and equivalence of $\mathcal{R}$-bilinear maps are defined as in the Sect. 14.2 and Sect. 14.6. The equivalence classes of $\mathcal{R}$-bilinear maps form a commutative semiring with respect to $\oplus, \otimes$ and the classes of $\langle 0\rangle, \langle 1\rangle$. (Again we define $\langle r\rangle = \mathcal{R}^r$ for $r \in \mathbb{N}$.) The notion of restriction is the same as in Sect. 14.3. The resulting relation $\leq$ is a preorder satisfying

$$\phi \leq \phi', \ \psi \leq \psi' \implies \phi \oplus \psi \leq \phi' \oplus \psi', \ \phi \otimes \psi \leq \phi' \otimes \psi',$$

and

$$\phi \sim \phi', \ \psi \sim \psi', \ \phi \leq \psi \implies \phi' \leq \psi',$$

where ϕ, ϕ', ψ, ψ' are $\mathcal{R}$-bilinear maps.

Next we generalize the operations of scalar restriction and extension, which are well-known for modules, to bilinear maps. (Cf. Bourbaki [69, Chap. I, §5].)

Let $\rho: S \to \mathcal{R}$ be a morphism of commutative rings such that $\mathcal{R}$ is a finitely generated free S-module via ρ. Then, given an $\mathcal{R}$-bilinear map $\phi: U \times V \to W$, we may view U, V, W as S-modules (which are finitely generated and free) and ϕ as an S-bilinear map, which we denote by $_S\phi: U \times V \to W$. We call $_S\phi$ the *scalar restriction* of ϕ induced by ρ. Scalar restriction commutes with $\oplus$, but not with $\otimes$ (cf. Ex. 15.4). The scalar restrictions of equivalent $\mathcal{R}$-bilinear maps are again equivalent. (Of course, we can define the scalar restriction of a tensor over $\mathcal{R}$ in an analogous way.)

Now let $\rho: S \to \mathcal{R}$ be an arbitrary morphism of commutative rings. ($\mathcal{R}$ does not need to be a finitely generated free S-module via ρ.) Given an S-bilinear map $\phi: U \times V \to W$, we may form the scalar extensions $U^{\mathcal{R}} := \mathcal{R} \otimes_S U$, $V^{\mathcal{R}} := \mathcal{R} \otimes_S V$, $W^{\mathcal{R}} := \mathcal{R} \otimes_S W$ and the $\mathcal{R}$-bilinear map $\phi^{\mathcal{R}}: U^{\mathcal{R}} \times V^{\mathcal{R}} \to W^{\mathcal{R}}$ sending $(a \otimes u, b \otimes v)$ to $ab \otimes \phi(u, v)$. We call $\phi^{\mathcal{R}}$ the *scalar extension* of ϕ induced by ρ. (Of course, we can define the scalar extension of a tensor over S in an analogous way.)

We remark that on associative S-algebras this yields the scalar extension in the usual sense. Since our modules are all free, we can construct the scalar extension in a simple way: let $(\phi_{ij\ell})$ be the coordinate tensor of the S-bilinear map $\phi: U \times V \to W$ with respect to bases $(u_i), (v_j), (w_\ell)$ of U, V, W (i.e., $\phi(u_i, v_j) = \sum_\ell \phi_{ij\ell} w_\ell$). Then $(\rho(\phi_{ij\ell}))$ is the coordinate tensor of $\phi^{\mathcal{R}}$ with respect to the bases $(1 \otimes u_i)$, $(1 \otimes v_j), (1 \otimes w_\ell)$ of $U^{\mathcal{R}}, V^{\mathcal{R}}, W^{\mathcal{R}}$.

We note that scalar extension commutes with direct sum and tensor product:

$$(\phi \oplus \psi)^{\mathcal{R}} = \phi^{\mathcal{R}} \oplus \psi^{\mathcal{R}}, \quad (\phi \otimes_S \psi)^{\mathcal{R}} = \phi^{\mathcal{R}} \otimes_{\mathcal{R}} \psi^{\mathcal{R}}$$

and takes equivalent bilinear maps to equivalent ones. Also, for an S-bilinear map ϕ we have

$$(15.14) \qquad _S(\phi^{\mathcal{R}}) \simeq \mathcal{R} \otimes_S \phi$$

where the ring $\mathcal{R}$ stands here for the S-bilinear map "multiplication in $\mathcal{R}$." If $\mathcal{K} \to S, S \to \mathcal{R}$ are two ring morphisms and χ is a $\mathcal{K}$-bilinear map, then

$$(\chi^S)^{\mathcal{R}} \simeq \chi^{\mathcal{R}}.$$

These properties follow easily from the corresponding facts about scalar extension of modules.

Let us discuss now the effect of a change of scalars with respect to the restriction order. If ϕ, ψ are $\mathcal{R}$-bilinear, then

$$(15.15) \qquad \phi \leq \psi \Rightarrow _S\phi \leq _S\psi,$$

and if ϕ, ψ are S-bilinear, then

$$(15.16) \qquad \phi \leq \psi \Rightarrow \phi^{\mathcal{R}} \leq \psi^{\mathcal{R}}.$$

(The proofs are obvious. Of course, for (15.15) we need to assume that $\mathcal{R}$ is a finitely generated free S-module via ρ.) We remark that in general $\phi^{\mathcal{R}} \leq \psi^{\mathcal{R}}$ does not imply $\phi \leq \psi$. To see this let $p \in S[T]$ be an irreducible separable polynomial of degree $n > 1$ over a field S and $\mathcal{R}$ be a splitting field of p. Then

$$(S[T]/(p))^{\mathcal{R}} \simeq \mathcal{R}[T]/(p) \simeq \mathcal{R}^n,$$

in particular $(S[T]/(p))^{\mathcal{R}} \leq \langle n \rangle^{\mathcal{R}}$, but

$$S[T]/(p) \not\leq \langle n \rangle$$

because $R(S[T]/(p)) \geq 2n - 1$, see the forthcoming Prop. (17.1). However, we can say the following (compare also Ex. 15.7).

(15.17) Proposition. *Let A be an algebra over a field k and ϕ, ψ be k-bilinear maps. Assume $\phi^A \leq \psi^A$. Then:*

(1) $\exists M \in \mathbb{N}' \; \forall N \in \mathbb{N}: \quad \phi^{\otimes N} \leq \langle M \rangle \otimes \psi^{\otimes N}.$
(2) *If k is algebraically closed, then we have $\phi \leq \psi$.*

Proof. Choosing bases for the k-spaces underlying ϕ and ψ, these two bilinear maps may be represented by coordinate tensors $(\phi_{ij\ell})$ and $(\psi_{i_1 j_1 \ell_1})$. The relation $\phi^A \leq \psi^A$ means that there exist elements $\alpha_{i_1 i}, \beta_{j_1 j}, \gamma_{\ell \ell_1} \in A$ such that

$$\phi_{ij\ell} = \sum_{i_1, j_1, \ell_1} \gamma_{\ell \ell_1} \psi_{i_1 j_1 \ell_1} \alpha_{i_1 i} \beta_{j_1 j}.$$

Replacing A by the subalgebra generated by the $\alpha_{i_1 i}, \beta_{j_1 j}, \gamma_{\ell \ell_1}$, we may assume that A is a finitely generated k-algebra. By Hilbert's Nullstellensatz (cf. Lang [316, Chap. III, §2]) there is a k-algebra morphism from A to a finite field extension K of k. By applying the scalar extension $A \to K$ to $\phi^A \leq \psi^A$ we obtain $\phi^K \leq \psi^K$. If k is algebraically closed, then $K = k$, so $\phi \leq \psi$, which shows the second statement.

In general we conclude from $\phi^K \leq \psi^K$ that $(\phi^{\otimes N})^K \leq (\psi^{\otimes N})^K$ for every $N \in \mathbb{N}'$. This implies by (15.15) that $_k\big((\phi^{\otimes N})^K\big) \leq {}_k\big((\psi^{\otimes N})^K\big)$, and hence, using (15.14), we obtain

$$\phi^{\otimes N} \leq K \otimes_k \phi^{\otimes N} \leq K \otimes_k \psi^{\otimes N} \leq \langle M \rangle \otimes_k \psi^{\otimes N}$$

where $M := R(K)$. $\qquad\square$

We remark that this proposition (and the subsequent corollary) may be viewed as a statement about autarky similar to Thm. (4.17). In the proof of both results Hilbert's Nullstellensatz is crucial.

(15.18) Corollary (Schönhage). *The exponent of matrix multiplication is invariant under scalar extensions: if $k \subseteq K$ is a field extension, then*

$$\omega(k) = \omega(K).$$

Proof. (15.16) implies immediately $\omega(K) \leq \omega(k)$. To prove the reverse inequality, let $\tau > \omega(K)$. By definition, there exists $m \in \mathbb{N}'$ such that

$$R(\langle m, m, m \rangle^K) \leq m^\tau,$$

that is

$$\langle m, m, m \rangle^K \leq \langle \lfloor m^\tau \rfloor \rangle^K.$$

Prop. (15.17) implies the existence of some $M \in \mathbb{N}'$ such that for all $N \in \mathbb{N}'$

$$\langle m^N, m^N, m^N \rangle \simeq \langle m, m, m \rangle^{\otimes N} \leq \langle M \lfloor m^\tau \rfloor^N \rangle.$$

For given $n \in \mathbb{N}'$ we choose $N \in \mathbb{N}$ such that $m^{N-1} \leq n < m^N$. Then

$$R(\langle n, n, n \rangle) \leq M m^\tau n^\tau,$$

and therefore $\omega(k) \leq \tau$. We have thus shown that $\omega(k) \leq \omega(K)$.   □

15.4 Degeneration and Border Rank

The concept of border rank is motivated by the fact that a tensor $t \in \mathbb{C}^{m \times n \times p}$ of rank r may be the limit of a sequence of tensors in $\mathbb{C}^{m \times n \times p}$ of rank strictly smaller than r. For instance, let t be the structural tensor of the $\mathbb{C}$-algebra $\mathbb{C}[T]/(T^m)$ with respect to the basis $1, T, \ldots, T^{m-1}$. We know from Prop. (14.48) and the forthcoming Thm. (17.14) that $R(t) = 2m - 1$. On the other hand, denote for $\theta \in \mathbb{C}^\times$ the structural tensor of $\mathbb{C}[T]/(T^m - \theta^m) \simeq \mathbb{C}^m$ with respect to the basis $1, T, \ldots, T^{m-1}$ by t_θ. Then $R(t_\theta) = m$ for all $\theta \in \mathbb{C}^\times$ and $\lim_{\theta \to 0} t_\theta = t$. We will express this fact by saying that the "border rank" $\underline{R}(t)$ of t is at most m.

However, we will focus on a purely algebraic concept of approximation which works over any field k, and only later give the connection to the topological interpretation. As earlier, where we defined the restriction order instead of studying simply the rank function, thus allowing the comparison of any bilinear maps, we will define a *degeneration order* which expresses that a bilinear map is an "approximate" restriction of another.

In the following k denotes an arbitrary field unless otherwise specified. In case $k = \mathbb{R}$ the reader may think of the indeterminate ε used in the sequel as an "infinitely small" element.

(15.19) Definition. Let $\phi: U \times V \to W$ and ψ be k-bilinear maps, $q \in \mathbb{N}'$ and ε be an indeterminate over k. We call ϕ a *degeneration of order q* of ψ, in symbols $\phi \trianglelefteq_q \psi$, if there exists a $k[\varepsilon]$-bilinear map $\phi' : U^{k[\varepsilon]} \times V^{k[\varepsilon]} \to W^{k[\varepsilon]}$ such that

$$\varepsilon^{q-1} \phi^{k[\varepsilon]} + \varepsilon^q \phi' \leq \psi^{k[\varepsilon]}.$$

ϕ is said to be a *degeneration* of ψ, in symbols $\phi \trianglelefteq \psi$, if $\phi \trianglelefteq_q \psi$ for some $q \in \mathbb{N}'$. The *border rank* $\underline{R}(\phi)$ of ϕ is the smallest $r \in \mathbb{N}$ such that $\phi \trianglelefteq \langle r \rangle$, i.e.,

$$\forall r \in \mathbb{N}: \quad \underline{R}(\phi) \leq r \iff \phi \trianglelefteq \langle r \rangle.$$ •

A bilinear map ϕ is a degeneration of order one of some bilinear map ψ if and only if ϕ is a restriction of ψ. (To see this apply scalar extension via the k-algebra morphism $k[\varepsilon] \to k, \varepsilon \mapsto 0$.) In particular the border rank is not greater than the rank: $\underline{R}(\phi) \leq R(\phi)$.

For tensors t_1, t_2 over k we analogously define the relations $t_1 \trianglelefteq_q t_2$, $t_1 \trianglelefteq t_2$, and the border rank. It is immediate from the definition that for a tensor t in

$U \otimes V \otimes W$ we have $t \trianglelefteq_q \langle r \rangle$ iff there exist $u_\rho \in U^{k[\varepsilon]}$, $v_\rho \in V^{k[\varepsilon]}$, $w_\rho \in W^{k[\varepsilon]}$ for $1 \leq \rho \leq r$ such that

$$\varepsilon^{q-1} t + \varepsilon^q t' = \sum_{\rho=1}^r u_\rho \otimes v_\rho \otimes w_\rho$$

for some $t' \in U^{k[\varepsilon]} \otimes V^{k[\varepsilon]} \otimes W^{k[\varepsilon]}$. Thus our definition is consistent with the one given in Sect. 15.2. In terms of coordinate tensors this reads as follows. Let $t \in k^{m \times n \times p}$ and associate with t the trilinear form

$$F := \sum_{i,j,\ell} t_{ij\ell} X_i Y_j Z_\ell$$

where $X_1, \ldots, X_m, Y_1, \ldots, Y_n, Z_1, \ldots, Z_p$ are indeterminates over $k(\varepsilon)$. Then $t \trianglelefteq_q \langle r \rangle$ iff there exist linear forms

$$f_\rho \in \langle X_1, \ldots, X_m \rangle_{k[\varepsilon]}, \quad g_\rho \in \langle Y_1, \ldots, Y_n \rangle_{k[\varepsilon]}, \quad h_\rho \in \langle Z_1, \ldots, Z_p \rangle_{k[\varepsilon]}$$

over $k[\varepsilon]$ such that

$$\varepsilon^{q-1} F + \varepsilon^q G = \sum_{\rho=1}^r f_\rho g_\rho h_\rho$$

for some trilinear form G in X_i, Y_j, Z_ℓ over $k[\varepsilon]$.

(15.20) Example. Assume that k has a primitive mth root of unity ζ. We want to show that

$$\underline{R}(k[T]/(T^m)) \leq m.$$

The trilinear form F associated with the structural tensor of $k[T]/(T^m)$ with respect to the basis $1, T, \ldots, T^{m-1}$ has the form

$$F = \sum X_i Y_j Z_{i+j}$$

where the sum is taken over all i, j with $0 \leq i < m$, $0 \leq j < m$ satisfying $i + j < m$. It is straightforward to check that

$$\varepsilon^{m-1} F + \varepsilon^{2m-1} \sum_{i+j \geq m} X_i Y_j Z_{i+j-m}$$

$$= \frac{1}{m} \sum_{\rho=0}^{m-1} \left(\sum_{i=0}^{m-1} \zeta^{\rho i} \varepsilon^i X_i \right) \left(\sum_{j=0}^{m-1} \zeta^{\rho j} \varepsilon^j Y_j \right) \left(\sum_{\ell=0}^{m-1} \zeta^{-\rho \ell} \varepsilon^{m-1-\ell} Z_\ell \right)$$

(use $\sum_{\rho=0}^{m-1} \zeta^{\rho i} = 0$ for $i \not\equiv 0$ mod m). Therefore, we have $k[T]/(T^m) \trianglelefteq_m \langle m \rangle$. (For another proof see Ex. 15.9.)   ●

(15.21) Example. Recall the example due to Bini et al. which has been discussed in Sect. 15.2. There we have shown that the border rank of the bilinear map $\boxvert \times \boxvert$ of the multiplication of a 2×2-matrix $A = (a_{ij})$ satisfying $a_{22} = 0$ with a full 2×2-matrix B is at most five, which implied $\underline{R}(\langle 3, 2, 2 \rangle) \leq 10$.   ●

The degeneration may also be characterized in the following way, which differs only slightly from the original definition.

(15.22) Lemma. *For k-bilinear maps* $\phi: U \times V \to W$ *and* ψ *the following statements are equivalent:*

(1) $\phi \unlhd \psi$.

(2) *There exists a* $k[\varepsilon]$-*bilinear map* $\phi' : U^{k[\varepsilon]} \times V^{k[\varepsilon]} \to W^{k[\varepsilon]}$ *such that*

$$\phi^{k(\varepsilon)} + \varepsilon(\phi')^{k(\varepsilon)} \unlhd \psi^{k(\varepsilon)}.$$

Proof. (1) $\Rightarrow$ (2): clear.

(2) $\Rightarrow$ (1): Let $\psi : U_1 \times V_1 \to W_1$ be a k-bilinear map satisfying (2). There exist $k(\varepsilon)$-bilinear maps $\alpha: U^{k(\varepsilon)} \to U_1^{k(\varepsilon)}$, $\beta: V^{k(\varepsilon)} \to V_1^{k(\varepsilon)}$ and $\gamma: W_1^{k(\varepsilon)} \to W^{k(\varepsilon)}$ such that for all $u \in U^{k(\varepsilon)}$, $v \in V^{k(\varepsilon)}$

$$(\phi^{k(\varepsilon)} + \varepsilon(\phi')^{k(\varepsilon)})(u, v) = \gamma\psi(\alpha(u), \beta(v)).$$

Choosing bases we can express this as

$$\phi_{ij\ell} + \varepsilon\phi'_{ij\ell} = \sum_{i_1, j_1, \ell_1} \gamma_{\ell\ell_1} \psi_{i_1 j_1 \ell_1} \alpha_{i_1 i} \beta_{j_1 j}$$

where $(\phi_{ij\ell})$, $(\phi'_{ij\ell})$, $(\psi_{i_1 j_1 \ell_1})$ are the corresponding coordinate tensors and $\alpha_{i_1 i}$, $\beta_{j_1 j}$, $\gamma_{\ell\ell_1} \in k(\varepsilon)$. The inclusion $k[\varepsilon] \hookrightarrow k[[\varepsilon]]$ yields an embedding of $k(\varepsilon)$ into the field of fractions $k((\varepsilon))$ of the power series ring $k[[\varepsilon]]$. We note that

$$k((\varepsilon)) = \{\varepsilon^{-e} p \mid e \in \mathbb{N}, \ p \in k[[\varepsilon]]\}.$$

Hence, by multiplying with a sufficiently high power ε^{q-1} we obtain

(A) $$\varepsilon^{q-1}\phi_{ij\ell} + \varepsilon^q \phi'_{ij\ell} = \sum \tilde{\gamma}_{\ell\ell_1} \psi_{i_1 j_1 \ell_1} \tilde{\alpha}_{i_1 i} \tilde{\beta}_{j_1 j},$$

where $\tilde{\alpha}_{i_1 i}, \tilde{\beta}_{j_1 j}, \tilde{\gamma}_{\ell\ell_1} \in k[[\varepsilon]]$. We may achieve that the $\tilde{\alpha}_{i_1 i}, \tilde{\beta}_{j_1 j}, \tilde{\gamma}_{\ell\ell_1}$ are polynomials in $k[\varepsilon]$ by cutting off these power series after order $q - 1$ and by replacing the $\phi'_{ij\ell}$ by some other polynomials. Eq. (A) may then be expressed in a coordinate free way as follows

$$\varepsilon^{q-1}\phi^{k[\varepsilon]} + \varepsilon^q \phi' \unlhd \psi^{k[\varepsilon]},$$

which means that ϕ is a degeneration of ψ. $\qquad\square$

From this lemma we get easily the following.

(15.23) Lemma. *Let* $t \in U \otimes V \otimes W$ *be concise. Then*

$$\underline{R}(t) \geq \max\{\dim U, \dim V, \dim W\}.$$

Proof. We may assume $t \in k^{m \times n \times p}$ is concise. By Lemma (15.22) there exists $t_1 \in k[\varepsilon]^{m \times n \times p}$ such that the tensor $\tilde{t} := t + \varepsilon t_1 \in k(\varepsilon)^{m \times n \times p}$ has rank at most $\underline{R}(t)$. It is therefore sufficient to argue that $\tilde{t}$ is concise. But, by Prop. (14.43), the tensor t is 1-concise iff the vectors

$$(t_{1j\ell}, \ldots, t_{mj\ell}) \quad (j, \ell) \in \underline{n} \times \underline{p}$$

generate k^m. This means that some $(m \times m)$-minor of the matrix $(t_{ij\ell})_{i,(j,\ell)}$ in $k^{m \times (n \times p)}$ does not vanish. Then the corresponding minor of $(\tilde{t}_{ij\ell})_{i,(j,\ell)}$ does not vanish either and $\tilde{t}$ is therefore 1-concise. Analogously one sees that $\tilde{t}$ is concise. $\qquad\square$

We are now going to derive Schönhage's example (15.12), i.e., we want to show that

$$\underline{R}(\langle e, 1, \ell \rangle \oplus \langle 1, (e-1)(\ell-1), 1 \rangle) = e\ell + 1.$$

for positive natural numbers e, ℓ.

Proof of (15.12). Matrix tensors are concise, thus the lower bound is a consequence of (15.23). Put $h := (e-1)(\ell-1)$. The trilinear form corresponding to $\langle e, 1, \ell \rangle \oplus \langle 1, h, 1 \rangle$ with respect to suitable bases is

$$F = \sum_{(i,j) \in \underline{e} \times \underline{\ell}} a_i b_j c_{ji} + \sum_{(i,j) \in \underline{e-1} \times \underline{\ell-1}} X_{ij} Y_{ij} Z$$

where the $a_i, b_j, c_{ji}, X_{ij}, Y_{ij}$ and Z are indeterminates over k. (The X and Y in the inner product have double indices for technical reasons.) Additionally we define for $(i, j) \in \underline{e} \times \underline{\ell}$

$$X_{i\ell} = 0, \qquad\qquad Y_{i\ell} = -\sum_{j=1}^{\ell-1} Y_{ij},$$
$$X_{ej} = -\sum_{i=1}^{e-1} X_{ij}, \qquad Y_{ej} = 0.$$

Then it is easy to check that (by cancellation of the constant and of all terms linear in ε)

$$\sum_{(i,j) \in \underline{e} \times \underline{\ell}} (a_i + \varepsilon X_{ij})(b_j + \varepsilon Y_{ij})(\varepsilon^2 c_{ji} + Z) - \left(\sum_{i=1}^{e} a_i \right)\left(\sum_{j=1}^{\ell} b_j \right) Z$$
$$= \varepsilon^2 F + \varepsilon^3 G$$

for some $G \in k[\varepsilon][a, b, c, X, Y, Z]$. Therefore

$$\langle e, 1, \ell \rangle \oplus \langle 1, h, 1 \rangle \trianglelefteq_3 \langle e\ell + 1 \rangle. \qquad\qquad \square$$

The next lemma summarizes some properties of $\trianglelefteq_q$ (compare (15.8)).

(15.24) Lemma. *Let $\phi, \phi_1, \chi, \psi, \psi_1$ be bilinear maps over k. Then*

(1) $\phi \trianglelefteq_q \psi \implies \phi \trianglelefteq_{q+1} \psi$,
(2) $\phi \trianglelefteq_p \chi, \chi \trianglelefteq_q \psi \implies \phi \trianglelefteq_{pq} \psi$,
(3) $\phi \trianglelefteq_q \psi, \phi_1 \trianglelefteq_{q_1} \psi_1 \implies \phi \oplus \phi_1 \trianglelefteq_{\max\{q,q_1\}} \psi \oplus \psi_1, \phi \otimes \phi_1 \trianglelefteq_{q+q_1-1} \psi \otimes \psi_1$,
(4) $\phi \trianglelefteq_q \psi, \phi \sim \phi_1, \psi \sim \psi_1 \implies \phi_1 \trianglelefteq_q \psi_1$.

Proof. (1) Clear.

(2) From Def. (15.19) we have with $k[\varepsilon]$-bilinear maps ϕ', χ'

$$\hat{\phi} := \varepsilon^{p-1} \phi^{k[\varepsilon]} + \varepsilon^p \phi' \leq \chi^{k[\varepsilon]}, \quad \varepsilon^{q-1} \chi^{k[\varepsilon]} + \varepsilon^q \chi' \leq \psi^{k[\varepsilon]}.$$

Replace in the second inequality ε by ε^p (i.e., apply scalar extension via $k[\varepsilon] \to k[\varepsilon], \varepsilon \mapsto \varepsilon^p$), obtaining

$$\varepsilon^{pq-p} \chi^{k[\varepsilon]} + \varepsilon^{pq} \chi' \leq \psi^{k[\varepsilon]}.$$

Let (α, β, γ) be a restriction from $\chi^{k[\varepsilon]}$ to $\hat{\phi}$ and put $\phi'' := \gamma \chi'(\alpha-, \beta-)$. Then (α, β, γ) is also a restriction of $\varepsilon^{pq-p}\chi^{k[\varepsilon]} + \varepsilon^{pq}\chi'$ to $\varepsilon^{pq-p}\hat{\phi} + \varepsilon^{pq}\phi''$. So

$$\varepsilon^{pq-1}\phi^{k[\varepsilon]} + \varepsilon^{pq}(\phi' + \phi'') \leq \varepsilon^{pq-p}\chi^{k[\varepsilon]} + \varepsilon^{pq}\chi' \leq \psi^{k[\varepsilon]},$$

which implies $\phi \trianglelefteq_{pq} \psi$.

(3) From Def. (15.19) we have with $k[\varepsilon]$-bilinear maps ϕ', ϕ_1'

(A) $$\varepsilon^{q-1}\phi^{k[\varepsilon]} + \varepsilon^q\phi' \leq \psi^{k[\varepsilon]}, \quad \varepsilon^{q_1-1}\phi_1^{k[\varepsilon]} + \varepsilon^{q_1}\phi_1' \leq \psi_1^{k[\varepsilon]}.$$

For the statement on the direct sum we may assume $q = q_1$ using (1). By taking direct sums we get

$$\varepsilon^{q-1}(\phi \oplus \phi_1)^{k[\varepsilon]} + \varepsilon^q(\phi' \oplus \phi_1') \leq (\psi \oplus \psi_1)^{k[\varepsilon]}$$

and hence $\phi \oplus \phi_1 \trianglelefteq_q \psi \oplus \psi_1$. By taking tensor products we obtain from (A)

$$\varepsilon^{q+q_1-2}(\phi \otimes \phi_1)^{k[\varepsilon]} + \varepsilon^{q+q_1-1}(\phi^{k[\varepsilon]} \otimes \phi_1' + \phi' \otimes \phi_1^{k[\varepsilon]} + \varepsilon\phi' \otimes \phi_1') \leq (\psi \otimes \psi_1)^{k[\varepsilon]}$$

and therefore $\phi \otimes \phi_1 \trianglelefteq_{q+q_1-1} \psi \otimes \psi_1$.

(4) By Prop. (14.51) we know that $\phi_1 \sim \phi$ implies $\phi_1 \leq \phi$. Furthermore, $\phi \leq \chi$, $\chi \trianglelefteq_q \psi$ implies $\phi \trianglelefteq_q \psi$ by (2), as $\phi \leq \chi$ is equivalent to $\phi \trianglelefteq_1 \chi$.   □

The previous lemma shows that the degeneration $\trianglelefteq$ and the border rank $\underline{R}$ are well defined on elements of the semiring $\mathcal{T}$ of tensor classes. It also implies that the most important properties of the degeneration can be neatly summarized as follows.

(15.25) Proposition. *The degeneration $\trianglelefteq$ is a preorder on $\mathcal{T}$ which is finer than the restriction order (i.e., $a \leq b \Rightarrow a \trianglelefteq b$). $\trianglelefteq$ induces the natural order on $\mathbb{N}$ and is compatible with addition and multiplication in the semiring $\mathcal{T}$:*

$$\forall a, a', b, b' \in \mathcal{T} : a \trianglelefteq b, \ a' \trianglelefteq b' \implies a + a' \trianglelefteq b + b', \ aa' \trianglelefteq bb'.$$

Moreover

$$\forall a, b \in \mathcal{T}, \pi \in S_3 : a \trianglelefteq b \implies \pi a \trianglelefteq \pi b.$$

For $a \in \mathcal{T}$ we have

$$\underline{R}(a) = \min\{r \in \mathbb{N} \mid a \trianglelefteq r\}.$$

The border rank function $\underline{R} : \mathcal{T} \to \mathbb{N}$ is subadditive and submultiplicative.

When k is an algebraically closed field it is possible to characterize the degeneration in terms of the Zariski topology. In this case the degeneration $\trianglelefteq$ is actually a partial order on $\mathcal{T}$. For a discussion of this we refer the reader to the Appendix of Chap. 20.

We know that $\phi \leq \psi$ implies $\phi \trianglelefteq \psi$ for k-bilinear maps ϕ, ψ. The reverse implication is false, nevertheless the following can be said. (Compare statement (15.7) in Sect. 15.2 where we discussed how to pass from an approximative to an exact algorithm.)

(15.26) Proposition. *Let ϕ, ψ be k-bilinear maps. Then*

$$\phi \trianglelefteq_q \psi \implies \phi \leq k[\varepsilon]/(\varepsilon^q) \otimes \psi \leq \langle q^2 \rangle \otimes \psi.$$

If $|k| \geq 2q - 2$, then even $\phi \leq \langle 2q - 1 \rangle \otimes \psi$ follows.

Proof. Let $\phi: U \times V \to W$ and assume $\phi \trianglelefteq_q \psi$. By applying scalar extension with respect to the canonical morphism $k[\varepsilon] \to \mathcal{R} := k[\varepsilon]/(\varepsilon^q)$ we conclude from Def. (15.19) that $\varepsilon^{q-1} \phi^{\mathcal{R}} \leq \psi^{\mathcal{R}}$. Scalar restriction with respect to $k \to \mathcal{R}$ yields (cf. (15.14), (15.15))

$$_k(\varepsilon^{q-1} \phi^{\mathcal{R}}) \leq {}_k(\psi^{\mathcal{R}}) \simeq \mathcal{R} \otimes_k \psi.$$

We are going to show that ϕ is a restriction of $_k(\varepsilon^{q-1} \phi^{\mathcal{R}})$: consider the k-linear map $\pi : \mathcal{R} \to k$, $\sum_{i=0}^{q-1} \lambda_i \varepsilon^i \mapsto \lambda_{q-1}$. It is easy to check that the diagram

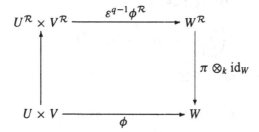

is commutative, hence $\phi \leq_k (\varepsilon^{q-1} \phi^{\mathcal{R}})$. Thus we may conclude that $\phi \leq \mathcal{R} \otimes_k \psi$. Moreover we know from Prop. (14.48) that $\mathcal{R} \leq \langle 2q - 1 \rangle$, provided $|k| \geq 2q - 2$. This proves the lemma. $\square$

As an application we prove the following result, which will be needed later on in Sect. 15.8.

(15.27) Lemma. *For a k-bilinear map $\phi \neq 0$ we have*

$$\limsup_{N \to \infty} \frac{1}{N} \log R(\phi^{\otimes N}) \leq \log \underline{R}(\phi).$$

Proof. Let $r := \underline{R}(\phi)$, say $\phi \trianglelefteq_q \langle r \rangle$. Then, by Lemma (15.24), we get $\phi^{\otimes N} \trianglelefteq_{Nq} \langle r^N \rangle$ for all N. From Prop. (15.26) we conclude $R(\phi^{\otimes N}) \leq (Nq)^2 r^N$. This clearly implies the assertion. $\square$

The sequence $N^{-1} \log R(\phi^{\otimes N})$ does actually have a limit which is called the *asymptotic rank* of ϕ, compare Ex. 15.24.

15.5 The Asymptotic Sum Inequality

We are now ready to prove Schönhage's asymptotic sum inequality (15.11), one of the main tools in proving upper estimates on the complexity of matrix multiplication. Let us recall its statement:

For $e_1, h_1, \ell_1, \ldots, e_s, h_s, \ell_s, r \in \mathbb{N}'$ we have

$$\underline{R}\left(\bigoplus_{i=1}^{s} \langle e_i, h_i, \ell_i \rangle\right) \le r \implies \sum_{i=1}^{s} (e_i h_i \ell_i)^{\omega/3} \le r.$$

Proof of (15.11). We prefer to work in the semiring $\mathcal{T}$ as only equivalence classes of tensors matter (cf. Sect. 14.6). The addition in $\mathcal{T}$ corresponds to the direct sum of tensors, whereas the multiplication corresponds to the tensor product. Put $r := \underline{R}(\sum_{i=1}^{s} \langle e_i, h_i, \ell_i \rangle)$. By definition we have for some $q \in \mathbb{N}'$ the inequality

$$\sum_{i=1}^{s} \langle e_i, h_i, \ell_i \rangle \trianglelefteq_q \langle r \rangle$$

in the semiring $\mathcal{T}$ with respect to the degeneration order. Taking Nth powers $(N \in \mathbb{N})$ we obtain by Lemma (15.24)(3)

$$\left(\sum_{i=1}^{s} \langle e_i, h_i, \ell_i \rangle\right)^{N} \trianglelefteq_{Nq} \langle r^{N} \rangle,$$

and this implies using Prop. (15.26)

$$\left(\sum_{i=1}^{s} \langle e_i, h_i, \ell_i \rangle\right)^{N} \le \langle (Nq)^2 r^{N} \rangle.$$

By performing the multinomial expansion in the semiring $\mathcal{T}$ we get

$$\left(\sum_{i=1}^{s} \langle e_i, h_i, \ell_i \rangle\right)^{N} = \sum_{\mu} \binom{N}{\mu} \prod_{i=1}^{s} \langle e_i, h_i, \ell_i \rangle^{\mu_i},$$

where the sum on the right-hand side is taken over all distributions $\mu \in \mathbb{N}^s$ with $\sum \mu_i = N$. From this we may conclude that

(A) $\qquad \binom{N}{\mu} \left\langle \prod_{i=1}^{s} e_i^{\mu_i}, \prod_{i=1}^{s} h_i^{\mu_i}, \prod_{i=1}^{s} \ell_i^{\mu_i} \right\rangle \le \langle (Nq)^2 r^{N} \rangle$

for every distribution $\mu \in \mathbb{N}^s$ with $\sum \mu_i = N$.

Now we come to the essential point of the proof. By the definition of ω there exists for every $\varepsilon > 0$ a constant $c_\varepsilon \in \mathbb{N}$ such that for all n

$$R(\langle n, n, n \rangle) \le c_\varepsilon n^{\omega + \varepsilon}.$$

Therefore, we have for every distribution $\mu \in \mathbb{N}^s$ with $\sum \mu_i = N$ that

$$\langle M_{\varepsilon, \mu}, M_{\varepsilon, \mu}, M_{\varepsilon, \mu} \rangle \le \left\langle c_\varepsilon \binom{N}{\mu} \right\rangle,$$

where we have set $M_{\varepsilon, \mu} := \lfloor \binom{N}{\mu}^{1/(\omega+\varepsilon)} \rfloor$. From (A) we deduce that

$$\langle e_\mu, h_\mu, \ell_\mu \rangle \leq \langle c_\varepsilon (Nq)^2 r^N \rangle,$$

where $e_\mu := M_{\varepsilon,\mu} \prod_{i=1}^s e_i^{\mu_i}$, $h_\mu := M_{\varepsilon,\mu} \prod_{i=1}^s h_i^{\mu_i}$, $\ell_\mu := M_{\varepsilon,\mu} \prod_{i=1}^s \ell_i^{\mu_i}$. This implies by Prop. (15.5) that

$$(e_\mu h_\mu \ell_\mu)^{\omega/3} \leq c_\varepsilon (Nq)^2 r^N,$$

hence (use $a/2 \leq \lfloor a \rfloor$ for $a \in [1, \infty)$, and $\binom{N}{\mu} \leq s^N$)

$$\binom{N}{\mu} \prod_{i=1}^s (e_i h_i \ell_i)^{\omega \mu_i/3} \leq 2^\omega s^{N\varepsilon/(\omega+\varepsilon)} c_\varepsilon (Nq)^2 r^N.$$

If we sum over all distributions $\mu \in \mathbb{N}^s$ with $\sum_i \mu_i = N$ we arrive at

$$\left(\sum_{i=1}^s (e_i h_i \ell_i)^{\omega/3} \right)^N \leq \binom{N+s-1}{s-1} 2^\omega s^{N\varepsilon/(\omega+\varepsilon)} c_\varepsilon (Nq)^2 r^N.$$

(Note that there are $\binom{N+s-1}{s-1}$ such distributions.) By taking Nth roots and letting N tend to infinity we conclude

$$\sum_{i=1}^s (e_i h_i \ell_i)^{\omega/3} \leq s^{\varepsilon/(\omega+\varepsilon)} r.$$

Since $\varepsilon > 0$ was chosen arbitrarily, the theorem is proved.   □

15.6 First Steps Towards the Laser Method

The goal of the next three sections is to develop a powerful method for gaining upper bounds on the exponent of matrix multiplication.

In the sequel $e_{ij\ell} \in k^{m \times n \times p}$ denotes a canonical basis vector, i.e., $(e_{ij\ell})_{i',j',\ell'} = \delta_{ii'}\delta_{jj'}\delta_{\ell\ell'}$. We may thus write a coordinate tensor $t \in k^{m \times n \times p}$ in the form $t = \sum_{i,j,\ell} t_{ij\ell} e_{ij\ell}$. The *support of* t is defined as the quadruple $(\underline{m}, \underline{n}, \underline{p}, \text{supp}\, t)$ where

$$\text{supp}\, t := \{(i, j, \ell) \in \underline{m} \times \underline{n} \times \underline{p} \mid t_{ij\ell} \neq 0\}.$$

This may be generalized as follows. Let $t \in U \otimes V \otimes W$ be a tensor over k and let D be a triple of direct sum decompositions of U, V, W into k-subspaces, i.e.,

(15.28) $D: \quad U = \bigoplus_{i \in I} U_i, \quad V = \bigoplus_{j \in J} V_j, \quad W = \bigoplus_{\ell \in L} W_\ell,$

where I, J, L are finite index sets. (We will also call D a *direct sum decomposition of* t.) Then we have decompositions

$$U \otimes V \otimes W = \bigoplus_{(i,j,\ell) \in I \times J \times L} U_i \otimes V_j \otimes W_\ell, \quad t =: \sum_{(i,j,\ell) \in I \times J \times L} t(i, j, \ell),$$

where $t(i, j, \ell) \in U_i \otimes V_j \otimes W_\ell$. We call

$$\text{supp}_D t := \{(i, j, \ell) \in I \times J \times L \mid t(i, j, \ell) \neq 0\}$$

the *D-support of t* and the nonzero $t(i, j, \ell)$ the *D-components of t*.

Our next topic is a combinatorial notion of degeneration.

(15.29) Definition. Let I, J, L be finite sets.

(1) A subset Δ of $I \times J \times L$ is called a *diagonal* iff the three projections $\Delta \to I$, $\Delta \to J$, $\Delta \to L$ are injective.

(2) A subset Ψ of $\Phi \subseteq I \times J \times L$ is called a *(combinatorial) degeneration* of Φ, $\Psi \trianglelefteq \Phi$, iff there exists functions $a : I \to \mathbb{Z}$, $b : J \to \mathbb{Z}$, $c : L \to \mathbb{Z}$ such that

$$\forall (i, j, \ell) \in \Psi \quad : \quad a(i) + b(j) + c(\ell) = 0$$
$$\forall (i, j, \ell) \in \Phi \setminus \Psi \quad : \quad a(i) + b(j) + c(\ell) > 0. \; \bullet$$

The relevance of diagonals for our purposes is expressed in the following fact. Let D be a direct sum decomposition of a tensor t and assume that $\Delta \subseteq \text{supp}_D t$ is a diagonal. Then we have

$$\sum_{(i,j,\ell) \in \Delta} t(i, j, \ell) \simeq \bigoplus_{(i,j,\ell) \in \Delta} t(i, j, \ell).$$

(Indeed, let $\Delta = \{(i_s, j_s, \ell_s) \mid s \in \underline{d}\}$ and assume w.l.o.g. that the three projections of Δ onto I, J, L are bijections. Then we have the direct sum decompositions $U = \oplus_{s=1}^d U_{i_s}$, $V = \oplus_{s=1}^d V_{j_s}$, $W = \oplus_{s=1}^d W_{\ell_s}$ such that $t(i_s, j_s, \ell_s) \in U_{i_s} \otimes V_{j_s} \otimes W_{\ell_s}$.)

The next proposition explains how the combinatorial degeneration is related to the degeneration of tensors.

(15.30) Proposition. *Assume D is a direct sum decomposition of a tensor t in $U \otimes V \otimes W$ and let $\Psi \subseteq \text{supp}_D t$. If Ψ is a degeneration of $\text{supp}_D t$, then*

$$\sum_{(i,j,\ell) \in \Psi} t(i, j, \ell) \trianglelefteq t,$$

where $t(i, j, \ell)$ denote the D-components of t.

Proof. Let $a : I \to \mathbb{Z}$, $b : J \to \mathbb{Z}$, $c : L \to \mathbb{Z}$ be such that $a(i) + b(j) + c(\ell) > 0$ for all $(i, j, \ell) \in \text{supp}_D t \setminus \Psi$ and $a(i) + b(j) + c(\ell) = 0$ for all $(i, j, \ell) \in \Psi$. We define a $k(\varepsilon)$-linear endomorphism α of $U^{k(\varepsilon)}$ by

$$\alpha\left(\sum_{i \in I} u_i\right) := \sum_{i \in I} \varepsilon^{a(i)} u_i$$

where $u_i \in U_i$. That is, with respect to a basis adapted to the given direct sum decomposition of U, α is described by a *diagonal matrix* whose entries in the ith block are all $\varepsilon^{a(i)}$. Analogously, we define $\beta \in \text{End}(V^{k(\varepsilon)})$ and $\gamma \in \text{End}(W^{k(\varepsilon)})$. Then we have

$$(\alpha \otimes \beta \otimes \gamma)t \quad = \quad \sum_{(i,j,\ell) \in \text{supp}_D t} \varepsilon^{a(i)+b(j)+c(\ell)} t(i,j,\ell)$$

$$= \quad \sum_{(i,j,\ell) \in \Psi} t(i,j,\ell) + \varepsilon t'$$

for some $t' \in (U \otimes V \otimes W)^{k[\varepsilon]}$. Hence the assertion follows by Lemma (15.22). $\square$

We leave it as an exercise to the reader to verify that the combinatorial degeneration defines a partial order on the set of subsets of $I \times J \times L$.

We may now explain the basic idea of the *laser method*. Remember that in Prop. (15.13) we obtained the bound $\omega < 2.55$ in the following way. We proved an upper bound on the border rank of a direct sum of matrix tensors, and then extracted asymptotic information from this by means of the asymptotic sum inequality. The new idea is to use tensors t which possess a direct sum decomposition D having the following two properties:

- All the D-components $t(i,j,\ell)$ are matrix tensors.
- A diagonal $\Delta \subseteq I \times J \times L$ of "large size" is a degeneration of the support $\text{supp}_D t$. (Large size will mean that $|\Delta| \geq c \cdot \min\{|I|, |J|, |L|\}$ for some constant $c > 0$ independent of t and D.)

If, moreover, an upper bound r on the border rank of t is known, then we have by Prop. (15.30)

$$\underline{R}\left(\bigoplus_{(i,j,\ell) \in \Delta} t(i,j,\ell) \right) \leq r.$$

The asymptotic sum inequality applied to this then yields an estimate of the exponent ω.

In fact, we will apply the above procedure to tensor powers $t^{\otimes N}$ of a tensor t. They have a direct sum decomposition which is inherited from a decomposition of t. To formalize this, let us focus for a moment on objects (I, J, L, Φ) where I, J, L are finite sets and $\Phi \subseteq I \times J \times L$. Two objects (I, J, L, Φ) and (I', J', L', Φ') are called *isomorphic* iff there are bijections $I \to I'$, $J \to J'$, $L \to L'$ such that the corresponding bijection $I \times J \times L \to I' \times J' \times L'$ sends Φ to Φ'. The *product* of (I, J, L, Φ) and (I', J', L', Φ') is defined as $(I \times I', J \times J', L \times L', \Psi)$ where Ψ is the image of $\Phi \times \Phi'$ under the natural bijection

$$(I \times J \times L) \times (I' \times J' \times L') \stackrel{\sim}{\to} (I \times I') \times (J \times J') \times (L \times L').$$

By abuse of notation, we often write Φ instead of (I, J, L, Φ) if I, J, L are clear from the context. The product of Φ and Φ' is then denoted by $\Phi \times \Phi'$. It is easy to check that

$$\text{supp}\langle e, h, \ell \rangle \quad = \quad \Big\{ ((i,j), (j,m), (m,i)) \in (\underline{e} \times \underline{h}) \times (\underline{h} \times \underline{\ell}) \times (\underline{\ell} \times \underline{e}) \ \Big| $$

$$i \in \underline{e}, j \in \underline{h}, m \in \underline{\ell} \Big\},$$

and

$$\text{supp}\langle e, h, \ell \rangle \times \text{supp}\langle e', h', \ell' \rangle = \text{supp}\langle ee', hh', \ell\ell' \rangle.$$

Assume now that $t \in U \otimes V \otimes W$ has a direct sum decomposition D as in (15.28). Let $t' \in U' \otimes V' \otimes W'$ have a direct sum decomposition

$$D': \quad U' = \bigoplus_{i' \in I'} U'_{i'}, \quad V' = \bigoplus_{j' \in J'} V'_{j'}, \quad W' = \bigoplus_{\ell' \in L'} W'_{\ell'}.$$

Then we get a direct sum decomposition $D \otimes D'$ of $U \otimes U'$, $V \otimes V'$, $W \otimes W'$ as follows

$$D \otimes D': \quad \begin{aligned} U \otimes U' &= \bigoplus_{(i,i') \in I \times I'} U_i \otimes U'_{i'}, \\ V \otimes V' &= \bigoplus_{(j,j') \in J \times J'} V_j \otimes V'_{j'}, \\ W \otimes W' &= \bigoplus_{(\ell,\ell') \in L \times L'} W_\ell \otimes W'_{\ell'}. \end{aligned}$$

It is obvious that

$$(t \otimes t')((i, i'), (j, j'), (\ell, \ell')) \simeq t(i, j, \ell) \otimes t(i', j', \ell'),$$

hence

$$\operatorname{supp}_{D \otimes D'} t \otimes t' = \operatorname{supp}_D t \times \operatorname{supp}_{D'} t'.$$

We also remark that if D is a direct sum decomposition of $t \in U_1 \otimes U_2 \otimes U_3$ and $\pi \in S_3$ is a permutation, then πD (defined in the natural way) is a direct sum decomposition of $\pi t \in U_{\pi^{-1}(1)} \otimes U_{\pi^{-1}(2)} \otimes U_{\pi^{-1}(3)}$. We have $\operatorname{supp}_{\pi D} \pi t = \pi(\operatorname{supp}_D t)$ and $(\pi t)(\pi(i_1, i_2, i_3)) = \pi(t(i_1, i_2, i_3))$ (of course, $\pi(i_1, i_2, i_3)$ denotes the permuted sequence $(i_{\pi^{-1}(1)}, i_{\pi^{-1}(2)}, i_{\pi^{-1}(3)})$).

To summarize, if D is a direct sum decomposition of t such that all D-components $t(i, j, \ell)$ are matrix tensors, then the direct sum decomposition $D^{\otimes N}$ of $t^{\otimes N}$ shares this property. Moreover, $\operatorname{supp}_{D^{\otimes N}} t^{\otimes N} = (\operatorname{supp}_D t)^N$.

The most difficult step in the application of the laser method is to identify classes of objects (I, J, L, Φ) which contain large diagonals as degenerations. Moreover, these classes should be closed under products, as we want to degenerate supports of the form $(\operatorname{supp}_D t)^N$. This combinatorial problem will be investigated systematically in the next section. Here we confine ourselves to show that the class of supports of cubic matrix tensors has the desired properties.

(15.31) Lemma. *There is a diagonal of size $\lceil 3h^2/4 \rceil$ which is a degeneration of* $\operatorname{supp}\langle h, h, h \rangle$.

Proof. Let $s \in \mathbb{N}$ and set for $i, j, \ell \in \underline{h}$

$$a(i, j) := (i - s)^2 + 2(i - s)j, \quad b(j, \ell) := j^2 + 2j\ell, \quad c(\ell, i) := \ell^2 + 2\ell(i - s).$$

Then $a(i, j) + b(j, \ell) + c(\ell, i) = (i + j + \ell - s)^2$ and therefore $a + b + c \geq 0$ on $\operatorname{supp}\langle h, h, h \rangle$. Moreover

$$\begin{aligned} &\left\{ ((i, j), (j, \ell), (\ell, i)) \in (\underline{h} \times \underline{h})^3 \, \Big| \, a(i, j) + b(j, \ell) + c(\ell, i) = 0 \right\} \\ &= \left\{ ((i, j), (j, \ell), (\ell, i)) \in (\underline{h} \times \underline{h})^3 \, \Big| \, i + j + \ell = s \right\}, \end{aligned}$$

which is a diagonal as two of the components i, j, ℓ determine the third one. It is straightforward to verify that this set has the maximal cardinality $\lceil 3h^2/4 \rceil$ for the value $s := \lceil 3h/2 \rceil + 1$ (cf. Ex. 15.18). $\qquad \square$

(15.32) Proposition (Strassen). *Let* $e, h, \ell, q \in \mathbb{N}'$. *If the tensor* t *possesses a direct sum decomposition* D *such that* $\mathrm{supp}_D\, t \simeq \mathrm{supp}\langle e, h, \ell \rangle$, *and such that all D-components of* t *are isomorphic to matrix tensors* $\langle m, n, p \rangle$ *satisfying* $mnp = q$ *(the m, n, p may vary), then* $(eh\ell)^2 q^\omega \le \underline{R}(t)^3$.

Proof. Let π denote the cyclic permutation (123) and let $N \in \mathbb{N}'$. The tensor $t_N := (t \otimes \pi t \otimes \pi^2 t)^{\otimes N}$ inherits the direct sum decomposition $D_N := (D \otimes \pi D \otimes \pi^2 D)^{\otimes N}$ from D. By our general considerations we have

$$
\begin{aligned}
\mathrm{supp}_{D_N}\, t_N &\simeq (\mathrm{supp}_D\, t \times \mathrm{supp}_{\pi D}\, \pi t \times \mathrm{supp}_{\pi^2 D}\, \pi^2 t)^N \\
&\simeq \mathrm{supp}\langle (eh\ell)^N, (eh\ell)^N, (eh\ell)^N \rangle.
\end{aligned}
$$

Moreover, the D-components of t_N are isomorphic to matrix tensors $\langle m, n, p \rangle$ satisfying $mnp = q^{3N}$. Lemma (15.31) shows the existence of a diagonal Δ_N of cardinality $\lceil 3(eh\ell)^{2N}/4 \rceil$ which is a degeneration of $\mathrm{supp}_{D_N}\, t_N$. Prop. (15.30) implies therefore

$$
\bigoplus_{(i,j,\ell) \in \Delta_N} t_N(i, j, \ell) \trianglelefteq t_N,
$$

and by the asymptotic sum inequality we get

$$
|\Delta_N| q^{N\omega} \le \underline{R}(t_N) \le \underline{R}(t)^{3N}.
$$

By taking Nth roots and letting $N \to \infty$ we obtain the desired statement. $\qquad\square$

(15.33) Corollary (Strassen, 1987). $\omega < 2.48$.

Proof. We apply Prop. (15.32) to the coordinate tensor

$$
t = \sum_{j=1}^{q} (e_{0jj} + e_{j0j}) \in k^{q+1} \otimes k^{q+1} \otimes k^q
$$

and the direct sum decomposition D

$$
D: \quad k^{q+1} = U_0 \oplus U_1, \quad k^{q+1} = V_0 \oplus V_1, \quad k^q =: W_0,
$$

where $U_0 = V_0 = k(1, 0, \ldots, 0)$, $U_1 = V_1 = \{\xi \in k^{q+1} \mid \xi_0 = 0\}$ (see Fig. 15.1). In particular $I = J = \{0, 1\}$, $L = \{0\}$. It is easy to check that

$$
\mathrm{supp}_D\, t = \{(0, 1, 0), (1, 0, 0)\} \simeq \mathrm{supp}\langle 1, 2, 1 \rangle,
$$

$$
t(0, 1, 0) \simeq \langle 1, 1, q \rangle, \quad t(1, 0, 0) \simeq \langle q, 1, 1 \rangle.
$$

Therefore the assumptions of Prop. (15.32) are fulfilled.

Let us determine the border rank of t. Consider the trilinear form

$$
F = \sum_{i,j,\ell} t_{ij\ell} X_i Y_j Z_\ell = \sum_{j=1}^{q} (X_0 Y_j Z_j + X_j Y_0 Z_j)
$$

corresponding to t. The identity

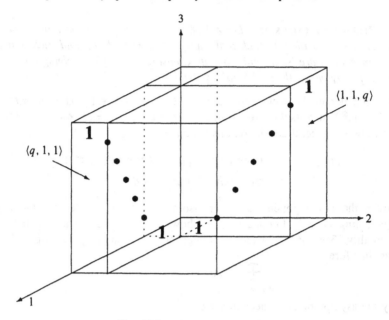

Fig. 15.1. $\mathrm{supp}_D\, t \simeq \mathrm{supp}\langle 1, 2, 1\rangle$.

$$\sum_{j=1}^{q}(X_0 + \varepsilon X_j)(Y_0 + \varepsilon Y_j)Z_j - X_0 Y_0\left(\sum_{j=1}^{q} Z_j\right) = \varepsilon F + \varepsilon^2 G$$

for some $G \in k[\varepsilon][X, Y, Z]$ shows that $\underline{R}(t) \le q+1$. (Note that even $\underline{R}(t) = q+1$ by Lemma (15.23), since t is concise.)

Prop. (15.32) yields now $4q^\omega \le (q + 1)^3$ and for $q = 5$ we obtain the desired estimate.   □

15.7* Tight Sets

I, J, L denote finite sets throughout this section.

The topic of this section is purely combinatorial. Our goal is to prove that the so-called tight subsets of a combinatorial cube $I \times J \times L$ may always be degenerated to a large diagonal. This crucial result will be needed for the further development of the laser method.

(15.34) Definition. A subset Φ of $I \times J \times L$ is called *tight* iff there are injective maps $\alpha: I \to \mathbb{Z}^r$, $\beta: J \to \mathbb{Z}^r$, $\gamma: L \to \mathbb{Z}^r$ for some $r \ge 1$ such that

$$\forall(i, j, \ell) \in \Phi : \alpha(i) + \beta(j) + \gamma(\ell) = 0.$$

We call Φ *b-tight* for $b \in \mathbb{N}'$ iff the above maps α, β, γ may be chosen such that their images are contained in $\{-b, -b+1, \ldots, b\}^r$.   •

In the above definition of tightness we may assume that $r = 1$. Indeed, we may replace the map $\alpha = (\alpha_1, \ldots, \alpha_r) \colon I \to \mathbb{Z}^r$ by the map $\tilde{\alpha} \colon I \to \mathbb{Z}$ defined by $\tilde{\alpha}(i) := \sum_{\rho=1}^r (2b+1)^\rho \alpha_\rho(i)$, and similarly for β and γ. (Of course, $|\tilde{\alpha}(i)|$ is not bounded by b.)

We leave the proof of the following remark as an exercise to the reader.

(15.35) Remark.
(1) The projection $\Phi \to I \times J$ of a tight subset $\Phi \subseteq I \times J \times L$ to the first two coordinates is injective.
(2) Subsets of tight sets are again tight.
(3) The support of a matrix tensor $\langle e, h, \ell \rangle$ is tight.
(4) If Φ and Φ' are b-tight, then so is their product.   •

The set Ψ_M defined below will be important in the proof of Thm. (15.39). As a word of warning, we mention that Ψ_M is not tight (cf. Ex. 15.20).

(15.36) Lemma. *Let $M \in \mathbb{N}'$. The subset*

$$\Psi_M := \left\{ (i, j, \ell) \in (\mathbb{Z}/M\mathbb{Z})^3 \,\middle|\, i + j + \ell = 0 \right\}$$

of $(\mathbb{Z}/M\mathbb{Z})^3$ contains a diagonal Δ of cardinality $|\Delta| \geq M/2$, which is a degeneration of Ψ_M.

Proof. W.l.o.g. we may prove the claim for the set

$$\Psi := \left\{ (i, j, \ell) \in (\mathbb{Z}/M\mathbb{Z})^3 \,\middle|\, i + j + \ell + 1 = 0 \right\}$$

obtained from Ψ_M by a permutation of the last coordinate. If we identify $\mathbb{Z}/M\mathbb{Z}$ with the set of representatives $I := \{0, 1, \ldots, M-1\} \subset \mathbb{Z}$, then we see that $\Psi = \Psi_1 \cup \Psi_2$ where

$$\Psi_1 := \{(i, j, \ell) \in I^3 \mid i + j + \ell = M - 1\},$$
$$\Psi_2 := \{(i, j, \ell) \in I^3 \mid i + j + \ell = 2M - 1\},$$

(the addition is meant to be in $\mathbb{Z}$). The subset

$$\Delta := \{(i, i, M - 1 - 2i) \mid 0 \leq i \leq (M-1)/2\}$$

of Ψ_1 is obviously a diagonal of cardinality $\lfloor (M+1)/2 \rfloor \geq M/2$.

Let us show that $\Delta \trianglelefteq \Psi$. We define the functions $a, b, c \colon I \to \mathbb{Z}$ by

$$a(i) = 4i^2, \quad b(j) := 4j^2, \quad c(\ell) := -2(M - 1 - \ell)^2$$

(the operations are in $\mathbb{Z}$). For all $(i, j, \ell) \in \Psi_1$ we have $a(i) + b(j) + c(\ell) = 2(i - j)^2 \geq 0$ with equality holding iff $(i, j, \ell) \in \Delta$. On the other hand, we have for all $(i, j, \ell) \in \Psi_2$

$$a(i) + b(j) + c(\ell) = 2(i - j)^2 + 4M(i + j) - 2M^2 \geq 2M^2 > 0.$$

This proves the claim.   □

We need two further auxiliary results.

(15.37) Remark. Let subsets $\Phi \subseteq I \times J \times L$, $\Phi' \subseteq I' \times J' \times L'$ and maps $A: I \to I'$, $B: J \to J'$, $C: L \to L'$ be given such that $F(\Phi) \subseteq \Phi'$, where $F: I \times J \times L \to I' \times J' \times L'$ is defined by $F(i, j, \ell) = (A(i), B(j), C(\ell))$. Then $\Psi' \trianglelefteq \Phi'$ implies $F^{-1}(\Psi') \cap \Phi \trianglelefteq \Phi$.   •

The obvious proof of this remark is left to the reader.

(15.38) Lemma. *Let Φ be a subset of the finite set $I \times J \times L$, and put*

$$\Pi := \left\{ \{(i, j, \ell), (i', j', \ell')\} \in \binom{\Phi}{2} \;\middle|\; i = i' \text{ or } j = j' \text{ or } \ell = \ell' \right\}.$$

Then there exist $I' \subseteq I$, $J' \subseteq J$, $L' \subseteq L$ such that

$$\Delta := (I' \times J' \times L') \cap \Phi$$

is a diagonal of cardinality at least $|\Phi| - |\Pi|$, which is a degeneration of Φ.

Proof. Consider the graph G whose set of vertices is Φ and whose set of edges is Π. Then G has at least $|\Phi| - |\Pi|$ connected components. Let $\Delta \subseteq \Phi$ be a set of representatives for the set of connected components of G and define $I' := \mathrm{pr}_1(\Delta)$, $J' := \mathrm{pr}_2(\Delta)$, $L' := \mathrm{pr}_3(\Delta)$, where pr_i denote the projections. Then (I', J', L', Δ) is a diagonal, and it is easy to see that $\Delta = (I' \times J' \times L') \cap \Phi$. To check that Δ is a degeneration of Φ, choose for a, b, c the indicator functions of the complement of I' in I, J' in J, L' in L, respectively.   □

The main theorem of this section goes back to a method of Coppersmith and Winograd [134]. The present variant is due to Strassen, who kindly allowed us to publish it here. Its proof is a fine example for the power of the probabilistic method in combinatorics. (For a systematic exposition of this method with numerous applications we refer to the book by Alon et al. [7].)

(15.39) Theorem. *Let $\Phi \subseteq I \times J \times L$ be a b-tight subset, $|I| \leq |J| \leq |L|$, and assume that the projections $p_I: \Phi \to I$, $p_J: \Phi \to J$, $p_L: \Phi \to L$ are surjective. Let $c \geq 1$ be such that $4(2b + 1)/9 \leq c|\Phi|/|I|$ and*

$$\max_{i \in I} |p_I^{-1}(i)| \leq c\frac{|\Phi|}{|I|}, \quad \max_{j \in J} |p_J^{-1}(j)| \leq c\frac{|\Phi|}{|J|}, \quad \max_{\ell \in L} |p_L^{-1}(\ell)| \leq c\frac{|\Phi|}{|L|}.$$

Then there exists a diagonal Δ in Φ of cardinality $|\Delta| \geq 2|I|/(27c)$, which is a degeneration of Φ.

Proof. Let $\alpha: I \to \mathbb{Z}^r$, $\beta: J \to \mathbb{Z}^r$, $\gamma: L \to \mathbb{Z}^r$ be injective maps which guarantee that Φ is b-tight. That is, $\alpha(i) + \beta(j) + \gamma(\ell) = 0$ for all $(i, j, \ell) \in \Phi$, and the images of α, β, γ are contained in $\{-b, -b + 1, \ldots, b\}^r$. Let $\alpha_1, \ldots, \alpha_r: I \to \mathbb{Z}$ denote the components of α, thus $\alpha(i) = (\alpha_1(i), \ldots, \alpha_r(i))$. Similarly, β_ρ, γ_ρ denote the components of β, γ, respectively.

Choose a prime number $M \geq 2b+1$ and elements $w_1, w_2, \ldots, w_{r+3}$ in the field $\mathbb{Z}/M\mathbb{Z}$. To these elements we assign the maps $A_w \colon I \to \mathbb{Z}/M\mathbb{Z}$, $B_w \colon J \to \mathbb{Z}/M\mathbb{Z}$, $C_w \colon L \to \mathbb{Z}/M\mathbb{Z}$ defined by

$$A_w(i) := \sum_{\rho=1}^{r} \overline{\alpha_\rho(i)} w_\rho + w_{r+1} - w_{r+2},$$

$$B_w(j) := \sum_{\rho=1}^{r} \overline{\beta_\rho(j)} w_\rho + w_{r+2} - w_{r+3},$$

$$C_w(\ell) := \sum_{\rho=1}^{r} \overline{\gamma_\rho(\ell)} w_\rho + w_{r+3} - w_{r+1},$$

where $m \mapsto \overline{m}$ denotes the residue class mapping $\mathbb{Z} \to \mathbb{Z}/M\mathbb{Z}$. For all $(i, j, \ell) \in \Phi$ we have $A_w(i) + B_w(j) + C_w(\ell) = 0$. Thus the map $F_w \colon I \times J \times L \to (\mathbb{Z}/M\mathbb{Z})^3$, $F_w(i, j, \ell) := (A_w(i), B_w(j), C_w(\ell))$, sends Φ into the subset

$$\Psi_M = \left\{ (x, y, z) \in (\mathbb{Z}/M\mathbb{Z})^3 \;\middle|\; x + y + z = 0 \right\}.$$

According to Lemma (15.36) there is a diagonal $D \trianglelefteq \Psi_M$ of size $|D| \geq M/2$. Hence by Rem. (15.37) the set $\Phi_w := F_w^{-1}(D) \cap \Phi$ is a degeneration of Φ.

For $d \in D$ we set $\Phi_w(d) := F_w^{-1}(d) \cap \Phi$. Thus we have a disjoint union $\Phi_w = \bigcup_{d \in D} \Phi_w(d)$. As D is a diagonal, even the projections $p_I(\Phi_w(d))$ of $\Phi_w(d)$, $d \in D$, are pairwise disjoint, and the same holds for p_J and p_L. From this we easily conclude that if $\Delta_d \trianglelefteq \Phi_w(d)$ are diagonals for all $d \in D$, then $\Delta := \bigcup_{d \in D} \Delta_d$ is a diagonal and a degeneration of Φ_w.

For $d \in D$ let

$$\Pi_w(d) := \left\{ \{(i, j, \ell), (i', j', \ell')\} \in \binom{\Phi_w(d)}{2} \;\middle|\; i = i' \text{ or } j = j' \text{ or } \ell = \ell' \right\}.$$

By Lemma (15.38) there exists a diagonal $\Delta_d \trianglelefteq \Phi_w(d)$ of size $|\Delta_d| \geq |\Phi_w(d)| - |\Pi_w(d)|$. To finish the proof it is therefore sufficent to show that M and the $w_\rho \in \mathbb{Z}/M\mathbb{Z}$ may be chosen such that

(A) $$S_w := \sum_{d \in D} (|\Phi_w(d)| - |\Pi_w(d)|) \geq \frac{2}{27c} |I|.$$

To achieve this, we think of $w_1, w_2, \ldots, w_{r+3}$ as independent random variables uniformly distributed over the field $\mathbb{Z}/M\mathbb{Z}$. (M will be specified later.) Then A_w, B_w, C_w become random maps and S_w is a random variable. It is sufficent to prove that the expected value of S_w satisfies $E(S_w) \geq 2|I|/(27c)$, since in that case, there must be a value w such that (A) holds.

For fixed $(i, j, \ell) \in I \times J \times L$ the random variables $w \mapsto A_w(i)$, $w \mapsto B_w(j)$, $w \mapsto C_w(\ell)$ are uniformly distributed over $\mathbb{Z}/M\mathbb{Z}$ and pairwise (stochastically) independent. For instance, $w \mapsto A_w(i)$ and $w \mapsto B_w(j)$ are independent because the $\mathbb{Z}/M\mathbb{Z}$–linear map

$$(\mathbb{Z}/M\mathbb{Z})^{r+3} \to (\mathbb{Z}/M\mathbb{Z})^2, \ w \mapsto (A_w(i), B_w(j))$$

is surjective.

We claim that $A_w(i)$, $A_w(i')$, $C_w(\ell)$ are independent if $i \neq i'$. For this we need to show that the $\mathbb{Z}/M\mathbb{Z}$-linear map

$$(\mathbb{Z}/M\mathbb{Z})^{r+3} \to (\mathbb{Z}/M\mathbb{Z})^3, \ w \mapsto (A_w(i), A_w(i'), C_w(\ell))$$

is surjective. Since $\mathbb{Z}/M\mathbb{Z}$ is a field, this is equivalent to the fact that the matrix

$$\begin{pmatrix} \overline{\alpha_1(i)} & \cdots & \overline{\alpha_r(i)} & 1 & -1 & 0 \\ \overline{\alpha_1(i')} & \cdots & \overline{\alpha_r(i')} & 1 & -1 & 0 \\ \overline{\gamma_1(\ell)} & \cdots & \overline{\gamma_r(\ell)} & -1 & 0 & 1 \end{pmatrix}$$

has rank three. If this rank were not three, then the first two rows of this matrix would be identical. As $|\alpha_\rho(i)|, |\alpha_\rho(i')| \leq b$ and $M \geq 2b+1$, this implied $\alpha_\rho(i) = \alpha_\rho(i')$ for all ρ, which contradicts the injectivity of α.

Let us calculate the expected value of $|\Phi_w(d)|$ for some $d = (x, y, z) \in D$:

$$\begin{aligned} E(|\Phi_w(d)|) &= \sum_{(i,j,\ell) \in \Phi} \Pr\{w \mid A_w(i) = x, B_w(j) = y, C_w(\ell) = z\} \\ &= \sum_{(i,j,\ell) \in \Phi} \Pr\{w \mid A_w(i) = x, B_w(j) = y\} = |\Phi| M^{-2}. \end{aligned}$$

(Note that $A_w(i) = x$ and $B_w(j) = y$ imply $C_w(\ell) = z$ for $(i, j, \ell) \in \Phi$, $(x, y, z) \in \Psi_M$. Moreover, $A_w(i)$, $B_w(j)$ are uniformly distributed and independent.)

To calculate the expected value of $|\Pi_w(d)|$ we consider for fixed $d = (x, y, z) \in D$ the set

$$\begin{aligned} U_w(d) &:= \left\{ \{(i, j, \ell), (i', j', \ell')\} \in \binom{\Phi_w(d)}{2} \mid \ell = \ell' \right\} \\ &= \left\{ \{(i, j, \ell), (i', j', \ell)\} \in \binom{p_L^{-1}(\ell)}{2} \mid A_w(i) = A_w(i') = x, \right. \\ &\qquad\qquad\qquad\qquad\qquad\qquad \left. B_w(j) = B_w(j') = y, C_w(\ell) = z \right\} \\ &= \left\{ \{(i, j, \ell), (i', j', \ell)\} \in \binom{p_L^{-1}(\ell)}{2} \mid A_w(i) = A_w(i') = x, \right. \\ &\qquad\qquad\qquad\qquad\qquad\qquad\qquad \left. C_w(\ell) = z \right\}. \end{aligned}$$

By the independence of $A_w(i)$, $A_w(i')$, $C_w(\ell)$ for $i \neq i'$ we see that

$$E(|U_w(d)|) = \sum_{\ell \in L} \frac{\zeta(\ell)(\zeta(\ell) - 1)}{2} M^{-3},$$

where $\zeta(\ell) := |p_L^{-1}(\ell)|$. Note that $\sum_{\ell \in L} \zeta(\ell) = |\Phi|$ and $\zeta(\ell) \leq c|\Phi|/|L|$ by assumption. Therefore, we obtain

$$E(|U_w(d)|) \le \frac{1}{2M^3} \sum_{\ell \in L} \zeta(\ell)^2 \le \frac{c|\Phi|^2}{2M^3|L|}.$$

By a similar reasoning for the other coordinates we get for $d \in D$

$$E(|\Pi_w(d)|) \le \frac{c|\Phi|^2}{2M^3} \left(\frac{1}{|I|} + \frac{1}{|J|} + \frac{1}{|L|} \right) \le \frac{3c|\Phi|^2}{2M^3|I|}$$

(recall that $|I| \le |J| \le |L|$). From this, $E(|\Phi_w(d)|) = |\Phi|M^{-2}$, and $|D| \ge M/2$ we derive that (cf. (A))

$$E(S_w) \ge \frac{|I|}{2c}(\lambda - \frac{3}{2}\lambda^2),$$

where $\lambda := c|\Phi|/(M \cdot |I|)$.

By Bertrand's postulate (cf. Chandrasekharan [106, p. 71]) we may choose the prime M such that

$$\frac{9}{4}\frac{c|\Phi|}{|I|} \le M \le \frac{9}{2}\frac{c|\Phi|}{|I|}$$

(which implies by assumption $M \ge 2b + 1$). This means $2/9 \le \lambda \le 4/9$, hence $\lambda - 3\lambda^2/2 \ge 4/27$. Altogether, we obtain $E(S_w) \ge 2|I|/(27c)$ which was to be shown. $\qquad\qquad\square$

15.8 The Laser Method

We continue our discussion of the laser method in Sect. 15.6 and prove a generalization of Prop. (15.32) based on Thm. (15.39).

We have encountered the entropy function already several times in this book. Let us recall its definition. If $P : I \to [0, 1]$ is a probability distribution on a finite set I, then

$$H(P) := - \sum_{P(i)>0} P(i) \log P(i)$$

is called the *entropy of P*. The function H, defined on the simplex of all probability distributions on I, is concave and attains its maximum value $\log|I|$ exactly for the uniform distribution on I. The entropy enters naturally into our discussion as the function describing the asymptotic growth of multinomial coefficients: there exists a sequence (ρ_N) with $\lim_{N\to\infty} \rho_N = 0$ such that for all $\mu: I \to \mathbb{N}$ satisfying $\sum_{i\in I} \mu_i = N$ we have

$$(15.40) \qquad \left| \frac{1}{N} \log \binom{N}{\mu} - H\left(\frac{\mu}{N}\right) \right| \le \rho_N.$$

(This follows by a straightforward calculation from Stirling's formula.) Let P be a probability distribution on $I \times J \times L$ where I, J, L are finite sets. The first marginal distribution $P_1: I \to [0, 1]$ of P is defined for all $i \in I$ by $P_1(i) := \sum_{(j,\ell)\in J\times L} P(i, j, \ell)$. The second and third marginal distributions are defined analogously.

Recall that $\mathcal{T}$ is the semiring defined in Sect. 14.6. It is useful to introduce the function

$$\zeta : \mathcal{T} \to [0, \infty), \quad \zeta(a) := (\dim U \dim V \dim W)^{1/6},$$

where (U, V, W, t) denotes a concise representative of the tensor class a. The function ζ is obviously multiplicative and we have $\zeta(\langle e, h, \ell \rangle) = (eh\ell)^{1/3}$. Using this function we may write the asymptotic sum inequality as follows:

$$\underline{R}\left(\bigoplus_{i=1}^{s} \langle e_i, h_i, \ell_i \rangle \right) \leq r \implies \sum_{i=1}^{s} \zeta(\langle e_i, h_i, \ell_i \rangle)^\omega \leq r.$$

We may now claim the announced generalization of Prop. (15.32).

(15.41) Theorem (Coppersmith and Winograd). *Let D denote a direct sum decomposition of a nonzero tensor $t \in U \otimes V \otimes W$ having the following properties:*

(1) *the D-support $\mathrm{supp}_D\, t$ of t is tight,*
(2) *all D-components $t(i, j, \ell)$ of t with respect to D are isomorphic to a matrix tensor.*

Then we have for every probability distribution P on $\mathrm{supp}_D\, t$

$$\min_{1 \leq m \leq 3} H(P_m) + \omega \sum_{\mathrm{supp}_D\, t} P(i, j, \ell) \log \zeta(t(i, j, \ell)) \leq \log \underline{R}(t).$$

Let us show first that Prop. (15.32) is an immediate consequence of the above theorem. Let D be a direct sum decomposition of $t \in U \otimes V \otimes W$ satisfying the assumptions of Prop. (15.32). Then for the symmetrized tensor $t' := t \otimes (123)t \otimes (132)t$ with the decomposition $D' := D \otimes (123)D \otimes (132)D$ we have $\mathrm{supp}_{D'}\, t' \simeq \mathrm{supp}\langle eh\ell, eh\ell, eh\ell \rangle$, which is tight, and $t'(i, j, \ell)$ is isomorphic to a matrix tensor with $\zeta(t'(i, j, \ell)) = q$ for all $(i, j, \ell) \in \mathrm{supp}_{D'}\, t'$. By choosing for P the uniform distribution on $\mathrm{supp}_{D'}\, t'$ we conclude from the above theorem that

$$\log(eh\ell)^2 + \omega \log q \leq \log \underline{R}(t'),$$

hence indeed $(eh\ell)^2 q^\omega \leq \underline{R}(t') \leq \underline{R}(t)^3$.

Proof of Thm. (15.41). Let $D: U = \bigoplus_{i \in I} U_i$, $V = \bigoplus_{j \in J} V_j$, $W = \bigoplus_{\ell \in L} W_\ell$ be a triple of direct sum decompositions for the tensor $t \in U \otimes V \otimes W$ satisfying the assumptions of the theorem. We assume that the support $\mathrm{supp}_D\, t$ is b-tight.

We choose a nonzero function $Q: \mathrm{supp}_D\, t \to \mathbb{N}$ and denote by N the sum of its values. The marginal distributions of Q are denoted by $\mu: I \to \mathbb{N}$, $\nu: J \to \mathbb{N}$, $\pi: L \to \mathbb{N}$. That is,

$$\mu(i) := \sum_{j,\ell} Q(i, j, \ell), \quad \nu(j) := \sum_{i,\ell} Q(i, j, \ell), \quad \pi(\ell) := \sum_{i,j} Q(i, j, \ell),$$

and obviously $\sum_i \mu(i) = \sum_j \nu(j) = \sum_\ell \pi(\ell) = N$. A sequence $x = (x_1, \ldots, x_N)$ in I^N is said to have the *distribution* μ iff the element $i \in I$ appears exactly at $\mu(i)$ positions in x, for all $i \in I$.

The $D^{\otimes N}$-support of the tensor $t^{\otimes N}$

$$(\text{supp}_D t)^N = \text{supp}_{D^{\otimes N}} t^{\otimes N} \subseteq I^N \times J^N \times L^N$$

is by Rem. (15.35) b-tight. We define in dependence on μ, ν, π

$$
\begin{aligned}
I_\mu &:= \{x \in I^N \mid x \text{ has distribution } \mu\}, \\
J_\nu &:= \{y \in J^N \mid y \text{ has distribution } \nu\}, \\
L_\pi &:= \{z \in L^N \mid z \text{ has distribution } \pi\}, \\
\Phi &:= (I_\mu \times J_\nu \times L_\pi) \cap (\text{supp}_D t)^N.
\end{aligned}
$$

Note that $|I_\mu| = \binom{N}{\mu}$, $|J_\nu| = \binom{N}{\nu}$, $|L_\pi| = \binom{N}{\pi}$. The set Φ is not empty as it contains any sequence in $(\text{supp}_D t)^N$ having the distribution Q. The projection $\Phi \to I_\mu$ is surjective and all its fibers have cardinality $|\Phi|/|I_\mu|$. (To see this note that S_N operates transitively on I_μ, and that the projection $\Phi \to I_\mu$ is S_N-equivariant.) Analogous facts hold for J_ν and L_π.

The $D^{\otimes N}$-components of $t^{\otimes N}$ are all matrix tensors: we have

$$t^{\otimes N}(x, y, z) \simeq \otimes_{\rho=1}^N t(x_\rho, y_\rho, z_\rho)$$

for all $(x, y, z) \in (\text{supp}_D t)^N \subseteq I^N \times J^N \times L^N$. We recall that

$$\zeta(t(i, j, \ell)) = (\dim U_i \dim V_j \dim W_\ell)^{1/6}$$

for $(i, j, \ell) \in \text{supp}_D t$ by the definition of the function ζ, as the $t(i, j, \ell)$ are all concise. The reason to restrict the situation to $I_\mu \times J_\nu \times L_\pi$ is the fact that $\zeta(t^{\otimes N}(x, y, z))$ has the same value for all $(x, y, z) \in \Phi$. Indeed, for $(x, y, z) \in \Phi$ we have

$$
\begin{aligned}
\zeta(t^{\otimes N}(x, y, z)) &= \prod_{\rho=1}^N \zeta(t(x_\rho, y_\rho, z_\rho)) = \prod_{\rho=1}^N (\dim U_{x_\rho} \dim V_{y_\rho} \dim W_{z_\rho})^{1/6} \\
\text{(A)} \qquad &= \prod_{i \in I} (\dim U_i)^{\mu(i)/6} \prod_{j \in J} (\dim V_j)^{\nu(j)/6} \prod_{\ell \in L} (\dim W_\ell)^{\pi(\ell)/6} \\
&= \prod_{(i,j,\ell) \in \text{supp}_D t} (\dim U_i \dim V_j \dim W_\ell)^{Q(i,j,\ell)/6} \\
&= \prod_{(i,j,\ell) \in \text{supp}_D t} \zeta(t(i, j, \ell))^{Q(i,j,\ell)}.
\end{aligned}
$$

We apply now Thm. (15.39) to the b-tight subset $\Phi \subseteq I_\mu \times J_\nu \times L_\pi$. This shows the existence of a diagonal Δ in Φ, which is a degeneration of Φ, such that

$$\text{(B)} \qquad |\Delta| \geq C \cdot \min\{|I_\mu|, |J_\nu|, |L_\pi|\} = C \cdot \min\left\{ \binom{N}{\mu}, \binom{N}{\nu}, \binom{N}{\pi} \right\},$$

where C is some constant only depending on b. (In fact, we may take $C = \min\{2/27, (6(2b+1))^{-1}\}$.) As trivially $\Phi \trianglelefteq (\text{supp}_D t)^N$, we have $\Delta \trianglelefteq (\text{supp}_D t)^N$, and thus by Prop. (15.30)

$$\bigoplus_{(x,y,z)\in\Delta} t^{\otimes N}(x, y, z) \trianglelefteq t^{\otimes N}.$$

The asymptotic sum inequality implies (using (A))

$$|\Delta| \prod_{\text{supp}_D t} \zeta(t(i, j, \ell))^{\omega Q(i,j,\ell)} \le \underline{R}(t^{\otimes N}),$$

or by taking logarithms

(C) $\quad \dfrac{1}{N} \log |\Delta| + \omega \displaystyle\sum_{\text{supp}_D t} \dfrac{1}{N} Q(i, j, \ell) \log \zeta(t(i, j, \ell)) \le \dfrac{1}{N} \log \underline{R}(t^{\otimes N}).$

Now let a probability distribution P on $\text{supp}_D t$ and a positive $\varepsilon > 0$ be given. By an easy continuity argument we see that for arbitrarily large natural numbers $N \ge \varepsilon^{-1}$ there is a function $Q: \text{supp}_D t \to \mathbb{N}$ with $\sum Q(i, j, \ell) = N$ such that

(D) $\quad \forall (i, j, \ell) \in \text{supp}_D t: \left| P(i, j, \ell) - \dfrac{1}{N} Q(i, j, \ell) \right| \le \varepsilon,$

(E)
$$\left| \frac{1}{N} \log \binom{N}{\mu} - H(P_1) \right| \le \varepsilon,$$
$$\left| \frac{1}{N} \log \binom{N}{\nu} - H(P_2) \right| \le \varepsilon,$$
$$\left| \frac{1}{N} \log \binom{N}{\pi} - H(P_3) \right| \le \varepsilon,$$

where μ, ν, π are the marginal distributions of Q. (For (E) use (15.40).) Moreover, by Lemma (15.27), we may assume that

(F) $\quad \dfrac{1}{N} \log \underline{R}(t^{\otimes N}) \le \log \underline{R}(t) + \varepsilon.$

Altogether, we deduce from (C) using (B), (D), (E), (F) that

$$\min_{1 \le m \le 3} H(P_m) + \omega \sum_{\text{supp}_D t} P(i, j, \ell) \log \zeta(t(i, j, \ell)) \le \log \underline{R}(t) + \varepsilon S,$$

where $S := 2 + \log C^{-1} + \omega \sum_{\text{supp}_D t} \log \zeta(t(i, j, \ell))$. The conclusion of the theorem follows by letting ε tend to zero. $\qquad\square$

For applying Thm. (15.41) the following remark is often useful. Its proof is left as an exercise to the reader.

(15.42) Remark. Let us make the assumptions of Thm. (15.41) with $I = J = L$, suppose further that $\text{supp}_D t$ is stable under permutations in S_3 and that the function

$$\text{supp}_D t \to \mathbb{N}, (i, j, \ell) \mapsto \zeta(t(i, j, \ell))$$

is constant on S_3-orbits of $\text{supp}_D t$. Then the left-hand side of the inequality in Thm. (15.41) is maximized at an S_3-invariant probability distribution P on $\text{supp}_D t$.

•

We give a first application of Thm. (15.41). For fixed $q \geq 2$ we consider the tensor

$$t^{(q)} = t = \sum_{i=1}^{q} (e_{0ii} + e_{i0i} + e_{ii0}) \in (k^{q+1})^{\otimes 3}.$$

As a direct sum decomposition D of t we choose $I = J = L = \{0, 1\}$ and

$$U_0 = V_0 = W_0 = k(1, 0, \ldots, 0),$$
$$U_1 = V_1 = W_1 = \{\xi \in k^{q+1} \mid \xi_0 = 0\}.$$

Then

$$\begin{aligned}
\operatorname{supp}_D t &= \{(0, 1, 1), (1, 0, 1), (1, 1, 0)\} \\
&= \{(i, j, \ell) \in \{0, 1\}^3 \mid i + j + \ell = 2\},
\end{aligned}$$

which is obviously tight, and

$$t(0, 1, 1) \simeq \langle 1, 1, q \rangle, \quad t(1, 0, 1) \simeq \langle q, 1, 1 \rangle, \quad t(1, 1, 0) \simeq \langle 1, q, 1 \rangle.$$

In order to give an upper bound on $\underline{R}(t)$ we focus on the trilinear form

$$F := \sum_{i.j.\ell} t_{ij\ell} X_i Y_j Z_\ell = \sum_{i=1}^{q} (X_0 Y_i Z_i + X_i Y_0 Z_i + X_i Y_i Z_0)$$

corresponding to t. The following identity

$$\begin{aligned}
&\sum_{i=1}^{q} \varepsilon^{-2} (X_0 + \varepsilon X_i)(Y_0 + \varepsilon Y_i)(Z_0 + \varepsilon Z_i) \\
&- \varepsilon^{-3} \left(X_0 + \varepsilon^2 \sum_{i=1}^{q} X_i \right) \left(Y_0 + \varepsilon^2 \sum_{i=1}^{q} Y_i \right) \left(Z_0 + \varepsilon^2 \sum_{i=1}^{q} Z_i \right) \\
&+ (\varepsilon^{-3} - q\varepsilon^{-2}) X_0 Y_0 Z_0 \\
&= F + \varepsilon G
\end{aligned}$$

holding in the polynomial ring $k(\varepsilon)[X, Y, Z]$ for some $G \in k[\varepsilon][X, Y, Z]$ shows that $\underline{R}(t) \leq q + 2$. We apply now Thm. (15.41). By Rem. (15.42) it is optimal to choose for P the uniform distribution on $\operatorname{supp}_D t$. Therefore

$$H(1/3, 2/3) + \frac{\omega}{3} \log q \leq \log(q + 2),$$

hence $\omega \leq \log_q(4(q + 2)^3/27)$. Setting $q = 8$ we obtain the following result.

(15.43) Corollary (Coppersmith and Winograd, 1987). $\omega < 2.41$.

(15.44) Remark. If the tensor $(t^{(q)})$ defined above had a border rank strictly smaller than $q + 2$, then our argument would yield the estimate $\omega \leq 2$ when choosing $q = 2$! However, as shown in Ex. 15.14(1), we have $\underline{R}(t^{(q)}) = q + 2$.

In Ex. 15.24(7) we will show that in Thm. (15.41) we may replace the border rank by the asymptotic rank $\underline{R}$. Therefore, if we had $\underline{R}(t^{(2)}) = 3$, then we could still conclude that $\omega = 2$. For an interesting approach to decide such questions we refer to Ex. 15.25. ●

We turn to a second application of Thm. (15.41). For fixed $q \geq 2$ we consider the tensor

$$t^{(q)} = t = \sum_{i=1}^{q}(e_{0ii} + e_{i0i} + e_{ii0}) + e_{0,0,q+1} + e_{0,q+1,0} + e_{q+1,0,0} \in (k^{q+2})^{\otimes 3}.$$

We define a direct sum decomposition D of t with $I = J = L = \{0, 1, 2\}$ by setting

$$
\begin{aligned}
U_0 = V_0 = W_0 &= k(1, 0, \dots, 0), \\
U_1 = V_1 = W_1 &= \{\xi \in k^{q+2} \mid \xi_0 = \xi_{q+1} = 0\}, \\
U_2 = V_2 = W_2 &= k(0, \dots, 0, 1).
\end{aligned}
$$

Then $\text{supp}_D t = \{(i, j, \ell) \in \{0, 1, 2\}^3 \mid i + j + \ell = 2\}$ is clearly tight, and

$$t(0, 1, 1) \simeq \langle 1, 1, q \rangle, \quad t(1, 0, 1) \simeq \langle q, 1, 1 \rangle, \quad t(1, 1, 0) \simeq \langle 1, q, 1 \rangle,$$
$$t(0, 0, 2) \simeq t(0, 2, 0) \simeq t(2, 0, 0) \simeq \langle 1, 1, 1 \rangle.$$

The identity

$$\sum_{i=1}^{q} \varepsilon^{-2}(X_0 + \varepsilon X_i)(Y_0 + \varepsilon Y_i)(Z_0 + \varepsilon Z_i)$$

$$- \varepsilon^{-3}\left(X_0 + \varepsilon^2 \sum_{i=1}^{q} X_i\right)\left(Y_0 + \varepsilon^2 \sum_{i=1}^{q} Y_i\right)\left(Z_0 + \varepsilon^2 \sum_{i=1}^{q} Z_i\right)$$

$$+ (\varepsilon^{-3} - q\varepsilon^{-2})(X_0 + \varepsilon^3 X_{q+1})(Y_0 + \varepsilon^3 Y_{q+1})(Z_0 + \varepsilon^3 Z_{q+1})$$

$$= \sum_{i=1}^{q}(X_0 Y_i Z_i + X_i Y_0 Z_i + X_i Y_i Z_0)$$

$$+ X_0 Y_0 Z_{q+1} + X_0 Y_{q+1} Z_0 + X_{q+1} Y_0 Z_0 + \varepsilon G$$

holding in the polynomial ring $k(\varepsilon)[X, Y, Z]$ for some $G \in k[\varepsilon][X, Y, Z]$ shows that $\underline{R}(t) \leq q + 2$. (Since t is concise we even have $\underline{R}(t) = q + 2$.) In applying Thm. (15.41) we may assume by Rem. (15.42) that the probability distribution P is S_3-invariant, say for $\beta \in [0, 1]$

$$
\begin{aligned}
P(0, 1, 1) = P(1, 0, 1) = P(1, 1, 0) &= (1 - \beta)/3, \\
P(0, 0, 2) = P(0, 2, 0) = P(2, 0, 0) &= \beta/3.
\end{aligned}
$$

We then obtain from Thm. (15.41)

$$H(\beta/3, (2 - 2\beta)/3, (1 + \beta)/3) + \frac{\omega}{2}(\beta \log 1 + (1 - \beta) \log q^{2/3}) \leq \log(q + 2),$$

hence

$$\omega \le 3 \, \frac{\log(q+2) - H(\beta/3, (2-2\beta)/3, (1+\beta)/3)}{(1-\beta)\log q}.$$

A numerical minimization shows that the right-hand side attains its minimum for $q = 6$ and $\beta \approx 0.048$ (cf. Ex. 15.23). This yields the following upper bound.

(15.45) Corollary (Coppersmith and Winograd, 1987). $\omega < 2.39$.

15.9* Partial Matrix Multiplication

In this section k is assumed to be an infinite field.

This section is devoted to a generalization of the asymptotic sum inequality due to Schönhage. The method developed here will be employed in the following section for proving an upper bound on the complexity of multiplying rectangular matrices. Both sections can be omitted at first reading.

Let $e, h, \ell \in \mathbb{N}'$ and $I \subseteq \underline{e} \times \underline{h}$, $J \subseteq \underline{h} \times \underline{\ell}$. We define the multiplication $\langle e, h, \ell \rangle_{I,J}$ of partially filled matrices of pattern (I, J) (partial matrix multiplication for short) as being the restriction of $\langle e, h, \ell \rangle$ to

$$\{a \in k^{e \times h} \mid a_{ij} = 0 \text{ for } (i, j) \notin I\} \times \{b \in k^{h \times \ell} \mid b_{j\ell} = 0 \text{ for } (j, \ell) \notin J\}$$

(cf. Fig. 15.2). Note that the bilinear map $\boxed{?} \times \boxed{}$ introduced in Sect. 15.2 is the

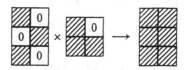

Fig. 15.2. $\langle 3, 2, 2 \rangle_{I,J}$, where $I = \{(1, 1), (2, 2), (3, 1)\}$, $J = \{(1, 1), (2, 1), (2, 2)\}$.

partial matrix multiplication $\langle 2, 2, 2 \rangle_{I,J}$ of pattern (I, J), where $J = \underline{2} \times \underline{2}$, and $I = J \setminus \{(2, 2)\}$.

Our aim is to show that upper bound estimates on the border rank of partial matrix multiplications also yield estimates on the exponent ω. It will be convenient to work in a more general, coordinate free setting. In the following all vector spaces are assumed to be finite dimensional.

Let E, H, L be k-spaces of dimensions e, h, ℓ, respectively. The bilinear map of composing linear endomorphisms defined as

$$\langle E, H, L \rangle \colon \quad \mathrm{Hom}(H, E) \times \mathrm{Hom}(L, H) \to \mathrm{Hom}(L, E), \quad (\phi, \psi) \mapsto \phi \circ \psi$$

is just a coordinate free version of the matrix multiplication map $\langle e, h, \ell \rangle$. Obviously, $\langle E, H, L \rangle$ and $\langle e, h, \ell \rangle$ are equivalent. In analogy to Prop. (14.26) we have

$$\langle E, H, L \rangle \otimes \langle E', H', L' \rangle \simeq \langle E \otimes E', H \otimes H', L \otimes L' \rangle,$$

if E', H', L' are further k-spaces.

Let $E = \oplus_{i \in \mu} E_i$, $H = \oplus_{j \in \nu} H_j$ be direct decompositions of k-spaces, μ, ν being finite index sets, and let $I \subseteq \mu \times \nu$. We define $\text{Hom}(H, E)_I$ as the subspace of those endomorphisms $\phi \in \text{Hom}(H, E)$ having the property that for all $j \in \nu$

$$\phi(H_j) \subseteq \bigoplus_{(i,j) \in I} E_i.$$

We note that if $E' = \oplus_{i' \in \mu'} E_{i'}$, $H' = \oplus_{j' \in \nu'} H_{j'}$ are further decompositions, $I' \subseteq \mu' \times \nu'$, then we have $E \otimes E' = \oplus_{(i,i') \in \mu \times \mu'} E_i \otimes E_{i'}'$, $H \otimes H' = \oplus_{(j,j') \in \nu \times \nu'} H_j \otimes H_{j'}'$ and

$$\text{Hom}(H, E)_I \otimes \text{Hom}(H', E')_{I'} \simeq \text{Hom}(H \otimes H', E \otimes E')_{I \times I'},$$

when we view $I \times I'$ as a subset of $(\mu \times \mu') \times (\nu \times \nu')$. Given a direct decomposition $L = \oplus_{\ell \in \pi} L_\ell$ of a third space and some $J \subseteq \nu \times \pi$, we may define the *partial matrix multiplication* $\langle E, H, L \rangle_{I,J}$ with the pattern (I, J) as the restriction of $\langle E, H, L \rangle$ to $\text{Hom}(H, E)_I \times \text{Hom}(L, H)_J$. It is clear that this notion coincides with the previous $\langle e, h, \ell \rangle_{I,J}$, if $E = k^e, H = k^h, L = k^\ell$ are decomposed in the canonical way into one dimensional subspaces. The tensor product of partial matrix multiplications is again a partial matrix multiplication. More specifically, we have

(15.46) $\langle E, H, L \rangle_{I,J} \otimes \langle E', H', L' \rangle_{I',J'} \simeq \langle E \otimes E', H \otimes H', L \otimes L' \rangle_{I \times I', J \times J'}.$

We leave the straightforward proof of this fact as an exercise to the reader. For instance we have

$$(\ulcorner \times \square)^{\otimes 3} = (\langle 2, 2, 2 \rangle_{I,J})^{\otimes 3} \simeq \langle 8, 8, 8 \rangle_{I_1, J_1},$$

where $J_1 = (\underline{2} \times \underline{2})^3$ and the set $I_1 = I^3$ can be illustrated as follows:

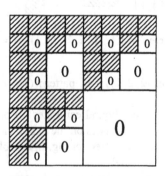

The next lemma shows that partial matrix multiplications of a certain pattern restrict to (full) matrix multiplications.

(15.47) Lemma. *Let $E = \oplus_{i \in \mu} E_i$, $H = \oplus_{j \in \nu} H_j$, $L = \oplus_{\ell \in \pi} L_\ell$ be direct decompositions of k-spaces and $I \subseteq \mu \times \nu$, $J \subseteq \nu \times \pi$. We require that*

$$M := \sum_{i:(i,j) \in I} \dim E_i, \quad P := \sum_{\ell:(j,\ell) \in J} \dim L_\ell$$

are independent of $j \in \nu$. Then $\langle k^M, H, k^P \rangle$ is a restriction of $\langle E, H, L \rangle_{I,J}$.

Proof. For $\alpha \in \mathrm{Hom}(E, k^M)$, $\beta \in \mathrm{Hom}(k^P, L)$ we define the linear maps

$$
\begin{aligned}
{}_\alpha \Delta &: \mathrm{Hom}(H, E)_I \to \mathrm{Hom}(H, k^M), && \phi \mapsto \alpha \circ \phi, \\
\Delta_\beta &: \mathrm{Hom}(L, H)_J \to \mathrm{Hom}(k^P, H), && \psi \mapsto \psi \circ \beta, \\
{}_\alpha \Delta_\beta &: \mathrm{Hom}(L, E) \to \mathrm{Hom}(k^P, k^M), && \chi \mapsto \alpha \circ \chi \circ \beta.
\end{aligned}
$$

The following diagram is commutative

$$
\begin{array}{ccc}
\mathrm{Hom}(H, E)_I \times \mathrm{Hom}(L, H)_J & \xrightarrow{\ \langle E, H, L \rangle_{I,J}\ } & \mathrm{Hom}(L, E) \\
\Big\downarrow {}_\alpha \Delta \times \Delta_\beta & & \Big\downarrow {}_\alpha \Delta_\beta \\
\mathrm{Hom}(H, k^M) \times \mathrm{Hom}(k^P, H) & \xrightarrow[\ \langle k^M, H, k^P \rangle\ ]{} & \mathrm{Hom}(k^P, k^M)
\end{array}
$$

Therefore, it suffices to show the existence of α, β such that ${}_\alpha \Delta$ and Δ_β are linear isomorphisms (cf. Rem. (14.30)).

There is some $\alpha \in \mathrm{Hom}(E, k^M)$ with the property that α restricted to $\oplus_{(i,j) \in I} E_i$ is an isomorphism, for all $j \in \nu$. (Namely, by choosing bases, this amounts to finding a matrix for which certain minors do not vanish. As k is infinite, this is certainly possible.) Let $\phi \in \mathrm{Hom}(H, E)_I$ be such that $\alpha \circ \phi = 0$. As $\phi(H_j)$ is contained in $\oplus_{(i,j) \in I} E_i$, and α is assumed to be injective on this subspace, we conclude that $\phi(H_j) = 0$, for all $j \in \nu$. Thus ${}_\alpha \Delta$ is injective. As $\dim \mathrm{Hom}(H, E)_I = M \dim H = \dim \mathrm{Hom}(H, k^M)$, we see that ${}_\alpha \Delta$ is an isomorphism. Analogously one shows the existence of a suitable β.   □

The following theorem generalizes Prop. (15.10).

(15.48) Theorem (Schönhage). *Let $E = \oplus_{i \in \mu} E_i$, $H = \oplus_{j \in \nu} H_j$, $L = \oplus_{\ell \in \pi} L_\ell$ be direct decompositions of k-spaces, $I \subseteq \mu \times \nu$, $J \subseteq \nu \times \pi$, and assume $\underline{R}(\langle E, H, L \rangle_{I,J}) \leq r$ for $r \in \mathbb{N}$. Then*

$$\left(\sum_{j \in \nu} m_j p_j \dim H_j \right)^{\omega/3} \leq r,$$

where $m_j := \sum_{(i,j) \in I} \dim E_i$, $p_j := \sum_{(j,\ell) \in J} \dim L_\ell$.

Proof. By Cor. (15.18) we may assume that k is infinite. There is some $q \in \mathbb{N}'$ such that $\langle E, H, L \rangle_{I,J} \trianglelefteq_q \langle r \rangle$. Using (15.46) we conclude for all $N \in \mathbb{N}'$

$$\langle E^{\otimes N}, H^{\otimes N}, L^{\otimes N} \rangle_{I^N, J^N} \simeq \left(\langle E, H, L \rangle_{I,J} \right)^{\otimes N} \trianglelefteq_{Nq} \langle r^N \rangle.$$

Let $\sigma: v \to \mathbb{N}$ with $\sum_{j \in v} \sigma(j) = N$ and define

$$v^\sigma := \{ j \in v^N \mid j \text{ has distribution } \sigma \}$$

and

$$H^\sigma := \bigoplus_{j \in v^\sigma} H_{j_1} \otimes \cdots \otimes H_{j_N}.$$

(Recall that $j \in v^N$ has distribution σ iff $|\{s \mid j_s = i\}| = \sigma(i)$ for all $i \in v$.) Moreover, we set $I^\sigma := I^N \cap (\mu^N \times v^\sigma)$ and $J^\sigma := J^N \cap (v^\sigma \times \pi^N)$. Then the partial matrix multiplication $\langle E^{\otimes N}, H^\sigma, L^{\otimes N} \rangle_{I^\sigma, J^\sigma}$ is a restriction of $\langle E^{\otimes N}, H^{\otimes N}, L^{\otimes N} \rangle_{I^N, J^N}$. Now observe that for all $j \in v^\sigma$

$$\dim \bigoplus_{(i,j) \in I^\sigma} E_{i_1} \otimes \cdots \otimes E_{i_N} = m_{j_1} \cdots m_{j_N} = m_1^{\sigma_1} \cdots m_v^{\sigma_v} =: M,$$

$$\dim \bigoplus_{(j,\ell) \in J^\sigma} L_{\ell_1} \otimes \cdots \otimes L_{\ell_N} = p_{j_1} \cdots p_{j_N} = p_1^{\sigma_1} \cdots p_v^{\sigma_v} =: P.$$

By construction, M and P are independent of $j \in v^\sigma$. Therefore, by Lemma (15.47), the partial matrix multiplication $\langle E^{\otimes N}, H^\sigma, L^{\otimes N} \rangle_{I^\sigma, J^\sigma}$ restricts to the matrix multiplication $\langle k^M, H^\sigma, k^P \rangle$. Thus we have

(15.49) $$\langle k^M, H^\sigma, k^P \rangle \trianglelefteq_{Nq} \langle r^N \rangle.$$

Prop. (15.10) implies $MP \dim H^\sigma \leq r^{3N/\omega}$. If we sum over all distributions σ and observe that

$$MP \dim H^\sigma = \binom{N}{\sigma} \prod_{j \in v} (m_j p_j \dim H_j)^{\sigma_j},$$

we get by the multinomial theorem

$$\left(\sum_{j \in v} m_j p_j \dim H_j \right)^N \leq \binom{N + |v| - 1}{|v| - 1} r^{3N/\omega}.$$

By taking Nth roots and letting $N \to \infty$ we obtain the desired statement. $\square$

(15.50) Example. In Sect. 15.2 was shown that $\underline{R}(\square \times \square) \leq 5$. By Thm. (15.48) this implies $6^{\omega/3} \leq 5$, hence $\omega < 2.70$. This improves the bound $\omega < 2.78$ obtained in Prop. (15.9) by applying Prop. (15.10).   •

We remark that Thm. (15.48) can be generalized to direct sums of partial matrix multiplications. For a statement and a proof of this we refer to Ex. 15.28.

15.10* Rapid Multiplication of Rectangular Matrices

It is clear that for positive natural numbers h, s the estimate $R(\langle h, h, s\rangle) \leq h^2 s$ holds. Therefore, if s depends on h in such a way that $s = s(h) = O(\log^2 h)$, then we have

$$R(\langle h, h, s(h)\rangle) = O(h^2 \log^2 h).$$

Astonishingly, we can prove that this estimate is also true for $s(h) = \lfloor h^\alpha \rfloor$, where α is a positive real number smaller than 0.17.

(15.51) Theorem (Coppersmith). *Over an infinite field k we have for $h \to \infty$*

$$\begin{aligned}
\underline{R}(\langle h, h, \lfloor h^\alpha \rfloor\rangle) &= O(h^2 \log h), \\
R(\langle h, h, \lfloor h^\alpha \rfloor\rangle) &= O(h^2 \log^2 h),
\end{aligned}$$

where $\alpha := \log 4/(5 \log 5) \approx 0.17$.

Proof. We consider the following partial matrix multiplication

$$\begin{pmatrix} a_{11} & a_{12} & a_{13} \\ 0 & a_{22} & a_{23} \end{pmatrix} \cdot \begin{pmatrix} b_{11} & b_{12} \\ b_{21} & 0 \\ b_{31} & 0 \end{pmatrix} = \begin{pmatrix} c_{11} & c_{12} \\ c_{21} & c_{22} \end{pmatrix},$$

let us call it ϕ. Since ϕ is 1-concise, we have $\underline{R}(\phi) \geq 5$. In fact, the border rank of ϕ equals 5, which is demonstrated by the following identity showing that $\phi \leq_3 \langle 5\rangle$: we have

$$\begin{aligned}
&(a_{11} + \varepsilon^2 a_{12})(b_{21} + \varepsilon^2 b_{11})c_{11} + (a_{11} + \varepsilon^2 a_{13})b_{31}(c_{11} - \varepsilon c_{21}) \\
&\quad + (a_{11} + \varepsilon^2 a_{22})(b_{21} - \varepsilon b_{12})c_{12} \\
&\quad\quad + (a_{11} + \varepsilon^2 a_{23})(b_{31} + \varepsilon b_{12})(c_{12} + \varepsilon c_{21}) - a_{11}(b_{21} + b_{31})(c_{11} + c_{12}) \\
&= \varepsilon^2 \Big(a_{11} b_{11} c_{11} + a_{11} b_{12} c_{21} + a_{12} b_{21} c_{11} \\
&\quad\quad + a_{13} b_{31} c_{11} + a_{22} b_{21} c_{12} + a_{23} b_{31} c_{12} \Big) + \varepsilon^3 F
\end{aligned}$$

with some $F \in k[\varepsilon][a, b, c]$. With respect to the direct decompositions

$$E = L = k^2 = k(1, 0) \oplus k(0, 1), \quad H = k^3 = k(1, 0, 0) \oplus (k(0, 1, 0) \oplus k(0, 0, 1))$$

the bilinear map ϕ is the partial matrix multiplication $\langle E, H, L\rangle_{I,J}$ with the pattern

$$I = \{(1, 1), (1, 2), (2, 2)\}, \quad J = \{(1, 1), (1, 2), (2, 1)\} \subseteq \{1, 2\}^2.$$

We refer now to the proof of Thm. (15.48). By statement (15.49) in this proof we have for any $N \in \mathbb{N}'$ and any $\sigma_1, \sigma_2 \in \mathbb{N}$ satisfying $\sigma_1 + \sigma_2 = N$

$$\left\langle 1^{\sigma_1} \cdot 2^{\sigma_2}, \binom{N}{\sigma_2} \cdot 1^{\sigma_1} \cdot 2^{\sigma_2}, 2^{\sigma_1} \cdot 1^{\sigma_2}\right\rangle \trianglelefteq_{3N} \langle 5^N\rangle.$$

By permuting the first two components and taking the tensor product we get by Lemma (15.24)(3)

$$\left\langle \binom{N}{\sigma_2} 4^{\sigma_2}, \binom{N}{\sigma_2} 4^{\sigma_2}, 4^{\sigma_1} \right\rangle \trianglelefteq_{6N} \left(5^{2N} \right),$$

which implies by Prop. (15.26)

(A) $$\left\langle \binom{N}{\sigma_2} 4^{\sigma_2}, \binom{N}{\sigma_2} 4^{\sigma_2}, 4^{\sigma_1} \right\rangle \leq \left(12N \cdot 5^{2N} \right).$$

We assume now that N is a multiple of five and choose $\sigma_1 = N/5$, $\sigma_2 = 4N/5$. Stirling's formula gives us the asymptotic estimate

$$\binom{N}{\sigma_2} 4^{\sigma_2} \sim c \cdot \frac{5^N}{\sqrt{N}} \quad (N \to \infty)$$

for some constant $c > 0$. Given a sufficiently large $h \in \mathbb{N}$, we choose N to be the minimal multiple of five satisfying $h \leq c5^N/(2\sqrt{N})$. Then it is a little exercise to show that (A) implies

$$R(\langle h, h, \lfloor h^\alpha \rfloor \rangle) \leq c'h^2 \log^2 h$$

for some positive constant c' and $\alpha := \log 4/(5 \log 5)$. (Note that $4^{\sigma_1} = (5^N)^\alpha$, and use that $h_1 = c5^N/\sqrt{N}$ implies $N \leq \log h_1$ for large enough N.) This proves the theorem. $\square$

15.11 Exercises

15.1. Let $A = (a_{ij})$, $B = (b_{ij})$ be 2×2-matrices over a ring S, and $C = (c_{ij}) := AB$. Check that the following scheme gives a bilinear computation of length seven for C with 15 additions or subtractions.

$$
\begin{array}{llll}
\xi_0 & := & a_{21} + a_{22} & \qquad \eta_0 & := & b_{12} - b_{11} \\
\xi_1 & := & \xi_0 - a_{11} & \qquad \eta_1 & := & b_{22} - \eta_0 \\
\xi_2 & := & a_{11} - a_{21} & \qquad \eta_2 & := & b_{22} - b_{12} \\
\xi_3 & := & a_{12} - \xi_1 & \qquad \eta_3 & := & \eta_1 - b_{21}
\end{array}
$$

$$
\begin{array}{llll}
P_1 & := & \xi_1 \cdot \eta_1 & \qquad P_5 & := & \xi_0 \cdot \eta_0 \\
P_2 & := & a_{11} \cdot b_{11} & \qquad P_6 & := & \xi_3 \cdot b_{22} \\
P_3 & := & a_{12} \cdot b_{21} & \qquad P_7 & := & a_{22} \cdot \eta_3 \\
P_4 & := & \xi_2 \cdot \eta_2 &
\end{array}
$$

$$
\begin{array}{llll}
c_{11} & = & P_3 + P_2 & \qquad c_{12} & = & Z_3 + P_6 \\
Z_1 & := & P_1 + P_2 & \qquad c_{21} & = & Z_2 - P_7 \\
Z_2 & := & Z_1 + P_4 & \qquad c_{22} & = & Z_2 + P_5 \\
Z_3 & := & Z_1 + P_5 &
\end{array}
$$

15.2. We present a generalization of Strassen's algorithm. Let $A = (a_{ij})$ and $B = (b_{ij})$ be two $h \times h$-matrices. With $3(h^2 - h)$ multiplications we compute the following products:

$$u_{ij} := (a_{ij} + a_{ii})b_{jj}, \quad v_{ij} := a_{ii}(b_{ij} - b_{jj}), \quad w_{ij} := (a_{ij} - a_{jj})\sum_s b_{js}$$

for distinct $i, j \in \underline{h}$. We further compute for pairwise distinct i, j, ℓ

$$p_{ij\ell} := a_{i\ell}b_{\ell j}, \quad q := (\sum_s a_{ss})(\sum_s b_{ss})$$

with $h(h-1)(h-2) + 1$ multiplications. Show that all entries of the product matrix $C = (c_{ij}) = AB$ can be obtained from these products by linear operations. This proves that $R(\langle h, h, h \rangle) \leq h^3 - h + 1$. In particular $R(\langle 2, 2, 2 \rangle) \leq 7$ and $R(\langle 3, 3, 3 \rangle) \leq 25$.

15.3. Let $A = (a_{ij})$ and $B = (b_{ij})$ be 3×3-matrices and $C = (c_{ij}) = AB$. Check the following identity showing that $R(\langle 3, 3, 3 \rangle) \leq 23$. (This is the best known upper bound on the rank of 3×3-matrix multiplication.) Let

$$
\begin{aligned}
m_1 &:= (a_{11} + a_{12} + a_{13} - a_{21} - a_{22} - a_{32} - a_{33})b_{22}, \\
m_2 &:= (a_{11} - a_{21})(-b_{12} + b_{22}), \\
m_3 &:= a_{22}(-b_{11} + b_{12} + b_{21} - b_{22} - b_{23} - b_{31} + b_{33}), \\
m_4 &:= (-a_{11} + a_{21} + a_{22})(b_{11} - b_{12} + b_{22}), \\
m_5 &:= (a_{21} + a_{22})(-b_{11} + b_{12}), \\
m_6 &:= a_{11}b_{11}, \\
m_7 &:= (-a_{11} + a_{31} + a_{32})(b_{11} - b_{13} + b_{23}), \\
m_8 &:= (-a_{11} + a_{31})(b_{13} - b_{23}), \\
m_9 &:= (a_{31} + a_{32})(-b_{11} + b_{13}), \\
m_{10} &:= (a_{11} + a_{12} + a_{13} - a_{22} - a_{23} - a_{31} - a_{32})b_{23}, \\
m_{11} &:= a_{32}(-b_{11} + b_{13} + b_{21} - b_{22} - b_{23} - b_{31} + b_{32}), \\
m_{12} &:= (-a_{13} + a_{32} + a_{33})(b_{22} + b_{31} - b_{32}), \\
m_{13} &:= (a_{13} - a_{33})(b_{22} - b_{32}), \\
m_{14} &:= a_{13}b_{31}, \\
m_{15} &:= (a_{32} + a_{33})(-b_{31} + b_{32}), \\
m_{16} &:= (-a_{13} + a_{22} + a_{23})(b_{23} + b_{31} - b_{33}), \\
m_{17} &:= (a_{13} - a_{23})(b_{23} - b_{33}), \\
m_{18} &:= (a_{22} + a_{23})(-b_{31} + b_{33}), \\
m_{19} &:= a_{12}b_{21}, \\
m_{20} &:= a_{23}b_{32}, \\
m_{21} &:= a_{21}b_{13}, \\
m_{22} &:= a_{31}b_{12}, \\
m_{23} &:= a_{33}b_{33}.
\end{aligned}
$$

Then

$$
\begin{aligned}
c_{11} &:= m_6 + m_{14} + m_{19}, \\
c_{12} &:= m_1 + m_4 + m_5 + m_6 + m_{12} + m_{14} + m_{15}, \\
c_{13} &:= m_6 + m_7 + m_9 + m_{10} + m_{14} + m_{16} + m_{18},
\end{aligned}
$$

$$c_{21} := m_2 + m_3 + m_4 + m_6 + m_{14} + m_{16} + m_{17},$$
$$c_{22} := m_2 + m_4 + m_5 + m_6 + m_{20},$$
$$c_{23} := m_{14} + m_{16} + m_{17} + m_{18} + m_{21},$$
$$c_{31} := m_6 + m_7 + m_8 + m_{11} + m_{12} + m_{13} + m_{14},$$
$$c_{32} := m_{12} + m_{13} + m_{14} + m_{15} + m_{22},$$
$$c_{33} := m_6 + m_7 + m_8 + m_9 + m_{23}.$$

15.4. Give an example showing that the scalar restriction of bilinear maps does not commute with $\otimes$.

15.5. Let ϕ, ψ be k-bilinear maps over an infinite field k and $K = k(X_1, \ldots, X_n)$ be a rational function field over k. Prove that $\phi^K \le \psi^K$ implies $\phi \le \psi$.

15.6.* We assume that k is an infinite field.

(1) Let $\phi: U \times U \to U$ be a bilinear map over k. Suppose we have $\phi(u, v) = \sum_{\rho=1}^{r} f_\rho(u) g_\rho(v) w_\rho$ for all $u, v \in U$, with $n \le r \le 3n$, $f_1, \ldots, f_r \in U^*$ in general position, $g_1, \ldots, g_r \in U^*$ also in general position, and $w_1, \ldots, w_r$ generating U. ($f_1, \ldots, f_r \in U^*$ are said to be in general position iff $(f_{\rho_1}, \ldots, f_{\rho_n})$ form a basis of U^* for every choice of $1 \le \rho_1 < \rho_2 < \ldots < \rho_n \le r$.) Prove that $\langle \lfloor (3n - r)/2 \rfloor \rangle$ is a restriction of ϕ.

(2) For a polynomial $p \in k[T]$ of degree $n \ge 1$ we have

$$\langle \lceil \tfrac{n}{2} \rceil \rangle \le k[T]/(p) \le \langle 2n - 1 \rangle.$$

15.7. Prove the following supplement to Prop. (15.17) by applying Ex. 15.6. We suppose that k is of characteristic zero.

(1) Let A be a k-algebra and ϕ, ψ be k-bilinear maps. Then $\phi^A \le \psi^A$ implies the existence of some $q \in \mathbb{N}'$ such that $\langle q \rangle \otimes \phi \le \langle 4q \rangle \otimes \psi$. (Hint: as $\operatorname{char} k = 0$, every finite field extension of k is simply generated.)

(2) If the additivity conjecture holds over k, then for any k-algebra A and k-bilinear map ϕ we have $R(\phi^A) \le R(\phi) \le 4R(\phi^A)$. (This statement might be helpful for disproving the additivity conjecture, at least for nonclosed ground-fields k.)

15.8. Let $A = (a_{ij})$, $B = (b_{ij})$ be 3×3-matrices with indeterminate entries, and let ε be a further indeterminate over k. Compute for $1 \le i, j \le 3$ the 21 products

$$u_{ii} := (a_{i1} + \varepsilon^2 a_{i2})(\varepsilon^2 b_{1i} + b_{2i}), \quad u_{ij} := (a_{i1} + \varepsilon^2 a_{j2})(b_{2i} - \varepsilon b_{1j}), \ (i \ne j),$$
$$v_{ii} := (a_{i1} + \varepsilon^2 a_{i3}) b_{3i}, \quad v_{ij} := (a_{i1} + \varepsilon^2 a_{j3})(b_{3i} + \varepsilon b_{1j}), \ (i \ne j),$$
$$w_i := a_{i1}(b_{2i} + b_{3i})$$

and define the 3×3-matrix $D = (d_{ij})$ by

$$d_{ij} := \varepsilon^{-2}(u_{ij} + v_{ij} - w_i) + \varepsilon^{-1}(v_{ji} - v_{jj}).$$

Check that $D = AB + \varepsilon R$, where R is some 3×3-matrix over $k[a, b]$. This shows $\underline{R}(\langle 3, 3, 3\rangle) \le 21$, which is the best known upper bound on the border rank of 3×3-matrix multiplication. Which estimate on the exponent do we get from this?

15.9. In Example (15.20) we showed that $k[T]/(T^m)$ has border rank m, assuming that k possesses a primitive mth root of unity. Give another proof of this fact which works over any field. Use that $k(\varepsilon)[T]/(f)$ with $f = T(T - \varepsilon) \cdots (T - \varepsilon^{m-1})$ has rank m and "approximates" $k[T]/(T^m)$.

15.10. Prove that for bilinear maps ϕ, ψ over k and $q \in \mathbb{N}'$ we have $\phi \unlhd_q \psi$ iff $\varepsilon^{q-1}\phi^{\mathcal{R}} \le \psi^{\mathcal{R}}$, where $\mathcal{R}$ denotes the k-algebra $k[\varepsilon]/(\varepsilon^q)$. Conclude that in Def. (15.19) the polynomial ring $k[\varepsilon]$ may be replaced by the ring of formal power series $k[[\varepsilon]]$.

15.11. We give a further characterization of the degeneration order. Let $R := k[[\varepsilon]]$ be the ring of formal power series in ε over k and $K := k((\varepsilon))$ its field of fractions. The group $G^K := \mathrm{GL}(m, K) \times \mathrm{GL}(n, K) \times \mathrm{GL}(p, K)$ acts on $K^{m \times n \times p} \simeq K^m \otimes K^n \otimes K^p$ via $(\alpha, \beta, \gamma) \cdot t := (\alpha \otimes \beta \otimes \gamma)(t)$. Further, the k-algebra morphism $R \to k$, $\sum_{i \ge 0} a_i \varepsilon^i \mapsto a_0$, induces a k-linear map $R^{m \times n \times p} \to k^{m \times n \times p}$, $t \mapsto t_{\varepsilon=0}$. Let $s, t \in k^{m \times n \times p}$. Prove that $s \unlhd t$ iff there exists some $g \in G^K$ such that $g \cdot t \in R^{m \times n \times p}$ and $(g \cdot t)_{\varepsilon=0} = s$.

15.12. Show that the equivalence in Lemma (15.22) remains true if we replace in (2) the restriction $\le$ by the degeneration $\unlhd$.

15.13.* In this exercise we discuss an obstruction for a bilinear map being a degeneration of another one. By a *filtration* $\mathcal{F}$ of a triple of k-spaces U, V, W we understand a triple $\mathcal{F} := ((U_i)_{i \in \underline{a}}, (V_j)_{j \in \underline{b}}, (W_\ell)_{\ell \in \underline{c}})$ of ascending sequences of subspaces of U, V, W, respectively. The crucial statement is as follows. Let $\phi, \psi \colon U \times V \to W$ be bilinear maps, $\phi \unlhd \psi$, and $\mathcal{F}$ be a filtration of U, V, W. Then there exists a filtration $\mathcal{F}' := ((U_i')_{i \in \underline{a}}, (V_j')_{j \in \underline{b}}, (W_\ell')_{\ell \in \underline{c}})$ of U, V, W such that for all $(i, j, \ell) \in \underline{a} \times \underline{b} \times \underline{c}$ we have $\dim U_i' = \dim U_i$, $\dim V_j' = \dim V_j$, $\dim W_\ell' = \dim W_\ell$ and

$$\phi(U_i' \times V_j') \subseteq W_\ell' \quad \text{provided} \quad \psi(U_i \times V_j) \subseteq W_\ell.$$

(1) Give a direct proof of this statement based on Prop. (8.54) and Ex. 15.11.
(2) Give another proof of the above statement for algebraically closed fields k based on the topological characterization of border rank in Thm. (20.24). Use the fact that the existence of subspaces $U' \subseteq U$, $V' \subseteq V$, $W' \subseteq W$ of prescribed dimensions satisfying $\psi(U' \times V') \subseteq W'$ is a closed condition on ψ. (Hint: Grassmann varieties!)

15.14. We give some applications of the previous Ex. 15.13.

(1) Let $\phi: U \times U \to U$ be a bilinear map such that $\underline{R}(\phi) = \dim U =: m$. Show that there exists a sequence $U_1 \subset U_2 \subset \ldots \subset U_m = U$ of subspaces such that $\dim U_i = i$ and $\dim\langle\phi(U_i \times U)\rangle \leq i$ for all i.

(2) Let A be a (not necessarily commutative) algebra with unit element satisfying $\underline{R}(A) = \dim A =: m$. Then there exists an ascending sequence of left ideals $L_1 \subset L_2 \subset \ldots \subset L_m = A$ of A satisfying $\dim L_i = i$.

(3) The tensor $t^{(q)} = \sum_{i=1}^{q}(e_{0ii} + e_{i0i} + e_{ii0}) \in (k^{q+1})^{\otimes 3}$ considered in the first application of Thm. (15.41) has border rank $q + 2$ iff $q > 1$.

(4) The bilinear map $\square \times \square$ considered in Sect. 15.2 has border rank 5.

15.15. We discuss an iterative procedure for computing the solution τ of the equation $f_1^\tau + \ldots + f_s^\tau = r$, where $1 < s < r < f_1 + \ldots + f_s$, the f_i being positive natural numbers not all equal to one.

(1) Prove that there is exactly one solution $\tau \in (0, 1)$.

(2) We define the sequence $\tau_0 := 1, \tau_1, \ldots$ recursively by

$$\tau_{j+1} := \tau_j \theta_j, \quad \text{where } \theta_j := \frac{\log r}{\log \sum_{i=1}^{s} f_i^{\tau_j}}.$$

Prove that (τ_j) converges monotonically towards the solution τ. Check also that we have linear convergence. (Hint: show that there is some constant $\gamma \in (0, 1)$ such that $\theta_j \leq \gamma$ for all j. Use the general inequality $(\sum_{i=1}^{s} u_i)^\theta < \sum_{i=1}^{s} u_i^\theta$ holding for $\theta \in (0, 1)$, $u_i > 0$, $s > 1$.)

15.16. Perform a numerical calculation to show that the best estimate on ω obtained from $(e\ell)^{\omega/3} + ((e - 1)(\ell - 1))^{\omega/3} \leq e\ell + 1$ is achieved for $e = \ell = 4$, namely $\omega < 2.55$.

15.17. In this exercise we show that for $h, \ell \in \mathbb{N}'$

$$\underline{R}(\langle 1, h, 2\ell\rangle \oplus \langle \ell, 2, h\rangle \oplus \langle 2h, \ell, 1\rangle) \leq 2(h + 1)(\ell + 2).$$

(1) The trilinear form corresponding to $\langle 1, h, \ell\rangle \oplus \langle \ell, 1, h\rangle \oplus \langle h, \ell, 1\rangle$ is

$$F := \sum_{i,j}(a_i y_{ij} w_j + u_j b_i z_{ij} + x_{ij} v_j c_i),$$

where the $a_i, b_i, c_i, u_j, v_j, w_j, x_{ij}, y_{ij}, z_{ij}$ are indeterminates and the summation runs over all $1 \leq i \leq h$, $1 \leq j \leq \ell$. Denote by a the sum over all a_i, by b the sum over all b_i, etc. Check the identity

$$\varepsilon^{12} \sum_{i,j} (a_i + \varepsilon^4 u_j + \varepsilon^6 x_{ij})(b_i + \varepsilon^2 v_j + \varepsilon^5 y_{ij})(c_i + \varepsilon^3 w_j + \varepsilon^4 z_{ij})$$

$$- \varepsilon^8 \sum_i (a_i + \varepsilon^8 u + \varepsilon^{10} x_i)(b_i + \varepsilon^6 v + \varepsilon^9 y_i)(c_i + \varepsilon^7 w + \varepsilon^8 z_i)$$

$$- \varepsilon^4 \sum_j (a + \varepsilon^8 u_j)(b + \varepsilon^6 v_j)(c + \varepsilon^7 w_j)$$

$$+ (a + \varepsilon^{12} u)(b + \varepsilon^{10} v)(c + \varepsilon^{11} w)$$

$$- (\ell \varepsilon^{12} - \varepsilon^8) \sum_i a_i b_i c_i - (1 - \ell \varepsilon^4) abc$$

$$= \varepsilon^{20} F + \varepsilon^{18} \sum_{i,j} a_i v_j z_{i,j} + O(\varepsilon^{21}).$$

(2) In order to cancel the ε^{18}-term we set up a second trilinear form

$$\overline{F} := \sum_{i,j} (a_i \overline{y}_{ij} \overline{w}_j + \overline{u}_j \overline{b}_i z_{ij} + \overline{x}_{ij} v_j \overline{c}_i)$$

involving the same a_i, v_j, z_{ij}, but different indeterminates $\overline{b}_i$, $\overline{c}_i$, $\overline{u}_j$, $\overline{w}_j$, $\overline{x}_{ij}$, $\overline{y}_{ij}$. Show that the tensor t corresponding to $F - \overline{F}$ has border rank at most $2(h + 1)(\ell + 2)$, and $t \simeq \langle 1, h, 2\ell \rangle \oplus \langle \ell, 2, h \rangle \oplus \langle 2h, \ell, 1 \rangle$.

(3) Which estimate on the exponent do we get from this? (Take $h = 5$ and $\ell = 11$.)

15.18. Verify that the set $\{(i, j, \ell) \in \underline{h}^3 \mid i + j + \ell = s\}$ has the maximal cardinality $\lceil 3h^2/4 \rceil$ for the value $s = \lceil 3h/2 \rceil + 1$.

15.19. Show that $\langle r \rangle \trianglelefteq \langle h, h, h \rangle$ for $r = \lceil 3h^2/4 \rceil$.

15.20. Prove that the set $\Psi_M = \{(x, y, z) \in (\mathbb{Z}/M\mathbb{Z})^3 \mid x + y + z = 0\}$ is not tight for $M \geq 2$.

15.21. Show that the entropy describes the asymptotic growth of the multinomial coefficients, as stated in (15.40). (Hint: use Stirling's formula $n! \sim \sqrt{2\pi n}\, (n/e)^n$.)

15.22. Prove Rem. (15.42). (Hint: use the concavity of the entropy function.)

15.23. For $q \in \mathbb{N}$, $q \geq 2$ define the function $f_q : [0, 1) \to \mathbb{R}$ by

$$f_q(\beta) := 3\, \frac{\log(q + 2) - H(\beta/3, (2 - 2\beta)/3, (1 + \beta)/3)}{(1 - \beta) \log q}.$$

Show that $\min_q \min_\beta f_q(\beta) \approx \min_\beta f_6(\beta) \approx f_6(0.048) \approx 2.39$ by a numerical minimization.

15.24.* This exercise is devoted to a discussion of the notion of asymptotic rank.

(1) Let (α_N) be a sequence of non-negative real numbers satisfying $\alpha_{N+M} \leq \alpha_N + \alpha_M$ for all $N, M \in \mathbb{N}$. Then $\lim_{N \to \infty} \alpha_N / N$ exists and equals $\inf_N \alpha_N / N$.

(2) The *asymptotic rank* $\underset{\sim}{R}(t)$ of a tensor t is defined as

$$\underset{\sim}{R}(t) := \lim_{N \to \infty} R(t^{\otimes N})^{1/N}.$$

By (1) this limit is well-defined. We have $\underline{R}(t) \leq \underset{\sim}{R}(t) \leq R(t)$ and $\underset{\sim}{R}(t^{\otimes m}) = \underset{\sim}{R}(t)^m$ for $m \in \mathbb{N}$. (Hint: Lemma (15.27).)

(3) $\underset{\sim}{R} : \mathcal{T} \to [0, \infty)$ is a subadditive and submultiplicative functional on the semiring $\mathcal{T}$ of tensor classes. $\underset{\sim}{R}$ is monotone w.r.t. $\leq$ and $\unlhd$ and it is S_3-invariant. (Hint: to show the subadditivity use that for all $a \in \mathcal{T}$ there is some monotonically increasing sequence $\varepsilon_N = o(N)$ of natural numbers satisfying $R(a^N) \leq \underset{\sim}{R}(a)^N 2^{\varepsilon_N}$.)

(4) $\underset{\sim}{R}(\langle 2, 2, 2 \rangle) = 2^\omega$.

(5) The asymptotic sum inequality (15.11) is true with $\underline{R}$ replaced by $\underset{\sim}{R}$. (Of course in its statement r may now be a real number.)

(6) Prove that $\underset{\sim}{R}\left(\sum_{i=1}^s \langle 2, 2, 2 \rangle^{N_i}\right) = \sum_{i=1}^s \underset{\sim}{R}(\langle 2, 2, 2 \rangle)^{N_i}$ by means of the asymptotic sum inequality.

(7) Thm. (15.41) remains true if we replace the border rank by the asymptotic rank.

15.25.* In this exercise we sketch how the conclusion "$\omega = 2$" could be deduced from some hypothetical combinatorial assumption. We work over the field $k = \mathbb{C}$.

(1) Consider the trilinear form

$$F := X_0 Y_1 Z_2 + X_0 Y_2 Z_1 + X_1 Y_0 Z_2 + X_2 Y_0 Z_1 + X_1 Y_2 Z_0 + X_2 Y_1 Z_0.$$

Show that for all $a, b \in \mathbb{C}^\times$ the tensor corresponding to

$$F + a^{-1} b^{-1} X_0 Y_0 Z_0 + a X_1 Y_1 Z_1 + b X_2 Y_2 Z_2$$

is isomorphic to $\mathbb{C}[T]/(T^3 - 1) \simeq \mathbb{C}^3$ and hence of rank 3.

(2) Show that the border rank of F equals 4. (For the lower bound use Ex. 15.14.)

(3) As in Rem. (15.44) argue that $\underset{\sim}{R}(F) = 3$ would imply $\omega = 2$.

(4) We say that a subset M of an abelian group G satisfies the *no three disjoint equivoluminous subsets property* iff for all pairwise disjoint subsets M_1, M_2, M_3 of M, not all empty, either

$$\sum_{m \in M_1} m \neq \sum_{m \in M_2} m \quad \text{or} \quad \sum_{m \in M_2} m \neq \sum_{m \in M_3} m.$$

Assume that $M \subseteq G$ has this property. We select two distinct characters $\chi, \psi : G \to \mathbb{C}$ of G. Let $t \in \mathbb{C}^{3 \times 3 \times 3}$ denote the coordinate tensor of the trilinear form F considered in (1). We know that there are $s_i \in \mathbb{C}^{3 \times 3 \times 3}$ for $0 \le i \le 2$ such that

$$t + \chi(m)^{-1}\psi(m)^{-1}s_0 + \chi(m)s_1 + \psi(m)s_2$$

has rank 3 for arbitrary $m \in M$. By taking the tensor product over all $m \in M$ and summing over all pairs of characters χ, ψ conclude that $R(t^{\otimes N}) \le |G|^2 3^{|M|}$. (Hint: the sum $\sum_\chi \chi(g)$ taken over all characters vanishes if $g \in G \setminus \{0\}$.)

(5) If there is a sequence of pairs (G_n, M_n) with the no three disjoint equivoluminous subsets property satisfying $|M_n| \to \infty$ and $|M_n|^{-1} \log |G_n| \to 0$ as $n \to \infty$, then $\underline{R}(t) = 3$. By (3) this would imply that $\omega = 2$ over $\mathbb{C}$.

15.26. Prove Fact (15.46) about the tensor product of partial matrix multiplications.

15.27. Apply Lemma (15.47) to the $(n \times n)$-partial matrix multiplication ϕ of pattern $(I, \underline{n} \times \underline{n})$, where $I := \{(i, i) \mid i \in \underline{n}\}$. Is ϕ equivalent to a full matrix multiplication?

15.28.* Let $E_\rho = \oplus_{i \in \mu_\rho} E_{\rho i}$, $H_\rho = \oplus_{j \in \nu_\rho} H_{\rho j}$, $L_\rho = \oplus_{\ell \in \pi_\rho} L_{\rho \ell}$ be direct decompositions of k-spaces, $I_\rho \subseteq \mu_\rho \times \nu_\rho$, $J_\rho \subseteq \nu_\rho \times \pi_\rho$ for $\rho \in \underline{s}$. Show that

$$\underline{R}\left(\sum_{\rho=1}^{s} \langle E_\rho, H_\rho, L_\rho \rangle_{I_\rho, J_\rho}\right) \le r$$

implies

$$\sum_{\rho=1}^{s} \left(\sum_{j \in \nu_\rho} m_{\rho j} p_{\rho j} \dim H_{\rho j}\right)^{\omega/3} \le r,$$

where $m_{\rho j} := \sum_{(i,j) \in I_\rho} \dim E_{\rho i}$, $p_{\rho j} := \sum_{(j,\ell) \in J_\rho} \dim L_{\rho \ell}$. (Hint: combine the proofs of Thm. (15.11) and (15.48).)

15.12 Open Problems

Problem 15.1. Determine $\underline{R}(\langle 2, 2, 2 \rangle)$. (We know that this border rank is either 6 or 7, cf. Cor. (19.14).)

Problem 15.2. We say that a concise tensor t of format (m, n, p) is of *minimal border rank* iff $\underline{R}(t) = \max\{m, n, p\}$. Such tensors are an important building stone in the construction of fast matrix multiplication algorithms. (The estimate on the exponent in Cor. (15.33) relies on the tensor of minimal border rank described in its proof; also in the deduction of Cor. (15.45) such a tensor is crucial.) Try to classify the tensors of minimal border rank!

Problem 15.3. Determine the exponent $\omega(k)$ of matrix multiplication.

Problem 15.4. Is the asymptotic spectrum Δ_m of matrix multiplication logarithmically convex? (Compare the Notes of this chapter.)

Problem 15.5. Is there a tensor of format (m, n, p) such that the asymptotic rank $\underline{R}(t)$ of t is strictly larger than $\max\{m, n, p\}$? (If this is not the case, then in particular $\omega(k) = 2$.)

15.13 Notes

We first briefly report on the history of upper estimates on the exponent ω (compare also Pan's survey [412], his book [413], and the survey by Strassen [510]). The investigation of fast matrix multiplication algorithms started with Strassen's discovery [493] in 1969 that $h \times h$-matrices can be multiplied using only $O(h^{2.81})$ arithmetic operations. For almost a decade after this discovery, no further progress was made. Then Pan [408, 410] got the ball rolling again in 1978 by showing that $\omega < 2.79$. One year later, Bini et al. [49] and Bini [45] introduced the new and powerful concept of border rank and obtained the (only marginally better) estimate $\omega < 2.78$. Schönhage [460] put this idea into effective use by his famous asymptotic sum inequality and proved $\omega < 2.55$. Shortly after, further improvements were achieved by Pan [411], Romani [443], and Coppersmith and Winograd [133] who showed $\omega < 2.50$. In 1987 Strassen [508] published his laser method by which he proved $\omega < 2.48$. His ideas were taken up and successfully extended in the same year by Coppersmith and Winograd [134] who achieved the present "world record" $\omega < 2.38$. (Note that in Sect. 15.8 for the sake of simplicity and clarity we only presented a proof of the slightly worse bound $\omega < 2.39$.)

Let us now proceed with detailed bibliographical remarks. Implicitly, Prop. (15.1) and (15.3) are already contained in Strassen's 1969 paper [493]. Prop. (15.5) is an immediate consequence of the invariance of tensor rank under the action of S_3, which was independently noticed by Pan [406] and Hopcroft and Musinski [253]. The concept of border rank was introduced by Bini et al. [49]. In that paper the authors concluded from the example $\langle 3, 2, 2 \rangle \trianglelefteq_2 \langle 10 \rangle$ that $h \times h$-matrices can be "approximatively" multiplied using only $O(h^{2.78})$ arithmetic operations. Bini [45] discovered that approximative bilinear algorithms (also called APA-algorithms) may be turned into exact ones (Eq. (15.7)) and thus deduced Prop. (15.10).

The asymptotic sum inequality – in the literature sometimes called the τ-theorem – has been discovered by Schönhage [460]. In the same paper he presented the remarkable Example (15.12) which shows that the border rank is not additive, and obtained the estimate on the exponent expressed in Prop. (15.13). Pan had developed a method, called trilinear aggregating, to construct tensors of small (border) rank, which he applied first in [408]. In [409], before Schönhage, he already had found an example which implied that the border rank is not additive, but in this paper Pan had not drawn this conclusion.

Our presentation of Sect. 15.3 and 15.4 closely follows Strassen article [508], from which we have also taken Prop. (15.17). However, Cor. (15.18) is due to Schönhage [460]. The notion of degeneration of bilinear maps was introduced into complexity theory by Strassen [508], while the concept of border rank (or approximative rank) had been described earlier by Bini et al. [49], Bini [45], and Schönhage [460]. Prop. (15.26) appears in Strassen [508] and is a straightforward generalization of Bini's result [45] (cf. Eq. (15.7)).

The laser method was introduced by Strassen in [508]. That paper contains all the results of Sect. 15.6. We remark that combinatorial degenerations correspond to degenerations by monomial matrices (which are essentially toric degenerations). (See [508, Thm. 6.1].) In [134] Coppersmith and Winograd strongly generalized Strassen's method. They obtained the estimates on the exponent in Cor. (15.43) and (15.45) by essentially proving Thm. (15.41). (For a more detailed presentation, see [135].) One of the key ingredients in Coppersmith and Winograd's generalization is the subtle technique of randomization which appears in the proof of Thm. (15.39). Our presentation of Sect. 15.7 and 15.8 is based on the paper [511] and a private communication by Strassen. The reader should note that the auxiliary result about "sequences free of three-term arithmetic progressions", which is needed in Coppersmith and Winograd's original proof, has been avoided.

The contents of Sect. 15.9 on the multiplication of partially filled matrices is due to Schönhage [460]. Thm. (15.51) on the rapid multiplication of rectangular matrices was discovered by Coppersmith [128].

We proceed by mentioning some further results. In the definition of the exponent $\omega(k)$ one may as well use the cost function c_+ which counts only the additions and subtractions (Lickteig [334]). The notion of approximative complexity makes not only sense for bilinear maps, one may as well consider arbitrary polynomials or rational functions. In fact, the notion had been originally introduced in this context (Strassen [499]). Griesser [206] generalized the known lower bounds for multiplicative complexity based on substitution, transcendence degree or algebraic geometric degree to the approximative complexity. Lickteig [334] employs the notion of approximative complexity to prove lower bounds. (Note that we used border rank in this chapter solely to find upper bounds.) For further results on border rank, resp. approximative complexity we refer to Notzaki [393], Bini et al. [50], Bini [44], Schnorr [451], Lickteig [330], Griesser [205], and Bini and Capovani [48].

An interesting by-product of Coppersmith and Winograd [133] is the discovery, that in the conclusion of the asymptotic sum inequality one always has strict inequality, unless in the trivial case where $e_i = h_i = \ell_i = 1$ for all i. (This result is sometimes expressed by saying that "ω is an infimum, not a minimum.") See also Strassen [509] for an interpretation of this result in terms of asymptotic spectra. In Rem. (15.44) we noted that if the asymptotic rank of the tensor $t^{(2)} = \sum_{i=1}^{2}(e_{0ii} + e_{i0i} + e_{ii0}) \in (k^3)^{\otimes 3}$ equaled three, then $\omega = 2$ would follow. Coppersmith [130] presented a combinatorial problem whose solution would imply $\underline{R}(t^{(2)}) = 3$ (see Ex. 15.25 and Coppersmith and Winograd [135]). Brockett and

Dobkin [79] have obtained a result related to Thm. (15.51), namely they showed that $R(\langle h, h, \lfloor \log h \rfloor \rangle) = h^2 + o(h^2)$ as $h \to \infty$. For some further investigations connected to Thm. (15.51) see also Lotti and Romani [342].

Finally, we give a short account of a recent approach by Strassen [509, 511] which sheds a new light on the topics discussed in this chapter. (Compare the survey [510].) In Ex. 15.24 we showed that the asymptotic rank $\underset{\sim}{R}: \mathcal{T} \to [0, \infty)$ is a subadditive and submultiplicative functional on the semiring $\mathcal{T}$ of tensor classes. In the same exercise we verified that $\underset{\sim}{R}\left(\sum_{i=1}^{s} \langle 2, 2, 2 \rangle^{N_i}\right) = \sum_{i=1}^{s} \underset{\sim}{R}(\langle 2, 2, 2 \rangle)^{N_i}$ as a consequence of the asymptotic sum inequality . In turns out that this equality holds for any $b \in \mathcal{T}$ in place of $\langle 2, 2, 2 \rangle$. Hence $\underset{\sim}{R}$ *behaves like the maximum functional* on the semiring of non-negative functions of some function ring. Strassen showed that there is a deeper reason for this. He introduced an asymptotic version of the degeneration order as follows. For $a, b \in \mathcal{T}$ we say that a is an *asymptotic degeneration* of b, $a \underset{\sim}{\unlhd} b$, iff there is some sequence $\varepsilon_N = o(N)$ of non-negative integers satisfying $a^N \unlhd b^N \cdot 2^{\varepsilon_N}$. The asymptotic degeneration $\underset{\sim}{\unlhd}$ is a preorder on $\mathcal{T}$ compatible with addition and multiplication. In Ex. 14.24 it is proved that $\mathcal{T}$ can be embedded uniquely into a commutative ring $\mathcal{B}$ such that $\mathcal{B} = \mathcal{T} - \mathcal{T}$. The surprising and crucial fact is now that $\underset{\sim}{\unlhd}$ allows an extension to all of $\mathcal{B}$, compatible with the ring operations. In this way, $\mathcal{B}$ becomes a preordered ring which, as it turns out, has an additional property which allows to bring in the structure theory of Stone, Kadison, and Dubois (see Becker and Schwartz [33]).

Strassen proved the following central result. For any subset $X \subseteq \mathcal{T}$ there is a compact space Δ and a ring morphism $\varphi: \mathbb{Z}[X] \to C(\Delta)$ (where $\mathbb{Z}[X]$ denotes the subring of $\mathcal{B}$ generated by X and $C(\Delta)$ the ring of continuous real functions on Δ) such that $\varphi(X)$ separates the points of Δ and such that for all $a, b \in \mathbb{Z}[X]$

$$a \underset{\sim}{\unlhd} b \iff (\varphi(a) \leq \varphi(b) \text{ pointwise on } \Delta).$$

An easy consequence of this is $\underset{\sim}{R}(a) = \max \varphi(a)$. The pair (Δ, φ) is unique up to canonical isomorphism and called an *asymptotic spectrum* of X. Actually, Δ can always be realized by a unique compact subset $\Delta(X)$ of $\mathbb{R}^X$ (in the product topology) such that $\varphi(x)$ is the coordinate projection with respect to x. $\Delta(X)$ is called *the asymptotic spectrum* of X. It can be shown that $\Delta_c := \Delta(\langle 2, 2, 2 \rangle) \subseteq \mathbb{R}$ is actually an asymptotic spectrum for all $\langle h, h, h \rangle$ via $\varphi(\langle h, h, h \rangle) = x^{\log h}|_{\Delta_c}$. The right endpoint of Δ_c is readily identified, namely $\max \Delta_c = \max_{\Delta_c} x = \max_{\Delta_c} \varphi(x) = \underset{\sim}{R}(\langle 2, 2, 2 \rangle) = 2^\omega$. This allows a transparent proof of the asymptotic sum inequality as follows (for simplicity, we confine ourselves to the square matrix multiplication):

$$\underset{\sim}{R}\left(\oplus_{i=1}^{s} \langle h_i, h_i, h_i \rangle\right) \leq r \implies \varphi\left(\sum_{i=1}^{s} \langle h_i, h_i, h_i \rangle\right) = \sum_{i=1}^{s} x^{\log h_i} \leq r \text{ on } \Delta_c.$$

Hence, as $2^\omega \in \Delta_c$, we get $\sum_{i=1}^{s} h_i^\omega \leq r$. As the asymptotic degeneration of tensor classes is completely described by their asymptotic spectrum, the computation of these objects is a problem of central importance! Note that asymptotic spectra also yield necessary conditions for the validity of degeneration relations.

What is known about the spectra of specific examples? One can show that $\Delta_c = [4, 2^\omega]$. Therefore $\omega = 2$ iff $\Delta_c = \{4\}$. Finding points in spectra thus means proving lower bounds. It can be shown that $\Delta_m := \Delta(\langle 2, 1, 1 \rangle, \langle 1, 2, 1 \rangle, \langle 1, 1, 2 \rangle) \subseteq \mathbb{R}^3$ is an asymptotic spectrum for all $\langle e, h, \ell \rangle$ via $\varphi(\langle e, h, \ell \rangle) = x_1^{\log e} x_2^{\log h} x_3^{\log \ell} |_{\Delta_m}$. Moreover, the convex hull Γ of the points $(1, 1, 0)$, $(1, 0, 1)$, $(0, 1, 1)$ is contained in $\log \Delta_m$ and $\log \Delta_m$ is star-shaped relative to Γ.

In [511] Strassen generalized this further by developing a method by which for many sets $X \subseteq T$ of tensor classes a singular simplex Γ lying in $\Delta(X)$ – the so-called support simplex – can actually be computed. He conjectures in [512] that the support simplex coincides with the asymptotic spectrum $\Delta(X)$ if X consists only of "tight" tensor classes. Strassen's conjecture implies that $\omega(k) = 2$, and this at least is believed by most experts today. His conjecture also implies that the statement of Problem 15.4 is true, and that Problem 15.5 has a negative answer for tight tensors. Strassen's approach involves ideas from convex analysis and is related to the formalism of statistical mechanics. For further investigations in this area we refer also to Bürgisser [98] and Tobler [518].

Ex. 15.1 is due to Winograd [554]. Ex. 15.2 is from Gastinel [182], Ex. 15.3 is due to Laderman [310], while Ex. 15.6, 15.7 and 15.12 are taken from Strassen [508]. Ex. 15.8, 15.15, 15.17 and 15.28 are due to Schönhage [460]. Ex. 15.13 is due to Strassen [511] and Bürgisser [98]. Ex. 15.24 is discussed in Strassen [509]. Ex. 15.25 is due to Coppersmith [130], see also Coppersmith and Winograd [135].

Problems 15.1 and 15.4 are asked in Strassen [510]. With regard to Problem 15.2 we remark that Coppersmith and Winograd [133, Rem. 6.2] have proved, among other things, the following interesting statement: a direct sum t of matrix tensors which has cubic format (n, n, n) is of minimal border rank iff $t \simeq \langle n \rangle$.

Chapter 16. Problems Related
to Matrix Multiplication

Fast matrix multiplication algorithms play a fundamental role in the design of algorithmic solutions to several problems in computational linear algebra. In this chapter we will show that the complexities of problems such as matrix inversion, computation of the determinant, *LUP*-decomposition, computing the characteristic polynomial, or orthogonal basis transform, are dominated asymptotically by the complexity of matrix multiplication. Though we shall restrict ourselves to the muliplicative complexity only, the algorithms we exhibit show that these upper bounds also hold for the total complexity. On the other hand, matrix multiplication can be reduced to special instances of these problems, which shows that – from a computational point of view – all these problems are asymptotically equivalent. To put this sort of reasoning into a formal framework, we shall begin by introducing the notion of an exponent for a certain type of problems. We then show, by exhibiting specific algorithms, that all the above problems (and some others) have the same exponent as matrix multiplication. In the last section of this chapter we show how fast matrix multiplication algorithms can be used to compute the transitive closure of a graph.

Unless stated otherwise, k denotes in this chapter an infinite field.

16.1 Exponent of Problems

In this chapter we discuss various computational problems in which matrix multiplication is involved and investigate them from the viewpoint of complexity theory. Let us first clarify what we mean by a problem. We define a *problem* Π formally as a relation $\Pi \subseteq k^m \times k^n$. Elements $\xi \in k^m$ are thought of as inputs, while elements $\eta \in k^n$ satisfying $(\xi, \eta) \in \Pi$ are interpreted as the solutions of the problem Π to input ξ. The set of inputs for which there exists a solution is called the *domain of definition* def Π of Π. Thus def Π is just the projection of Π onto the first coordinate. As an example consider the matrix multiplication problem $MaMu_n$ defined by

$$MaMu_n := \{((A, B), C) \in (k^{n \times n})^2 \times k^{n \times n} \mid AB = C\}.$$

This is the graph of a map defined on $(k^{n \times n})^2$, thus for any input (A, B) there is exactly one solution. Later on we will encounter problems which do not have this property.

Let $\Pi \subseteq k^m \times k^n$ be a problem. A partial function $f: k^m \supseteq \operatorname{def} f \to k^n$ is said to *solve* Π if and only if graph $f \subseteq \Pi$ and $\operatorname{def} f = \operatorname{def} \Pi$. To define and investigate the complexity of these problems we use the model of computation trees, presented in Sect. 4.4. Recall that $\Omega := k^c \cup k \cup \{+, -, *, /\}$ is the set of operational symbols and $P = \{=, \leq\}$ or $P = \{=\}$ is the set of relational symbols, according to whether k is an ordered field or not. Let $\Pi \subseteq k^m \times k^n$ be a problem, and $J := \{(k; a_1, \ldots, a_m) \mid (a_1, \ldots, a_m) \in \operatorname{def} \Pi\}$. Then any partial function $f = (f_1, \ldots, f_n): k^m \supseteq \operatorname{def} f \to k^n$ solving Π defines a collection (φ, π) for J, where $\pi := \{J\}$ and $\varphi(k; a) := (k; f_1(a), \ldots, f_n(a))$ for $(k; a) \in J$. We define $C(f) := C^*(\varphi, \pi)$ and call

$$C(\Pi) := \min\{C(f) \mid f \text{ a partial function solving } \Pi\}.$$

the *(multiplicative) complexity* of Π. Note that we are only dealing with "pure computational problems" in the sense of Chap. 4 (see p. 117).

(16.1) Definition. Let $\Pi = (\Pi_1, \Pi_2, \ldots)$ be a sequence of problems. Then

$$\omega(\Pi) := \inf\{\gamma \mid C(\Pi_n) = O(n^\gamma)\}$$

is called the *exponent* of the sequence Π.   ●

In the sequel we shall see that for the sequence $MaMu = (MaMu_n)_{n \geq 1}$ the above definition of the exponent is consistent with the one of Sect. 15.1. For this, and also for further applications in this chapter, we shall need the following result whose proof is similar to that of Prop. (4.26) and is left to the reader. (Cf. Ex. 16.1.)

(16.2) Lemma. *Let $\Pi \subseteq k^m \times k^n$ be a problem whose domain of definition is Zariski dense in k^m. Then there exist elements $g_1, \ldots, g_n$ in the rational function field $K := k(X_1, \ldots, X_m)$ such that $(\xi, (g_1(\xi), \ldots, g_n(\xi))) \in \Pi$ for all ξ in some Zariski dense subset of $\operatorname{def} \Pi$ and*

$$C(\Pi) \geq L_K(g_1, \ldots, g_n).$$

This lemma gives a lower bound for the complexity of a problem in terms of the straight-line complexity of rational functions from which we know that they satisfy certain relations. The clue will be to exploit this information in specific situations.

As a first simple application let us show that $\omega(MaMu_n)$ equals the exponent ω of matrix multiplication as defined in Sect. 15.1.

Indeed, let $n \geq 1$, $A = (a_{ij})$ and $B = (b_{ij})$ be $n \times n$-matrices over k with indeterminate entries, and $D := AB$. Set $R := k[a_{ij}, b_{ij} \mid 1 \leq i, j \leq n]$ and $K := \operatorname{Quot}(R)$. Obviously R is the coordinate ring of the irreducible affine variety $\operatorname{def} MaMu_n$. By Lemma (16.2) we have $C(MaMu_n) \geq L_K(D)$. On the other side, since the entries of D are quadratic forms we have $L_K(D) = L_R(D)$ by Thm. (7.1).

Any straight line program for computing AB without divisions obviously can be interpreted as a computation tree for $MaMu_n$, hence $C(MaMu_n) \leq L_R(D)$. We thus obtain

$$C(MaMu_n) = L_R(D) = L(\langle n, n, n \rangle).$$

This implies our assertion $\omega = \omega(MaMu_n)$.

16.2 Triangular Inversion

One of the fundamental problems in computational linear algebra is that of matrix inversion. For $n \geq 1$ we define

$$Inv_n := \{(A, B) \in \mathrm{GL}(n, k) \times \mathrm{GL}(n, k) \mid AB = I_n\}.$$

The problem sequence Inv is then defined by $Inv = (Inv_1, Inv_2, \ldots)$. Following Winograd, we consider the restricted problem

$$\widetilde{Inv}_{3n} := \left\{ \left(\begin{pmatrix} I_n & A & 0 \\ 0 & I_n & B \\ 0 & 0 & I_n \end{pmatrix}, \begin{pmatrix} I_n & -A & AB \\ 0 & I_n & -B \\ 0 & 0 & I_n \end{pmatrix} \right) \,\middle|\, A, B \in k^{n \times n} \right\}$$

to see that $C(MaMu_n) \leq C(\widetilde{Inv}_{3n}) \leq C(Inv_{3n})$. Hence $\omega \leq \omega(Inv)$. Later in Sect. 16.4 we shall prove that $\omega(Inv) \leq \omega$, thereby obtaining $\omega(Inv) = \omega$. In the remaining of this section we study a restricted inversion problem, namely that of inverting triangular matrices. For $n \geq 1$ we define

$$TInv_n := \{(A, B) \in \mathrm{GL}(n, k) \times \mathrm{GL}(n, k) \mid A, B \text{ upper triangular}, AB = I_n\}.$$

and set $TInv := (TInv_1, TInv_2, \ldots)$. (We discuss here the case of upper triangular matrices only; lower triangular matrices can be handled analogously.)

(16.3) Proposition. *We have $\omega(TInv) = \omega$.*

Proof. As the previously defined problem $\widetilde{Inv}_{3n}$ is actually a subset of $TInv_{3n}$, we obtain $\omega \leq \omega(TInv)$. To show the other inequality let A be a nonsingular upper triangular $(2m \times 2m)$-matrix over k. Partition A into $m \times m$-blocks A_{ij}, $i, j = 1, 2$. Then A_{11} and A_{22} are invertible, upper triangular, and

$$A^{-1} = \begin{pmatrix} A_{11}^{-1} & -A_{11}^{-1} A_{12} A_{22}^{-1} \\ 0 & A_{22}^{-1} \end{pmatrix}.$$

This shows that if $n = 2^\ell$, then

$$C(TInv_{2^\ell}) \leq 2C(TInv_{2^{\ell-1}}) + 2C(MaMu_{2^{\ell-1}}).$$

If $\gamma > \omega$ we have $C(MaMu_{2^\ell}) = O(2^{\gamma \ell})$. Solving the above recursion with Ex. 2.14 shows that $C(TInv_{2^\ell}) = O(2^{\gamma \ell})$. Since the restriction to matrices of 2-power size is irrelevant, we obtain $\omega(TInv) \leq \omega$. $\qquad \square$

16.3 *LUP*-decomposition

The *LUP*-decomposition of a matrix $A \in k^{m \times n}$ consists of a triple $(L, U, P) \in k^{m \times m} \times k^{m \times n} \times k^{n \times n}$ such that $A = LUP$, L is lower triangular with 1's in the diagonal, U is upper triangular, and P is a permutation matrix. The importance of the *LUP*-decomposition comes from the fact that many algorithms in linear algebra are based on this decomposition. Examples are the computation of the determinant and the inverse of a nonsingular matrix, as well as the computation of approximations of the eigenvalues of a complex matrix. The formal definition of the *LUP*-decomposition problem is as follows:

$$
\begin{aligned}
LUP_{m,n} \ = \ & \{(A, (L, U, P)) \in k^{m \times n} \times (k^{m \times m} \times k^{m \times n} \times k^{n \times n}) \mid \\
& m \leq n, A = LUP, \\
& L \text{ lower triangular with 1's in the diagonal,} \\
& U \text{ upper triangular, } P \text{ permutation matrix}\}.
\end{aligned}
$$

We say that a matrix has an *LUP*-decomposition if it is in $\mathrm{def} LUP_{m,n}$ for some m and n. Not every matrix needs to have an *LUP*-decomposition. (Take for example $\left(\begin{smallmatrix} 0 & 0 \\ 1 & 1 \end{smallmatrix}\right)$.) Note that the *LUP*-decomposition of a matrix may not be unique. (Consider the example $\left(\begin{smallmatrix} 1 & 2 \\ 0 & 3 \end{smallmatrix}\right)$.) Below we shall prove that any $m \times n$-matrix with $m \leq n$ having full rank m has an *LUP*-decomposition. In doing so we give a recursive procedure to compute such matrices L, U, and P.

(16.4) Theorem. *Let m and n be positive integers, $m \leq n$, and $A \in k^{m \times n}$ have rank m. Then $A \in \mathrm{def}\ LUP_{m,n}$.*

Proof. We prove by induction on m that A has an *LUP*-decomposition $A = LUP$ such that the first m columns of U are linearly independent.

For the induction start "$m = 1$" note that by assumption the ith column of A is nonzero for some $1 \leq i \leq n$. Let P be the permutation matrix corresponding to the transposition $(1, i)$. We set $L := 1$, $U := AP$.

Now let us proceed with the induction step. We may suppose that $m > 1$. Choose some p with $1 \leq p \leq m - 1$ and decompose A as

$$
A = \begin{pmatrix} A_1 \\ A_2 \end{pmatrix} = \begin{pmatrix} A_{11} & A_{12} \\ A_{21} & A_{22} \end{pmatrix},
$$

where $A_1 \in k^{p \times n}$, $A_2 \in k^{(m-p) \times n}$, $A_{11} \in k^{p \times p}$, $A_{12} \in k^{p \times (n-p)}$, $A_{21} \in k^{(m-p) \times p}$, and $A_{22} \in k^{(m-p) \times (n-p)}$. Since A_1 has full rank p, it has an *LUP*-decomposition by the induction hypothesis: $A_1 = L_1 U_1 P_1$. Decompose U_1 and $A_2 P_1^{-1}$ into blocks as $U_1 = (\tilde{U}_1 \mid B)$ and $A_2 P_1^{-1} = (C \mid D)$ where $\tilde{U}_1, C \in k^{p \times p}$ and $\tilde{U}_1$ is upper triangular. By the induction hypothesis $\tilde{U}_1$ is invertible. Setting $F := D - C\tilde{U}_1^{-1} B$ we have

$$
A = \begin{pmatrix} L_1 & 0 \\ C\tilde{U}_1^{-1} & I_{m-p} \end{pmatrix} \begin{pmatrix} \tilde{U}_1 & B \\ 0 & F \end{pmatrix} P_1.
$$

F has rank $m - p$ since A has rank m and hence, by the induction hypothesis, F has an *LUP*-decomposition $F = L_2 U_2 P_2$, where the first $(m - p)$ columns of U_2 are linearly independent. Let $\tilde{P} := I_p \oplus P_2$. ($\tilde{P}$ is an $n \times n$-permutation matrix.) Then we easily verify that

$$A = \underbrace{\left(\begin{array}{cc} L_1 & 0 \\ C\tilde{U}_1^{-1} & L_2 \end{array} \right)}_{=:L} \underbrace{\left(\begin{array}{cc} \tilde{U}_1 & BP_2^{-1} \\ 0 & U_2 \end{array} \right)}_{=:U} \underbrace{(\tilde{P} P_1)}_{=:P}.$$

Clearly, L is lower triangular, U is upper triangular, and the first m columns of U are linearly independent. ☐

The above proof provides us with a recursive procedure to compute an *LUP*-decomposition of an $m \times n$-matrix having full rank. We consider the sequence of problems

$$LUP := (LUP_{n,n})_{n \geq 1}.$$

(16.5) Theorem (Bunch and Hopcroft). *We have* $\omega(LUP) \leq \omega$.

Proof. We regard the procedure described in the proof of Thm. (16.4) as a computation tree and count the number of nonscalar multiplications/divisions in this tree. If $n = 1$, we have no nonscalar operations. If $n > 1$, then we need to compute an *LUP*-decomposition of A_1 with cost $C(LUP_{p,n})$, compute $\tilde{U}_1^{-1}$ with cost $C(TInv_p)$, compute $C\tilde{U}_1^{-1}$ with cost $C(MaMu_p)$, compute $(C\tilde{U}_1^{-1})B$ with cost $\leq \lceil \frac{n-p}{p} \rceil C(MaMu_p) \leq \lceil \frac{n}{p} \rceil C(MaMu_p)$, and finally, compute an *LUP*-decomposition of F with cost $C(LUP_{m-p,n})$. (Note that multiplication with a permutation matrix, inversion of a permutation matrix, additions, and subtractions are for free in our model.) We thus obtain

(A) $\qquad C(LUP_{2m,n}) \leq 2C(LUP_{m,n}) + C(TInv_m) + (n/m + 1)C(MaMu_m).$

For the sake of simplicity we assume in the following that $n = 2^\ell$ is a power of 2. Let $\gamma > \omega$. Applying Prop. (16.3), we obtain from (A) for any $q \in \underline{\ell}$

$$C(LUP_{2^q,n}) \leq 2C(LUP_{2^{q-1},n}) + c2^{(\ell-q)}2^{q\gamma}$$

for some constant c (not depending on q). Using Ex. 2.14 we obtain for $q \in \underline{\ell}$ the inequality

$$C(LUP_{2^q,2^\ell}) \leq \tilde{c} \cdot 2^\ell \cdot 2^{(\gamma-1)q}$$

for some constant $\tilde{c}$. In particular, the assertion follows by setting $q = \ell$. ☐

16.4 Matrix Inversion and Determinant

The *LUP*-decomposition described in the previous section allows to show that *Inv* has the same exponent as *MaMu*.

(16.6) Proposition (Bunch and Hopcroft). $\omega(Inv) = \omega$.

Proof. In view of the results in Sect. 16.2 we only need to show that $\omega(Inv) \leq \omega$. First note that

$$C(Inv_n) \leq C(LUP_{n.n}) + 2C(TInv_n) + C(MaMu_n).$$

Indeed, given an invertible $n \times n$-matrix A we compute an *LUP*-decomposition $A = LUP$. Then $A^{-1} = P^{-1}U^{-1}L^{-1}$. (Multiplication with a permutation matrix is free of charge in our model.) Now Prop. (16.3) and Thm. (16.5) imply the result.   □

One of the important problems in computational linear algebra is that of computing the determinant of a square matrix. For $n \geq 1$ we define

$$Det_n := \left\{(A, a) \in \mathrm{GL}(n, k) \times k \mid a = \det A\right\}$$

and $Det = (Det_1, Det_2, \ldots)$. In the following we show that $\omega(Det) = \omega$. As a by-product, we also obtain $\omega(LUP) = \omega$. Our proof is based on the differential methods developed in Chap. 7.

(16.7) Theorem. *We have* $\omega(LUP) = \omega(Det) = \omega$.

Proof. The proof proceeds in several steps. In the first we show that for all n

(A) $C(Det_n) \leq C(LUP_{n.n}) + n.$

Using Thm. (16.5) this gives

$$\omega(Det) \leq \omega.$$

Let $A \in \mathrm{GL}(n, k)$ be given. We compute an *LUP*-decomposition $A = LUP$. Computing $\det P$ is free of cost in our model. The determinant of L is 1 and that of U can be computed by multiplying the diagonal entries of U and this requires n multiplications. This proves (A).

In the second step we derive a lower bound for $\omega(Det)$. Assume w.l.o.g. that $n = 2m$. Let $A = (a_{ij})$ be an $n \times n$-matrix with indeterminate entries over k, $R := k[a_{ij} \mid 1 \leq i, j \leq n]$, and $K := \mathrm{Quot}(R)$. By Lemma (16.2) we have

$$C(Det_n) \geq L_K(\det A).$$

Now we use Cor. (7.9) (which is a consequence of the derivative inequality (7.7)) to obtain

$$L_K(\det A) \geq \frac{1}{3}(L_K(A^{-1}) - n^2).$$

We claim that

(B) $$L_K(A^{-1}) \geq C(MaMu_m).$$

Let Γ be a straight-line program of c_*-length $L_K(A^{-1})$ executable on $(K; A)$ which computes A^{-1} on this input. Ex. 4.4 shows that for Zariski almost all $\xi \in k^{n \times n}$ the program Γ is executable on $(\mathcal{O}_\xi; A)$, where $\mathcal{O}_\xi \subset K$ denotes the local ring of ξ. Hence, as $GL(n, k)$ is open in $k^{n \times n}$, there exists $\xi \in GL(n, k)$ such that $A^{-1} \in \mathcal{O}_\xi$ and $L_K(A^{-1}) = L_{\mathcal{O}_\xi}(A^{-1})$. Let $I := I_n$. The k-algebra isomorphism $\mathcal{O}_\xi \to \mathcal{O}_I$, $A \mapsto A\xi$ shows that we may w.l.o.g. assume that $\xi = I$. (We can compute A^{-1} from $(\xi A)^{-1} = A^{-1}\xi^{-1}$ at no cost, as we do not count linear operations in our complexity model and ξ is a scalar matrix.) Noting that $L_{\mathcal{O}_I}(A^{-1}) = L_{\mathcal{O}_0}((I - A)^{-1})$, we obtain

$$L_K(A^{-1}) = L_{\mathcal{O}_\xi}(A^{-1}) = L_{\mathcal{O}_0}((I - A)^{-1}).$$

The Taylor expansion of $(I - A)^{-1}$ at 0 is $(I - A)^{-1} = I + A + A^2 + \cdots$. We now apply Thm. (7.1) with $d = 2$ to get $L_{\mathcal{O}_0}((I - A)^{-1}) \geq L_R(A^2)$. Partition A into $m \times m$-blocks A_{ij}, $i, j \in \{1, 2\}$, and define the substitution

$$\psi : R \to R, \quad \psi(A) := \begin{pmatrix} 0 & A_{12} \\ A_{21} & 0 \end{pmatrix}.$$

Obviously we have $\psi(A)^2 = \begin{pmatrix} A_{12}A_{21} & 0 \\ 0 & A_{21}A_{12} \end{pmatrix}$. We obtain

$$C(MaMu_m) = L_R(A_{12}A_{21}) \leq L_R(\psi(A)^2) \leq L_R(A^2) \leq L_K(A^{-1}),$$

which proves claim (B). Combining all the above results we get

$$C(MaMu_{n/2}) \leq 3C(Det_n) + n^2 \leq 3C(LUP_{n,n}) + n^2 + 3n,$$

which implies $\omega \leq \omega(Det) \leq \omega(LUP)$, provided $\omega > 2$. The case where $\omega = 2$ is settled in Ex. 16.4. $\square$

16.5* Transformation to Echelon Form

For many applications such as solving systems of linear equations or computing the rank of a matrix it is desirable to transform a matrix into echelon form. Recall that an $m \times n$-matrix $A = (a_{ij})$ is said to be in echelon form if there exist integers $r \geq 0$, $1 \leq j_1 < j_2 < \cdots < j_r \leq n$ such that $a_{ij} = 0$ for $i > r$ or ($i \leq r$ and $j < j_i$), and $a_{ij_i} \neq 0$ for $i \leq r$. Obviously, r must be the rank of A.

The problem of transforming a matrix into echelon form is given by

$$Ech_{m,n} := \{(A, (U, S)) \in k^{m \times n} \times (SL(n, k) \times k^{m \times n}) \mid S \text{ in echelon form}, UA = S\}.$$

The aim of this section is to describe a recursive procedure which will give an upper bound for $C(Ech_{m,n})$ in terms of $C(MaMu_n)$ if $n \leq 2m$. Besides being of interest in its own, this result will be needed later in Sect. 16.6 when dealing with the computation of the characteristic polynomial.

We will need to study three auxiliary problems Π_1, Π_2, Π_3 described below. Informally, Π_2 describes the problem of transforming a square matrix to upper triangular form; our description of a fast algorithm for Π_2 calls for the problem Π_1. Finally, we show how to transform an upper triangular matrix to echelon form (problem Π_3).

For $1 \leq n \leq m \leq 2n$ we define

$$\Pi_1^{m,n} := \left\{((A, R), (U, R')) \in (k^{(m-n)\times n} \times k^{n\times n}) \times (SL(m, k) \times k^{n\times n}) \mid \right.$$
$$\left. R, R' \text{ upper triangular, } U\binom{A}{R} = \binom{R'}{0}\right\}.$$

Further, for $n \geq 1$ we set

$$\Pi_2^n := \left\{(A, (U, R)) \in k^{n\times n} \times (SL(n, k) \times k^{n\times n}) \mid \right.$$
$$\left. R \text{ upper triangular, } UA = R\right\},$$

and

$$\Pi_3^n := \left\{(R, (U, S)) \in k^{n\times n} \times (SL_n(k) \times k^{n\times n}) \mid \right.$$
$$\left. R \text{ upper triangular, } S \text{ in echelon form, } UR = S\right\}.$$

Below we will sketch recursive algorithms for the above auxiliary problems to get upper bounds on their complexities in terms of the complexity of matrix multiplication. We leave the details of our description to the reader.

(16.8) Proposition (Schönhage). *Let* $\gamma > \omega$. *Then* $C(\Pi_1^{m,n}) = O(n^\gamma)$ *for* $1 \leq n \leq m \leq 2n$.

Proof. (Sketch) For simplicity we assume that $n = 2^\ell$ is a power of 2 and set $m = 2n$. If $(m, n) = (2, 1)$, we can compute an appropriate U easily. Suppose now that $\ell \geq 1$. Divide A and R into $2^{\ell-1} \times 2^{\ell-1}$-blocks as

$$A = \begin{pmatrix} A_1 & A_2 \\ A_3 & A_4 \end{pmatrix}, \quad R = \begin{pmatrix} R_1 & A_5 \\ 0 & R_2 \end{pmatrix}.$$

The following diagram shows how to obtain U and R' from A and R:

$$\begin{pmatrix} A_1 & A_2 \\ A_3 & A_4 \\ R_1 & A_5 \\ 0 & R_2 \end{pmatrix} \xrightarrow{1} \begin{pmatrix} A_1 & A_2 \\ R_3 & A_6 \\ 0 & A_7 \\ 0 & R_2 \end{pmatrix} \xrightarrow{2} \begin{pmatrix} R_4 & A_8 \\ 0 & A_9 \\ 0 & A_7 \\ 0 & R_2 \end{pmatrix} \xrightarrow{3} \begin{pmatrix} R_4 & A_8 \\ 0 & A_9 \\ 0 & R_5 \\ 0 & 0 \end{pmatrix}$$

$$\xrightarrow{4} \begin{pmatrix} R_4 & A_8 \\ 0 & R_6 \\ 0 & 0 \\ 0 & 0 \end{pmatrix} =: \begin{pmatrix} R' \\ 0 \end{pmatrix}.$$

In Step **1** we apply the algorithm recursively to $\binom{A_3}{R_1}$ to obtain U_1 and R_3 with $U_1\binom{A_3}{R_1} = \binom{R_3}{0}$ and set $\binom{A_6}{A_7} := U_1\binom{A_4}{A_5}$. In Step **2** we apply the algorithm recursively to $\binom{A_1}{R_3}$ to obtain U_2 and R_4 and set $\binom{A_8}{A_9} := U_2\binom{A_2}{A_6}$. In Step **3** we apply the algorithm to $\binom{A_7}{R_2}$ to get U_3 and R_5, and, finally, in Step **4** we apply the algorithm to $\binom{A_9}{R_5}$ to get U_4 and R_6. Altogether we have

$$
\begin{pmatrix} I_{2^{\ell-1}} & \\ U_4 & \\ & I_{2^{\ell-1}} \end{pmatrix}
\begin{pmatrix} U_2 & \\ & \\ & U_3 \end{pmatrix}
\begin{pmatrix} I_{2^{\ell-1}} & \\ U_1 & \\ & I_{2^{\ell-1}} \end{pmatrix}
\begin{pmatrix} A \\ R \end{pmatrix} = \begin{pmatrix} R' \\ 0 \end{pmatrix}.
$$

Let $\gamma > \omega$ and set $x_\ell := C(\Pi_1^{m,n})$. Then we have with a suitable constant $c > 0$

$$
x_\ell \le 4x_{\ell-1} + c2^{\gamma\ell},
$$

and Ex. 2.14 yields the result.   □

(16.9) Proposition. *We have* $\omega(\Pi_2) \le \omega$.

Proof. (Sketch) For simplicity we assume that $n = 2^\ell$ is a power of 2. If $\ell = 0$, nothing has to be computed, thus suppose that $\ell \ge 1$. We divide A into $2^{\ell-1} \times 2^{\ell-1}$-blocks as $A = \binom{A_1\,A_2}{A_3\,A_4}$ and consider the diagram

$$
\begin{pmatrix} A_1 & A_2 \\ A_3 & A_4 \end{pmatrix} \xrightarrow{\ \mathbf{1}\ }
\begin{pmatrix} A_1 & A_2 \\ R_1 & A_5 \end{pmatrix} \xrightarrow{\ \mathbf{2}\ }
\begin{pmatrix} R_2 & A_6 \\ 0 & A_7 \end{pmatrix} \xrightarrow{\ \mathbf{3}\ }
\begin{pmatrix} R_2 & A_6 \\ 0 & R_3 \end{pmatrix} =: R.
$$

Here, Step **1** is the recursive application of this algorithm to A_3 which yields U_1 and R_1, $A_5 := U_1 A_4$; Step **2** is the application of the algorithm for $\Pi_1^{2^\ell.2^{\ell-1}}$ to $\binom{A_1}{R_1}$ which gives U_2 and R_2, and $U_2\binom{A_2}{A_5} =: \binom{A_6}{A_7}$. We proceed with Step **3** which is the recursive application of this algorithm to A_7 which gives U_3 and R_3. It is straightforward to verify that the matrix

$$
U := \begin{pmatrix} I_{2^{\ell-1}} & 0 \\ 0 & U_3 \end{pmatrix} U_2 \begin{pmatrix} I_{2^{\ell-1}} & 0 \\ 0 & U_1 \end{pmatrix}
$$

together with R is a solution of Π_2^n for input A. An analysis of this procedure yields the result.   □

(16.10) Proposition. *We have* $\omega(\Pi_3) \le \omega$.

Proof. (Sketch) For simplicity we assume that $n = 2^\ell$. The case $\ell = 0$ is trivial, thus suppose $\ell \ge 1$. We decompose R into $2^{\ell-1} \times 2^{\ell-1}$-blocks as $R = \binom{R_1\,A_1}{0\,R_2}$ and consider the following diagram:

$$
\begin{pmatrix} R_1 & A_1 \\ 0 & R_2 \end{pmatrix} \xrightarrow{\ \mathbf{1}\ }
\begin{pmatrix} S_1 & A_2 \\ 0 & A_3 \\ 0 & R_2 \end{pmatrix} \xrightarrow{\ \mathbf{2}\ }
\begin{pmatrix} S_1 & A_2 \\ 0 & R_3 \\ 0 & 0 \end{pmatrix} \xrightarrow{\ \mathbf{3}\ }
\begin{pmatrix} S_1 & A_2 \\ 0 & S_2 \\ 0 & 0 \end{pmatrix} =: S.
$$

Step 1 is the recursive application of this algorithm to R_1 which yields U_1 and $S_1 \in k^{\mathrm{rk}(R_1) \times 2^{\ell-1}}$; we set $\binom{A_2}{A_3} := U_1 A_1$. Step 2 is the application of the algorithm for $\Pi_1^{(2^\ell - \mathrm{rk}(R_1)), 2^{\ell-1}}$ to $\binom{A_3}{R_2}$ which yields U_2 and R_3; finally, Step 3 is the recursive application of this algorithm to R_3 which gives U_3 and S_2. It is easy to verify that S together with

$$
U := \begin{pmatrix} I_{\mathrm{rk}(R_1)} & 0 & 0 \\ 0 & U_3 & 0 \\ 0 & 0 & I_{\mathrm{cork}(R_1)} \end{pmatrix} \begin{pmatrix} I_{\mathrm{rk}(R_1)} & 0 \\ 0 & U_2 \end{pmatrix} \begin{pmatrix} U_1 & 0 \\ 0 & I_{2^{\ell-1}} \end{pmatrix}
$$

is a solution of Π_3^n for input R. An analysis of this procedure yields the result. $\square$

Now we can prove that transformation to echelon form is not harder than matrix multiplication.

(16.11) Proposition. *Let $\gamma > \omega$. Then we have*

(1) $C(Ech_{m,n}) = O(m^\gamma)$ *for* $n \le m$.
(2) *If* $n \ge m$, *then there exists* $0 \le \varepsilon \le m$ *such that* $C(Ech_{m,n}) \le C(Ech_{m,m}) + C(Ech_{\varepsilon, n-m})$.

Proof. (1) Concatenating an $m \times n$-matrix with a zero matrix we see that $C(Ech_{m,n}) \le C(Ech_{m,m})$. If A is an $m \times m$-matrix we transform it to upper triangular form via the algorithm for Π_2^m given in the proof of Prop. (16.9) and then transform the upper triangular matrix to echelon form via the algorithm for Π_3^m, as given in the proof of Prop. (16.10).

(2) Let $A \in k^{m \times n}$. Divide A as $A = (A_1 | A_2)$, where $A_1 \in k^{m \times m}$ and $A_2 \in k^{m \times (n-m)}$ and consider the following sequence of transformations:

$$
A = (A_1 | A_2) \xrightarrow{\ 1\ } \begin{pmatrix} S_1 & A_3 \\ 0 & A_4 \end{pmatrix} \xrightarrow{\ 2\ } \begin{pmatrix} S_1 & A_3 \\ 0 & S_2 \end{pmatrix} =: S.
$$

Here, Step 1 is the recursive application of the procedure in (1) to A_1 which gives S_1 and U_1. We are done if $n = m$; if $n > m$ we proceed with Step 2 which is the recursive application of this algorithm to the $(m - \mathrm{rk}(S_1)) \times (n - m)$-matrix A_4 to obtain S_2. Now we put $\varepsilon := m - \mathrm{rk}(S_1)$. $\square$

(16.12) Theorem (Keller-Gehrig). *Let $\gamma > \omega$ and let n_m be a sequence of integers satisfying $n_m = O(m)$. Then $C(Ech_{m,n_m}) = O(m^\gamma)$.*

Proof. We set $n := n_m$. Multiple use of Prop. (16.11)(2) gives a sequence $m = \varepsilon_0 \ge \varepsilon_1 \ge \cdots \ge \varepsilon_\ell \ge 0$ such that $\sum_{i=0}^{\ell-1} \varepsilon_i \le n$ and $C(Ech_{m,n}) \le \sum_{i=0}^{\ell} C(Ech_{\varepsilon_i, \varepsilon_i})$. (Proceed as follows: if $m = 0$ or $m > n$ then set $\ell := 0$. Otherwise there exists $0 \le \varepsilon_1 \le m$ such that $C(Ech_{m,n}) \le C(Ech_{m,m}) + C(Ech_{\varepsilon_1, n-m})$. If $\varepsilon_1 = 0$ or $\varepsilon_1 \ge n - m$, set $\ell := 1$, etc.) Since $\sum_{i=0}^{\ell} \varepsilon_i \le n + \varepsilon_\ell \le n + m$ we obtain from Prop. (16.11)(1) the estimate $C(Ech_{m,n}) = O((m + n)^\gamma)$. (Note that $\sum_{i=0}^{\ell} \varepsilon_i^\gamma \le (\sum_{i=0}^{\ell} \varepsilon_i)^\gamma$.) Since $n = O(m)$, we are done. $\square$

A matrix A is said to be in *rank normal form* if $A = \begin{pmatrix} I_r & 0 \\ 0 & 0 \end{pmatrix}$, where $r = \operatorname{rk} A$. We define

$$RNF_{m,n} := \left\{ (A, (S, T)) \in k^{m \times n} \times (\mathrm{GL}(m, k) \times \mathrm{GL}(n, k)) \mid SAT \text{ is in rank normal form} \right\}.$$

The computation of the rank normal form of a matrix can, in principle, be reduced to that of the echelon form as the proof of the following proposition shows.

(16.13) Proposition. *If $\gamma > \omega$ and n_m is a sequence of integers satisfying $n_m = O(m)$, then $C(RNF_{m,n}) = O(m^{\gamma})$.*

Proof. Let $m, n, \in \mathbb{N}'$ and $\mu := \min\{m, n\}$. It is sufficient to show that

$$C(RNF_{m,n}) \leq C(Ech_{m,n}) + C(TInv_{\mu}) + \left\lceil \frac{n}{\mu} \right\rceil C(MaMu_{\mu}).$$

Let $A \in k^{m \times n}$. We compute U such that $UA = B$ is in echelon form. Then we compute (with zero cost) a permutation matrix P such that $BP = \begin{pmatrix} R & C \\ 0 & 0 \end{pmatrix}$ with an invertible ($\operatorname{rk} A \times \operatorname{rk} A$)-matrix R. We set $S := U$. We have $T := P \begin{pmatrix} R^{-1} & -R^{-1}C \\ 0 & I \end{pmatrix}$, where I is the $(n - \operatorname{rk} A) \times (n - \operatorname{rk} A)$-identity matrix. Note that $R^{-1}C$ can be computed from R^{-1} and C with cost $\leq \lceil n/\mu \rceil C(MaMu_{\mu})$ and that triangular inversion is monotone with respect to matrix size. $\qquad \square$

16.6* The Characteristic Polynomial

In Sect. 16.4 we saw that computing the determinant is about as hard as matrix multiplication. In this section we shall see that even the problem of computing *all* coefficients of the characteristic polynomial of a matrix has the same exponent as matrix multiplication.

Let X be an indeterminate over the field k. For $n \geq 1$ we define

$CharPol_n :=$

$$\left\{ (A, (a_0, \ldots, a_{n-1})) \in k^{n \times n} \times k^n \mid \det(X I_n - A) = X^n + \sum_{j=0}^{n-1} a_j X^j \right\}$$

and $CharPol = (CharPol_1, CharPol_2, \ldots)$. We have $\omega(Det) \leq \omega(CharPol)$. Hence, if $\omega > 2$, then $\omega(CharPol) \geq \omega$ by Thm. (16.7). To prove "$\leq$" we present in this section an algorithm for computing the characteristic polynomial.

We begin with introducing some notation. A matrix $F \in k^{n \times n}$ is said to be in Frobenius (normal) form if

$$F = \begin{pmatrix} 0 & & & -a_0 \\ 1 & \ddots & & \vdots \\ & \ddots & 0 & -a_{n-2} \\ & & 1 & -a_{n-1} \end{pmatrix} \in k^{n \times n},$$

for some $a_0, \ldots, a_{n-1} \in k$. Note that the characteristic polynomial of F equals $X^n + \sum_{j=0}^{n-1} a_j X^j$. If an invertible matrix T is known such that $T^{-1}AT$ is in Frobenius form, we can obtain the characteristic polynomial of A with one inversion and two multiplications of $n \times n$-matrices. Unfortunately such a T does not always exist.

Let us fix a notation: if $A_1, \ldots, A_m$ are matrices with an equal number of rows we denote by $(A_1|A_2| \cdots |A_m)$ the matrix obtained by concatenating the columns of $A_1, \ldots, A_m$.

If we can find $v \in k^n$ such that $T := (v|Av| \ldots |A^{n-1}v)$ is regular, then $T^{-1}AT$ is in Frobenius form. If A is generic, any nonzero $v \in k^n$ will serve, see Ex. 16.6. To compute T in this case, we proceed as follows: let $\ell := \lceil \log n \rceil - 1$. We first compute $A = A^{2^0}, A^{2^1}, \ldots, A^{2^\ell}$. Then we successively compute

$$
\begin{aligned}
Av &= Av, \\
(A^3 v | A^2 v) &= A^2 (Av|v), \\
&\ \ \vdots & &\ \ \vdots \\
(A^{2 \cdot 2^\ell - 1} v | \cdots | A^{2^\ell} v) &= A^{2^\ell}(A^{2^\ell - 1} v | \cdots | v).
\end{aligned}
$$

We can compute in this way T and hence the coefficients of the characteristic polynomial in time $O(n^\gamma \log n)$, where $\gamma > \omega$. For generic A however, we have to modify this simple idea.

We denote by $(e_1, \ldots, e_n)$ the standard basis of k^n. Further, we impose a $k[X]$-module structure on k^n by setting $Xv := Av$ for $v \in k^n$. For $v_1, \ldots, v_m \in k^n$ we write $\langle v_1, \ldots, v_m \rangle$ for the k-space generated by $v_1, \ldots, v_m$, while $\langle v_1, \ldots, v_m \rangle_{k[X]}$ denotes the $k[X]$-module generated by $v_1, \ldots, v_m$. For $B \in k^{\ell \times \ell}$ we define colsp(B) as the k-space generated by the columns of B and by col(B) the number ℓ of columns of B. By definition we set colsp$(B) = \emptyset$ if col$(B) = 0$. The following lemma shows that if A is an arbitrary matrix, then there exists U such that $U^{-1}AU$ is an upper block triangular matrix, whose diagonal blocks are in Frobenius form.

(16.14) Lemma. *For every $A \in k^{n \times n}$ there exists a unique sequence of non-negative integers $(m_1, \ldots, m_n)$ and a unique matrix $U = U(A)$ satisfying*

(1) $U = (U_1| \cdots |U_n)$, $U_j \in k^{n \times m_j}$ for $j \in \underline{n}$,
(2) $\forall j \in \underline{n}$: $U_j = (e_j|Ae_j| \cdots |A^{m_j-1}e_j)$, if $m_j \neq 0$,
(3) *For all $j \in \underline{n}$ the columns of $(U_1| \cdots |U_j)$ form a k-basis of $\langle e_1, \ldots, e_j \rangle_{k[X]}$.*

Moreover, we have

$$U^{-1}AU = \begin{pmatrix} F_1 & & & * \\ & F_2 & & \\ & & \ddots & \\ & & & F_n \end{pmatrix},$$

where $F_j \in k^{m_j \times m_j}$ is in Frobenius form for $j \in \underline{n}$ and the entries below the block diagonal are zero.

Proof. Existence: We recursively define the sequence $0 = m_0, m_1, \ldots, m_n$ and $\emptyset = U_0, U_1, \ldots, U_n$ by setting for $j \geq 1$

$$m_j := \min\{t \mid A^t e_j \in \text{colsp}(U_1 | \cdots | U_{j-1} | e_j | \cdots | A^{t-1} e_j)\}$$

and $U_j := (e_j | Ae_j | \cdots | A^{m_j-1} e_j)$. Properties (1)-(3) are verified quickly.

Uniqueness is easy and left to the reader. The assertion on the structure of $U^{-1}AU$ can be seen by regarding conjugation with U as a base change in k^n. □

Once we have computed $U = U(A)$, we can obtain the characteristic polynomial of A by one inversion (namely U^{-1}), two multiplications of $n \times n$-matrices to obtain $U^{-1}AU$, and an additional $O(n^2)$ multiplications corresponding to the multiplication of the characteristic polynomials of the F_i's. (According to Cor. (2.15) the multiplication of these polynomials can be accomplished more efficiently, but we do not need that in the sequel.) Before proceeding further with our discussions let us first consider an example.

(16.15) Example. Consider the matrix

$$A := \begin{pmatrix} 0 & -1 & -1 & -1 & -1 \\ 1 & 2 & 0 & 0 & 0 \\ 0 & 0 & 2 & -1 & 0 \\ 0 & 0 & 1 & 3 & 1 \\ 0 & 0 & -1 & 0 & 1 \end{pmatrix}$$

over the field $\mathbb{Q}$ of rational numbers. We want to compute the sequence $(m_i)_{1 \leq i \leq 5}$ and the matrix $U = U(A)$. We have $Ae_1 = e_2$ and $A^2 e_1 = -e_1 + 2e_2 \in \langle e_1, Ae_1 \rangle$. This gives $m_1 = 2$ and $U_1 = (e_1 | Ae_1) = (e_1 | e_2)$. Since $e_2 \in \text{colsp}(U_1)$, we obtain $m_2 = 0$. Further, $Ae_3 = -e_1 + 2e_3 + e_4 - e_5$, $A^2 e_3 = -2e_1 - e_2 + 3e_3 + 4e_4 - 3e_5$, $A^3 e_3 = -3e_1 - 4e_2 + 2e_3 + 15e_4 - 6e_5$. We see that $Ae_3 \notin \text{colsp}(U_1 | e_3)$, $A^2 e_3 \notin \text{colsp}(U_1 | e_3 | Ae_3)$, but $A^3 e_3 \in \text{colsp}(U_1 | e_3 | Ae_3 | A^2 e_3)$. Hence, $U_3 = (e_3 | Ae_3 | A^2 e_3)$ and $m_3 = 3$. Clearly, $m_4 = m_5 = 0$. Now let $U = (U_1 | U_3)$. This gives

$$U^{-1}AU = \begin{pmatrix} 0 & -1 & 0 & 0 & -3 \\ 1 & 2 & 0 & 0 & 2 \\ 0 & 0 & 0 & 0 & 8 \\ 0 & 0 & 1 & 0 & -12 \\ 0 & 0 & 0 & 1 & 6 \end{pmatrix}.$$

Thus $(X^2 - 2X + 1)(X^3 - 6X^2 + 12X - 8) = (X - 1)^2 (X - 2)^3$ is the characteristic polynomial of the matrix A. •

The rest of this section is devoted to the presentation of an algorithm for computing $U(A)$ which is faster than the procedure described in the above example.

Let $r := \lfloor \log n + 1 \rfloor$, i.e., r is the smallest integer greater than $\log n$. We successively construct a sequence $V_0, V_1, \ldots, V_r$ of regular $n \times n$-matrices such that $V_r = U(A)$; for each i we construct a sequence of non-negative integers $m_{i1}, \ldots, m_{in}$ such that $V_i = (V_{i1} | \cdots | V_{in})$, $V_{ij} \in k^{n \times m_{ij}}$, and $V_{ij} = (e_j | A e_j | \cdots | A^{m_{ij}-1} e_j)$ if $m_{ij} \neq 0$.

For $i = 0$ we set $m_{0j} := 1$, $V_{0j} := e_j$, i.e., $V_0 := I_n$. Now assume that $V_{i-1,j}$ has already been constructed for some $i \geq 1$. If $m_{i-1,j} = 0$ for some $j \in \underline{n}$, we set $m_{ij} := 0$. Otherwise, let

$$
W_{ij} := \begin{cases} V_{i-1,j} & \text{if } \mathrm{col}(V_{i-1,j}) < 2^{i-1}, \\ (V_{i-1,j} | A^{2^{i-1}} V_{i-1,j}) & \text{otherwise,} \end{cases}
$$

and $W_i := (W_{i1} | \cdots | W_{in})$. (To compute W_i we just need to multiply $A^{2^{i-1}}$ and V_{i-1}.) Now we define

$$
m_{ij} := \min \{ t \mid t < 2^i, \, A^t e_j \in \mathrm{colsp}(W_{i1} | \cdots | W_{i,j-1} | e_j | \cdots | A^{t-1} e_j) \} \cup \{ 2^i \}
$$

and set $V_{ij} := (e_j | A e_j | \cdots | A^{m_{ij}-1} e_j)$. Note that V_{ij} may be empty (if $m_{ij} = 0$), $m_{ij} \leq 2^i$, and if $0 < m_{ij} < 2^i$, then the m_{ij}th column of W_{ij} is the first column which is linearly dependent on the preceding columns of W_{ij} and all columns of the matrices $W_{i1}, \ldots, W_{i,j-1}$. Also note that the columns of $V_i := (V_{i1} | \cdots | V_{in})$ are linearly independent. The m_{ij} and V_{ij}, $j \in \underline{n}$ may be computed by transforming W_i to echelon form.

The following lemma shows that $V_r = U(A)$.

(16.16) Lemma. *For $j \in \underline{n}$ and $0 \leq i \leq r - 1$ we have*

$$
\langle \langle e_1, \ldots, e_{j-1} \rangle_{k[X]}, \mathrm{colsp}(V_{ij}) \rangle \subseteq \langle e_1, \ldots, e_j \rangle_{k[X]}
$$

with equality holding if $\mathrm{col}(V_{ij}) < 2^i$.

Before proving the lemma let us see that it implies $V_r = U(A)$: indeed, $\mathrm{col}(V_{rj}) \leq n < 2^r$ for all $j \in \underline{n}$. Hence, using the lemma and applying induction on j we see that $\mathrm{colsp}(V_{i1} | \cdots | V_{ij}) = \langle e_1, \ldots, e_j \rangle_{k[X]}$. Thus, V_r and $(\mathrm{col}(V_{r1}), \ldots, \mathrm{col}(V_{rn}))$ satify the conditions of Lemma (16.14) and this implies $V_r = U$ by uniqueness.

Proof of Lemma (16.16). Since $V_{ij} = (e_j | A e_j | \cdots | A^{m_{ij}-1} e_j)$, we have $\mathrm{colsp}(V_{ij}) \subseteq \langle e_j \rangle_{k[X]}$, hence "$\subseteq$" follows. Let us now prove the "$=$" part. Let $L_j := \langle e_1, \ldots, e_j \rangle_{k[X]}$. By the construction of the W_i it is obvious that $\mathrm{colsp}(W_{i1} | \cdots | W_{ij}) \subseteq L_j$ for all i, j. Let $m_{ij} = \mathrm{col}(V_{ij}) < 2^i$. The construction of the V_{ij} implies that

$$
A^{m_{ij}} e_j \in \langle W_{i1} | \cdots | W_{i,j-1} | e_j | \cdots | A^{m_{ij}-1} e_j \rangle,
$$

hence $A^{m_{ij}} e_j \equiv f(A) e_j \mod L_{j-1}$ for some polynomial $f \in k[X]$ of degree less than m_{ij}. We infer that for all $\ell \in \mathbb{N}$ there exists some polynomial $f_\ell \in k[X]$ of degree less than m_{ij} such that $A^\ell e_j \equiv f_\ell(A) e_j \mod L_{j-1}$. ($f_\ell$ is the remainder of the division of X^ℓ by $X^{m_{ij}} - f(X)$.) The assertion follows. $\square$

The algorithm for the computation of the characteristic polynomial of a matrix $A \in k^{n \times n}$ is now as follows.

(1) Compute $r := \lfloor \log n + 1 \rfloor$ and the matrices $A^2, \ldots, A^{2^{r-1}}$.
(2) Set $V_0 := I_n$; for $i := 1$ to r do Steps (3) and (4).
(3) Compute W_i defined above by multiplying $A^{2^{i-1}}$ and V_{i-1}.
(4) Transform W_i to echelon form and compute V_i.
(5) Set $U := V_r$. Compute U^{-1}.
(6) Compute $U^{-1}AU$. This matrix has now the form as in Lemma (16.14).
(7) Compute the product of the characteristic polynomials of the F_i.

An analysis of this algorithm gives us the estimate

$$C(CharPol_n) \le (2r + 1)C(MaMu_n) + rC(Ech_{n,2n}) + C(Inv_n) + \beta n^2$$

for all $n \ge 1$, where β is a constant taking care of the time for the multiplication of the characteristic polynomials of the F_i. By Thm. (16.12) and Prop. (16.6) we conclude $\omega(CharPol) \le \omega$. Taking into account $C(Det_n) \le C(CharPol_n)$ and Thm. (16.7) we obtain the main theorem of this section.

(16.17) Theorem (Keller-Gehrig). *We have* $\omega(CharPol) = \omega$.

16.7* Computing a Basis for the Kernel

This section is devoted to showing that computing the kernel of a square matrix is about as hard as matrix multiplication.

For $n \ge 1$ we define the problem

$$Ker_n := \left\{ (A, B) \in k^{n \times n} \times \bigsqcup_{i=0}^{n} k^{n \times i} \;\middle|\; \right.$$

$$\left. B \in k^{n \times rk(B)}, \, rk(A) + rk(B) = n, \, AB = 0 \right\},$$

and $Ker := (Ker_1, Ker_2, \ldots)$. Note that $(A, B) \in Ker_n$ iff the columns of B form a basis for the kernel of A. It is easy to give a relative upper bound for the complexity of Ker_n in terms of the complexity of matrix multiplication: given the matrix $A \in k^{n \times n}$, we transform it to echelon form with cost $C(Ech_{n,n})$. Then, using a standard procedure, we can compute a basis for the kernel of A with at most n further multiplications/divisions. We thus have

$$C(Ker_n) \le C(Ech_{n,n}) + n.$$

This implies $\omega(Ker) \le \omega$ by Thm. (16.12). The following gives a lower bound for $C(Ker_n)$ in terms of the complexity of matrix multiplication.

(16.18) Theorem (Bürgisser et al.). $C(Ker_n) \geq C(MaMu_{\lfloor n/4 \rfloor})$ *for all* $n \geq 4$. *In particular*, $\omega(Ker) = \omega$.

Proof. We may assume w.l.o.g. that $n = 4m$ for some $m \in \mathbb{N}$. (Note that $C(Ker_n) \geq C(Ker_{n-1})$.) Consider the restricted problem

$$\widetilde{Ker}_n := Ker_n \cap \left(\left\{ \begin{pmatrix} \xi & \eta \\ 0 & 0 \end{pmatrix} \; \middle| \; \xi, \eta \in k^{2m \times 2m} \right\} \times \bigsqcup_{i=0}^{n} k^{n \times i} \right).$$

Let $X = (X_{ij})$ and $Y = (Y_{ij})$ denote $2m \times 2m$-matrices whose entries are indeterminates over k. We put $R := k[X_{ij}, Y_{ij} \mid i, j \in \underline{2m}]$ and $K := \text{Quot}(R)$. Application of Lemma (16.2) to $\widetilde{Ker}_n$ yields the existence of $B \in K^{4m \times 2m}$ such that $\text{rk}(B) = 2m$, $(X|Y)B = 0$, and $L_K(B) \leq C(Ker_n)$. From these conditions it easily follows that B can be written in the form

$$B = \begin{pmatrix} X^{-1}U \\ -Y^{-1}U \end{pmatrix}$$

for some $U \in \text{GL}(2m, K)$. We therefore have

$$L_K(X^{-1}U, Y^{-1}U) \leq C(Ker_n).$$

As in the proof of Thm. (16.7) we deduce that there exist $\xi, \eta \in \text{GL}(2m, k)$ such that

$$X, Y, U \in \text{GL}(2m, \mathcal{O}_{(\xi,\eta)}), \quad \text{and} \quad L_{\mathcal{O}_{(\xi,\eta)}}(X^{-1}U, Y^{-1}U) = L_K(X^{-1}U, Y^{-1}U),$$

where $\mathcal{O}_{(\xi,\eta)} \subset K$ is the local ring of (ξ, η). Replacing U by $UU(\xi, \eta)^{-1}$ we may assume that $U(\xi, \eta) = I_{2m}$. Application of the isomorphism

$$\varphi \colon \mathcal{O}_{(\xi,\eta)} \to \mathcal{O}_{(I_{2m}, I_{2m})}, \quad \varphi(X) := X\xi, \quad \varphi(Y) := Y\eta,$$

shows that we may assume w.l.o.g. that $\xi = \eta = I_{2m}$. (Note that we do not count linear operations; see also the proof of Thm. (16.7).) Furthermore

$$L_{\mathcal{O}_{(I_{2m}, I_{2m})}}(X^{-1}U, Y^{-1}U) = L_{\mathcal{O}_{(0,0)}}((I_{2m} - X)^{-1}V, (I_{2m} - Y)^{-1}V),$$

where V denotes the image of U under the variable substitution $X \mapsto I_{2m} - X$, $Y \mapsto I_{2m} - Y$. We use now Thm. (7.1) with $d = 2$ to get

$$L_R(X^2 + XV^{(1)} + V^{(2)}, Y^2 + YV^{(1)} + V^{(2)})$$
$$\leq L_{\mathcal{O}_{(0,0)}}((I_{2m} - X)^{-1}V, (I_{2m} - Y)^{-1}V),$$

where $V = I_{2m} + V^{(1)} + V^{(2)} + \cdots$ is the Taylor expansion of V at $(0,0)$. The left-hand side of the above equation can be estimated from below by the quantity $L_R(X^2 - Y^2 + (X - Y)V^{(1)})$. We decompose X and Y into blocks as

$$X = \begin{pmatrix} X^{11} & X^{12} \\ X^{21} & X^{22} \end{pmatrix}, \quad Y = \begin{pmatrix} Y^{11} & Y^{12} \\ Y^{21} & Y^{22} \end{pmatrix},$$

with $X^{ij}, Y^{ij} \in R^{m \times m}$. The entries of X^{ij} and Y^{ij} are pairwise different indeterminates over k. Now we use the linear substitution

$$\psi: R \rightarrow R, \quad \psi(X) := \begin{pmatrix} 0 & X^{12} \\ X^{21} & X^{22} \end{pmatrix}, \quad \psi(Y) := \begin{pmatrix} 0 & X^{12} \\ 0 & 0 \end{pmatrix}$$

and obtain

$$\psi(X^2 - Y^2 + (X - Y)V^{(1)}) = \begin{pmatrix} X^{12}X^{21} & X^{12}X^{22} \\ P & Q \end{pmatrix}$$

for some $P, Q \in R^{m \times m}$. Therefore

$$C(MaMu_m) = L_R(X^{12}X^{21}) \leq C(Ker_n). \qquad \square$$

16.8* Orthogonal Basis Transform

Throughout this section we assume that k is a field of characteristic $\neq 2$.

Let A be a symmetric matrix over k. It is well-known that there exists an invertible matrix S over k such that SAS^{T} is diagonal. The problem to compute such an S from A is formalized as

$$OgB_n := \{(A, S) \in k^{n \times n} \times GL(n, k) \mid A \text{ symmetric}, SAS^{\mathsf{T}} \text{ diagonal}\},$$

$OgB := (OgB_1, OgB_2, \ldots)$. In this section we will prove that $\omega(OgB) \geq \omega$, provided $\omega > 2$. We proceed as follows: we first relate the problem to that of computing the trace of the inverse of a symmetric matrix with indeterminate entries. Multiple use of differential methods from Chap. 7 yields then a lower bound in terms of the complexity of computing the square of a symmetric matrix with indeterminate entries over k. By applying a suitable substitution we finally obtain a lower bound in terms of the complexity of matrix multiplication.

We start with an auxiliary result the proof of which is left as an exercise to the reader (see Ex. 16.2).

(16.19) Lemma. *Let $A = (a_{ij})$ be a symmetric matrix with indeterminate entries over k. Then the distinct entries of A^{-2} are algebraically independent over k.*

(16.20) Theorem (Bürgisser et al.). *The sequence OgB satisfies*

$$C(OgB_n) \geq \frac{1}{6^2} C(MaMu_{\lfloor n/4 \rfloor}) - 4n^2 - n.$$

In particular, $\omega(OgB) \geq \omega$, provided $\omega > 2$.

Proof. We may w.l.o.g. assume that $n = 4m$ for some $m \in \mathbb{N}$. Let $A = (a_{ij})$ denote a symmetric $n \times n$-matrix with indeterminate entries over k. Put $R := k[a_{ij} \mid 1 \leq i \leq j \leq n]$ and $K := \mathrm{Quot}(R)$. By Lemma (16.2) there exists a matrix $S \in \mathrm{GL}(n, K)$ such that $D := SAS^\top$ is diagonal and $L_K(S) \leq C(OgB_n)$. Denote that diagonal entries of D by $d_1, \ldots, d_n$ and note that

$$
\begin{pmatrix} d_1 \\ \vdots \\ d_n \end{pmatrix} = D \begin{pmatrix} 1 \\ \vdots \\ 1 \end{pmatrix} = S\left(A\left(S^\top \begin{pmatrix} 1 \\ \vdots \\ 1 \end{pmatrix}\right)\right).
$$

We see that D can be computed from A and S with $2n^2$ multiplications. Further, we have $\mathrm{Tr}(A^{-1}) = \mathrm{Tr}(S^\top(D^{-1}S))$. Hence, since the trace of the product of two $n \times n$-matrices can be computed with no more than n^2 multiplications, we have

$$
L_K(\mathrm{Tr}(A^{-1})) \leq L_K(S) + 4n^2 + n.
$$

(Computing D^{-1} requires n divisions.) Now let ε be an indeterminate over K and $V = (v_{ij}) \in k^{n \times n}$ be symmetric. The expansion of $\mathrm{Tr}((A + \varepsilon V)^{-1})$ in $K[[\varepsilon]]$ yields

$$
\mathrm{Tr}((A + \varepsilon V)^{-1}) = \mathrm{Tr}(A^{-1}) + \varepsilon \sum_{i \leq j} \frac{\partial \, \mathrm{Tr}(A^{-1})}{\partial a_{ij}} v_{ij} + O(\varepsilon^2).
$$

On the other hand, we have (geometric series expansion)

$$
\mathrm{Tr}((A + \varepsilon V)^{-1}) = \mathrm{Tr}(A^{-1}) - \varepsilon \, \mathrm{Tr}(A^{-1} V A^{-1}) + O(\varepsilon^2).
$$

A comparison of the two terms yields

$$
\frac{\partial \, \mathrm{Tr}(A^{-1})}{\partial a_{ij}} = \begin{cases} -2(A^{-2})_{ij}, & \text{if } i \neq j, \\ -(A^{-2})_{ij}, & \text{otherwise.} \end{cases}
$$

From the derivative inequality (7.7) we deduce

$$
\frac{1}{3} L_K(A^{-2}) \leq L_K(\mathrm{Tr}(A^{-1})) \leq L_K(S) + 4n^2 + n.
$$

By Lemma (16.19) we have a well defined k-algebra morphism $K \to K, A \mapsto A^{-2}$, which shows that $L_K(A^4 \mid A^{-2}) \leq L_K(A^{-2})$. We thus obtain

$$
L_K(A^4) \leq L_K(A^4 \mid A^{-2}) + L_K(A^{-2}) \leq 2L_K(A^{-2})
$$

Now we apply Thm. (7.1) with $d = 4$ to get $L_R(A^4) \leq \binom{4}{2} L_K(A^4)$. Let $\mathcal{O}_I$ denote the local ring of $I := I_n$. Then we have $L_R(A^4) \geq L_{\mathcal{O}_I}(A^4) = L_{\mathcal{O}_0}((I - A)^4)$. Now we use Thm. (7.1) again with $d = 2$ together with the expansion

$$
(I - A)^4 = I - 4A + 6A^2 + \cdots
$$

to obtain $L_R(A^4) \geq L_R(A^2)$. Summarizing, we have

$$C(OgB_n) \geq \frac{1}{6^2} L_R(A^2) - 4n^2 - n.$$

We proceed further by dividing the matrix A into $m \times m$-blocks $A_{ij} \in R^{m \times m}$ and defining the substitution

$$\psi: R \to R, \quad \psi(A) := \begin{pmatrix} 0 & 0 & A_{13} & 0 \\ 0 & 0 & A_{23} & 0 \\ A_{13}^{\top} & A_{23}^{\top} & 0 & 0 \\ 0 & 0 & 0 & 0 \end{pmatrix}.$$

We obtain

$$\psi(A)^2 = \begin{pmatrix} A_{13}A_{13}^{\top} & A_{13}A_{23}^{\top} & 0 & 0 \\ A_{23}A_{13}^{\top} & A_{23}A_{23}^{\top} & 0 & 0 \\ 0 & 0 & W & 0 \\ 0 & 0 & 0 & 0 \end{pmatrix},$$

where $W = A_{13}^{\top}A_{13} + A_{23}^{\top}A_{23}$. Hence

$$C(MaMu_m) = L_R(A_{13}A_{23}^{\top}) \leq L_R(A^2).$$

This gives the assertion.   □

In the rest of this section we will derive a relative upper bound for $C(OgB_n)$. We denote by Σ_n the space of symmetric $n \times n$-matrices and by Σ_n^{ℓ} the set of those $A \in \Sigma_n$ which are decomposable as $A = \begin{pmatrix} A_1 & B \\ B^{\top} & C \end{pmatrix}$ with regular $A_1 \in \Sigma_{\ell}$. We will need several preparatory lemmas.

(16.21) Lemma. Let $\gamma > \omega$. For $1 \leq \ell \leq n - 1$ there exists a computation tree $T_{n,\ell}^{(1)}$ which, on input $A = \begin{pmatrix} A_1 & B \\ B^{\top} & C \end{pmatrix} \in \Sigma_n^{\ell}$, computes some $S \in GL(n, k)$ satisfying $SAS^{\top} = \begin{pmatrix} A_1 & 0 \\ 0 & A_2 \end{pmatrix} \in \Sigma_n^{\ell}$. Moreover, $C(T_{n,\ell}^{(1)}) = O(n^{\gamma})$.

Proof. It is enough to compute

$$S := \begin{pmatrix} I_{\ell} & 0 \\ -B^{\top}A_1^{-1} & I_{n-\ell} \end{pmatrix}.$$ □

For giving an upper bound for $C(OgB_n)$ we may restrict ourselves to invertible matrices, as the following lemma shows.

(16.22) Lemma. Let $\gamma > \omega$. For any $n \geq 1$ there exists a computation tree $T_n^{(2)}$ which on input $A \in \Sigma_n$ computes a matrix $S \in GL(n, k)$ such that $SAS^{\top} = \begin{pmatrix} A_1 & 0 \\ 0 & 0 \end{pmatrix}$, where A_1 is regular. Moreover, $C(T_n^{(2)}) = O(n^{\gamma})$.

Proof. Let S_1 be a matrix which transforms A to echelon form and P be a permutation matrix such that $S_1AP = \begin{pmatrix} R & B \\ 0 & 0 \end{pmatrix}$, where $R \in k^{rk(A) \times rk(A)}$ is upper triangular of full rank, and $B \in k^{rk(A) \times (n - rk(A))}$. There is a permutation matrix Q such that $S := QPS_1$ satisfies $SAS^{\top} = \begin{pmatrix} A_1 & 0 \\ 0 & 0 \end{pmatrix}$. Computation of P and Q can be performed with $O(n^2)$ queries. Now apply Thm. (16.12).   □

(16.23) Lemma. *Let* $\gamma > \omega$. *Suppose that* $n = 2\ell \geq 2$. *There exists a computation tree* $T_n^{(3)}$ *with the following property: on input* $A = \begin{pmatrix} 0 & A_1 \\ A_1^\top & A_2 \end{pmatrix}$ *with* $A_1 \in GL(\ell, k)$, $T_n^{(3)}$ *computes* $S \in GL(n, k)$ *such that* $SAS^\top \in \Sigma_n^\ell$. *Further,* $C(T_n^{(3)}) = O(n^\gamma)$.

Proof. By Lemma (16.22) we can compute with cost $O(n^\gamma)$ a regular matrix U such that $U A_2 U^\top = \begin{pmatrix} A_3 & 0 \\ 0 & 0 \end{pmatrix}$, A_3 regular of size t. Let

$$S_1 = \begin{pmatrix} (A_1 U^\top)^{-1} & 0 \\ 0 & U \end{pmatrix}.$$

(S_1 can be computed with cost $O(n^\gamma)$.) Then we have

$$S_1 A S_1^\top = \begin{pmatrix} 0 & 0 & I & 0 \\ 0 & 0 & 0 & I \\ I & 0 & A_3 & 0 \\ 0 & I & 0 & 0 \end{pmatrix} =: B,$$

where the identity matrices in the first and third row have size t, whereas the identity matrices in the second and fourth row are of size $n - t$. It is easy to give $S_2 \in GL(n, k)$ such that

$$S_2 B S_2^\top = \begin{pmatrix} A_3 & 0 & I & 0 \\ 0 & 2I & 0 & I \\ I & 0 & 0 & 0 \\ 0 & I & 0 & 0 \end{pmatrix}.$$

Now set $S := S_2 S_1$.   □

(16.24) Lemma. *Let* $\gamma > \omega$. *For* $1 \leq \ell n - 1$ *there exists a computation tree* $T_{n,\ell}^{(4)}$ *which computes on input* $A \in \Sigma_n \cap GL(n, k)$ *some* $S \in GL(n, k)$ *such that* $SAS^\top \in \Sigma_n^\ell$ *in time* $C(T_{n,\ell}^{(4)}) = O(n^\gamma)$.

Proof. Let $A = \begin{pmatrix} A_1 & A_2 \\ A_3 & A_4 \end{pmatrix} \in \Sigma_n$, $A_1 \in \Sigma_\ell$. By Lemma (16.22) we may assume that A is regular. Now we proceed as follows:

$$\begin{pmatrix} A_1 & A_2 \\ A_2^\top & A_4 \end{pmatrix} \xrightarrow{1} \begin{pmatrix} B_1 & 0 & B_2 \\ 0 & 0 & \\ B_2^\top & A_4 \end{pmatrix} \xrightarrow{2} \begin{pmatrix} B_1 & 0 & 0 \\ 0 & 0 & C_1 \\ 0 & C_1^\top & C_2 \end{pmatrix} \xrightarrow{3}$$

$$\begin{pmatrix} B_1 & 0 & 0 & 0 \\ 0 & 0 & D_1 & D_2 \\ 0 & D_1^\top & D_3 & D_4 \\ 0 & D_2^\top & D_4^\top & D_6 \end{pmatrix} \xrightarrow{4} \begin{pmatrix} B_1 & 0 & 0 & 0 \\ 0 & F_1 & F_2 & F_3 \\ 0 & F_2^\top & & F_4 \\ 0 & F_3^\top & & \end{pmatrix} \in \Sigma_n^\ell.$$

Step **1** describes the application of Lemma (16.22) to A_1. Step **2** is an application of Lemma (16.21). Since A and B_1 are regular, we may find a permutation matrix P such that $C_1 P = (D_1 | D_2)$ with *regular* D_1. In step **3** this transformation is performed. Finally, Step **4** is the application of Lemma (16.23) to the matrix $\begin{pmatrix} 0 & D_1 \\ D_1^\top & D_3 \end{pmatrix}$.   □

Now we are in a position to prove the main theorem of this section.

(16.25) Theorem (Bürgisser et al.). $\omega(OgB) = \omega$ *provided* $\omega > 2$.

Proof. Given $A \in \Sigma_n^\ell$ we use the computation tree $T_n^{(2)}$ introduced in Lemma (16.22) to compute S such that $SAS^T = \left(\begin{smallmatrix} A_1 & 0 \\ 0 & 0 \end{smallmatrix}\right)$, with regular $A_1 \in \Sigma_r$ ($r = $ rk A). We use the computation tree $T_{r,\lceil r/2\rceil}^{(4)}$ in Lemma (16.24) to transform A_2 to $A_3 = \left(\begin{smallmatrix} B_1 & B_2 \\ B_2^\top & B_3 \end{smallmatrix}\right)$ with $B_1 \in GL(\lceil r/2\rceil, k)$. We apply the computation tree $T_{r,\lceil r/2\rceil}^{(1)}$ of Lemma (16.21) to transform A_1 to $A_2 = \left(\begin{smallmatrix} C_1 & 0 \\ 0 & C_2 \end{smallmatrix}\right)$. Now we apply our algorithm recursively to C_1 and C_2. The matrix S can be obtained by multiplying the intermediate transformation matrices. Let $\gamma > \omega$. Using Lemmas (16.21)–(16.24), an analysis of this algorithm gives

$$C(OgB_n) \leq O(n^\gamma) + 2C(OgB_{\lceil r/2\rceil}) \leq O(n^\gamma) + 2C(OgB_{\lceil n/2\rceil}),$$

hence $C(OgB_n) = O(n^\gamma)$. This implies $\omega(OgB) \leq \omega$. The other inequality follows from Thm. (16.20). $\qquad\square$

(16.26) Remark. It can be shown that $C(OgB_n) \geq cn^2$ for some constant c, see the Notes. This implies $\omega(OgB) = \omega$ without further assumptions on ω. $\qquad\bullet$

16.9* Matrix Multiplication and Graph Theory

Matrix multiplication algorithms can be applied to some problems in graph theory, such as computing the transitive closure. This is the content of this section.

We start with defining Boolean matrices. We introduce on the set $\{0, 1\}$ the operations $+$ (OR) and $*$ (AND) given by

$+$	0	1
0	0	1
1	1	1

$*$	0	1
0	0	0
1	0	1

Clearly, $(\{0, 1\}, +, *)$ is a commutative semiring. A matrix $A \in \{0, 1\}^{n \times n}$ is called an $n \times n$ *Boolean matrix*. The Boolean sum $+$ and the Boolean product $*$ of two Boolean matrices $A = (a_{ij})$ and $B = (b_{ij})$ is defined by $A + B := C = (c_{ij})$ and $A * B := D = (d_{ij})$ where for all $i, j \in \underline{n}$

$$c_{ij} = a_{ij} + b_{ij}, \qquad d_{ij} = \sum_{\mu=1}^{n} a_{i\mu} * b_{\mu j}.$$

It is easily seen that also $(\{0, 1\}^{n \times n}, +, *)$ is a commutative semiring.

(16.27) Remark. Let A, B, and C be $n \times n$ Boolean matrices and $I = I_n$ be the $n \times n$ identity matrix regarded as a Boolean matrix. Then we have:

(1) $A + A = A$, $I * A = A * I = A$.
(2) For any $\ell \geq 1$ we have $(I + A)^{*\ell} = I + A + A^{*2} + \cdots + A^{*\ell}$, where for a Boolean matrix A^{*i} denotes the i-fold Boolean power of A.   •

The entries of a power of a Boolean matrix can be characterized in the following way.

(16.28) Lemma. *Let A be an $n \times n$ Boolean matrix and $\ell \geq 1$. Then*

$$(A^{*\ell})_{ij} = 1 \iff \exists \mu_1, \ldots, \mu_{\ell-1} \in \underline{n}: \quad a_{i\mu_1} = a_{\mu_1\mu_2} = \cdots = a_{\mu_{\ell-1}j} = 1.$$

Proof. This is a consequence of the formula

$$(A^{*\ell})_{ij} = \sum_{\mu_1=1}^{n} \cdots \sum_{\mu_{\ell-1}=1}^{n} a_{i\mu_1} * a_{\mu_1\mu_2} * \cdots * a_{\mu_{\ell-1}j}. \qquad \square$$

Let $G = (V, E)$ be a directed graph and $\hat{G} = (V, \hat{E})$ be its transitive closure, i.e., $(v, w) \in \hat{E}$ iff there exist $u_1, \ldots, u_{\ell-1}$ such that

$$(v, u_1), (u_1, u_2), \ldots, (u_{\ell-1}, w) \in E.$$

Let $n = |V|$ and $A = (a_{ij})$ be the $n \times n$ adjacency matrix of G. Then the adjacency matrix of $\hat{G}$ is given by the following formula.

(16.29) Proposition. *The adjacency matrix $\hat{A}$ of the transitive closure $\hat{G}$ of G is given by*

$$\hat{A} = (I_n + A)^{*n} = I_n + A + A^{*2} + \cdots + A^{*n}.$$

Proof. We prove by induction on i: the graph $G_i = (V, E_i)$ given by the adjacency matrix $(I_n + A)^{*i}$ satisfies: $(v, w) \in E_i$ iff there exists $j \leq i$ and $u_1, \ldots, u_{j-1} \in V$ such that $(v, u_1), \ldots, (u_{j-1}, w) \in E_i$. The assertion is obvious for $i = 1$. Now note that $(I_n + A)^{*(i+1)} = (I_n + A)^{*i} + A^{*(i+1)}$. Applying the foregoing lemma we get the assertion. Therefore $(I_n + A)^{*n}$ is the adjacency matrix of $\hat{G}$. $\square$

By abuse of notation, we call for any $n \times n$ Boolean matrix A the matrix $(I_n + A)^{*n}$ its transitive closure and denote it by $\hat{A}$. For computing the transitive closure we need to know how to multiply Boolean matrices.

(16.30) Proposition. *Let n be a positive integer, $p > n$ be a prime, A, B be $n \times n$ Boolean-matrices. Let $(c_{ij}) = C = A * B$ and $D = (d_{ij}) = AB \in \mathbb{F}_p^{n \times n}$, where in the latter case A and B are regarded as matrices over $\mathbb{F}_p$. Then we have for all $i, j \in \underline{n}$*

$$c_{ij} = 0 \iff d_{ij} = 0.$$

Proof. Note that $d_{ij} = \sum_{\mu=1}^{n} a_{i\mu}b_{\mu j}$. Consider for a moment $a_{i\mu}$ and $b_{\mu j}$ as integers. Then

$$0 \leq \sum_{\mu=1}^{n} a_{i\mu}b_{\mu j} \leq p - 1.$$

Hence, $d_{ij} = 0$ iff for all $\mu \in \underline{n}$ we have $a_{i\mu}b_{\mu j} = 0$ and this is the case iff $c_{ij} = 0$. $\square$

The foregoing proposition immediately implies the following.

(16.31) Theorem (Fischer and Meyer). *Suppose that for all $m \geq 1$ there exists a computation tree T_m over $\mathbb{Z}$ which computes $MaMu_m$ with total cost $O(m^\gamma)$, and denote by B_m the minimum number of bit-operations sufficient to compute the transitive closure of an $m \times m$ Boolean matrix. Then $B_m = o(m^\gamma (\log m)^3)$.*

Proof. For any positive integer m let $p = p_n$ be the smallest prime greater than m. By Prop. (16.30) and (16.29) the transitive closure of an $m \times m$ Boolean matrix may be computed by computing $(I_m + A)^m$ over $\mathbb{F}_p$. The latter requires $O(m^\gamma \log m)$ arithmetic operations in $\mathbb{F}_p$ using square and multiply. Each $\mathbb{F}_p$-operation requires at most $o(\log^2 p)$ bit-operations by Schönhage and Strassen [466]. Since by Bertrand's postulate (see [106]) we have $p \leq 2m + 2$, we are done. $\square$

16.10 Exercises

16.1. Prove Lemma (16.2)

16.2. Let k be a field of characteristic $p \geq 0$ and let $A = (a_{ij})$ be a symmetric $n \times n$-matrix whose distinct entries are algebraically independent over k. Show that for $\ell \in \mathbb{Z} \setminus p\mathbb{Z}$ the entries of A^ℓ are algebraically independent. (Hint: consider the Taylor expansion of A at $I = I_n$: $(I - A)^\ell = I - \ell A + O(A^2)$. Now use Ex. 8.5.)

16.3. (Inversion of generic matrices) Let $A = (a_{ij})$ be an $n \times n$-matrix whose entries are pairwise different indeterminates over k. Let K denote the quotient field of $R := k[a_{ij} \mid 1 \leq i, j \leq n]$ and put $M(n) := L_K(A^{-1})$.

(1) Suppose that $n = 2m$ and decompose A into $m \times m$-blocks A_{ij}, $1 \leq i, j \leq 2$. Let $\Delta := A_{22} - A_{21} A_{11}^{-1} A_{12} \in K^{n \times n}$. Show that

$$A = \begin{pmatrix} I_m & 0 \\ A_{21} A_{11}^{-1} & I_m \end{pmatrix} \begin{pmatrix} A_{11} & 0 \\ 0 & \Delta \end{pmatrix} \begin{pmatrix} I_m & A_{11}^{-1} A_{12} \\ 0 & I_m \end{pmatrix}.$$

Using this identity develop a recursive procedure for computing A^{-1} if n is a power of 2.
(2) Show that if $C(MaMu_n) = O(n^\gamma)$, then $M(n) = O(n^\gamma)$.

16.4. Prove that $\omega(LUP) \geq \omega(Det) \geq 2$. (Hint: compare Thm. (16.7). Use the dimension bound (4.12) and the derivative inequality (7.7).)

16.5. We keep the notation of the previous exercise.

(1) Show that there exist $0 \neq \alpha \in k$, $C \in k^{n \times 1}$, and $D \in k^{1 \times n}$ such that

$$L\left(\begin{pmatrix} \alpha & D \\ C & A \end{pmatrix}^{-1} \right) \leq M(n + 1).$$

(2) Show that $M(n) \leq M(n+1)$ by considering the factorization

$$\begin{pmatrix} 1 & 0 \\ -D/\alpha & I_n \end{pmatrix} \begin{pmatrix} \alpha & C \\ D & A \end{pmatrix} \begin{pmatrix} 1 & -C/\alpha \\ 0 & I_n \end{pmatrix} = \begin{pmatrix} \alpha & 0 \\ 0 & A - (DC/\alpha) \end{pmatrix}.$$

16.6. Let A be an $n \times n$-matrix whose entries are algebraically independent over the field k. Let $v \in k^n$ be nonzero. Show that $(v \mid Av \mid \cdots \mid A^{n-1}v)$ is regular.

16.7. For an $n \times n$-matrix A we denote by $\mathrm{Tr}(A)$ its trace. Let A be an $n \times n$-matrix whose entries a_{ij} are algebraically independent over a field k. Further, $R := k[a_{ij} \mid 1 \leq i, j \leq n]$ and $K := \mathrm{Quot}(R)$. Define $T(n) := L_K(\mathrm{Tr}(A^{-1}))$.

(1) Show that $L_k(A^4) \leq 6T(3n)$ and hence $L_R(A^4) \leq 36T(3n)$. (Hint: look at the proof of Thm. (16.20).)
(2) Let $n = 3m$. Show that $C(MaMu_m) \leq 36T(n)$ by partitioning A into $m \times m$-blocks A_{ij}, $1 \leq i, j \leq 3$ and considering the substitution

$$\psi: R \to R, \quad A \mapsto \begin{pmatrix} 0 & A_{12} & 0 \\ 0 & 0 & A_{23} \\ I_m & 0 & 0 \end{pmatrix}.$$

(3) Show that $C(MaMu_n) \leq 828T(n) + O(n^2)$. (Hint: use the inequality $C(MaMu_{3m}) \leq 23C(MaMu_m)$, see Laderman [310], or Ex. 15.3.)

16.8. Show that $T(n)$ is nondecreasing. (Hint: proceed as in Ex. 16.5 and note that

$$\mathrm{Tr}(A^{-1}) = \frac{1}{\det(A)} \sum_{i=1}^{n} \det A_i,$$

where A_i is the matrix obtained from A by deleting the ith row and the ith column.)

16.9.* For $n \geq 1$ we define

$$3\text{-}Cpr_n := \left\{ ((A_1, A_2, A_3), (B_1, B_2)) \in (k^{n \times n})^3 \times (k^{n \times n})^2 \mid A_1 A_2 A_3 = B_1 B_2 \right\}$$

and $3\text{-}Cpr := (3\text{-}Cpr_1, 3\text{-}Cpr_2, \ldots)$. In this exercise we prove that $\omega(3\text{-}Cpr) = \omega$ provided $\omega > 2$.

(1) Show that $\omega(3\text{-}Cpr) \leq \omega$.
(2) Let $A = (a_{ij})$, $B = (b_{ij})$, and $C = (c_{ij})$ be $n \times n$-matrices with indeterminate entries over k and let $K := k(a_{ij}, b_{ij}, c_{ij} \mid 1 \leq i, j \leq n)$. Show that $L_K(\mathrm{Tr}(ABC)) \leq C(3\text{-}Cpr_n) + n^2$. (Hint: show that there exist $U, V \in K^{n \times n}$ such that $L_K(UV) \leq C(3\text{-}Cpr_n)$.)
(3) Prove that $C(MaMu_n) \leq 3C(3\text{-}Cpr_n) + 3n^2$. This gives $\omega(3\text{-}Cpr) = \omega$ provided $\omega > 2$. (Hint: First prove that $\partial \, \mathrm{Tr}(ABC)/\partial a_{ij} = (BC)_{ji}$, where $A = (a_{ij})$. Then apply the derivative inequality.)

16.10.* For many applications in linear algebra it is desirable to find a base change such that a given linear map has a sparse representation matrix with respect to the new bases. Let $c \geq 1$ be a real number. We call an $n \times n$-matrix c-sparse if it has at most cn nonzero entries. For an integer $n \geq 1$ we define $Spr(c)_n := Spr_n$ by

$$Spr_n := \{(A, (S, T, B)) \in k^{n \times n} \times (GL(n, k)^2 \times k^{n \times n}) \mid$$
$$B = SAT, B \text{ is } c\text{-sparse}\}.$$

As usual, $Spr := (Spr_1, Spr_2, \ldots)$.

(1) Show that def $Spr_n = k^{n \times n}$ for all $c \geq 1$.
(2) Show that $\omega(Spr) \leq \omega$.
(3) Suppose that $n = 3m$ and let $A = (a_{ij})$ be an $n \times n$-matrix with indeterminate entries over k. Let $R := k[a_{ij} \mid 1 \leq i, j \leq n]$. Show that $L_R(\mathrm{Tr}(A^3)) \leq 3C(Spr_n) + 6n^2 + 10cn^2$. (Hint: let $K := \mathrm{Quot}(R)$. There exist $S, T, B \in GL(n, \mathcal{O}_0)$, $\mathcal{O}_0$ being the local ring of 0, such that $B = S(E - A)T$, B is c-sparse, and $L_{\mathcal{O}_0}(S, T, B) \leq C(Spr_n)$. Let $S = S^{(0)} + S^{(1)} + \cdots$, $T = T^{(0)} + T^{(1)} + \cdots$, and $B = B^{(0)} + B^{(1)} + \cdots$ denote the Taylor expansions in 0 of S, T, and B, respectively. Note that the matrices $B^{(j)}$ are also c-sparse. Show that

$$L_R(S^{(2)}, S^{(3)}, T^{(2)}, T^{(3)}) \leq 3L_{\mathcal{O}_0}(S, T, B).$$

Further, prove that $\mathrm{Tr}(A^3) = \sum_{i+j+\ell=3} \mathrm{Tr}(T^{(i)}(B^{-1})^{(j)} S^{(\ell)})$. The products $(B^{-1})^{(j)} S^{(\ell)}$ for $j + \ell \leq 3$ can be computed from $B^{(2)}$, $B^{(3)}$, $S^{(2)}$, and $S^{(3)}$ with only $10cn^2$ multiplications.)
(4) Subdivide A into $m \times m$ blocks $A_{ij} \in R^{m \times m}$ and consider the substitution

$$\psi : R \to R, \quad \psi(A) := \begin{pmatrix} 0 & A_{12} & 0 \\ 0 & 0 & A_{23} \\ A_{13} & 0 & 0 \end{pmatrix}.$$

Show that $\psi(\mathrm{Tr}(A^3)) = 3\,\mathrm{Tr}(A_{12}A_{23}A_{31})$.
(5) Show that $C(MaMu_m) \leq 3L_R(\mathrm{Tr}(A_{12}A_{23}A_{31}))$. (Use the derivative inequality.)
(6) Prove that $\omega(Spr) = \omega$ provided $\omega > 2$.

16.11. (Evaluation of a polynomial at a matrix) Let $g \in k[X]$ be a polynomial of degree ≥ 2, $n \in \mathbb{N}'$, and $\Pi_{g,n}$ be the problem defined by

$$\Pi_{g,n} := \{(A, B) \in k^{n \times n} \times k^{n \times n} \mid B = g(A)\}.$$

(1) Show that $C(\Pi_{g,n}) \leq 2C(MaMu_n)\sqrt{\deg g}$. (Hint: use Prop. (9.1).)
(2) Show that $C(\Pi_{g,n}) = O(C(MaMu_n)\sqrt{n} + \deg g)$. (Hint: compute first the characteristic polynomial of A.)

(3) Let k be an infinite field. Show that $C(MaMu_n) \leq C(\Pi_{g,3n})$. (Hint: Consider the matrix

$$A := \begin{pmatrix} I_n & A_1 & 0 \\ 0 & cI_n & A_2 \\ 0 & 0 & I_n \end{pmatrix}$$

where c is some element in k. What is A^j?)

For $A = (a_{ij}) \in k^{n \times n}$ let $F_1, \ldots, F_n$ denote the linear forms defined by $F_i := \sum_{j=1}^n a_{ij} X_j$. By Thm. (13.1) we know that the linear complexity $L_\infty(A)$ of A, up to a factor ≤ 3, equals $L_{k(X)}^{tot}(F_1, \ldots, F_n)$. The aim of the next three exercises is to give fast algorithms for solving linear equations defined by matrices of low linear complexity.

16.12. Let V be a k-space. A sequence $v = (v_1, v_2, \ldots) \in V^{\mathbb{N}'}$ is called *linearly generated* iff there exists a nonzero $f = \sum_{i=0}^n f_i X^i \in k[X]$ such that $\sum_{i=0}^n f_i v_{i+j} = 0$ for all $j \geq 0$; f is called a *generating polynomial* of v. A generating polynomial of least degree is called a *minimum polynomial* of v and is denoted by f_v. (Cf. Ex. 3.9.)

(1) Show that if $\varphi: V \to W$ is a morphism of k-spaces, then the sequence $\varphi(v) := (\varphi(v_1), \varphi(v_2), \ldots)$ is linearly generated and $f_{\varphi(v)}$ divides f_v.
(2) Let $A \in GL(n, k)$, $b \in k^n$, and $f_{A,b} = \sum_{i=0}^m f_i X^i$ be a minimum polynomial of $(b, Ab, A^2b, \ldots)$. Show that $f_0 \neq 0$ and $A^{-1}b = -(\sum_{i=1}^m f_i A^{i-1}b)/f_0$. Hence, once $f_{A,b}$ is known, one can compute the solution $x \in k^n$ of $Ax = b$ with a total cost of $O(mL_\infty(A)) = O(nL_\infty(A))$ arithmetic operations.
(3) Design an algorithm that computes $f_{A,b}$ with $O(nM(n)\log n)$ arithmetic operations, where $M(n)$ is defined as in Sect. 2.1. (Hint: use Ex. 3.9.)

16.13.
(1) Let $n \in \mathbb{N}'$, $f \in k[X]$ be a polynomial of degree n. $W_f := k[X]_{<n}$ is a $k[X]$-module via $X \circ g := Xg \bmod f$ for $g \in W_f$. Let λ be the k-linear form on W_f defined by $\lambda(g_0 + \ldots + g_{n-1}X^{n-1}) := g_{n-1}$. For $g \in W_f$ show that f is a minimum polynomial of the sequence $(\lambda(g), \lambda(X \circ g), \lambda(X^2 \circ g), \ldots)$ iff $GCD(f, g) = 1$.
(2) Let $k = \mathbb{F}_q$. Show that the proportion of those $g \in W_f$ such that $GCD(g, f) = 1$ equals $\Phi(f) := \prod_{i=1}^r (1 - q^{-d_i})$, where $d_1, \ldots, d_r$ are the degrees of the distinct irreducible factors of f, and that $\Phi(f) \geq 1/(6\lceil \log_q 2 \deg f \rceil)$. (Hint: first show that if $\deg f \leq q + q^2 + \ldots + q^\ell$, then $\Phi(f) \geq 1/(6\ell)$.)

16.14. (Wiedemann's algorithm)
(1) Let V be a $k[X]$-module and for $v \in V$ let $\langle v \rangle_{k[X]}$ denote the cyclic $k[X]$-module generated by v. Suppose that the sequence $(v, Xv, X^2v, \ldots)$ is linearly generated and has minimum polynomial f_v. Show that $\langle v \rangle_{k[X]}$ and W_{f_v} are isomorphic $k[X]$-modules and the set of isomorphisms between them is in bijection to the set of those $g \in W_{f_v}$ such that $GCD(f_v, g) = 1$.

(2) Let $k = \mathbb{F}_q$, and for $A \in k^{n \times n}$, $b \in k^n$ let $f_{A,b}$ be as defined in Ex. 16.12(2). Show that the proportion of those linear forms η of k^n such that $s := (\eta(b), \eta(Ab), \ldots)$ has minimum polynomial $f_{A,b}$ is at least $1/(6\lceil \log_q 2n \rceil)$. Hence, if η is chosen uniformly at random, then the probability that s has minimum polynomial $f_{A,b}$ is at least $1/(6\lceil \log_q 2n \rceil)$.

(3) Design a randomized algorithm that computes the solution x of a nonsingular system $Ax = b$ of n linear equations in n unknowns with $O(nL_\infty(A) + M(n) \log n)$ arithmetic operations, and which has an error probability at most $(1 - 1/(6\lceil \log_q 2n \rceil))$.

The aim of the next three exercises is to provide a different analysis for the probabilistic algorithm of the last exercise.

16.15. Let $X_1, \ldots, X_n$ be indeterminates over k, $F \in k[X_1, \ldots, X_n]$ be nonzero, and $S \subseteq k$ be finite. Show that

$$\frac{|\{(c_1, \ldots, c_n) \in S^n \mid F(c_1, \ldots, c_n) \neq 0\}|}{|S|^n} \geq 1 - \frac{\deg F}{|S|}.$$

(Hint: use induction on n.)

16.16. Let $(a_0, a_1, \ldots) \in k^{\mathbb{N}}$ be a linearly generated sequence with a minimum polynomial of degree m. For $\mu \geq 1$ let the Toeplitz matrix T_μ be defined by

$$T_\mu := \begin{pmatrix} a_{\mu-1} & a_{\mu-2} & \cdots & a_0 \\ a_\mu & a_{\mu-1} & \cdots & a_1 \\ \vdots & \vdots & & \vdots \\ a_{2\mu-2} & a_{2\mu-3} & \cdots & a_{\mu-1} \end{pmatrix}.$$

Prove that T_μ is singular for $\mu > m$ and nonsingular for $\mu = m$. (Hint: use Ex. 3.7(4).)

16.17. For $\ell = (\ell_1, \ldots, \ell_n) \in k^{1 \times n}$, $b \in k^{n \times 1}$, and $A \in k^{n \times n}$ let $f_{A,b}^\ell$ denote a minimum polynomial of the sequence $(\ell b, \ell Ab, \ell A^2 b, \ldots)$. Further let $f_{A,b}$ denote a minimum polynomial of $(b, Ab, \ldots)$ and $S \subseteq k$ be a finite subset. Show that

$$\frac{|\{\ell = (\ell_1, \ldots, \ell_n) \in S^n \mid f_{A,b}^\ell = f_{A,b}\}|}{|S|^n} \geq 1 - \frac{\deg f_{A,b}}{|S|}.$$

16.11 Open Problems

Problem 16.1. Consider the problem of solving a linear system of size n, formally defined as follows

$$SLS_n := \{((A, b), v) \in (GL(n, k) \times k^n) \times k^n \mid Av = b\}.$$

By Prop. (16.6) we have $\omega(SLS) \leq \omega$ for the corresponding problem sequence $SLS = (SLS_n)_{n \geq 1}$. Decide whether these exponents are equal!

The following open problems deal with "decision problems," for which the notion of exponent can be defined in analogy to the case of pure computational problems as discussed in Sect. 16.1.

Problem 16.2. Consider the problem of deciding whether a given matrix $A \in k^{n \times n}$ is regular. Is the exponent of the corresponding problem sequence equal to the exponent of matrix multiplication?

Problem 16.3. Consider the problem of verifying matrix multiplication, that is, to decide for given $A, B, C \in k^{n \times n}$ whether $AB = C$. Is the exponent of the corresponding problem sequence equal to the exponent of matrix multiplication?

Problem 16.4. Let $O(n, k) := \{A \in k^{n \times n} \mid AA^\top = I_n\}$ be the orthogonal group of matrices of size n. Note that each matrix in $O(n, k)$ has the determinant plus or minus one. Consider the problem of determining the sign of the determinant of a given orthogonal matrix. Does the corresponding problem sequence have the exponent of matrix multiplication?

16.12 Notes

The interest in the reduction of problems in computational linear algebra to matrix multiplication goes back to Strassen [493]. Based on his algorithm for multiplying 2×2-matrices he designed in this paper fast algorithms for generic inversion, computing the determinant, or solving generic systems of linear equations; in the terminology used in this chapter this means that the exponents of these problems are majorized by that of matrix multiplication. Schönhage [456] continued these investigations by deriving upper bounds for the QR-decomposition, for unitary transformation of arbitrary complex matrices to upper Hessenberg form, and for unitary triangularization of Hermitian matrices in terms of the complexity of matrix multiplication. Later, Bunch and Hopcroft [97] invented fast algorithms for the LUP-decomposition of nonsingular matrices based on fast matrix multiplication; as a by-product, they showed that computing the inverse of a nonsingular matrix is at most as hard as matrix multiplication. Fast matrix multiplication algorithms also serve as the basis of efficient algorithmic solutions of problems from other areas, such as the recognition of context-free languages (Valiant [524]), or computing the transitive closure of Boolean matrices (Fischer and Meyer [173] and Sect. 16.9).

The proof of $C(MaMu_n) \leq C(Inv_{3n})$ in Sect. 16.2 is from Bunch and Hopcroft [97]; the authors give the credit for the idea to Winograd. The triangular inversion algorithm is a special case of Strassen's algorithm for generic inversion [493]. Our presentation follows Bunch and Hopcroft [97]. Schönhage [456] has independently shown that inversion of any nonsingular matrix over an ordered field can be done with $O(n^\gamma)$ steps, provided $C(MaMu_n) = O(n^\gamma)$. (Compare also Borodin and Munro [65, p. 50] and Aho et al. [1, Ex. 6.25]). The *LUP*-decomposition-algorithm as well as the assertion of Prop. (16.6) in Sect. 16.4 is due to Bunch and Hopcroft [97]. See also [1, Chap. 6.4].

The major step in deriving a relative lower bound for *Det* in Sect. 16.4 is the application of the derivative inequality (7.7) and is due to Baur and Strassen [32]. The problem of inverting matrices has also been considered in other settings. For instance, Kalorkoti [279] has shown that even the computation of the trace of the inverse of a generic matrix is about as hard as matrix multiplication (Ex. 16.7). The results of Sect. 16.5–16.6 are due to Keller-Gehrig [298]; our presentation follows this paper rather closely. However, the algorithm for the problem $\Pi_1^{m,n}$ in Sect. 16.5 is originally due to Schönhage [456], where he uses a modification of this algorithm to obtain the QR-decomposition of complex matrices.

Sect. 16.7 and 16.8 follow Bürgisser et al. [100] where the authors also derive absolute lower bounds for the problems *OgB*, *Spr*, and *3-Cpr*. This shows that these problems have the same exponent as matrix multiplication, even when $\omega = 2$. The results in Sect. 16.9 are due to Fischer and Meyer [173]. A more efficient algorithm based on matrix multiplication has been exhibited by Munro [388]. For other transitive closure algorithms see Purdom [428] and the so-called "Four Russians algorithm" of Arlazarov et al. [11]. See also Aho et al. [1, Chap. 6.6] and Manber [347] and the references therein.

Ex. 16.3 and 16.5 are from Strassen [493, 498]. Ex. 16.7 and 16.8 are from Kalorkoti [279]. Ex. 16.9 and 16.10 are from Bürgisser et al. [100]. Ex. 16.11(1) is from Paterson and Stockmeyer [415], while parts (2) and (3) of that exercise follow Giesbrecht [195]. Ex. 16.12 to 16.14 are due to Wiedemann [549]. In that paper it is also shown how to extend these methods to solve singular systems of linear equations, and to compute minimum, as well as characteristic polynomials of matrices. *Wiedemann's algorithm*, as his randomized algorithm for solving sparse systems of linear equations is called, is especially useful for sieve-based integer factoring methods or solving the discrete logarithm problem in finite fields; for the former see the book of Cohen [117] or the article of Lenstra and Lenstra [326], for the latter the papers of Coppersmith [129], Odlyzko [400], Coppersmith et al. [132], and LaMacchia and Odlyzko [313]. In particular, see the last reference for further algorithms which deal with solving sparse systems of linear equations. The analysis of Wiedemann's algorithm given in Ex. 16.17 is due to Kaltofen and Saunders [290]. For the design of these exercises the lecture notes of Kaltofen [285] have been very helpful. Ex. 16.15 is due to Schwarz [468]; see also Zippel [573]. For a block version of Wiedemann's algorithm see Coppersmith [131] and Kaltofen [287].

Problem 16.1 is from Strassen [510]. With respect to Problem 16.2 we remark that Lickteig has an affirmative answer for the related problem to decide whether the determinant of a given matrix equals one. (Compare the Notes of Chap. 11.)

Many of the algorithms presented in this chapter have more efficient randomized and parallel versions. For these topics we refer the reader to the book of Ja'Ja [268], the papers of von zur Gathen [186], Eberly [153], Giesbrecht [195], Kaltofen and Pan [288, 289], and the references therein.

Chapter 17. Lower Bounds
for the Complexity of Algebras

We discuss methods for proving lower bounds for the multiplicative complexity of bilinear maps; as usual, we shall focus on multiplication maps of (associative) algebras and on matrix multiplication. The first section discusses various lower bounds, the highlight being Lafon and Winograd's theorem [311] which gives a general lower bound for the multiplicative complexity of matrix multiplication. The methods of Sect. 17.1 will be unified in the second section where we prove Alder and Strassen's theorem [6], a lower bound for the multiplicative complexity of associative algebras in terms of their dimension and their number of maximal two-sided ideals. The question naturally arises for which algebras this lower bound is sharp. In Sect. 17.3 we investigate division algebras and characterize all those which have minimal multiplicative complexity. Finally, in Sect. 17.4 we give a complete characterization, due to de Groote and Heintz [220], of all commutative algebras which have minimal rank.

Unless otherwise specified, k denotes any field, a k-space is always assumed to be of finite dimension, and a k-algebra is assumed to be associative and finite-dimensional.

17.1 First Steps Towards Lower Bounds

In this section we introduce some basic techniques for proving lower bounds for the rank and multiplicative complexity of a bilinear map by means of several examples. In the next section we will put these investigations into a formal framework. We are mainly interested in k-algebras and matrix multiplication maps $\langle m, n, p \rangle$ (of not necessarily quadratic size).

It is a rule of thumb that proving non-trivial lower bounds for the rank of a bilinear map ϕ is generally easier than proving lower bounds for the multiplicative complexity. Think for example of the conciseness argument: if ϕ is concise and of format (m, n, p), then $R(\phi) \geq \max\{m, n, p\}$. However, we have not yet encountered a similar assertion for the multiplicative complexity. While it is easy to prove that 3-conciseness of ϕ yields $L(\phi) \geq p$ (the proof is exactly the same as for the rank), it is not totally trivial to deduce from the 1- or 2-conciseness of

ϕ the inequality $L(\phi) \geq m$ or $L(\phi) \geq n$. We will study the relation between the conciseness of a bilinear map and lower bounds for its multiplicative complexity below in Lemma (17.4) (separation lemma).

We begin with one of the simplest lower bound proofs for the rank. A k-algebra D in which every nonzero element is invertible is called a *division algebra* over k. For example, every field extension of k is a division algebra over k. (However, division algebras do not need to be commutative; see, e.g., Example (17.32).)

(17.1) Proposition. *For a division algebra D of dimension n over k we have $R(D) \geq 2n - 1$.*

Proof. Let $(f_1, g_1, w_1; \ldots; f_r, g_r, w_r)$ be a bilinear computation for D. Choose $0 \neq e \in \cap_{i=1}^{n-1} \ker f_i$. Then we have

$$D = eD \subseteq \sum_{i=n}^{r} kw_i.$$

The left-hand side is of dimension n, so $r - n + 1 \geq n$ which shows $r \geq 2n - 1$. $\square$

We note in passing the following corollary which, for example, implies that the algorithm for the multiplication of complex numbers given in Chap. 1 is an optimal bilinear computation.

(17.2) Corollary. *If $K \supseteq k$ is a finite simple extension of degree n and $|k| \geq 2n-2$, then $R(K) = 2n - 1$.*

(For the upper bound apply Prop. (14.48)(1).) We now aim at proving the same lower bound as in the above proposition for the multiplicative complexity of a division algebra. It is easy to see that the above argument cannot be used for this purpose. However, we can use another type of conciseness arguments which will be prepared in the following. Let U_1 be a subspace of the k-space U. The linear forms $\lambda_1, \ldots, \lambda_s \in U^*$ on U are said to *separate the points of U_1* iff $U_1 \cap (\cap_{i=1}^{s} \ker \lambda_i) = 0$. The proof of the following remark is left as an exercise (cf. Ex. 17.1).

(17.3) Remark. $\lambda_1, \ldots, \lambda_s$ separate the points of U_1 if and only if the restrictions of $\lambda_1, \ldots, \lambda_s$ to U_1 generate the dual space of U_1. Then there are $1 \leq i_1 < i_2 < \cdots < i_q \leq s$, $q := \dim U_1$, such that $\lambda_{i_1}, \ldots, \lambda_{i_q}$ restricted to U_1 form a basis of U_1^*. Thus, the zeroset $\cap_{j=1}^{q} \ker \lambda_{i_j}$ of $\lambda_{i_1}, \ldots, \lambda_{i_q}$ is a complement of U_1 in U. •

In the examples we study in the sequel we shall often encounter the following situation: we start from a bilinear map $\phi \in \mathrm{Bil}(U, V; W)$, construct a linear endomorphism π of $U \times V$ and carry on with the investigation of $L(\phi \circ \pi)$. In general, $\phi \circ \pi$ is not bilinear but only quadratic. However, it is a quadratic map of special type; namely, it is a map from $U \times V$ to W. In analogy to bilinear maps, we define the left- and the right-kernel of a quadratic map $\phi: U \times V \to W$ as

$$\text{lker}(\phi) \ := \ \{x \in U \mid \forall u \in U, v \in V : \phi(x+u, v) = \phi(u, v)\}$$
$$\text{rker}(\phi) \ := \ \{y \in V \mid \forall u \in U, v \in V : \phi(u, y+v) = \phi(u, v)\}.$$

It is easily checked that the left- and the right-kernel of ϕ are subspaces of U and V, respectively. We call a quadratic map 1-*concise* if its left-kernel is zero, 2-*concise* if its right-kernel is zero, 3-*concise* if the span of its image is W, and concise if it is i-concise for $i = 1, 2, 3$. The next lemma discusses the relationship between the conciseness of a quadratic map and lower bounds for its multiplicative complexity in a general setting.

(17.4) Separation Lemma. *Let* $(f_i, g_i, w_i)_{i \leq \ell}$ *be a quadratic computation for the quadratic map* $\phi : U \times V \to W$ *and let* U_1 *be a subspace of* U *such that* $U_1 \cap \text{lker}(\phi) = 0$. *Then there exists* $(\lambda_1, \ldots, \lambda_\ell)$ *in* $\prod_{i=1}^{\ell} \{f_i, g_i\}$ *such that* $\lambda_1, \ldots, \lambda_\ell$ *separate the points of* $U_1 \times 0$. *In particular,* $L(\phi) \geq \dim U_1$. *An analogous assertion holds with the roles of* U *and* V *interchanged.*

Proof. Let $U_2 \subseteq U_1$ be a maximal subspace of U_1 such that there exists $(\lambda_1, \ldots, \lambda_\ell) \in \prod_{i=1}^{\ell} \{f_i, g_i\}$ separating the points of $U_2 \times 0$. W.l.o.g. we may assume that $f_1, \ldots, f_q$ form upon restriction a basis of $(U_2 \times 0)^*$, where $q := \dim U_2$. By Rem. (17.3) the subspace $E := \cap_{i \leq q} \ker f_i$ is a complement of $U_2 \times 0$ in $U \times V$. If $U_2 \neq U_1$ there exists $0 \neq x \in U_1$ such that $(x, 0) \in E \cap (U_1 \times 0)$. Note that $x \notin U_2$. We claim that

$$\forall i > q : \quad f_i(x, 0) = g_i(x, 0) = 0.$$

Otherwise, there is some $j > q$ and some $\lambda \in \{f_j, g_j\}$ such that $f_1, \ldots, f_q, \lambda$ separate the points of $(U_2 \oplus kx) \times 0$, which is a contradiction to the maximality of U_2. Let $(u, v) \in U \times V$. There exists $u' \in U_2$ such that

$$f_1(u + u', v) = \cdots = f_q(u + u', v) = 0.$$

Hence we obtain

$$\phi(x + u + u', v) - \phi(u + u', v)$$
$$= \sum_{i=1}^{\ell} \big(f_i(u + u', v) + f_i(x, 0)\big)\big(g_i(u + u', v) + g_i(x, 0)\big) w_i$$
$$\quad - \sum_{i=1}^{\ell} f_i(u + u', v) g_i(u + u', v) w_i$$
$$= \sum_{i=1}^{\ell} f_i(u + u', v) g_i(x, 0) w_i = 0.$$

Since u and v were arbitrary, we get the contradiction $0 \neq x \in U_1 \cap \text{lker}(\phi) = 0$. $\square$

(17.5) Corollary. *Let U_1, U_2, and U_3 be k-spaces and $\phi \in \text{Bil}(U_1, U_2; U_3)$. Then the i-conciseness of ϕ implies that $L(\phi) \geq \dim U_i$.*

To prove 1- or 2-conciseness of a quadratic map we shall frequently use the following.

(17.6) Remark. Let $\phi: U \times V \to W$ be a quadratic map such that $\phi(U, 0) = 0$. Then $y \in \text{rker}(\phi)$ implies that $\phi(U, y) = 0$. Indeed, $y \in \text{rker}(\phi)$ implies that for all $u \in U$ we have $\phi(u, y) = \phi(u, 0) = 0$.   •

Now we can prove a result analogous to Prop. (17.1) for the multiplicative complexity.

(17.7) Proposition (Fiduccia and Zalcstein). *For a division algebra D of dimension n over k we have $L(D) \geq 2n - 1$.*

Proof. Let $(f_1, g_1, w_1; \ldots; f_\ell, g_\ell, w_\ell)$ be a quadratic computation for D and $T \subset D$ be an $(n-1)$-dimensional subspace. Since D is concise, we may assume w.l.o.g. by the separation lemma (17.4) that $f_1, \ldots, f_{n-1}$ separate the points of $T \times 0$. Let π be the projection onto $\cap_{i=1}^{n-1} \ker f_i$ along $T \times 0$. It suffices to show that $\mu \circ \pi$ is 2-concise, where we have denoted by μ the multiplication in D. Suppose that there is a nonzero $b \in \text{rker}(\mu \circ \pi)$. Since $(\mu \circ \pi)(D, 0) = 0$, Rem. (17.6) implies that $(\mu \circ \pi)(D, b) = 0$. Hence, $D \cdot b \subseteq T \cdot b$, which is a contradiction, since $\dim(D \cdot b) = \dim D = n$. (Note that for all $a, b \in D$ there exists $t \in T$ such that $\pi(a, b) = (a + t, b)$.)   □

We now proceed with lower bound proofs for matrix multiplication. Recall Strassen's algorithm for multiplying 2×2-matrices (see Chap. 1 or Ex. 14.5). Is this algorithm optimal or does there exist a quadratic computation of length less than 7? At the end of this section we will give an answer to this question. For the moment, we discuss a simpler problem, namely, that of deriving a lower bound for the *rank* of matrix multiplication. More generally, we give in the subsequent proposition a lower bound for the rank of any *simple algebra*. Recall that a nonzero algebra A is called simple iff its only two-sided ideals are 0 and A. For instance, every division algebra is simple. Also, it is easy to show, using elementary rules of matrix multiplication, that any matrix algebra is simple (see Ex. 17.4). A right (left) ideal I of an algebra A is called minimal if the only right (left) ideal properly contained in I is 0. It is called maximal if the only right (left) ideal properly containing I is A itself. We shall need the following facts. (See Ex. 17.5.)

(17.8) Fact. Let A be a simple k-algebra. Then we have:

(1) Any two minimal left (right) ideals of A have the same dimension.
(2) If L_1 is a minimal and L_2 a maximal left (right) ideal of the algebra A, then $\dim L_1 + \dim L_2 = \dim A$.
(3) For any minimal left (right) ideal L_1 of A there exists a complementary maximal left (right) ideal L_2 of A (that is, $L_1 \oplus L_2 = A$).
(4) No non-zero right (left) ideal of A is contained in a proper left (right) ideal of A.   •

(17.9) Proposition. *If A is a simple k-algebra then $R(A) \geq 2 \dim A - 1$.*

Proof. (Baur) Let $n := \dim A$ and $(f_1, g_1, w_1; \ldots; f_r, g_r, w_r)$ be an optimal bilinear computation for A. Assume that $r < 2n - 1$.

Since A is concise, we may w.l.o.g. assume that $f_1, \ldots, f_n$ form a basis of A^*. Since $r < 2n - 1$, there exists $0 \neq b \in A$ such that $g_n(b) = \cdots = g_r(b) = 0$. We thus have $Ab = \sum_{i=1}^{n-1} f_i(A) g_i(b) w_i \subseteq \sum_{i=1}^{n-1} kw_i$, which shows that Ab is a proper left ideal of A; hence $Ab \subseteq L$ for a maximal left ideal L. Let m denote the dimension of any minimal left ideal of A (see Fact (17.8)(1)). Since $\dim Ab \geq m$ we may w.l.o.g. assume that $g_1(b) \cdots g_m(b) \neq 0$. We infer that

$$\sum_{i=1}^{m-1} kw_i \subset Ab \subseteq L,$$

since $f_1, \ldots, f_n$ form a basis of A^*. The linear forms $g_m, \ldots, g_r$ generate A^*: otherwise, they have a common zero $y \neq 0$ and we obtain $Ay \subseteq \sum_{i=1}^{m-1} kw_i$ which contradicts $\dim Ay \geq m$. We may thus, after reordering the indices $m, m+1, \ldots, r$ assume that $g_m, \ldots, g_{n-1}$ form upon restriction a basis of L^*. (Note that $\dim L = n - m$ by Fact (17.8)(2).) This property of the g_i implies that for any $y \in A$ there exists $c \in L$ such that $g_j(c) = g_j(y)$, for $m \leq j < n$. Since $r < 2n - 1$, there exists $0 \neq a \in A$ such that $f_n(a) = \cdots = f_r(a) = 0$. Hence we obtain $\forall y \in A \; \exists c \in L: \quad ay - ac = a(y - c) \in \sum_{i=1}^{m-1} kw_i \subseteq L$. Thus we have $ay \in L$ for all $y \in A$, hence $aA \subseteq L$, a contradiction to Fact (17.8)(4). $\qquad \square$

(17.10) Corollary (Hopcroft and Kerr, Winograd). *The rank of $k^{2 \times 2}$ is equal to 7. Hence, Strassen's algorithm for multiplying 2×2-matrices is an optimal bilinear computation.*

Let us now proceed with lower bound proofs for the multiplicative complexity of matrix multiplication $\langle m, n, p \rangle$.

(17.11) Proposition. *We have $L(\langle m, n, p \rangle) \geq (m + p - 1)n$, for positive integers m, n, p. In particular, $L(\langle n, n, n \rangle) \geq 2n^2 - n$.*

Proof. The proof is similar to the above lower bound proof for division algebras: let $(f_1, g_1, w_1; \ldots; f_\ell, g_\ell, w_\ell)$ be a quadratic computation for $\langle m, n, p \rangle$ and $U := \{(a_{ij}) \in k^{m \times n} \mid \forall j \in \underline{n}: a_{1j} = 0\}$. Since $\langle m, n, p \rangle$ is concise by Prop. (14.41), we may w.l.o.g. assume by the separation lemma (17.4) that $f_1, \ldots, f_{(m-1)n}$ separate the points of $U \times 0$. Let π be the projection of $k^{m \times n} \times k^{n \times p}$ onto $\bigcap_{i=1}^{(m-1)n} \ker f_i$ along $U \times 0$. It suffices to show that $\langle m, n, p \rangle \circ \pi$ is 2-concise. Let b be an element of $\mathrm{rker}(\langle m, n, p \rangle \circ \pi)$. Since $(\langle m, n, p \rangle \circ \pi)(k^{m \times n}, 0) = 0$, we obtain by Rem. (17.6) $(\langle m, n, p \rangle \circ \pi)(k^{m \times n}, b) = 0$, which implies

$$k^{m \times n} \cdot b \subseteq U \cdot b \subseteq \{(c_{ij}) \in k^{m \times p} \mid \forall j \in \underline{p}: c_{1j} = 0\}.$$

Hence $k^{1 \times n} \cdot b = 0$. We conclude that $b = 0$. $\qquad \square$

Unfortunately, the above proposition yields $L(\langle 2, 2, 2 \rangle) \geq 6$ instead of the desired 7. The next theorem gives the best general lower bound for $L(\langle m, n, p \rangle)$ known at the present time. (See however Sect. 18.4.)

(17.12) Theorem (Lafon and Winograd). *For integers $m, n, p \geq 2$ we have*

$$L(\langle m, n, p \rangle) \geq (m + p)n + p - n - 1.$$

Proof. We begin with introducing some notation:

- We denote by $E_{i,j}$ the matrix whose (i, j)-entry is 1 and which has zeros elsewhere. Whenever we use this notation, the format of $E_{i,j}$ will be clear from the context.
- If $\{f_i \mid i \in I\}$ are linear forms on $k^{m \times n} \times k^{n \times p}$ which separate the points of a subspace $U \times V$ of dimension $|I|$, then we call the projection along $U \times V$ onto $\bigcap_{i \in I} \ker f_i$ simply the *corresponding projection*.

We shall work with the following direct sum decompositions.

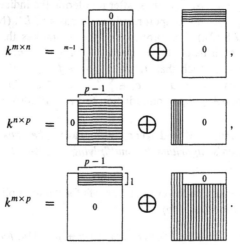

We further consider the decomposition $k^{m \times p} = U_1 \oplus U_2$, where

$$U_1 := \bigoplus_{i=1}^{p} k E_{1,i}, \qquad U_2 := \bigoplus_{i=2}^{m} \bigoplus_{j=1}^{p} k E_{i,j}.$$

Now let $(f_i, g_i, w_i)_{i \leq \ell}$ be a quadratic computation for $\phi := \langle m, n, p \rangle$. The proof proceeds in several steps.

Step 1. Let τ be the projection from $k^{m \times p}$ onto U_1 along U_2; i.e., for $w \in k^{m \times p}$ we have $\tau(w) = Dw$ where $D := \text{diag}(1, 0, \ldots, 0) \in k^{p \times p}$. We first show that w.l.o.g.

(A) $\qquad \tau(w_1) = E_{1,2}, \ldots, \tau(w_{p-1}) = E_{1,p}, \tau(w_p) = E_{1,1}.$

Since ϕ is 3-concise, $w_1, \ldots, w_\ell$ generate $k^{m \times p}$. Hence we may w.l.o.g assume that $\tau(w_1), \ldots, \tau(w_p)$ form a basis of U_1. Let $T \in \text{GL}(p, k)$ be the transition matrix from the basis $(E_{1,1}, \ldots, E_{1,p})$ of U_1 to the basis $(\tau(w_1), \ldots, \tau(w_p))$, i.e., if $T = (t_{ij})_{1 \le i, j \le p}$, then $Dw_i = \tau(w_i) = \sum_{j=1}^{p} t_{ij} E_{1,j} = E_{1,i} T$ for $i \in \underline{p}$. We thus obtain $E_{1,i} = Dw_i T^{-1} = \tau(w_i T^{-1})$. Now we construct a new quadratic computation for ϕ satisfying (A) in the following way: for $a \in k^{m \times n}$, $b \in k^{n \times p}$ let $\tilde{f}_i(a, b) := f_i(a, bT)$, $\tilde{g}_i(a, b) := g_i(a, bT)$, and $\tilde{w}_i := w_i T^{-1}$ for all $i \le \ell$. Since $ab = abTT^{-1} = \sum_{i \le \ell} f_i(a, bT) g_i(a, bT) w_i T^{-1}$, $(\tilde{f}_i, \tilde{g}_i, \tilde{w}_i)_{i \le \ell}$ is a quadratic computation for ϕ. A permutation of the indices yields now a quadratic computation satisfying (A).

We have $\sum_{i=1}^{p-1} kw_i \oplus \boxed{} = k^{m \times p}$. (Take images under τ and use (A).) Let β be the projection along $\sum_{i=1}^{p-1} kw_i$ onto $\boxed{}$. We claim that

(B) $\forall x \in \boxed{}$:   (1, 1)-entry of $\beta(x) = 0$.

For the proof note that for all $x \in k^{m \times p}$ we have $\beta(x) = x + \sum_{i=1}^{p-1} \lambda_i w_i$ for some $\lambda_i \in k$ (depending on x). Since the (1, 1)-entries of x and $\tau(x)$ coincide, we obtain for all $x \in \boxed{}$ by taking images under τ and using (A)

$$(1, 1)\text{-entry of } \beta(x) = (1, 1)\text{-entry of } (x + \sum_{i=1}^{p-1} \lambda_i E_{1,i+1}) = 0.$$

Step 2. Now we show that the bilinear map $\psi_1 := \beta \circ \phi$ is 2-concise. Take $b = (b_{v\sigma}) \in k^{n \times p}$ such that $\beta(k^{m \times n} \cdot b) = 0$. Let $v \in \underline{n}$ and $\sigma \in \underline{p}$. Then $E_{2,v} \cdot b \in U_2 \subset \boxed{}$; hence $\beta(E_{2,v} \cdot b) = E_{2,v} \cdot b$, and

$$0 = (2, \sigma)\text{-entry of } \beta(E_{2,v} \cdot b) = (2, \sigma)\text{-entry of } E_{2,v} \cdot b = b_{v\sigma}.$$

Since v and σ are arbitrary, we obtain $b = 0$. This establishes the 2-conciseness of ψ_1. Since ψ_1 has the quadratic computation $(f_i, g_i, \beta(w_i))_{p \le i \le \ell}$, we may w.l.o.g. assume by the separation lemma (17.4) that $\{f_i \mid p \le i < p + (p-1)n\}$ separate the points of $0 \times \boxed{}$. (Note that $\boxed{}$ has dimension $(p-1)n$.) Let π_1 be the corresponding projection. Then $(f_i \circ \pi_1, g_i \circ \pi_1, \beta(w_i))_{p+(p-1)n \le i \le \ell}$ is a quadratic computation for $\psi_2 := \psi_1 \circ \pi_1$.

Step 3. We show that $\mathrm{lker}(\psi_2) \cap \boxed{} = 0$. Choose $a = (a_{\mu v}) \in \boxed{} \cap \mathrm{lker}(\psi_2)$. Hence $\psi_2(a, k^{n \times p}) = \psi_2(0, k^{n \times p}) = 0$. Let $\mu \in \underline{m}$ and $v \in \underline{n}$ be arbitrary and consider the $(\mu, 1)$-entry of $\psi_2(a, E_{v,1}) = \psi_1 \circ \pi_1(a, E_{v,1})$. There exists $l \in \boxed{}$ such that $\pi_1(a, E_{v,1}) = (a, E_{v,1} + l)$. Since $a \cdot (E_{v,1} + l) \in U_2 \subset \boxed{}$, we have $0 = \psi_1(a, E_{v,1} + \ell) = a(E_{v,1} + \ell)$ and hence

$$0 = (\mu, 1)\text{-entry of } \psi_2(a, E_{v,1}) = (\mu, 1)\text{-entry of } a(E_{v,1} + l) = a_{\mu v}.$$

We obtain $a = 0$, which shows the assertion about the left-kernel of ψ_2. By the separation lemma (17.4) we may w.l.o.g. assume that the linear forms $f_i \circ \pi_1$, $p + (p-1)n \le i < p + (p-1)n + (m-1)n$, separate the points of $\quad \times\, 0$.

(Note that $\dim \quad = (m-1)n$.) Let π_2 be the corresponding projection and α be the projection of $k^{m\times p}$ onto $kE_{1,1}$ along $\oplus_{(i,j)\neq(1,1)} kE_{i,j}$. (That is, $\alpha(a)$ is the $(1,1)$-entry of a.) Then $\psi_3 := \alpha \circ \psi_2 \circ \pi_2$ has the quadratic computation

$$(f_i \circ \pi_1 \circ \pi_2,\ g_i \circ \pi_1 \circ \pi_2,\ \alpha \circ \beta(w_i))_{p+(p-1)n+(m-1)n \le i \le \ell}.$$

We claim that $\psi_3(0, k^{n\times p}) = 0$. To this end, let $b \in k^{n\times p}$. Then there exist $r \in \quad$ and $l \in \quad$ such that $(\pi_1 \circ \pi_2)(0, b) = (r, b+l)$. Hence, $\psi_3(0, b) = \alpha \circ \beta(r(b+l))$. Since $r(b+l) \in U_2$, we have $\beta(r(b+l)) = r(b+l)$ and $\psi_3(0, b) = \alpha(r(b+l)) = 0$.

Step 4. Now we show that $\mathrm{lker}(\psi_3) \cap \quad = 0$. Let $a = (a_{\mu\nu}) \in \quad \cap \mathrm{lker}(\psi_3)$. Then $\psi_3(a, k^{n\times p}) = \psi_3(0, k^{n\times p}) = 0$. Let $\nu \in \underline{n}$ be arbitrary. There exist $r \in \quad$ and $l \in \quad$ such that

$$\begin{aligned}
0 &= \psi_3(a, E_{\nu,1}) = (1,1)\text{-entry of } \beta((a+r)(E_{\nu,1}+l)) \\
&= (1,1)\text{-entry of } \beta(aE_{\nu,1} + al),
\end{aligned}$$

the last equality being a consequence of $r(E_{\nu,1}+l) \in U_2 \subset \quad$. Since $aE_{\nu,1} = a_{1\nu}E_{1,1} \in \quad$ and since $al \in \quad$, we have by (B)

$$0 = \psi_3(a, E_{\nu,1}) = a_{1\nu}E_{1,1}.$$

Since ν was arbitrary, we obtain $a = 0$, which shows the assertion about the right-kernel of ψ_3. Now we have $n \le L(\psi_3) \le \ell - p - (p-1)n - (m-1)n + 1$, which proves the theorem. $\square$

(17.13) Remarks.
(1) The above theorem shows that $L(\langle 2, 2, 2\rangle) = 7$. Hence, Strassen's algorithm for multiplying 2×2-matrices is even an optimal quadratic computation.
(2) If m, n, or p equal 1, then $L(\langle m, n, p\rangle) = mnp$ by the conciseness of $\langle m, n, p\rangle$ and Cor. (17.5).
(3) If $AB = C$, then $B^\top A^\top = C^\top$. This shows that $L(\langle m, n, p\rangle) = L(\langle p, n, m\rangle)$. (This gives, e.g., $L(\langle 5, 4, 3\rangle) \ge 32$ instead of $L(\langle 5, 4, 3\rangle) \ge 30$.)
(4) For later use in the next section we remark that the proof of Thm. (17.12) shows the following: if $(f_i, g_i, w_i)_{i \le \ell}$ is a quadratic computation for $\langle m, n, p\rangle$, then there is a $(p-1)$-dimensional subspace W of $k^{m\times p}$ (equal to $\sum_{i=1}^{p-1} kw_i$ in the above proof) such that, after permuting some of the f_i with g_i, the set $\{f_i \mid w_i \notin W\}$ separates the points of $\quad \times k^{n\times p}$. (Cf. Ex. 17.2.)

(5) With some minor modifications the proof of Thm. (17.12) can be used to obtain the following stronger result: we decompose $k^{m \times p} = (V_1 \oplus V_2) \oplus V_3$, where $V_1 = \oplus_{i=1}^{m} kE_{i,1}$, $V_2 = \oplus_{i=1}^{2} \oplus_{j=2}^{p} kE_{i,j}$, and $V_3 := \oplus_{i=3}^{m} \oplus_{j=2}^{p} kE_{i,j}$. Let π be the projection of $k^{m \times p}$ along V_3 onto $V_1 \oplus V_2$. Then we even have $L(\pi \circ \langle m, n, p \rangle) \geq (m + p)n + p - n - 1$. (Details are left to the reader.) This result may suggest that the lower bound of Thm. (17.12) is sharp only for some small values of m, n, p.

(6) The following generalization of Thm. (17.12) is also true: if t is a positive integer and $m_i, n_i, p_i, i = 1, \ldots, t$, are integers ≥ 2, then

$$L(\langle m_1, n_1, p_1 \rangle \oplus \cdots \oplus \langle m_t, n_t, p_t \rangle) \geq \sum_{i=1}^{t} \Big((m_i + p_i)n_i + p_i - n_i - 1 \Big).$$

(The proof is, up to some obvious modifications, exactly as above. Details are left to the reader; see also the proof of Prop. (17.23).) In particular, if $A = \oplus_{i=1}^{t} k^{m_i \times m_i}$, then $L(A) \geq 2 \dim A - t$. By a theorem of Wedderburn [316, XVII] any semisimple algebra A over an algebraically closed field is isomorphic to $\oplus_{i=1}^{t} k^{m_i \times m_i}$ for some integer t (which necessarily equals the number of maximal two-sided ideals of A) and some positive integers $m_1, \ldots, m_t$. Hence, $L(A) \geq 2 \dim A -$number of maximal two-sided ideals of A. In the next section we show that this theorem is valid for any associative algebra over any field. ●

17.2 Multiplicative Complexity of Associative Algebras

The aim of this section is to prove the following theorem.

(17.14) Theorem (Alder and Strassen). *For an associative k-algebra A of dimension n with exactly t maximal two-sided ideals we have $L(A) \geq 2n - t$.*

Note that Prop. (17.7) and Thm. (17.12) (for $m = n = p$) are special cases of this theorem, since division- and matrix algebras are simple, hence only have one maximal two-sided ideal. Further examples are provided in the exercises and in the rest of this chapter.

For the proof of this theorem we shall put the conciseness arguments used in the previous section into a formal framework. The main tool for the proof of lower bounds for a bilinear map $\phi \in \mathrm{Bil}(U, V; W)$ in the last section was the separation lemma (17.4). We used it in the following way: starting from a quadratic computation $(f_i, g_i, w_i)_{i \leq \ell}$ for ϕ we used the separation lemma to prove that, after interchanging some f_i with g_i, the set $\{f_i \mid w_i \notin W_1\}$ separates the points of $U_1 \times V_1$ for some subspaces $U_1 \subseteq U$, $V_1 \subseteq V$, and $W_1 \subseteq W$ (see also Rem. (17.13)(4)). This gives rise to the following notation.

(17.15) Notation. Let U, V, W be k-spaces, $\phi \in \mathrm{Bil}(U, V; W)$ a bilinear map and $\gamma = (f_1, g_1, w_1; \ldots; f_\ell, g_\ell, w_\ell)$ a quadratic computation for ϕ of length r. We call a triple (U_1, V_1, W_1) of k-subspaces $U_1 \subseteq U$, $V_1 \subseteq V$, $W_1 \subseteq W$ *separable by* γ iff some sequence of linear forms contained in the product $\prod_\rho \{f_\rho, g_\rho\}$ over all $\rho \in \underline{\ell}$ satisfying $w_\rho \notin W_1$ separates the points of $U_1 \times V_1$. We denote by $S(\phi, \gamma)$ the set of all triples which are separable by γ. If ϕ is the multiplication map of an associative algebra A, we write $S(A, \gamma)$ instead of $S(\phi, \gamma)$.   •

(17.16) Remark. Let $\phi \in \mathrm{Bil}(U, V; W)$, γ be a quadratic computation of ϕ and $U_2 \subseteq U_1 \subseteq U$, $V_2 \subseteq V_1 \subseteq V$, $W_2 \subseteq W_1 \subseteq W$ be subspaces. Then:

(1) $(0, 0, W_1) \in S(\phi, \gamma)$.
(2) If $(U_1, V_1, W_1) \in S(\phi, \gamma)$, then $(U_2, V_2, W_2) \in S(\phi, \gamma)$.
(3) In our new terminology the separation lemma (17.4) states the following: if ϕ is 1-concise, then $(U, 0, 0) \in S(\phi, \gamma)$. Similarly, if ϕ is 2-concise, then $(0, V, 0) \in S(\phi, \gamma)$.   •

The reason why to study separable triples of subspaces is explained in the following lemma for which we need another useful concept. The *quotient* $\phi/U_1 \times V_1$ of a bilinear map $\phi \in \mathrm{Bil}(U, V; W)$ by subspaces U_1, V_1 of U, V, respectively, is defined as the bilinear map $U/U_1 \times V/V_1 \to W/\widetilde{W}$ sending $(u + U_1, v + V_1)$ to $\phi(u, v) + \widetilde{W}$, where $\widetilde{W} := \langle \phi(U_1 \times V) + \phi(U \times V_1) \rangle$. (This bilinear map is well-defined.) It is easy to see that $L(\phi/U_1 \times V_1) \le L(\phi)$. The concept of separable triples allows to sharpen this inequality.

(17.17) Lemma. *Let* $\phi \in \mathrm{Bil}(U, V; W)$, $\gamma = (f_1, g_1, w_1; \ldots; f_\ell, g_\ell, w_\ell)$ *be a quadratic computation for* ϕ *of length* ℓ *and* $(U_1, V_1, W_1) \in S(\phi, \gamma)$ *be a triple of subspaces separable by* γ. *Then we have for any linear endomorphism* $\pi\colon W \to W$ *satisfying* $W_1 \subseteq \ker \pi$

$$L\left(\pi \circ \phi/U_1 \times V_1\right) + \dim(U_1 \times V_1) + |\{\rho \in \underline{\ell} \mid w_\rho \in W_1\}| \le \ell.$$

In particular, if $W_1 = 0$, *we may choose* $\pi = \mathrm{id}$ *and obtain*

$$L\left(\phi/U_1 \times V_1\right) + \dim(U_1 \times V_1) \le \ell.$$

Proof. Let $(\lambda_1, \ldots, \lambda_s) \in \prod_{\{\rho \mid w_\rho \notin W_1\}} \{f_\rho, g_\rho\}$ be separating the points of $U_1 \times V_1$. Then there are $1 \le i_1 < i_2 < \ldots < i_q \le s$, $q := \dim(U_1 \times V_1)$ such that $\lambda_{i_1}, \ldots, \lambda_{i_q}$ restricted to $U_1 \times V_1$ form a basis of $(U_1 \times V_1)^*$; thus the zeroset E of $\lambda_{i_1}, \ldots, \lambda_{i_q}$ is a complement of $U_1 \times V_1$ in $U \times V$ (see Rem. (17.3)). Let ψ denote the restriction of ϕ to E. Then we obviously have

(A) $L(\pi \circ \psi) \le \ell - \dim(U_1 \times V_1) - |\{\rho \in \underline{\ell} \mid w_\rho \in W_1\}|.$

Putting $\widetilde{W} := \langle (\pi \circ \phi)(U_1 \times V) + (\pi \circ \phi)(U \times V_1) \rangle$ we have the canonical commutative diagram

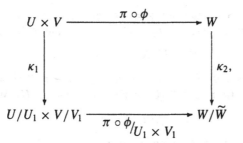

κ_1 and κ_2 being the canonical mappings. This yields

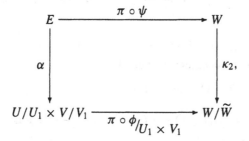

where α is the restriction of κ_1 to E. By our assumptions, the linear map α is an isomorphism, thus ${}^{\pi \circ \phi}/_{U_1 \times V_1} = \kappa_2 \circ \pi \circ \psi \circ \alpha^{-1}$ and therefore $L\left({}^{\pi \circ \phi}/_{U_1 \times V_1}\right) \leq L(\pi \circ \psi)$ by Rem. (14.5). Using (A) this proves the lemma.

$\square$

In our applications of the previous lemma we let W_1 be the k-space generated by some of the w_i. Then, if (U_1, V_1, W_1) is a triple separable by a quadratic computation for a bilinear map ϕ, we deduce that $L(\phi) \geq \dim U_1 + \dim V_1 + \dim W_1$. Our aim is thus to find separable triples for which the sum of the dimensions of its components is as large as possible. The subsequent lemma is the key tool for proving that certain triples are separable by a quadratic computation. It is a generalization of the separation lemma (17.4) and plays here the same role as that lemma did in the last section.

(17.18) Extension Lemma. *Let $\phi \in \mathrm{Bil}(U, V; W)$, γ be a quadratic computation for ϕ and $U_1 \subseteq U$, $V_1 \subseteq V$, $W_1 \subseteq W$ be subspaces, and assume that $U_2 \subseteq U_1$ is a* maximal *subspace satisfying $(U_2, V_1, W_1) \in S(\phi, \gamma)$. Then, if $U_2 \neq U_1$, there exists some $u \in U_1 \setminus U_2$ such that*

$$\phi(u, V) \subseteq \langle \phi(U_1 \times V_1) \rangle + W_1.$$

An analogous statement holds for the roles of U and V interchanged.

Using this lemma we may "extend" a given triple $(U_2, V_1, W_1) \in S(\phi, \gamma)$ to some new triple $(U_1, V_1, W_1) \in S(\phi, \gamma)$, $U_1 \supset U_2$, if the above inclusion does not hold for any $u \in U_1 \setminus U_2$.

Proof. (U_2, V_1, W_1) is separable by $\gamma =: (f_1, g_1, w_1; \ldots; f_\ell, g_\ell, w_\ell)$. Further, we may w.l.o.g. assume that $w_1, \ldots, w_p \in W_1$, $w_{p+1}, \ldots, w_\ell \notin W_1$, and that

$$E \oplus (U_2 \times V_1) = U \times V,$$

where E is the zeroset of $\{f_{p+1}, \ldots, f_q\}$, $\dim U_2 = q - p$. If $U_2 \neq U_1$, then there exists some nonzero $(u, v) \in E \cap (U_1 \times V_1)$. Note that $u \notin U_2$. We claim that

$$\forall i > q: \quad f_i(u, v) = g_i(u, v) = 0.$$

Otherwise, there would exist some $j > q$ and some $\lambda \in \{f_j, g_j\}$ such that $f_{p+1}, \ldots, f_q, \lambda$ separate the points of $(U_2 \oplus ku) \times V_1$, which is a contradiction to the maximality of U_2. Let $y \in V$ be arbitrary. There exists $(s, t) \in U_2 \times V_1$ such that

$$f_{p+1}(s, t + y) = \cdots = f_q(s, t + y) = 0.$$

We obtain the following congruences mod W_1:

$$
\begin{aligned}
\phi(u &+ s, v + t + y) - \phi(s, t + y) \\
&\equiv \sum_{i=p+1}^{\ell} \big(f_i(s, t + y) + f_i(u, v)\big)\big(g_i(s, t + y) + g_i(u, v)\big)w_i \\
&\quad - \sum_{i=p+1}^{\ell} f_i(s, t + y)g_i(s, t + y)w_i \\
&= \sum_{i=p+1}^{\ell} f_i(s, t + y)g_i(u, v)w_i \\
&\equiv 0.
\end{aligned}
$$

Hence, the bilinearity of ϕ implies

$$\phi(u, y) \equiv -\phi(u + s, v) - \phi(u, t) \bmod W_1.$$

Since $y \in V$ was arbitrary, the proof is complete. $\square$

In the subsequent proposition we give some examples for triples $(U_1, V_1, 0)$ which are separated by *all* quadratic computations of a bilinear map ϕ. We shall need a concept from the theory of rings. The intersection of all maximal left ideals of a ring A is called the *radical* of A and is denoted by $\mathrm{Rad}(A)$. A ring is called *semisimple* iff its radical is 0. We have the following facts whose proofs can be found in any standard textbook such as [140, 259, 316].

(17.19) Fact. Let A be a ring.

(1) $\mathrm{Rad}(A)$ is a two-sided ideal of A and $A/\mathrm{Rad}(A)$ is semisimple.
(2) $\mathrm{Rad}(A)$ is nilpotent, i.e., there exists $n \in \mathbb{N}'$ such that $\mathrm{Rad}(A)^n = 0$. Hence, every element of $\mathrm{Rad}(A)$ is nilpotent.
(3) $A/\mathrm{Rad}(A)$ and A have the same number of maximal two-sided ideals.
(4) A semisimple algebra is a direct product of simple algebras.   •

(17.20) Proposition. (1) $(\mathrm{Rad}(A), A, 0) \in S(A, \gamma)$ *for all quadratic computations γ of an associative k-algebra A.*

(2) *Let A be a simple associative k-algebra and R be a maximal right ideal of A. Then $(R, A, 0) \in S(A, \gamma)$ for all quadratic computations γ of A.*

Proof. (1) Clearly, $(0, \mathrm{Rad}(A), 0) \in S(A, \gamma)$ by Rem. (17.16). By reverse induction on i we prove that for all $i \in \mathbb{N}$

$$(\mathrm{Rad}(A)^i, \mathrm{Rad}(A), 0) \in S(A, \gamma).$$

(For $i = 0$ we get the desired assertion.) For large i the statement is clear by Fact (17.19)(2). Let $i > 0$ and let the subspace U_2 be maximal with respect to the conditions

$$\mathrm{Rad}(A)^i \subseteq U_2 \subseteq \mathrm{Rad}(A)^{i-1}, \quad (U_2, \mathrm{Rad}(A), 0) \in S(A, \gamma).$$

If $U_2 = \mathrm{Rad}(A)^{i-1}$ we are done. Otherwise, by the extension lemma, there is some $a \in \mathrm{Rad}(A)^{i-1} \setminus U_2$ such that

$$aA \subseteq \langle \mathrm{Rad}(A)^{i-1} \mathrm{Rad}(A) \rangle = \mathrm{Rad}(A)^i,$$

which is a contradiction to $a \notin \mathrm{Rad}(A)^i$.

(2) We have $(R, 0, 0) \in S(A, \gamma)$. Let $V_2 \subseteq A$ be a maximal subspace satisfying $(R, V_2, 0) \in S(A, \gamma)$. If $V_2 \neq A$, then by the extension lemma there exists some $b \in A \setminus V_2$ such that

$$Ab \subseteq \langle RA \rangle = R.$$

This, however, contradicts Fact (17.8)(4). $\qquad\square$

From Lemma (17.17), Prop. (17.20)(1), and Rem. (17.16)(2) we immediately obtain the following.

(17.21) Corollary. *For an associative k-algebra A we have*

$$L(A) \geq L(A/\mathrm{Rad}(A)) + 2 \dim \mathrm{Rad}(A).$$

This corollary and Fact (17.19) show that for proving Thm. (17.14) we may confine ourselves to the case of semisimple algebras. Note the similarity between the proofs of Prop. (17.20)(2) and (17.11). More generally, using the same argument as in the proof of Prop. (17.11), we can prove that for a simple k-algebra A of dimension n we have $L(A) \geq 2n - m$, where m denotes the dimension of minimal right ideals of A (see Ex. 17.6.) We aim now at improving this bound.

(17.22) Proposition. $L(A) \geq 2 \dim A - 1$ *for every simple k-algebra A.*

Proof. We first show that

(A) $\qquad\qquad\qquad (A, 0, W_1) \in S(A, \gamma)$

for any subspace W_1 of A of dimension less than m, where m denotes the dimension of any minimal right (or left) ideal of A, see Fact (17.8)(1).

Suppose the contrary were true and let $U_2 \subset A$ be a maximal subspace such that $(U_2, 0, W_1)$ is separated by γ. (U_2 exists by Rem. (17.16)(1).) By our assumption U_2 is properly contained in A and the extension lemma implies the existence of some $a \in A \setminus U_2$ such that $aA \subseteq W_1$. But aA is a nonzero right ideal, thus $\dim aA \geq m$ which is a contradiction.

Let L_1 be a minimal left ideal of A and L_2 be a complementary maximal left ideal of A (cf. Fact (17.8)(3)). Let τ be the projection of A onto L_1 along L_2. Since A is concise we may assume w.l.o.g. that $\tau(w_1), \ldots, \tau(w_m)$ form a basis of L_1. Setting $W_1 := kw_1 + \ldots + kw_{m-1}$ we have $W_1 \cap L_2 = 0$. By Lemma (17.17) it is sufficient to prove that $(A, L_2, W_1) \in S(A, \gamma)$, since then

$$2 \dim A - 1 = 2 \dim A - m + (m - 1) \leq L(A).$$

For proving this we consider the proper right ideal

$$R := (\tau(W_1) : L_1) := \{a \in A \mid aL_1 \subseteq \tau(W_1)\}.$$

We are going to show now that

$$(R, L_2, W_1) \in S(A, \gamma).$$

By (A) we know that $(R, 0, W_1) \in S(A, \gamma)$. Let $V_2 \subseteq L_2$ be a maximal subspace satisfying $(R, V_2, W_1) \in S(A, \gamma)$. If $V_2 \neq L_2$, then the extension lemma implies the existence of some $b \in L_2 \setminus V_2$ such that $Ab \subseteq \langle RL_2 \rangle + W_1$. Since $Ab \subseteq L_2$ and $L_2 \cap W_1 = 0$, we conclude that $Ab \subseteq \langle RL_2 \rangle$, hence the nonzero left ideal Ab is contained in the proper right ideal R, which contradicts Fact (17.8)(4). Finally we prove that

$$(A, L_2, W_1) \in S(A, \gamma).$$

Let U_2 be a maximal subspace of A containing R such that $(U_2, L_2, W_1) \in S(A, \gamma)$ and assume $U_2 \neq A$. The extension lemma shows the existence of some $a \in A \setminus U_2$ such that $aA \subseteq \langle AL_2 \rangle + W_1 = L_2 + W_1$. Therefore $aL_1 \subseteq \tau(W_1)$, which implies $a \in (\tau(W_1) : L_1) = R$, a contradiction. $\square$

If the additivity conjecture (see p. 360) were true, the proof of Thm. (17.14) would be complete. Now the point is that we do not need the additivity conjecture since the above proof can be generalized to the case of semisimple algebras as well.

(17.23) Proposition. *Let A and B be k-algebras and assume A to be simple. Then*

$$L(A \times B) \geq 2 \dim A - 1 + L(B).$$

Proof. Let m denote the dimension of the minimal left ideals of A and let γ be an optimal quadratic computation of A. Let L_1 be a minimal left ideal of A and L_2 be a complementary maximal left ideal of A (cf. Fact (17.8)(3)). Denote by τ the projection along $L_2 \times B$ onto $L_1 \times 0$. We may w.l.o.g. assume that $\tau(w_1), \ldots, \tau(w_m)$ form a basis of $L_1 \times 0$. Then

$$(kw_1 + \ldots + kw_m) \oplus (L_2 \times B) = A \times B;$$

we denote the projection onto $L_2 \times B$ according to the above direct decomposition by π. Put $W_1 := kw_1 + \ldots + kw_{m-1}$. We claim that

$$(A \times 0, L_2 \times 0, W_1) \in S(A \times B, \gamma).$$

This follows from the statement $(A, L_2, \mathrm{pr}_1(W_1)) \in S(A, \tilde{\gamma})$ demonstrated in the proof of the previous proposition; here pr_1 is the projection of $A \times B$ onto the first factor and $\tilde{\gamma}$ is the quadratic computation

$$\tilde{\gamma} = \left(f_1 \circ \iota, g_1 \circ \iota, \mathrm{pr}_1(w_1), \ldots, f_r \circ \iota, g_r \circ \iota, \mathrm{pr}_1(w_r) \right)$$

for A, where ι is the injection $A \hookrightarrow A \times B, a \mapsto (a, 0)$.

We may therefore conclude from Lemma (17.17) that

$$L\left(\pi \circ \phi_{/(A \times 0) \times (L_2 \times 0)} \right) + 2 \dim A - 1 \leq L(A \times B),$$

where ϕ denotes the multiplication map of $A \times B$. Let ψ denote the restriction of ϕ to $(A \times B) \times (L_2 \times B)$. Then obviously

$$L\left(\pi \circ \psi_{/(A \times 0) \times (L_2 \times 0)} \right) \leq L\left(\pi \circ \phi_{/(A \times 0) \times (L_2 \times 0)} \right).$$

On the other hand, $\pi \circ \psi = \psi$ as $\mathrm{im}\,\psi = L_2 \times B$. Moreover $\psi_{/(A \times 0) \times (L_2 \times 0)}$ is isomorphic to the multiplication map of B. Summarizing, we have shown that indeed

$$L(B) + 2 \dim A - 1 \leq L(A \times B). \qquad \square$$

Now we are finally in a position to show our main theorem.

Proof of Thm. (17.14). Let A be an associative k-algebra having exactly t maximal two-sided ideals. Then there exist simple algebras $A_1, \ldots, A_t$ such that $A/\mathrm{Rad}(A) \simeq A_1 \times \cdots \times A_t$ (cf. Fact (17.19)). Hence

$$
\begin{aligned}
L(A) \;\geq\;& L(A/\mathrm{Rad}(A)) + 2 \dim \mathrm{Rad}(A) && \text{(by Cor. (17.21))} \\
\geq\;& \sum_{i=1}^{t} (2 \dim A_i - 1) + 2 \dim \mathrm{Rad}(A) && \text{(by Prop. (17.23)} \\
& && \text{using induction)} \\
=\;& 2 \dim A - t. \quad \square
\end{aligned}
$$

(17.24) Remark. Although the bound in Thm. (17.14) is sharp, it can be shown that the rank of "almost all" algebras is quadratic in their dimension. See Ex. 17.20.

•

17.3* Multiplicative Complexity of Division Algebras

Unless otherwise specified, k will denote a field and D will be a division algebra of dimension n over k.

By Prop. (17.7) we have $L(D) \geq 2n - 1$. Our first aim will be the characterization of those division algebras for which $L(D) = 2n - 1$ holds. The following theorem reduces this problem to the investigation of division algebras for which $R(D) = 2n - 1$.

(17.25) Theorem (Feig). *Let D be a division algebra over k of dimension n with $L(D) = 2n - 1$. Then any optimal quadratic computation $(f_i, g_i, w_i)_{i \leq 2n-1}$ for D is essentially bilinear, i.e., after interchanging some f_ρ with g_ρ we have for all $a, b \in D$: $f_\rho(a, b) = f_\rho(a, 0)$, $g_\rho(a, b) = g_\rho(0, b)$. In particular $R(D) = 2n - 1$.*

Proof. (Baur) For $\rho \leq 2n - 1$ we define the linear forms f_ρ', f_ρ'' on D by $f_\rho(a, b) = f_\rho'(a) + f_\rho''(b)$ for all $a, b \in D$. Analogously we define g_ρ' and g_ρ''. We want to show that after interchanging some f_ρ with g_ρ we have $f_1'' = g_\rho' = 0$. Note that $f_1', \ldots, f_{2n-1}'$ or $f_1'', \ldots, f_{2n-1}''$ generate D^*: otherwise there would exist $a, b \in D$, both nonzero, such that $ab = 0$, a contradiction. W.l.o.g. we may assume that $f_1', \ldots, f_n'$ form a basis of D^*. Let $U_i := \cap_{\rho \leq i} \ker f_\rho'$ for $i = 1, \ldots, 2n - 1$. We first show that $U_n = 0 \times D$, thereby proving $f_1'' = \cdots = f_n'' = 0$. Since $U_n \oplus (D \times 0) = D \times D$ it is sufficient to show that $\pi(U_n) = 0$ where π is the first projection of $D \times D$. Suppose on the contrary that $\pi(U_n) \neq 0$. Since the U_i are nested, we have

(A) $\forall\, i \geq n: \qquad U_i \cap (D \times 0) = 0 \times 0.$

Now $\pi(U_{2n-1}) = 0$. (Suppose $(a, b) \in U_{2n-1}$, $a \neq 0$. By the definition of U_{2n-1} we obtain $ab = 0$, hence $b = 0$. But then (A) implies $a = 0$, a contradiction.) Since $\dim U_i / U_{i+1} \leq 1$ and $\pi(U_n) \neq 0$ by assumption, there exists $p \geq n$ such that $\pi(U_p)$ is one-dimensional, i.e., $\pi(U_p) = ke$ for some $0 \neq e \in D$. Note that $d := \dim U_p \geq 2n - p$. We can find a basis of the form $((e, b_1), \ldots, (e, b_d))$ for U_p. Since $p \geq n$, (A) implies that $b_1, \ldots, b_d$ are linearly independent and so are $eb_1, \ldots, eb_d$. ($\sum_i \lambda_i b_i = 0$ implies $(\sum_i \lambda_i e, 0) \in U_p$, hence $\sum_i \lambda_i e = 0$; this in turn shows that $\sum_i \lambda_i (e, b_i) = 0$, which implies $\lambda_1 = \cdots = \lambda_d = 0$.) But the definition of U_p gives

$$\forall\, i \in \underline{d}: \qquad eb_i = \sum_{\rho = p+1}^{2n-1} f_\rho(e, b_i) g_\rho(e, b_i) w_\rho \in \langle w_{p+1}, \ldots, w_{2n-1} \rangle.$$

Hence, $2n - p \leq \dim U_p \leq 2n - 1 - p$. This yields the desired contradiction to $\pi(U_n) \neq 0$.

The above argument yields: if $f_{\rho_1}', \ldots, f_{\rho_n}'$ form a basis of D^* for some $1 \leq \rho_1 < \ldots < \rho_n \leq 2n - 1$, then $f_{\rho_1}'', \ldots, f_{\rho_n}'' = 0$. This, together with the optimality of the computation shows that for all $1 \leq \rho \leq 2n - 1$

$$f_\rho' \neq 0 \iff f_\rho'' = 0.$$

Analogously, one shows: $g'_\rho \neq 0 \Leftrightarrow g''_\rho = 0$. Thus we obtain disjoint sets F, G, H, and I such that $F \cup G \cup H \cup I = \{1, \ldots, 2n-1\}$ and

$$
\begin{aligned}
F &= \{\rho \mid f''_\rho = g''_\rho = 0\} \\
G &= \{\rho \mid f''_\rho = g'_\rho = 0\} \\
H &= \{\rho \mid f'_\rho = g'_\rho = 0\} \\
I &= \{\rho \mid f'_\rho = g''_\rho = 0\}.
\end{aligned}
$$

For all $\rho \in I$ we interchange f_ρ with g_ρ. Then, for all $a, b \in D$ we have

$$
ab = \sum_{\rho \in F} f'_\rho(a) g'_\rho(a) w_\rho + \sum_{\rho \in G} f'_\rho(a) g''_\rho(b) w_\rho + \sum_{\rho \in H} f''_\rho(b) g''_\rho(b) w_\rho.
$$

Note that for $\rho \in F$ we have $f_\rho(a, c) = f'_\rho(a)$ for all $c \in D$. Hence, by setting $b = 0$ in the above equation, we obtain

$$
\forall a, c \in D: \quad 0 = \sum_{\rho \in F} f'_\rho(a) g'_\rho(a) w_\rho = \sum_{\rho \in F} f_\rho(a, c) g_\rho(a, c) w_\rho.
$$

The optimality of the computation immediately yields $F = \emptyset$. Analogously, by setting $a = 0$ we obtain $H = \emptyset$, hence the assertion. $\qquad\square$

(17.26) Remark. The above proof can actually be applied to the (slightly more general) case of a bilinear map $\phi \in \mathrm{Bil}(U, V; W)$ which satisfies $L(\phi) = \dim U + \dim V - 1$ and has no zero divisors, i.e., $\phi(a, b) = 0$ implies $a = 0$ or $b = 0$. $\qquad\bullet$

The next theorem gives a lower bound for the rank of an arbitrary division algebra. In the following we set $[D : k(a)] := \dim_k D - \dim_k k(a)$ for $a \in D$.

(17.27) Theorem (Baur). *We have $R(D) \geq 2n - 2 + [D : k(a)]$ for a suitable element $a \in D$.*

Proof. Let $(f_1, g_1, w_1; \ldots; f_r, g_r, w_r)$ be a bilinear computation for D. Since D is concise, we may w.l.o.g. assume that $f_1, \ldots, f_n$ form a basis of D^*. Let $x_1, \ldots, x_n$ be the corresponding dual basis of D. Let $\tilde{f}_\rho(a) := f_\rho(x_n a)$, $\tilde{g}_\rho(b) := g_\rho(b)$ and $\tilde{w}_\rho := x_n^{-1} w_\rho$. Then $ab = x_n^{-1}(x_n a)b = \sum_{\rho \leq r} \tilde{f}_\rho(a) \tilde{g}_\rho(b) \tilde{w}_\rho$. For all $i, j \in \underline{n}$ we have $\tilde{f}_i(x_n^{-1} x_j) = f_i(x_j) = \delta_{ij}$. Thus $(x_n^{-1} x_1, \ldots, x_n^{-1} x_n)$ is the dual basis of $(\tilde{f}_1, \ldots, \tilde{f}_n)$. Hence, replacing f_ρ by $\tilde{f}_\rho$ if necessary, we may assume that $x_n = 1$. Now $g_n, \ldots, g_r$ generate D^*; otherwise there would exist a common zero $0 \neq b \in D$ of $g_n, \ldots, g_r$ for which we would have $x_n b = b = 0$, a contradiction. W.l.o.g. we may assume that $g_n, \ldots, g_{2n-1}$ form a basis of D^*. Let $y_n, \ldots, y_{2n-1}$ be the corresponding dual basis. Defining $m := [D : k(x_1)]$, we may, after a possible reordering of the terms, assume that $y_n, \ldots, y_{n+m-1}$ form a basis of the $k(x_1)$-vector space D. Furthermore, we have for any $n \leq i \leq n + m - 1$:

$$
x_1 y_i \in kw_1 + kw_i + \sum_{\rho \geq 2n} kw_\rho,
$$

$$
y_i = x_n y_i \in kw_n + kw_i + \sum_{\rho \geq 2n} kw_\rho.
$$

We conclude that

$$\sum_{i=n}^{n+m-1} k y_i + \sum_{i=n}^{n+m-1} k x_1 y_i \subseteq k w_1 + \sum_{i=n}^{n+m-1} k w_i + \sum_{i \geq 2n} k w_i.$$

For $n \geq 2$ the elements $y_n, \ldots, y_{n+m-1}, x_1 y_n, \ldots, x_1 y_{n-m+1}$ are k-linearly independent, hence we obtain $2m \leq 1 + m + r - (2n - 1)$, i.e., $r \geq m + 2n - 2$. For $n = 1$ the assertion of the theorem is clear. $\square$

(17.28) Remarks.
(1) The proof of the above theorem shows even the following: there is a k-basis $\{a_1, \ldots, a_{n-1}, 1\}$ of D such that $R(D) \geq 2n - 2 + [D{:}k(a_i)]$ for all $1 \leq i \leq n - 1$.
(2) Thm. (17.27) is also true for a local k-algebra D, i.e., a k-algebra with a unique maximal ideal; see Ex. 17.11.   ●

(17.29) Theorem (Fiduccia and Zalcstein, Feig, de Groote). *If k has at least $2n - 2$ elements, then the following assertions are equivalent:*

(1) $L(D) = 2n - 1$,
(2) $R(D) = 2n - 1$,
(3) D *is a simple field extension of k.*

Proof. Note that (1) implies (2) by Thm. (17.25) and (2) implies (3) by Thm. (17.27). For showing that (3) implies (1), note that $L(D) \geq 2n - 1$ by Prop. (17.7) and $L(D) \leq R(D) \leq 2n - 1$ by Prop. (14.48)(3). $\square$

(17.30) Remark. It can be shown that $R(D) \geq 2n$ if D is a field extension of degree n of a finite field with less than $2n - 2$ elements, see the Notes.   ●

An interesting application of Thm. (17.27) is given below. We briefly recall the notion of a *central division algebra*. A division algebra D is called *central* over k iff k is the center of D, where we have identified k with the subring $\{\kappa \cdot 1 \mid \kappa \in k\}$ of D. It is well-known that the dimension n of any division algebra over k is a square number and that maximal subfields of D, i.e., maximal field extensions of k contained in D, have dimension $\sqrt{n}$ over k (see, e.g., Huppert [259, Satz 14.6]).

(17.31) Corollary. *We have $R(D) \geq 2n - 2 + \sqrt{n}$ for central division algebras D over k.*

(17.32) Example. We can now determine the multiplicative complexity of the 4-dimensional $\mathbb{R}$-algebra $\mathbb{H}$ of real quaternions. $\mathbb{H}$ has an $\mathbb{R}$-basis $\{1, i, j, k\}$ with multiplication defined by

$$i^2 = j^2 = k^2 = -1, \ ij = k = -ji, \ jk = i = -kj, \ ki = j = -ik.$$

If

$$(x_0 + i x_1 + j x_2 + k x_3)(y_0 + i y_1 + j y_2 + k y_3) = (f_0 + i f_1 + j f_2 + k f_3),$$

one verifies that

$$
\text{(A)} \quad
\begin{aligned}
f_0 &= x_0 y_0 - x_1 y_1 - x_2 y_2 - x_3 y_3, \\
f_1 &= x_0 y_1 + x_1 y_0 + x_2 y_3 - x_3 y_2, \\
f_2 &= x_0 y_2 - x_1 y_3 + x_2 y_0 + x_3 y_1, \\
f_3 &= x_0 y_3 + x_1 y_2 - x_2 y_1 + x_3 y_0.
\end{aligned}
$$

It is not difficult to show that $\mathbb{H}$ is a central division algebra over $\mathbb{R}$, see Ex. 17.12. We thus obtain from Cor. (17.31) $L(\mathbb{H}) \geq 8$. To obtain an upper bound for $L(\mathbb{H})$, we consider the group algebra $A := \mathbb{R}[C_2 \times C_2]$, where C_2 is the cyclic group of order 2. We have

$$
C_2 \times C_2 = \{ \underbrace{(1,1)}_{:=1}, \underbrace{(1,-1)}_{:=\tilde{i}}, \underbrace{(-1,1)}_{:=\tilde{j}}, \underbrace{(-1,-1)}_{:=\tilde{k}} \},
$$

with component-wise multiplication. If

$$
(\tilde{x}_0 + \tilde{i}\tilde{x}_1 + \tilde{j}\tilde{x}_2 + \tilde{k}\tilde{x}_3)(\tilde{y}_0 + \tilde{i}\tilde{y}_1 + \tilde{j}\tilde{y}_2 + \tilde{k}\tilde{y}_3) = (\tilde{f}_0 + \tilde{i}\tilde{f}_1 + \tilde{j}\tilde{f}_2 + \tilde{k}\tilde{f}_3),
$$

we obtain for $\tilde{f}_0, \ldots, \tilde{f}_3$ a similar scheme as (A), with minus signs replaced by plus signs. We then get

$$
\begin{aligned}
f_0 &= -\tilde{f}_0 + 2 x_0 y_0, \\
f_1 &= \tilde{f}_1 - 2 x_3 y_2, \\
f_2 &= \tilde{f}_2 - 2 x_1 y_3, \\
f_3 &= \tilde{f}_3 - 2 x_2 y_1.
\end{aligned}
$$

This shows that $L(\mathbb{H}) \leq L(\mathbb{R}[C_2 \times C_2]) + 4$. To compute $L(\mathbb{R}[C_2 \times C_2])$ we use

$$
\mathbb{R}[C_2 \times C_2] \simeq \mathbb{R}[x]/(x^2 - 1) \otimes_{\mathbb{R}} \mathbb{R}[x]/(x^2 - 1) \simeq \mathbb{R}^4,
$$

as $\mathbb{R}$-algebras. Hence $L(\mathbb{R}[C_2 \times C_2]) = 4$, which implies $L(\mathbb{H}) \leq 8$.   •

(17.33) Corollary. *Let* $\operatorname{char} k = p > 0$ *and* K *be a field extension of* k *of degree* n *such that for every* $a \in K$ *we have* $a^p \in k$. *Then* $R(K) \geq 2n - 2 + n/p$.

Proof. The assumptions imply that the minimal polynomial of every $a \in K$ divides $x^p - a^p$. Hence $[k(a):k] \leq p$, which shows that $[K:k(a)] \geq n/p$. Application of Thm. (17.27) yields the result.   □

(17.34) Remark. An example for a field extension $K \supset k$, $\operatorname{char} k = p > 0$, such that $a^p \in k$ for all $a \in K$ is given by the purely inseparable extension $\mathbb{F}_p(x) \supset \mathbb{F}_p(x^p)$.   •

17.4* Commutative Algebras of Minimal Rank

Throughout this section k is an infinite field.

By Thm. (17.14) the multiplicative complexity $L(A)$ of an associative k-algebra A of dimension n satisfies $L(A) \geq 2n - t$ where t is the number of maximal two-sided ideals of A. In the last section we saw that this inequality is sharp for the class of simple field extensions of k. The question which naturally arises is that of a characterization of all algebras for which Thm. (17.14) is sharp. Unfortunately a complete answer to this question is not known at the present time. Even a complete characterization of all algebras which have the minimum possible rank (rather than multiplicative complexity) seems to be very difficult. However, if we restrict ourselves to certain classes of algebras, such a characterization can be within reach. The aim of this section is to do this for the class of commutative k-algebras. We begin with a definition.

(17.35) Definition. An n-dimensional k-algebra A with exactly t maximal two-sided ideals is said to be of *minimal rank*, iff $R(A) = 2n - t$.   •

Let us first study some examples. We have already seen that simple field extensions of k are of minimal rank. These algebras are isomorphic to $k[X]/(p)$ for some irreducible polynomial $p \in k[X]$. More generally, we can prove that *simply generated algebras* are of minimal rank. These are algebras of the type $k[w]$. For such an algebra to be finite dimensional, w must satisfy an equation $f(w) = 0$ for some non-constant polynomial $f \in k[X]$; if f is of minimal degree, it is easily seen that $k[w]$ is isomorphic to $k[X]/(f)$. If $f = p_1^{\nu_1} \cdots p_t^{\nu_t}$ is the prime factor decomposition of f in $k[X]$, we have by the Chinese remainder theorem $k[X]/(f) \simeq k[X]/(p_1^{\nu_1}) \oplus \cdots \oplus k[X]/(p_t^{\nu_t})$; the residue class rings $k[X]/(p_i^{\nu_i})$ are *local* algebras; hence t is the number of maximal two-sided ideals of $k[X]/(f)$. We deduce from Thm. (17.14) that $R(k[X]/(f)) \geq 2n - t$, where $n := \deg f$. We even have equality by Prop. (14.48)(2), hence $k[X]/(f)$ is of minimal rank.

Let us proceed with another type of commutative algebras which will later be proved to be of minimal rank. A k-algebra A is called a *generalized null algebra* if there exist nilpotent elements $w_1, \ldots, w_s \in A$ such that $w_i w_j = 0$ for $i \neq j$ and $A = k[w_1, \ldots, w_s]$. Some basic properties of these algebras are summarized in the following.

(17.36) Lemma. *Let A be a generalized null algebra, $A = k[w_1, \ldots, w_s]$, with nonzero and nilpotent elements $w_1, \ldots, w_s$ satisfying $w_i w_j = 0$ for $i \neq j$. Further let $n_i := \max\{n \mid w_i^n \neq 0\}$. Then we have:*

(1) *A is commutative.*
(2) *A is a local algebra with maximal ideal $Aw_1 + \cdots + Aw_s$.*
(3) *A is an epimorphic image of*

$$k[X_1, \ldots, X_s]/(X_1^{n_1+1}, \ldots, X_s^{n_s+1}, X_i X_j \mid i, j \in \underline{s}, i \neq j).$$

(4) *Suppose that* $d := \dim\langle w_1^{n_1}, \ldots, w_s^{n_s}\rangle_k = \dim\langle w_1^{n_1}, \ldots, w_d^{n_d}\rangle_k$. *Then*

$$\{1\} \cup \{w_i^{j_i} \mid i \in \underline{s}, j_i = 1, \ldots, n_i - v_i\}$$

is a k-basis for A, where $v_1, \ldots, v_d = 0, v_{d+1}, \ldots, v_s = 1$. *In particular, A has dimension* $1 + \sum_{i=1}^{s} n_i - (s - d)$.

Proof. (1) Obvious.

(2) $\mathfrak{m} := Aw_1 + \cdots + Aw_s$ is maximal since $A/\mathfrak{m} \simeq k$ as k-algebras. It is also nilpotent by definition of the w_i. Hence there exists $n \in \mathbb{N}$ such that $\mathfrak{m}^n = 0$. Further, if $a \in A \setminus \mathfrak{m}$, then a is a scalar multiple of $1 - m$ for some $m \in \mathfrak{m}$. Since $(1 - m)(1 + m + \cdots + m^{n-1}) = 1 - m^n = 1$, we conclude that $a \in A^{\times}$.

(3) The kernel of the substitution morphism $k[X_1, \ldots, X_s] \to A$, $X_i \mapsto w_i$ for $i \in \underline{s}$ contains $(X_1^{n_1+1}, \ldots, X_s^{n_s+1}, X_i X_j \mid i \neq j)$.

(4) Obviously the given elements generate A as a k-space. To show that they are linearly independent, suppose that

$$a_0 + \sum_{i=1}^{s} \sum_{j_i=1}^{n_i - v_i} a_{ij_i} w_i^{j_i} = 0.$$

Since $w_i w_j = 0$ for $i \neq j$, we obtain for all i, by successive multiplication of the above equation with $w_i^{n_i}, \ldots, w_i^{v_i}$, the equalities $a_0 = 0, a_{i1} = 0, \ldots, a_{ij_i} = 0$. $\square$

Generalized null algebras are of minimal rank, as the following shows.

(17.37) Lemma. *If A is a generalized null algebra of dimension n, then* $R(A) = 2n - 1$, *i.e., A is of minimal rank.*

Proof. We use Lemma (17.36)(4). Let

$$\alpha := a_0 + \sum_{i=1}^{s} \sum_{j_i=1}^{n_i - v_i} a_{ij_i} w_i^{j_i},$$

$$\beta := b_0 + \sum_{i=1}^{s} \sum_{j_i=1}^{n_i - v_i} b_{ij_i} w_i^{j_i}.$$

It is easily verified that

$$\alpha\beta = -(s - 1)a_0 b_0 + \sum_{i=1}^{s} \left(a_0 + \sum_{j_i=1}^{n_i - v_i} a_{ij_i} w_i^{j_i} \right) \left(b_0 + \sum_{j_i=1}^{n_i - v_i} b_{ij_i} w_i^{j_i} \right).$$

Hence, multiplication in A can be accomplished as follows: Compute first $a_0 b_0$ and then the s products

$$\left(a_0 + \sum_{j_i=1}^{n_i - v_i} a_{ij_i} X^{j_i} \right) \left(b_0 + \sum_{j_i=1}^{n_i - v_i} b_{ij_i} X^{j_i} \right),$$

where X is an indeterminate over k. By Ex. 14.17 there exists a bilinear computation of length $1 + \sum_{i=1}^{s} 2(n_i - v_i) = 2n - 1$ which computes these products. $\square$

The main theorem of this section says that a commutative algebra of minimal rank is built up from simply generated algebras and generalized null algebras.

(17.38) Theorem (de Groote and Heintz). *A commutative algebra is of minimal rank if and only if it is isomorphic to a product of simply generated algebras and generalized null algebras.*

Let us give an outline of the proof. First we discuss the structure of a commutative algebra. It turns out that these algebras are built up from simpler blocks, namely, from local algebras. Then we derive in Lemma (17.42) a certain normal form for optimal bilinear computations for commutative algebras of minimal rank. With this lemma we can reduce the statement to the case of local algebras. Then we use Lemma (17.42) again to reduce the problem to a result on local algebras, see Lemma (17.44). This last lemma establishes the assertion.

By applying the Chinese remainder theorem, we see that the commutative algebra $k[X]/(f)$, f a non-constant polynomial in $k[X]$, is a product of local algebras. This is a special case of the following general structure theorem for commutative algebras. For a proof the reader is referred to, e.g., Atiyah and Macdonald [12].

(17.39) Fact. Every finite dimensional commutative algebra A is isomorphic to a direct product of local algebras $A_1, \ldots, A_t$: $A \simeq A_1 \times \cdots \times A_t$. Up to the ordering and isomorphisms this decomposition is unique.   ●

Note that the number t above is necessarily the number of maximal ideals of A. In the sequel A will always denote a commutative algebra which we identify with the direct product $A_1 \times \cdots \times A_t$ according to the above fact.

For $A = A_1 \times \cdots \times A_t$, we denote by $p_i: A \to A_i$ the projection of A onto A_i. We will use the following easy observations very often in the sequel.

(17.40) Remarks. Let $A = A_1 \times \cdots \times A_t$ be a decomposition of the commutative algebra A into local algebras.

(1) $A^\times = A_1^\times \times \cdots \times A_t^\times$.
(2) Let $e_i = (0, \ldots, 0, 1_{A_i}, 0, \ldots, 0)$ where the unit element 1_{A_i} of A_i is in the ith position. Then $e_i e_j = \delta_{ij} e_i$, δ being the Kronecker delta, $e_1 + \cdots + e_t = 1_A$.
(3) If $\varepsilon_1, \ldots, \varepsilon_t \in A$ are such that $p_i(\varepsilon_j) = 0$ for $i \neq j$ and $\sum_{i=1}^{t} \varepsilon_i = 1$, then $\varepsilon_i = e_i$ for all $i = 1, \ldots, t$.
(4) If $x \in A$ is such that $e_i x \in kx$, then $x \in Ae_i = A_i$.   ●

(17.41) Lemma. *Let $A = A_1 \times \cdots \times A_t$ be a decomposition of A into local algebras. Further let $a_1, \ldots, a_s \in A$ be such that for every $i \in \underline{t}$ there exists $j \in \underline{s}$ satisfying $p_i(a_j) \in A_i^\times$. Then there exist $\lambda_1 \ldots, \lambda_s \in k$ such that $\sum_{j=1}^{s} \lambda_j a_j \in A^\times$.*

Proof. Let $\mathfrak{m}_i$ denote the maximal ideal of A_i. For each $i \in \underline{t}$ consider the proper k-subspaces $V_i := \{(\lambda_1, \ldots, \lambda_s) \in k^s \mid p_i(\sum_{j=1}^{s} \lambda_j a_j) \in \mathfrak{m}_i\} \subset k^s$. Since k is infinite, the union $\cup_{i=1}^{t} V_i$ is properly contained in k^s. Now Rem. (17.40)(1) implies the assertion.   □

The next lemma plays a fundamental role for the proof of Thm. (17.38). It asserts that any optimal bilinear computation for a commutative algebra of minimal rank can be transformed into a bilinear computation with certain properties by means of a sequence of transformations. These are as follows: let

$$\gamma := (f_1, g_1, w_1; \ldots; f_r, g_r, w_r)$$

be a bilinear computation for A. If π is any permutation in the symmetric group S_r, *permutation of γ by π* yields a bilinear computation $(f_{\pi\rho}, g_{\pi\rho}, w_{\pi\rho})_{\rho \leq r}$ of length r. If for $\rho \in \underline{r}$ the elements $\alpha_\rho, \beta_\rho, \delta_\rho \in k$ are such that $\alpha_\rho \beta_\rho \delta_\rho = 1$, then *scaling γ by* $(\alpha_\rho, \beta_\rho, \delta_\rho)_{\rho \leq r}$ yields another bilinear computation $(\alpha_\rho f_\rho, \beta_\rho g_\rho, \delta_\rho w_\rho)_{\rho \leq r}$ of length r for A. Similarly, if $x, y, z \in A^\times$, then

$$\sum_i f_i(a) g_i(b) w_i = ab = x(x^{-1}ay)(y^{-1}bz)z^{-1}$$

$$= \sum_i f_i(x^{-1}ay) g_i(y^{-1}bz) x w_i z^{-1}$$

$$=: \sum_i \overline{f}_i(a) \overline{g}_i(b) \overline{w}_i,$$

hence *the linear transformation of γ by the triple* (x, y, z) yields a bilinear computation $(\overline{f}_\rho, \overline{g}_\rho, \overline{w}_\rho)_{\rho \leq r}$ of length r for A.

(17.42) Lemma. *Let $A = A_1 \times \cdots \times A_t$ be a decomposition of the n-dimensional commutative algebra A of minimal rank $2n - t$ into local algebras. Then any optimal bilinear computation for A can be transformed by means of permutations, scalings, and linear transformations into a bilinear computation $(f_i, g_i, w_i)_{i \leq 2n-t}$ satisfying*

(1) $(w_1, \ldots, w_n)$ *is a basis of A and $(f_1, \ldots, f_n)$ is the corresponding dual basis,*
(2) $(w_{n-t+1}, \ldots, w_{2n-t})$ *is a basis of A and $(g_{n-t+1}, \ldots, g_{2n-t})$ is the corresponding dual basis,*
(3) $w_{n-t+i} = e_i = (0, \ldots, 0, 1_{A_i}, 0, \ldots, 0)$ *for all $i \in \underline{t}$.*

Proof. We transform an optimal bilinear computation $(f_i, w_i, g_i)_{i \leq r}$ for A by means of permutations, scalings, and linear transformations to a computation satisfying (1)-(3).

(i) (Permutation) Since A is concise, $f_1, \ldots, f_{2n-t}$ generate A^*, and after permuting the f_i's we may assume that $(f_1, \ldots, f_n)$ form a basis of A^*. Let $(x_1, \ldots, x_n)$ be the corresponding dual basis of A.

(ii) (Permutation and scaling) For each $i \leq t$ the elements $p_i(x_1), \ldots, p_i(x_n)$ generate A_i, hence there exists for all $i \leq t$ an x_{j_i} such that $p_i(x_{j_i}) \in A_i^\times$. ($A_i$ is local.) After an appropriate permutation we may w.l.o.g. assume that $\{x_{n-s+1}, \ldots x_n\} = \{x_{j_1}, \ldots, x_{j_t}\}$. By Lemma (17.41) $\sum_{j=n-s+1}^n \lambda_j x_j \in A^\times$ for some $\lambda_j \in k$. By decreasing s if necessary, we may assume that $\lambda_j \neq 0$ for all j. By scaling the original computation by $(\lambda_\rho, 1, \lambda_\rho^{-1})_{\rho \leq r}$ (set $\lambda_j = 1$ for $j \leq n - s$ or $j > n$), we may assume that $\sum_{j=n-s+1}^n x_j \in A^\times$.

(iii) ($s = t$) Let $x := \sum_{j=n-s+1}^{n} x_j$. By (ii) the linear mapping $b \mapsto xb$ is an isomorphism and its image lies in $\langle w_{n-s+1}, \ldots, w_{2n-t} \rangle$. Hence $s = t$ and $(g_{n-t+1}, \ldots, g_{2n-t})$ is a basis of A^*.

(iv) (Permutation) By permuting the indices $n - t + 1, \ldots, n$ if necessary, we may assume that $p_i(x_{n-t+i}) \in A_i^\times$ for $i \in \underline{t}$.

(v) (Permutation and scaling) Let $(y_{n-t+1}, \ldots, y_{2n-t})$ denote the basis of A dual to $(g_{n-t+1}, \ldots, g_{2n-t})$. As in (ii)-(iv) we can transform the bilinear computation in such a way that $p_i(y_{n-t+i}) \in A_i^\times$ for $i \in \underline{t}$. However, by permuting the index set $\{n - t + 1, \ldots, 2n - t\}$ the linear forms $f_{n-t+1}, \ldots, f_{2n-t}$ will possibly be replaced by new linear forms $\tilde{f}_{n-t+1}, \ldots, \tilde{f}_{2n-t}$. As in (iii) we deduce that $(f_1, \ldots, f_{n-t}, \tilde{f}_{n-t+1}, \ldots, \tilde{f}_n)$ is a basis for A^*. For the moment let $(\tilde{x}_1, \ldots, \tilde{x}_n)$ denote the corresponding dual basis. Then $\tilde{x}_{n-t+i} y_{n-t+j} = 0$ for all $i, j \le t, i \ne j$, and since $\langle \tilde{x}_{n-t+1}, \ldots, \tilde{x}_n \rangle = \bigcap_{\rho \le n-t} \ker f_\rho = \langle x_{n-t+1}, \ldots, x_n \rangle$, we have

(A) $\langle \tilde{x}_{n-t+1}, \ldots, \tilde{x}_n \rangle \cap A^\times \ne \emptyset.$

Noting that $p_j(y_{n-t+j}) \in A_j^\times$, we deduce $p_j(\tilde{x}_{n-t+i}) = 0$ for all $i, j \le t, i \ne j$. Hence (A) implies $p_i(\tilde{x}_{n-t+i}) \in A_i^\times$ for $i \le t$. Thus, by replacing x_{n-t+i} by $\tilde{x}_{n-t+i}$ for $i \in \underline{t}$ if necessary, we may in addition to (i)-(iv) assume that $p_i(y_{n-t+i}) \in A_i^\times$.

(vi) The equality $x_{n-t+i} y_{n-t+j} = 0$ for $i \ne j, i, j \le t$ together with (iv) and (v) implies $p_j(x_{n-t+i}) = 0 = p_i(y_{n-t+j})$.

(vii) (Linear transformation) Let $x := \sum_{i=n-t+1}^{n} x_i$ and $y := \sum_{j=n-t+1}^{n} y_j$. Then by (iv) and (v) we have $x, y \in A^\times$. After a linear transformation of the bilinear computation with $(x^{-1}, 1, y)$ we may assume $x = y = 1$. Now (vi) and Lemma (17.40)(3) imply $x_{n-t+i} = e_i = y_{n-t+i}$ for $i \le t$. We have by Lemma (17.40)(2) $e_i = e_i e_i = x_{n-t+i} y_{n-t+i} = w_{n-t+i}$ for $i \le t$. This establishes (3).

(viii) (Scaling) For $\rho \le n - t$ we have by (vii) $x_\rho = x_\rho \cdot 1 = x_\rho \sum_{j=n-t+1}^{n} y_j$ $= g_\rho(1) w_\rho$. Hence $g_\rho(1) \ne 0$. Scaling by $(1, c_\rho^{-1}, c_\rho)_{\rho \le r}$, where $c_\rho = g_\rho(1)$ for $1 \le \rho \le n - t$ and $c_\rho = 1$ for $\rho \ge n - t + 1$, we obtain a bilinear algorithm satisfying (i)-(vii) and $x_\rho = w_\rho$ for $\rho \le n - t$. This establishes (1). Analogously, after appropriate scaling, we may w.l.o.g. assume that $y_\rho = w_\rho$ for $\rho \ge n + 1$, hence (2) is also established.   □

This lemma allows to reduce the proof of Thm. (17.38) to the case of local algebras.

(17.43) Corollary. *Let $A = A_1 \times \cdots \times A_t$ be a decomposition of the commutative algebra A into a product of local algebras A_i, $i = 1, \ldots, t$. Then A is of minimal rank if and only if A_i is of minimal rank for all $1 \le i \le t$.*

Proof. Suppose that all A_i are of minimal rank $2n_i - 1$. Then $2n - t \le R(A) \le \sum_{i=1}^{t} (2n_i - 1) = 2n - t$. Hence A is of minimal rank.
For the proof of the converse, suppose that $A = \sum_{\rho \le 2n-t} f_\rho \otimes g_\rho \otimes w_\rho$ where f_ρ, g_ρ and w_ρ satisfy the conditions (1)-(3) of Lemma (17.42). (Note the identification of A with its structural tensor.) We shall first prove

(A) $\qquad\qquad \forall \rho \le 2n - t \ \exists \ \text{unique } i: \qquad e_i w_\rho = w_\rho.$

The uniqueness is clear, since $w_\rho \ne 0$ by the optimality of the computation. As for the existence, note first that the statement is trivial for $n - t < \rho \le n$. Suppose that $\rho > n$ and let $i \in \underline{t}$. We then have

$$
e_i w_\rho = \underbrace{\sum_{j=1}^{n-t} \underbrace{f_j(e_i)}_{=0} \, g_j(w_\rho)w_j}_{\text{by (1), (3) of Lemma (17.42)}} + \sum_{j=n-t+1}^{n} f_j(e_i) \underbrace{g_j(w_\rho)}_{=\delta_{j\rho}} \, w_j
$$

$$
= f_\rho(e_i)w_\rho \in k w_\rho.
$$

Since $0 \ne w_\rho = 1 \cdot w_\rho = \sum_{i=1}^{t} e_i w_\rho$, this implies the existence of i such that $e_i w_\rho = w_\rho$. The case $\rho \le n - t$ can be proved analogously. This proves (A). Let $J_i := \{\rho \mid e_i w_\rho = w_\rho\}$. By (A) the set $\{1, \ldots, 2n - t\}$ is the disjoint union of $J_1, \ldots, J_t$. For $i \le t$ let $\iota_i \colon A_i \to A_1 \times \cdots \times A_t$ be defined by $\iota_i(a) := (0, \ldots, 0, a, 0, \ldots, 0)$, where a is in the ith position. Clearly, ι_i is a right inverse of p_i for all i, i.e., $p_i \circ \iota_i = \mathrm{id}_{A_i}$. For all $a, b \in A_i$

$$
ab = p_i(\iota_i(a)\iota_i(b)) = \sum_{\rho \in J_i} (f_\rho \circ \iota_i)(a)(g_\rho \circ \iota_i)(b)p_i(w_\rho),
$$

which shows that $R(A_i) \le |J_i|$. Hence we have

$$
2n - t = \sum_{i=1}^{t}(2n_i - 1) \le \sum_{i=1}^{t} R(A_i) \le \sum_{i=1}^{t} |J_i| = 2n - t,
$$

which yields $R(A_i) = 2n_i - 1$ for all $i \le t$. $\qquad\qquad\qquad\qquad\qquad\square$

Now let A be a local algebra of minimal rank $2n - 1$. We can apply Lemma (17.42) with $t = 1$. Then $w_n = 1$ and for all $i < n < j$ we have

$$
w_i w_j = \sum_{\rho=1}^{2n-1} f_\rho(w_i)g_\rho(w_j)w_\rho = g_i(w_j)w_i + f_j(w_i)w_j \in k w_i + k w_j.
$$

So for the proof of Thm. (17.38) in case of local algebras we only need the following result.

(17.44) Lemma. *Suppose that $(x_1, \ldots, x_{n-1}, 1)$ and $(y_1, \ldots, y_{n-1}, 1)$ are k-bases for the commutative local k-algebra A such that $x_i y_j \in k x_i + k y_j$ for all $i, j = 1, \ldots, n - 1$. Then A is either simply generated or a generalized null algebra.*

Proof. Suppose A is not simply generated. By assumption, for all i, j there exist $a_{ij}, b_{ij} \in k$ such that $x_i y_j = b_{ij} x_i + a_{ij} y_j$, hence

(A) $\qquad\qquad (x_i - a_{ij})(y_j - b_{ij}) = a_{ij} b_{ij}.$

Note that

(B) $\qquad\qquad a_{ij} b_{ij} \ne 0 \ \Rightarrow \ k[x_i] = k[y_j],$

since the assumption implies $k[x_i] = k[x_i - a_{ij}] = k[(x_i - a_{ij})^{-1}] = k[y_j]$ (see Ex. 17.13). In the sequel we prove a couple of assertions which will ultimately imply the lemma.

Claim 1. *For all i there exists j such that $a_{ij}b_{ij} = 0$.* Otherwise (B) implies that there exists i such that for all j we have $k[y_j] = k[x_i]$, which is a contradiction since A is not simply generated.

Claim 2. *For all i there exists $\alpha_i \in k$ such that $(x_i - \alpha_i) \in \mathrm{Rad}(A)$ (=maximal ideal of A).* Note that by Claim 1 there exists j such that $(x_i - a_{ij})(y_j - b_{ij}) = 0$. The second factor is nonzero, as 1, y_j are linearly independent by assumption. Thus, setting $\alpha_i := a_{ij}$, $(x_i - \alpha_i)$ is a non-unit, hence in $\mathrm{Rad}(A)$. *Similarly, there exists for all j an element $\beta_j \in k$ such that $(y_j - \beta_j) \in \mathrm{Rad}(A)$.*

Claim 3. *W.l.o.g. $x_i, y_j \in \mathrm{Rad}(A)$ for all i, j.* Otherwise, we replace x_i by $\xi_i := x_i - \alpha_i$, and y_j by $\eta_j := y_j - \beta_j$. Clearly, $(1, \xi_1, \ldots, \xi_{n-1})$ form a basis of A, as do $(1, \eta_1, \ldots, \eta_{n-1})$, and (A) implies

$$\xi_i \eta_j = (x_i - \alpha_i)(y_j - \beta_j) \in kx_i + ky_j + k = k\xi_i + k\eta_j + k.$$

Since $\mathrm{Rad}(A) \cap k = 0$ and $\xi_i, \eta_j \in \mathrm{Rad}(A)$, we obtain $\xi_i \eta_j \in k\xi_i + k\eta_j$.

Claim 4. *$(x_1, \ldots, x_{n-1})$ and $(y_1, \ldots, y_{n-1})$ form a basis of* $\mathrm{Rad}(A)$. Obviously, the k-span of these elements is a subset of $\mathrm{Rad}(A)$ by Claim 3 and is of dimension $n - 1$. Since $\mathrm{Rad}(A)$ is of dimension at most $n - 1$, the assertion follows.

Claim 5. *For all i and j we have either $x_i y_j = 0$ or $k[x_i] = k[y_j]$.* If $a_{ij}b_{ij} \neq 0$, then $k[x_i] = k[y_j]$ by (B). If $a_{ij}b_{ij} = 0$, we have $a_{ij} = 0$: otherwise $(x_i - a_{ij}) = -a_{ij}^{-1}(1 - a_{ij}^{-1}x_i) \in A^\times$ (since $x_i \in \mathrm{Rad}(A)$) and thus $y_j = b_{ij} \in \mathrm{Rad}(A) \cap k = 0$, a contradiction. Similarly $b_{ij} = 0$. Hence $x_i y_j = 0$.

Claim 6. *For all i and j we have either $x_i x_j = 0$ or $k[x_i] = k[x_j]$.* Let i and j be fixed. Two cases can occur: either $x_i y_\ell = 0$ for all ℓ or there exists ℓ such that $x_i y_\ell \neq 0$. In the first case we obtain by Claim 4 $x_i \mathrm{Rad}(A) = 0$, thus $x_i x_j = 0$. In the second case Claim 5 implies that $k[x_i] = k[y_\ell]$. If $k[x_j] = k[y_\ell]$ we are done. Otherwise, $x_j y_\ell = 0$ by Claim 5. Then we obtain from $x_i \in k[y_\ell]$ the identity $x_i = \alpha_m y_\ell^m + \cdots + \alpha_1 y_\ell$ for some $\alpha_1, \ldots, \alpha_m \in k$ (since $x_i, y_\ell \in \mathrm{Rad}(A)$, no constant term occurs). Hence $x_i x_j = x_j x_i = x_j(\alpha_m y_\ell^m + \cdots + \alpha_1 y_\ell) = 0$.

Now we define an equivalence relation $\sim$ on $\{x_1, \ldots, x_{n-1}\}$ by $x_i \sim x_j$ iff $k[x_i] = k[x_j]$. Let $w_1, \ldots, w_s$ be a set of representatives of the $\sim$-classes. Then $A = k[x_1, \ldots, x_{n-1}] = k[w_1, \ldots, w_s]$, $w_i w_j = 0$ for $i \neq j$ by Claim 6, and $w_i \in \mathrm{Rad}(A)$ by Claim 3. Thus A is a generalized null algebra. $\square$

(17.45) Remark. Thm. (17.38) shows that the additivity conjecture is true for the class of local algebras of minimal rank.   •

17.5 Exercises

17.1. Prove Rem. (17.3).

17.2. Let $V = U_1 \oplus U_2$ be a direct sum decomposition of a finite dimensional k-space V. Let $f_1, \ldots, f_p \in V^*$, $p := \dim U_1$, separate the points of U_1. Let π be the projection of V along $\cap_{i=1}^{p} \ker f_i$ onto U_1. Let $g_1, \ldots, g_q \in V^*$, $q := \dim U_2$, be such that $g_1 \circ \pi, \ldots, g_q \circ \pi$ separate the points of U_2. Prove that $f_1, \ldots, f_p, g_1, \ldots, g_q$ form a basis of V^*.

17.3. Prove Rem. (17.13)(5).

17.4. Show that matrix algebras over (skew) fields are simple.

17.5. Prove Fact (17.8).

17.6. Let A be a simple associative k-algebra of dimension n and suppose that minimal left ideals of A have dimension m. Prove that $L(A) \geq 2n - m$ without using Thm. (17.14).

17.7. Let k be an infinite field, $f \in k[X]$ be a polynomial over k with t distinct prime factors in $k[X]$, and $A = k[X]/(f)$, where (f) denotes the ideal of $k[X]$ generated by f. Show that $L(A) = 2 \deg f - t$.

17.8.
(1) Prove Rem. (17.26).
(2) Let $\Phi_k^{m,n}$ be the polynomial multiplication map as defined in Sect. 14.5 and assume that k is a field having at least $m + n - 2$ elements. Show that $\Phi_k^{m,n}$ satisfies the assumptions of Rem. (17.26).
(3) An $n \times n$-matrix $A = (a_{ij})$ is called a *Toeplitz matrix* if $a_{ij} = a_{i-1,j-1}$ for $2 \leq i, j \leq n$. Let U be the $(2n - 1)$-dimensional k-space of all Toeplitz matrices over k, $V = W := k^{n \times 1}$. Let the bilinear map $\phi \in \mathrm{Bil}(U, V; W)$ be defined by $\phi(A, b) := Ab$. Show that if $k = \mathbb{Q}$, then ϕ has no zero-divisors and ϕ is concise.
(4) Show that $R(\phi) = 2n - 1 = L(\phi)$ and that every quadratic algorithm for ϕ is essentially bilinear if $k = \mathbb{Q}$. (Hint: Show that there is a permutation π of S_3 such that $\pi\phi = \Phi_k^{n,n}$.)

17.9.* Let k be a field, $n \in \mathbb{N}'$, and $\Phi_k^{m,n}$ be the polynomial multiplication map. Show that $R(\Phi_k^{m,n}) = m + n - 1$ iff k has at least $m + n - 2$ elements. (Hint: Consider the 3-slices of $\Phi_k^{m,n}$.)

17.10.

(1) Let ϕ be the bilinear map assigning to $(a, b) \in k^{m \times n} \times k^{n \times p}$ the components of the first column and first row of the product matrix ab. Prove that $L(\phi) \geq (m + p - 1)n$ if $m, p \geq 1$. (Hint: see the proof of Prop. (17.11).)

(2) Conclude that $R(\Box \times \Box) = 6$ for the example in Sect. 15.2.

17.11. (Cf. Rem. (17.28)(2).) Let A be a local k-algebra of dimension n. Show that there exists $a \in A$ such that $R(A) \geq 2n - 2 + \dim_k A - \dim_k k(a)$. (Hint: Combine the proofs of Thm. (17.27) and Lemma (17.41).)

17.12. Prove that $\mathbb{H}$ is a central division algebra over $\mathbb{R}$.

17.13. Let A be a commutative k-algebra and $a \in A^\times$. Show that $k[a] = k[a^{-1}]$.

17.14. Let A be an n-dimensional local k-algebra. Show the equivalence of the following assertions:

(1) $R(A) \leq 2n - 1 + d$.

(2) There exist a pair of bases $(1, x_2, \ldots, x_n)$ and $(1, y_2, \ldots, y_n)$ of A, sequences $(X_1, \ldots, X_n)$, $(Y_1, \ldots, Y_n)$, $(Z_1, \ldots, Z_n)$ in A, and $\xi_{ij}, \eta_{ij}, \xi_i^{(\delta)}, \eta_i^{(\delta)} \in k$ such that for all i, j we have

$$x_i y_j = \xi_{ij} X_i + \eta_{ij} Y_j + \sum_{1 \leq \delta \leq d} \xi_i^{(\delta)} \eta_i^{(\delta)} Z_\delta.$$

For the next three exercises we assume familiarity with Wedderburn's theorem on the structure of semisimple k-algebras. (See, e.g., Lang [316, XVII].)

17.15. Let A denote an n-dimensional k-algebra. Prove the following:

(1) $R(A) \geq L(A) \geq n$.

(2) $R(A) = n$ implies $A \simeq k^n$.

(3) $L(A) = n$ iff $R(A) = n$.

(4) Use (3) to prove $R(T_2) = 4$, where T_2 denotes the k-algebra of upper triangular 2×2 matrices over k.

17.16. Let k be an algebraically closed field, $n \in \mathbb{N}'$, A be an n-dimensional k-algebra and T_2 as in the previous exercise. Show the equivalence of the following assertions.

(1) $R(A) = n + 1$,

(2) $L(A) = n + 1$,

(3) $A / \operatorname{Rad}(A) \simeq k^{n-1}$,

(4) $A \simeq k[X]/(X^2) \times k^{n-2}$ or $A \simeq T_2 \times k^{n-3}$.

(Hint: (3) $\Rightarrow$ (4): $\mathrm{Rad}(A) = ku$ for some $u \in A$, $u^2 = 0$. Let ρ_u and λ_u be the right, resp. left multiplication with u. Then $I := \ker \rho_u \cap \ker \lambda_u$ is a two-sided ideal of A of dimension $n - 1$ or $n - 2$. If $\dim_k I = n - 1$, use the central primitive idempotents of A to find a k-basis $(e_0, \ldots, e_{n-1})$ of A such that $e_i e_j = \delta_{ij} e_j$ for $i, j \geq 1$, $e_i e_0 = e_0 e_i = 0$ for all $i \geq 2$, $e_0^2 = 0$, and $e_0 e_1 = e_1 e_0 = e_0$. Deduce that $A \simeq k[X]/(X^2) \times k^{n-2}$. In case $\dim_k I = n - 2$ find a k-basis $(e_0, \ldots, e_{n-1})$ of A such that $e_0^2 = 0$, $e_0 e_1 = e_0$, $e_1 e_0 = 0$, $e_0 e_2 = 0$, $e_2 e_0 = e_0$, $e_0 e_i = e_i e_0$ for $i \geq 3$, and $e_i e_j = \delta_{ij} e_j$ for $i, j \geq 1$. Deduce that $A \simeq T_2 \times k^{n-3}$.)

17.17. Let k be any field which is not algebraically closed and A be a semisimple k-algebra of dimension n. Show that $L(A) = n + 1$ iff $A \simeq k_1 \times k^{n-2}$, where k_1 is a quadratic field extension of k.

17.18. Let D_{2n} be the dihedral group of order $2n$ and $\mathbb{C} D_{2n}$ denote the group ring of D_{2n} over $\mathbb{C}$. Show that

$$L(\mathbb{C} D_{2n}) = \begin{cases} \frac{1}{2}(7n - 3) & \text{for } n \text{ odd,} \\ \frac{1}{2}(7n - 6) & \text{for } n \text{ even.} \end{cases}$$

(Hint: compute the irreducible matrix representations of $\mathbb{C} D_{2n}$, see also Chap. 13.)

17.19. Let A be the group algebra of the symmetric group S_3 over $\mathbb{R}$, (e_1, e_2, e_3) be the standard basis of $\mathbb{R}^3$, and U be the subspace of $\mathbb{R}^3$ generated by $e_1 + e_2 + e_3$.

(1) The linear extension of the S_3-action on (e_1, e_2, e_3) given by

$$\pi e_i := e_{\pi(i)}$$

for $\pi \in S_3$, $i \in \underline{3}$, turns $\mathbb{R}^3$ into an A-left module.
(2) $\mathbb{R}^3 = U \oplus U^\perp$ is a direct decomposition of $\mathbb{R}^3$ into a direct sum of simple A-modules.
(3) $A \simeq \mathbb{R}^2 \times \mathbb{R}^{2 \times 2}$.
(4) $L(A) = 9$.

17.20. (Existence of algebras of high rank) Let $Alg_n \subset k^{n \times n \times n}$ denote the subset of the structural tensors of all n-dimensional associative k-algebras having as unit element the first standard basis vector e_1 of k^n. We assume that k is algebraically closed.

(1) Prove that Alg_n is Zariski closed in $k^{n \times n \times n}$, hence it is a subvariety.
(2) Let E, S be a subspaces of k^n of dimensions $m, n - 1 - m$, respectively such that $k^n = ke_1 \oplus E \oplus S$. For a bilinear map $\phi: E \times E \to S$ define $\Phi \in \mathrm{Bil}(k^n, k^n; k^n)$ by

$$\Phi(\lambda e_1 + a + s, \lambda' e_1 + a' + s') := \lambda \lambda' e_1 + (\lambda a' + \lambda' a) + (\lambda s' + \lambda' s + \phi(a, a')),$$

where $a, a' \in E$, $s, s' \in S$, and $\lambda, \lambda' \in k$. Show that Φ turns k^n into an n-dimensional k-algebra A_Φ with unit element e_1.

(3) The morphism $E^* \otimes E^* \otimes S \to Alg_n$ given by $\phi \mapsto A_\phi$ is an injective morphism of algebraic varieties. Use this to show that Alg_n contains a component C of dimension $\geq m^2(n-1-m)$. Deduce that $\dim C \geq n^3/9$.

(4) There exists $A \in Alg_n$ such that the border rank $\underline{R}(A)$ of A satisfies $\underline{R}(A) > n^2/27$. (Hint: Let X_r denote the Zariski closure of the set of all tensors in $k^{n \times n \times n}$ of border rank r. Then $\dim X_r \leq r(3n-2)$, see Prop. (20.4))

(5) Show that there exists a Zariski open subset of the component C of Alg_n described in (3) such that $R(A) > n^2/27$ for all k-algebras A belonging to this set.

17.21. The *small isotropy group* $\Gamma^0(\phi)$ of a bilinear map $\phi \in \mathrm{Bil}(U, V; W)$ is defined as the set of all $(\alpha, \beta, \gamma) \in GL(U) \times GL(V) \times GL(W)$ satisfying $\gamma \circ \phi = \phi \circ (\alpha \times \beta)$. We assume that our ground field k is algebraically closed. Show the following:

(1) $\Gamma^0(\phi)$ is a closed subgroup of $GL(U) \times GL(V) \times GL(W)$.

(2) Let A be an associative k-algebra (not necessarily commutative). Let $\mathrm{Aut}(A) < GL(A)$ denote the group of algebra automorphisms of A. Show that $\Gamma^0(A)$ is a semi-direct product of $\mathrm{Aut}(A)$ and $(A^\times)^2$, the latter being a normal subgroup. (Hint: see the proof of Prop. (14.13).)

(3) Describe both the automorphism group and the small isotropy group of $k^{n \times n}$. What are their dimensions? (Hint: you may apply the Skolem-Noether theorem, cf. Pierce [417, p. 230].)

17.22. Prove that Thm. (17.46) in the Notes is indeed a generalization of Thm. (17.38).

17.23. Let k be an algebraically closed field and A be a basic k-algebra of minimal rank. (For the definition of basic k-algebras see the Notes of this chapter.) Then the following holds.

(1) $\mathrm{Rad}(A)$ satisfies Condition (3) in Thm. (17.46).

(2) If I is a two-sided ideal of A, then A/I is of minimal rank.

17.24.* Let k be an algebraically closed field and denote by $T_n(k)$ the k-algebra of upper triangular $n \times n$ matrices. Then $T_n(k)$ is of minimal rank if and only if $n = 2$. Furthermore, $R(T_3(k)) = 10$, hence the trivial 3×3 matrix multiplication algorithm is optimal in $T_3(k)$. (Hint: use without proof Thm. (17.46) given in the Notes of this chapter, and the previous exercise.)

17.6 Open Problems

Problem 17.1. Determine the rank $r := R(\langle 3, 3, 3 \rangle)$. (It is known that $17 \leq r \leq 23$ by Prop. (17.9) and Ex. 15.3. Over $\mathbb{F}_2$ one even has $r \geq 18$, see Sect. 18.4.)

Problem 17.2. Improve the lower bound of Thm. (17.12).

Problem 17.3. Decide whether the following analogue of Thm. (17.14) is correct:

$$\underline{R}(A) \geq 2 \dim A - \dim \operatorname{Rad}(A) - t.$$

Problem 17.4. Generalize Thm. (17.25) in the following sense: find all k-algebras for which optimal quadratic computations are essentially bilinear.

Problem 17.5. Determine the algebras of minimal rank.

17.7 Notes

The arguments we have used for proving lower bounds in this chapter are essentially based on the substitution method discussed in Chap. 6. Since the beginning of the 1970's there have been numerous efforts to prove nontrivial lower bounds for the rank and multiplicative complexity of bilinear maps. One of the important achievements was Lafon and Winograd's [311] result, presented here as Thm. (17.12). Prior to this discovery, Brockett and Dobkin [80] had proved Thm. (17.12) for the rank instead of multiplicative complexity. In 1981 Alder and Strassen [6] unified all the previous lower bounds for the multiplicative complexity of different associative algebras in a single result, Thm. (17.14) in Sect. 17.2. Their proof of this theorem is the quintessence of all lower bound proofs based on linear independence arguments.

Which associative algebras do realize the lower bound of Alder and Strassen's theorem? De Groote [218] began studying this problem by characterizing all division algebras of minimal rank; his result is the contents of the implication (2) $\Rightarrow$ (3) in Thm. (17.29). These investigations were carried on further by de Groote and Heintz [220] who gave a complete characterization of commutative algebras of minimal rank, see Thm. (17.38). In a further paper, Heintz and Morgenstern [243] gave a complete characterization of all basic algebras of minimal rank, see below. A classification of all associative algebras of minimal rank is, however, still unknown.

Let us now proceed with some remarks on the history of lower bounds for the complexity of algebras prior to the proof of Thm. (17.12) by Lafon and Winograd. In 1971 Winograd [553] proved that $L(\mathbb{C}) = 3$, where $\mathbb{C}$ is regarded as an $\mathbb{R}$-algebra. In the same year Hopcroft and Kerr [252] and Winograd [553] showed that $L(k^{2\times 2}) = 7$, thereby proving the optimality of Strassen's algorithm for 2×2-matrix multiplication as a quadratic computation. Winograd [554] has also shown that among the bilinear algorithms of length 7 Strassen's algorithm is not optimal with respect to additions/multiplications by exhibiting an algorithm with 7 multiplications and 15 additions/subtractions, cf. Ex. 15.1. Probert [426] has shown that any bilinear algorithm of length 7 for the multiplication of 2×2-matrices requires at least 15 additions/subtractions; hence Winograd's algorithm is optimal both with respect to multiplications and additions/subtractions. In 1977 it was shown by Fiduccia and Zalcstein [171] that $L(A) \geq 2 \dim A - 1$ for a division algebra A, with equality holding if A is a simple field extension. This is the contents of the implication (3) $\Rightarrow$ (1) in Thm. (17.29). In the same paper the authors proved that $L(k[X]/(f)) \leq 2 \deg f - m$, where f is a univariate polynomial over k with m distinct prime factors, and conjectured that this upper bound is in fact optimal. This conjecture was proved independently by several authors, see Winograd [557] and Bini and Capovani [47]. (Cf. Ex. 17.7.) Dobkin [144], de Groote [215], and Howell and Lafon [256] have independently proved that $L(\mathbb{H}) = 8$.

Thm. (17.14) of Alder and Strassen has been generalized by Hartmann [230]. For an associative k-algebra A and an A-module M, he derives lower bounds for the multiplicative complexity $L(A, M)$ of the multiplication map $A \times M \to M$. In particular, if $M = A$, he recovers Alder and Strassen's theorem. The case where $A = K$ is a k-division algebra and $M = K^n$ had been previously considered by Auslander and Winograd [19]. For extensions of Hartmann's result see Bshouty [91].

The proof of Prop. (17.9) in Sect. 17.1 has been taken from Baur [29]. There are a variety of isolated results on the complexity of different quadratic and bilinear maps, which we have not included in this section due to lack of space. For instance, Dobkin and van Leeuwen [147] have shown that the multiplicative complexity of computing products of pure real quaternions is 6. (For the lower bound for this problem the authors refer to Howell and Lafon [256].) Another example is van Leeuwen's [319] quadratic computation of length 5 for squaring 2×2-matrices which he has also proved optimal (see also Ex. 14.18).

The arrangements of the statements and the proofs in Sect. 17.3 and 17.4 have been taken from Strassen's lectures [503]. Thm. (17.25) is due to Feig [159]. We have traded his original proof for a simplified version due to Baur (unpublished, see [503]). Thm. (17.27) which has greatly simplified the characterization of division algebras of minimal rank in Thm. (17.29) is also an unpublished result of Baur. It has been taken from Strassen's lectures [503]. Winograd [559] and de Groote [218] have shown that the rank of a simple field extension of k of degree n is $2n - 1$ iff k has at least $2n - 2$ elements. For extensions of these results see Chudnovsky and Chudnovsky [109, 110], Shokrollahi [473, 474, 475, 476], and Sect. 18.5. The proof of de Groote and Heintz's theorem (17.38) presented in Sect. 17.4 follows Strassen's lectures [503] almost word by word. A generalization of this theorem to certain non-commutative algebras has been achieved by Büchi and Clausen [96], see also Büchi [95]. Their result has itself been generalized by Heintz and Morgenstern [243], see below.

A problem related to Ex. 17.7 is that of the multiplication of two polynomials modulo a third polynomial. This problem has been investigated by Winograd [562]. All the minimal algorithms for this problem have been classified by Averbuch et al. [21, 22, 20]. Ex. 17.8 is from Feig [159], Ex. 17.14 from Strassen's lectures [503], Ex. 17.9 from de Groote [219], Ex. 17.18 from Alder and Strassen [6]. Ex. 17.20 is from Büchi [95]; he gives the credit for the construction of A_ϕ to Guerino Mazzola. Ex. 17.22-17.24 are from Heintz and Morgenstern [243].

If the rank of an algebra is known, it is natural to ask about the set of its optimal bilinear computations. The small isotropy group of the algebra (see Ex. 17.21) acts on this set. De Groote [216] has shown that this action is transitive for $A = k^{2 \times 2}$ which means that there is essentially one unique bilinear algorithm for the multiplication in $k^{2 \times 2}$. On the other hand, field extensions generally possess higher dimensional orbit varieties of optimal algorithms, see Winograd [557] and de Groote [216, 217]. For the varieties of optimal algorithms for commutative algebras of minimal rank see Fellmann [167, 168]; for algorithm varieties of $\mathfrak{sl}(2, k)$ see Mirwald [367].

Very little is known about the problem of inversion (of units) and division (by units) in finite dimensional associative algebras. (These problems are not represented by quadratic maps.) Alt and van Leeuwen [8] have shown that complex inversion has multiplicative complexity 4 (see also Ex. 6.16), and Lickteig [332, 333] has proved that the multiplicative complexity of complex division (as well as that of solving a linear system of size 2) is 6.

Thm. (17.38) has been generalized by Heintz and Morgenstern [243] as follows. An algebra A is called *basic* if $A/\operatorname{Rad}(A)$ is a finite product of division algebras, or equivalently, if for each maximal two-sided ideal $\mathfrak{m}$ of A, the residue class ring $A/\mathfrak{m}$ is a division algebra. Observe that commutative algebras are basic. For a basic algebra A we define the left ideal L_A and the right ideal R_A by

$$L_A := \{x \in \operatorname{Rad}(A) \mid x \operatorname{Rad}(A) = 0\}, \quad R_A := \{x \in \operatorname{Rad}(A) \mid \operatorname{Rad}(A)x = 0\}.$$

Hence, L_A and R_A are the left and right annihilator of $\operatorname{Rad}(A)$ in $\operatorname{Rad}(A)$, respectively. If A is of dimension n and $A/\operatorname{Rad}(A)$ is a p-fold product of division algebras, we say that A belongs to the class $\mathcal{M}_k$ if there exists a pair of bases $\Sigma = ((x_1, \ldots, x_n), (y_1, \ldots, y_n))$ of A which satisfies the following properties:

(1) $x_\nu y_\mu \in kx_\nu + ky_\mu$ for each $1 \le \nu, \mu \le n$;
(2) $x_\pi = y_\pi$ for $1 \le \pi \le p$ and $x_1, \ldots, x_p$ are mutually orthogonal idempotents of A. Furthermore, $1 = x_1 + \cdots + x_p$, $x_\pi y_\mu \in ky_\mu$, and $x_\nu y_\pi \in kx_\nu$ for $1 \le \pi \le p$, $1 \le \nu \le \mu \le n$.

As a generalization of Thm. (17.38) Heintz and Morgenstern [243] prove the following.

(17.46) Theorem (Heintz and Morgenstern). *Let k be an algebraically closed field and A be a basic k-algebra. Then the following three conditions are equivalent:*

(1) *A is of minimal rank,*
(2) *$A \in \mathcal{M}_k$,*
(3) *There exist $\omega_1, \ldots, \omega_m \in \operatorname{Rad}(A)$ such that*

$$\operatorname{Rad}(A) = L_A + A\omega_1 A + \cdots + A\omega_m A = R_A + A\omega_1 A + \cdots + A\omega_m A,$$

and $\omega_i \omega_j = 0$ for $1 \le i \ne j \le m$.

If A is a Lie algebra over a field k and $\phi: A \times A \to A$ is the corresponding bilinear map (called multiplication in the sequel), we define the *multiplicative complexity*, resp. *rank, of A* as the multiplicative complexity, resp. rank, of ϕ. (Do not confuse this notion of rank with the one used in the theory of Lie algebras as the dimension of a Cartan subalgebra of A.) The interest in the rank of Lie algebras is partly due to a theorem of Strassen [498, Korollar 4] according to which the complexity of multiplication in a connected linear algebraic group is estimated from below, up to a constant factor, by the rank of the corresponding Lie algebra. The following results are due to de Groote and Heintz [221]. (See also de Groote et al. [222].)

(17.47) Theorem (de Groote and Heintz). *For the series of classical simple Lie algebras over the field $\mathbb{C}$ of complex numbers one has:*

(1) $L(\mathfrak{sl}(l+1, \mathbb{C})) \geq 2 \dim_{\mathbb{C}} \mathfrak{sl}(l+1, \mathbb{C}) - 2l$,

(2) $L(\mathfrak{o}(2l+1, \mathbb{C})) \geq 2 \dim_{\mathbb{C}} \mathfrak{o}(2l+1, \mathbb{C}) - 4l + 2$,

(3) $L(\mathfrak{sp}(2l, \mathbb{C})) \geq 2 \dim_{\mathbb{C}} \mathfrak{sp}(2l, \mathbb{C}) - 4l + 2$,

(4) $L(\mathfrak{o}(2l, \mathbb{C})) \geq 2 \dim_{\mathbb{C}} \mathfrak{o}(2l, \mathbb{C})) - 4l + 4$.

(5) *Let* $\mathfrak{g} \in \{\mathfrak{sl}(l+1, \mathbb{C}), \mathfrak{o}(2l+1, \mathbb{C}), \mathfrak{o}(2l, \mathbb{C})\}$ *and* $\mathfrak{h}$ *be a Cartan subalgebra of* $\mathfrak{g}$. *Then* $R(\mathfrak{g}) \geq 2 \dim_{\mathbb{C}} \mathfrak{g} - \dim_{\mathbb{C}} \mathfrak{h} = 2 \dim_{\mathbb{C}} \mathfrak{g} - l$ *with equality holding if and only if* $\mathfrak{g} = \mathfrak{sl}(2, \mathbb{C})$.

In each of the cases of the above theorem, l is the dimension of a Cartan subalgebra of the corresponding Lie algebra.

Chapter 18. Rank over Finite Fields and Codes

Although the bilinear complexity of a bilinear map ϕ over a finite field may not be the minimum number of multiplications and divisions necessary for computing ϕ, the study of such maps gives some insight into the problem of computing a bilinear map over the ring of integers of a global field, such as the ring $\mathbb{Z}$ of integers: any bilinear computation defined over $\mathbb{Z}$ (that is, whose coefficients belong to $\mathbb{Z}$) gives via reduction of constants modulo a prime p a bilinear computation over the finite field $\mathbb{F}_p$. In this chapter we introduce a relationship observed by Brockett and Dobkin [80] between the rank of bilinear maps over a finite field and the theory of linear error-correcting codes. More precisely, we associate to any bilinear computation of length r of a bilinear map over a finite field a linear code of block length r; the dimension and minimum distance of this code depend only on the bilinear map and not on the specific computation. The question about lower bounds for r can then be stated as the question about the minimum block length of a linear code of given dimension and minimum distance. This question has been extensively studied by coding theorists; we use their results to obtain linear lower bounds for different problems, such as polynomial and matrix multiplication. In particular, following Bshouty [85, 86] we show that the rank of $n \times n$-matrix multiplication over $\mathbb{F}_2$ is $\frac{5}{2}n^2 - o(n^2)$. In the last section of this chapter we discuss an interpolation algorithm on algebraic curves due to Chudnovsky and Chudnovsky [110]. Combined with a result on algebraic curves with many rational points over finite fields, this algorithm yields a linear upper bound for $R(\mathbb{F}_{q^n}/\mathbb{F}_q)$ for fixed q.

18.1 Linear Block Codes

In this section we briefly review the fundamental concepts of the theory of linear error-correcting codes. For an introduction to this theory we recommend the classic book of MacWilliams and Sloane [345], or van Lint [337].

In the sequel, $\mathbb{F}_q$ denotes the finite field with q elements for a prime power q. We will consider $\mathbb{F}_q^n$ as an n-dimensional vector space over the field $\mathbb{F}_q$.

A k-dimensional subspace C of $\mathbb{F}_q^n$ is called a *linear block code* of *dimension* k and *block length* n, an $[n, k]_q$-*code* for short. Elements of C are called *codewords*. $\mathbb{F}_q^n$ is equipped with the *Hamming metric* $d_H(x, y) := |\{i \mid x_i \neq y_i\}|$; $\mathrm{wgt}(x) := d_H(x, 0)$ is called *the Hamming weight of x*; if C is nonzero, then

$d(C) := \min\{\text{wgt}(c) \mid 0 \neq c \in C\}$ is called *the minimum distance of C*. An $[n, k]_q$-code with minimum distance d is called an $[n, k, d]_q$-*code*. Unless stated otherwise, we assume in this chapter that all the codes appearing are nonzero.

There are several inequalities relating the parameters n, k, d, and q of a linear code. Let C be an $[n, k, d]_q$-code. If $d \geq 2e + 1$ for some natural number e, the Hamming spheres of radius e around the codewords $c \in C$, i.e., the sets $\{u \in \mathbb{F}_q^n \mid d_H(u, c) \leq e\}$ are disjoint. This readily yields *the sphere packing bound*:

$$\sum_{i=0}^{e} \binom{n}{i}(q - 1)^i \leq q^{n-k}.$$

A lower bound for n in terms of k, d, and q is provided by the following *Griesmer bound* (cf. Ex. 18.2):

$$n \geq \sum_{i=0}^{k-1} \left\lceil \frac{d}{q^i} \right\rceil.$$

One of the most interesting open problems in coding theory is that of determining all n, k, d, and q for which there exists an $[n, k, d]_q$-code. The problem is solved, once the quantity

$$N_q[k, d] := \min\{n \mid \exists \, [n, k, d]_q\text{-code}\}$$

is known for all k, d, and q. (One can blow up the block length of any linear code without changing the other parameters, by appending zeros to the codewords.) The Griesmer bound for example implies that $N_q[k, d] \geq \sum_{i=0}^{k-1} \lceil \frac{d}{q^i} \rceil$. The determination of all values of the function N_q seems to be outside the range of the methods available at the present time. However, it is possible to make assertions about $N_q[k, d]$ for large k and d. To be more specific, we define for any prime power q the function $\alpha_q \colon [0, 1] \to [0, 1]$ by

$$\alpha_q(\delta) := \sup\{\rho \mid \exists \text{ sequence } (C_i)_{i \geq 1} \text{ of different } [n_i, k_i, d_i]_q\text{-codes}$$
$$\text{with } \lim_{i \to \infty}(d_i/n_i, k_i/n_i) = (\delta, \rho)\}.$$

The function α_q is monotonically decreasing and satisfies $\alpha_q(x) = 0$ for $x \geq \frac{q-1}{q}$. (See Ex. 18.8.) The subsequent lemma shows that upper estimates for the function α_q yield lower bounds for N_q. Before stating this lemma we introduce the following useful convention.

(18.1) Convention. Let $(a_n)_{n \geq 1}$ be a sequence of real numbers and f, g be functions from $\mathbb{N}$ to $\mathbb{R}$. We write $a_n \geq g(n) - o(f(n))$ if there exists a sequence $(\varepsilon_n)_{n \geq 1}$ such that $(\varepsilon_n/f(n)) \to 0$ for $n \to \infty$, and $a_n \geq g(n) - \varepsilon_n$ for all n.   •

(18.2) Lemma. *Let R_q be a continuous, monotonically decreasing function in the interval $[0, \frac{q-1}{q}]$ such that $R_q(x) \geq \alpha_q(x)$ for every $x \in [0, \frac{q-1}{q}]$ and $R_q(\frac{q-1}{q}) = 0$. Let ξ be the unique solution of $R_q(\xi) = \xi$ in the interval $(0, \frac{q-1}{q})$. Then $N_q[d, d] \geq d/\xi - o(d)$.*

Proof. For each d let C_d be an $[N_q[d, d], d, d]_q$-code and denote the upper limit $\limsup_{d\to\infty} d/N_q[d, d]$ by δ. This implies that $d/N_q[d, d] \leq \delta - o(1)$, hence $N_q[d, d] \geq d/\delta - o(d)$. Now we obtain the assertion: the assumptions imply $\delta \leq \alpha_q(\delta) \leq R_q(\delta)$, hence $\delta \leq \xi$ since R_q is assumed to be monotonically decreasing.   □

The exact behavior of the function α_q is unknown. (Compare however Ex. 18.8.) But there exist upper estimates for α_q which satisfy the assumptions of Lemma (18.2). Recall that the binary entropy function $H(x)$ is defined by

$$H(x) := -x \log x - (1 - x) \log(1 - x),$$

on $[0, 1]$.

(18.3) Theorem. (1) (McEliece et al.) *We have* $H(\frac{1}{2} - \sqrt{x(1 - x)})) \geq \alpha_2(x)$ *for all* $x \in [0, \frac{1}{2}]$.
(2) (Plotkin) *For every prime power q we have* $1 - \frac{q}{q-1}x \geq \alpha_q(x)$ *for* $x \in [0, \frac{q-1}{q}]$.

18.2 Linear Codes and Rank

The connection of rank and coding theory is given by the theorem below. One can attach to every bilinear computation Γ of a bilinear map ϕ a linear code whose block length is the length of Γ, but its dimension and minimum distance only depend on ϕ. It is possible in this way to connect the problem of determining the rank of a bilinear problem with the problem of determining the minimal length of a code with given dimension and minimum distance, i.e., the value $N_q[k, d]$.

(18.4) Theorem (Brockett and Dobkin). *Let U_1, U_2, and U_3 be $\mathbb{F}_q$-spaces and $\phi \in \mathrm{Bil}(U_1, U_2; U_3)$. Further, let*

$$d_1 := \min\{\dim \phi(u_1, U_2) \mid 0 \neq u_1 \in U_1\},$$
$$d_2 := \min\{\dim \phi(U_1, u_2) \mid 0 \neq u_2 \in U_2\},$$
$$d_3 := \min\{R(\lambda \circ \phi) \mid 0 \neq \lambda \in U_3^*\}.$$

Then $R(\phi) \geq N_q[\dim U_i, d_i]$, provided ϕ is i-concise, for any $i \in \underline{3}$.

Proof. Let ϕ be represented as $\phi = \sum_{i=1}^r f_i \otimes g_i \otimes w_i \in U_1^* \otimes U_2^* \otimes U_3$. We define the following vector space morphisms:

$$\gamma_1 \in \mathrm{Hom}(U_1, \mathbb{F}_q^r), \quad \gamma_1(a) := (f_1(a), \ldots, f_r(a)),$$
$$\gamma_2 \in \mathrm{Hom}(U_2, \mathbb{F}_q^r), \quad \gamma_1(b) := (g_1(b), \ldots, g_r(b)),$$
$$\gamma_3 \in \mathrm{Hom}(U_3^*, \mathbb{F}_q^r), \quad \gamma_3(\lambda) := (\lambda(w_1), \ldots, \lambda(w_r)).$$

For $i \in \underline{3}$, the image of γ_i is a linear code C_i over $\mathbb{F}_q$ and the i-conciseness of ϕ implies that γ_i is injective. Hence it remains to compute the minimum distance of these codes. We only show that the minimum distance of C_1 is at least d_1, the other cases being similar. Since $\phi(u_1, U_2) \subseteq \sum_{i, f_i(u_1) \neq 0} \mathbb{F}_q w_i$, we obtain $d_1 \leq \dim \phi(u_1, U_2) \leq \mathrm{wgt}(\gamma_1(u_1))$ for all $0 \neq u_1 \in U_1$, hence the minimum distance of C_1 is at least d_1. But this implies that there exists an $[r, \dim U_1, d_1]_q$-code, see Ex. 18.4. $\qquad\square$

(18.5) Remark. Let $\phi \in \mathrm{Bil}(U, V; W)$ and suppose that for all $0 \neq u \in U$ we have $\phi(u, V) = W$. Then $R(\lambda \circ \phi) = \dim U$ for all $0 \neq \lambda \in W^*$. $\bullet$

18.3 Polynomial Multiplication over Finite Fields

In this section we investigate the rank of the polynomial multiplication map $\Phi_k^{n,n} : k[X]_{<n} \times k[X]_{<n} \to k[X]_{<2n-1}$ from Sect. 14.5. Recall that $R(\Phi_k^{n,n}) \geq 2n-1$ with equality holding if k is an infinite field. Hence it remains to study the case where $k = \mathbb{F}_q$ is a finite field. We write Φ_q^n for $\Phi_q^{n,n}$.

(18.6) Theorem (Kaminski). $R(\Phi_q^n) \geq \max\{N_q[\ell, 2n - \ell] \mid n \leq \ell < 2n\}$.

Proof. Let $p \in \mathbb{F}_q[x]$ be irreducible of degree ℓ where $n \leq \ell < 2n$. Consider the following commutative diagram

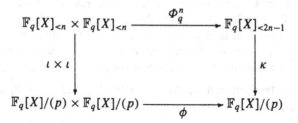

where ϕ is the multiplication in the field $\mathbb{F}_q[x]/(p)$, ι is the residue class map on $\mathbb{F}_q[x]$ restricted to $\mathbb{F}_q[x]_{<n}$, and κ is the residue class mapping restricted to $\mathbb{F}_q[x]_{<2n-1}$; ι is injective and κ is surjective since $n \leq \ell < 2n$. Let $\lambda \in (\mathbb{F}_q[x]/(p))^*$. Then $R(\lambda \circ \kappa \circ \Phi_q^n) \geq R(\lambda \circ \phi) - 2(\ell - n)$. (This follows from the following fact: if U, V, W, and X are k-spaces and $\varphi : U \to V$, $\sigma : V \to W$, and $\tau : W \to X$ are k-space morphisms, then $\mathrm{rk}(\tau \circ \sigma \circ \varphi) \geq \mathrm{rk}(\varphi) - \mathrm{cork}(\sigma) - \mathrm{cork}(\tau)$.) By Rem. (18.5) we have $R(\lambda \circ \phi) = \ell$ since $\mathbb{F}_q[x]/(p)$ is a field. Hence we obtain $R(\lambda \circ \kappa \circ \Phi_q^n) \geq 2n - \ell$. The bilinear map $\kappa \circ \Phi_q^n$ is 3-concise since κ is surjective, thus Thm. (18.4) implies $R(\kappa \circ \Phi_q^n) \geq N_q[\ell, 2n - \ell]$. The rest follows from the trivial observation $R(\Phi_q^n) \geq R(\kappa \circ \Phi_q^n)$. $\qquad\square$

(18.7) Example. Thm. (18.6) and the sphere packing bound imply $R(\Phi_2^4) \geq N_2[5, 3] \geq 9$. On the other hand, straightforward generalization of an optimal bilinear computation for Φ_2^2 implies that $R(\Phi_2^4) \leq 9$. Hence, $R(\Phi_2^4) = 9$. $\bullet$

(18.8) Corollary (Brown and Dobkin). $R(\Phi_2^n) \geq 3.527n - o(n)$.

Proof. Applying Thm. (18.6) with $\ell = n$ we obtain $R(\Phi_2^n) \geq N_2[n, n]$. We use the criterion of Lemma (18.2) with $R_2(x) := \mathrm{H}(\frac{1}{2} - \sqrt{x(1-x)})$ from Thm. (18.3)(1). Solving $R_2(\xi) = \xi$ numerically for $\xi \in (0, \frac{1}{2})$ we obtain $\xi \approx 0.283477$, hence by Lemma (18.2) we have $N_2[n, n] \geq n/\xi - o(n) \geq 3.527n - o(n)$. $\square$

For other values of q we may apply the bound of Thm. (18.3)(2).

(18.9) Corollary (Lempel et al.). *We have* $R(\Phi_q^n) > (2 + \frac{1}{q-1})n - o(n)$.

Proof. Again we apply Thm. (18.6) with $\ell = n$ to obtain $R(\Phi_q^n) \geq N_q[n, n]$. We use the criterion of Lemma (18.2) with $R_q(x) = 1 - \frac{q}{q-1}x$ from Thm. (18.3)(2) and get $R_q(\xi) = \xi$ for $\xi = (q-1)/(2q-1)$, hence the assertion. $\square$

Using another type of argument we can improve upon the last corollary. The rest of this section is devoted to the proof of the following theorem.

(18.10) Theorem (Baur, Bshouty and Kaminski). $R(\Phi_q^n) \geq 3n - o(n)$.

In the sequel, we will denote $\mathbb{F}_q[X]_{<n}$ by P_n. We need some preliminaries.

(18.11) Lemma. *Let* $u_1, \dots, u_s \in P_n$, *and* $\deg u_1 = n - 1 \geq 1$. *Further let* $\deg \mathrm{GCD}(u_1, \dots, u_s) = m$. *Then* $\dim \sum_{i=1}^{s} P_n u_i \geq 2n - m - 1$.

Proof. Let $g := \mathrm{GCD}(u_1, \dots, u_s)$, $(g) = g \cdot \mathbb{F}_q[X]$ be the principal ideal of g in $\mathbb{F}_q[X]$, and $h \in (g) \cap P_{2n-1}$. By the extended Euclidean algorithm there exist $h_1, \dots, h_s \in \mathbb{F}_q[X]$ such that $h = \sum_{i=1}^{s} h_i u_i$. Further, for every $i = 2, \dots, s$ there exist $q_i, r_i \in \mathbb{F}_q[X]$, $\deg r_i < n - 1$, such that $h_i = q_i u_1 + r_i$. We then have

$$h = \sum_{i=2}^{s} r_i u_i + \left(h_1 + \sum_{i=2}^{s} q_i u_i \right) u_1.$$

Since both $\sum_{i=2}^{s} r_i u_i$ and h belong to P_{2n-1}, and $\deg u_1 = n - 1$, we have that $(h_1 + \sum_{i=2}^{s} q_i u_i) \in P_n$ and thus $h \in \sum_{i=1}^{s} P_n u_i$. This implies

$$(g) \cap P_{2n-1} \subseteq \sum_{i=1}^{s} P_n u_i.$$

But since $\deg g = m$, the $(2n - m - 1)$-dimensional $\mathbb{F}_q$-space $\sum_{i=0}^{2n-2-m} \mathbb{F}_q X^i g$ is contained in the left-hand side of the above inclusion and we obtain the assertion.

$\square$

(18.12) Lemma. *For any basis* $u_1, \dots, u_n$ *of* P_n *there exists a set S contained in* $\{1, \dots, n\}$ *with the following properties.*

- $|S| \leq \frac{2n-2}{\log_q \log_q n} + 1$,
- $\deg \mathrm{GCD}(u_i \mid i \in S) \leq \log_q n$.
- *There exists* $i \in S$ *such that* $\deg u_i = n - 1$.

Proof. We inductively define pairwise different indices $i_1, \ldots, i_j, \ldots, i_s \in \underline{n}$, and set $S := \{i_j \mid j \in \underline{s}\}$. We may assume w.l.o.g. that $\deg u_1 = n - 1$ and set $i_1 := 1$.

Suppose that we have already constructed $i_1, \ldots, i_j$. If the degree of $h_j := \mathrm{GCD}(u_{i_1}, \ldots, u_{i_j})$ is $\leq \log_q(n)$, set $s := j$. Otherwise there exists an irreducible polynomial $p \in \mathbb{F}_q[X]$ such that p divides h_j to the power e and $\deg p^e \geq \frac{1}{2} \log_q \deg h_j > \frac{1}{2} \log_q \log_q n =: d$. To see this, note that if $f \in \mathbb{F}_q[X]$, $f = p_1^{v_1} \cdots p_r^{v_r}$ is the prime factor decomposition of f, and $\deg p_i^{v_i} \leq \delta$ for some δ and all i, then

$$\deg f \leq r\delta \leq \delta|P_{\delta-1}| = \delta q^\delta \leq q^{2\delta}.$$

Hence $\delta \geq \frac{1}{2} \log_q \deg f$. Choose i_{j+1} such that $u_{i_{j+1}}$ is not divisible by p. Since $u_1, \ldots, u_n$ is a basis, this is always possible.

We have by our construction $\deg h_s \leq \log_q n$ and

$$0 \leq \deg h_j \leq \deg h_{j-1} - d \leq \ldots \leq \deg h_1 - (j-1)d = (n-1) - (j-1)d$$

for all $j \in \underline{s}$. This implies $s \leq (n-1)/d + 1$. $\square$

Proof of Thm. (18.10). Let $f_i, g_i \in P_n^*$ and $w_i \in P_n$ be such that $\Phi_q^n = \sum_{i=1}^r f_i \otimes g_i \otimes w_i$. By the conciseness of Φ_q^n we may w.l.o.g. assume that $g_1, \ldots, g_n$ form a basis of P_n^*. Let $u_1, \ldots, u_n$ be the corresponding dual basis of P_n. There exists by Lemma (18.12) a subset S of $\{1, \ldots, n\}$ with the properties stated above. W.l.o.g. we may assume that $S = \{1, \ldots, s\}$. We then have

$$\sum_{i=1}^s P_n u_i \subseteq \sum_{i=1}^s \mathbb{F}_q w_i + \sum_{i=n+1}^r \mathbb{F}_q w_i.$$

The dimension of the left-hand side is $\geq 2n - \log_q n - 1$ by Lemma (18.11). We thus obtain $2n - \log_q n - 1 \leq s + r - n$, i.e.,

$$r \geq 3n - \log_q n - \frac{2n-2}{\log_q \log_q n} - 2 = 3n - o(n).$$ $\square$

18.4* Matrix Multiplication over Finite Fields

In this section we shall derive a lower bound for the rank of $\mathbb{F}_2^{m \times m}$ using the Griesmer bound. Let us first examine what happens if we apply the coding theory bound (18.4) directly. If $0 \neq A \in \mathbb{F}_2^{m \times m}$, then $A \cdot \mathbb{F}_2^{m \times m}$ is a non-vanishing right ideal in $\mathbb{F}_2^{m \times m}$. The dimension of a minimal right ideal of $\mathbb{F}_2^{m \times m}$ equals m. Hence $R(\mathbb{F}_2^{m \times m}) \geq N_2[m^2, m]$. Estimating $N_2[m^2, m]$ by the Griesmer-bound, we see that this method does not even yield the lower bound of $2m^2 - 1$ of Thm. (17.14). The next proposition is the key observation for obtaining better lower bounds.

(18.13) Proposition. *Let A be an associative $\mathbb{F}_q$-algebra and U be an $\mathbb{F}_q$-subspace of A such that $U \setminus 0 \subseteq A^\times$. Let ϕ be the restriction of multiplication in A to $U \times A$. Then $R(A) \geq R(\phi) \geq N_q[\dim U, \dim A]$.*

Proof. $\phi \in \text{Bil}(U, A; A)$ is 1-concise. Since the nonzero elements of U are invertible, left multiplication by such an element is an isomorphism. Hence Thm. (18.4) yields the assertion. ☐

(18.14) Corollary. $R(\mathbb{F}_q^{n \times n}) \geq N_q[n, n^2]$.

Proof. Consider the $\mathbb{F}_q$-algebra morphism (regular representation)

$$\mathbb{F}_{q^n} \to \text{End}_{\mathbb{F}_q}(\mathbb{F}_{q^n}) \simeq \mathbb{F}_q^{n \times n}, \quad a \mapsto (b \mapsto ab).$$

Since $\mathbb{F}_{q^n}$ is a field, the image of $\mathbb{F}_{q^n}^{\times}$ is in $\text{GL}(n, \mathbb{F}_q)$. Now apply Prop. (18.13). ☐

It is easily seen that any $\mathbb{F}_q$-subspace U of $\mathbb{F}_q^{n \times n}$ such that $U \setminus 0 \subseteq \text{GL}(n, \mathbb{F}_q)$ has dimension $\leq n$. This follows from $U \cap I = 0$ for any proper right ideal I of $\mathbb{F}_q^{n \times n}$ and the fact that maximal right ideals have dimension $n^2 - n$.

For the rest of this section we assume that $q = 2$, since for other values of q the method described does not yield interesting lower bounds.

(18.15) Corollary. $\mathbb{F}_2^{n \times n}$ *is not of minimal rank for* $n \geq 5$, *which means* $R(\mathbb{F}_2^{n \times n}) > 2n^2 - 1$ *for* $n \geq 5$.

Proof. We have $R(\mathbb{F}_2^{n \times n}) \geq N_2[n, n^2] \geq \sum_{i=0}^{n-1} \lceil n^2 2^{-i} \rceil$ by Cor. (18.14) and the Griesmer bound. The right-hand side of this inequality is easily shown to be greater than $2n^2 - 1$ for $n \geq 5$. ☐

(18.16) Remark. It is known that $R(\langle 3, 3, n \rangle / \mathbb{F}_2) \geq \lceil 91n/16 \rceil$. (Compare the Notes.) Putting $n = 3$ yields $R(\mathbb{F}_2^{3 \times 3}) \geq 18$. However, it is not known whether $\mathbb{F}_2^{4 \times 4}$ is of minimal rank. ●

Combining Prop. (18.13) with the techniques of Sect. 17.1, one can prove good asymptotic lower bounds which we are going to discuss in the following. Let s and t be positive integers. First we want to study the algebra $\mathbb{F}_2^{st \times st}$. We identify $\mathbb{F}_2^{st \times st}$ with $(\mathbb{F}_2^{s \times s})^{t \times t}$. Let V be the set of all matrices $(V_{ij})_{1 \leq i, j \leq t} \in (\mathbb{F}_2^{s \times s})^{t \times t}$ such that $V_{ij} = 0$ for $i \leq j$. Further let U be the set of all matrices $(U_{ij})_{1 \leq i, j \leq t}$ such that $U_{ij} = 0$ for $i \neq j$, and $U_{11} = \cdots = U_{tt} \in \mathbb{F}_{2^s}$, where $\mathbb{F}_{2^s}$ is regarded as a subspace of $\mathbb{F}_2^{s \times s}$ via the regular representation (see the proof of Cor. (18.14)). We have the following pictorial description of U and V.

We first observe some obvious properties of U and V. (Proofs are left to the reader.)

(18.17) Remark. Notation being as above we have:

(1) $U \cap V = 0$.
(2) $\dim V = s^2 \binom{t}{2}$, $\dim U = s$.

(3) Any element of $(U \oplus V) \setminus V$ is invertible.

(4) If π is any projection of $U \oplus V$ along V, then $\pi(a)$ is invertible for any nonzero element a of U.   ●

Now consider the bilinear mapping ψ which is defined as the restriction of the multiplication in $\mathbb{F}_2^{st \times st}$ to $(U \oplus V) \times \mathbb{F}_2^{st \times st}$. Apparently we have $r := R(\psi) \leq R(\mathbb{F}_2^{st \times st})$. Let $\psi = \sum_{i=1}^{r} f_i \otimes g_i \otimes w_i \in (U \oplus V)^* \otimes (\mathbb{F}_2^{st \times st})^* \otimes \mathbb{F}_2^{st \times st}$. Since ψ is 1-concise, we may w.l.o.g. assume that $f_1, \ldots, f_\ell$ form upon restriction a basis of V^*, where $\ell := \dim V = s^2 \binom{t}{2}$ by Lemma (18.17)(2). Let π be the projection of $U \oplus V$ onto $\cap_{i=1}^{\ell} \ker f_i$ along V and $\phi := \psi \circ (\pi \times \mathrm{id})$. ($\pi$ exists by Rem. (17.3).) Since $f_1 \circ \pi = \cdots = f_\ell \circ \pi = 0$, we have $R(\phi) \leq R(\psi) - \ell \leq R(\mathbb{F}_2^{st \times st}) - \ell$. Further, ϕ is the restriction of the multiplication in $\mathbb{F}_2^{st \times st}$ to $\pi(U) \times \mathbb{F}_2^{st \times st}$. By Lemma (18.17)(2),(4) and Prop. (18.13) we get $R(\phi) \geq N_2[s, s^2 t^2]$. Applying the Griesmer bound we obtain

$$
\begin{aligned}
R(\mathbb{F}_2^{st \times st}) \;&\geq\; N_2[s, s^2 t^2] + s^2 \binom{t}{2} \\
\text{(A)} \qquad &\geq\; \sum_{i=0}^{s-1} \left\lceil \frac{s^2 t^2}{2^i} \right\rceil + s^2 \binom{t}{2} \\
&\geq\; \frac{5}{2} s^2 t^2 \left(1 - \frac{4}{5 \cdot 2^s} - \frac{1}{5t} \right).
\end{aligned}
$$

(18.18) Theorem (Bshouty). *We have $R(\mathbb{F}_2^{n \times n}) \geq \frac{5}{2} n^2 - o(n^2)$.*

Proof. For every n let $s = s_n$ be such that $s_n = o(n)$ and $s_n \to \infty$ for $n \to \infty$. Further let t and r be defined by $n = ts + r$ where $0 \leq r \leq s - 1$. The obvious injection $\mathbb{F}_2^{st \times st} \to \mathbb{F}_2^{n \times n}$ of $\mathbb{F}_2$-spaces shows that $\mathbb{F}_2^{st \times st}$ is a restriction of $\mathbb{F}_2^{n \times n}$ and hence $R(\mathbb{F}_2^{st \times st}) \leq R(\mathbb{F}_2^{n \times n})$. Since $s^2 t^2 = n^2 - 2nr + r^2 = n^2 - o(n^2)$, (A) yields the result.   □

(18.19) Remark. The above technique can also be used to obtain lower bounds for multiplication of matrices of arbitrary (not necessarily square) size. Compare the Notes and Ex. 18.11.   ●

18.5* Rank of Finite Fields

Throughout this section we assume familiarity with the theory of algebraic function fields of one variable. A good reference for this theory is the monograph of Stichtenoth [488].

Let $k \subseteq K$ be an extension of fields and $n := [K : k]$. We already know that $R(K/k) = 2n - 1$ iff $|k| \geq 2n - 2$, see Rem. (17.30). In this section we shall study $R(\mathbb{F}_{q^n}/\mathbb{F}_q)$ for fixed q and n tending to ∞. More precisely we shall prove the following.

(18.20) Theorem (Chudnovsky and Chudnovsky). *For every prime power q there exists a constant c_q such that $R(\mathbb{F}_{q^n}/\mathbb{F}_q) \le c_q n$.*

The method we shall employ is an interpolation technique on algebraic curves defined over $\mathbb{F}_q$. Let us fix some notation. We denote an algebraic function field K of one variable (AF1 for short) with constant field $\mathbb{F}_q$ by $K/\mathbb{F}_q$. The set of prime divisors (also called *places*) of degree m of $K/\mathbb{F}_q$ is denoted by $\mathbf{P}_m(K/\mathbb{F}_q)$ and the (additively written) group of divisors of $K/\mathbb{F}_q$ by $\mathbf{D}(K/\mathbb{F}_q)$. The linear space of a divisor A is denoted by $\mathcal{L}(A)$. For $Q \in \mathbf{P}_m(K/\mathbb{F}_q)$ we denote by ev_Q the evaluation map at Q, i.e., if $\mathcal{O}_Q$ is the local ring of Q and $\mathcal{M}_Q$ its maximal ideal, then ev_Q is the canonical map $\mathcal{O}_Q \to \mathcal{O}_Q/\mathcal{M}_Q \simeq \mathbb{F}_{q^m}$. If Q is of degree one and $f \in \mathcal{O}_Q$, we also write $f(Q)$ for $\mathrm{ev}_Q(f)$.

We begin with an auxiliary result which guarantees the existence of prime divisors of degree n of $K/\mathbb{F}_q$ under some mild conditions.

(18.21) Lemma. *Let $K/\mathbb{F}_q$ be an AF1 of genus g and n be an integer satisfying $n \ge 2\log_q g + 6$. Then there exists a prime divisor of degree n of $K/\mathbb{F}_q$.*

The proof of this lemma uses the *Hasse-Weil theorem* and involves some easy manipulations of the ζ-function of $K/\mathbb{F}_q$ (cf. Ex. 18.12). A central step in the proof of Thm. (18.20) is the following interpolation algorithm.

(18.22) Proposition (Chudnovsky and Chudnovsky). *Let $K/\mathbb{F}_q$ be an AF1 of genus g, $n \ge 2\log_q g + 6$, and assume that there exist at least $4g + 2n$ prime divisors of degree one of $K/\mathbb{F}_q$. Then we have $R(\mathbb{F}_{q^n}/\mathbb{F}_q) \le 3g + 2n - 1$.*

Proof. Let $m := n + 2g - 1$ and P be a prime divisor of degree one of $K/\mathbb{F}_q$. By Lemma (18.21) there exists $Q \in \mathbf{P}_n(K/\mathbb{F}_q)$. The linear space $\mathcal{L}(mP)$ of mP is contained in the valuation ring of Q; hence ev_Q is defined on $\mathcal{L}(mP)$ and gives a homomorphism $\kappa \colon \mathcal{L}(mP) \to \mathbb{F}_{q^n}$ of $\mathbb{F}_q$-spaces. Its kernel is $\mathcal{L}(mP - Q)$ and thus has dimension $m - n - g + 1$ by the Riemann-Roch theorem. Since $\mathcal{L}(mP)$ has dimension $m - g + 1$ by Riemann-Roch, we deduce that κ is surjective. Now let $\ell := 4g + 2n - 1$ and $P_1, \ldots, P_\ell \in \mathbf{P}_1(K/\mathbb{F}_q)$ be different from P, and consider the evaluation map $\gamma \colon \mathcal{L}(2mP) \to \mathbb{F}_q^\ell$, $f \mapsto (f(P_1), \ldots, f(P_\ell))$. Its kernel is $\mathcal{L}(2mP - \sum_{i=1}^\ell P_i) = 0$, since $2m < \ell = 2m + 1$. We may w.l.o.g. assume that $P_1, \ldots, P_r$, $r := \dim \mathcal{L}(2mP) = 2m - g + 1 = 3g + 2n - 1$, are such that $\gamma' \colon \mathcal{L}(2mP) \to \mathbb{F}_q^r$, $f \mapsto (f(P_1), \ldots, f(P_r))$ is an isomorphism of $\mathbb{F}_q$-spaces. We have thus the following commutative diagram:

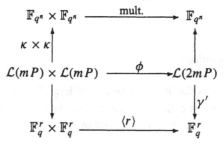

where ϕ is the restriction of the multiplication map of $K/\mathbb{F}_q$. κ is surjective, hence $\mathbb{F}_{q^n} \le \phi$; γ' is injective (even an isomorphism), hence $\phi \le \langle r \rangle$. (Cf. Lemma (14.10).) We thus obtain $\mathbb{F}_{q^n} \le \phi \le \langle r \rangle = \langle 3g + 2n - 1 \rangle$. $\square$

(18.23) Remark. The evaluation map γ used in the above proof is also used to construct geometric Goppa codes. See Ex. 18.13.   •

To derive Thm. (18.20) from the last proposition it is obvious that we have to look for AF1's which have many prime divisors of degree one relative to their genus. Such function fields have been studied in connection with geometric Goppa codes. The proof of the following theorem can be found in Garcia and Stichtenoth [180].

(18.24) Theorem (Garcia and Stichtenoth). *Let q be a prime power, X_1 be an indeterminate over $\mathbb{F}_{q^2}$, and $K_1 := \mathbb{F}_{q^2}(X_1)$. For $m \ge 1$ let $K_{m+1} := K_m(Z_{m+1})$, where Z_{m+1} satisfies the Artin-Schreier equation $Z_{m+1}^q + Z_{m+1} = X_m^{q+1}$ and $X_m := Z_m/X_{m-1} \in K_m$ (for $m \ge 2$). Then $K_m/\mathbb{F}_{q^2}$ has genus g_m given by*

$$
g_m = \begin{cases}
q^m + q^{m-1} - q^{\frac{m+1}{2}} - 2q^{\frac{m-1}{2}} + 1 & \text{if } m \equiv 1 \bmod 2, \\[2mm]
q^m + q^{m-1} - \frac{1}{2}q^{\frac{m}{2}+1} - \frac{3}{2}q^{\frac{m}{2}} - q^{\frac{m}{2}-1} + 1 & \text{if } m \equiv 0 \bmod 2,
\end{cases}
$$

and $|\mathbb{P}_1(K_m/\mathbb{F}_{q^2})| \ge (q^2 - 1)q^{m-1} + 2q \ge (q-1)g_m$.

Proof of Thm. (18.20). By Prop. (14.23), the rank is submultiplicative. Thus we may suppose that q is a square ≥ 49 and that n is larger than some integer $n_0(q)$ depending on q. Let g_m be defined as in Thm. (18.24), m be the minimum integer such that $n \le g_m$, and suppose that n is large enough so that the inequalities $n \ge 2\log_q g_m + 6$ and $g_m \le 2\sqrt{q}g_{m-1} < 2\sqrt{q}n$ hold. By Thm. (18.24) there exists an AF1 $K_m/\mathbb{F}_q$ of genus g_m such that $|P_1(K_m/\mathbb{F}_q)| \ge (\sqrt{q}-1)g_m \ge 6g_m \ge 4g_m + 2n$. Hence, Prop. (18.22) implies that $R(\mathbb{F}_{q^n}/\mathbb{F}_q) \le 3g_m + 2n - 1 < 2(3\sqrt{q}+1)n$ and we are done. $\square$

(18.25) Remark. The constants in the proof above are far from optimal. See the Notes for a further discussion.   •

18.6 Exercises

18.1. Prove that d_H is indeed a metric.

18.2. An $[n, k]_q$-code C is the image of an embedding $\mathbb{F}_q^k \to \mathbb{F}_q^n$. A representation matrix of this embedding is called a *generator matrix* $G \in \mathbb{F}_q^{k \times n}$ for C. The rows of G form a basis for C. Let d be the minimum distance of C.

(1) Show that (after a permutation of the coordinates) C has a generator matrix of the form

$$G = \left(\begin{array}{c|c} 0 \cdots 0 & * \cdots * \\ \hline G' & * \\ & * \end{array} \right).$$

(2) Prove that if m is a positive integer and $u, v \in \mathbb{F}_q^m$ have weight m, then there exists $\lambda \in \mathbb{F}_q \setminus 0$ such that $\lambda u + v$ has weight $\leq m\frac{q-2}{q-1}$.

(3) Use (2) to show that the code C' with the generator matrix G' as in (1) has minimum distance $\geq \lceil \frac{d}{q} \rceil$.

(4) Prove the Griesmer bound.

18.3. Let C be an $[n, k, d]_q$-code with generator matrix $G \in \mathbb{F}_q^{k \times n}$. Any matrix $H \in \mathbb{F}_q^{n \times (n-k)}$ of full rank such that $G \cdot H = 0$ is called a *parity check matrix* for C. Let H be a parity check matrix for C. Show that any $d - 1$ columns of H are linearly independent and that there exist d linearly dependent columns of H. Deduce the *Singleton bound* $d + k \leq n + 1$.

18.4. Let C be an $[n, k, d]_q$-code.

(1) There exists an $[m, k, d]_q$-code for every $m \geq n$.

(2) There exists an $[n - \delta, k, d - \delta]$-code for any $0 \leq \delta \leq d - 1$. (Hint: Delete an appropriate coordinate in every codeword.)

(3) There exists an $[n - \kappa, k - \kappa, d]_q$-code for every $0 \leq \kappa \leq k - 1$. (Hint: Fix a coordinate position i and consider the subcode consisting of codewords with vanishing ith coordinate.)

18.5. Let $r = R(\Phi_q^n)$.

(1) For any $1 \leq v \leq n$ there exists an $[r - n + v, v, n]_q$-code.

(2) We have $r \geq (2 + \frac{1}{q-1})n - (v + \frac{q}{q-1}nq^{-v})$.

(3) Show that the right-hand side of (2) is maximum if $v = \lceil \log_q an \rceil$, where $a = q/((q-1)\log_q e)$ and e is the basis of the natural logarithm.

18.6. Let $\phi \in \mathrm{Bil}(\mathbb{F}_q^\ell, \mathbb{F}_q^m; \mathbb{F}_q^n)$. Then

$$R(\phi) \geq \left\lceil \frac{1}{q^{\ell-1}(q-1)} \sum_{u \in \mathbb{F}_q^\ell} \dim \phi(u, \mathbb{F}_q^m) \right\rceil.$$

Interpret the result in terms of matrices. (Hint: use Ex. 18.7 and Thm. (18.4).)

18.7. (Plotkin bound) Prove that $d \leq nq^{k-1}(q - 1)/(q^k - 1)$ for an $[n, k, d]_q$-code C. (Hint: consider the $(q^k - 1) \times n$-matrix consisting of all the nonzero codewords of C.)

18.8.* For any nonzero $[n, k, d]_q$-code C we define $P(C) := (d/n, k/n) \in [0, 1]^2$. Let Σ_q be the set of all $P(C)$ where C runs over the set of all nonzero linear codes over $\mathbb{F}_q$, and S_q be the set of limit points of Σ_q. Let $(\delta_0, R_0) \in S_q$, $\delta_0 R_0 \neq 0$.

(1) $S_q = \{(\delta, R) \in [0, 1]^2 \mid R \leq \alpha_q(\delta)\}$. (For the definition of α_q see p. 490.)
(2) Show that $\{(\delta, \frac{R_0}{1-\delta_0}(1 - \delta)) \mid 0 \leq \delta \leq \delta_0\} \subset S_q$. (Hint: use Ex. 18.4(2).)
(3) Show that $\{(\frac{\delta_0}{1-R_0}(1 - R), R) \mid 0 \leq R \leq R_0\} \subset S_q$.
(4) Show that $\alpha_q(\delta) = 0$ for $\delta \geq \frac{q-1}{q}$. (Hint: use Ex. 18.7 and (2).)
(5) Show that α_q is monotonically decreasing on $[0, \frac{q-1}{q}]$.
(6) Show that α_q is continuous.

For solving (4)–(6) the following picture can be helpful:

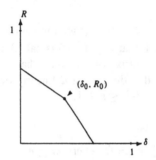

18.9. (The Gilbert-Varshamov bound) For a prime power q, $n \in \mathbb{N}'$, and $r \in \mathbb{N}$ define $V_q(n, r) := \sum_{i=0}^{r} \binom{n}{i}(q - 1)^i$ and $\theta := q^{-1}(q - 1)$.

(1) For $0 \leq \delta \leq \theta$ show that

$$\binom{n}{\lfloor \delta n \rfloor}(q - 1)^{\lfloor \delta n \rfloor} \leq V_q(n, \lfloor \delta n \rfloor) \leq (1 + \lfloor \delta n \rfloor)\binom{n}{\lfloor \delta n \rfloor}(q - 1)^{\lfloor \delta n \rfloor},$$

and deduce from this that $\lim_{n \to \infty} \log_q(V_q(n, \lfloor \delta n \rfloor)) = H_q(\delta)$, where $H_q(x)$ is the q-ary entropy function defined by $H_q(x) := x \log_q(q - 1) - x \log_q(x) - (1 - x) \log_q(1 - x)$ for $0 < x \leq \theta$, and $H_q(0) := 0$.
(2) Show that if $n, k, d \in \mathbb{N}'$ satisfy $V_q(n, d - 1) < q^{n-k+1}$, then there exists an $[n, k, d]_q$-code.
(3) Show that $\alpha_q(\delta) \geq 1 - H_q(\delta)$ for $0 \leq \delta \leq \theta$.

18.10.* For any prime power q define the function g_q recursively by $g_q(2) := g_q(3) - 1 := 2$, and $g_q(n) := 2q g_q(\lceil \log_q(2n) \rceil)$ for $n \geq 4$.

(1) Prove that g_q is non-decreasing and unbounded, but grows slower than

$$\overbrace{\log \circ \log \circ \cdots \circ \log}^{m \text{ times}}$$

for any fixed number m of iterations.

(2) Prove that $R(\mathbb{F}_{q^n}) \le ng_q(n)$. (Hint: let $t := \lceil \log_q(2n) \rceil$. Prove first

$$R(\mathbb{F}_{q^n}) \le R(\Phi_q^n) \le R(k[X]/(X^{(q^t-1)} - 1)).$$

Now use induction on n, the factorization of $X^{(q^t-1)} - 1$ over $\mathbb{F}_q$, and the Chinese remainder theorem.)

18.11. (Lower bound for the multiplication of rectangular matrices)

(1) Let m, n, p be positive integers. Show that for every $0 \le \mu \le m - 1$, $R(\langle m, n, p \rangle) \ge R(\langle m - \mu, n, p \rangle) + \max\{n\mu, p\mu\}$.

(2) Show that for every $1 \le \ell \le m$ we have

$$R(\langle m, n, p \rangle / \mathbb{F}_2) \ge \binom{\ell}{2} \left\lfloor \frac{m}{\ell} \right\rfloor^2 + (m - n)p + 2p\ell \left\lfloor \frac{m}{\ell} \right\rfloor - \frac{1}{2^{\ell-1}} \left\lfloor \frac{m}{\ell} \right\rfloor.$$

(Hint: use the method of Sect. 18.4 to show that $R(\langle m, m, p \rangle / \mathbb{F}_2)$ is greater than or equal to $\binom{\ell}{2} \lfloor \frac{m}{\ell} \rfloor^2 + N_2[p, p\ell \lfloor \frac{m}{\ell} \rfloor]$.)

Throughout the remaining exercises we assume familiarity with the theory of algebraic function fields of one variable.

18.12. In this exercise we give a proof of Lemma (18.21). For a positive integer n let $\mathbb{P}_n(K/\mathbb{F}_q)$ denote the set of prime divisors of degree n of $K/\mathbb{F}_q$.

(1) Let $\Omega := K\mathbb{F}_{q^n}$ be the constant extension of K by $\mathbb{F}_{q^n}$. Show that

$$|\mathbb{P}_1(\Omega/\mathbb{F}_{q^n})| = \sum_{d|n} |\mathbb{P}_d(K/\mathbb{F}_q)| d,$$

where the sum is over all positive divisors of n.

(2) Show that

$$|\mathbb{P}_n(K/\mathbb{F}_q)| \ge \frac{1}{n} \left(q^n - q^{n/2}(4g + q) \right).$$

(Hint: let g be the genus of $K/\mathbb{F}_q$. The Hasse-Weil theorem [233, 546] implies that there are complex numbers $w_1, \ldots, w_{2g}$ of absolute value $\sqrt{q}$ such that $|\mathbb{P}_1(K\mathbb{F}_{q^n}/\mathbb{F}_{q^n})| = q^n + 1 - \sum_{i=1}^{2g} w_i^n$ for any positive integer n. Use (1) and the Möbius inversion formula to deduce that

$$n|\mathbb{P}_n(K/\mathbb{F}_q)| = \sum_{d|n} \mu\left(\frac{n}{d}\right) \left(q^d - \sum_{i=1}^{2g} w_i^d \right).)$$

(3) Prove Lemma (18.21).

18.13.* (Codes beyond the Gilbert-Varshamov bound) Let q be a prime power.

(1) Let $K/\mathbb{F}_q$ be an AF1 of genus g, $P, P_1, \ldots, P_n$ be different prime divisors of degree 1 of $K/\mathbb{F}_q$, and m be an integer satisfying $m < n$. Let $\gamma: \mathcal{L}(mP) \to \mathbb{F}_q^n$,

$f \mapsto (f(P_1), \ldots, f(P_n))$. The image C of γ is called a *geometric Goppa code*. Using the Riemann-Roch theorem show that C is an $[n, k, d]_q$-code with $k \geq m - g + 1$ and $d \geq n - m$. Deduce that there exists an $[n, k, d]_q$-code, where $k + d = n - g + 1$.

(2) Let $A(q) := \limsup_{K/\mathbb{F}_q} N(K)/g(K)$, where the limit is taken over all AF1's $K/\mathbb{F}_q$, $N(K)$ is the number of prime divisors of degree one and $g(K)$ is the genus of $K/\mathbb{F}_q$. (Since $|q + 1 - N(K)| \leq 2g(K)\sqrt{q}$ by the Hasse-Weil theorem, the limit exists and is $\leq 2\sqrt{q}$.) Further let $\gamma_q := A(q)^{-1}$. Show that $\alpha_q(\delta) \geq 1 - \gamma_q - \delta$ for $0 \leq \delta \leq 1 - \gamma_q$.

(3) Using Thm. (18.24) show that $\gamma_q \leq (\sqrt{q} - 1)^{-1}$ if q is a square.

(4) Let q be a square ≥ 49. Show that the equation $H_q(\delta) - \delta = (\sqrt{q} - 1)^{-1}$ has two different solutions $0 < \delta_1 < \delta_2 < 1$, and that $\alpha_q(\delta) > 1 - H_q(\delta)$ for $\delta_1 < \delta < \delta_2$.

18.7 Open Problems

Problem 18.1. Determine $R(\mathbb{F}_{q^n})$ for all q and n. Similarly, determine the rank of $\Phi_k^{m,n}$ for all finite k and all m, n.

Problem 18.2. Determine all values of q and n for which $\mathbb{F}_q^{n \times n}$ is of minimal rank. (Of course we may assume $n > 2$.)

18.8 Notes

The connection between the rank of bilinear maps and error-correcting codes discussed in Sect. 18.2 has been first observed by Brockett and Dobkin [80] in 1978. Since then this connection has been exploited by several authors to give lower bounds for bilinear maps over finite fields. One of the most interesting applications in this direction is Bshouty's [85, 86] result on matrix multiplication presented in Sect. 18.4 for square matrices and in Ex. 18.11 for matrices of arbitrary format. The Griesmer bound presented in Sect. 18.1 is due to Griesmer [204]; the first bound in Thm. (18.3) is called the MRRW-bound (or "the bound of four"); it is due to Mc Eliece et al. [358]. The second bound is called the Plotkin-bound and is due to Plotkin [423].

Thm. (18.6) and Example (18.7) are due to Kaminski [292], and Cor. (18.8) is due to Brown and Dobkin [81]. Cor. (18.9) appears in Lempel et al. [324]. Thm. (18.10) has been found independently by Baur (unpublished) and Bshouty and Kaminski [93]. The $o(n)$ in the latter paper is smaller than the one given here. The proof we have presented is due to Baur and has been taken from [29]. In [93] the authors also prove that the rank of the multiplication of polynomials of degree n over $\mathbb{F}_q$ is $3n + 1 - \lfloor q/2 \rfloor$ if $q/2 < n \leq q + 1$. In a further paper [294] the same authors prove a lower bound of $\frac{5}{2}n - o(n)$ for the multiplicative complexity of polynomial multiplication over an arbitrary finite field. (Note that for the computation of bilinear forms over finite fields divisions may help, since equality might not hold in Prop. (14.1), see Problem 14.1.) The assertion $R(\langle 3, 3, n\rangle/\mathbb{F}_2) \geq \lceil\frac{91}{16}n\rceil$ in Sect. 18.4 (cf. Rem. (18.16)) is due to Ja'Ja and Takche [270]. For other results in this direction see Ja'Ja [266].

It has long been an open problem whether $R(\mathbb{F}_{q^n})$ is bounded from above by a linear function of n (cf. Thm. (18.20)). This question was settled by Chudnovsky and Chudnovsky [110] and most of the results in Sect. 18.5 (apart from Thm. (18.24)) are due to them. More precisely, the authors prove in [110] that for fixed q there exists a constant d_q such that $R(\mathbb{F}_{q^n}/\mathbb{F}_q) \leq d_q n$ for infinitely many n. The first rigorous proof of Thm. (18.20) has been given by Shparlinski et al. [478]; here, the authors also deduce better constants than those given here by using Shimura curves. They show for instance that one may take $c_q = 2(1 + (\sqrt{q} - 2)^{-1})$ if q is a square ≥ 9, $c_q = 6(1 + (q-2)^{-1})$ for arbitrary prime powers $q > 2$, and $c_2 = 27$. Also, if q is a prime or the square of a prime, one can use reductions of classical modular curves to derive constants independent of q; see for instance Shokranian and Shokrollahi [472] for more details. The quantity $A(q)$ defined in Ex. 18.13 plays here an important role. Drinfeld and Vlădut [149] proved that $A(q) \leq \sqrt{q} - 1$. Ihara [262] and Tsfasman et al. [521] showed that $A(q^2) = q - 1$. Their proofs require deep results from algebraic geometry; outlines of these proofs can be found in Tsfasman and Vlădut [522]. A very elegant and rather elementary proof of $A(q^2) = q - 1$ is given by Garcia and Stichtenoth's [180] theorem (18.24), which can be proved by studying ramifications in Artin-Schreier extensions. Using class field theory, Serre [469] has proved that $A(q) > 0$ for any prime power q; however, the exact value of $A(q)$ is still unknown if q is not a square. Using a modification of the interpolation algorithm in Prop. (18.22) and applying it to elliptic function fields, Shokrollahi [476] has shown that $R(\mathbb{F}_{q^n}/\mathbb{F}_q) = 2n$ if $\frac{1}{2}q + 1 < n < \frac{1}{2}(q + 1 + \varepsilon(q))$, where $\varepsilon(q)$ is a constant depending on q; for instance, if q is a square, one has $\varepsilon(q) = 2\sqrt{q}$. For other results in this direction see Shokrollahi [473, 475].

Ex. 18.5 is from Lempel et al. [324]. Its result has been generalized by Bshouty [89] in the following sense: if Φ_f denotes the bilinear map corresponding to the multiplication modulo f of two arbitrary polynomials of degree $< \deg f$ over $\mathbb{F}_q$, then $R(\Phi_f) \geq (2 + 1/(q-1))\deg f - o(\deg f)$. Ex. 18.6 has been observed by several authors; see, e.g., Ja'Ja [266], Mirwald and Schnoor [369, 370], Mirwald [368]. Ex. 18.8 has been taken from Tsfasman and Vlădut [522]. The continuity of the function α_q (Ex. 18.8(6)) is due to Manin [349]. The Gilbert-Varshamov bound (Ex. 18.9) is due to Gilbert [196], Varshamov [534], and Sacks [446]. We have taken the arrangement of the statements from van Lint [337]. Ex. 18.10 is from Grigoriev [207, 208]; a similar result has been independently obtained by Lempel et al. [324]. Ex. 18.11 is from Bshouty [85]. For a solution to Ex. 18.12 see Shokrollahi [474, Satz 7.5]. Geometric Goppa codes (Ex. 18.13(1)) have been discovered by Goppa [200]. There exist several books and monographs on this topic; see for instance Goppa [201], Moreno [379], Tsfasman and Vlădut [522], or Stichtenoth [488]. The existence of sequences of linear codes beyond the Gilbert-Varshamov bound is due to Tsfasman et al. [521].

Lempel and Winograd [325] use the connection between linear codes and bilinear computations over finite fields to establish a new approach to the theory of linear codes. For an application of coding theory to the investigation of the maximal rank of an (m, n, p)-tensor of unbalanced format see von zur Gathen [190].

Chapter 19. Rank of 2-Slice and 3-Slice Tensors

It is easy to compute the rank of a matrix: one transforms it to echelon form for example via Gaussian elimination; the rank can then be read off this form. In the first part of this chapter we show that similar results also hold for a pair of matrices. Following the Weierstraß-Kronecker theory, we define certain invariants for pairs of matrices and give a formula, due to Grigoriev [207, 208] and Ja'Ja' [265, 266] for the rank of such a pair in terms of these invariants. For triples of matrices, resp. 3-slice tensors, no such formula is known. In the second part of this chapter we present some results due to Strassen [505]. We prove a lower bound for the border rank of 3-slice tensors. Moreover, we show that for the format $(n, n, 3)$, $n \geq 3$ odd, the complement of the set of tensors of maximal border rank is a hypersurface and we explicitly determine its equation. For this we will also rely on a result which will be obtained in Chap. 20.

19.1 The Weierstraß-Kronecker Theory

In this and the following section we assume familiarity with the Weierstraß-Kronecker theory of canonical forms for pairs of matrices. Throughout this section X and Y denote indeterminates over a field k.

In the sequel we will collect the main definitions and facts of the Weierstraß-Kronecker theory without proofs. For a detailed account of this theory the reader may consult Gantmakher [179].

We associate to a pair $(A, B) \in (k^{m \times n})^2$ the *pencil* $\mathfrak{A} := XA + YB$ *of type* (m, n). In the sequel we denote pencils by capital gothic letters like $\mathfrak{A}$, $\mathfrak{B}$, etc. Also, we freely mix up pairs of matrices with their corresponding pencils.

Two pencils (A, B) and $(\tilde{A}, \tilde{B})$ are called *equivalent* if there exist regular matrices $P \in k^{m \times m}$ and $Q \in k^{n \times n}$ such that $(PAQ^{-1}, PBQ^{-1}) = (\tilde{A}, \tilde{B})$. We will introduce in this section a canonical representative of the equivalence class of a pencil $\mathfrak{A}$, the *canonical form* of $\mathfrak{A}$.

Let $\mathfrak{A} = XA + YB$ have rank r, i.e., r is the maximal order of a non-vanishing minor of $\mathfrak{A}$. (Here, rank stands for the usual rank, i.e., the rank of $\mathfrak{A}$ considered as a matrix in $k[X, Y]$.) For $i \in \underline{r}$ let $\Delta_i = \Delta_i(X, Y)$ be the greatest common divisor of the minors of order i of $\mathfrak{A}$. For every $i \in \underline{r}$, Δ_i divides Δ_{i+1}. The non-constant among the homogeneous polynomials

$$d_1 := \frac{\Delta_r}{\Delta_{r-1}}, \dots, d_{r-1} := \frac{\Delta_2}{\Delta_1}, d_r := \Delta_1,$$

are called the *invariant divisors* of $\mathfrak{A}$. They are uniquely determined up to a constant factor. Further, $d_{i+1} | d_i$ for $1 \le i < r$, see Gantmakher [179, Chap. 6.3]. Let $d_1, \dots, d_s$ be the invariant divisors of $\mathfrak{A}$. For $i \in \underline{s}$, each d_i can be decomposed as a product

$$d_i(X, Y) = Y^{e_i} \prod_{j=1}^{r_i} p_{i,j}(X, Y)^{v_{i,j}},$$

where $e_i \ge 0$, $v_{i,j} > 0$, and the $p_{i,j}(X, 1)$ are irreducible polynomials in $k[X]$ of positive degree. The Y^{e_i}, $i \in \underline{s}$, are called the *infinite elementary divisors* while the $p_{i,j}^{v_{i,j}}$ are called the *finite elementary divisors* of the pencil $\mathfrak{A}$ over k. Together they form the set of *elementary divisors* of $\mathfrak{A}$ over k. (Note that the invariant divisors of a pencil remain the same when switching to an extension field of k while the elementary divisors may change.) We call a pencil $\mathfrak{A} = XA + YB$ of type (m, n) *regular* if $m = n$ and $\det \mathfrak{A} \ne 0$. The pencil is called *singular* otherwise.

From the definition it is clear that the determinant $\det \mathfrak{A}$ of a regular pencil $\mathfrak{A} = xA + yB$ equals the product of its invariant divisors. As $\det A$ is the coefficient of X^n in $\det \mathfrak{A}$, we see that $\mathfrak{A}$ has no infinite elementary divisors iff $\det A \ne 0$.

Let us introduce the following convenient assumption.

(19.1) Assumption. *For the rest of this section we assume that the occurring pencils do not have infinite elementary divisors.*

The canonical representatives of the equivalence class of a regular pencil are built up from certain blocks of simple type. These are as follows. Let $n \ge 1$, $p = T^n + p_{n-1}T^{n-1} + \dots + p_0 \in k[T]$, and $a \in k$. We define

$$F_p := \begin{pmatrix} 0 & & & -p_0 \\ 1 & \ddots & & \vdots \\ & \ddots & 0 & -p_{n-2} \\ & & 1 & -p_{n-1} \end{pmatrix} \in k^{n \times n}, \quad \mathfrak{F}_p := XI_n + YF_p,$$

and

$$J_{n,a} := \begin{pmatrix} a & 1 & & \\ & \ddots & \ddots & \\ & & \ddots & 1 \\ & & & a \end{pmatrix} \in k^{n \times n}, \quad \mathfrak{J}_{n,a} := XI_n + YJ_{n,a},$$

where as usual I_n denotes the $n \times n$-identity matrix. $\mathfrak{F}_p$ is called the *(Frobenius) companion block* of p, while $\mathfrak{J}_{n,a}$ is called the *(upper) Jordan block of $(X+aY)^n$*. We leave it to the reader to verify that the only invariant divisors of $\mathfrak{F}_p$ and $\mathfrak{J}_{n,a}$ are $X^n + \sum_{i=0}^{n-1}(-1)^{n-i}p_i X^i Y^{n-i}$ and $(X + aY)^n$, respectively. (See Ex. 19.3; for the matrices F_p compare also Sect. 16.6.)

For two pencils $\mathfrak{A}$ and $\mathfrak{B}$ we define their *direct block sum* $\mathfrak{A} \boxplus \mathfrak{B}$ by the block-diagonal-pencil $\begin{pmatrix} \mathfrak{A} & 0 \\ 0 & \mathfrak{B} \end{pmatrix}$. (The unusual notation $\boxplus$ rather than $\oplus$ will be explained in the next section.) We say that $\mathfrak{A}$ is in the *first canonical form* if it is the direct block sum of the companion blocks of its invariant divisors. $\mathfrak{A}$ is said to be in the *second canonical form* if it is the direct block sum of the companion blocks of its elementary divisors.

(19.2) Theorem. *Two regular pencils are equivalent if and only if they have the same invariant divisors. Any regular pencil is equivalent to a pencil in first (or second) canonical form.*

For a proof of this theorem see Gantmakher [179, Thm. 12.3].

A pencil $\mathfrak{A}$ is said to be in the *third canonical form* if it is a direct sum of the Jordan blocks corresponding to its completely reducible elementary divisors and the companion blocks of the rest of its elementary divisors. (At this point our terminology differs slightly from that of [179].) Ex. 19.3 and Thm. (19.2) imply that *any regular pencil is equivalent to a pencil in the third canonical form.* Obviously, pencils in the first, second, or third canonical form are uniquely determined by their invariant divisors up to a permutation of the blocks. (This justifies the phrase "the first normal form" instead of "a first normal form.")

For studying the general case we need to introduce two further types of elementary blocks. Let $\varepsilon, \eta \geq 1$. We define the $(\varepsilon + 1) \times \varepsilon$-pencil $\mathfrak{C}_\varepsilon$ and the $\eta \times (\eta + 1)$-pencil $\mathfrak{R}_\eta$ by

$$\mathfrak{C}_\varepsilon := \begin{pmatrix} X & & & \\ Y & X & & \\ & Y & \ddots & \\ & & \ddots & X \\ & & & Y \end{pmatrix}, \quad \mathfrak{R}_\eta := \begin{pmatrix} X & Y & & & \\ & X & Y & & \\ & & \ddots & \ddots & \\ & & & X & Y \end{pmatrix}.$$

$\mathfrak{C}_\varepsilon$ is called a *minimal column block of index* ε, while $\mathfrak{R}_\eta$ is called a *minimal row block of index* η. We have the following theorem, whose proof can be found in Gantmakher [179, Thm. 12.5] for the case of infinite fields and in Mirwald [368] for finite fields.

(19.3) Theorem (Weierstraß, Kronecker). *For every pencil $\mathfrak{A}$ of type (m, n) there exist uniquely determined nonnegative integers $c, \ell, r, m_0, n_0, 0 \leq \varepsilon_1 \leq \ldots \leq \varepsilon_c, 0 \leq \eta_1 \leq \ldots \leq \eta_r$, and a regular pencil $\mathfrak{B}$ of type (ℓ, ℓ), unique up to equivalence, such that $\mathfrak{A}$ is equivalent to the following direct sum:*

$$\mathfrak{C}_{\varepsilon_1} \oplus \ldots \oplus \mathfrak{C}_{\varepsilon_c} \oplus \mathfrak{R}_{\eta_1} \oplus \ldots \oplus \mathfrak{R}_{\eta_r} \oplus \mathfrak{B} \oplus 0^{m_0 \times n_0}.$$

A pencil of the form described in the above theorem is said to be in *(Weierstraß-Kronecker) canonical form,* and $\mathfrak{B}$ is called its *regular kernel.* Using Thm. (19.2) we may assume that $\mathfrak{B}$ is in the first, second, or third canonical form. Accordingly, in generalization of the previous definitions, a pencil in canonical form is said to be in the *first, second,* or *third (Weierstraß-Kronecker) canonical form* if the same is true for its regular kernel. The numbers $\varepsilon_1, \ldots, \varepsilon_c$ are called the *minimal column indices* and the numbers $\eta_1, \ldots, \eta_r$ the *minimal row indices* of the pencil $\mathfrak{A}$.

19.2 Rank of 2-Slice Tensors

Throughout this section k denotes an algebraically closed field.

In this section we will compute the rank of a 2-slice tensor, or equivalently, of a bilinear map $\phi \in \mathrm{Bil}(k^m, k^n; k^2)$. With respect to canonical bases, ϕ can be represented by a pair (A, B) of matrices in $k^{m \times n}$, or equivalently, by a pencil $\mathfrak{A}$ of type (m, n). We shall often change our point of view and regard bilinear maps as pencils, or tensors, or vice-versa. In particular, we can talk about the *rank* $R(\mathfrak{A}) := R(\phi)$ of the pencil $\mathfrak{A}$, or we can say that $\mathfrak{A}$ is *i-concise* when the same holds for ϕ; i-conciseness of a pencil $\mathfrak{A} = XA + YB$ can be rephrased by linear algebraic properties of A and B, see Ex. 19.9. In particular, regular pencils are concise and more generally, a pencil in the canonical form given in Thm. (19.3) is concise if and only if $m_0 = n_0 = 0$.

Let $\mathfrak{A} = XA + YB$ be a regular pencil of type (n, n). By a linear transformation of k^2 we may always ensure that assumption (19.1) is satisfied. Indeed, as k is infinite, there exists a regular matrix $\begin{pmatrix} a & b \\ c & d \end{pmatrix}$ such that $\det(aA + bB) \neq 0$. This matrix transforms the pencil $\mathfrak{A}$ to $X(aA + bB) + Y(cA + dB)$, which has no infinite elementary divisors. Moreover the linear transformation does not change the rank.

The fact that we do not distinguish between bilinear maps and their pencils explains the unusual notation of "direct *block* sum $\boxplus$" since the direct sum of bilinear maps has a different meaning, see Sect. 14.2. However the direct block sum is a restriction of the direct sum as is easily seen.

We can now employ the theory of canonical forms of pencils introduced in the last section to compute $R(\mathfrak{A})$.

(19.4) Theorem (Grigoriev, Ja'Ja'). *Let $\mathfrak{A}$ be a pencil of type (m, n) in canonical form without infinite elementary divisors. Let $\varepsilon_1, \ldots, \varepsilon_c$ be the minimal column indices, $\eta_1, \ldots, \eta_r$ be the minimal row indices, and the pencil $\mathfrak{B}$ of type (ℓ, ℓ) be the regular kernel of $\mathfrak{A}$. Further let $\delta(\mathfrak{B})$ denote the number of invariant divisors of $\mathfrak{B}$ which are not squarefree. Then we have*

$$R(\mathfrak{A}) = \sum_{i=1}^{c} (\varepsilon_i + 1) + \sum_{j=1}^{r} (\eta_j + 1) + \ell + \delta(\mathfrak{B}).$$

Let us illustrate this theorem by means of an example.

(19.5) Example. Consider the regular pencil $\mathfrak{A} = \bigoplus_{i=1}^{3} \bigoplus_{j=1}^{s_i} \mathfrak{J}_{n_{ij}, a_i}$, where

	n_{i1}	n_{i2}	n_{i3}	n_{i4}	n_{i5}	n_{i6}
a_1	7	5	4	3	1	1
a_2	10	8	9	9	1	
a_3	13	11	1	8		

$\mathfrak{A}$ has six invariant divisors $d_1, \ldots, d_6$ satisfying $d_{i+1} | d_i$. To determine the d_i make each row of the above table weakly decreasing from left to right. Then the resulting columns encode the d_i as follows (check this!):

$$
\begin{aligned}
d_1 &= (X + a_1 Y)^7 (X + a_2 Y)^{10} (X + a_3 Y)^{13}, \\
d_2 &= (X + a_1 Y)^5 (X + a_2 Y)^9 (X + a_3 Y)^{11}, \\
d_3 &= (X + a_1 Y)^4 (X + a_2 Y)^9 (X + a_3 Y)^8, \\
d_4 &= (X + a_1 Y)^3 (X + a_2 Y)^8 (X + a_3 Y), \\
d_5 &= (X + a_1 Y)(X + a_2 Y), \\
d_6 &= (X + a_1 Y).
\end{aligned}
$$

Four of the invariant divisors are not squarefree, hence $\delta(\mathfrak{A}) = 4$. This gives $R(\mathfrak{A}) = \sum_{i,j} n_{ij} + 4 = 95$.   •

The next lemma prepares for the proof of the above theorem.

(19.6) Lemma. *Let m, n, p, q be positive integers, and $\mathfrak{A}_{11}, \mathfrak{A}_{12}, \mathfrak{A}_{22}$ be pencils of type $(m, p), (m, q), (n, q)$ respectively. Further let the pencil $\mathfrak{A}$ have the block decomposition $\mathfrak{A} = \left(\begin{smallmatrix} \mathfrak{A}_{11} & \mathfrak{A}_{12} \\ 0 & \mathfrak{A}_{22} \end{smallmatrix} \right)$. Then we have:*

(1) *If $\mathfrak{A}_{22}$ is 1-concise, then $R(\mathfrak{A}) \geq R(\mathfrak{A}_{11}) + n$. (In particular, this is the case when $\mathfrak{A}$ is 1-concise.)*
(2) *$R(\mathfrak{A}_{11} \boxplus \mathfrak{A}_{22}) \leq R(\mathfrak{A}_{11}) + R(\mathfrak{A}_{22})$.*
(3) *If $\mathfrak{A}_{22}$ is concise and $R(\mathfrak{A}_{22}) = \max\{n, q\}$, then equality holds in (2).*

Proof. (1) Let $\phi = \sum_{i=1}^r f_i \otimes g_i \otimes w_i \in (k^m \oplus k^n)^* \otimes (k^p \oplus k^q)^* \otimes k^2$ be the tensor corresponding to $\mathfrak{A}$, where $r := R(\mathfrak{A})$. The assumption implies that we may assume w.l.o.g. that $f_1, \ldots, f_n$ form upon restriction a basis of $(k^n)^*$. Let π be the projection onto $\bigcap_{i=1}^n \ker f_i$ along k^n. Then $\phi \circ (\pi \times \mathrm{id})$ and $\mathfrak{A}_{11}$ agree on $k^m \times k^p$, hence $\mathfrak{A}_{11}$ is a restriction of $\phi \circ (\pi \times \mathrm{id})$ and the latter has rank not greater than $r - n$.
(2) Note that $\mathfrak{A}_{11} \boxplus \mathfrak{A}_{22}$ is a restriction of the direct sum $\mathfrak{A}_{11} \oplus \mathfrak{A}_{22}$. Now apply Prop. (14.23)(1).
(3) Follows from (1) and (2).   □

(19.7) Remark. In general, equality does not hold in Lemma (19.6)(2). Consult Ex. 19.10.   •

The first step in proving Thm. (19.4) consists of showing that it is valid for $\mathfrak{A} = \mathfrak{F}_p$.

(19.8) Proposition. *Thm. (19.4) is valid for $\mathfrak{A} = \mathfrak{F}_p$.*

Proof. Let n denote the degree of p. Since $\mathfrak{F}_p$ is 1-concise we obtain $R(\mathfrak{F}_p) \geq n$ by Rem. (14.38). Let us now apply the criterion of Prop. (14.45): $R(\mathfrak{F}_p) = n$ if and only if there exist $n \times n$-matrices U, V and $n \times n$-diagonal matrices D_1 and D_2 such that $I_n = U D_1 V$ and $F_p = U D_2 V$. We may assume that $D_1 = I_n$ (otherwise replace U by $U D_1$). Hence, $R(\mathfrak{F}_p) = n$ is equivalent to the diagonalizability of

F_p. Since p is the only invariant divisor of $\mathfrak{F}_p$ the latter condition is equivalent to $\delta(\mathfrak{F}_p) = 0$. (Consider the third canonical form of $\mathfrak{F}_p$.) It remains now to prove that $R(\mathfrak{F}_p) \leq n + 1$. Since k is infinite, there exists $c = c(T) = \sum_{i=0}^{n-1} c_i T^i$ such that $p + c$ has n different zeros in k, i.e., $\delta(\mathfrak{F}_{p+c}) = 0$. We have

$$
\mathfrak{F}_p = \mathfrak{F}_{p+c} + \begin{pmatrix} 0 & \cdots & c_0 Y \\ 0 & \cdots & c_1 Y \\ \vdots & \ddots & \vdots \\ 0 & \cdots & c_{n-2} Y \\ 0 & \cdots & c_{n-1} Y \end{pmatrix}.
$$

The assertion follows, since $R(\mathfrak{F}_{p+c}) = n$ and the rank of the second summand on the right-hand side is 1. $\qquad\square$

The last proposition allows computing the rank of minimal column and row blocks.

(19.9) Proposition. *Let $\varepsilon, \eta \geq 1$. Then $R(\mathfrak{C}_\varepsilon) = \varepsilon + 1$, $R(\mathfrak{R}_\eta) = \eta + 1$.*

Proof. $\mathfrak{C}_\varepsilon$ is 1-concise, hence its rank is $\geq \varepsilon + 1$. Let $p = p(T) = T^\varepsilon + \sum_{i=0}^{\varepsilon-1} p_i T^i$ have ε different zeros in k. Then we have

$$
\mathfrak{C}_\varepsilon = \left(\begin{array}{c|c} \mathfrak{F}_p \\ \hline 0 \ \cdots \ 0 \end{array} \right) + \begin{pmatrix} 0 & \cdots & 0 & p_0 Y \\ \vdots & \ddots & \vdots & \vdots \\ 0 & \cdots & 0 & p_{\varepsilon-2} Y \\ 0 & \cdots & 0 & p_{\varepsilon-1} Y \\ 0 & \cdots & 0 & Y \end{pmatrix}.
$$

Since $R(\mathfrak{F}_p) = \varepsilon$ by the previous proposition and the rank of the second summand is obviously 1, we obtain $R(\mathfrak{C}_\varepsilon) \leq \varepsilon + 1$. The assertion for $\mathfrak{R}_\eta$ follows by transposing. $\qquad\square$

In view of Lemma (19.6)(3) and the previous proposition we may suppose for the proof of Thm. (19.4) that $\mathfrak{A}$ is a regular pencil of type (n, n). In this case we deduce from Prop. (19.8) and Lemma (19.6)(2) that

$$
R(\mathfrak{A}) \leq n + \delta(\mathfrak{A}).
$$

(Consider the first normal form of A.) Let $\ell := \delta(\mathfrak{A})$. If $\ell = 0$, then $R(\mathfrak{A}) = n$, since $\mathfrak{A}$ is concise. Otherwise, if $p_r | p_{r-1} | \cdots | p_1$ denote the invariant divisors of $\mathfrak{A}$, we contend that there exists $a \in k$ such that $p_1, \ldots, p_\ell$ are divisible by $(X + aY)^2$, while $p_{\ell+1}, \ldots, p_r$ are not. (Recall that k is algebraically closed.)

Considering the third canonical form of $\mathfrak{A}$ we infer that there exist positive integers $n_1, \ldots, n_\ell$, $n_i \geq 2$ for all i, and a regular pencil $\mathfrak{A}'$ such that $\mathfrak{A}$ is equivalent to

$$
\mathfrak{J}_{n_1.a} \boxplus \cdots \boxplus \mathfrak{J}_{n_\ell.a} \boxplus \mathfrak{A}'.
$$

We may w.l.o.g. assume that $\mathfrak{A}$ itself has the above form. For proving Thm. (19.4) we need to show that

$$R(\mathfrak{A}) \geq n + \ell = \text{Number of rows of } \mathfrak{A} + \ell.$$

The first step will be the reduction to the case where $\mathfrak{A}' = 0$. We use Lemma (19.6)(1) to deduce from $R(\mathfrak{J}_{n_1,a} \boxplus \cdots \boxplus \mathfrak{J}_{n_\ell,a}) \geq \sum_{i=1}^{\ell} n_i + \ell$ the inequality

$$R(\mathfrak{A}) \geq R(\mathfrak{J}_{n_1,a} \boxplus \cdots \boxplus \mathfrak{J}_{n_\ell,a}) + \text{Number of rows of } \mathfrak{A}' \geq n + \ell.$$

In the second step we reduce further to the case where $n_1 = \cdots = n_l = 2$. Namely, by Lemma (19.6)(1) we have $R(\mathfrak{B} \boxplus \mathfrak{J}_{m,a}) \geq R(\mathfrak{B} \boxplus \mathfrak{J}_{m-1,a}) + 1$ for any $m \geq 2$ and any pencil $\mathfrak{B}$. Repeating this argument we obtain

$$R(\mathfrak{J}_{n_1,a} \boxplus \cdots \boxplus \mathfrak{J}_{n_\ell,a}) \geq R(\mathfrak{J}_{2,a} \boxplus \cdots \boxplus \mathfrak{J}_{2,a}) + \sum_{i=1}^{\ell}(n_i - 2).$$

Thus, the proof of Thm. (19.4) is complete once we have shown the following.

(19.10) Proposition. *Let ℓ be a positive integer. Then Thm. (19.4) is valid if $\mathfrak{A}$ is a direct block sum of ℓ copies of $\mathfrak{J}_{2,a}$, i.e., $R(\boxplus_{i=1}^{\ell} J_{2,a}) = 2\ell + \ell = 3\ell$.*

Proof. An appropriate change of bases replaces each pair $(I_2, J_{2,a})$ by $(I_2, J_{2,a} - aI_2)$; we may thus suppose that $a = 0$.

Let $\varphi_1, \varphi_2 \in \text{Bil}(k^{2\ell}, k^{2\ell}; k)$ be the bilinear forms corresponding to $I_{2\ell}$ and a direct sum of ℓ copies of $J_{2,0}$, respectively. Further, let $\psi = (\varphi_1, \varphi_2)$ be the bilinear map corresponding to $\mathfrak{A}$ with respect to the canonical bases in $k^{2\ell}$ and k^2, and let

$$U_1 := \langle e_1, e_3, \dots, e_{2l-1} \rangle, \quad U_2 := \langle e_2, e_4, \dots, e_{2l} \rangle,$$

where $(e_1, \dots, e_\ell)$ is the canonical basis in $k^{2\ell}$. We have

(A) $\varphi_1(U_2, U_1) = 0, \quad R(\varphi_1) = \text{rk } I_{2\ell} = 2\ell$

(B) $\varphi_2(U_1, U_1) = \varphi_2(U_2, U_2) = \varphi_2(U_2, U_1) = 0, \quad R(\varphi_2) = \ell \cdot \text{rk } J_{2,0} = \ell.$

Now let $\psi = \sum_{i=1}^{r} f_i \otimes g_i \otimes w_i \in (k^{2\ell})^* \otimes (k^{2\ell})^* \otimes k^2$, where $r := R(\psi)$. Since ψ is concise, we may assume that $f_1, \dots, f_\ell$ form upon restriction a basis of U_2^*. Let π be the projection along U_2 onto $\cap_{i=1}^{l} \ker f_i$. It suffices to show that $\psi \circ (\pi \times \text{id})$ is 2-concise.

To this end, suppose that $v \in k^{2\ell}$ is such that for all $u \in k^{2\ell}$ we have $\psi(\pi(u), v) = 0$. Then $\varphi_2(\pi(u), v) = 0$ for all $u \in k^{2\ell}$. Suppose that $u = u_1 + u_2$ and $v = v_1 + v_2$ are the decompositions of u and v with respect to the decomposition $k^{2\ell} = U_1 \oplus U_2$. Using (B) we obtain

$$0 = \varphi_2(\pi(u), v) = \varphi_2(u_1, v_2).$$

Since $R(\varphi_2) = \ell$, φ_2 is 2-concise, hence $v_2 = 0$. Now, using (A), we have $0 = \varphi_1(\pi(u), v) = \varphi_1(\pi(u), v_1) = \varphi_1(u_1, v_1)$, which implies $v_1 = 0$, since φ_1 is 2-concise.   □

(19.11) Remark. There exist formulas for the rank of pairs of matrices over *any* field. See the Notes for a further discussion.   •

19.3* Rank of 3-Slice Tensors

We assume some familiarity with the content of Chap. 20. In particular, the proof of the last theorem of this section will rely on Prop. (20.20). k denotes an algebraically closed field.

In the previous section we completely determined the rank of a 2-slice tensor in terms of its Weierstraß-Kronecker canonical form. For 3-slice tensors no such classification is known. However, we have the following lower bound result, which even holds for the border rank.

(19.12) Theorem (Strassen). *Let* $t = [t_{ij\ell}] \in k^{n \times n \times 3}$ *be a 3-slice tensor with the 3-slices*

$$A := [t_{ij1}]_{i,j}, \quad B := [t_{ij2}]_{i,j}, \quad C := [t_{ij3}]_{i,j} \in k^{n \times n}$$

such that $\det A \neq 0$. *Then the border rank of* t *satisfies*

$$\underline{R}(t) \geq n + \frac{1}{2} \mathrm{rk}(BA^{-1}C - CA^{-1}B).$$

Proof. Let us show first that it is sufficient to prove the lower bound for the rank. Put $S_r := \{t \in k^{n \times n \times 3} \mid R(t) \leq r\}$ for $r \in \mathbb{N}$. By the topological characterization of border rank (cf. Thm. (20.3)) the Zariski closure X_r of S_r equals $\{t \in k^{n \times n \times 3} \mid \underline{R}(t) \leq r\}$. If the lower bound is true for the rank, then we have for all $r \in \mathbb{N}$

$$S_r \subseteq \left\{t \mid \det A \neq 0, \; n + \frac{1}{2} \mathrm{rk}(BA^{-1}C - CA^{-1}B) \leq r\right\} \cup \left\{t \mid \det A = 0\right\}.$$

This implies

$$S_r \subseteq \left\{t \mid n + \frac{1}{2} \mathrm{rk}(B\mathrm{adj}\,(A)\,C - C\mathrm{adj}\,(A)\,B) \leq r\right\} \cup \left\{t \mid \det A = 0\right\},$$

where $\mathrm{adj}\,(A) := \det A \cdot A^{-1}$ denotes the adjoint of A. (Note that the entries of $\mathrm{adj}\,(A)$ are polynomials in the entries of A.) As the right-hand side of this inclusion is Zariski closed, the left-hand side S_r may be replaced by its closure X_r. The set X_r is irreducible, hence it is contained in one member of the union, but certainly not in $\{t \mid \det A = 0\}$ if $r \geq n$. Noting that $\det A \neq 0$ implies $\underline{R}(t) \geq n$ (cf. Lemma (15.23)) the lower bound of the theorem for the border rank follows.

Let us show now for $\det A \neq 0$ that

(A) $$R(t) \geq n + \frac{1}{2} \mathrm{rk}(BA^{-1}C - CA^{-1}B).$$

By replacing t with the isomorphic tensor with slices I_n, $A^{-1}B$, $A^{-1}C$ we may assume w.l.o.g. that A is the identity matrix I_n. Let $r := R(t)$, i.e.,

(B) $$\forall i, j, \ell : \quad t_{ij\ell} = \sum_{\rho=1}^{r} \eta_{\rho i} \zeta_{\rho j} \theta_{\rho\ell}$$

for some $r \times n$-matrices $H := [\eta_{\rho i}]$, $Z := [\zeta_{\rho j}]$, and some $r \times 3$-matrix $[\theta_{\rho \ell}]$. If we introduce the three diagonal matrices

$$T_\ell := \mathrm{diag}(\theta_{1\ell}, \theta_{2\ell}, \ldots, \theta_{r\ell})$$

for $\ell = 1, 2, 3$, then we can express condition (B) as (compare Prop. (14.45))

(C) $\qquad H^\mathsf{T} T_1 Z = I_n, \quad H^\mathsf{T} T_2 Z = B, \quad H^\mathsf{T} T_3 Z = C.$

Suppose first that T_1 is invertible. By replacing Z with $T_1 Z$ and T_ℓ with $T_\ell T_1^{-1}$ ($\ell = 1, 2, 3$) we may assume w.l.o.g. that $T_1 = I_r$. Now we augment Z to an invertible $r \times r$-matrix $\hat{Z}$ by attaching $r - n$ columns to the right-hand side orthogonal to the rows of H^T. Then $\hat{Z}^{-1}$ is obtained from H^T by attaching $r - n$ rows to the bottom. If we define

$$\hat{B} := \hat{Z}^{-1} T_2 \hat{Z}, \quad \hat{C} := \hat{Z}^{-1} T_3 \hat{Z},$$

and write $\hat{B}$, $\hat{C}$ in block form according to the subdivision $r = n + (r - n)$ as

$$\hat{B} = \begin{pmatrix} B_{11} & B_{12} \\ B_{21} & B_{22} \end{pmatrix}, \quad \hat{C} = \begin{pmatrix} C_{11} & C_{12} \\ C_{21} & C_{22} \end{pmatrix},$$

then we see from (C) that $B_{11} = B$, $C_{11} = C$. Now $\hat{B}\hat{C} - \hat{C}\hat{B} = 0$, in particular

$$BC + B_{12}C_{21} - CB - C_{12}B_{21} = 0.$$

Therefore

$$\mathrm{rk}(BC - CB) = \mathrm{rk}(C_{12}B_{21} - B_{12}C_{21}) \leq 2(r - n)$$

since C_{12} and B_{12} have only $r - n$ columns. Thus we have shown (A) under the additional hypothesis that T_1 is invertible.

We complete the proof of (A) by induction on $s := |\{\rho \mid \theta_{\rho 1} = 0\}|$. The start "$s = 0$" has just been demonstrated, so assume $s > 0$, say $\theta_{r1} = 0$. Define the coordinate tensor $\tilde{t} \in k^{n \times n \times 3}$ by

$$\tilde{t}_{ij\ell} = \sum_{\rho=1}^{r-1} \eta_{\rho i} \zeta_{\rho j} \theta_{\rho \ell}$$

and denote its slices by $\tilde{A}$, $\tilde{B}$, $\tilde{C}$. Then $\tilde{A} = E_n$ and by the induction hypothesis we have $\mathrm{rk}(\tilde{B}\tilde{C} - \tilde{C}\tilde{B}) \leq 2(r - 1 - n)$. Moreover $B = \tilde{B} + \theta_{r2} D$, $C = \tilde{C} + \theta_{r3} D$ where $D := [\eta_{ri}\zeta_{rj}]_{i,j}$ is a matrix of rank one. Thus

$$BC - CB = \tilde{B}\tilde{C} - \tilde{C}\tilde{B} + (\theta_{r3}\tilde{B} - \theta_{r2}\tilde{C})D + D(\theta_{r2}\tilde{C} - \theta_{r3}\tilde{B})$$

and therefore

$$\mathrm{rk}(BC - CB) \leq \mathrm{rk}(\tilde{B}\tilde{C} - \tilde{C}\tilde{B}) + 2 \leq 2(r - n).$$

The theorem is now completely proved. $\qquad \square$

We proceed with two applications of this theorem. Let Λ be a finite dimensional (associative) k-algebra with unit element and M be a finite dimensional Λ-module. By identifying M with its structural bilinear map $\Lambda \times M \to M$ we may speak of the border rank $\underline{R}(M)$ of M. In particular, the border rank $\underline{R}(\Lambda)$ of the algebra Λ is defined.

(19.13) Corollary. *Let Λ be a finite dimensional k-algebra with unity and M be an n-dimensional Λ-module. Let $\ell_a \colon M \to M, x \mapsto ax$ denote the left multiplication by $a \in \Lambda$. Then we have for all $b, c \in \Lambda$*

$$\underline{R}(M) \geq n + \frac{1}{2}\,\mathrm{rk}(\ell_{bc-cb}).$$

Proof. We may w.l.o.g. assume that $1, b, c$ are linearly independent. Let Λ_0 denote their linear hull. The bilinear map $\phi \colon \Lambda_0 \times M \to M, (a, x) \to ax$ is a restriction of the structural map of M. Therefore, $\underline{R}(\phi) \leq \underline{R}(M)$ (compare Sect. 15.4). The 1-slices of the coordinate tensor of ϕ with respect to the basis $(1, b, c)$ and some basis of M are just the matrices corresponding to ℓ_1, ℓ_b, ℓ_c. Thus we have by Thm. (19.12)

$$\underline{R}(\phi) \geq n + \frac{1}{2}\,\mathrm{rk}(\ell_b\ell_c - \ell_c\ell_b),$$

which implies the assertion.   □

(19.14) Corollary. *Let $m \geq 2$. Then*

$$\underline{R}(k^{m\times m}) \geq 3m^2/2, \quad \underline{R}(T_m) \geq m(3m+1)/4,$$

where T_m denotes the k-algebra of upper triangular $m \times m$-matrices over k.

Proof. For $\Lambda = k^{m\times m}$ choose $b = \mathrm{diag}(\lambda_1, \ldots, \lambda_m)$ with λ_i pairwise distinct, and $c \in k^{m\times m}$ with $c_{ij} = 1$ if $i + j = m + 1$, $c_{ij} = 0$ otherwise. Then $bc - cb$ is invertible and $\mathrm{rk}(\ell_{bc-cb}) = m^2$. The assertion follows from Cor. (19.13).

For $\Lambda = T_m$ choose $b = \mathrm{diag}(\lambda_1, \ldots, \lambda_m)$ with λ_i pairwise distinct, and $c \in T_m$ with $c_{ij} = 1$ if $j = i + 1$, $c_{ij} = 0$ otherwise. Then it is easy to see that $\mathrm{rk}(\ell_{bc-cb}) = m(m-1)/2$, hence the assertion follows from Cor. (19.13).   □

For the remainder of this section we assume familiarity with the notation introduced in Sect. 20.1 and with the statement of Prop. (20.20). In particular we recall that $X_r(f)$ denotes the secant variety consisting of the tensors of format f which have border rank at most r. Are the formats of 3-slice tensors good?

Let A, B, C be $n \times n$-matrices with indeterminate entries and define the rational function

(19.15) $$F_n := (\det A)^2 \cdot \det(BA^{-1}C - CA^{-1}B).$$

We consider F_n as a rational function on $k^{n\times n\times 3}$ by identifying A, B, C with the 3-slices of a tensor t in $k^{n\times n\times 3}$. Thm. (19.12) implies that $\underline{R}(t) \geq 3n/2$ for tensors t satisfying $\det A \neq 0$ and $F_n(t) \neq 0$. Since there exists such a tensor (check this!), we see that the typical border rank $\underline{R}(f_n)$ of the format $f_n := (n, n, 3)$ satisfies $\underline{R}(f_n) \geq 3n/2$. However, for odd $n \geq 3$, we have $\lceil 3n^2/(2n+1) \rceil = (3n-1)/2 < 3n/2$, thus the format f_n is not good!

In the following, we will prove $\underline{R}(f_n) = (3n+1)/2$ and show that the complement $X_{(3n-1)/2}(f_n)$ of the set of tensors of maximal border rank is a hypersurface with irreducible generator F_n.

(19.16) Lemma. *Let $n \geq 3$. F_n is an irreducible polynomial which is homogeneous of degree n in the entries of A as well as in the entries of B and C. Moreover, F_n is a relative invariant with respect to the action of $\mathrm{GL}(n, k) \times \mathrm{GL}(n, k) \times \mathrm{GL}(3, k)$ on the space of tensors of format $(n, n, 3)$ represented by the slices A, B, C: we have for all $S, T \in \mathrm{GL}(n, k)$*

$$F_n(SAT, SBT, SCT) = (\det S \det T)^3 F_n(A, B, C),$$

and for all $R = [r_{ij}] \in \mathrm{GL}(3, k)$

$$F_n(\tilde{A}, \tilde{B}, \tilde{C}) = (\det R)^n F_n(A, B, C),$$

where

$$\begin{aligned}
\tilde{A} &= r_{11}A + r_{12}B + r_{13}C, \\
\tilde{B} &= r_{21}A + r_{22}B + r_{23}C, \\
\tilde{C} &= r_{31}A + r_{32}B + r_{33}C.
\end{aligned}$$

Proof. We have

$$\begin{aligned}
F_n(A, B, C) &= (\det A)^2 (\det C)^2 \det(C^{-1}BA^{-1} - A^{-1}BC^{-1}) \\
&= (\det C)^2 \det(AC^{-1}B - BC^{-1}A) = F_n(C, A, B),
\end{aligned}$$

hence F_n is invariant under cyclic permutations of A, B, C. Using this and the fact that F_n is a polynomial in B and C, we see that F_n must be a polynomial. Moreover, F_n is homogeneous of degree n in the entries of C, hence as well in the entries of A and B.

The behavior of F_n under transformations in $\mathrm{GL}(n, k) \times \mathrm{GL}(n, k)$ is clear. The behavior with respect to transformations in $\mathrm{GL}(3, k)$ is easily checked for diagonal matrices, for permutation matrices, and for the matrices ($\lambda \in k$)

$$R = \begin{pmatrix} 1 & 0 & 0 \\ \lambda & 1 & 0 \\ 0 & 0 & 1 \end{pmatrix}.$$

Since the group $\mathrm{GL}(3, k)$ is generated by such matrices, the statement follows.

Let us prove the irreducibility of F_n. First note that $F_n \neq 0$. (Substitute for instance for A the identity matrix, for B a diagonal matrix with pairwise distinct eigenvalues and for C the permutation matrix with respect to the cycle $(12 \ldots n)$.) Suppose we had a nontrivial decomposition $F = GH$. Then G and H are homogeneous in each of the set of entries of A, B, and C, respectively. Now we substitute for A the identity matrix I_n and for C a diagonal matrix $\Lambda = \text{diag}(\lambda_1, \ldots, \lambda_n)$ where the $\lambda_i \in k$ are pairwise distinct. Then

$$G(I_n, B, \Lambda) \cdot H(I_n, B, \Lambda) = F(I_n, B, \Lambda) = \det[(\lambda_i - \lambda_j)b_{ij}]_{ij}.$$

The latter determinant (call it D) is irreducible. Namely, since D is linear in each row (and each column), all the variables of each row (or column) appear in only one of the factors D_1, D_2 of a decomposition $D = D_1 D_2$. This is only possible if one of the factors is constant (use $n \geq 3$).

Therefore either $G(I_n, B, \Lambda)$ or $H(I_n, B, \Lambda)$ are constant. Thus, as G and H are homogeneous in B, either G or H have degree zero in B. By symmetry, the same holds with respect to A and C. So w.l.o.g. $G = G(A)$ is a polynomial in A and $H = H(B, C)$ is a polynomial in B, C. Then

$$G(A)H(B, C) = F(A, B, C) = (-1)^n F(B, A, C) = (-1)^n G(B)H(A, C),$$

and we obtain for some $\xi \in k^\times$

$$F(A, B, C) = \xi G(A)G(B)G(C),$$

yielding the contradiction $0 = F(A, A, A) = \xi G(A)^3$. Thus F_n is indeed irreducible. $\qquad\square$

(19.17) Theorem (Strassen). *Let $n \geq 3$ be odd and f_n be the format $(n, n, 3)$. Then the complement $X_{(3n-1)/2}(f_n)$ of the set of tensors of format f_n which have maximal border rank is an irreducible hypersurface in $k^{n \times n \times 3}$ with irreducible generator F_n. The format f_n is not good and its typical rank equals $\underline{R}(f_n) = (3n + 1)/2$.*

Proof. Put $r_n := (3n - 1)/2$. By Thm. (19.12) we have

$$X_{r_n}(f_n) \cap \{t \mid \det A \neq 0\} \subseteq \{t \mid F_n(t) = 0\},$$

implying $X_{r_n}(f_n) \subseteq \{t \mid F_n(t) = 0\}$. Lemma (19.16) shows that the zeroset of F_n is an irreducible hypersurface with irreducible generator F_n. On the other hand, we will show in Prop. (20.20) that $\dim X_{r_n}(f_n) \geq 3n^2 - 1$. (The machinery of Sect. 20.3 will be introduced only for proving this.) Hence $X_{r_n}(f_n)$ coincides with the zeroset of F_n. It follows immediately that $\underline{R}(f_n) = r_n + 1$. Further, the format f_n is not good since $\lceil 3n^2/(2n + 1) \rceil = (3n - 1)/2 < r_n$ as we have already observed. $\qquad\square$

19.4 Exercises

19.1. Let k be an infinite field, $\bar{k}$ be its algebraic closure, and $A, B \in k^{n \times n}$. Prove that there exists $T \in \text{GL}(n, \bar{k})$ such that $T^{-1}AT = B$ iff there exists $U \in \text{GL}(n, k)$ such that $U^{-1}AU = B$.

19.2. Show that the set of elementary divisors of a pencil $\mathfrak{A}$ uniquely determines its invariant divisors up to multiplication by elements of the underlying field. (Hint: recall that if $d_1, \ldots, d_r$ are the invariant divisors of a pencil, then $d_{i+1} | d_i$ for $1 \leq i < r$.)

19.3. Show that $X^n + \sum_{i=0}^{n-1}(-1)^{n-i} p_i X^i Y^{n-i}$ and $(X + aY)^n$ are the only invariant divisors of $\mathfrak{F}_p$ and $\mathfrak{J}_{n,a}$, respectively. What are the elementary divisors of these pencils?

19.4. Show that if p and q are coprime polynomials, then $\mathfrak{F}_{pq}$ is equivalent to $\mathfrak{F}_p \boxplus \mathfrak{F}_q$. What happens if p and q are not coprime? (Hint: what are the elementary divisors of $\mathfrak{F}_p \boxplus \mathfrak{F}_q$?)

19.5. Let $p \in k[T]$ be of degree $n \geq 1$. Show that $k[T]/(p) \to k^{n \times n}$, T mod $p \mapsto F_p$ is an injective morphism of k-algebras.

19.6. Let p be a separable polynomial over k. Show that $k[F_p]$ is the set of all matrices commuting with F_p. (Hint: F_p is diagonalizable over an extension of k.)

19.7. Let $p(T) = \prod_{i=1}^{r}(T + a_i)^{v_i}$. Show that $\mathfrak{F}_p$ is equivalent to the direct block sum of $\mathfrak{J}_{v_i, a_i}$ for $i \in \underline{r}$.

19.8. Show that the set of all regular pencils of type (n, n) with rank n is Zariski open in $(k^{n \times n})^2$.

19.9. The *left kernel* lker A of an $m \times n$-matrix A is defined as the set of all $u \in k^m$ such that $u^{\top} A = 0$. The *right kernel* rker is defined similarly.

(1) $\mathfrak{A} = XA + YB$ is 1-concise if and only if lker $A \cap$ lker $B = 0$.
(2) $\mathfrak{A}$ is 2-concise if and only if rker $A \cap$ rker $B = 0$.
(3) $\mathfrak{A}$ is 3-concise if and only if A and B are linearly independent.
(4) Companion blocks, Jordan blocks, minimal row blocks, and minimal column blocks are concise. The same holds for direct block sums of any of these blocks.

19.10. Show that in Lemma (19.6)(2) equality does not hold in general. (Hint: use Ex. 19.4.)

19.11. Let n be a positive integer and k be a field having at least n elements.

(1) Show that $R(\mathfrak{C}_n) = R(\mathfrak{R}_n) = n + 1$.
(2) If $n \geq 2$ then $R(\mathfrak{J}_{n,a}) = n + 1$ for any $a \in k$.

19.12. Show that the maximum possible rank of a pencil of type (n, n) over an algebraically closed field k is $\lfloor 3n/2 \rfloor$.

19.13. Give a bilinear computation of length 3 for the pair

$$X_1 Y_1 + X_2 Y_2 + X_3 Y_3, \quad X_2 Y_1 + X_2 Y_3 + X_3 Y_2$$

of bilinear forms over a field with at least 3 elements.

19.14. Give a bilinear computation of length 4 for the pair

$$X_1 Y_1 + X_2 Y_2 + X_3 Y_3, \quad X_1 Y_2 + X_2 Y_3$$

over an field with at least 3 elements.

19.15. Show that the rank of the bilinear map given by the matrix multiplication

$$\begin{pmatrix} X_1 & X_2 & & \\ & \ddots & \ddots & \\ & & \ddots & X_2 \\ & & & X_1 \end{pmatrix} \begin{pmatrix} Y_1 \\ \vdots \\ Y_n \end{pmatrix}$$

is $n + 1$ over any field with at least n elements. (Hint: consider $\pi \phi$ for some $\pi \in S_3$, where ϕ is the bilinear map of the problem.)

19.16. Show that the rank of the $\mathbb{R}$-bilinear map $\mathbb{C} \times \mathbb{C}^n \to \mathbb{C}^n$ which sends the tuple $(x, z_1, \ldots, z_n)$ of complex numbers to the tuple $(x z_1, \ldots, x z_n)$ equals $3n$.

19.17. Let $t_i \in k^{n_i \times n_i \times 3}$ be a 3-slice tensor with the 3-slices $A_i, B_i, C_i \in k^{n \times n}$, $i = 1, 2$. We define the direct block sum $t_1 \boxplus t_2 \in k^{(n_1+n_2) \times (n_1+n_2) \times 3}$ of t_1 and t_2 as being the 3-slice tensor with 3-slices $A_1 \oplus A_2$, $B_1 \oplus B_2$, $C_1 \oplus C_2$ (compare Sect. 19.1).

(1) Prove that $F_{n_1+n_2}(t_1 \boxplus t_2) = F_{n_1}(t_1) F_{n_2}(t_2)$. ($F_n$ is defined in (19.15).)
(2) Assume that n_1 and n_2 are odd. Show that if t_1, t_2 are of maximal border rank for the formats $(n_1, n_1, 3)$, $(n_2, n_2, 3)$, respectively, then $t_1 \boxplus t_2$ is of maximal border rank for the format $(n_1 + n_2, n_1 + n_2, 3)$.

19.18.* Assume $\operatorname{char} k = 0$ and let $sl(2, k)$ denote the 3-dimensional Lie algebra of 2×2-matrices with trace zero.

(1) For $m \in \mathbb{N}$ we may give the space $k[X, Y]_{(m)}$ of homogeneous bivariate polynomials of degree m the structure of an $sl(2, k)$-module $V(m)$ by setting

$$(z \cdot f) := -(z_{11} \partial_1 f + z_{21} \partial_2 f) X - (z_{12} \partial_1 f + z_{22} \partial_2 f) Y$$

for $z \in sl(2, k)$, $f \in k[X, Y]_{(m)}$. One can show that $V(m)$ is a simple $sl(2, k)$-module, and that any finite dimensional $sl(2, k)$-module is isomorphic to a direct sum of such modules. (See Humphreys [258, §7].) Consider the basis $(v_i)_{0 \le i \le m}$ of $V(m)$, where $v_i := (-1)^i \binom{m}{i} X^i Y^{m-i}$, and let

$$h := \begin{pmatrix} 1 & 0 \\ 0 & -1 \end{pmatrix}, \quad x := \begin{pmatrix} 0 & 1 \\ 0 & 0 \end{pmatrix}, \quad y := \begin{pmatrix} 0 & 0 \\ 1 & 0 \end{pmatrix}$$

be the standard basis of $sl(2, k)$. Verify that for $0 \leq i \leq m$

$$h \cdot v_i = (m - 2i)v_i, \quad x \cdot v_i = (m + 1 - i)v_{i-1}, \quad y \cdot v_i = (i + 1)v_{i+1}$$

($v_{-1} := v_{m+1} := 0$). By computing the 3-slices A, B, C of the bilinear map $sl(2, k) \times V(m) \to V(m), (z, f) \mapsto z \cdot f$ with respect to the above bases prove that $F_{m+1}(A, B, C) \neq 0$ if $m > 0$. (Caution: $\det A = 0$ if m is even!) Conclude that $\underline{R}(V(m)) = \lceil 3(m + 1)/2 \rceil$ if $m > 0$.

(2) Prove that $\underline{R}(M) = \lceil \frac{3}{2} \dim M \rceil$ for any $sl(2, k)$-module M not containing $V(0)$. (Use Ex. 19.17.) This shows that coordinate tensors of M are of maximal border rank for the format $(\dim M, \dim M, 3)$.

19.5 Notes

The rank of a pair of matrices, or equivalently of $(m, n, 2)$-tensors, was first studied by Grigoriev [208, 207] over algebraically closed fields. These investigations were carried on and generalized by Ja'Ja' [265, 266] by relaxing the conditions on the ground field. Finally, Teichert [515] proved fomulas for the rank of a pair of matrices over any ground field, see below.

The lower bound on the border rank of 3-slice tensors in Thm. (19.12), as well as the explicit determination of the generator of the hypersurface consisting of the tensors of maximal border rank in Thm. (19.17), are due to Strassen [505]. We have followed his paper in Sect. 19.3 quite closely. Lickteig [330] has improved Cor. (19.14) to $\underline{R}(k^{m \times m}) \geq 3m^2/2 + m/2 - 1$. (See also Griesser [205].)

For a detailed account of the Weierstraß-Kronecker theory the reader may consult Gantmakher [179]. Mirwald [368] gives a detailed description of methods for computing the various canonical forms of a pair of matrices.

Thm. (19.4) has been proved by Grigoriev [208, 207] and independently by Ja'Ja [265]. Prop. (19.8) and (19.9) are from Ja'Ja' [265]. Our presentation follows Teichert [515], where it is also shown that Thm. (19.4) is valid over an arbitrary infinite field if we let $\delta(\mathfrak{B})$ denote the number of invariant divisors of $\mathfrak{B}$ which are not decomposable into a product of non-associated linear factors. (We call two linear forms non-associated if their quotient is not constant.) For finite fields we have the following result, also due to Teichert [515]: let $\mathfrak{A}$ be a pencil of format (m, n) in normal form without infinite elementary divisors. Let $\varepsilon_1, \ldots, \varepsilon_c$ be the minimal column indices, $\eta_1, \ldots, \eta_r$ be the minimal row indices, and the pencil $\mathfrak{B} = XC + XD$ of format (ℓ, ℓ) be the regular kernel of $\mathfrak{A}$. Further let $\delta(\mathfrak{B})$ be the number of invariant divisors of $\mathfrak{B}$ which are not decomposable into a product of non-associated linear factors in k ($\delta(\mathfrak{B}) := 0$ if $\ell = 0$). Then we have

$$R(\mathfrak{A}/\mathbb{F}_q) = \max(\alpha, \beta + \lceil \gamma \rceil),$$

where $\alpha = \sum_{i=1}^{c}(\varepsilon_i + 1) + \sum_{j=1}^{r}(\eta_j + 1) + \ell + \delta(\mathfrak{B})$, $\beta = \sum_{\varepsilon_i \leq q}(\varepsilon_i + 1) + \sum_{\eta_j \leq q}(\eta_j + 1)$, and $\gamma = q^{-1}[(q+1)(\varepsilon + \eta) + \mathrm{rk}(C) + \sum_{\lambda \in \mathbb{F}_q} \mathrm{rk}(\lambda C + D)]$ with $\varepsilon := \sum_{\varepsilon_j > q} \varepsilon_j$ and $\eta := \sum_{\eta_j > q} \eta_j$. Special cases of this general theorem have been proved earlier by Ja'Ja [266]. For a connection between the rank of pencils over $\mathbb{F}_2$ and the multiplicative complexity of pairs of Boolean quadratic forms, see Mirwald and Schnorr [370].

Ex. 19.11 is from Ja'Ja [265, 266]. Ex. 19.12 is from Grigoriev [208]. See Ja'Ja [265] for Ex. 19.15. Ex. 19.16 has also been obtained by Auslander et al. [17] using different methods. In fact, they showed that under some conditions on a bilinear map ϕ the additivity conjecture is true for $\phi \oplus \cdots \oplus \phi$. For the bilinear map of the multiplication in quadratic extension fields these conditions are fulfilled. Ex. 19.18 is taken from Strassen [505].

Chapter 20. Typical Tensorial Rank

The typical rank $\underline{R}(f)$ of a format f is the rank of Zariski almost all tensors of that format. Following Strassen [505] and Lickteig [331] we determine the asymptotic growth of the function $\underline{R}$ and determine its value for some special formats. In particular, we consider the formats of 3-slice tensors and prove a result needed in Chap. 19. The problem amounts to determining the dimension of higher secant varieties to Segre varieties. We achieve this by computing the dimension of the tangent space to these varieties, for which some machinery is developed. In the appendix we give a topological characterization of the border rank due to Alder [5], on which all our investigations in this chapter are based.

Throughout this chapter, k denotes an algebraically closed field.

20.1 Geometric Description

A *format* is a triple $f = (f_1, f_2, f_3)$ of positive natural numbers. To a format f we assign the tensor product of k-vector spaces of dimensions f_1, f_2, f_3

$$T(f) := U_1 \otimes U_2 \otimes U_3 \quad (= k^{f_1 \times f_2 \times f_3});$$

for definiteness we assume $U_i = k^{f_i}$.

(20.1) Lemma. *The subset of nonzero triads*

$$S(f) := \{u_1 \otimes u_2 \otimes u_3 \mid u_i \in U_i\} \setminus \{0\}$$

is a smooth and irreducible Zariski closed subset of $T(f)$ of dimension $f_1 + f_2 + f_3 - 2$. Its tangent space at a point $t = u_1 \otimes u_2 \otimes u_3 \in S(f)$ is given by

$$U_1 \otimes u_2 \otimes u_3 + u_1 \otimes U_2 \otimes u_3 + u_1 \otimes u_2 \otimes U_3.$$

Proof. The morphism of projective spaces

$$\psi \colon \mathbb{P}^{f_1-1} \times \mathbb{P}^{f_2-1} \times \mathbb{P}^{f_3-1} \to \mathbb{P}^{f_1 f_2 f_3 -1}$$

induced by the tensor product map $(u_1, u_2, u_3) \mapsto u_1 \otimes u_2 \otimes u_3$ is called the Segre embedding. It is well-known (and easily verified directly) that the image $\mathrm{im}\,\psi$

is closed and that ψ is an isomorphism onto im ψ (cf. Hartshorne [232, p. 13]). Hence im ψ is smooth and of dimension $f_1 + f_2 + f_3 - 3$. Now note that $S(f)$ is just the affine cone (without top 0) of im ψ.

To prove the assertion about the tangent space we observe that the subspace $U_1 \otimes u_2 \otimes u_3$ is contained in $S(f)$, hence it is also contained in the tangent space of $S(f)$ at $t = u_1 \otimes u_2 \otimes u_3$. Arguing in the same way with the other components, we see that

$$T := U_1 \otimes u_2 \otimes u_3 + u_1 \otimes U_2 \otimes u_3 + u_1 \otimes u_2 \otimes U_3$$

is contained in the tangent space of $S(f)$ at t. It is readily verified that $\dim T = f_1 + f_2 + f_3 - 2 = \dim S(f)$. Hence the assertion follows by a comparison of dimensions. $\qquad\square$

We call $S(f)$ the *Segre variety* of the format f. The image of the summation map

$$\sigma_r(f) : S(f)^r \to T(f), \quad (t_1, \ldots, t_r) \mapsto \sum_{\rho=1}^{r} t_\rho,$$

denoted by $S_r(f)$, consists of the tensors in $T(f)$ of rank $\leq r$. Obviously $S_1(f) = S(f)$. We will call the Zariski closure of $S_r(f)$ the $(r-1)$th *secant variety* of the format f.

(20.2) Remark. Let $Y \subseteq \mathbb{P}^n$ be a closed subset and $\ell \in \mathbb{N}'$. In algebraic geometry one studies the *secant variety* $S_\ell(Y) \subseteq \mathbb{P}^n$, which is defined as the union of the secant ℓ-planes to Y (cf. Harris [228]). In this terminology, the variety $X_r(f)$ is just the affine cone of the secant variety $S_{r-1}(Y)$, where Y is the image of the Segre imbedding $\mathbb{P}^{f_1-1} \times \mathbb{P}^{f_2-1} \times \mathbb{P}^{f_3-1} \to \mathbb{P}^{f_1 f_2 f_3 - 1}$. This justifies our naming of $X_r(f)$. $\qquad\bullet$

The notion of border rank, which was introduced in Def. (15.19) in a purely algebraic way, can also be characterized topologically. The next theorem expresses this important insight. A proof can be found in Appendix 20.6.

(20.3) Theorem (Alder). *The secant variety $X_r(f)$ is irreducible and consists of all tensors in $T(f)$ having border rank at most r, i.e.,*

$$X_r(f) = \{t \in T(f) \mid \underline{R}(t) \leq r\}.$$

We denote the *maximal possible border rank* of a tensor of format f by $\underline{R}(f)$. By the previous theorem we have

$$\underline{R}(f) = \max\{\underline{R}(t) \mid t \in T(f)\} = \min\{r \mid X_r(f) = T(f)\}.$$

In the subsequent proposition we will see that $\underline{R}(f)$ is the rank (and also the border rank) of Zariski almost all tensors of format f. We thus call $\underline{R}(f)$ the *typical tensorial rank* of the format f. Our goal is to find $\underline{R}(f)$ for given formats, or more generally, to determine the dimensions of the secant varieties $X_r(f)$. The subsequent proposition summarizes some of the properties of $\underline{R}$ and the secant varieties $X_r(f)$.

(20.4) Proposition. (1) $\underline{R}(f)$ *is monotonic with respect to each component of* f *and invariant under permutations of the three components.*

(2) *The chain of algebraic sets* $\{0\} = X_0(f) \subset X_1(f) \subset \ldots \subset X_{\underline{R}(f)}(f) = T(f)$ *is strictly ascending.*

(3) *For all* t *in some nonempty Zariski open subset of* $T(f)$ *we have* $R(t) = \underline{R}(t) = \underline{R}(f)$.

(4) $\dim X_r(f) \leq \min\{r(f_1 + f_2 + f_3 - 2), f_1 f_2 f_3\}$.

(5) $f_1 f_2 f_3/(f_1 + f_2 + f_3 - 2) \leq \underline{R}(f) \leq \min\{f_1 f_2, f_1 f_3, f_2 f_3\}$.

Proof. (1) Clear by definition.

(2) We have $S_{r-1}(f) + S(f) = S_r(f)$. Since the addition of tensors is continuous and $X_r(f)$ is closed we see that $X_{r-1}(f) + S(f) \subseteq X_r(f)$. Therefore, if $X_{r-1}(f) = X_r(f)$, we conclude by induction on $m \geq 1$

$$X_{r-1}(f) + S_m(f) \subseteq X_{r-1}(f).$$

Hence $X_{r-1}(f) = T(f)$.

(3) Let $r = \underline{R}(f)$. The constructible set $S_r(f)$ contains a nonempty subset U which is open in $X_r(f) = T(f)$ (cf. Ex. 4.19). Since $\dim X_{r-1}(f) < \dim T(f)$ by (2), the set $U \setminus X_{r-1}(f)$ is not empty. However, $R(t) = \underline{R}(t) = r$ for all $t \in U \setminus X_{r-1}(f)$.

(4) The inequality follows from the definition of $X_r(f)$ and $\dim S(f) = f_1 + f_2 + f_3 - 2$.

(5) The upper bound is trivial and the lower bound is a consequence of (4). $\square$

(20.5) Example.

(1) Let $f = (m, n, 1)$. Then $T(f)$ can be identified with the space of $m \times n$ matrices over k,

$$S_r(f) = X_r(f) = \{A \in k^{m \times n} \mid \operatorname{rk} A \leq r\},$$

and $\underline{R}(f) = \min\{m, n\}$.

(2) Let $f = (n, n, 2)$. Almost all pencils of type (n, n) are equivalent to a pencil of the form $\mathfrak{A} = XI_n + YD$, where D is a nonsingular diagonal matrix (cf. Chap. 19). Since $R(\mathfrak{A}) = n$, we conclude that $\underline{R}(n, n, 2) = n$. In the next section we will again obtain this by a different method, which does not rely on the normal form of 2-slice tensors. $\bullet$

(20.6) Remark. We could also consider the *maximal rank* $R(f)$ *of format* f

$$R(f) := \max\{R(t) \mid t \in T(f)\} = \min\{r \mid S_r(f) = T(f)\}.$$

This quantity is more difficult to investigate than $\underline{R}(f)$. We certainly have $\underline{R}(f) \leq R(f)$. This inequality can be strict, so typical and maximal rank do not need to be the same (cf. Ex. 20.5). For more information about $R(f)$ consult the Notes. $\bullet$

We introduce the following notation

$$Q(f) := \frac{f_1 f_2 f_3}{f_1 + f_2 + f_3 - 2}$$

for a format f.

(20.7) Definition. A format f is called *good* iff for all r the inequality in Prop. (20.4)(4) is an equality. A format f is called *perfect* iff it is good and additionally $Q(f) \in \mathbb{N}$.   ●

(20.8) Remark. By Prop. (20.4)(5) we always have $Q(f) \le \underline{R}(f)$. If the format f is good, then $\underline{R}(f) = \lceil Q(f) \rceil$.   ●

The goal of this chapter is to prove the following theorem.

(20.9) Theorem (Strassen, Lickteig). (1) *Every format $f \in 2(\mathbb{N}')^3$ for which $Q(f)/\max f$ is an integer is perfect. For example, $(n, n, n + 2)$ is perfect if $6 \mid n$.*
(2) *We have for every format f*

$$\underline{R}(f) \sim Q(f) \quad \text{as min } f \to \infty.$$

Let us briefly outline the main idea of the proof. The problem amounts to determining the dimensions of the secant varieties $X_r(f)$. How can we do this? The image of the differential of the morphism $\sigma = \sigma_r(f): S(f)^r \to T(f)$ in $t = (t_1, \ldots, t_r) \in S(f)^r$ is the vector space

$$\sum_{\rho=1}^{r} (U_1 \otimes u_2^\rho \otimes u_3^\rho + u_1^\rho \otimes U_2 \otimes u_3^\rho + u_1^\rho \otimes u_2^\rho \otimes U_3),$$

where $t_\rho = u_1^\rho \otimes u_2^\rho \otimes u_3^\rho$. It is a known fact that $\dim X_r(f)$ equals the dimension of this vector space, if $t \in S(f)^r$ is sufficiently generally chosen and $\operatorname{char} k = 0$ (compare Lemma (20.10)). Thus it is enough to determine the dimension of this vector space for a generic t. This, however, turns out to be a quite difficult task. We will attack this problem in the next section after developing some machinery.

20.2 Upper Bounds on the Typical Rank

We introduce the following notation. For a triad $t = u_1 \otimes u_2 \otimes u_3 \in S(f)$ we denote by

$$T(f) \diamond_1 t, \quad T(f) \diamond_2 t, \quad T(f) \diamond_3 t$$

the subspaces

$$U_1 \otimes u_2 \otimes u_3, \quad u_1 \otimes U_2 \otimes u_3, \quad u_1 \otimes u_2 \otimes U_3$$

of $T(f)$, respectively. By Lemma (20.1) the sum of these three subspaces is just the tangent space of $S(f)$ in t.

(20.10) Lemma. (1) *The image of the differential of $\sigma_r(f)$ in a point $t = \sum_\rho t_\rho$ in $S(f)^r$ is given by*

$$\sum_{\rho=1}^r \left(T(f) \diamond_1 t_\rho + T(f) \diamond_2 t_\rho + T(f) \diamond_3 t_\rho \right).$$

(2) *If there is a point $t \in S(f)^r$ such that*

$$\dim \sum_{\rho=1}^r \left(T(f) \diamond_1 t_\rho + T(f) \diamond_2 t_\rho + T(f) \diamond_3 t_\rho \right) \geq d,$$

then $\dim X_r(f) \geq d$. The reverse is also true if char $k = 0$.

Proof. Write $S := S(f)$, $X := X_r(f)$, $\sigma := \sigma_r(f)$. The tangent space of S^r in t is given by $T_t S^r = \bigoplus_{\rho=1}^r T_{t_\rho} S$. Since σ is the restriction to S^r of the linear map $T(f)^r \to T(f)$, $(v_1, \ldots, v_r) \mapsto \sum_i v_i$, we obtain for the image of the differential $d_t \sigma$ of σ in t

$$\operatorname{im} d_t \sigma = \sum_{\rho=1}^r T_{t_\rho} S,$$

which shows the first assertion.

The dimension of $\operatorname{im} d_t \sigma$ is a Zariski lower semicontinuous function of t. (See the subsequent considerations below.) Therefore, if $\dim \operatorname{im} d_t \sigma \geq d$ for some t, this inequality also holds for some t' such that $\sigma(t')$ is a smooth point of X. As $\operatorname{im} d_{t'} \sigma \subseteq T_{\sigma(t')} X$ we conclude

$$d \leq \dim \operatorname{im} d_{t'} \sigma \leq \dim T_{\sigma(t')} X = \dim X.$$

If char $k = 0$, we may even achieve by a suitable choice of t' that the differential of σ in t' is surjective, which shows the reverse direction. $\qquad\square$

We are now going to introduce some symbolic calculus for determining the dimensions of vector spaces which are of a slightly more general form than the ones occuring in the last lemma.

In the following we call a 4-tuple $s := (s_0; s_1, s_2, s_3) \in \mathbb{N}^4$ a *configuration*. If

$$(t; x, y, z) \in S(f)^{s_0} \times S(f)^{s_1} \times S(f)^{s_2} \times S(f)^{s_3} =: S(f)^{|s|},$$

we denote by $\Sigma_f(t; x, y, z)$ the following subspace of $T(f)$

$$\Sigma_f(t; x, y, z) := \sum_{\rho=1}^{s_0} \left(T(f) \diamond_1 t_\rho + T(f) \diamond_2 t_\rho + T(f) \diamond_3 t_\rho \right)$$

(20.11)
$$+ \sum_{\alpha=1}^{s_1} T(f) \diamond_1 x_\alpha + \sum_{\beta=1}^{s_2} T(f) \diamond_2 y_\beta + \sum_{\gamma=1}^{s_3} T(f) \diamond_3 z_\gamma.$$

First we observe that the map

$$S(f)^{|s|} \longrightarrow \mathbb{N}, \quad (t; x, y, z) \mapsto \dim \Sigma_f(t; x, y, z)$$

is Zariski lower semicontinuous, i.e., the sets $\{(t; x, y, z) \mid \dim \Sigma_f(t; x, y, z) > r\}$ are Zariski open for all $r \in \mathbb{N}$. (This follows easily from the determinantal criterion for linear dependence, cf. Ex. 20.2.) We denote the maximum value of the above map by $d(s, f)$ and call it the *dimension of the configuration s in the format f*.

Note that, by the semicontinuity, $d(s, f)$ is also the generic value of the above map. The following dimension estimate holds

$$d(s, f) \leq \min\{s_0(f_1 + f_2 + f_3 - 2) + s_1 f_1 + s_2 f_2 + s_3 f_3, f_1 f_2 f_3\}.$$

(20.12) Definition. A configuration s is said to *fill a format* f, $s \succ f$, iff $d(s, f) = f_1 f_2 f_3$. The configuration s is said to *fill f exactly*, $s \asymp f$, iff

$$d(s, f) = s_0(f_1 + f_2 + f_3 - 2) + s_1 f_1 + s_2 f_2 + s_3 f_3 = f_1 f_2 f_3. \qquad \bullet$$

So $s \succ f$ means that $\Sigma_f(t; x, y, z) = T(f)$ for Zariski almost all $(t; x, y, z)$ in $S(f)^{|s|}$. The configuration s fills f exactly iff the right-hand side of (20.11) is a direct sum decomposition of $T(f)$ for almost all $(t; x, y, z)$.

We list some of the properties of the relations $\succ$ and $\asymp$.

(20.13) Lemma. (1) *The relations $\succ$ and $\asymp$ are invariant under simultaneous permutation of the components of f and the last three components of s.*
(2) *If $S \geq s$, $f \geq F$ componentwise, then $s \succ f$ implies $S \succ F$.*
(3) *$(r; 0, 0, 0) \succ f \implies \underline{R}(f) \leq r$.*
(4) *$(r; 0, 0, 0) \asymp f \implies f$ perfect and $\underline{R}(f) = r$.*

Proof. (1) Clear.

(2) The implication $s \succ f \Rightarrow S \succ f$ is obvious. It remains to prove that $s \succ f$ implies $s \succ F$, where w.l.o.g. $F = (f_1, f_2, f_3 - 1)$. Let $T(f) = U_1 \otimes U_2 \otimes U_3$, $T(F) = U_1 \otimes U_2 \otimes U_3'$, where $U_3' \subseteq U_3$ is a subspace of codimension one. Choose a surjective linear map $\pi: U_3 \to U_3'$. The linear map $\mathrm{id} \otimes \mathrm{id} \otimes \pi: T(f) \to T(F)$ induces a map $S(f)^{|s|} \to S(F)^{|s|}$. Let $(\tilde{t}; \tilde{x}, \tilde{y}, \tilde{z})$ be the image of $(t; x, y, z) \in S(f)^{|s|}$ under this map. It is straightforward to verify that

$$(\mathrm{id} \otimes \mathrm{id} \otimes \pi)\Sigma_f(t; x, y, z) = \Sigma_F(\tilde{t}; \tilde{x}, \tilde{y}, \tilde{z}).$$

This shows the implication $s \succ f \Rightarrow s \succ F$.

(3) By Lemma (20.10)(2) the relation $(r; 0, 0, 0) \succ f$ implies $X_r(f) = T(f)$, hence $\underline{R}(f) \leq r$.

(4) Assume $(r; 0, 0, 0) \asymp f$. By (3) and the definition we have $\dim X_r(f) = r(f_1 + f_2 + f_3 - 2) = f_1 f_2 f_3$, in particular $Q(f) = r \in \mathbb{N}$. It remains to show that the format f is good. Let $r' < r$. The morphism $X_{r'}(f) \times S(f)^{r-r'} \to T(f)$, $(t_1, t_2) \mapsto t_1 + t_2$ is dominant, hence

$$r \dim S = \dim T(f) \leq \dim X_{r'}(f) + (r - r') \dim S(f).$$

Therefore $\dim X_{r'}(f) = r'(f_1 + f_2 + f_3 - 2)$, and we see that f is good. $\qquad \square$

Our strategy to establish a relation $s_+ \succ f_+$ will be to split up f_+ into smaller formats f, f' and s_+ into smaller configurations s, s' according to certain rules, and then to derive $s_+ \succ f_+$ from the relations $s \succ f$, $s' \succ f'$. (For $\asymp$ we proceed analogously.) This is expressed in the following splitting lemma I.

(20.14) Splitting Lemma I. *If we have*

$$
\begin{aligned}
s &= (r; a, b, c + r'), & f &= (m, n, p), \\
s' &= (r'; a', b', c + r), & f' &= (m, n, p'), \\
s_+ &= (r + r'; a + a', b + b', c), & f_+ &= (m, n, p + p'),
\end{aligned}
$$

then $d(s_+, f_+) \geq d(s, f) + d(s', f')$. In particular,

$$
\begin{aligned}
s \succ f, \ s' \succ f' &\Rightarrow s_+ \succ f_+, \\
s \asymp f, \ s' \asymp f' &\Rightarrow s_+ \asymp f_+.
\end{aligned}
$$

Similar rules hold when the first or second component of the format are distinguished.

Proof. Let $T(f) = U \otimes V \otimes W$, $T(f') = U \otimes V \otimes W'$, and finally, let $T(f_+) = U \otimes V \otimes (W \oplus W')$. Observe that for a triad $\tau \in S(f)$

(A)
$$
\begin{aligned}
T(f_+) \diamond_i \tau &= T(f) \diamond_i \tau = T(f') \diamond_i \tau & \text{for } i = 1, 2, \\
T(f_+) \diamond_3 \tau &= T(f) \diamond_3 \tau + T(f') \diamond_3 \tau.
\end{aligned}
$$

Now choose

$$
\begin{aligned}
t &\in S(f)^r, & x &\in S(f)^a, & y &\in S(f)^b, & z &\in S(f)^c, \\
t' &\in S(f')^{r'}, & x' &\in S(f')^{a'}, & y' &\in S(f')^{b'}.
\end{aligned}
$$

Let $tt' \in S(f_+)^{r+r'}$ denote the concatenation of t and t'. Using (A) we obtain

$$
\Sigma_{f_+}(tt'; xx', yy', z) =
$$

$$
\sum_{\rho=1}^{r} \left(T(f) \diamond_1 t_\rho + T(f) \diamond_2 t_\rho + T(f) \diamond_3 t_\rho + T(f') \diamond_3 t_\rho \right)
$$

$$
+ \sum_{\rho=1}^{r'} \left(T(f') \diamond_1 t'_\rho + T(f') \diamond_2 t'_\rho + T(f') \diamond_3 t'_\rho + T(f) \diamond_3 t'_\rho \right)
$$

$$
+ \sum_{\alpha=1}^{a} T(f) \diamond_1 x_\alpha + \sum_{\alpha=1}^{a'} T(f') \diamond_1 x'_\alpha
$$

$$
+ \sum_{\beta=1}^{b} T(f) \diamond_2 y_\beta + \sum_{\beta=1}^{b'} T(f') \diamond_2 y'_\beta
$$

$$
+ \sum_{\gamma=1}^{c} \left(T(f) \diamond_3 z_\gamma + T(f') \diamond_3 z_\gamma \right).
$$

This means

$$\Sigma_{f_+}(tt'; xx', yy', z) = \Sigma_f(t; x, y, zt') \oplus \Sigma_{f'}(t'; x', y', zt).$$

(The sum is direct, as the first summand is contained in $T(f)$, the second summand is contained in $T(f')$, and $T(f) \cap T(f') = 0$.) As $\dim \Sigma_f$ and $\dim \Sigma_{f'}$ depend semicontinuously on their arguments, we infer that

$$d(s_+, f_+) \geq d(s, f) + d(s', f').$$

(In fact, we may find t, t', x, x', y, y', z such that $\dim \Sigma_f(t; x, y, zt') = d(s, f)$ and $\dim \Sigma_{f'}(t'; x', y', zt) = d(s', f')$ hold *simultaneously*!)

This proves $s \succ f$, $s' \succ f' \Rightarrow s_+ \succ f_+$. In order to settle the other implication, one only needs to verify that

$$
\begin{aligned}
& r(m + n + p - 2) + am + bn + (c + r')p \\
+\ & r'(m + n + p' - 2) + a'm + b'n + (c + r)p' \\
=\ & (r + r')(m + n + p + p' - 2) + (a + a')m + (b + b')n + c(p + p'). \quad \square
\end{aligned}
$$

This lemma allows to derive successively "more complicated" relations $s \succ f$ starting with simpler ones, as exemplified in the following lemma.

(20.15) Lemma. *For all $a, b, c, d \in \mathbb{N}$ the following relations hold:*

(1) $(1; 0, 0, 0) \asymp (1, 1, a)$, $(0; 0, 0, 1) \asymp (1, 1, a)$ *if $a > 0$,*
(2) $(0; bc, 0, 0) \asymp (a, b, c)$ *if $abc > 0$,*
(3) $(0; bc, ad, 0) \asymp (a, b, c + d)$ *if $ab(c + d) > 0$,*
(4) $(1; ab, 0, 0) \asymp (1, a + 1, b + 1)$,
(5) $(a; b, 0, 0) \asymp (a, 2, a + b)$ *if $a > 0$,*
(6) $(2a; 0, ab, 0) \asymp (2a + b, 2, 2a)$ *if $a > 0$,*
(7) $(2a; 0, 2ab + 2ac + 4bc, 0) \asymp (2a + 2b, 2, 2a + 2c)$ *if $(a + b)(a + c) > 0$,*
(8) $(2ad; 0, 0, 2a(b + c - d + 1) + 4bc) \asymp (2a + 2b, 2a + 2c, 2d)$
    if $(a + b)(a + c)d > 0$ and $a(b + c - d + 1) + 2bc \geq 0$,
(9) $(2ad; 0, 0, 0) \succ (2a + 2b, 2a + 2c, 2d)$
    if $(a + b)(a + c)d > 0$ and $a(b + c - d + 1) + 2bc \leq 0$.

(20.16) Remark. Part five of the above lemma implies $(a; 0, 0, 0) \asymp (a, 2, a)$. Using Lemma (20.13)(4) we again conclude that the formats $(a, 2, a)$ of 2-slice tensors are perfect.   ●

Proof of Lemma (20.15). (1) Clear.

(2) We proceed by induction on bc. The start $bc = 1$ is given by (1) (after a permutation). For the inductive step we may assume that $c > 1$. By the inductive hypothesis we have

$$(0; b(c - 1), 0, 0) \asymp (a, b, c - 1), \quad (0; b, 0, 0) \asymp (a, b, 1).$$

Splitting lemma I implies $(0; bc, 0, 0) \asymp (a, b, c)$.

(3) By (2) we may assume that $cd > 0$. Then (2) gives us

$$(0; bc, 0, 0) \asymp (a, b, c), (0; 0, ad, 0) \asymp (a, b, d).$$

Splitting lemma I implies $(0; bc, ad, 0) \asymp (a, b, c + d)$.

(4) By (1) we may assume that $a > 0$. Then (1) and (3) give us

$$(1; 0, 0, 0) \asymp (1, 1, b + 1), (0; ab, 1, 0) \asymp (1, a, b + 1).$$

Hence, by splitting lemma I with the second component distinguished, we get that $(1; ab, 0, 0) \asymp (1, a + 1, b + 1)$.

(5) We proceed by induction on a. The start $a = 1$ follows from (4). For the step we have by the inductive hypothesis and (4)

$$(a-1; b+1, 0, 0) \asymp (a-1, 2, a-1+b+1), (1; b+a-1, 0, 0) \asymp (1, 1+1, a+b).$$

Hence, by splitting lemma I $(a; b, 0, 0) \asymp (a, 2, a + b)$.

(6) We proceed by induction on b. The start $b = 0$ is covered by (5). For the step we have by the inductive hypothesis and (3)

$$(2a; 0, a(b - 1), 0) \asymp (2a + b - 1, 2, 2a), (0; 2a, a, 0) \asymp (1, 2, a + a).$$

Hence, by splitting lemma I $(2a; 0, ab, 0) \asymp (2a + b, 2, 2a)$.

(7) We may assume that $a > 0$ since by (3) $(0; 0, 4bc, 0) \asymp (2b, 2, 0 + 2c)$. We also may assume that $bc > 0$ as by (6)

$$(2a; 0, 2ab, 0) \asymp (2a + 2b, 2, 2a), (2a; 0, 2ac, 0) \asymp (2a, 2, 2a + 2c).$$

By (5) and (3) we have

$$(2a; 0, 0, 0) \asymp (2a, 2, 2a), (0; 0, 2ac, 2a) \asymp (a + a, 2, 2c).$$

Thus splitting lemma I implies $(2a; 0, 2ac, 0) \asymp (2a, 2, 2a+2c)$. (3) gives us also $(0; 2a, 2ab+4bc, 0) \asymp (2b, 2, a + (a+2c))$. By applying splitting lemma I again we obtain

$$(2a; 0, 2ab + 2ac + 4bc, 0) \asymp (2a + 2b, 2, 2a + 2c).$$

(8) We proceed by induction on d: the start $d = 1$ is covered by (7). Inductively we assume

$$(2a(d - 1); 0, 0, 2a(b + c - d + 2) + 4bc) \asymp (2a + 2b, 2a + 2c, 2(d - 1))$$

and apply splitting lemma I to this and to the start concluding

$$(2ad; 0, 0, 2a(b + c - d + 1) + 4bc) \asymp (2a + 2b, 2a + 2c, 2d).$$

(9) For fixed a, b, c choose d_0 such that

$$2a \geq 2a(b + c - d_0 + 1) + 4bc > 0.$$

We are going to show by induction on $e \geq 1$ that

(A) $(2a(d_0 + e); 0, 0, 0) \succ (2a + 2b, 2a + 2c, 2(d_0 + e))$.

From (8) we conclude

$$(2ad_0; 0, 0, 2a(b + c - d_0 + 1) + 4bc) \asymp (2a + 2b, 2a + 2c, 2d_0).$$

Hence, by Lemma (20.13)(2),

$$(2ad_0; 0, 0, 2a) \succ (2a + 2b, 2a + 2c, 2d_0).$$

Analogously we obtain for all $e \geq 1$

(B) $(2a; 0, 0, 2a(d_0 + e - 1)) \succ (2a + 2b, 2a + 2c, 2)$.

From these two relations (the second with $e = 1$) we get by splitting lemma I the induction start

$$(2a(d_0 + 1); 0, 0, 0) \succ (2a + 2b, 2a + 2c, 2(d_0 + 1)).$$

Let $e > 1$. By the induction hypothesis and Lemma (20.13) we have

$$(2a(d_0 + e - 1); 0, 0, 2a) \succ (2a + 2b, 2a + 2c, 2(d_0 + e - 1)).$$

Together with (B) and splitting lemma I this implies the assertion (A).   □

Proof of Thm. (20.9). Let $f \in 2(\mathbb{N}')^3$ be such that $1 \leq f_1 \leq f_2 \leq f_3$. Further, let $a := \lceil f_1 f_2/(f_1 + f_2 + f_3 - 2) \rceil$. Define the natural numbers

$$b := f_1/2 - a, \ c := f_2/2 - a, \ d := f_3/2.$$

A calculation shows that

(A) $2a(b + c - d + 1) + 4bc = f_1 f_2 - a(f_1 + f_2 + f_3 - 2) \leq 0$.

From Lemma (20.15)(9) we may conclude that the configuration $(af_3; 0, 0, 0)$ fills the format f, hence by Lemma (20.13)

(B) $$\underline{R}(f) \leq af_3 = \left\lceil \frac{f_1 f_2}{f_1 + f_2 + f_3 - 2} \right\rceil f_3.$$

If, additionally, $f_1 f_2/(f_1 + f_2 + f_3 - 2)$ is an integer, then we have equality in (A) and Lemma (20.15)(8) implies that $(af_3; 0, 0, 0) \asymp f$. Hence, by Lemma (20.13), the format f is perfect.

To prove the asymptotic statement we may w.l.o.g. assume that $f \in 2(\mathbb{N}')^3$ and $1 \leq f_1 \leq f_2 \leq f_3$ (use Prop. (20.4)(1)). From (B) and Rem. (20.8) we get

$$1 \leq \underline{R}(f)/Q(f) \leq 1 + f_3/Q(f)$$

implying the asymptotic assertion if $f_3/(f_1 f_2) \to 0$ as $f_1 \to \infty$. However, if $f_3/(f_1 f_2)$ is bounded from below by a positive constant, then we can use Prop. (20.4)(5) and obtain

$$1 \leq \underline{R}(f)/Q(f) \leq 1 + (f_1 + f_2 - 2)/f_3 \leq 1 + O(1/f_1)$$

as $f_1 \to \infty$.   □

20.3* Dimension of Configurations in Formats

We present here an extension of the method described in the previous section to compute the dimension $d(s, f)$ of a configuration s in a format f. Our goal is to supply a proof of Prop. (20.20) which states that for formats $f = (n, n, 3)$ of 3-slice tensors the codimension of the secant variety $X_r(f)$ in $T(f)$ is at most one, where $r = (3n - 1)/2$, n odd. This result was used in Chap. 19. Our discussion is of a fairly technical nature and may be skipped by the reader at first reading.

In the following we are going to consider *pairs of configurations* (s, s') and *pairs of formats* (f, f'), where we always assume that $s \geq s'$, $f \geq f'$ component-wise. We think of $T(f')$ as being a subspace of $T(f)$. The map

$$S(f)^{|s-s'|} \times S(f')^{|s'|} \longrightarrow N$$
$$((t; x, y, z), (t'; x', y', z')) \longmapsto \dim(\Sigma_f(tt'; xx', yy', zz') + T(f'))/T(f')$$

is Zariski lower semicontinuous. (This can be shown as in Ex. 20.2.) We denote its maximum value by $d((s, s'), (f, f'))$ and call it the *dimension of (s, s') in (f, f')*. By the semicontinuity, it is also the generic value of the above map.

(20.17) Lemma. (1) $(0; 3, 3, 0) \asymp (2, 2, 3)$.
(2) *Let $s = (3; 1, 1, 0)$, $s' = (0; 1, 1, 0)$, $f = (3, 3, 3)$, $f' = (1, 1, 3)$. Then we have $d((s, s'), (f, f')) = 24$.* •

Proof. The proof is by direct verification.
(1) Let (e_1, e_2), (f_1, f_2), (g_1, g_2, g_3) be bases of U, V, W, respectively. Some computation shows that

$$U \otimes V \otimes W =$$
$$U \otimes f_2 \otimes g_2 + U \otimes (f_1 + f_2) \otimes (g_1 + g_2) + U \otimes (f_1 - f_2) \otimes (g_1 + g_3)$$
$$+ e_2 \otimes V \otimes g_3 + e_1 \otimes V \otimes (g_1 - g_2) + (e_1 + e_2) \otimes V \otimes (g_2 + g_3).$$

(2) Let (e_1, e_2, e_3), (f_1, f_2, f_3), (g_1, g_2, g_3) be bases of U, V, W, respectively. We may identify $T(f) = U \otimes V \otimes W$, $T(f') = e_1 \otimes f_1 \otimes W$. Some computation shows that

$$\sum_{\rho=1}^{3} \sum_{i=1}^{3} T(f) \diamond_i t_\rho + U \otimes f_1 \otimes (g_2 + g_3) + e_1 \otimes V \otimes g_1 + T(f') = T(f),$$

where $t_1 = e_2 \otimes f_2 \otimes g_2$, $t_2 = e_3 \otimes f_3 \otimes g_3$, $t_3 = (e_1 + e_2 + e_3) \otimes (f_1 + f_2 + f_3) \otimes g_1$. □

(20.18) Lemma. *We have $d(s, f) \geq d(s', f') + d((s, s'), (f, f'))$.*

Proof. Let $\pi: T(f) \to T(f)/T(f')$ denote the canonical projection. For elements $u = (t; x, y, z)$ in $S(f)^{|s-s'|}$ and $u' = (t'; x', y', z')$ in $S(f')^{|s'|}$ we define $uu' := (tt'; xx', yy', zz')$. As $\Sigma_{f'}(u')$ is contained in $T(f') = \ker \pi$, we get

$$\dim \Sigma_f(uu') \geq \dim \pi(\Sigma_f(uu')) + \dim \Sigma_{f'}(u').$$

By the semicontinuity we know that the sets

$$\{u' \mid \dim \Sigma_{f'}(u') = d(s', f')\}, \quad \{uu' \mid \dim \pi(\Sigma_f(uu')) = d((s, s'), (f, f'))\}$$

are nonempty open subsets of $S(f')^{|s'|}$ and $S(f)^{|s-s'|} \times S(f')^{|s'|}$, respectively. Hence there exist u, u' such that $\Sigma_{f'}(u') = d(s', f')$ and $\dim \pi(\Sigma_f(uu')) = d((s, s'), (f, f'))$ hold simultaneously. As $d(s, f) \geq \dim \Sigma_f(uu')$, the assertion follows.   □

In the previous section we computed lower bounds for $d(s_+, f_+)$ by the following strategy: we split up f_+, s_+ into smaller formats f, f' and configurations s, s' according to splitting lemma I (20.14) and achieved $d(s_+, f_+) \geq d(s, f) + d(s', f')$. We will proceed here in an analogous way. However, as we are dealing with pairs of formats, respectively configurations, the splitting rule is now more complicated.

(20.19) Splitting Lemma II. *If we have*

$$
\begin{aligned}
(s_+, s'_+) &= \big((r + a + b + c; 0, 0, 0), \\
&\qquad (r' + a + b + c; 0, 0, 0)\big),
\end{aligned}
$$

$$
\begin{aligned}
(f_+, f'_+) &= \big((m + \overline{m}, n + \overline{n}, p), \\
&\qquad (m' + \overline{m}, n' + \overline{n}, p)\big),
\end{aligned}
$$

$$(s, s') = \big((r; b, a, 0), (r'; b, a, 0)\big),$$

$$(f, f') = \big((m, n, p), (m', n', p)\big),$$

$$g = (m - m', \overline{n}, p),$$

$$h = (\overline{m}, n - n', p),$$

then

$$d((s_+, s'_+), (f_+, f'_+)) \geq$$
$$d((s, s'), (f, f')) + d((0; a + c, r - r', 0), g) + d((0; r - r', b + c, 0), h).$$

Proof. Let $U_1, U_2, U_3, V_1, V_2, V_3$, and W be k-spaces of dimensions $\overline{m}, m', m - m', \overline{n}, n', n - n', p$, respectively and write $g' := (m', \overline{n}, p)$, $h' := (\overline{m}, n', p)$, $\overline{f} := (\overline{m}, \overline{n}, p)$. We may assume $T(f) = (U_2 \oplus U_3) \otimes (V_2 \oplus V_3) \otimes W$ and

$$T(f') = U_2 \otimes V_2 \otimes W, \quad T(g) = U_3 \otimes V_1 \otimes W, \quad T(h) = U_1 \otimes V_3 \otimes W,$$
$$T(g') = U_2 \otimes V_1 \otimes W, \quad T(h') = U_1 \otimes V_2 \otimes W, \quad T(\overline{f}) = U_1 \otimes V_1 \otimes W.$$

By decomposing the corresponding subspaces we see that for $x \in S(g')$

$$\sum_{i=1}^{3} T(f_+) \diamond_i x = \sum_{i=1}^{3} T(g') \diamond_i x + T(\overline{f}) \diamond_1 x + T(g) \diamond_1 x + T(f) \diamond_2 x,$$

hence

$$\sum_{i=1}^{3} T(f_+) \diamond_i x \equiv T(g) \diamond_1 x + T(f) \diamond_2 x \mod T(f_+').$$

Analogously for $y \in S(h')$, $z \in S(\overline{f})$

$$\sum_{i=1}^{3} T(f_+) \diamond_i y \equiv T(f) \diamond_1 y + T(h) \diamond_2 y \mod T(f_+'),$$

$$\sum_{i=1}^{3} T(f_+) \diamond_i z \equiv T(g) \diamond_1 z + T(h) \diamond_2 z \mod T(f_+').$$

Moreover, if $t = (u_2 + u_3) \otimes (v_2 + v_3) \otimes w \in S(f)$, $u_i \in U_i$, $v_j \in V_j$, then

$$\sum_{i=1}^{3} T(f_+) \diamond_i t \equiv \sum_{i=1}^{3} T(f) \diamond_i t + T(h) \diamond_1 t_h + T(g) \diamond_2 t_g \mod T(f_+'),$$

where $t_g := u_3 \otimes (v_2 + v_3) \otimes w$, $t_h := (u_2 + u_3) \otimes v_3 \otimes w$. Now choose

(A) $t \in S(f)^{r-r'}$, $t' \in S(f')^{r'}$, $x \in S(g')^a$, $y \in S(h')^b$, $z \in S(\overline{f})^c$.

We conclude from the above observations that

$$\Sigma_{f_+}(tt'xyz; 0, 0, 0) \equiv$$
$$\Sigma_f(tt'; y, x, 0) + \Sigma_g(0; xz, t_g, 0) + \Sigma_h(0; t_h, yz, 0) \mod T(f_+'),$$

where $t_g := ((t_1)_g, \ldots, (t_{r-r'})_g)$, $t_h := ((t_1)_h, \ldots, (t_{r-r'})_h)$. It is possible to make the choice (A) such that

$$\dim(\Sigma_f(tt'; y, x, 0) + T(f'))/T(f') = d((s, s'), (f, f')),$$
$$\dim \Sigma_g(0; xz, t_g, 0) = d((0; a+c, r-r', 0), g),$$
$$\dim \Sigma_h(0; t_h, yz, 0) = d((0; r-r', b+c, 0), h)$$

simultaneously hold! The assertion follows as

$$d((s_+, s_+'), (f_+, f_+')) \geq \dim(\Sigma_{f_+}(tt'xyz; 0, 0, 0) + T(f_+'))/T(f_+').$$

$\square$

(20.20) Proposition (Strassen). *For odd $n \geq 3$ we have* $\dim X_{r_n}(f_n) \geq 3n^2 - 1$, *where $f_n := (n, n, 3)$, $r_n := (3n-1)/2$.*

(20.21) Remark. In Sect. 19.3 it was shown that $X_{r_n}(f_n) \neq T(f_n)$, thus the above proposition implies that $X_{r_n}(f_n)$ is a hypersurface in $T(f_n)$.   •

Proof of Prop. (20.20). Put $s_n := (r_n; 0, 0, 0)$. By Lemma (20.10) it is enough to prove that $d(s_n, f_n) \geq 3n^2 - 1$. We show this by induction on n.

Let $n = 3$. By splitting lemma I we have

$$d((4; 0, 0, 0), (3, 3, 3)) \geq d((3; 1, 0, 0), (2, 3, 3)) + d((1; 3, 0, 0); (1, 3, 3)).$$

Moreover, $d((3; 1, 0, 0), (2, 3, 3)) = 18$ as $(3; 0, 0, 0) \times (2, 3, 3)$, see (20.15)(5). By direct verification or from (20.15)(4) we get $d((1; 3, 0, 0), (1, 3, 3)) = 8$. Therefore $d((4; 0, 0, 0), (3, 3, 3)) \geq 26$.

Assume $n \geq 5$. We conclude from Lemma (20.18) that

$$d(s_n, f_n) \geq d(s_{n-2}, f_{n-2}) + d((s_n, s_{n-2}), (f_n, f_{n-2})).$$

By the induction hypothesis it is therefore sufficient to prove that

(A) $$d_n := d((s_n, s_{n-2}), (f_n, f_{n-2})) \geq 3(n^2 - (n-2)^2).$$

We show (A) by induction on $n \geq 5$. Let us verify the induction step first. Splitting lemma II (with $a = b = 0$, $c = 3$) implies

$$d_n \geq d_{n-2} + 2d((0; 3, 3, 0), (2, 2, 3)).$$

By Lemma (20.17)(1) we have $d((0; 3, 3, 0), (2, 2, 3)) = 12$. Therefore, we get $d_n \geq d_{n-2} + 24$, and statement (A) follows by the induction hypothesis.

It remains to verify that $d_5 \geq 48$. By splitting lemma II (with $a = b = 1$, $c = 2$) we obtain

$$d_5 \geq d((s, s'), (f, f')) + 2d((0; 3, 3, 0), (2, 2, 3)),$$

where $s = (3, 1, 1, 0)$, $s' = (0; 1, 1, 0)$, $f = (3, 3, 3)$, $f' = (1, 1, 3)$. Using Lemma (20.17)(2) we conclude that in fact $d_5 \geq 24 + 2 \cdot 12 = 48$.   □

20.4 Exercises

20.1. Let $A \in k[X_1, \ldots, X_n]^{m \times p}$ be a matrix over a polynomial ring. Prove that the map

$$k^n \to \mathbb{N}, \; \xi \mapsto \mathrm{rk}\, A(\xi)$$

is Zariski lower semicontinuous.

20.2. Prove that the map $S(f)^{|s|} \to \mathbb{N}$, $(t; x, y, z) \mapsto \dim \Sigma_f(t; x, y, z)$ is Zariski lower semicontinuous.

20.3. Let $f = (m, n, 1)$, $m \leq n$, and assume $\operatorname{char} k = 0$. Using Lemma (20.10) prove that $\dim X_r(f) = r(m + n - r)$ for $0 \leq r \leq m$. (In fact, this is true for any characteristic.) Conclude that the format f is not good if $m \geq 2$.

20.4. Let $f = (m, n, 2)$, $m \leq n$. Prove by the method of this chapter that $\underline{R}(f) = \min\{n, 2m\}$. Give an alternative proof using the normal form of pencils discussed in Chap. 19.

20.5.* Let $m \leq n$. Prove that

$$R(m, n, 2) = \min\{2m, m + \lfloor n/2 \rfloor\}.$$

Conclude that $\underline{R}(m, n, 2) < R(m, n, 2)$ if $\lceil n/2 \rceil < m$. (Hint: use Thm. (19.4).)

20.6. Let $f = (m, n, p)$ and assume $(m - 1)(n - 1) + 1 \leq p \leq mn$. Show that

(1) $\underline{R}(f) = p$.
(2) f is perfect if $p = (m - 1)(n - 1) + 1$.
(3) f is not good if $p > (m - 1)(n - 1) + 2$.

20.7. A configuration s is said to be *of full dimension in a format* f, $s \prec f$, iff $d(s, f) = s_0(f_1 + f_2 + f_3 - 2) + s_1 f_1 + s_2 f_2 + s_3 f_3$. Obviously $s \asymp f$ iff $s \prec f$ and $s \succ f$. Prove the following statements.

(1) If $S \leq s$ and $f \leq F$ componentwise, then $s \prec f$ implies $S \prec F$. (Hint: use splitting lemma I.)
(2) $(r; 0, 0, 0) \prec f$ implies that $\dim X_\rho(f) = \rho \dim S(f)$ for all $\rho \leq r$.
(3) A format f is good iff $(\lfloor Q \rfloor; 0, 0, 0) \prec f$ and $(\lceil Q \rceil; 0, 0, 0) \succ f$, where $Q := f_1 f_2 f_3 / (f_1 + f_2 + f_3 - 2)$.

20.8. Prove that $\underline{R}(n, n, 3) = (3n+1)/2$ for odd $n \geq 3$ without using the technique developed in Sect. 20.3. Show in the same way that $\underline{R}(n, n, 3) = 3n/2$ if n is even. (Hint: prove that $(0; r - 3, 3, 0) \succ (2, n - 2, 3)$, where $r = (3n + 1)/2$ if n is odd, and where $r = 3n/2$ if n is even.)

20.9. Show that the format $(n, n, 4)$ is good if $n \geq 3$. In particular, $\underline{R}(n, n, 4) = 2n - 1$. Deduce from $\underline{R}(4, 4, 4) = 7$ a nonconstructive proof for the existence of fast matrix multiplication algorithms! (Hint: show that $(n - 1; 0, 0, n) \asymp (n, n, 2)$ for $n \geq 3$. The case where n is odd is easier.)

20.10.* Let $f_n = (n, n, 3)$, $n \geq 3$ odd. Proceeding exactly as in the proof of Prop. (20.20) show that for $0 \leq r \leq (3n - 3)/2$ we have

$$\dim X_r(f_n) = r(2n + 1).$$

20.5 Open Problems

Problem 20.1. Determine all good formats!

Problem 20.2. In Sect. 19.3 we found the irreducible generator F_n of the vanishing ideal of $X_r(f)$ for the format $f = (n, n, 3)$, where $r = (3n-1)/2$, n odd. Construct generators of $X_r(f)$ for other formats! The representation theory of the symmetric, resp. full linear group might be helpful for this.

20.6* Appendix: Topological Degeneration

In this appendix we want to prove that the degeneration order for bilinear maps, which was introduced in Sect. 15.4 in a purely algebraic way, can also be characterized in terms of the Zariski topology. This will also give us the topological description of the notion of border rank expressed in Thm. (20.3). The proof of this characterization relies on some deeper facts from commutative algebra and algebraic geometry. We recall that the field k is always assumed to be algebraically closed.

(20.22) Definition. Let s, t be tensors over k, $s \in U \otimes V \otimes W$. The tensor s is said to be a *topological degeneration* of t, $s \trianglelefteq_{\text{top}} t$, iff s lies in the Zariski closure of the subset $\{s' \in U \otimes V \otimes W \mid s' \leq t\}$ of tensors which are a restriction of t. •

Let us first formulate this in a somewhat different way. The linear group $G := \text{GL}(U) \times \text{GL}(V) \times \text{GL}(W)$ acts rationally on $U \otimes V \otimes W$ via $g \cdot t := (\alpha \otimes \beta \otimes \gamma)(t)$, where $g = (\alpha, \beta, \gamma) \in G$ and $t \in U \otimes V \otimes W$. The orbits $G \cdot t$ are the isomorphism classes of tensors living in $U \otimes V \otimes W$. Let $s, t \in U \otimes V \otimes W$, $E := \text{End}(U) \times \text{End}(V) \times \text{End}(W)$. Since

$$\{s' \in U \otimes V \otimes W \mid s' \leq t\} = \{(\alpha \otimes \beta \otimes \gamma)(t) \mid (\alpha \otimes \beta \otimes \gamma) \in E\},$$

and G is dense in E, we see that

$$(20.23) \qquad\qquad s \trianglelefteq_{\text{top}} t \iff s \in \overline{G \cdot t},$$

where the closure is taken with respect to the Zariski topology. Thus, the topological degenerations of t are just the elements in the closure of the orbit of t under the above action of the group G.

Our goal is to prove the following theorem.

(20.24) Theorem (Strassen). *Topological degeneration and degeneration coincide, i.e., $s \trianglelefteq_{\text{top}} t$ iff $s \trianglelefteq t$ for tensors s, t over an algebraically closed field k.*

We note that Alder's theorem (20.3) is an immediate consequence of this result. Another implication is the following.

(20.25) Corollary. *The degeneration order is a partial order on the semiring T of tensor classes (compare Prop. (15.25)).*

Proof. We need to show that for all tensors $s, t \in U \otimes V \otimes W$

$$s \leq_{\mathrm{top}} t, \ t \leq_{\mathrm{top}} s \ \Rightarrow \ s \simeq t.$$

Assume $s \leq_{\mathrm{top}} t$ and $s \not\simeq t$. Then, by (20.23), $s \in \overline{G \cdot t} \setminus G \cdot t$. Since this set is stable under the action of G we infer that $G \cdot s \subseteq \overline{G \cdot t} \setminus G \cdot t$. However, it is a well-known fact that $G \cdot t$ is open in its closure (cf. Humphreys [258, p. 60]). Hence we obtain $\overline{G \cdot s} \subseteq \overline{G \cdot t} \setminus G \cdot t$. Therefore, t is not a topological degeneration of s. $\qquad\square$

Our proof of Thm. (20.24) is based on three preparatory lemmas.

(20.26) Lemma. *Let X be an irreducible affine variety of dimension $n \geq 1$, $U \subseteq X$ be a nonempty open subset and $x \in X$. Then there exists an irreducible curve $C \subseteq X$ containing x and meeting U.*

Proof. W.l.o.g. $x \notin U$. By the Noether normalization theorem (cf. Lang [316, Chap. X, §4] or Matsumura [356, §33, Lemma 2]) there exists a finite morphism $\pi \colon X \to k^n$. Let $y' \in k^n \setminus \pi(X \setminus U)$ and C' be the line through $\pi(x)$ and y'. The going-down theorem (cf. Matsumura [356, §9, Thm. 9.4]) tells us that there exists a curve $C \subseteq X$ containing x such that $\pi(C) = C'$. Hence $C \cap U \neq \emptyset$. $\qquad\square$

(20.27) Lemma. *Let $\varphi \colon X \to Y$ be a morphism of affine varieties and let y be in $\overline{\mathrm{im}\, \varphi} \setminus \mathrm{im}\, \varphi$. Then there exist irreducible curves $X_1 \subseteq X$, $Y_1 \subseteq Y$ with $y \in Y_1$ and $\overline{\varphi(X_1)} = Y_1$.*

Proof. We may assume that φ is dominant, Y is irreducible, and $\dim Y \geq 1$. (Replace Y by an irreducible component of $\overline{\mathrm{im}\, \varphi}$ containing y.) Let $U \subseteq \mathrm{im}\, \varphi$ be nonempty and open in Y. By Lemma (20.26) there exists an irreducible curve $Y_1 \subseteq Y$ containing y and meeting U. Let X_2 be a component of $\varphi^{-1}(Y_1)$ dominating Y_1. Another application of Lemma (20.26) yields the existence of an irreducible curve $X_1 \subseteq X_2$ which is not contained in a fiber of φ. Thus $\overline{\varphi(X_1)} = Y_1$. $\qquad\square$

In the following $A(X)$ denotes the coordinate ring of an affine variety X. $R := k[[\varepsilon]]$ stands for the ring of formal power series in the indeterminate ε, and $K := k((\varepsilon))$ denotes its field of fractions.

(20.28) Lemma. *Let $\varphi \colon X \to Y$ be a morphism of affine varieties and $y \in \overline{\mathrm{im}\, \varphi}$. Then there exists a commutative diagram*

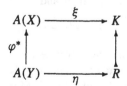

where ξ, η are k-algebra morphisms, the upgoing arrow on the right is the inclusion, and η sends the maximal ideal of y into the maximal ideal of R.

Proof. We may suppose that $y \notin \text{im } \varphi$. By Lemma (20.27) we may assume that X and Y are irreducible curves and φ is dominant. In the following we view $A(Y)$ as a subring of $A(X)$. The function field $k(X)$ is a finite algebraic extension of $k(Y)$; let B denote the integral closure of $A(Y)$ in $k(X)$. Then B is integrally closed with field of fractions $k(X)$, and B is a finite $A(Y)$-module (cf. Zariski and Samuel [571, Chap. V, §4, Thm. 9]). In particular B is a one-dimensional regular noetherian domain (cf. Matsumura [356, §11, Thm. 11.2]. Let $m \subset B$ be the maximal ideal lying over the maximal ideal $m_y \subset A(Y)$ of y (such an m exists by the going-up theorem, cf. Matsumura [356, §9, Thm. 9.4]). The completion $\hat{B}_m$ of the regular local ring B_m is by the Cohen structure theorem (cf. Zariski and Samuel [572, §12, Cor.]) isomorphic to a power series ring R. We thus have a commutative diagram of k-algebra morphisms

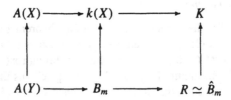

and the composition of the two morphisms of the bottom row sends m_y into the maximal ideal of R.   □

Proof of Thm. (20.24). Let $U = k^m$, $V = k^n$, $W = k^p$, $U \otimes V \otimes W = k^{m \times n \times p}$ for concreteness, and identify the group G with $\text{GL}(m, k) \times \text{GL}(n, k) \times \text{GL}(p, k)$. The k-algebra morphism $R \to k$, $\sum_{i \geq 0} a_i \varepsilon^i \mapsto a_0$, applied componentwise, induces a k-linear map

$$R^{m \times n \times p} \to k^{m \times n \times p}, \; t \mapsto t_{\varepsilon=0}.$$

We have an action of the group $G^K := \text{GL}(m, K) \times \text{GL}(n, K) \times \text{GL}(p, K)$ on $K^{m \times n \times p}$ defined analogously as the action of G. Let $s, t \in k^{m \times n \times p}$. By Ex. 15.11 we know that $s \trianglelefteq t$ iff there exists $g \in G^K$ such that

$$g \cdot t \in R^{m \times n \times p}, \; (g \cdot t)_{\varepsilon=0} = s.$$

Assume $s \trianglelefteq t$. Then there exists $g = (\alpha, \beta, \gamma) \in G^K$ such that $g \cdot t = s + \varepsilon s'$ for some $s' \in R^{m \times n \times p}$. By multiplying with a sufficiently high power of ε we may assume that all entries of α, β, γ lie in R and

(A) $g \cdot t = \varepsilon^h s + \varepsilon^{h+1} s'$

for some $h \in \mathbb{N}$. If we cut off the power series $\alpha_{ii_1}, \beta_{jj_1}, \gamma_{\ell\ell_1}$ after order h we obtain a new $g = (\alpha, \beta, \gamma)$ having entries in the polynomial ring $k[\varepsilon]$ and such that (A) still holds with some new $s' \in k[\varepsilon]^{m \times n \times p}$. Moreover, by adding the multiple of the unit matrix $\varepsilon^{h+1} I_m$ to α, $\varepsilon^{h+1} I_n$ to β and $\varepsilon^{h+1} I_p$ to γ we may achieve that $\det \alpha \det \beta \det \gamma \neq 0$. Therefore $\det \alpha(\theta) \det \beta(\theta) \det \gamma(\theta) \neq 0$ for all $\theta \in k \setminus \mathcal{E}$, where $\mathcal{E}$ is a finite subset of exceptional values. ($\alpha(\theta)$ arises from α by substituting ε with θ.) Hence $\{\theta^{-h} g(\theta) \cdot t \mid \theta \in k \setminus \mathcal{E}, \theta \neq 0\}$ is a subset of the orbit $G \cdot t$ of t which contains s in its Zariski closure. Thus we have shown that $s \trianglelefteq_{\text{top}} t$.

Assume now that $s \trianglelefteq_{\text{top}} t$. By assumption s lies in the closure of the image of the the orbit map $\varphi \colon G \to k^{m \times n \times p}$, $g \mapsto g \cdot t$. By Lemma (20.28) there is a commutative diagram of k-algebra morphisms

$$
\begin{array}{ccc}
A(G) & \xrightarrow{\;\xi\;} & K \\[2pt]
{\scriptstyle \varphi^*}\big\uparrow & & \big\uparrow \\[2pt]
A(k^{m \times n \times p}) & \xrightarrow[\;\eta\;]{} & R
\end{array}
$$

where η sends the maximal ideal of s into the maximal ideal of R. Note that

$$
\begin{aligned}
A(k^{m \times n \times p}) &= k[Y_{ij\ell} \mid i \leq m, \ j \leq n, \ \ell \leq p], \\
A(G) &= k[X^1_{ii_1}, X^2_{jj_1}, X^3_{\ell\ell_1}, (\det X^1 \det X^2 \det X^3)^{-1} \mid \\
& \qquad i, i_1 \leq m, \ j, j_1 \leq n, \ \ell, \ell_1 \leq p],
\end{aligned}
$$

where the $X^\nu_{ii_1}$ and $Y_{ij\ell}$ denote indeterminates over k. We define

$$
\alpha_{ii_1} := \xi(X^1_{ii_1}), \quad \beta_{jj_1} := \xi(X^2_{jj_1}), \quad \gamma_{\ell\ell_1} := \xi(X^3_{\ell\ell_1}), \quad u_{ij\ell} := \eta(Y_{ij\ell}).
$$

It is straightforward to check that $g := (\alpha, \beta, \gamma) \in G^K$, $u \in R^{m \times n \times p}$, $u_{\varepsilon=0} = s$ and $g \cdot t = u$. This shows that $s \trianglelefteq t$ and proves the theorem. $\qquad\square$

20.7 Notes

The lower bound $Q(f) \leq R(f)$ for the maximal rank $R(f)$ of a format f appeared first in Brockett [78], Grigoriev [208] and Howell [255]. Atkinson and Stephens [16] obtained the upper bound $R(m, n, p) \leq m + \lfloor p/2 \rfloor n$ for $m \leq n$. For the typical rank this was subsequently improved by Atkinson and Lloyd [14] to $\underline{R}(m, n, p) \leq \lceil p/2 \rceil n$. With the direct methods used by these authors, an asymptotically sharp determination of the typical rank seems to be impossible. A breakthrough here was independently achieved by Strassen [505] and Lickteig [331] who introduced the idea of computing upper bounds on the typical rank via linearization, as well as the splitting technique. The main result of this chapter, Thm. (20.9), is due to them. Our presentation of the splitting technique, in particular the symbolic calculus introduced in Sect. 20.3, is based on Lickteig's work [329, 331]. Prop. (20.20), which forms a part of Thm. (19.17), was first proved by Strassen [505] by a different method than the one we presented in that section. We remark that Lickteig [331] exactly determined the typical rank for cubic formats. He showed that $\underline{R}(n, n, n) = \lceil Q(f) \rceil$ if $n \neq 3$, whereas $\underline{R}(3, 3, 3) = 5$.

The topological characterization of the border rank in Thm. (20.3) is due to Alder [5]. Its generalization, Thm. (20.24), was shown by Strassen [508]. Our presentation of the proof of this result follows Strassen [508] and Kraft [308, III.2.3]. In fact, this proof goes even back to Hilbert [248]. Lehmkuhl and Lickteig [323] have carried over Thm. (20.3) to the case of real closed fields (in fact, their proof shows that also Thm. (20.24) carries over). They further give explicit upper bounds on the order of degenerations. It is not known whether Cor. (20.25) is true over arbitrary fields k (compare, however, Bürgisser [98]).

For results on the maximal rank of unbalanced formats (i.e., formats (m, n, p) or $(m, n, mn - p)$ with p "small") we refer to Atkinson and Lloyd [15], Bshouty [88] and von zur Gathen [190] (over finite fields).

The result of Ex. 20.4 is in Bini [44]. Ex. 20.5 is mentioned in Atkinson and Stephens [16]. For Ex. 20.6 compare also Bini [46]. Ex. 20.7 is essentially contained in Strassen [505]. Ex. 20.9 and 20.10 are from Lickteig [329]. For a result on the typical rank of associative algebras due to Büchi [95] see Ex. 17.20. Problems 20.1 and 20.2 are posed in Strassen [510].

Part V

Complete Problems

Chapter 21. P Versus NP:
A Nonuniform Algebraic Analogue

One of the most important topics in structural complexity theory is the question, whether nondeterministic polynomial time for Turing machine computations is more powerful than deterministic polynomial time. Some insight into Cook's hypothesis "$P \neq NP$" is provided by polynomial time reductions which allow to single out the NP-complete problems as the hardest problems in NP, see Cook [124], Karp [297], and Levin [328]. This chapter discusses Valiant's nonuniform algebraic analogue of NP-completeness [526, 529] which grew out of his studies of enumeration problems culminating in the theory of #P-completeness. The first section introduces Valiant's algebraic complexity classes **VP** and **VNP**, as well as p-projections as an algebraic analogue of polynomial time reductions. The elements of these classes are certain families of multivariate polynomials over some fixed field k. Valiant's hypothesis "**VP** $\neq$ **VNP**" gains interest by his theorem stating that the family $PER = (PER_n)$ of generic permanents is **VNP**-complete over any field of characteristic different from 2. The proof of this fundamental result, which is the main goal of this chapter, is based on an alternative characterization of **VNP** in terms of the expression or formula size of polynomials, see Sect. 21.2. In the third section it is shown that every polynomial of expression size u is a projection of PER_{2u+2}. Sect. 21.4 presents a simplified proof of Valiant's theorem based on a generalized Laplace expansion theorem. In the last section we prove Brent's theorem [72] which describes the close connection between depth and expression size, as well as results of Hyafil [260] and Valiant, Skyum, Berkowitz, and Rackoff [530] estimating the depth of a polynomial in terms of its degree and complexity. On the basis of these results the extended Valiant hypothesis can be formulated in purely algebraic terms: for any fixed positive constant c there is no possibility of writing all generic permanents as $PER_n = DET_{m(n)}(A)$, where A is an $m(n) \times m(n)$ matrix over $k \cup \{X_{\mu\nu} | \mu, \nu \in \underline{n}\}$ and $m(n) = 2^{O(\log^c n)}$.

21.1 Cook's Versus Valiant's Hypothesis

Let us first recall some basic terminology. An *alphabet* Σ is a finite set with at least two elements, $\Sigma^* := \bigcup_{n \geq 0} \Sigma^n$ denotes the set of all words of finite length over Σ, and $\Sigma^+ := \bigcup_{n \geq 1} \Sigma^n$ is the set of all nonempty words. Every subset A of Σ^* is called a *language* over Σ. Such an A specifies a *decision problem*: given

$x \in \Sigma^*$ one has to decide whether or not x belongs to A, and complexity theory adds to this the question how hard it is to get the correct answer?

One of the fundamental problems in the theory of computability is to separate *tractable* and *intractable* problems. It is widely believed that the class **P** of decision problems that can be solved in polynomial time by a deterministic Turing machine is a good approximation of the informal notion of tractability or feasibility. This class is compared with the class **NP** of those decision problems $A \subseteq \Sigma^*$, where membership of $x \in \Sigma^*$ to A has a short *proof* that can be *verified* in polynomial time. The keen interest in the class **NP** results from the fact that a number of computational problems that come up frequently in practice like the traveling salesman problem, the integer linear programming problem, the Hamiltonian circuit problem, the satisfiability problem SAT, and many others, belong to this class. Moreover, the above mentioned problems are **NP**-complete, i.e., in a sense they are the hardest problems in **NP**. So the "global" question "**P** $\neq$ **NP**?" is equivalent to the "local" question "$SAT \notin$ **P**?". In the sequel we will briefly recall the theory of **NP**-completeness. (For a detailed discussion we refer to Garey and Johnson's book [181].) In doing so, we will also prepare for a nonuniform algebraic analogue of that theory.

A function $t: \mathbb{N} \to \mathbb{N}$ is called *polynomially bounded* (or *p-bounded*), iff it is majorized by a polynomial.

(21.1) Definition. A language A over the alphabet Σ belongs to the class **P** iff the corresponding characteristic function $\chi_A: \Sigma^* \to \{0, 1\}$ is computable by a deterministic Turing machine M in polynomial time, i.e., there exists a p-bounded function $t: \mathbb{N} \to \mathbb{N}$ such that for all $n \in \mathbb{N}$ and for all $x \in \Sigma^n$, M computes $\chi_A(x)$ in at most $t(n)$ steps.   ●

We are going to motivate a nonuniform algebraic analogue of this definition. W.l.o.g. we may assume in the last definition that $\Sigma = \{0, 1\}$. Then the above Turing machine M can be viewed as a machine computing a family $(f_n)_{n \geq 0}$ of Boolean functions, $f_n: \{0, 1\}^n \to \{0, 1\}$ being the indicator of $\Sigma^n \cap A$ in Σ^n. This family corresponds to the partition $A = \sqcup_{n \geq 0}(\Sigma^n \cap A)$. Sometimes it is more natural to use the parameter n in a more liberal way: consider for example the language A of all invertible matrices over the field $\mathbb{F}_2$. This language is partitioned naturally as $A = \sqcup_{n \geq 1} A_n$, where A_n is the set of all invertible matrices of size n. Notice that the Boolean function corresponding to A_n is now a function $f_n: \{0, 1\}^{n^2} \to \{0, 1\}$. As the class of p-bounded functions is closed under composition, this does not matter: $(f_n)_{n \geq 1}$ is computable in time polynomial in n^2 iff it is computable in time polynomial in n. Next notice that every n-ary Boolean function f_n is represented by a polynomial $F_n \in \mathbb{F}_2[X_1, \ldots, X_n]$ of degree at most 1 in each indeterminate; in particular, $\deg F_n \leq n$. The family (f_n) is uniformly described by the Turing machine M. In the sequel we shall study families (f_n) of multivariate polynomials over a field k, but in contrast to the discrete case, we do not require that this family is uniformly describable. (This corresponds to the class **P**/poly consisting of all families (f_n) of Boolean functions that can be computed by a polynomially sized family of Boolean circuits, see, e.g., Balcázar et al. [23].)

Now we are well-prepared to understand Valiant's definition of a nonuniform algebraic analogue of **P**. For a multivariate polynomial f, we denote the minimum number of indeterminates involved in f by $v(f)$. If $v(f) = m$ then w.l.o.g. we may assume that $f \in k[X_1, \ldots, X_m]$. Sometimes we write X instead of $X_1, \ldots, X_m$.

(21.2) Definition. Let k be a field, $X_1, X_2, \ldots$ indeterminates over k.

(1) A sequence $f = (f_n)_{n \geq 1}$ of multivariate polynomials over k is called a *p-family* (of polynomials over k) iff the functions $n \mapsto v(f_n)$ and $n \mapsto \deg f_n$ are *p*-bounded.

(2) A *p*-family $f = (f_n)_{n \geq 1}$ is said to be *p-computable* iff the function that maps n to $L^{\text{tot}}_{k(X)}(f_n \mid X)$ is *p*-bounded.

(3) **VP = VP**(nonuniform; k) denotes Valiant's class of *p*-computable families over k.   •

Two complexity measures L and L' defined for polynomials are *polynomially related* iff for every *p*-family $f = (f_n)$ of polynomials over k, the function $n \mapsto L(f_n)$ is *p*-bounded iff $n \mapsto L'(f_n)$ is *p*-bounded.

In the above definition $L^{\text{tot}}_{k(X)}$ can be replaced by any polynomially related complexity measure. For the rest of this chapter we shall therefore work with more convenient straight-line programs.

(21.3) Definition. For $f \in A := k[X_1, \ldots, X_{v(f)}]$ we define

$$L(f) := L_k(f)$$

as the minimum $1_{k \cup \{*, +\}}$-length of a division- and subtraction free straight-line program which computes f on input $(A; X)$.   •

(21.4) Proposition. *L and $L^{\text{tot}}_{k(X)}$ are polynomially related.*

Proof. Recall from Chap. 4 that we denote by k the set of all scalar multiplications by elements of k, and by k^c the set of all λ^c, where λ^c is the 0-ary operation of taking the constant $\lambda \in k$. Let $c := 1_{k \cup \{+, -, *\}}$ and let $L'(f)$ denote the minimum c-length of a division free straight-line program Γ which computes f on input $(A; X)$. Ex. 7.2 shows that $L^{\text{tot}}_{k(X)}$ and L' are polynomially related. In Γ we can replace each subtraction by a scalar multiplication and an addition. Hence, L and L' are polynomially related.   □

(21.5) Convention. *For the rest of this chapter straight-line programs are always assumed to be division- and subtraction free and the length of a straight-line program is always its $1_{k \cup \{*, +\}}$-length.*   •

(21.6) Example. The following *p*-families belong to **VP**:

(1) $SUM := (SUM_n)_{n \geq 1}$, where $SUM_n := X_1 + \ldots + X_n$.
(2) $PROD := (PROD_n)_{n \geq 1}$, where $PROD_n := X_1 \cdots X_n$.
(3) $POWSUM := (POWSUM_n)_{n \geq 1}$, where $POWSUM_n := \sum_{i=1}^{n} X_i^n$.

(4) $DET := (DET_n)_{n \geq 1}$, where $DET_n := \det(X_{ij})_{1 \leq i,j \leq n}$ and $X_{ij} := X_{(i-1)n+j}$. The fact that $DET \in$ **VP** follows by Gaussian elimination.   •

We come back to the discrete setting.

(21.7) Definition. A language $A \subseteq \Sigma^*$ belongs to the class **NP**, iff there exist a p-bounded $t: \mathbb{N} \to \mathbb{N}$ and a language $B \subseteq (\Sigma \sqcup \{\#\})^*$ belonging to **P** such that

$$\forall n \in \mathbb{N} \ \forall x \in \Sigma^n \ (x \in A \Leftrightarrow \exists e \in \Sigma^{t(n)} : x\#e \in B).$$ •

Let $x \in A \cap \Sigma^n$ and $e \in \Sigma^{t(n)}$ such that $x\#e \in B$. Then e is a *short proof* for the fact that $x \in A$, and the possibility of deciding $x\#e \in B$ in polynomial time can be interpreted as an *efficient verification* of the fact that x belongs to A. In terms of characteristic functions the equivalence in the above definition can be rewritten in the following way:

$$\chi_A(x) = \bigvee_{e \in \Sigma^{t(n)}} \chi_B(x\#e).$$

Replacing the disjunction over e by a sum over e leads to the subsequent nonuniform algebraic analogue of **NP**.

(21.8) Definition.
(1) A p-family $f = (f_n)$ of polynomials over k is said to be *p-definable* iff there exists a family $g = (g_n) \in$ **VP** such that for all $n \in \mathbb{N}'$, $t(n) := v(g_n) - v(f_n) \geq 0$ and $f_n(X) = \sum_{e \in \{0,1\}^{t(n)}} g_n(X, e)$, if $t(n) > 0$, and $f_n(X) = g_n(X)$, if $t(n) = 0$. (Here, $X := (X_1, \ldots, X_{v(f_n)})$.)
(2) **VNP** = **VNP**(nonuniform; k) denotes Valiant's class of all p-definable families over k.   •

Obviously, **VP** $\subseteq$ **VNP**.

The theory of **NP**-completeness is based on the concept of reduction. We briefly recall the definition of polynomial reducibility.

(21.9) Definition. Let $A_1 \subseteq \Sigma_1^*$ and $A_2 \subseteq \Sigma_2^*$ be languages.

(1) A_1 is said to be *polynomially reducible* to A_2 (for short: $A_1 \leq_p A_2$), iff there exists a function $f: \Sigma_1^* \to \Sigma_2^*$, computable in polynomial time, such that for all $x \in \Sigma_1^* : x \in A_1 \Leftrightarrow f(x) \in A_2$.
(2) A_1 and A_2 are *p-equivalent* ($A_1 \equiv_p A_2$), iff $A_1 \leq_p A_2$ and $A_2 \leq_p A_1$.   •

Obviously, p-equivalence is an equivalence relation and $\leq_p$ induces a partial ordering on the $\equiv_p$-classes. Furthermore, if $A \in$ **NP** and $B \leq_p A$, then $B \in$ **NP**. The following definition specifies the $\leq_p$-hardest problems in **NP**.

(21.10) Definition. A language $A \subseteq \Sigma^*$ is *NP-complete*, iff $A \in$ **NP** and $B \leq_p A$, for all $B \in$ **NP**.   •

Obviously, if A is **NP**-complete, then **P** = **NP** iff $A \in$ **P**. It is not obvious that **NP**-complete problems do exist. In a seminal 1971 paper, S. Cook [124] proved that the satisfiability problem *SAT* is **NP**-complete. Recall that *SAT* is (a suitable encoding of) the problem to decide for a Boolean formula F, which is the conjunction of finitely many clauses, whether there is a satisfying truth assignment for F. (A clause is a disjunction of Boolean variables x_i or its negations $\overline{x_i}$.)

(21.11) Cook's Theorem. *SAT is NP-complete.*

For a proof of this theorem see Garey and Johnson's book [181]. Meanwhile, there are thousands of problems recognized as NP-complete, see for instance [181] and Johnson's survey article [277].

Algebraic analogues of reducibility and completeness are provided by the subsequent definitions.

(21.12) Definition.
(1) $f \in k[X_1, \ldots, X_n]$ is a *projection* of $g \in k[X_1, \ldots, X_m]$, iff there are $a_1, \ldots, a_m \in k \cup \{X_1, \ldots, X_n\}$ such that $f = g(a_1, \ldots, a_m)$.
(2) Let $f = (f_n)$ and $g = (g_n)$ be *p*-families of polynomials over k. Then f is called a *p-projection* of g (for short: $f \preceq_p g$), iff there is a *p*-bounded function t such that for every n, f_n is a projection of $g_{t(n)}$. (In this case we sometimes write $f \preceq_p^t g$.) •

Note that $f \preceq_p^t g$ forces $v(f_n) \leq v(g_{t(n)})$, for all $n \geq 1$.

(21.13) Remark.
(1) The relation $\preceq_p$ is transitive; more precisely, $f \preceq_p^t g$ and $g \preceq_p^s h$ implies $f \preceq_p^{sot} h$.
(2) The classes **VP** and **VNP** are closed under *p*-projections. •

(21.14) Definition. A family $g = (g_n)$ of polynomials over k is said to be **VNP-complete**, iff $g \in$ **VNP** and $f \preceq_p g$, for all $f \in$ **VNP**. •

Let us look for promising candidates of **VNP**-complete problems. Pursuing further the above analogy we start with a suitable NP-complete problem and define with its help a family of polynomials. We illustrate this procedure by an example. Consider the Hamiltonian cycle problem. Here, the task is to decide whether for a graph G with points $1, 2, \ldots, n$ a cycle $\sigma \in S_n$ of length n exists such that $(i, \sigma(i))$ is an edge of G, for all $1 \leq i \leq n$. If we ask for the number of all such cycles then the answer is a polynomial expression in the entries of the $n \times n$ adjacency matrix (a_{ij}) of G. More precisely, the corresponding family $HC = (HC_n)$ of Hamiltonian cycle polynomials

$$HC_n := \sum_{\sigma: \, n\text{-cycle}} \prod_{i=1}^n X_{i\sigma(i)},$$

is the *enumerator* of that problem since HC_n evaluated at the adjacency matrix of G gives (over the rationals) the number of *all* Hamiltonian cycles in G. Obviously, G has a Hamiltonian cycle iff $HC_n(a_{ij}) > 0$. Thus a fast procedure for the evaluation of the polynomial family $HC = (HC_n)$ would yield a fast decision procedure for the Hamiltonian cycle problem. Many other NP-complete problems possess natural enumerators, and it is plausible that these do not differ essentially from each other in their computational complexity. It is a surprising fact, proved by Valiant [528], that there are problems *in* **P** whose corresponding enumerators are about as hard as typical enumerators of NP-complete problems. The most striking example is the problem of deciding whether a bipartite graph has a perfect matching. According to M. Hall's algorithm [226] this problem is in **P**. The corresponding enumerator is the family $PER = (PER_n)$ of generic permanents, where

$$PER_n := \sum_{\sigma \in S_n} \prod_{i=1}^{n} X_{i\sigma(i)}.$$

In the rest of this chapter we will mainly be concerned with the family PER.

(21.15) Proposition. *PER and HC belong to* **VNP**.

Proof. Let $X = (X_{ij})$ and $Y = (Y_{\ell m})$ be two $n \times n$ matrices over k with $2n^2$ indeterminates X_{ij} and $Y_{\ell m}$, $1 \le i, j, \ell, m \le n$.

$PER \in$ **VNP**: we define the polynomial g_n by

$$g_n(X, Y) := \underbrace{\left(\underbrace{\prod_{i,j,\ell,m} (1 - Y_{ij} Y_{\ell m})}_{=:\alpha_n(Y)} \right) \cdot \underbrace{\left(\prod_{i=1}^{n} \sum_{j=1}^{n} Y_{ij} \right)}_{=:\beta_n(Y)} \cdot \underbrace{\left(\prod_{i=1}^{n} \sum_{j=1}^{n} X_{ij} Y_{ij} \right)}_{=:\mu_n(X,Y)}}_{=:\gamma_n(Y)},$$

where the product in α_n is over all $1 \le i, j, \ell, m \le n$ such that $i = \ell$ iff $j \ne m$. Note that $g = (g_n)_{n \ge 1} \in$ **VP** since $v(g_n) = 2n^2$, $\deg g_n = O(n^3)$, and $L^{\text{tot}}_{k(X,Y)}(g_n) = O(n^3)$. So it suffices to prove that for all $e \in \{0, 1\}^{n \times n}$ the following statements are valid:

(A) $\gamma_n(e) \ne 0$ iff e is a permutation matrix.
(B) $\gamma_n(e) \in \{0, 1\}$.
(C)   If $\gamma_n(e) \ne 0$, then $\mu_n(X, e) = \prod_{i=1}^{n} X_{i\sigma(i)}$, where σ denotes the permutation corresponding to e.
(D) $PER_n = \sum_{e \in \{0,1\}^{n \times n}} g_n(X, e)$.

For proving (A), note that $\alpha_n(e) \ne 0$ iff every row and every column of e contains at most one 1. If $\alpha_n(e) \ne 0$, then $\beta_n(e) \ne 0$ iff every row of e contains at least one 1. Altogether, $\gamma_n(e) = \alpha_n(e)\beta_n(e) \ne 0$ iff e is a permutation matrix. Now (B)-(D) follow easily and thus $PER \in$ **VNP**.

$HC \in$ **VNP**: Let e be a permutation matrix corresponding to the permutation $\pi \in S_n$. Then π is an n-cycle iff $\pi^q(1) \ne 1$, for all $1 \le q < n$. In terms of matrices this means $(e^q)_{11} \ne 1$, for all $1 \le q < n$. Let

$$\varepsilon_n(Y) := (1 - Y_{11})(1 - (Y^2)_{11}) \cdots (1 - (Y^{n-1})_{11})$$

and $e \in \{0, 1\}^{n \times n}$ be a permutation matrix. Then $\varepsilon_n(e) = 1$ if e corresponds to an n-cycle, otherwise, $\varepsilon_n(e) = 0$. If $C_1(Z)$ denotes the first column of the matrix Z, then $C_1(Y^j) = Y \cdot C_1(Y^{j-1})$, and an easy induction shows that $L^{\text{tot}}_{k(Y)}(\varepsilon_n(Y)) = O(n^3)$. Thus with the notation of the first part of our proof and

$$G_n(X, Y) := \gamma_n(Y)\varepsilon_n(Y)\mu_n(X, Y)$$

we obtain $G := (G_n) \in \mathbf{VP}$. Furthermore, $G_n(X, e) = \prod_{i=1}^{n} X_{i\sigma(i)}$, if $e \in \{0, 1\}^{n \times n}$ corresponds to an n-cycle σ, and $G_n(X, e) = 0$ for all other $e \in \{0, 1\}^{n \times n}$. Altogether, $HC_n = \sum_{e \in \{0,1\}^{n \times n}} G_n(X, e)$. Thus $HC \in \mathbf{VNP}$. $\qquad\qquad \square$

(21.16) Remark.

(1) $PER = DET$, if char $k = 2$. Hence $PER \in \mathbf{VP}$ in characteristic 2, see Example (21.6)(4).

(2) The fastest known algorithm to compute the permanent is based on Ryser's formula: $PER_n = \mathrm{per}(X) = \sum_J (-1)^{|J|} \prod_{i=1}^{n} (\sum_{j \in \underline{n} \setminus J} X_{ij})$, where the sum is over all proper subsets J of $\underline{n}$, see Ex. 21.1. This formula shows that $L(PER_n) = O(n 2^n)$. (Check this!)   •

The following nonuniform algebraic analogue of Cook's theorem gives some evidence of the intrinsic difficulty of computing permanents.

(21.17) Valiant's Theorem. *The family HC is* **VNP***-complete. The same is true for the family PER, if* char $k \neq 2$.

Cook's and Valiant's theorem support the following two fundamental conjectures.

(21.18) Cook's Hypothesis. $P \neq NP$.

(21.19) Valiant's Hypothesis. $\mathbf{VP} \neq \mathbf{VNP}$ *over any field.*

In the subsequent sections we concentrate on a detailed proof of the second claim of Valiant's theorem. For proofs of the first claim the reader is referred to Valiant [526] and von zur Gathen [187]; see also Ex. 21.8. The remaining part of the present section outlines a proof of the **VNP**-completeness of *PER*.

In a first step, see the next section, we give an alternative characterization of **VNP** involving the expression size of polynomials. The set of *arithmetic expressions* (or *formulas*) over $I := k \cup \{X_1, \ldots, X_n\}$ is inductively defined as follows: every element in I is an expression; furthermore, if φ_1 and φ_2 are expressions, then so is $(\varphi_1 \circ \varphi_2)$, for $\circ \in \{+, *\}$. The *size* $E(\varphi)$ of an expression φ is the number of $+$ and $*$ used to build it. Every expression φ obviously represents a unique polynomial $\mathrm{val}(\varphi) \in k[X_1, \ldots, X_n]$. The *expression size* $E(f)$ of $f \in k[X_1, \ldots, X_n]$ is the smallest size of a formula φ with $\mathrm{val}(\varphi) = f$.

Note that an arithmetic expression for the polynomial f may be viewed as a special straight-line program, where intermediate results can be used only once in the sequel; in particular,

$$L(f) \leq E(f).$$

However, $L(f)$ and $E(f)$ might differ substantially. A number of examples supporting this claim can be constructed on the basis of the following.

(21.20) Lemma. $E(f) \geq \deg(f) - 1$, *for every nonzero* $f \in k[X_1, \ldots, X_n]$.

Proof. The claim is proved by induction on $E(f)$. The start being clear, assume $E(f) \geq 1$. Then an optimal arithmetic expression φ for f is of the form $\varphi = (\varphi_1 \circ \varphi_2)$ with $\circ \in \{+, *\}$ and optimal arithmetic expressions φ_1 and φ_2 for $f_1 := \mathrm{val}(\varphi_1)$

and $f_2 := \text{val}(\varphi_2)$; thus $E(f) = E(f_1) + E(f_2) + 1$. By optimality, $f_1 \cdot f_2 \neq 0$. Let d, d_1, and d_2 be the degrees of f, f_1, and f_2, respectively. Then $d \leq d_1 + d_2$ and by induction we get $E(f) = E(f_1) + E(f_2) + 1 \geq d_1 - 1 + d_2 - 1 + 1 \geq d - 1. \square$

Consider for example the family $f = (f_n)$ with $f_n := X_1^{2^n}$. Then $L(f_n) = n$, but $E(f_n) = 2^n - 1$. Thus $n \mapsto L(f_n)$ is p-bounded whereas $n \mapsto E(f_n)$ is exponential in n. Note however that f does not belong to **VP** as $n \mapsto \deg f_n$ is not p-bounded. The question is whether such a substantial difference is also possible if we restrict ourselves to families in **VP**.

(21.21) Definition.
(1) A p-family $g = (g_n)$ over k is called *p-expressible* iff $n \mapsto E(g_n)$ is p-bounded.
(2) $\mathbf{VP}_e = \mathbf{VP}_e(\text{nonuniform}; k)$ denotes Valiant's class of all p-expressible families over k.
(3) $\mathbf{VNP}_e = \mathbf{VNP}_e(\text{nonuniform}; k)$ denotes Valiant's class of all families $f = (f_n)$ over k such that there exists a p-expressible family g with $v(g_n) - v(f_n) =: t(n) \geq 0$ and $f_n(X) = \sum_{e \in \{0,1\}^{t(n)}} g_n(X, e)$, if $t(n) > 0$, and $f_n(X) = g_n(X)$, if $t(n) = 0$. (Again, $X := (X_1, \ldots, X_{v(f_n)})$.) ●

Obviously, $\mathbf{VP}_e \subseteq \mathbf{VP} \subseteq \mathbf{VNP}$ and $\mathbf{VNP}_e \subseteq \mathbf{VNP}$. Besides Valiant's hypothesis $\mathbf{VP} \neq \mathbf{VNP}$ a prominent conjecture is that $\mathbf{VP}_e \neq \mathbf{VP}$. Surprisingly, as shown in the next section, the corresponding "nondeterministic" classes coincide:

$$\mathbf{VNP}_e = \mathbf{VNP}.$$

In the second step we show in Sect. 21.3 that every polynomial of expression size u is both a projection of DET_{2u+2} and PER_{2u+2}. Now let $f \in \mathbf{VNP} = \mathbf{VNP}_e$. Then there exists $g \in \mathbf{VP}_e$ such that $f_n(X) = \sum_e g_n(X, e)$. Hence, there is a matrix A_n over $k \cup \{X_i, Y_j | i, j\}$ of size $N = 2E(g_n) + 2$ such that $g_n(X, Y) = \text{per}(A_n)$. Under the assumption that $\text{char } k \neq 2$ it is shown in a third step in Sect. 21.4 that the matrix A_n can be modified to a matrix A'_n over $k \cup \{X_i | i\}$ of size $N' \leq 10N$ such that $\text{per}(A'_n) = \sum_e g_n(X, e) = f_n(X)$. Thus f_n is a projection of $PER_{20E(g_n)+20}$, which shows the **VNP**-completeness of PER.

21.2 p-Definability and Expression Size

The goal of this section is to prove that $\mathbf{VNP}_e = \mathbf{VNP}$. To this end we first reformulate our claim.

(21.22) Lemma. $\mathbf{VNP}_e = \mathbf{VNP}$ iff $\mathbf{VP} \subseteq \mathbf{VNP}_e$.

Proof. $\Rightarrow$: This follows from $\mathbf{VNP}_e = \mathbf{VNP} \supseteq \mathbf{VP}$.

$\Leftarrow$: As $\mathbf{VNP}_e \subseteq \mathbf{VNP}$ it remains to prove that $\mathbf{VNP} \subseteq \mathbf{VNP}_e$. Let $f \in \mathbf{VNP}$. Then $f_n(X) = \sum_e g_n(X, e)$, for a suitable $g \in \mathbf{VP}$. By our assumption, there exists $h \in \mathbf{VP}_e$ such that $g_n(Y) = \sum_{e'} h_n(Y, e')$, for all n. Hence $f_n(X) = \sum_{e,e'} h_n(X, e, e')$. $\square$

The main tool in the proof of **VP** $\subseteq$ **VNP$_e$** are homogeneous straight-line programs.

(21.23) Definition. A straight-line program is called *homogeneous* iff its result sequence on input $(k[X]; X)$ consists of homogeneous polynomials. For a finite subset F of $k[X]$ we denote by $H(F)$ the minimum length of a homogeneous straight-line program computing all the homogeneous parts of each $f \in F$. •

(21.24) Remark. Let $b_1, \ldots, b_r$ be the result sequence of a homogeneous straight-line program Γ. Then $b_i b_j \neq 0$ for $i \leq_\Gamma j$ implies $\deg b_i \leq \deg b_j$. (Here, $<_\Gamma$ denotes the partial ordering on the vertices of the directed acyclic multigraph corresponding to Γ.) •

The following lemma is a variant of results in Chap. 7.

(21.25) Lemma. *If $f \in k[X]$ is of degree d with h nonzero homogeneous parts then $L(f) - h + 1 \leq H(f) \leq (d+1)^2 L(f)$.*

Proof. Obviously, $L(f) \leq H(f) + h - 1$. To prove the other claim let $\Gamma = (\Gamma_1, \ldots, \Gamma_r)$ be a straight-line program which computes f optimally with α additions, μ multiplications, and σ multiplications by scalars (thus $\alpha + \mu + \sigma = L(f)$). Let (b_i) be the corresponding result sequence. If $\Gamma_i = (+; p, q)$, $\Gamma_j = (*; s, t)$, and $\Gamma_\ell = (\lambda; p)$, then the corresponding homogeneous parts of degree δ read as follows: $b_i^{(\delta)} = b_p^{(\delta)} + b_q^{(\delta)}$, $b_j^{(\delta)} = \sum_{u=0}^{\delta} b_s^{(u)} b_t^{(\delta-u)}$, and $b_\ell^{(\delta)} = \lambda b_p^{(\delta)}$. As in addition the constants are free, we see that $\{b_i^{(\delta)} | 1 \leq i \leq r, 0 \leq \delta \leq d\}$ can be computed by a homogeneous straight-line program in at most $\alpha(d+1) + \mu \sum_{\delta=0}^{d} ((\delta+1)+\delta) + \sigma = \alpha(d+1) + \mu(d+1)^2 + \sigma \leq (\alpha + \mu + \sigma)(d+1)^2$ steps. $\square$

Now we can prove the main result of this section.

(21.26) Theorem. **VNP$_e$** = **VNP**.

Proof. According to Lemma (21.22) it suffices to show that **VP** $\subseteq$ **VNP$_e$**. Let $f = (f_n) \in$ **VP**. For proving $f \in$ **VNP$_e$** we can w.l.o.g. assume by Lemma (21.25) that all f_n are homogeneous; furthermore we can restrict ourselves to homogeneous straight-line programs. So let $f_n \in k[X_1, \ldots, X_m]$, $m = v(f_n)$, be homogeneous of degree d. We are going to construct recursively a family $g \in$ **VP$_e$** such that $f_n(X) = \sum_e g_n(X, e)$, for all n. In a first step we describe and analyse the type of recursion. To this end let $P(C, d, m)$ denote the set of all homogeneous polynomials f in $k[X_1, \ldots, X_m]$ with $\deg f \leq d$ and $H(f) \leq C$. Furthermore define $A(C, d, m)$ as

$$\max_f \min\{E(h) \mid h \text{ homogeneous}, v(h) \leq v(f) + 2C + 4, f(X) = \textstyle\sum_e h(X, e)\},$$

where the maximum is over all $f \in P(C, d, m)$. Obviously, $A(C, 1, m) \leq 2m$ and $A(C, 2, m) \leq 3(\binom{m}{2} + m) - 1 \leq 3m^2$. We claim that the theorem is a consequence of the following recursion:

$$A(C, d, m) \leq 3A\big(4C + \lfloor d/2 \rfloor, \lfloor d/2 \rfloor + 1, m + C\big) + \gamma C^2,$$

where γ is a suitable constant. To prove this claim, put $(C_0, d_0, m_0) := (C, d, m)$, and for $i \geq 1$ let

$$(C_i, d_i, m_i) := \left(4C_{i-1} + \lfloor d_{i-1}/2 \rfloor, \lfloor d_{i-1}/2 \rfloor + 1, m_{i-1} + C_{i-1}\right).$$

Then an easy induction shows that for $2 \leq 2^{\ell-1} < d_0 \leq 2^\ell$ the following holds for all $i \leq \ell$: $d_i \leq 2^{\ell-i} + 1$; $d_\ell = 2$, $C_i \leq 4C_{i-1} + 2^{\ell-i} \leq 4^i C_0 + 2^{\ell-i}\frac{8^i-1}{7}$, $m_i \leq m_0 + \sum_{j=0}^{i-1} C_j$, and $A(C_0, d_0, m_0) \leq 3^\ell A(C_\ell, 2, m_\ell) + \gamma \sum_{j=0}^{\ell-1} C_j^2$. Hence if C, d, and m are p-bounded functions in n, then so is the solution to the above recurrence. Hence all what remains to do is to deduce the above recurrence.

Let $\Gamma = (\Gamma_1, \ldots, \Gamma_r)$ be an optimal homogeneous straight-line program computing $f_n \in P(C, d, m)$. W.l.o.g. Γ_j is of the form $\Gamma_j = (\omega_j; \alpha_j, \beta_j)$ or $\Gamma_j = (\lambda_j^c)$. Let (b_i) be the corresponding result sequence, and let d_i denote the degree of b_i.

Claim 1. If $J := \{j \in \underline{r} \mid \Gamma_j = (*; \alpha_j, \beta_j), \ d_j > d/2 \geq d_{\alpha_j}, d_{\beta_j}\}$, then there exists a homogeneous polynomial $S \in k[X_1, \ldots, X_m, E_0, E_j \mid j \in J]$ of degree $\lceil d/2 \rceil$ such that

(A) $H(S) \leq H(f_n) + \lceil d/2 \rceil - 2$.
(B) $f_n(X) = \sum_{i \in J} b_i(X) S(X, e_0 + e_i)$, where e_i denotes the indicator function of $i \in J \cup \{0\}$.

Claim 1 will be proved by modifying Γ to a homogeneous straight-line program Γ' of length $H(f_n) + \lceil d/2 \rceil - 2$ that on input $(k[X, E]; X, E)$ produces a result sequence (b_i') with $d_i' := \deg b_i'$ such that for all $j \in \underline{r}$

(C)$_j$ $b_j = b_j'$, if $d_j \leq d/2$,
(D)$_j$ $b_j = \sum_{i \in J} b_i(X) b_j'(X, e_0 + e_i)$ and $d_j' = d_j - \lfloor d/2 \rfloor$, if $d_j > d/2$.

In particular, with $S := b_r'$ and the fact that $f_n = b_r$ we see that (A) and (B) hold.

Γ' is defined as follows. (In the sequel it will be convenient to deviate from the usual indexing of the instructions.) First of all we let Γ_{0j}' compute E_0^j, for $2 \leq j \leq \lceil d/2 \rceil - 1$. (This is possible in $\lceil d/2 \rceil - 2$ steps.) For $j \in \underline{r} \setminus J$ we put $\Gamma_j' := \Gamma_j$, and for $j \in J$ we let Γ_j' compute $E_0^{d_j - \lfloor d/2 \rfloor - 1} E_j$. For $\Gamma' = (\Gamma_{02}', \ldots, \Gamma_{0\lceil d/2 \rceil-1}', \Gamma_1', \ldots, \Gamma_r')$, Claim (C)$_j$ holds obviously for all j. We prove (D)$_j$ for all j by induction on the partial ordering $<_\Gamma$ on the vertices of the directed acyclic multigraph corresponding to Γ. To prove the start we have to show that (D)$_j$ holds for all elements of J. So let $j \in J$. Then $b_j'(X, E) = E_0^{d_j - \lfloor d/2 \rfloor - 1} E_j$, $b_j'(X, e_0 + e_i) = \delta_{ij}$, for all $i \in J$, and b_j' is of degree $d_j - \lfloor d/2 \rfloor$. This settles the start. Now let $d_j > d/2$, but $j \notin J$. By the induction hypothesis, (C)$_\ell$ and (D)$_\ell$ are valid for all $\ell <_\Gamma j$.

Case 1. $\Gamma_j = (+; \alpha, \beta)$.
Since Γ is homogeneous we have $d_\alpha = d_\beta > d/2$, and as (D)$_\alpha$ and (D)$_\beta$ hold we obtain

$$b_j = b_\alpha + b_\beta = \sum_{i \in J} b_i(X) \left(b_\alpha'(X, e_0 + e_i) + b_\beta'(X, e_0 + e_i) \right)$$

$$= \sum_{i \in J} b_i(X)\underbrace{(b'_\alpha + b'_\beta)}_{=b'_j}(X, e_0 + e_i).$$

Finally, one easily checks that $d'_j = d_j - \lfloor d/2 \rfloor$.

Case 2. $\Gamma_j = (*; \alpha, \beta)$.

As $d_j > d/2$, but $j \notin J$ we have w.l.o.g. $d_\alpha > d/2 \geq d_\beta$. Hence, by (C)$_\beta$ and (D)$_\alpha$ and the fact that $b_\beta = b'_\beta$, we obtain

$$\begin{aligned} b_j &= b_\alpha * b_\beta = \sum_{i \in J} b_i(X) b'_\alpha(X, e_0 + e_i) b_\beta(X) \\ &= \sum_{i \in J} b_i(X)\underbrace{(b'_\alpha * b'_\beta)}_{=b'_j}(X, e_0 + e_i). \end{aligned}$$

Furthermore, $d'_j = d'_\alpha + d_\beta = (d_\alpha - \lfloor d/2 \rfloor) + d_\beta = d_j - \lfloor d/2 \rfloor$ which proves (D)$_j$.

Claim 2. Let $J' := \{\alpha_j, \beta_j \mid j \in J\}$. Then the polynomial $T(X, V) := \sum_{i \in J'} b_i V_i V_0^{\lfloor d/2 \rfloor - d_i}$ is homogeneous of degree $\lfloor d/2 \rfloor + 1$ and $H(T) \leq H(f_n) + 3|J'| + \lfloor d/2 \rfloor - 3$. Furthermore, $b_i(X) = T(X, v_0 + v_i)$ for all $i \in J'$, where v_i is the indicator function of $i \in J' \cup \{0\}$.

In fact, a homogeneous straight-line program Γ'' with the properties proving Claim 2 is obtained by deleting from Γ every instruction Γ_i with $d_i > d/2$, adding further instructions that compute all powers V_0^i, for $2 \leq i < \lfloor d/2 \rfloor$, as well as all instructions that multiply each b_i, $i \in J'$, with $V_i V_0^{\lfloor d/2 \rfloor - d_i}$ to get the intermediate result z_i, and finally by adding further instructions to sum z_i for $i \in J'$.

Now consider the homogeneous polynomial $R(Z, V, V', E)$ defined by

$$Z V_0 V'_0 E_0 \sum_{j \in J} V_{\alpha_j} V'_{\beta_j} E_j \prod_{i \in J' \setminus \{\alpha_j\}} (Z - V_i) \prod_{i \in J' \setminus \{\beta_j\}} (Z - V'_i) \prod_{\rho \in J \setminus \{j\}} (Z - E_\rho).$$

Note that R evaluated at a binary vector (z, v, v', e) of length $1 + 2(|J'| + 1) + (|J| + 1)$ gives 0 or 1. It is 1 iff $z = 1$ and there exists a (unique) index $j \in J$ such that $e = e_0 + e_j$, $v = v_0 + v_{\alpha_j}$, and $v' = v_0 + v_{\beta_j}$. Hence

(E) $\qquad f_n(X) = \sum_{(z,v,v',e)} T(X, v) T(X, v') S(X, e) R(z, v, v', e).$

Recall that we are looking for a family $g \in \text{VP}_e$ such that $f_n(X) = \sum_e g_n(X, e)$, for all n. We have already found a factor of g_n, namely R, which obviously has expression size at most $\gamma H(f_n)^2$, for some constant γ. The next step in the recursion processes $T(X, V)$ and $S(X, E)$. With (A), Claim 2, and $|J| + |J'| \leq H(f_n)$, we see that $4H(f_n) + \lfloor d/2 \rfloor$ is both an upper bound for $H(S)$ and $H(T)$. Thus using $J \neq \emptyset \neq J'$ Eq. (E) yields the stated recursion for $A(C, d, m)$. $\qquad \square$

21.3 Universality of the Determinant

In this section we show that the methods of (multi)linear algebra are, at least in principle, sufficient to devise feasible algorithms whenever they exist.

(21.27) Theorem. *Let* $f \in k[X_1, \ldots, X_n]$ *have expression size u. Then* f *is both a projection of* DET_{2u+2} *and a projection of* PER_{2u+2}.

In particular, every family in $\mathbf{VP}_e$ is a p-projection of both DET and PER. However, the universality of the permanent is pointless under this computational aspect, due to the notorious lack of feasible computations for the permanent. A more significant interpretation of this theorem will be given in the last section.

For the proof of the above theorem we need a fact from multilinear algebra.

(21.28) Lemma. *Let* R *be a commutative ring. For* $i = 1, 2$ *let* $A_i \in R^{d_i \times d_i}$ *be an upper triangular matrix with 1's on the diagonal,* $\alpha_i = (\alpha_{i1}, \alpha_{i2}, \ldots) \in R^{1 \times d_i}$ *and* $\beta_i = (\beta_{i1}, \beta_{i2}, \ldots)^\top \in R^{d_i \times 1}$. *Then for the matrices*

$$M_1 := \begin{pmatrix} \alpha_1 & 0 \\ A_1 & \beta_1 \end{pmatrix}, M_2 := \begin{pmatrix} \alpha_2 & 0 \\ A_2 & \beta_2 \end{pmatrix}, M := \begin{pmatrix} \alpha_1 & \alpha_2 & 0 \\ A_1 & 0 & \beta_1 \\ 0 & A_2 & \beta_2 \end{pmatrix},$$

we have

$$\det(M) = (-1)^{d_2} \det(M_1) + (-1)^{d_1} \det(M_2)$$

and

$$\mathrm{per}(M) = \mathrm{per}(M_1) + \mathrm{per}(M_2).$$

Proof. Let $M[i|j]$ denote the matrix obtained by deleting in $M = (m_{ab})$ row i and column j. Then the Laplace expansions w.r.t. column j read as follows:

(A)
$$\det(M) = \sum_i (-1)^{i+j} m_{ij} \det(M[i|j])$$

(B)
$$\mathrm{per}(M) = \sum_i m_{ij} \mathrm{per}(M[i|j]).$$

We prove our first claim by induction on d_2. The start $d_2 = 1$ is similar to the step. So we only prove the induction step. Let $d_2 \geq 2$. We apply (A) to column $d_1 + 1$ of M and observe that M has at most two nonzero entries in that column (namely in row 1 and row $d_1 + 2$); hence

$$\begin{aligned} \det(M) &= (-1)^{1+d_1+1} \alpha_{21} \det(M[1|d_1 + 1]) \\ &\quad + (-1)^{d_1+2+d_1+1} \det(M[d_1 + 2|d_1 + 1]). \end{aligned}$$

As $M' := M[d_1 + 2|d_1 + 1]$ is of the same type as M but now corresponding to $M_1' := M_1$ and $M_2' := M_2[2|1]$ with parameters d_1 and $d_2 - 1$, respectively, we get by the induction hypothesis

$$\det(M[d_1 + 2|d_1 + 1]) = (-1)^{d_2-1} \det(M_1) + (-1)^{d_1} \det(M_2[2|1]).$$

Combining this with $\det(M[1|d_1 + 1]) = \det(M_2[1|1])$ yields

$$\det(M) = (-1)^{d_2} \det(M_1) + (-1)^{d_1} \underbrace{\left(\alpha_{21} \det(M_2[1|1]) - \det(M_2[2|1])\right)}_{= \det(M_2) \text{ by (A)}}.$$

This proves our first claim. Using (B), the second claim can be shown analogously. □

Proof of Thm. (21.27). Let $\mathcal{E}$ be the set of expressions over $I = k \cup \{X_1, \ldots, X_n\}$. We will define a mapping $\mu \colon \mathcal{E} \to \bigcup_{s \geq 1} I^{s \times s}$ with the following properties, for all $\varphi \in \mathcal{E}$:

(A) $\mathrm{val}(\varphi) = \det(\mu(\varphi))$.
(B) If φ has expression size u, then $\mu(\varphi)$ has size $s \times s$ with $s = 2u + 2$.
(C) There exist $A \in I^{(s-1) \times (s-1)}$, $\alpha \in I^{1 \times (s-1)}$, $\beta \in I^{(s-1) \times 1}$, with $s = 2u + 2$ as in (B), such that A is upper triangular with ones on the diagonal and

(D) $\mu(\varphi)$ has in each column at most one entry which is an indeterminate. The last column contains no indeterminate.

(The first part of the theorem follows from (A) and (B); (C) and (D) will be needed for technical reasons.) The definition of μ proceeds by induction along the construction of φ.

Case 1. $(u = 0)$ Let $\varphi \in I$.
Then $\mu(\varphi) := \left(\begin{smallmatrix} \varphi & 0 \\ 1 & 1 \end{smallmatrix}\right) \in I^{2 \times 2}$ satisfies (A)–(D).

Case 2. $\varphi = (\varphi_1 * \varphi_2)$.
For $i \in \{1, 2\}$ let u_i denote the expression size of φ_i, and define $\mu(\varphi)$ as follows:

Then (C) and (D) are satisfied. Furthermore, $u = u_1 + u_2 + 1$ is the size of φ and, by the induction hypothesis, the size s of $\mu(\varphi)$ equals $(2u_1 + 2) + (2u_2 + 2) = 2u + 2$. This proves (B). Since $\mu(\varphi)$ is block triangular, (A) is clear too.

Case 3. $\varphi = (\varphi_1 + \varphi_2)$.

By the induction hypothesis, $M_1 := \mu(\varphi_1)$ and $M_2 := \mu(\varphi_2)$ (of even sizes $s_1 = 2u_1 + 2$ and $s_2 = 2u_2 + 2$, respectively) satisfy the assumptions of Lemma (21.28) with odd parameters $d_1 = s_1 - 1$ and $d_2 = s_2 - 1$, respectively. Then the corresponding M satisfies $\det(M) = -\det(M_1) - \det(M_2) = -\mathrm{val}(\varphi)$. We now get $\mu(\varphi)$ by adding to M a last row and a last but one column consisting of zeros except a one at their intersection. Then $\det(\mu(\varphi)) = -\det(M) = \mathrm{val}(\varphi)$, and (A) holds. Properties (C) and (D) are clear, and the size of $\mu(\varphi)$ is $s + 1 = 2u_1 + 2u_2 + 4 = 2u + 2$. This proves the universality of DET. The universality of PER is shown in a similar way. $\qquad\square$

The size of the matrix $\mu(\varphi)$ is not the best possible. For an improvement see Ex. 21.7. As we are primarily interested in a streamlined proof of Valiant's theorem, we pause with our discussion of the universality of the determinant. After having finished the proof of the completeness of PER, we will continue our discussion in Sect. 21.5.

21.4 Completeness of the Permanent

In this section we prove the **VNP**-completeness of the family PER assuming that $\mathrm{char}\, k \neq 2$. By Prop. (21.15) we know that $PER \in \mathbf{VNP}$. Now let $f \in \mathbf{VNP} = \mathbf{VNP}_e$. Then there exists $g \in \mathbf{VP}_e$ such that $f_n(X) = \sum_e g_n(X, e)$, for all n. Let $m = v(f_n)$ and $t = v(g_n) - v(f_n)$. It will be convenient to put $X = (X_1, \ldots, X_m)$ and $Y = (Y_1, \ldots, Y_t) := (X_{m+1}, \ldots, X_{m+t})$. By Thm. (21.27) there is a matrix A over $k \cup \{X_1, \ldots, X_m\} \cup \{Y_1, \ldots, Y_t\}$ of size $N = 2E(g_n) + 2$ such that $g_n(X, Y) = \mathrm{per}(A)$. Moreover, by property (D) in the proof of that theorem we can assume that A has in each column at most one entry which is an indeterminate. Thus the **VNP**-completeness of PER will follow from the subsequent result.

(21.29) Theorem. *Suppose that* $\mathrm{char}\, k \neq 2$ *and let* $A = A(X, Y)$ *be an* $N \times N$ *matrix over* $k \cup \{X_1, \ldots, X_m, Y_1, \ldots, Y_t\}$ *having in each column at most one entry not in* k. *Put* $g(X, Y) := \mathrm{per}(A)$. *Then a matrix* A' *of size* $N' \leq 10N$ *over the set* $k \cup \{X_1, \ldots, X_m\}$ *can be constructed such that* $\mathrm{per}(A') = \sum_{e \in \{0,1\}^t} g(X, e)$.

Proof. W.l.o.g. we may assume that each Y_i, $i \in \underline{t}$, occurs in A with multiplicity $\mu_i \geq 1$. The matrix A' will have a block and a fine structure. The only nonzero blocks of A' are in the first block row, in the first block column and along the block diagonal. The following figure illustrates the block structure for $t = 3$:

$$A' = \begin{pmatrix} A_0 & \mathcal{Y}_{01} & \mathcal{Y}_{02} & \mathcal{Y}_{03} \\ \mathcal{Y}_{10} & \mathcal{Y}_1 & 0 & 0 \\ \mathcal{Y}_{20} & 0 & \mathcal{Y}_2 & 0 \\ \mathcal{Y}_{30} & 0 & 0 & \mathcal{Y}_3 \end{pmatrix}.$$

The fine structure of the matrix $A'[i] := \begin{pmatrix} A_0 & \mathcal{Y}_{0i} \\ \mathcal{Y}_{i0} & \mathcal{Y}_i \end{pmatrix}$ reads as follows:

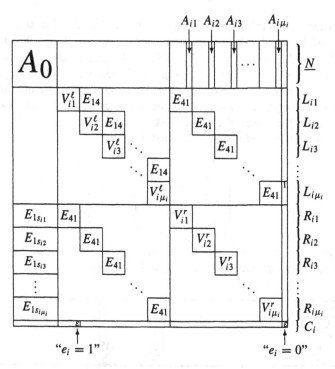

$$A_{i1} \quad A_{i2} \quad A_{i3} \qquad A_{i\mu_i}$$

Some comments are necessary. Substituting in $A = A(X, Y)$ each occurrence of a Y_i, $i \in \underline{t}$, by $\delta \in \{0, 1\}$ we obtain a matrix A_δ; in particular, $A_0 := A(X, 0, \ldots, 0)$ and $A_1 := A(X, 1, \ldots, 1)$. E_{ij} denotes a rectangular matrix of suitable size with a one at position (i, j) and zeros elsewhere. For $i \in \underline{t}$, $s_{i1} < \ldots < s_{i\mu_i}$ denote the indices of all columns of A containing an entry equal to Y_i, and $A_{i1}, \ldots, A_{i\mu_i}$ denote the corresponding columns in A_1. As $\mathrm{char}\, k \neq 2$, the value $\varepsilon = \varepsilon_i := 4^{-2\mu_i}$ is defined. As we will see later, the left and right ε in the last row of $A'[i]$ control the summation over $e_i = 1$ and $e_i = 0$, respectively. C_i is the singleton containing the row index of the ith control row. All $V_{i\alpha}^\ell$ and all $V_{i\alpha}^r$ are equal to the *Valiant matrix*

$$V = \begin{pmatrix} 0 & 1 & -1 & -1 \\ 1 & -1 & 1 & 1 \\ 0 & 1 & 1 & 2 \\ 0 & 1 & 3 & 0 \end{pmatrix}.$$

This matrix has some nice properties which will play a key role in the proof of Thm. (21.29). Namely, if $V[R|C]$ denotes V but with rows $r \in R$ and columns $c \in C$ removed, then the following hold:

$$\mathrm{per}(V) = \mathrm{per}(V[1|1]) = \mathrm{per}(V[4|4]) = \mathrm{per}(V[1, 4|1, 4]) = 0$$

and

$$\mathrm{per}(V[1|4]) = \mathrm{per}(V[4|1]) = 4.$$

We proceed with further comments on the matrix A'. According to our convention, E_{14} and E_{41} are 4×4 matrices, whereas $E_{15_{ij}}$ is a $4 \times N$ matrix. Altogether, the size N' of A' is equal to $N + \sum_{i=1}^{t}(8\mu_i + 1)$, and by the properties of the matrix A we know that $t \leq \mu_1 + \ldots + \mu_t \leq N$; hence $N' \leq 10N$. L_{ij} and R_{ij} denote 4-sets of row indices specified later. Rectangles which are not marked by letters are zero everywhere. The same is true for the entries in the four 5×5 block matrices outside the E and V block entries.

We illustrate the construction of the matrix A' for

$$A = \begin{pmatrix} Y_1 & 2 & 3 & 4 & 5 \\ 6 & Y_2 & X_1 & 7 & 8 \\ 9 & 10 & 11 & Y_1 & 12 \\ 13 & 14 & 15 & 16 & 17 \\ 18 & 19 & 20 & 21 & 22 \end{pmatrix}.$$

Here, $m = 1$, $t = 2$, $\mu_1 = 2$, $s_{11} = 1$, $s_{12} = 4$, $\mu_2 = 1$, and $s_{21} = 2$. Thus A' is equal to the following matrix in which $\varepsilon_1 = 4^{-4}$ and $\varepsilon_2 = 4^{-2}$.

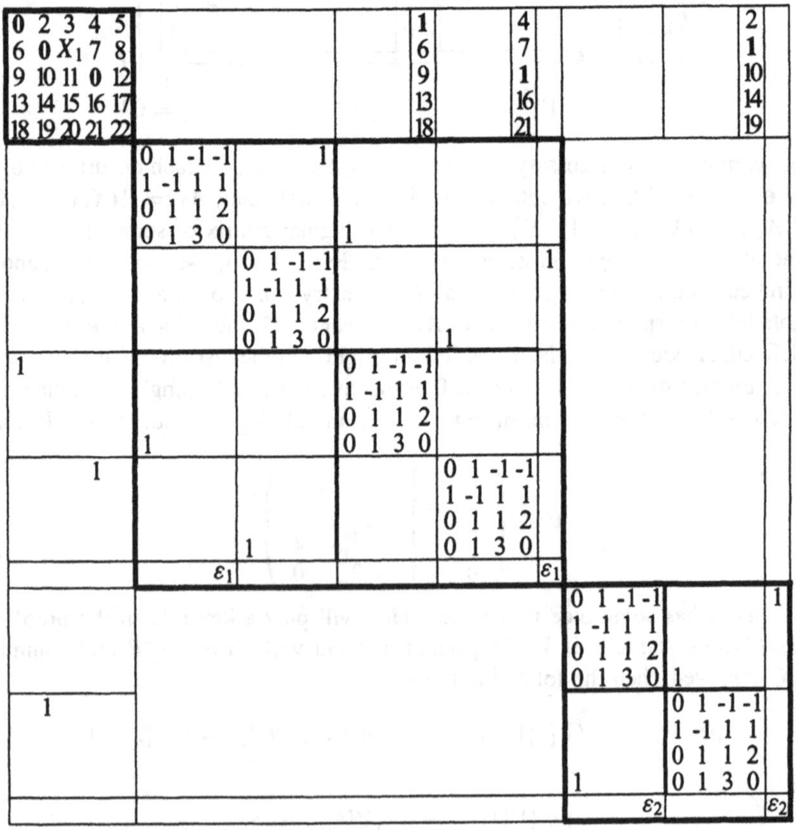

$$
\begin{array}{|ccccc|c|c|c|c|c|c|c|}
\hline
\begin{matrix}0 & 2 & 3 & 4 & 5\\ 6 & 0 & X_1 & 7 & 8\\ 9 & 10 & 11 & 0 & 12\\ 13 & 14 & 15 & 16 & 17\\ 18 & 19 & 20 & 21 & 22\end{matrix} & & & & & \begin{matrix}1\\6\\9\\13\\18\end{matrix} & \begin{matrix}4\\7\\1\\16\\21\end{matrix} & & & & \begin{matrix}2\\1\\10\\14\\19\end{matrix} \\
\hline
& \begin{matrix}0&1&-1&-1\\1&-1&1&1\\0&1&1&2\\0&1&3&0\end{matrix} & 1 & & & & & & & \\
\hline
\end{array}
$$

(The following is the full block matrix A' as displayed.)

$0\,2\,3\,4\,5$ $6\,0\,X_1\,7\,8$ $9\,10\,11\,0\,12$ $13\,14\,15\,16\,17$ $18\,19\,20\,21\,22$					$\begin{matrix}1\\6\\9\\13\\18\end{matrix}$	$\begin{matrix}4\\7\\1\\16\\21\end{matrix}$				$\begin{matrix}2\\1\\10\\14\\19\end{matrix}$
	$\begin{matrix}0&1&-1&-1\\1&-1&1&1\\0&1&1&2\\0&1&3&0\end{matrix}$	1								
		$\begin{matrix}0&1&-1&-1\\1&-1&1&1\\0&1&1&2\\0&1&3&0\end{matrix}$		1		1				
1			$\begin{matrix}0&1&-1&-1\\1&-1&1&1\\0&1&1&2\\0&1&3&0\end{matrix}$							
	1			1	$\begin{matrix}0&1&-1&-1\\1&-1&1&1\\0&1&1&2\\0&1&3&0\end{matrix}$					
		ε_1		1		ε_1				
						$\begin{matrix}0&1&-1&-1\\1&-1&1&1\\0&1&1&2\\0&1&3&0\end{matrix}$		1		1
1							$\begin{matrix}0&1&-1&-1\\1&-1&1&1\\0&1&1&2\\0&1&3&0\end{matrix}$			
						1	ε_2		ε_2

A main ingredient in the proof of Thm. (21.29) is the following version of Laplace expansion.

(21.30) Laplace Expansion Theorem. *Let R be a commutative ring, Z a finite totally ordered set, and $A \in R^{Z \times Z}$. Then for any partition $Z_1 \sqcup \ldots \sqcup Z_q$ of Z we have*

$$\text{per}(A) = \sum_{(S_1,\ldots,S_q)} \prod_{j=1}^{q} \text{per}(A \downarrow Z_j \times S_j),$$

where the summation is over all partitions $S_1 \sqcup \ldots \sqcup S_q = Z$ with $|S_j| = |Z_j|$, for all j, and where $A \downarrow Z_j \times S_j$ denotes the restriction of $A: Z \times Z \to R$ to $Z_j \times S_j$.

Proof. Viewing $\text{Sym}(Z_1) \times \ldots \times \text{Sym}(Z_q)$ in the obvious way as a subgroup of the symmetric group $\text{Sym}(Z)$, we have the following left coset decomposition

$$\text{Sym}(Z) = \bigsqcup_{\sigma \in \text{Sh}} \sigma(\text{Sym}(Z_1) \times \ldots \times \text{Sym}(Z_q)),$$

where $\text{Sh} := \text{Sh}(Z_1, \ldots, Z_q) := \{\sigma \in \text{Sym}(Z) \mid \forall j : \sigma \downarrow Z_j \text{ increasing}\}$ denotes the set of all $(Z_1, \ldots, Z_q)$-*shuffle permutations* of Z. Now let $A = (a_{ij}) \in R^{Z \times Z}$. Then

$$\text{per}(A) = \sum_{\pi \in \text{Sym}(Z)} \prod_{i \in Z} a_{i,\pi(i)} = \sum_{\sigma \in \text{Sh}} \sum_{(\rho_j) \in \prod_j \text{Sym}(Z_j)} \prod_{j=1}^{q} \prod_{i \in Z_j} a_{i,\sigma\rho_j(i)}$$

$$= \sum_{\sigma \in \text{Sh}} \underbrace{\left(\sum_{\rho_1 \in \text{Sym}(Z_1)} \prod_{i \in Z_1} a_{i,\sigma\rho_1(i)} \right)}_{= \text{per}(A \downarrow Z_1 \times \sigma(Z_1))} \cdots \underbrace{\left(\sum_{\rho_q \in \text{Sym}(Z_q)} \prod_{i \in Z_q} a_{i,\sigma\rho_q(i)} \right)}_{= \text{per}(A \downarrow Z_q \times \sigma(Z_q))}.$$

As $\sigma \mapsto (\sigma(Z_1), \ldots, \sigma(Z_q))$ establishes a one-to-one correspondence between all $(Z_1, \ldots, Z_q)$-shuffles of Z and all partitions $S_1 \sqcup \ldots \sqcup S_q = Z$ with $|S_j| = |Z_j|$, for all j, our claim follows. $\square$

We proceed with the proof of Thm. (21.29) by performing a Laplace expansion of $\text{per}(A')$ corresponding to the partition

$$\{1, \ldots, N'\} = \underline{N} \sqcup \bigsqcup_{i=1}^{t} \left(\left(\bigsqcup_{j=1}^{\mu_i} L_{ij} \sqcup R_{ij} \right) \sqcup C_i \right).$$

Here, $L_{ij} = \{L_{ij1}, L_{ij2}, L_{ij3}, L_{ij4}\}$, $R_{ij} = \{R_{ij1}, R_{ij2}, R_{ij3}, R_{ij4}\}$, and $C_i = \{c_i\}$ denote the sets of row and column indices of A' corresponding to V_{ij}^{ℓ}, V_{ij}^{r}, and ε_i, respectively. (Thus $c_i = N + \sum_{j=1}^{i}(8\mu_j + 1)$, and with $c_0 = N$ we have $L_{ijp} = c_{i-1} + 4(j-1) + p$, and $R_{ijp} = c_{i-1} + 4(\mu_i + j - 1) + p$.) According to the Laplace expansion theorem we obtain

(A)
$$\text{per}(A') = \sum \text{per}(A' \downarrow \underline{N} \times \underline{N}^*) \prod_{i=1}^{t} \text{per}(A' \downarrow C_i \times C_i^*) \cdot$$

$$\prod_{i=1}^{t} \prod_{j=1}^{\mu_i} \text{per}(A' \downarrow L_{ij} \times L_{ij}^*) \, \text{per}(A' \downarrow R_{ij} \times R_{ij}^*),$$

where the sum is over all partitions

(B)
$$\underline{N}' = \underline{N}^* \sqcup \coprod_{i=1}^{t} \left(\left(\coprod_{j=1}^{\mu_i} L_{ij}^* \sqcup R_{ij}^* \right) \sqcup C_i^* \right),$$

with $|\underline{N}^*| = N$, $|L_{ij}^*| = |R_{ij}^*| = 4$, and $|C_i^*| = 1$, for all i, j.

We now fix such a partition as described in (B) whose corresponding summand in (A) is nonzero. As $\prod_{i=1}^{t} \operatorname{per}(A' \downarrow C_i \times C_i^*) \neq 0$ we see that $C_i^* = C_i$ or $C_i^* = \{L_{i14}\}$ for all $i \in \underline{t}$. We are going to show that the partition in (B) can be recovered from $(C_1^*, \ldots, C_t^*)$. To this end, we note that both $A' \downarrow L_{ij} \times \underline{N}'$ and $A' \downarrow R_{ij} \times \underline{N}'$ contain exactly six nonzero columns; those four of the Valiant matrix and the two unit vectors $(1, 0, 0, 0)^\top$ and $(0, 0, 0, 1)^\top$ in columns denoted by ℓ_{ij1} and ℓ_{ij4} (resp. r_{ij1} and r_{ij4}). Thus we already know that

$$L_{ij}^* \subset L_{ij} \sqcup \{\ell_{ij1}, \ell_{ij4}\}, \quad |L_{ij}^*| = 4,$$

and

$$R_{ij}^* \subset R_{ij} \sqcup \{r_{ij1}, r_{ij4}\}, \quad |R_{ij}^*| = 4.$$

As $A' \downarrow (\underline{N}' \setminus L_{ij}) \times \{L_{ij2}, L_{ij3}\}$ and $A' \downarrow (\underline{N}' \setminus R_{ij}) \times \{R_{ij2}, R_{ij3}\}$ are zero matrices, we obtain the additional restrictions

$$\{L_{ij2}, L_{ij3}\} \subset L_{ij}^* \quad \text{and} \quad \{R_{ij2}, R_{ij3}\} \subset R_{ij}^*.$$

We also have $L_{ij}^* \neq L_{ij}$, for otherwise $A' \downarrow L_{ij} \times L_{ij} = V$ and $\operatorname{per}(V) = 0$. Furthermore, $\{\ell_{ij1}, \ell_{ij4}\} \not\subseteq L_{ij}^*$, since otherwise $L_{ij}^* = \{\ell_{ij1}, L_{ij2}, L_{ij3}, \ell_{ij4}\}$ and $\operatorname{per}(A' \downarrow L_{ij} \times L_{ij}^*) = \operatorname{per}(V[1, 4|1, 4]) = 0$. Finally, neither $\{L_{ij1}, \ell_{ij4}\}$ nor $\{L_{ij4}, \ell_{ij1}\}$ are subsets of L_{ij}^*, for otherwise $\operatorname{per}(A' \downarrow L_{ij} \times L_{ij}^*) = \operatorname{per}(V[4|4]) = 0$ or $\operatorname{per}(A' \downarrow L_{ij} \times L_{ij}^*) = \operatorname{per}(V[1|1]) = 0$. Together with $\ell_{ij1} = L_{i,j+1,4}$ and $\ell_{ij4} = R_{ij1}$, for all $j < \mu_i$, and $\ell_{i,\mu_i,1} = c_i$, $\ell_{i,\mu_i,4} = R_{i,\mu_i,1}$ we see that the following alternatives $(\mathbf{L})_{ij0}$ or $(\mathbf{L})_{ij1}$ hold for all $i \leq t$ and $j \leq \mu_i$

$(\mathbf{L})_{ij0}$ $\qquad\qquad L_{ij}^* = \{L_{ij2}, L_{ij3}, L_{ij4}, R_{ij1}\},$

$(\mathbf{L})_{ij1}$ $\qquad\qquad L_{ij}^* = \{L_{ij1}, L_{ij2}, L_{ij3}, L_{i,j+1,4}\}$, if $1 \leq j < \mu_i$, and
$\qquad\qquad\qquad L_{ij}^* = \{L_{ij1}, L_{ij2}, L_{ij3}, c_i\}$, if $j = \mu_i$.

Observing that $r_{ij1} = s_{ij}$ and $r_{ij4} = L_{ij1}$, for all $j \leq \mu_i$, one sees with a similar argument that the following alternatives $(\mathbf{R})_{ij0}$ and $(\mathbf{R})_{ij1}$ hold for all $i \leq t$ and $j \leq \mu_i$

$(\mathbf{R})_{ij0}$ $\qquad\qquad R_{ij}^* = \{R_{ij2}, R_{ij3}, R_{ij4}, L_{ij1}\},$

$(\mathbf{R})_{ij1}$ $\qquad\qquad R_{ij}^* = \{R_{ij1}, R_{ij2}, R_{ij3}, s_{ij}\}.$

In all cases we have $\operatorname{per}(A' \downarrow L_{ij} \times L_{ij}^*) = 4 = \operatorname{per}(A' \downarrow R_{ij} \times R_{ij}^*)$. Now row c_i controls which of the alternatives applies. More precisely, we shall show that

(C)$_0$ $\qquad\qquad\qquad$ $C_i^* = C_i \Rightarrow \forall j \le \mu_i :$ **(L)**$_{ij0} \wedge$ **(R)**$_{ij0}.$

(C)$_1$ $\qquad\qquad\qquad$ $C_i^* = \{L_{i14}\} \Rightarrow \forall j \le \mu_i :$ **(L)**$_{ij1} \wedge$ **(R)**$_{ij1}.$

To prove **(C)**$_0$ note that $C_i^* = C_i$ implies that **(L)**$_{i\mu_i 0}$ is valid. Furthermore, for $1 < j \le \mu_i$ the following domino properties hold

(D) $\qquad\qquad\qquad$ **(L)**$_{ij0} \Rightarrow$ **(L)**$_{i,j-1,0}$ and **(L)**$_{i,j-1,1} \Rightarrow$ **(L)**$_{ij1}.$

(In fact, **(L)**$_{ij0}$ and **(L)**$_{i,j-1,1}$ cannot hold simultaneously, for otherwise $L_{ij4} \in L_{ij}^* \cap L_{i,j-1}^* = \emptyset$.) Thus **(L)**$_{ij0}$ holds for all $1 \le j \le \mu_i$. To complete the proof of **(C)**$_0$ observe that **(L)**$_{ij0}$ implies **(R)**$_{ij0}$. (In fact, the validity of both **(L)**$_{ij0}$ and **(R)**$_{ij1}$ would yield $R_{ij1} \in R_{ij}^* \cap L_{ij}^* = \emptyset$, which is absurd.)

In a similar way we prove **(C)**$_1$. $C_i^* = \{L_{i14}\}$ implies the validity of **(L)**$_{i11}$, and (D) implies that **(L)**$_{ij1}$ holds, for every $1 \le j \le \mu_i$. As **(L)**$_{ij1} \wedge$ **(R)**$_{ij0}$ would give the contradiction $L_{ij1} \in L_{ij}^* \cap R_{ij}^* = \emptyset$, we see that **(L)**$_{ij1} \Rightarrow$ **(R)**$_{ij1}$, which completes the proof of **(C)**$_1$. Combining **(C)**$_0$ and **(C)**$_1$ we obtain

$$\underline{N}^* = \bigcup_{C_i^* = \{L_{i14}\}} \{R_{ij4} \mid 1 \le j \le \mu_i\} \cup \left(\underline{N} \setminus \bigcup_{C_i^* = \{L_{i14}\}} \{s_{ij} \mid 1 \le j \le \mu_i\} \right).$$

Thus if we put $e_i = 0$, if $C_i^* = C_i$ and $e_i = 1$, if $C_i = \{L_{i14}\}$, see our illustration of $A'[i]$, then $A' \downarrow \underline{N} \times \underline{N}^*$ is up to reordering of the columns equal to $A(X, e_1, \ldots, e_t)$. As

$$\underbrace{\prod_{i=1}^{t} \operatorname{per}(A' \downarrow C_i \times C_i^*)}_{=4^{-2\mu_i}} \cdot \prod_{i=1}^{t} \prod_{j=1}^{\mu_i} \underbrace{\operatorname{per}(A' \downarrow L_{ij} \times L_{ij}^*)}_{=4} \underbrace{\operatorname{per}(A' \downarrow R_{ij} \times R_{ij}^*)}_{=4} = 1,$$

we see that $\operatorname{per}(A') = \sum_{e \in \{0,1\}^t} \operatorname{per}(A(X, e)) = \sum_e g(X, e)$, which completes the proof of Thm. (21.29). $\qquad\qquad\qquad\qquad\qquad\qquad\qquad\qquad\square$

21.5* The Extended Valiant Hypothesis

In the previous sections we have seen that $\mathbf{VNP}_e = \mathbf{VNP}$; furthermore we have specified **VNP**-complete families in **VNP**. On the other hand there is the fundamental conjecture that $\mathbf{VP}_e$ is a proper subclass of **VP**. In this section we will work with p-families whose expression sizes resp. complexities are allowed to grow quasi-polynomially, see the definition below. This growth might be much faster than that of any polynomial, but is much slower than that of any exponential function $n \mapsto 2^{n^\varepsilon}$, for $\varepsilon > 0$. It will turn out that the corresponding complexity classes coincide. Furthermore, we shall show that under a more generous kind of reduction the family DET is complete in this complexity class. This enables us to formulate the extended Valiant hypothesis on the one hand in terms of complexity theory and on the other hand purely in terms of algebra.

(21.31) Definition.
(1) A function $t: \mathbb{N} \to \mathbb{N}$ is called *quasi-polynomially bounded* (*qp*-bounded), if there exists a positive constant c such that $t(n) \le n^{O(\log^c n)}$.
(2) A *p*-family $f = (f_n)$ over k is called *qp-computable* (resp. *qp-expressible*) iff $n \mapsto L(f_n)$ (resp. $n \mapsto E(f_n)$) is *qp*-bounded.
(3) **VQP** = **VQP**$(k;$ nonuniform$)$ (resp. **VQP**$_e$ = **VQP**$_e(k;$ nonuniform$)$) denotes Valiant's class of all *qp*-computable (resp. *qp*-expressible) families over k. •

Obviously, **VP** $\subseteq$ **VQP** and **VP**$_e$ $\subseteq$ **VQP**$_e$. The following conjecture generalizes the hypothesis **VNP** $\setminus$ **VP** $\ne \emptyset$.

(21.32) Extended Valiant Hypothesis. **VNP** $\setminus$ **VQP** $\ne \emptyset$ *over any field*.

The goal of this section is to show that this conjecture is equivalent to the statement: **VNP**$_e$ $\setminus$ **VQP**$_e$ $\ne \emptyset$ over any field. As we already know that **VNP**$_e$ = **VNP**, it remains to show the following.

(21.33) Theorem. **VQP**$_e$ = **VQP** *over any field*.

This theorem will result from interrelations between several complexity measures: the expression size $E(f)$, the complexity $L(f)$, and the depth $D(f)$ of a polynomial f. Let us briefly recall the notion of depth, introduced in Sect. 4.1. Every straight-line program $\Gamma = (\Gamma_1, \ldots, \Gamma_r)$ with $\Gamma_i = (\omega_i; i', i'')$ defines a directed acyclic multigraph: its set of nodes is $\{i \in \mathbb{Z} \mid -n < i \le r\}$ and for $1 \le i \le r$ it has edges $i' \to i, i'' \to i$. The length $D(\Gamma)$ of the longest directed path in this multigraph is called the *depth* of Γ. The *depth* $D(f) = D_k(f)$ of $f \in k[X_1, \ldots, X_n]$ is the smallest depth of a straight-line program that computes f from $X_1, \ldots, X_n$. There is also a notion of depth for an arithmetic expression φ over I: if $\varphi \in I$ then its depth is 0; if $\varphi = (\varphi_1 \circ \varphi_2)$ then depth$(\varphi) := \max\{\text{depth}(\varphi_1), \text{depth}(\varphi_2)\} + 1$. Every expression φ defines a binary tree T_φ as follows. If $\varphi \in I$ then T_φ has one node labeled φ; if $\varphi = (\varphi_1 \circ \varphi_2)$ then T_{φ_1} and T_{φ_2} are the left and right subtrees of T_φ, respectively, and the root of T_φ has label $\circ$. An expression ψ is a *subexpression* of φ iff T_ψ is a subtree of T_φ. Obviously, the depth of T_φ equals the depth of φ. The *expression depth (formula depth)* $T(f) = T_k(f)$ of $f \in k[X_1, \ldots, X_n]$ is the smallest depth of a formula φ with val$(\varphi) = f$.

(21.34) Remark. For all nonzero $f \in k[X_1, \ldots, X_n]$ the following statements hold:

(1) $D(f) \le L(f) \le E(f)$.
(2) $D(f) = T(f)$.
(3) $D(f) \ge \log \deg f$.   •

The following two theorems show much closer relationships between the depth, the formula size, and the complexity of a polynomial. Their proofs are based on certain program transformations.

(21.35) Theorem (Brent). *If $f \in k[X_1, \ldots, X_n]$ is of degree $d \geq 2$, then*

$$\log(E(f) + 1) \leq D(f) \leq \frac{2}{\log \varepsilon} \log(E(f)) + 1,$$

where $\varepsilon := (1 + \sqrt{5})/2$ is the ratio of the golden mean.

Proof. For proving the lower bound, let φ be an expression with $\mathrm{val}(\varphi) = f$ and $D(f) = \mathrm{depth}(\varphi) = T(f)$, see Rem. (21.34)(2). Let $|T_\varphi|$ denote the number of leaves of the tree corresponding to φ. Then an easy induction shows that $|T_\varphi| = E(\varphi) + 1$. Hence

$$E(f) + 1 \leq E(\varphi) + 1 = |T_\varphi| \leq 2^{\mathrm{depth}(\varphi)} = 2^{D(f)},$$

which proves the lower bound.

To prove the upper bound, let φ be an expression with $\mathrm{val}(\varphi) = f$ and $e := E(\varphi) = E(f)$. We may assume that $E(\varphi) \geq 2$. Let q be any fixed real number with $1 < q \leq E(\varphi)$. Consider the tree T_φ corresponding to φ and associate to each node of T_φ the expression size of the corresponding subexpression. Starting at the root of T_φ and following in each step a largest subexpression we will come to a node v, where both subexpressions β and γ have size $\leq q$. Replacing the subexpression corresponding to v by a new variable Y defines an expression α:

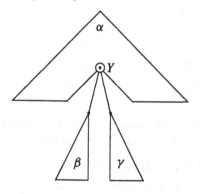

Thus for every real q satisfying $1 < q \leq E(\varphi)$, there exist expressions α, β, γ over $k \cup \{X_1, \ldots, X_n, Y\}$ and $\circ \in \{+, *\}$ such that

(A) Y occurs in α.
(B) β and γ are subexpressions of φ.
(C) Substituting Y by $(\beta \circ \gamma)$ transforms α into φ.
      (for short: $\varphi = \alpha(Y \leftarrow \beta \circ \gamma)$)
(D) $E(\beta), E(\gamma) \leq q$.
(E) $q < E(\beta \circ \gamma)$.

We now prove the upper bound for $D(f)$ by induction on $e := E(\varphi)$. For $e \leq 3$, the claim is easily verified. So let $e \geq 4$. Apply (A)–(E) with $q := \varepsilon^{-2}e$. (As $e \geq 4$, we have $1 < q < e$.) Then $E(\beta), E(\gamma) \leq q < e$, and since $\varepsilon^2 = \varepsilon + 1$ and $\varepsilon^{-1} = \varepsilon - 1$, we get

$$E(\alpha) \;=\; E(\varphi) - E(\beta \circ \gamma)$$
$$< \; e - q = e(1 - \varepsilon^{-2}) = \varepsilon^{-1}e < e.$$

Hence the induction hypothesis applies to α, β, and γ and gives

$$\max\{D(\mathrm{val}(\beta)), D(\mathrm{val}(\gamma))\} \le \frac{2}{\log\varepsilon}\log(\varepsilon^{-2}e) + 1$$

and

$$D(\mathrm{val}(\alpha)) \le \frac{2}{\log\varepsilon}\log(\varepsilon^{-1}e) + 1.$$

Now

$$D(\mathrm{val}(\beta \circ \gamma)) \le 1 + \max\{D(\mathrm{val}(\beta)), D(\mathrm{val}(\gamma))\}.$$

Interpreting $\mathrm{val}(\alpha)$ as a polynomial in Y with coefficients in $k[X_1, \ldots, X_n]$, we can write

$$\mathrm{val}(\alpha) = a_0 + a_1 Y,$$

for suitable $a_i \in k[X_1, \ldots, X_n]$. Let $\alpha_0 := \alpha(Y \leftarrow 0)$. Then $\mathrm{val}(\alpha_0) = a_0$ and

$$D(a_0) \le D(\mathrm{val}(\alpha)).$$

Let $\circ_1 \to \circ_2 \to \ldots \circ_m \to Y$ be the path from the root of T_α to Y. Replace Y by 1 and cancel every $\circ_i = +$.

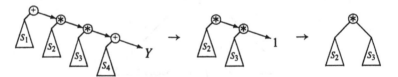

This gives an expression α_1 with $\mathrm{val}(\alpha_1) = a_1$ and $D(\mathrm{val}(\alpha_1)) \le D(\mathrm{val}(\alpha))$. Since $f = \mathrm{val}(\varphi) = \mathrm{val}(\alpha_0) + \mathrm{val}(\alpha_1) * \mathrm{val}(\beta \circ \gamma)$, we get

$$
\begin{aligned}
D(f) \;&\le\; 2 + \max\{D(\mathrm{val}(\alpha_0)), D(\mathrm{val}(\alpha_1)), D(\mathrm{val}(\beta \circ \gamma))\} \\
&\le\; 2 + \max\{D(\mathrm{val}(\alpha)), D(\mathrm{val}(\beta \circ \gamma))\} \\
&\le\; 2 + \max\left\{\frac{2}{\log\varepsilon}\log e - 1, \; \frac{2}{\log\varepsilon}\log e - 2\right\} \\
&\le\; \frac{2}{\log\varepsilon}\log e + 1,
\end{aligned}
$$

which was to be shown. $\square$

The fact that $D = \Theta(\log E)$ supports the conjecture that $\mathbf{VP}_e \ne \mathbf{VP}$, for $\mathbf{VP}_e = \mathbf{VP}$ would imply that for every $f \in \mathbf{VP}$ (in particular for $f = DET$) there is a constant c with $D(f_n) \le c \log n$, for all n. However, this seems very unlikely to be true.

As a final tool for proving $\mathbf{VQP}_e = \mathbf{VQP}$ we shall estimate the depth of a polynomial in terms of its degree and its complexity.

(21.36) Theorem (Hyafil, Valiant et al.). *Let f be an n-variate polynomial of degree $d \geq 1$ over k. Then*

$$D(f) \leq c\big(\log(dL(f))\log d + \log n\big),$$

for some universal constant c. Moreover, f can be computed by a straight-line program of length $O\big(d^6 L(f)^3\big)$ and depth $O\big(\log(dL(f))\log d + \log n\big)$.

Proof. W.l.o.g. we may assume that $f(0) = 0$. Let $\Gamma = (\Gamma_1, \ldots, \Gamma_r)$ be an optimal homogeneous straight-line program that on input $(k[X]; X)$ computes the homogeneous parts of f. Let $b = (b_j)_{j \in N}$ with $N := \{j \mid -n < j \leq r\}$ be the corresponding result sequence. As Γ is optimal and $f(0) = 0$ there is no instruction of the form $\Gamma_j = (\lambda^c)$. We partition $\underline{r}$ as $\underline{r} = A \sqcup M \sqcup S$, where $\Gamma_j = (+; j', j'')$, for $j \in A$, $\Gamma_j = (*; j', j'')$, for $j \in M$, and $\Gamma_j = (\lambda_j; j')$, for $j \in S$. By Lemma (21.25) we can assume that $r \leq (d+1)^2 L(f)$. Furthermore, we can w.l.o.g. assume that the following hold (see also Rem. (21.24)).

(A) Each b_j is nonzero and homogeneous of degree d_j, and $i \leq_\Gamma j$ implies
     $d_i \leq d_j \leq d$ for all i, j.
(B) $d_{j'} \geq d_{j''}$ for all $j \in A \sqcup M$.

Let $B = B(\Gamma, b) = (b_{ij})$ be an upper triangular matrix in $k[X_1, \ldots, X_n]^{N \times N}$ with ones on the main diagonal whose columns B_j are defined recursively in the following way (e_j denotes the indicator of $j \in N$):

$$\begin{aligned}
j \in N, j \leq 0 \;&\Rightarrow\; B_j := e_j, \\
\Gamma_j = (\lambda_j; j') \;&\Rightarrow\; B_j := e_j + \lambda_j \cdot B_{j'}, \\
\Gamma_j = (+; j', j'') \;&\Rightarrow\; B_j := e_j + B_{j'} + B_{j''}, \\
\Gamma_j = (*; j', j'') \;&\Rightarrow\; B_j := e_j + b_{j''} \cdot B_{j'}.
\end{aligned}$$

Before proceeding with the proof we illustrate the construction of the matrix B by an example. The following describes a straight-line program $\Gamma = (\Gamma_1, \ldots, \Gamma_5)$ and its result sequence $(b_i)_{-2 \leq i \leq 5} = (X_1, X_2, X_3, b_0, b_1, \ldots, b_5)$ on input $(k[X]; X)$:

$$\begin{aligned}
\Gamma_1 &= (\lambda; -2) & b_1 &= \lambda X_1 \\
\Gamma_2 &= (+; -1, 0) & b_2 &= X_2 + X_3 \\
\Gamma_3 &= (\mu; 2) & b_3 &= \mu b_2 = \mu X_2 + \mu X_3 \\
\Gamma_4 &= (+; 1, 3) & b_4 &= b_1 + b_3 = \lambda X_1 + \mu X_2 + \mu X_3 \\
\Gamma_5 &= (*; 4, 2) & b_5 &= b_4 \cdot b_2 = (\lambda X_1 + \mu X_2 + \mu X_3)(X_2 + X_3)
\end{aligned}$$

Hence $B_j = e_j$ for $-2 \leq j \leq 0$ and

$$\begin{aligned}
B_1 &= e_1 + \lambda B_{-2}, & B_2 &= e_2 + B_{-1} + B_0, & B_3 &= e_3 + \mu B_2, \\
B_4 &= e_4 + B_1 + B_3, & B_5 &= e_5 + b_2 \cdot B_4;
\end{aligned}$$

this yields

$$
B = \begin{array}{c}
\begin{array}{cccccccc}
B_{-2} & B_{-1} & B_0 & B_1 & B_2 & B_3 & B_4 & B_5
\end{array} \\
\begin{array}{c}
X_1 \\ X_2 \\ X_3 \\ b_1 \\ b_2 \\ b_3 \\ b_4 \\ b_5
\end{array}
\left(
\begin{array}{cccccccc}
1 & & & \lambda & & & \lambda & \lambda b_2 \\
& 1 & & & 1 & \mu & \mu & \mu b_2 \\
& & 1 & & 1 & \mu & \mu & \mu b_2 \\
& & & 1 & & & 1 & b_2 \\
& & & & 1 & \mu & \mu & \mu b_2 \\
& & & & & 1 & 1 & b_2 \\
& & & & & & 1 & b_2 \\
& & & & & & & 1
\end{array}
\right)
\end{array}
$$

We continue with the proof and mention some properties of the matrix B:

(C) $B = (b_{ij})$ is an upper triangular matrix with $b_{ii} = 1$ for all $i \in N$. Furthermore, $b_{ij} \neq 0$ implies $i \leq_r j$.

The fact that B is an upper triangular matrix with ones on all diagonal positions follows by construction. We prove the second statement by induction on j. W.l.o.g. $j \geq 1$, $i < j$ and $b_{ij} \neq 0$. Suppose first that $j \in A$. Then $b_{ij} = b_{ij'} + b_{ij''} \neq 0$, hence $b_{ij'} \neq 0$ or $b_{ij''} \neq 0$. By induction we know that $i \leq_r j'$ or $i \leq_r j''$. Combining this with $j' \leq_r j$ and $j'' \leq_r j$ we get $i \leq_r j$. If $j \in M$, then $b_{ij} = b_{j''}b_{ij'} \neq 0$; hence $b_{ij'} \neq 0$. Again by induction one knows that $i \leq_r j'$, and as $j' \leq_r j$ we also have $i \leq_r j$. The case $j \in S$ is handled analogously. This completes the proof of (C).

(D)   Each nonzero b_{ij} is a homogeneous polynomial of degree $d_j - d_i$.

We proceed by induction on j. W.l.o.g. $j \geq 1$, $i < j$, and $b_{ij} \neq 0$. If $j \in S$ then $b_{ij} = \lambda_j b_{ij'}$ and $d_j = d_{j'}$. By induction, $b_{ij'}$ is homogeneous of degree $d_{j'} - d_i = d_j - d_i$. Hence the same is true for b_{ij}. In the remaining cases we have $b_{ij} = b_{ij'} + b_{ij''}$ or $b_{ij} = b_{ij'} \cdot b_{j''}$. By induction we know that $b_{ij'}$ and $b_{ij''}$ are homogeneous or zero, $b_{ij'} \neq 0$ implies $\deg b_{ij'} = d_{j'} - d_i$, and $b_{ij''} \neq 0$ implies $\deg b_{ij''} = d_{j''} - d_i$. If $j \in A$, then by (B) $b_{j'}b_{j''} \neq 0$ and $d_{j'} = d_{j''} = d_j$; hence b_{ij} is homogeneous as well, and $\deg b_{ij} = d_j - d_i$. If $j \in M$, then b_{ij} is homogeneous of degree $\deg(b_{ij'}) + d_{j''} = d_{j'} - d_i + d_{j''} = d_j - d_i$. This proves (D).

Next we introduce $\leq_r$-antichains that will help to describe small depth computations. For every $a \geq 1$ we define

$$\Gamma_b(a) := \{ t \in \underline{r} \mid t \in M; d_{t'}, d_{t''} \leq a < d_t \}.$$

For every a, $\Gamma_b(a)$ is a $\leq_r$-antichain that is involved in the computation of the b_i and b_{ij} in the following way.

(E)   For $a \geq 1$ and $i, j \leq r$ with $d_i \leq a < d_j$ the following hold:

$$b_{ij} = \sum_{t \in \Gamma_b(a)} b_{it}b_{tj} \quad \text{and} \quad b_j = \sum_{t \in \Gamma_b(a)} b_t b_{tj}.$$

First note that $i \neq j$. As $d_j > a \geq 1$ we also have $j \geq 1$. Keeping a, Γ, and i fixed, we prove the claims by induction on j. The start being clear, we focus on the induction step.

Case 1. $\Gamma_j = (+; j', j'')$.
On the one hand $b_j = b_{j'} + b_{j''}$ and $d_j = d_{j'} = d_{j''}$ and on the other hand j', $j'' < j$. Thus

$$b_{ij} = b_{ij'} + b_{ij''} = \sum_{t \in \Gamma_b(a)} b_{it}(b_{tj'} + b_{tj''}) = \sum_{t \in \Gamma_b(a)} b_{it} b_{tj},$$

and

$$b_j = b_{j'} + b_{j''} = \sum_{t \in \Gamma_b(a)} b_t(b_{tj'} + b_{tj''}) = \sum_{t \in \Gamma_b(a)} b_t b_{tj}.$$

Case 2. $\Gamma_j = (*; j', j'')$.
We distinguish two subcases.

Case 2.1. $d_{j'} \leq a$.
By (B) $d_{j''} \leq d_{j'}$, hence $j \in \Gamma_b(a)$. As $\Gamma_b(a)$ is an antichain, we have $b_{tj} = 0$ for all $t \in \Gamma_b(a) \setminus \{j\}$. Thus, as $b_{jj} = 1$,

$$b_{ij} = \underbrace{b_{ij}b_{jj}}_{t=j} + \sum_{t \in \Gamma_b(a)\setminus\{j\}} b_{it} \underbrace{b_{tj}}_{=0}, \quad \text{and} \quad b_j = \underbrace{b_j b_{jj}}_{t=j} + \sum_{t \in \Gamma_b(a)\setminus\{j\}} b_t \underbrace{b_{tj}}_{=0}.$$

Case 2.2. $a < d_{j'} \leq d_j$.
Then $B_j = e_j + b_{j''} B_{j'}$, in particular, $b_{tj} = b_{tj'} b_{j''}$, for all $t < j$. Applying the induction hypothesis to j' we obtain

$$b_{ij} = b_{ij'} b_{j''} = b_{j''} \sum_{t \in \Gamma_b(a)} b_{it} b_{tj'} = \sum_{t \in \Gamma_b(a)} b_{it} b_{tj}.$$

Our second claim in this subcase can be shown in a similar way. As the case $\Gamma_j = (\lambda_j; j')$ can be handled analogously, (E) follows.

(F)   There exists a homogeneous straight-line program Γ' of length $r' = O(r^3)$ and depth $D' = O(\log(r) \cdot \log(d))$, which computes the homogeneous parts of f. (Recall that r denotes the length of Γ.)

W.l.o.g. all $X_1, \ldots, X_n$ occur in f. Thus $r \geq n - 1$. We are going to construct Γ' in $\lceil \log d \rceil$ stages. Each stage will contribute at most $2 + \lceil \log r \rceil$ to the depth of the final program Γ'. (Thus $D' \leq \lceil \log d \rceil (2 + \lceil \log r \rceil)$.)
    Stage 0. Compute all b_j and b_{ij} of degree $\leq 2^0 = 1$.
All these b_j and b_{ij} are linear forms $\sum_{\nu=1}^{n} a_\nu X_\nu$ in n indeterminates or constants. Thus Stage 0 can be accomplished in depth

$$1 + \lceil \log n \rceil \leq 1 + \lceil \log(r + 1) \rceil \leq 2 + \lceil \log r \rceil.$$

    Stage $\delta + 1$. Compute all b_j and b_{ij}, whose degrees are in the interval $(2^\delta, 2^{\delta+1}]$. (By (A) and (D), we are done after $\lceil \log d \rceil$ stages.) We first concentrate on the b_j. Let $2^{\delta+1} \geq d_j > 2^\delta =: a$. By (E) we have

(G) $b_j = \sum_{t \in \Gamma_b(2^\delta)} b_t b_{tj} = \sum_{t \in \Gamma_b(2^\delta)} b_{t'} b_{t''} b_{tj}$

and by (D) and the definition of $\Gamma_b(2^\delta)$, the three polynomials $b_{t'}$, $b_{t''}$ and b_{tj} all have degree $\leq 2^\delta$ for every $t \in \Gamma_b(2^\delta)$. Hence these polynomials have already been computed in previous stages. Thus those b_j can be computed in additional depth at most

$$2 + \lceil \log(|\Gamma_b(2^\delta)|) \rceil \leq 2 + \lceil \log r \rceil.$$

Next we consider the b_{ij}. Let $2^{\delta+1} \geq \deg b_{ij} = d_j - d_i > 2^\delta$. Put $a := 2^\delta + d_i$. As $d_i \leq a < d_j$, we obtain by (E)

(H) $b_{ij} = \sum_{t \in \Gamma_b(a)} b_{it} b_{tj} = \sum_{t \in \Gamma_b(2^\delta)} b_{t''} b_{it'} b_{tj}.$

Now both $b_{it'}$ and b_{tj} are of degree $\leq 2^\delta$ (check this!), hence have been computed earlier. However, $b_{t''}$ might have a larger degree, say $d_{t'} \geq d_{t''} > 2^{\delta+1}$. We claim that in this case $b_{t''} b_{it'} b_{tj} = 0$. In fact, if $b_{it'} \neq 0$ and $d_{t''} > 2^{\delta+1}$, then $d_{t'} \geq d_i$, hence $d_t = d_{t'} + d_{t''} > d_i + 2^{\delta+1} \geq d_j$, thus $b_{tj} = 0$. Altogether, the b_{ij} in question can be computed with additional depth $\leq 2 + \lceil \log r \rceil$ and the resulting straight-line program Γ' satisfies the depth requirements stated in (F). (G), (H), and the fact that $r \geq n - 1$ imply that $r' = O(r^3)$. This proves (F). Finally, combining Lemma (21.25) and (F) our claims follow.   □

Proof of Thm. (21.33).   To show the only nontrivial inclusion **VQP** $\subseteq$ **VQP**$_e$, let $f = (f_n) \in$ **VQP** and $d_n := \deg f_n$. Then $n \mapsto L(f_n)$ is qp-bounded. By the last theorem we know that $D(f_n) \leq c(\log(d_n L(f_n)) \log d_n + \log n)$. As $D = \Theta(\log E)$, a straightforward computation shows that also $n \mapsto E(f_n)$ is qp-bounded. Thus $f \in$ **VQP**$_e$, which completes the proof of Thm. (21.33).   □

Thm. (21.36) has another interesting corollary. For its formulation, we define the following nonuniform algebraic analogue of the complexity class **NC**, known as Nick's class (Pippenger [420], Cook [125]).

(21.37) Definition. For $d \in \mathbb{N}'$, **VNC**^{d} = **VNC**$^d(k;$ nonuniform$)$ denotes Valiant's class of all p-families $f = (f_n)$ of polynomials over k such that there is a sequence (Γ^n) of straight-line programs, Γ^n computing f_n, such that $n \mapsto \text{length}(\Gamma^n)$ is p-bounded and depth$(\Gamma^n) = O(\log^d n)$.   •

Obviously, **VNC**$^1 \subseteq$ **VNC**$^2 \subseteq \ldots \subseteq$ **VP**. Surprisingly, this hierarchy collapses, as is seen by the last theorem.

(21.38) Corollary. VP = **VNC**^{2} *for every field.*

The analogous statement **P** = **NC**^{2} in the discrete setting is not known to be true. Next we discuss a completeness result for the class **VQP**.

(21.39) Definition.

(1) Let $f = (f_n)$ and $g = (g_n)$ be p-families over k. Then f is a *qp-projection* of g iff there exists a qp-bounded function t such that for every n the polynomial f_n is a projection of $g_{t(n)}$.

(2) A family g is **VQP**-*complete* iff $g \in$ **VQP** and every $f \in$ **VQP** is a qp-projection of g. •

VQP is closed under qp-projections, as is shown by a straightforward computation.

(21.40) Corollary. *DET is* **VQP**-*complete.*

Proof. We already know that $DET \in$ **VP** $\subseteq$ **VQP**. Now let $f \in$ **VQP** $=$ **VQP**$_e$. Then $n \mapsto E(f_n)$ is qp-bounded and by Thm. (21.27) we know that f_n is a projection of $DET_{2E(f_n)+2}$. Thus f is a qp-projection of DET. □

By Cor. (21.38) we know that DET_n can be computed by a straight-line program of length polynomial in n and depth $O(\log^2 n)$. (For a direct proof of this fact consult Ex. 21.3.) Together with $D = \Theta(\log E)$ this yields the estimate

$$E(DET_n) = 2^{O(\log^2 n)}.$$

This should be compared with the best known upper bound for the expression size of the permanent: $E(PER_n) = O(n^2 2^n)$, see Ex. 21.1. Thus finally, we can state the following purely algebraic equivalent to the extended Valiant hypothesis in characteristic different from two.

(21.41) Extended Valiant Hypothesis. *PER is not a qp-projection of DET unless* char $k = 2$.

The extended Valiant hypothesis is true iff there is no constant c such that $E(PER_n) = 2^{O(\log^c n)}$. The problem of deriving the permanent from the determinant by substitution is classical. The best result known so far says that PER_n is not a projection of DET_m if $m < \sqrt{2}n$. So there is much left for research in Algebraic Complexity Theory.

21.6 Exercises

21.1. Prove that $\mathrm{per}(X) = \sum_J (-1)^{|J|} \prod_{i=1}^n (\sum_{j \in \underline{n} \setminus J} X_{ij})$, where the sum is over all proper subsets J of $\underline{n}$. Deduce that $E(PER_n) = O(n^2 2^n)$ and $L(PER_n) = O(n2^n)$. (Hint: use the principle of inclusion and exclusion.)

21.2. Let $\chi_n := \chi(A) = \det(T I_n - A)$ denote the characteristic polynomial of the generic n by n matrix $A = (a_{ij})$ over k. Give a direct proof of $(\chi_n) \in$ **VNC**^{2}. (Hint: consider for $r \in \underline{n}$ the lower right submatrix $A_r = (a_{ij})_{r \le i, j \le n}$ of size $R := n+1-r$, and put $d_r := \det(I_R - T A_r)$, $d_{n+1} := 1$. Then $\chi(A) = T^n d_1(T^{-1})$. Write $(b_{ij}^{(r)}) := (I_R - T A_r)^{-1} =: I_R + \sum_{\ell \ge 1} A_{r\ell} T^\ell \in K[[T]]^{R \times R}$, where $K := k(a_{ij}|i, j)$. Prove that $b_{rr}^{(r)} = d_{r+1}/d_r$ and $\prod_{r \in \underline{n}} b_{rr}^{(r)} = 1/d_1$. Now calculate $A_{r\ell}$ for all $r, \ell \in \underline{n}$, then $d_1^{-1} \bmod T^{n+1}$, and finally, $d_1 \bmod T^{n+1}$ by a Newton iteration.)

21.3. Give a direct proof of $DET \in \mathbf{VNC}^2$ by describing a sequence of straight-line programs $(\Gamma^n)_n$, where Γ^n computes DET_n in $O(n^4)$ steps with depth $O(\log^2 n)$.

21.4. Give a direct proof of $HC \in \mathbf{VNP}_e$.

21.5. Let $f = (f_n)$ with $f_n = \sum_{e \in \{0,1\}^n} g_n(e) X^e \in k[X_1, \ldots, X_n]$ and $g_n(e) \in \{0, 1\}$, for all n and e. Suppose that there exists a polynomial-time deterministic Turing machine which for every n on input $e \in \{0, 1\}^n$ computes $g_n(e)$. Prove that $f \in \mathbf{VNP}$.

21.6. For $n \in \mathbb{N}$ let $SUM_n := \sum_{i=1}^n X_i$ and $PROD_n := \prod_{i=1}^n X_i$. Show that the families SUM and $PROD$ are not p-projections of each other.

21.7.* In this exercise we show that every $f \in k[X_1, \ldots, X_n]$ of expression size u is both a projection of DET_{u+3} and of PER_{u+3}.
This can be shown as follows. Every arithmetic expression φ over $I = k \cup \{X_1, \ldots, X_n\}$ defines a sixtuple $G(\varphi) = (V, E, s, t, \lambda, \varepsilon)$, where (V, E) is a DAG with one input node s and one output node t in which every path from s to t has a length which is congruent $\varepsilon \in \{0, 1\}$ modulo 2. Furthermore, $\lambda \colon E \to I$ is a weight function. The definition of $G(\varphi)$ is recursively in the construction of φ. If $\varphi \in I$ then $G(\varphi) = (\{1, 2\}, \{(1, 2)\}, 1, 2, (1, 2) \mapsto \varphi, 1)$. In case $\varphi = (\varphi_1 \circ \varphi_2)$ with formulas φ_1, φ_2 and $\circ \in \{+, *\}$ we distinguish four cases in defining $G = G(\varphi) = (V, E, s, t, \lambda, \varepsilon)$. Let $G_i = G(\varphi_i) = (V_i, E_i, s_i, t_i, \lambda_i, \varepsilon_i)$, for $i = 1, 2$. In all cases we start with the disjoint union of G_1 and G_2 and identify certain sources and sinks (this will be indicated by $\equiv$) and add sometimes a new edge (always weighted one).
Case 1: $\circ = *$. In this case $s := s_1$, $t_1 \equiv s_2$, $t := t_2$, and $\varepsilon := \varepsilon_1 + \varepsilon_2 \bmod 2$.
In the remaining cases $\circ = +$.
Case 2: $\varphi_1, \varphi_2 \in I$. In this case add a new source s as well as two edges (s, s_1) and (s, s_2). Finally, put $\varepsilon := 0$.
In the remaining cases we assume that at least one φ_i does not belong to I.
Case 3: $\varepsilon_1 = \varepsilon_2$. In this case $s := s_1 \equiv s_2$, $t := t_1 \equiv t_2$, and $\varepsilon := \varepsilon_1 = \varepsilon_2$.
Case 4: $\varepsilon_1 \neq \varepsilon_2$. In this case add a new edge (t_2, t_1) and put $s := s_1 \equiv s_2$, $t := t_1$, $\varepsilon := \varepsilon_1$.
    Show that G has at most $u + 3$ nodes. Let $G(s, t)$ denote the set of all paths from s to t in G. Prove that $\mathrm{val}(\varphi) = \sum_{\pi \in G(s,t)} \lambda(\pi)$, where $\lambda(\pi)$ denotes the product of the weights of all edges involved in π. Now let φ be a formula for f of size $u = E(f)$. Modify $G(\varphi) = (V, E, s, t, \lambda, \varepsilon)$ to $G' = (V, E', \lambda')$ as follows. For all $v \in V \setminus \{s, t\}$ add self-loops (v, v) of weight one to E. If $\varepsilon = 1$ identify s and t, if $\varepsilon = 0$ add an edge (t, s) of weight one. Show that the cycle covers of G' all have positive sign and furthermore they are in one-to-one correspondence with the elements of $G(s, t)$. Finally prove the following: If $\Phi \colon V \times V \to I$ equals λ' on E' and vanishes outside E', then $f = \mathrm{val}(\varphi) = \mathrm{per}(\Phi) = \det(\Phi)$.

21.8.* Prove that HC is **VNP**-complete over any field. (Hint: use Ex. 21.7 and give a detailed proof of Valiant's sketch in [526] along the lines of our proof of the **VNP**-completeness of PER.)

21.9.* In analogy to the discrete case, where problems in **P** form a tiny fraction of all computable problems, this exercise shows that the complexity of n-variate polynomials of degree n is typically exponential in n. Let k be an infinite field. For $d, n \in \mathbb{N}$ let $k_{n,d} \subseteq k[X_1, \ldots, X_n]$ denote the vector space of all polynomials of degree at most d in n indeterminates over k. Thus $p := \dim(k_{n,d}) = \binom{d+n}{d}$. Prove the following statements:

(1) $L(f) \leq 2p - 2$, for all $f \in k_{n,d}$.
(2) $L(f) \geq \frac{3}{2}(p - 1)$, for Zariski almost all $f \in k_{n,d}$.
(3) We have $d + 1 \leq L(f) \leq 2d$ for almost all univariate polynomials f of degree d.
(4) We have $L(f) \geq 2^{2n}/(7n^2)$ for almost all n-variate polynomials f of degree $n \geq 2$.

(Hints: to prove (1) show that $L(n, d) := \max\{L(f) \mid f \in k_{n,d}\} \leq 2p - 2$ by writing $f \in k_{n,d}$ as $f = f(0) + \sum_{i=1}^n X_i F_i$ with $F_i \in k[X_1, \ldots, X_i]$ of degree at most $d - 1$. To show (2) imitate the proof of the transcendence degree bound (5.9).)

21.10. Prove the statements in Rem. (21.34).

21.11.* Prove the following characterization of **VNP** over k: $f \in$ **VNP** iff there exists a p-expressible family g such that f is a p-projection of the family h defined by $h_n(X) := \sum_{e \in \{0,1\}^m} g_n(e) X^e$, where $m := v(g_n)$.

21.12. Define $f_n \in k[X_1, \ldots, X_n]$ to be zero when n is not a power of 4, and otherwise inductively by $f_1 = X_1$ and

$$f_n = f_{n/4}(X_1, \ldots, X_{n/4}) * f_{n/4}(X_{n/4+1}, \ldots, X_{n/2})$$
$$+ f_{n/4}(X_{n/2+1}, \ldots, X_{3n/4}) * f_{n/4}(X_{3n/4+1}, \ldots, X_n).$$

Thus f_n is the polynomial computed by the complete binary tree with n leaves and alternating layers of $*$ and $+$. Show that the family $f = (f_n)$ is **VP**$_e$-complete, i.e., $f \in$ **VP**$_e$ and every $g \in$ **VP**$_e$ is a p-projection of f.

The next three exercises describe an alternative way for estimating the depth of a polynomial in terms of its degree and its complexity.

21.13. Let $\Gamma = (\Gamma_1, \ldots, \Gamma_r)$ be a homogeneous straight-line program which on input $(k[X]; X)$ produces the result sequence (b_i). If any of the instructions is replaced by the trivial instruction (0^c) while the other instructions remain unchanged, then the new straight-line program Γ' is homogeneous as well and the corresponding result sequence (b_i') (on the same input) satisfies $\deg(b_i') = \deg(b_i)$, for all $i \in \underline{r}$. (Convention: the zero polynomial is homogeneous of every degree.)

21.14. Let $f \in k[X]$ be homogeneous of degree $d \geq 2$. Show that there exist homogeneous polynomials $p_1, \ldots, p_s, q_1, \ldots, q_s$ for some $s \leq H(f)$, satisfying

(1) $f = \sum_{i=1}^{s} p_i q_i$,
(2) $\deg(p_i q_i) = d$ and $\deg(p_i), \deg(q_i) \in [\frac{1}{3}d, \frac{2}{3}d]$, for all $i \leq s$.
(3) $H(p_i) \leq H(f) - i$ and $H(q_i) \leq 7(H(f) - i + 1)$, for all $i \leq s$.

(Hint: let $\Gamma = (\Gamma_1, \ldots, \Gamma_r)$ be an optimal homogeneous straight-line program computing f on input $(k[X]; X)$. Let (b_i) be the corresponding result sequence. Then $b_r = f$. As Γ is homogeneous, there exists an index m such that Γ_m is a multiplication instruction and $\deg(b_m) \in [\frac{1}{3}d, \frac{2}{3}d]$. Let Γ' be equal to Γ except that Γ_m is replaced by the trivial instruction (0^c). Let (b_i') denote the result sequence corresponding to Γ' on the same input. Note that $b_m' = 0$ and $b_i' = b_i$ for all i with $i \not\geq_\Gamma m$. Prove the following claim: there exist polynomials $B_1, \ldots, B_r$ such that for $i \in \underline{r}$ the subsequent statements hold.

(A)$_i$ B_i is a homogeneous polynomial of degree $\deg(b_i) - \deg(b_m)$, moreover
$b_i = b_i' + b_m B_i$, and $H(\{b_j, b_j', B_j | j \leq_\Gamma i\}) \leq 7i$.

The statements in (A)$_i$ are trivially true for all $i \not\geq_\Gamma m$ (why?). Now consider those i with $i \geq_\Gamma m$ and prove (A)$_i$ by induction on the length of the longest path from m to i in the multigraph corresponding to Γ.)

21.15. On the basis of the last exercises give an alternative proof of the fact that there is a universal constant c such that $D(f) \leq c(\log(dL(f))\log d + \log n)$, for every n-variate polynomial f of degree d.

In the remaining exercises we consider algebraic computations which are not allowed to rely on the commutativity of multiplication. Thus formally we will be working in the k-algebra $k\{X\} := k\{X_1, \ldots, X_n\}$ of polynomials in the non-commuting indeterminates $X_1, \ldots, X_n$. The polynomial ring $k[X] := k[X_1, \ldots, X_n]$ is an epimorphic image of this algebra. On the other hand, $k[X]$ may be viewed as a k-subspace of $k\{X\}$ by mapping each monomial in $k[X]$ to its lexicographically smallest companion. In this way DET_n and PER_n will be viewed as elements of $k\{X_{ij}\}$ which are homogeneous of degree n. As multiplication in $k\{X\}$ is not commutative, care has to be taken when defining the formula size $E\{f\}$, the complexity $L\{f\}$ (w.r.t. $\Omega = \{+, *\}$), and the depth $D\{f\}$ of $f \in k\{X\}$. The goal of the remaining exercises is to deduce exponential lower bounds for both $E\{PER_n\}$ and $E\{DET_n\}$.

21.16. For $f \in k\{X\}$ define $E\{f\}$, $L\{f\}$, and $D\{f\}$. Show that $E(f) \leq E\{f\}$, $L(f) \leq L\{f\}$, and $D(f) \leq D\{f\}$, for $f \in k[X]$. Construct a polynomial f for which the above inequalities are strict.

We need one further complexity measure closely related to digraphs whose edge weights are Linear Forms and whose vertices are partitioned into Levels. More precisely, a (homogeneous) LFL-Program of degree d is a directed acyclic graph with one source and one sink. The vertices of this graph are partitioned into levels *numbered from 0 to d, where edges may only go from level i to level i + 1. The source (resp. sink) is the only vertex at level 0 (resp. d). Each edge is labeled with a homogeneous linear form $\sum_j c_j X_j \in k\{X\}$. The size of an LFL-program is the number of vertices. An LFL-program computes an element of $k\{X\}$ in an obvious way: the sum over all paths from the source to the sink of the product of all linear forms in the order of their occurrence in the path. Obviously, every LFL-program of degree d computes a homogeneous element of $k\{X\}$ of degree d. The minimal size $B\{f\}$ of an LFL-program computing $f \in k\{X\}$ is called the LFL-complexity of f.*

21.17.* If $f \in k\{X_1, \ldots, X_n\}$ is homogeneous of degree d, then the following holds:

(1) $B\{f\} \le d(E\{f\} + 1)$.
(2) $L\{f\} = O(nB\{f\}^2)$.
(3) $D\{f\} = O(\log B\{f\} \log d)$.
(4) $E\{f\} \le 2^{D\{f\}}$.

(Hints: (1) Let φ be an optimal formula for f. Use φ to construct a *non-homogeneous* LFL-program of size at most $E\{f\} + 1$ computing f, i.e. an LFL-program where the nodes are not required to be partitioned into levels, and where each edge may be labeled by a constant or a variable. In a second stage convert this non-homogeneous LFL-program into a homogeneous one by partitioning the function computed at each vertex to its homogeneous components.

(2+3): Note that an LFL-program of degree d with n_i vertices at level i can be described by a sequence $(B_1, \ldots, B_d)$ of matrices B_i of size $n_{i-1} \times n_i$ with entries in $\{\sum_{j=1}^n c_j X_j | c_j \in k\}$ such that $(f) = B_1 \cdots B_d$.)

21.18.* The LFL-complexity of a homogeneous polynomial $f \in k\{X_1, \ldots, X_n\}$ of degree d can be characterized as follows. Fix n and let M_δ, for $0 \le \delta \le d$, denote the set of all monic monomials of degree δ in $k\{X\}$. (Thus $|M_\delta| = n^\delta$.) To f and δ we let $N_\delta(f) \in k^{M_\delta \times M_{d-\delta}}$ denote the matrix which has at position (m_1, m_2) the coefficient of the monomial $m_1 * m_2$ in f. Prove that $B\{f\} = \sum_{\delta=0}^d \mathrm{rk}(N_\delta(f))$. (Hints: let $(B_1, \ldots, B_d)$ describe an optimal LFL-program for f, see the previous exercise, B_δ of size $n_{\delta-1} \times n_\delta$. For $0 \le \delta \le d$ define two matrices $L_\delta \in k^{M_\delta \times n_\delta}$ and $R_\delta \in k^{n_\delta \times M_{d-\delta}}$ as follows: $L_\delta[m_1, v]$ is the coefficient of the monomial m_1 of $(B_1 \ldots B_\delta)[1, v]$ and similarly, $R_\delta[v, m_2]$ is the coefficient of the monomial m_2 of $(B_{\delta+1} \ldots B_d)[v, m_2]$. Prove that $N_\delta(f) = L_\delta R_\delta$. As $\mathrm{rk}(N_\delta(f)) \le \mathrm{rk}(L_\delta) \le n_\delta$ it follows that $\sum_\delta \mathrm{rk}(N_\delta(f)) \le \sum_\delta n_\delta = B\{f\}$. To prove the other inequality show the following: if $\mathrm{rk}(L_\delta) < n_\delta$ or $\mathrm{rk}(R_\delta) < n_\delta$, for some δ, then one can build a smaller LFL-program that computes f. Thus both L_δ and R_δ have full rank n_δ; hence $\mathrm{rk}(N_\delta(f)) = \mathrm{rk}(L_\delta R_\delta) = \mathrm{rk}(L_\delta) = n_\delta$.)

21.19. With the notation of the previous exercise show that for $0 \le \delta \le n$ the following holds: $\mathrm{rk}(N_\delta(PER_n)) = \mathrm{rk}(N_\delta(DET_n)) = \binom{n}{\delta}$. Furthermore prove that $E\{PER_n\} \ge 2^{\Omega(n)}$ and $E\{DET_n\} \ge 2^{\Omega(n)}$.

21.20. Show that $L\{DET_n\} = O(n4^n)$ and $L\{PER_n\} = O(n4^n)$.

21.7 Open Problems

Problem 21.1. Is $VNC^1 \neq VNC^2$?

Problem 21.2. Is $VP_e \neq VP$?

Problem 21.3. Is DET **VP**-complete in **VP**?

Problem 21.4. Is Valiant's hypothesis true?

Problem 21.5. Is the extended Valiant hypothesis true?

Problem 21.6. Is **VQP** a proper subclass of **VNP**?

21.8 Notes

The significance of the complexity class **P** as a good approximation of what is "efficiently solvable" was first pointed out by Cobham [116] and Edmonds [154]. The theory of **NP**-completeness has been developed by Cook [124] and Karp [297], see also Levin [328]. For a detailed account see Garey and Johnson [181]; for an overview see Johnson's survey article [277]. The nonuniform algebraic analogue of this theory, presented in this chapter, goes back to Valiant [527, 529]. All results of this chapter are due to him, unless otherwise stated. Writing this chapter has been greatly facilitated by von zur Gathen's tutorial [187]. The terms "Cook's hypothesis" and "Valiant's hypothesis" were coined by Strassen [507]. Jerrum [275] has proved the **VNP**-completeness of the family of enumerators for (nonperfect) matching in two-dimensional lattice graphs for char $k \neq 2$, as well as for several families of multivariate polynomials connected to graph reliability problems.

The topic discussed in this chapter had its origin in Valiant's study of search and counting problems. Let t be a p-bounded function and $R \subset \Sigma^+ \times \Sigma^+$ a relation such that $R(x, y) \Rightarrow |y| \leq t(|x|)$. Furthermore, assume that $\{x\#y \mid R(x, y)\} \in \mathbf{P}$. Then $\{x \mid \exists y : R(x, y)\}$ is called a (polynomially bounded) *search problem* and the function that maps every x to the binary encoding of the number of all y with $R(x, y)$ is called the corresponding *counting problem*. The class **NP** is just the class of all those search problems. Every $L \in \mathbf{NP}$ defines a counting problem #L, and #**P** denotes the class of all counting problems that can be computed by a polynomial-time bounded *counting Turing machine*, for details see Valiant [528]. A counting problem B is called #**P**-*complete* iff $B \in \#\mathbf{P}$ and there are polynomial-time Turing reductions to it for all problems in **#P**. It is known that the counting versions of most complete problems in **NP** are **#P**-complete, see Simon [483], Valiant [528], and Johnson [277]. (For more examples of **#P**-complete problems, see Dyer and Frieze [151], Provan [427], and Jerrum [276].) Valiant [527] made the surprising discovery that there are problems in **P** whose corresponding counting problems are **#P**-complete. The most striking example is the perfect matching problem in bipartite graphs. The corresponding counting problem, which is just the problem of computing permanents of zero-one matrices over the integers, is nonetheless **#P**-complete. If one performs this computation in $\mathbb{Z}/m\mathbb{Z}$, then Valiant's results in [527] tell us that two cases have to be distinguished: if m is a power of two then the permanent of an integer matrix mod 2^ℓ can be computed in $O(n^{4\ell-3})$ steps (thus for fixed ℓ this bound is polynomial in n); on the other hand, if m is not an exact power of two, the complexity of computing the permanent of a 0-1 matrix mod m is **UP**-hard (i.e., a polynomial time algorithm for it would imply that any single-valued function whose graph is easy to decide, is itself easy to compute).

Valiant [529] presents a number of equivalent definitions for the algebraic complexity class **VNP**, see also von zur Gathen [187]. Thm. (21.26) and Ex. 21.11 give two of these alternative characterizations. In his original proof of the universality of the determinant, Valiant used the more compact construction sketched in Ex. 21.7. Our proof of the universality follows von zur Gathen [187]. His construction combined with a modification of Valiant's original proof of the **VNP**-completeness of the permanent has led to a simplified proof in Sect. 21.4. Thm. (21.35) is due to Brent [72]. His construction is *off-line* for it starts to work after completely knowing the expression, which needs to be transformed. For an *on-line* construction see Miller and Reif [365]. The first statement in Thm. (21.36) is due to Hyafil [260], the second claim is due to Valiant et al. [530]. For an on-line version of this result see Miller et al. [364].

The problem of deriving the permanent from the determinant is a classical mathematical problem. Szegö [514], answering a question posed by Pólya [424], showed that for $n \geq 3$ there is no sign matrix (ε_{ij}) of size n such that $PER_n = \det(\varepsilon_{ij}X_{ij})$. Marcus and Minc [350] showed that for $n \geq 3$ there are no n^2 linear forms f_{pq} in the X_{ij} such that $PER_n = \det(f_{pq})$, and von zur Gathen [188], proved that PER_n is not the projection of DET_m for $m < \sqrt{2}n$. Determinants and permanents are the extreme cases of the following notion. Let $\chi: S_n \to \mathbb{C}$ denote an irreducible character of the symmetric group S_n. Then $d_\chi := \sum_{\sigma \in S_n} \chi(\sigma) \prod_{i=1}^n X_{i\sigma(i)}$ is the *immanent* corresponding to χ. If ι and ε denote the trivial and the alternating character of S_n, respectively, then $PER_n = d_\iota$ and $DET_n = d_\varepsilon$. The complexity of immanents is discussed in Hartmann [229].

The formula for the permanent in Ex. 21.1 is due to Ryser [445]. Ex. 21.2 follows Chistov [108]. Specific algorithms solving Ex. 21.3 may be found in Csanky [136], Borodin et al. [66], Berkowitz [40], and Chistov [108]. Hints to solve Ex. 21.4 can be found in Appendix 2 of Valiant [529]. Ex. 21.5, 21.6 and 21.7 are due to Valiant [526]. An alternative solution to Ex. 21.8 may be found in von zur Gathen [187]. Ex. 21.9 follows von zur Gathen [187]. Ex. 21.11 is due to Valiant [529], and Ex. 21.12 is from Fich et al. [158]. Ex. 21.13–21.15 follow Hyafil [260]. Ex. 21.16–21.20 are due to Nisan [392]. The upper bounds in Ex. 21.20 can both be improved to $O(n2^n)$, see Nisan [392]. Compared to Nisan's exponential lower bound for the expression size of DET in the noncommutative setting, we only know that $E(DET_n) = \Omega(n^3)$. This has been shown by Kalorkoti [278] with a transcendence degree argument.

It is an interesting and important question whether the classes **VP** and **VNP** are closed under some natural mathematical operations, such as GCD, factorization, derivative. For a detailed discussion of this question see von zur Gathen [187] and Kaltofen [282, 283]. The above question is closely related to a fundamental question in computer algebra: how should one represent polynomials when one has to manipulate them? Several data structures for representing multivariate polynomials have been suggested: the dense representation, the sparse representation, formulas, and straight-line programs. In theory, the use of straight-line programs is the most powerful approach, see Kaltofen [282, 283]. In the sparse representation, von zur Gathen and Kaltofen [191] have shown that irreducible factors may have a length which is more than polynomial in the size of the input polynomial. Von zur Gathen [185] and Kaltofen [283] have shown the theoretical feasibility of the straight-line program approach, by solving standard problems of symbolic manipulation (like testing for irreducibility, GCD's or factoring) in this data structure in random polynomial time. Freeman et al. [176] report on an implementation of a computer algebra system based on this approach. For a recent approach to use straight-line programs for solving problems in elimination theory, we refer to Giusti et al. [197, 198] and the references given there.

Bibliography

We use the following abbreviations.

FOCS   IEEE Annual Smposium on Foundations of Computer Science
LNCS   Lecture Notes in Computer Science
STOC   ACM Annual Symposium on the Theory of Computing

[1] A.V. Aho, J.E. Hopcroft, and J.D. Ullman: The Design and Analysis of Computer Algorithms. Addison-Wesley, 1974

[2] M. Aigner: Combinatorial Theory. Volume 234 of Grundlehren der mathematischen Wissenschaften. Springer, 1979

[3] A.G. Akritas: A new method for computing polynomial greatest common divisors and polynomial remainder sequences. Num. Math. 52:119–127, 1988

[4] A.G. Akritas: Elements of Computer Algebra with Applications. John Wiley & Sons, New York 1989

[5] A. Alder: Grenzrang und Grenzkomplexität aus algebraischer und topologischer Sicht. PhD thesis, Zürich University 1984

[6] A. Alder and V. Strassen: On the algorithmic complexity of associative algebras. Theoret. Comp. Sc. 15:201–211, 1981

[7] N. Alon, J.H. Spencer, and P. Erdös: The probabilistic method. John Wiley & Sons, 1992

[8] H. Alt and J. van Leeuwen: Complexity of basic complex operations. Computing 27:205–215, 1981

[9] A. Antoniou: Digital filters: analysis and design. McGraw-Hill, New York 1979

[10] T.M. Apostol: Introduction to Analytic Number Theory. Undergraduate Texts in Mathematics. Springer, 1979

[11] V.L. Arlazarov, E.A. Dinic, M.A. Kronrod, and I.A. Faradzev: On economical construction of the transitive closure of a directed graph. Dokl. Akad. Nauk USSSR 194:487–488, 1970. (In Russian.) English translation in Soviet Math. Dokl. 11:1209–1210

[12] M.F. Atiyah and I.G. Macdonald: Introduction to Commutative Algebra. Addison-Wesley, London 1969

[13] M.D. Atkinson, editor: Computational Group Theory. Academic Press, 1984. Proceedings of the London Mathematical Society Symposium on Computational Group Theory

[14] M.D. Atkinson and S. Lloyd: Bounds on the ranks of some 3-tensors. Lin. Alg. Appl. 31:19–31, 1980

[15] M.D. Atkinson and S. Lloyd: The ranks of $m \times n \times (mn - 2)$-tensors. SIAM J. Comp. 12:611–615, 1983

[16] M.D. Atkinson and N.M. Stephens: On the maximal multiplicative complexity of a family of bilinear forms. Lin. Alg. Appl. 27:1–8, 1979

[17] L. Auslander, E. Feig, and S. Winograd: Direct sums of bilinear algorithms. Lin. Alg. Appl. 38:175–192, 1981

[18] L. Auslander, E. Feig, and S. Winograd: Abelian semi-simple algebras and algorithms for the discrete Fourier transform. Adv. Appl. Math. 5:31–55, 1984

[19] L. Auslander and S. Winograd: The multiplicative complexity of certain semilinear systems defined by polynomials. Adv. Appl. Math. 1:257–299, 1980

[20] A. Averbuch, N.H. Bshouty, and M. Kaminski: A classification of quadratic algorithms for multiplying polynomials of small degree over finite fields. J. Algorithms 13:577–588, 1992

[21] A. Averbuch, Z. Galil, and S. Winograd: Classification of all the minimal bilinear algorithms for computing the coefficients of the product of two polynomials modulo a polynomial. Theoret. Comp. Sc. 58:17–56, 1988. Part I: the algebra $G[u]/\langle Q(u)^{\ell}\rangle$

[22] A. Averbuch, Z. Galil, and S. Winograd: Classification of all the minimal bilinear algorithms for computing the coefficients of the product of two polynomials modulo a polynomial. Theoret. Comp. Sc. 86:143–203, 1991. Part II: the algebra $G[u]/\langle u^n\rangle$

[23] J. L. Balcázar, J. Díaz, and J. Gabarró: Structural Complexity I. Springer, 1988

[24] U. Baum: Existence and efficient construction of Fast Fourier Transforms for supersolvable groups. Comp. Compl. 1:235–256, 1991

[25] U. Baum and M. Clausen: Some lower and upper complexity bounds for the generalized Fourier transforms and their inverses. SIAM J. Comp. 20:451–459, 1991

[26] U. Baum and M. Clausen: Computing irreducible representations of supersolvable groups. Mathematics of Computation 63(207):351–359, 1994

[27] U. Baum, M. Clausen, and B. Tietz: Improved upper complexity bounds for the Discrete Fourier Transform. Applicable Algebra in Engineering, Communication and Computing 2:35–43, 1991

[28] W. Baur: On the algebraic complexity of rational iteration procedures. Theoret. Comp. Sc. 88:313–324, 1988

[29] W. Baur: Algebraische Berechnungskomplexität. Lectures at the department of mathematics of the Konstanz University, 1989

[30] W. Baur, 1994: private communication

[31] W. Baur and M.O. Rabin: Linear disjointness and algebraic complexity. In Logic and Algorithmic: An international Symposium held in honor of Ernst Specker, pages 35–46. Monogr. No. 30 de l'Enseign. Math., 1982

[32] W. Baur and V. Strassen: The complexity of partial derivatives. Theoret. Comp. Sc. 22:317–330, 1983

[33] E. Becker and N. Schwartz: Zum Darstellungssatz von Kadison-Dubois. Archiv Math. 40:421–428, 1983

[34] T. Becker and V. Weispfenning: Gröbner Bases, a Computational Approach to Commutative Algebra, volume 141 of GTM. Springer, 1993

[35] E.G. Belaga: Evaluation of polynomials of one variable with preliminary processing of the coefficients. In A.A. Lyapunov, editor, Problems of Cybernetics, volume 5, pages 1–13. Pergamon Press, 1961

[36] R.E. Bellman: Addition chains of vectors. Amer. Math. Monthly 70:765, 1963

[37] M. Ben-Or: Lower bounds for algebraic computation trees. In Proc. 15th ACM STOC, Boston, pages 80–86, 1983

[38] M. Ben-Or: Algebraic computation trees in characteristic $p > 0$. In Proc. 34th ACM FOCS, Boston, pages 534–539, 1994

[39] R. Benedetti and J.-J. Risler: Real Algebraic and Semi-Algebraic Sets. Hermann, 1990

[40] S. Berkowitz: On computing the determinant in small parallel time using a small number of processors. Inf. Proc. Letters 18:147–150, 1984

[41] E.R. Berlekamp: Algebraic Coding Theory. Mc Graw-Hill, New York 1968

[42] A.S. Besicovitch: On the linear independence of fractional powers of integers. J. London Math. Soc. 15:3–6, 1940

[43] T. Beth: Verfahren der schnellen Fourier-Transformation. Teubner, 1984

[44] D. Bini: Border rank of $(p \times q \times 2)$-tensors and the optimal approximation of a pair of bilinear forms. Number 85 in LNCS, pages 98–108. Springer, 1980

[45] D. Bini: Relation between exact and approximate bilinear algorithms. Applications. Calcolo 17:87–97, 1980

[46] D. Bini: Border rank of $m \times n \times (mn - q)$ tensors. Lin. Alg. Appl. 79:45–51, 1986

[47] D. Bini and M. Capovani: Lower bounds of the complexity of linear algebras. Inf. Proc. Letters 9:46–47, 1979

[48] D. Bini and M. Capovani: Tensor rank and border rank of band Toeplitz matrices. SIAM J. Comp. 2:252–258, 1987

[49] D. Bini, M. Capovani, G. Lotti, and F. Romani: $O(n^{2.7799})$ complexity for matrix multiplication. Inf. Proc. Letters 8:234–235, 1979

[50] D. Bini, G. Lotti, and F. Romani: Approximate solution for the bilinear form computation problem. SIAM J. Comp. 9:692–697, 1980

[51] D. Bini and V. Pan: Polynomial and Matrix Computations—Volume 1: Fundamental Algorithms. Progress in Theoretical Computer Science. Birkhäuser, Boston 1994

[52] A. Björner: Subspace arrangements. In Proc. of 1st European Congress of Mathematics, pages 211–227, Paris, 1992. Birkhäuser

[53] A. Björner: Nonpure shellability, f-vectors, subspace arrangements and complexity. 1994. to appear

[54] A. Björner and L. Lovász: Linear decision trees, subspace arrangements and Möbius functions. J. Amer. Math. Soc. 7(3):677–706, 1994

[55] A. Björner, L. Lovász, and A.C. Yao: Linear decision trees: volume estimates and topological bounds. In Proc. 24th ACM STOC, pages 171–177, 1992

[56] A. Björner, M. Las Vergnas, B. Sturmfels, N. White, and G.M. Ziegler: Oriented Matroids. Cambridge University Press, 1992

[57] L.I. Bluestein: A linear filtering approach to the computation of the discrete Fourier transform. IEEE Trans. Electroacoustics 18:451–455, 1970

[58] L. Blum, F. Cucker, M. Shub, and S. Smale: Complexity and Real Computation. Springer. to appear

[59] L. Blum, M. Shub, and S. Smale: On a theory of computation and complexity over the real numbers. Bull. Amer. Math. Soc. 21:1–46, 1989

[60] A. Blumer, A. Ehrenfeucht, D. Haussler, and M. Warmuth: Classifying learnable geometric concepts with the Vapnik-Chervonenkis dimension. In Proc. 18th ACM STOC, Berkeley, volume 35, pages 273–282, 1986

[61] J. Bochnak, M.Coste, and M.F. Roy: Géometrie algébrique réelle, volume 12 of Ergebnisse der Mathematik und ihrer Grenzgebiete, 3. Folge. Springer, 1987

[62] B. Bollobás: Graph Theory. Volume 63 of Graduate Texts in Mathematics. Springer, 1979

[63] A. Borodin and S. Cook: On the number of additions to compute specific polynomials. SIAM J. Comp. 5:146–157, 1976

[64] A. Borodin and R. Moenck: Fast modular transforms. J. Comp. Syst. Sci. 8:366–386, 1974

[65] A. Borodin and I. Munro: The Computational Complexity of Algebraic and Numeric Problems. American Elsevier, 1975

[66] A. Borodin, J. von zur Gathen, and J. Hopcroft: Fast parallel matrix and GCD computations. Inf. Control 52:241–256, 1982

[67] A. Borodin: Horner's rule is uniquely optimal. In Z. Kohavi and A. Paz, editors, Theory of Machines and Computations, pages 45–58. Academic Press, 1971

[68] A. Borodin: On the number of arithmetics required to compute certain functions. In J.F. Traub, editor, Complexity of sequential and parallel numerical algorithms, pages 149–180. Academic Press, 1973

[69] N. Bourbaki: Elément de Mathématique. Algèbre, volume I. Hermann, 1970

[70] N. Bourbaki: Elément de Mathématique. Topologie générale. Hermann, 1971

[71] A. Brauer: On addition chains. Bull. Amer. Math. Soc. 45:736–739, 1939
[72] R.P. Brent: The complexity of multiprecision arithmetic. In Proc. Seminar on Compl. of Comp. Problem Solving, Brisbane, pages 126–165, 1975
[73] R.P. Brent: Multiple-precision zero-finding methods and the complexity of elementary function evaluation. In J.F. Traub, editor, Analytic Computational Complexity, pages 151–176. Academic Press, New York 1975
[74] R.P. Brent, F.G. Gustavson, and D.Y.Y. Yun: Fast solution of Toeplitz systems of equations and computation of Padé approximants. J. Algorithms 1:259–295, 1980
[75] R.P. Brent and H.T. Kung: Fast algorithms for manipulating formal power series. J. ACM 25:581–595, 1978
[76] R.P. Brent and J.F. Traub: On the complexity of composition and generalized composition of power series. SIAM J. Comp. 9(1):54–66, 1980
[77] L. Bröcker: Characterization of basic semi-algebraic sets. Technical report, Münster University 1987
[78] R.W. Brockett: On the generic degree of a 3-tensor. unpublished manuscript, Harvard University 1976
[79] R.W. Brockett and D. Dobkin: On the number of multiplications required for matrix multiplication. SIAM J. Comp. 5:624–628, 1976
[80] R.W. Brockett and D. Dobkin: On the optimal evaluation of a set of bilinear forms. Lin. Alg. Appl. 19:207–235, 1978
[81] M. Brown and D. Dobkin: An improved lower bound on polynomial multiplication. IEEE Trans. Comp. C-29(5):337–340, 1980
[82] W.S. Brown: On Euclid's algorithm and the computation of polynomial greatest common divisors. J. ACM 18:478–504, 1971
[83] W.S. Brown: The subresultant prs algorithm. ACM Transaction on Mathematical Software 4:237–249, 1978
[84] W.S. Brown and J.F. Traub: On Euclid's algorithm and the theory of subresultants. J. ACM 18:505–514, 1971
[85] N.H. Bshouty: A lower bound for matrix multiplication. In Proc. 29th FOCS, pages 64–67, 1988
[86] N.H. Bshouty: A lower bound for matrix multiplication. SIAM J. Comp. 18:759–765, 1989
[87] N.H. Bshouty: On the extended direct sum conjecture. In Proc. 21st ACM STOC, pages 177–185, 1989
[88] N.H. Bshouty: Maximal rank of $(m \times n \times (mn - k))$-tensors. SIAM J. Comp. 19:467–471, 1990
[89] N.H. Bshouty: A lower bound for the multiplication of polynomials modulo a polynomial. Inf. Proc. Letters 41:321–326, 1992
[90] N.H. Bshouty: On the direct sum conjecture in the straight-line model. In ESA, First Ann. European Symp. Bad Honnef, Germany, number 726 in LNCS, pages 85–96, 1993
[91] N.H. Bshouty: On the complexity of bilinear forms over associative algebras. SIAM J. Comp. 23:815–833, 1994
[92] N.H. Bshouty: Multiplicative complexity of direct sums of quadratic systems. 1995
[93] N.H. Bshouty and M. Kaminski: Multiplication of polynomials over finite fields. SIAM J. Comp. 19:452–456, 1990
[94] B. Buchberger, G. Collins, and R. Loos, editors: Computer Algebra. Symbolic and Algebraic Computation. Springer, second edition, 1982
[95] W. Büchi: Über eine Klasse von Algebren minimalen Ranges. PhD thesis, Zürich University 1984
[96] W. Büchi and M. Clausen: On a class of primary algebras of minimal rank. Lin. Alg. Appl. 69:249–268, 1985

[97] J. Bunch and J. Hopcroft: Triangular factorization and inversion by fast matrix multiplication. Math. Comp. 28:231–236, 1974

[98] P. Bürgisser: Degenerationsordnung und Trägerfunktional bilinearer Abbildungen. PhD thesis, Konstanz University 1990

[99] P. Bürgisser: Decision complexity of generic complete intersections. Research Report 8578-CS, Institut für Informatik der Universität Bonn 1992

[100] P. Bürgisser, M. Karpinski, and T. Lickteig: Some computational problems in linear algebra as hard as matrix multiplication. Comp. Compl. 1:131–155, 1991

[101] P. Bürgisser, M. Karpinski, and T. Lickteig: On randomized semialgebraic decision complexity. J. Compl. 9:231–251, 1993

[102] P. Bürgisser and T. Lickteig: Verification complexity of linear prime ideals. J. Pure Appl. Alg. 81:247–267, 1992

[103] P. Bürgisser, T. Lickteig, and M. Shub: Test complexity of generic polynomials. J. Compl. 8:203–215, 1992

[104] S. Cabay and D.-K. Choi: Algebraic computations of scaled Padé fractions. SIAM J. Comp. 15:243–270, 1986

[105] D.G. Cantor and E. Kaltofen: On fast multiplication of polynomials over arbitrary algebras. Act. Inf. 28:693–701, 91

[106] K. Chandrasekharan: Introduction to Analytic Number Theory, volume 148 of Grundlehren der mathematischen Wissenschaften. Springer, 1968

[107] U. Cheng: On the continued fraction and Berlekamp's algorithm. IEEE Trans. Inform. Theory, pages 541–544, 1984

[108] A.L. Chistov: Fast parallel calculation of the rank of matrices over a field of arbitrary characteristic. In Fundamentals of Computation Theory, number 199 in LNCS, pages 63–69. Springer, 1985

[109] D.V. Chudnovsky and G.V. Chudnovsky: Algebraic complexities and algebraic curves over finite fields. Proc. Natl. Acad. Sci. USA 84:1739–1743, 1987

[110] D.V. Chudnovsky and G.V. Chudnovsky: Algebraic complexity and algebraic curves over finite fields. J. Compl. 4:285–316, 1988

[111] K.L. Clarkson: New applications of random sampling in computational geometry. Discrete Comput. Geom. 2:195–222, 1987

[112] M. Clausen: Beiträge zum Entwurf schneller Spektraltransformationen. Habilitationsschrift, Universität Karlsruhe 1988

[113] M. Clausen: Fast generalized Fourier transforms. Theoret. Comp. Sc. 67:55–63, 1989

[114] M. Clausen: Multivariate polynomials, standard tableaux, and representations of symmetric groups. J. Symb. Comp. 11:485–522, 1991

[115] M. Clausen and U. Baum: Fast Fourier Transforms. BI-Wissenschaftsverlag, Mannheim 1993

[116] A. Cobham: The intrinsic computational difficulty of functions. In Y. Bar-Hillel, editor, Proc. 1964 Intern. Congress for Logic Methodology and Philosophy of Science, pages 24–30. North-Holland, Amsterdam 1964

[117] H. Cohen: A course in computational algebraic number theory, volume 138 of GTM. Springer, 1993

[118] P.J. Cohen: Decision procedures for real and p-adic fields. Comm. Pure and Appl. Math. 22:131–151, 1969

[119] G.E. Collins: Subresultants and reduced polynomial remainder sequences. J. ACM 14:128–142, 1967

[120] G.E. Collins: The calculation of multivariate polynomial resultants. J. ACM 18:515–532, 1971

[121] G.E. Collins and R. Loos: Real zeros of polynomials. In B. Buchberger, G. Collins, and R. Loos, editors, Computer Algebra. Symbolic and Algebraic Computation, pages 83–94. Springer, second edition, 1982

[122] S. Cook and C. Rackoff: Time bounded random access machines. J. Comp. Syst. Sci. 7:354–375, 1973

[123] S.A. Cook: On the minimum computation time of functions. PhD thesis, Harvard 1966

[124] S.A. Cook: The complexity of theorem proving procedures. In Proc. 3rd ACM STOC, pages 151–158, 1971

[125] S.A. Cook: A taxonomy of problems with fast parallel algorithms. Inf. Control 64:2–22, 1985

[126] J.W. Cooley, P.A.W. Lewis, and P.D. Welch: History of the fast Fourier transform. Proc. IEEE 55:1675–1677, 1967

[127] J.W. Cooley and J.W. Tukey: An algorithm for the machine calculation of complex Fourier series. Math. Comp. 19:297–301, 1965

[128] D. Coppersmith: Rapid multiplication of rectangular matrices. SIAM J. Comp. 11:467–471, 1982

[129] D. Coppersmith: Fast evaluation of logarithms in fields of characteristic two. IEEE Trans. Inform. Theory IT-30:587–594, 1984

[130] D. Coppersmith: How not to multiply matrices. Technical report, IBM Research Report RC, 1988

[131] D. Coppersmith: Solving homogeneous linear equations over $GF(2)$ via block Wiedemann algorithm. Math. Comp. 62:333–350, 1994

[132] D. Coppersmith, A. Odlyzko, and R. Schroeppel: Discrete logarithms in $GF(p)$. Algorithmica 1:1–15, 1986

[133] D. Coppersmith and S. Winograd: On the asymptotic complexity of matrix multiplication. SIAM J. Comp. 11:482–492, 1982

[134] D. Coppersmith and S. Winograd: Matrix multiplications via arithmetic progression. In Proc. 19th ACM STOC, pages 1–6, 1987

[135] D. Coppersmith and S. Winograd: Matrix multiplication via arithmetic progressions. J. Symb. Comp. 9:251–280, 1990

[136] L. Csanky: Fast parallel matrix inversion algorithms. SIAM J. Comp. 5:618–623, 1976

[137] F. Cucker, M. Karpinski, P. Koiran, T. Lickteig, and K. Werther: On real Turing machines that toss coins. In Proc. 27th ACM STOC, Las Vegas, pages 335–342, 1995

[138] F. Cucker and T. Lickteig: Nash trees and Nash complexity. In Proceedings of the 1995 AMS-SIAM Summer Seminar in Park City, Utah, Lectures in Applied Mathematics. AMS, 1995

[139] C.W. Curtis and I. Reiner: Representation Theory of Finite Groups and Associative Algebras. John Wiley & Sons, New York 1962

[140] C.W. Curtis and I. Reiner: Methods of Representation Theory. John Wiley & Sons, New York 1990

[141] S.R. Czapor and K.O. Geddes: A comparison of algorithms for the symbolic computation of Padé approximants. In J. Fitch, editor, EUROSAM 84, volume 174 of LNCS, pages 248–259, 1984

[142] C.A. Desoer and E.S. Kuh: Basic Circuit Theory. McGraw-Hill, New York 1969

[143] P. Diaconis and D. Rockmore: Efficient computation of the Fourier transform on finite groups. J. of the AMS 3:297–332, 1990

[144] D. Dobkin: On the Arithmetic Complexity of a Class of Arithmetic Computations. Aiken Comput. Lab., Harvard University 1973

[145] D. Dobkin and R.J. Lipton: Multidimensional searching problems. SIAM J. Comp. 5(2):181–186, 1976

[146] D. Dobkin and R.J. Lipton: A lower bound of $\frac{1}{2}n^2$ on linear search programs for the knapsack problem. J. Comp. Syst. Sci. 16:413–417, 1978

[147] D. Dobkin and J. van Leeuwen: The complexity of cross products. Inf. Proc. Letters 4(6):149–154, 1976

[148] J.L. Dornstetter: On the equivalence between Berlekamp's and Euclid's algorithms. IEEE Trans. Inform. Theory 33:428–431, 1987

[149] V.G. Drinfeld and S.G. Vlădut: Number of points of an algebraic curve. Func. Anal. 17:53–54, 1983

[150] R.M. Dudley: Central limit theorems for empirical measures. Annals of Probability 6:899–929, 1978

[151] M.E. Dyer and A.M. Frieze: On the complexity of computing the volume of a polyhedron. SIAM J. Comp. 17(5):967–974, 1988

[152] W. Eberly: Very fast polynomial arithmetic. SIAM J. Comp. 18:955–976, 1989

[153] W. Eberly: On efficient band matrix arithmetic. In Proc. 33rd FOCS, pages 457–463, 1992

[154] J. Edmonds: Paths, trees, and flowers. Canad. J. Math. 17:449–467, 1965

[155] H.M. Edwards: Fermat's Last Theorem. Number 50 in GTM. Springer, 1977

[156] P. Erdös: Remarks on number theory. III. On addition chains. Acta Arithmetica 6:77–81, 1960

[157] J. Eve: The evaluation of polynomials. Numerische Mathematik 6:17–21, 1964

[158] F. Fich, J. von zur Gathen, and C. Rackoff: Complete families of polynomials. Manuscript, 1986

[159] E. Feig: On certain systems of bilinear forms whose minimal division-free algorithms are all bilinear. J. Algorithms 2:261–281, 1981

[160] E. Feig: Certain systems of bilinear forms whose minimal algorithms are all quadratic. J. Algorithms 4:137–149, 1983

[161] E. Feig: Minimal algorithms for bilinear forms may have divisions. J. Algorithms 4:81–84, 1983

[162] E. Feig, L. Auslander, and S. Winograd: On the multiplicative complexity of the discrete Fourier transform. Adv. Appl. Math. 5:87–109, 1984

[163] E. Feig and S. Winograd: On the direct sum conjecture. In Proc. 22nd FOCS, pages 91–94, 1981

[164] E. Feig and S. Winograd: The multiplicative complexity of the discrete Fourier transform. Adv. Appl. Math. 5:87–109, 1984

[165] E. Feig and S. Winograd: On the direct sum conjecture. Lin. Alg. Appl. 63:193–219, 1984

[166] W. Feller: An introduction to probability theory and its applications, volume 1. John Wiley & Sons, 1968

[167] A. Fellmann: Optimal algorithms for finite dimensional simply generated algebras. In Algebraic Algorithms and Error-Correcting Codes, number 229 in LNCS, pages 288–295, 1985

[168] A. Fellmann: Algorithmenvarietäten der kommutativen Algebren minimalen Ranges. PhD thesis, Universität Frankfurt 1988

[169] C.M. Fiduccia: Fast matrix multiplication. In Proc. 3rd ACM STOC, pages 45–49, 1971

[170] C.M. Fiduccia: Polynomial evaluation via the division algorithm: the fast Fourier transform revisited. In Proc. 4th ACM STOC, pages 88–93, 1972

[171] C.M. Fiduccia and Y. Zalcstein: Algebras having linear multiplicative complexities. J. ACM 24:311–331, 1977

[172] L. Finkelstein and W.M. Kantor, editors: Groups and Computation, volume 11 of DIMACS Series in Discrete Mathematics and Theoretical Computer Science. AMS, 1993

[173] M.J. Fischer and A.R. Meyer: Boolean matrix multiplications and transitive closure. In Conference Record, IEEE 12th Annual Symposium on Switching and Automata Theory, pages 129–131, 1971

[174] N. Fitchas, A. Galligo, and J. Morgenstern: Precise sequential and parallel complexity bounds for quantifier elimination over algebraically closed fields. J. Pure Appl. Alg. 67:1–14, 1990

[175] R. Fleischer: Decision trees: Old and new results. In Proc. 25th ACM STOC, pages 468–477, 1993

[176] T.S. Freeman, G. Imirzian, and E. Kaltofen: A system for manipulating polynomials given by straight-line programs. Technical Report 86-15, Dept. of Computer Science, Rensselaer Polytechnic Institute, Troy, NY, September 1986

[177] W. Fulton: Intersection Theory. Springer, 1984

[178] O. Gabber and Z. Galil: Explicit constructions of linear superconcentrators. J. Comp. Syst. Sci. 22:407–420, 1981

[179] F.R. Gantmakher: Matrizenrechnung I/II. Springer, Berlin 1958/59

[180] A. Garcia and H. Stichtenoth: A tower of Artin-Schreier extensions of function fields attaining the Drinfeld-Vlădut bound. Invent. Math. 121:211–222, 1995

[181] M.R. Garey and D.S. Johnson: Computers and Intractability: A Guide to the theory of NP-completeness. W.H. Freeman and Company, New York 1979

[182] N. Gastinel: Sur le calcul de produits de matrices. Num. Math. 17:222–229, 1971

[183] J. von zur Gathen: Hensel and Newton methods in valuation rings. Math. Comp. 42:637–661, 1984

[184] J. von zur Gathen: Parallel algorithms for algebraic problems. SIAM J. Comp. 13:802–824, 1984

[185] J. von zur Gathen: Irreducibility of multivariate polynomials. J. Comp. Syst. Sci. 31:225–264, 1985

[186] J. von zur Gathen: Parallel arithmetic computations: a survey. In Proc. 12th FOCS, Bratislava, volume 233 of LNCS, pages 93–112, 1986

[187] J. von zur Gathen: Feasible arithmetic computations: Valiant's hypothesis. J. Symb. Comp. 4:137–172, 1987

[188] J. von zur Gathen: Permanent and determinant. Lin. Alg. Appl. 96:87–100, 1987

[189] J. von zur Gathen: Algebraic complexity theory. Ann. Review of Comp. Sci. 3:317–347, 1988

[190] J. von zur Gathen: Maximal bilinear complexity and codes. Lin. Alg. Appl. 144:49–61, 1991

[191] J. von zur Gathen and E. Kaltofen: Factoring sparse multivariate polynomials. J. Comp. Syst. Sci. 31:265–287, 1985

[192] J. von zur Gathen and V. Strassen: Some polynomials that are hard to compute. Theoret. Comp. Sc. 11:331–336, 1980

[193] C.F. Gauß: Nachlaß Werke, volume III, chapter Theoria Interpolationis Methodo Nova Tractata, page 265. Göttingen, 1876

[194] K.O. Geddes, S.R. Czapor, and G. Labahn: Algorithms for Computer Algebra. Kluwer Academic Publisher, Boston 1992

[195] M. Giesbrecht: Fast algorithms for matrix normal forms. In Proc. 33rd FOCS, pages 121–130, 1992

[196] E.N. Gilbert: A comparison of signaling alphabets. Bell Syst. Techn. Jnl. 31:504–522, 1952

[197] M. Giusti, J. Heintz, J.E. Morais, J. Morgenstern, and L.M. Pardo: Straight-line Programs in Geometric Elimination Theory. J. Pure Appl. Alg., 1995. to appear

[198] M. Giusti, K. Hägele, J. Heintz, J.L. Montaña, and L.M. Pardo: Lower Bounds for Diophantine Approximations. In Proc. MEGA 96, 1996. to appear

[199] H. Goldstine: A History of Numerical Analysis from the 16th Through the 19th Century. Springer, 1977

[200] V.D. Goppa: Codes on algebraic curves. Sov. Math. Dokl. 24:170–172, 1981

[201] V.D. Goppa: Geometry and Codes. Kluwer Academic Publishers, 1988

[202] W.B. Gragg: The Padé table and its relation to certain algorithms of numerical analysis. SIAM Rev. 14:1–62, 1972

[203] R.L. Graham: An efficient algorithm for determining the convex hull of a finite planar set. Inf. Proc. Letters 1:132–133, 1972

[204] J.H. Griesmer: A bound for error-correcting codes. IBM J. Res. Develop. 4:532–542, 1960

[205] B. Griesser: A lower bound for the border rank of bilinear map. Calcolo 23(2):105–114, 1986

[206] B. Griesser: Lower bounds for the approximative complexity. Theoret. Comp. Sc. 46:329–338, 1986

[207] D.Yu. Grigoriev: Multiplicative complexity of a pair of bilinear forms and the polynomial multiplication. Number 64 in LNCS, pages 250–256. Springer, 1978

[208] D.Yu. Grigoriev: Some new bounds on tensor rank. LOMI preprint E-2-78, 1978

[209] D.Yu. Grigoriev: Notes of the scientific seminars of LOMI 118:25–82, 1982

[210] D.Yu. Grigoriev: Lower bounds in algebraic complexity. Notes of scientific Seminars of LOMI, 1982

[211] D.Yu. Grigoriev and M. Karpinski: Computability of the additive complexity for algebraic circuits with root extracting. SIAM J. Comp., 1993

[212] D.Yu. Grigoriev and M. Karpinski: Lower bounds on complexity of testing membership to a polygon for algebraic and randomized decision trees. Technical Report TR-93-042, Int. Comp. Sc. Inst., Berkeley 1993

[213] D.Yu. Grigoriev, M. Karpinski, and N. Vorobjov: Complexity lower bounds on testing membership to a polyhedron by algebraic decision trees. In Proc. 26th ACM STOC, pages 635–644, 1994

[214] D.Yu. Grigoriev, M. Karpinski, and N. Vorobjov: Improved lower bound on testing membership to a polyhedron by algebraic decision trees. In Proc. 36th STOC, pages 258–265, 1995

[215] H.F. de Groote: On the complexity of quaternion multiplication. Inf. Proc. Letters 3:177–179, 1975

[216] H.F. de Groote: On varieties of optimal algorithms for the computation of bilinear mappings II: optimal algorithms for 2×2-matrix multiplication. Theoret. Comp. Sc. 7:127–148, 1978

[217] H.F. de Groote: On varieties of optimal algorithms for the computation of bilinear mappings I: The isotropy group of bilinear mapping. Theoret. Comp. Sc. 7:1–24, 1978

[218] H.F. de Groote: Characterization of division algebras of minimal rank and the structure of their algorithm varieties. SIAM J. Comp. 12:101–117, 1983

[219] H.F. de Groote: Lectures on the Complexity of Bilinear Problems. Number 245 in LNCS. Springer, 1986

[220] H.F. de Groote and J. Heintz: Commutative algebras of minimal rank. Lin. Alg. Appl. 55:37–68, 1983

[221] H.F. de Groote and J. Heintz: A lower bound for the bilinear complexity of some semisimple Lie algebras. In Proc. AAECC 3, number 229 in LNCS, pages 211–227, 1986

[222] H.F. de Groote, J. Heintz, S. Möhler, and H. Schmidt: On the complexity of Lie algebras. In Proc. Fundamentals of Computation Theory, number 278 in LNCS, pages 172–179, 1986

[223] B. Grünbaum: Convex Polytopes. Wiley, 1967

[224] Håstad, J.: Tensor rank is NP-complete. In Automata, Languages, and Programming: Proceedings of the 16th International Colloquium, number 372 in LNCS, pages 451–460, 1989

[225] W. Habicht: Eine Verallgemeinerung des Sturmschen Wurzelzählverfahrens. Commentarii Mathematici Helvetici 21:99–116, 1948

[226] M. Hall: An algorithm for distinct representatives. Amer. Math. Monthly 63:716–717, 1956

[227] W. Hansen: Zum Scholz-Brauerschen Problem. Crelles J. Reine Angew. Math. 202:129–136, 1959

[228] J. Harris: Algebraic Geometry: A First Course. Volume 133 of Graduate Texts in Mathematics. Springer, New York 1992

[229] W. Hartmann: On the complexity of immanents. Linear and Multilinear Algebra 18:127–140, 1985

[230] W. Hartmann: On the multiplicative complexity of modules over associative algebras. SIAM J. Comp. 14:383–395, 1985

[231] W. Hartmann and P. Schuster: Multiplicative complexity of some rational functions. Theoret. Comp. Sc. 10:53–61, 1980

[232] R. Hartshorne: Algebraic Geometry. Volume 52 of Graduate Texts in Mathematics. Springer, 1977

[233] H. Hasse: Beweis des Analogons der Riemannschen Vermutung für die Artinschen und F.K. Schmidtschen Kongruenzzetafunktionen in gewissen elliptischen Fällen. Nachr. der math. Gesellschaft zu Göttingen, pages 253–262, 1933

[234] H. Hasse: Number Theory. Number 229 in Grundlehren der mathematischen Wissenschaften. Springer, 1980

[235] D. Haussler and E. Welzl: Epsilon-nets and simplex range queries. Discrete Comput. Geom. 2:127–151, 1987

[236] A. Hearn: Non-modular computation of polynomial GCD's using trial division. In Proceedings of Eurosam 79, number 72 in LNCS, pages 227–239, 1979

[237] M.T. Heideman: Multiplicative Complexity, Convolution, and the DFT. Springer, New York 1988

[238] M.T. Heideman, D.H. Johnson, and C.S. Burrus: Gauss and the history of the fast Fourier transform. Archive for History of Exact Sciences 34(3):265–277, 1985

[239] J. Heintz: Definability and fast quantifier elimination in algebraically closed fields. Theoret. Comp. Sc. 24:239–277, 1983

[240] J. Heintz: On polynomials with symmetric Galois group which are easy to compute. Theoret. Comp. Sc. 47:99–105, 1986

[241] J. Heintz: On the computational complexity of polynomials and bilinear mappings. A survey. In L.Huguet and A. Poli, editors, Applied Algebra, Algebraic Algorithms and Error-Correcting Codes, number 356 in LNCS, pages 269–300. Springer, 1989

[242] J. Heintz and G. Matera, 1994: private communication at the 1994 Oberwolfach Conference on Complexity Theory

[243] J. Heintz and J. Morgenstern: On associative algebras of minimal rank. In Proc. AAECC 2, number 228 in LNCS, pages 1–24, 1986

[244] J. Heintz and J. Morgenstern: On the intrinsic complexity of elimination theory. Journal of Complexity 9:471–498, 93

[245] J. Heintz, T. Recio, and M.F. Roy: Algorithms in real algebraic geometry and applications to computational geometry. DIMACS Series in Discrete Mathematics and Theoretical Computer Science 6:137–163, 1991

[246] J. Heintz and C.P. Schnorr: Testing polynomials which are hard to compute. In Logic and Algorithmic: An international Symposium held in honor of Ernst Specker, pages 353–363. Monogr. No. 30 de l'Enseign. Math., 1982

[247] J. Heintz and M. Sieveking: Lower bounds for polynomials with algebraic coefficients. Theoret. Comp. Sc. 11:321–330, 1980

[248] D. Hilbert: Über die vollen Invariantensysteme. Math. Ann. 42:313–373, 1893

[249] M.D. Hirsch: Lower bounds for the non-linear complexity of algebraic computation trees with integer inputs. Comp. Compl. 1:257–268, 1991

[250] M.D. Hirsch: Applications of topology to lower bound estimates in computer science. In From Topology to Computation: Proceedings of the SMALEFEST, pages 395–418. Springer, 1993

[251] M.W. Hirsch: Differential Topology. Number 33 in GTM. Springer, 1976

[252] J. Hopcroft and L. Kerr: On minimizing the number of multiplications necessary for matrix multiplication. SIAM J. Appl. Math. 20:30–36, 1971

[253] J. Hopcroft and J. Musinski: Duality applied to the complexity of matrix multiplication and other bilinear forms. SIAM J. Comp. 2:159–173, 1973

[254] E. Horowitz: A fast method for interpolation using preconditioning. Inf. Proc. Letters 1:157–163, 1972

[255] T.D. Howell: Global properties of tensor rank. Lin. Alg. Appl. 22:9–23, 1978

[256] T.D. Howell and J.C. Lafon: The complexity of quaternion product. Tech-rep. 75-245, Dept. Comp. Sc., Cornell University 1975

[257] D.A. Huffman: A method for the construction of minimum-redundancy codes. Proc. of IRE 40:1098–1101, 1952

[258] J.E. Humphreys: Linear Algebraic Groups. Volume 21 of Graduate Texts in Mathematics. Springer, 1975

[259] B. Huppert: Endliche Gruppen, volume I. Volume 134 of Grundlehren der mathematischen Wissenschaften. Springer, Berlin 1967

[260] L. Hyafil: On the parallel evaluation of multivariate polynomials. SIAM J. Comp. 8:120–123, 1979

[261] L. Hyafil and J.P. van de Wiele: On the additive complexity of specific polynomials. Inf. Proc. Letters 4:45–47, 1975

[262] Y. Ihara: Some remarks on the number of points of an algebraic curve over finite fields. J. Fac. Sci. Tokio 28:721–724, 1981

[263] I.M. Isaacs: Character Theory of Finite Groups. Number 69 in Pure and Applied Mathematics. Academic Press, 1976

[264] N. Jacobson: Basic Algebra, volume II. W.H. Freeman and Company, 1989

[265] J. Ja'Ja: Optimal evaluation of pairs of bilinear forms. SIAM J. Comp. 8:443–462, 1979

[266] J. Ja'Ja: Computation of bilinear forms over finite fields. J. ACM 27:822–830, 1980

[267] J. Ja'Ja: On the complexity of bilinear forms with commutativity. SIAM J. Comp. 9:713–728, 1980

[268] J. Ja'Ja: An Introduction to Parallel Algorithms. Addison Wesley, Reading, Massachusetts 1992

[269] J. Ja'Ja and T. Gonzalez: On the complexity of computing bilinear forms with {0, 1} constants. J. Comp. Syst. Sci. 20:77–95, 1980

[270] J. Ja'Ja and J. Takche: Improved lower bounds for some matrix multiplication problems. Inf. Proc. Letters 21:123–127, 1985

[271] J. Ja'Ja and J. Takche: On the validity of the direct sum conjecture. SIAM J. Comp. 15:1004–1020, 1986

[272] G.D. James and A. Kerber: The Representation Theory of the Symmetric Group. Volume 16 of Encyclopedia of Mathematics and its Applications. Addison-Wesley, 1981

[273] G.J. Janusz: Primitive idempotents in group algebras. Proc. Am. Math. Soc. 17:520–523, 1966

[274] J.W. Jaromczyk: An extension of Rabin's complete proof concept. In J. Gruska and M. Chytill, editors, Math. Foundations of Comp. Sci. 1981, number 118 in LNCS, pages 321–326. Springer, 1981

[275] M. Jerrum: On the Complexity of Evaluating Multivariate Polynomials. PhD thesis, Aiken Comput. Lab., Dept. of Computer Science, University of Edinburgh 1981

[276] M. Jerrum: Two-Dimensional Monomer-Dimer Systems are Computationally Intractable. J. Stat. Phys. 48:121–134, 1987. Erratum in: J. Stat. Phys. 59:1087–1088, 1990

[277] D.S. Johnson: A catalog of complexity classes. In J. van Leeuwen, editor, Handbook of Theoretical Computer Science, volume A, chapter 2, pages 61–161. Elsevier Science Publishers B. V., 1990

[278] K. Kalorkoti: A lower bound for the formula size of rational functions. SIAM J. Comp. 14:678–687, 1985

[279] K. Kalorkoti: The trace invariant and matrix multiplication. Theoret. Comp. Sc. 59:277–286, 1988

[280] K. Kalorkoti: Inverting polynomials and formal power series. SIAM J. Comp. 22:552–559, 1993

[281] E. Kaltofen: Polynomial factorization. In B. Buchberger, G.E. Collins, , and R. Loos, editors, Computer Algebra. Symbolic and Algebraic Computation, pages 95–113. Springer, second edition, 1982

[282] E. Kaltofen: Uniform closure properties of p-computable functions. In Proc. 18th ACM STOC, pages 330–337, 1987

[283] E. Kaltofen: Greatest common divisors of polynomials given by straight-line programs. J. ACM, pages 231–264, 1988

[284] E. Kaltofen: Polynomial factorization 1982–1986. In D.V. Chudnovsky and R.D. Jenks, editors, Computer in Mathematics, pages 285–300. Marcel Dekker, 1990

[285] E. Kaltofen: Efficient solution of sparse linear systems. Lecture Notes, Dept. Comput. Sci., Rensselaer Polytech. Inst., Troy, New York 1992

[286] E. Kaltofen: Polynomial factorization 1987–1991. In I. Simon, editor, Proc. Latin '92, number 583 in LNCS, pages 294–313. Springer, 1992

[287] E. Kaltofen: Analysis of Coppersmith's block Wiedemann algorithm for the parallel solution of sparse linear systems. Math. Comp. 64(210):777–806, 1995

[288] E. Kaltofen and V. Pan: Processor efficient parallel solution of linear systems over an abstract field. In Proc. 3rd Ann. ACM Symp. Parallel Algor. Architecture, pages 180–191. ACM Press, 1991

[289] E. Kaltofen and V. Pan: Processor-efficient parallel solution of linear systems II: the positive characteristic and singular cases. In Proc. 33rd FOCS, pages 714–723, 1992

[290] E. Kaltofen and B.D. Saunders: On Wiedemann's method of solving sparse linear systems. In Proc. AAECC-9, number 539 in LNCS, pages 29–38. Springer, 1991

[291] E. Kaltofen and V. Shoup: Subquadratic-time factoring of polynomials over finite fields. In Proc. 27th STOC, pages 398–406, 1995

[292] M. Kaminski: A lower bound for polynomial multiplication. Theoret. Comp. Sc. 40:319–322, 1985

[293] M. Kaminski: An algorithm for polynomial multiplication that does not depend on the ring constants. J. Algorithms 8:137–147, 1988

[294] M. Kaminski and N.H. Bshouty: Multiplicative complexity of polynomial multiplication over finite fields. J. ACM 36:150–170, 1989

[295] M. Kaminski, D.G. Kirkpatrick, and N.H. Bshouty: Addition requirements for matrix and transposed matrix products. J. Algorithms 9:354–364, 1988

[296] A. Karatsuba and Y. Ofman: Multiplication of multidigit numbers on automata. Soviet Physics Doklady 7:714–716, 1963

[297] R.M. Karp: Reducibility among combinatorial problems. In R.E. Miller and J.W. Thatcher, editors, Complexity of Computer Computations, pages 85–104. Plenum Press, New York 1972

[298] W. Keller-Gehrig: Fast algorithms for the characteristic polynomial. Theoret. Comp. Sc. 36:309–317, 1985

[299] J.L. Kelley: General Topology. Number 27 in GTM. Springer, 1955

[300] A.G. Khovanskii: On a class of systems of transcendental equations. Soviet Math. Dokl. 22(3):762–765, 1980

[301] A.G. Khovanskii: Fewnomials. Number 88 in Translations of Mathematical Monographs. American Mathematical Society, 1991

[302] J.H. Kingston: Algorithms and Data Structures. International Computer Science Series. Addison-Wesley, Sydney 1990

[303] D.E. Knuth: Evaluation of polynomials by computers. Comm. ACM 6:595–599, 1962

[304] D.E. Knuth: The Art of Computer Programming, volume 2: semi-numerical algorithms. Addison-Wesley, 1969

[305] D.E. Knuth: The analysis of algorithms. In Actes du congrès international des Mathématiciens, Nice, volume 3, pages 269–274, 1970

[306] D.E. Knuth: The Art of Computer Programming, volume 1: fundamental algorithms. Addison-Wesley, 1973

[307] D.E. Knuth: The Art of Computer Programming, volume 2: semi-numerical algorithms. Addison-Wesley, second edition, 1981

[308] H. Kraft: Geometrische Methoden der Invariantentheorie. Number D1 in Aspects of Mathematics. Vieweg, 1984

[309] H.T. Kung: On computing reciprocals of power series. Num. Math. 22:341–348, 1974

[310] J. Laderman: A noncommutative algorithm for multiplying 3×3-matrices using 23 multiplications. Bull. Amer. Math. Soc. 82:180–182, 1976

[311] J.C. Lafon and S. Winograd: A lower bound for the multiplicative complexity of the product of two matrices. Centre de Calcul de L'Esplanade, U.E.R. de Mathematique, Univ. Louis Pasteur, Strasbourg, France 1978

[312] J.L. Lagrange: Journal de l'école polytechnique, II(VIII), 1812

[313] B.A. LaMacchia and A.M. Odlyzko: Solving large sparse linear systems over finite fields. In Advances in Cryptology: CRYPTO '90, number 537 in LNCS, pages 109–133. Springer, 1991

[314] S. Lang: Diophantine Geometry. Number 11 in Interscience Tracts in Pure and Applied Mathematics. Interscience Publishers, 1962

[315] S. Lang: Real Analysis. Addison-Wesley, second edition, 1983

[316] S. Lang: Algebra. Addison-Wesley, second edition, 1984

[317] S. Lang: Algebra. Addison-Wesley, third edition, 1993

[318] J. van Leeuwen: An extension of Hansen's theorem for star chains. Crelles J. Reine Angew. Math. 295:202–207, 1977

[319] J. van Leeuwen: Squaring a 2×2-matrix. EATCS Bulletin 9:11–13, 1979

[320] J. van Leeuwen and P. van Emde Boas: Some elementary proofs of lower bounds in complexity theory. Lin. Alg. Appl. 19:63–80, 1978

[321] J. van Leeuwen (ed.): Handbook of Theoretical Computer Science, volume A. Elsevier Science Publishers B. V., 1990

[322] D.H. Lehmer: Euclid's algorithm for large numbers. Amer. Math. Monthly 45:227–233, 1938

[323] T. Lehmkuhl and T. Lickteig: On the Order of Approximation in Approximative Triadic Decompositions of Tensors. Theoret. Comp. Sc. 69:1–14, 1989

[324] A. Lempel, G. Seroussi, and S. Winograd: On the complexity of multiplication in finite fields. Theoret. Comp. Sc. 22:285–296, 1983

[325] A. Lempel and S. Winograd: A new approach to error-correcting codes. IEEE Trans. Inform. Theory 23:503–508, 1977

[326] A.K. Lenstra and H.W. Lenstra Jr: Algorithms in number theory. In J. van Leeuwen, editor, Handbook of Theoretical Computer Science, volume A, chapter 12, pages 673–715. Elsevier Science Publishers B. V., 1990

[327] A. Lev: On large subgroups of finite groups. J. Algebra 152:434–438, 1992

[328] L.A. Levin: Universal sorting problems. Problems of Information Transmission 9:265–266, 1973

[329] T. Lickteig: Untersuchungen zum Tensorrangproblem. PhD thesis, Univ. Konstanz 1982

[330] T. Lickteig: A Note on Border Rank. Inf. Proc. Letters 18:173–178, 1984

[331] T. Lickteig: Typical Tensorial Rank. Linear Algebra Appl., pages 95–120, 1985

[332] T. Lickteig: Gaussian elimination is optimal for solving linear equations in dimension two. Inf. Proc. Letters 22:277–279, 1986

[333] T. Lickteig: The computational complexity of the division in quadratic extension fields. SIAM J. Comp. 16:278–311, 1987

[334] T. Lickteig: On semialgebraic decision complexity. Technical Report TR-90-052, Int. Comp. Sc. Inst., Berkeley 1990. Habilitationsschrift, Universität Tübingen

[335] T. Lickteig: Semi-algebraic decision complexity, the real spectrum, and degree. J. Pure Appl. Alg. 110:131–184, 1996

[336] S. Linnainmaa: Taylor expansion of the accumulated rounding error. BIT 16:146–160, 1976

[337] J.H. van Lint: Introduction to Coding Theory. Volume 86 of Graduate Texts in Mathematics. Springer, 1982

[338] J.D. Lipson: Chinese remainder and interpolation algorithms. In Proc. 2nd Symp. on Symbolic and Algebraic Manipulation, pages 372–391, 1971

[339] R.J. Lipton: Polynomials with 0-1 coefficients that are hard to evaluate. In Proc. 16th FOCS, pages 6–10, 1975

[340] R.J. Lipton and L.J. Stockmeyer: Evaluation of polynomials with super-preconditioning. J. Comp. Syst. Sci. 16:124–139, 1978

[341] R. Loos: Generalized polynomial remainder sequences. In B. Buchberger, G.E. Collins, and R. Loos, editors, Computer Algebra, Symbolic and Algebraic Computation, number 4 in Computing Supplementum, pages 115–137. Springer, 1982

[342] G. Lotti and F. Romani: On the asymptotic complexity of rectangular matrix multiplication. Theoret. Comp. Sc. 23:171–185, 1983

[343] A. Lubotzky, R. Philips, and P. Sarnak: Ramanujan graphs. Combinatorica 8:261–277, 1988

[344] K. Lux and H. Pahlings: Computational aspects of representation theory of finite groups. In G. Michler and C.M. Ringel, editors, Representation Theory of Finite Groups and Finite-Dimensional Algebras, volume 95 of Progress in Mathematics, pages 37–64. Birkhäuser, 1991

[345] F.J. MacWilliams and N.J.A. Sloane: The Theory of Error-Correcting Codes. North-Holland, 1988

[346] L. Mahé: Une demonstration élementaire du théoreme de Bröcker-Scheiderer. C. R. Acad. Sci. Paris 309:613–616, 1989

[347] U. Manber: Introduction to Algorithms: A Creative Approach. Addison-Wesley, Reading, Massachusetts 1989

[348] U. Manber and M. Tompa: Probabilistic, nondeterministic, and alternating decision trees. In Proc. 14th ACM STOC, pages 234–244, 1982

[349] Yu.I. Manin: What is the maximum number of points on a curve over $\mathbb{F}_2$? J. Fac. Sci. Tokio 28:715–720, 1981

[350] M. Marcus and H. Minc: On the relation between the determinant and the permanent. Illinois J. Math. 5:376–381, 1961

[351] G.A. Margulis: Explicit constructions of concentrators. Probl. Inf. Trans. 10:325–332, 1975

[352] G.A. Margulis: Explicit group-theoretical constructions of combinatorial schemes and their application to the design of expanders and superconcentrators. Probl. Inf. Trans. 24:39–46, 1988

[353] D.K. Maslen and D.N. Rockmore: Generalized FFTs - a survey of some recent results. DIMACS workshop on Groups and Computation, 1995

[354] D.K. Maslen and D.N. Rockmore: Separation of variables and the computation of Fourier transforms on finite groups, I. manuscript, 1995

[355] J.L. Massey: Shift register synthesis and BCH decoding. IEEE Trans. Inform. Theory 15:122–127, 1969

[356] H. Matsumura: Commutative Ring Theory. Cambridge University Press 1986

[357] R.J. McEliece: The Theory of Information and Coding. Volume 3 of Encyclopedia of Mathematics and its Applications. Addison-Wesley, 1977

[358] R.J. McEliece, E.R. Rodemich, H.C. Rumsey Jr. and L.R. Welch: New upper bounds on the rate of a code via the Delsarte-MacWilliams inequalities. IEEE Trans. Inform. Theory 23:157–166, 1977

[359] S. Meiser: Point location in arrangements of hyperplanes. Information and Computation 106:286–303, 1993

[360] F. Meyer auf der Heide: A polynomial linear search algorithm for the n-dimensional knapsack problem. J. ACM 31:668–676, 1984

[361] F. Meyer auf der Heide: Simulating probabilistic by deterministic algebraic computation trees. Theoret. Comp. Sc. 41:325–330, 1985

[362] F. Meyer auf der Heide: Fast algorithms for n-dimensional restrictions of hard problems. J. ACM 35:740–747, 1988

[363] F. Meyer auf der Heide: On genuinely time bounded computations. In Proc. STACS 89, number 349 in LNCS, pages 1–16. Springer, 1989

[364] G.L. Miller, V. Ramachandran, and E. Kaltofen: Efficient parallel evaluation of straight-line code and arithmetic circuits. SIAM J. Comp. 17(4):686–695, 1988

[365] G.L. Miller and J.H. Reif: Parallel tree contraction and its applications. In Proc. 26th FOCS, pages 478–489, 1985

[366] J. Milnor: On the Betti numbers of real varieties. In Proc. AMS, volume 15, pages 275–280, 1964

[367] R. Mirwald: The algorithmic structure of $sl(2, k)$. In Proc. AAECC 3, number 229 in LNCS, pages 274–287, 1986

[368] R. Mirwald: Rang von Matrizenpaaren über $\mathbb{Z}_2$ und Komplexität von Paaren Boolescher quadratischer Formen. PhD thesis, Frankfurt University 1991

[369] R. Mirwald and C.P. Schnorr: The multiplicative complexity of Boolean quadratic forms. In Proc. 28th FOCS, pages 141–150, 1987

[370] R. Mirwald and C.P. Schnorr: The multiplicative complexity of Boolean quadratic forms. Theoret. Comp. Sc. 102:307–328, 1992

[371] R. Mirwald and C.P. Schnorr: The multiplicative complexity of Boolean quadratic forms: A survey. In M.S. Paterson, editor, Boolean Function Complexity: Selected papers from LMS Symposium, Durham, July 1990, number 169 in London Mathematical Society, Lecture Notes Series, pages 95–108. Cambridge University Press, 1992

[372] B. Mishra: Algorithmic Algebra. Texts and Monographs in Computer Science. Springer, 1993

[373] R. Moenck and A. Borodin: Fast modular transforms via division. In Proc. 13th Annual Symposium on Switching & Automata Theory, pages 90–96, 1972

[374] R.T. Moenck: Fast computation of GCDs. In Proc. 5th ACM STOC, pages 142–151, 1973

[375] J.L. Montaña, J.E. Morais, and L.M. Pardo: Lower bounds for arithmetic networks II: Sum of Betti numbers. Applicable Algebra in Engineering, Communication and Computing 7:41–51, 1996

[376] J.L. Montaña and L.M. Pardo: Lower bounds for arithmetic networks. Applicable Algebra in Engineering, Communication and Computing 4:1–24, 1993

[377] J.L. Montaña, L.M. Pardo, and T. Recio: The non-scalar model of complexity in computational geometry. In C. Traverso and T. Mora, editors, Proc. MEGA'90, number 94 in Progress in Mathematics, pages 347–362. Birkhäuser, 1991

[378] J.L. Montaña, L.M. Pardo, and T. Recio: A note on Rabin's width of a complete proof. Comp. Compl. 4:12–36, 1994

[379] C.J. Moreno: Algebraic Curves over Finite Fields with Applications to Coding Theory. Cambridge University Press, 1990

[380] J. Morgenstern: Algorithmes linéaires tangents et complexité. C. R. Acad. Sc. Paris 277:367–369, 1973

[381] J. Morgenstern: Note on a lower bound of the linear complexity of the fast Fourier transform. J. ACM 20:305–306, 1973

[382] J. Morgenstern: Complexité linéaire de calcul. Thèse, Univ. de Nice 1978

[383] J. Morgenstern: How to compute fast a function and all its derivatives. SIGACT News 16/4:60–62, 1985

[384] M. Morgenstern: Ramanujan diagrams. SIAM J. Disc. Math. 7(4):560–570, 1994

[385] J. Moses and D.Y.Y. Yun: The EZGCD algorithm. In Proceedings of the ACM Annual Conference, pages 159–166, Atlanta 1973

[386] T.S. Motzkin: Evaluation of polynomials. Bull. Am. Soc. 61:163, 1955

[387] D. Mumford: Algebraic Geometry I: Complex Projective Varieties. Volume 221 of Grundlehren der mathematischen Wissenschaften. Springer, 1976

[388] I. Munro: Efficient determination of the transitive closure of a directed graph. Inf. Proc. Letters 1:56–58, 1971

[389] M.V. Mihaĭljuk: On the complexity of calculating the elementary symmetric functions over finite fields. Sov. Math. Dokl. 20:170–174, 1979

[390] I. Newton: Philosophiae Naturalis Principia Mathematica, volume III. first edition, 1687

[391] Ngọc-Minh Lê: On Voronoi diagrams in the l_p-metric in higher dimensions. In P. Enjalbert, E.W. Mayr, and K.W. Wagner, editors, Proc. STACS 94, number 775 in LNCS, pages 711–722, 1994

[392] N. Nisan: Lower bounds for non-commutative computation. In Proc. 23rd ACM STOC, pages 410–418, 1991

[393] A. Nozaki: A note on the complexity of approximative evaluation of polynomials. Inf. Proc. Letters 9:73–75, 1979

[394] H.J. Nussbaumer: Fast polynomial transform algorithms for digital convolutions. IEEE trans. ASSP 28:205–215, 1980

[395] O.A. Oleĭnik: Estimates of the Betti numbers of real algebraic hypersurfaces. Math. Sb. (N.S.) 28 (70):635–640, 1951

[396] O.A. Oleĭnik and I.B. Petrovskii: On the topology of real algebraic surfaces. Izv. Akad. Nauk SSSR 13:389–402, 1949

[397] U. Oberst: The fast Fourier transform. Publ. Centro Matematico Vito Volterra, Universita di Roma II, 79, 1991

[398] U. Oberst: Galois theory and the fast Gelfand transform. Publ. Centro Matematico Vito Volterra, Universita di Roma II, 99:1–34, 1992

[399] U. Oberst: The optimal fast Gelfand, Fourier and Hartley transforms. In Oberwolfach conference: "Large Discrete Systems", 1995

[400] A.M. Odlyzko: Discrete logarithms in finite fields and their cryptographic significance. In Advances in Cryptology: Proceedings of Eurocrypt '84, number 209 in LNCS, pages 224–314. Springer, 1985

[401] J. Olivos: On vectorial addition chains. J. Algorithms 2:13–21, 1981

[402] P. Orlik and H. Terao: Arrangements of Hyperplanes. Number 300 in Grundlehren der mathematischen Wissenschaften. Springer, 1992

[403] A.M. Ostrowski: On two problems in abstract algebra connected with Horner's rule. In Studies in Mathematics and Mechanics presented to Richard von Mises, pages 40–48. Academic Press, New York 1954

[404] V.Ya. Pan: Certain schemes for the evaluation of polynomials with real coefficients. Problemy Kibernetiki 5:17–29, 1961

[405] V.Ya. Pan: Methods for computing values of polynomials. Russ. Math. Surv. 21:105–136, 1966

[406] V.Ya. Pan: On schemes for the evaluation of products and inverses of matrices. Uspekhi Matematicheskikh Nauk 5(167):249–250, 1972

[407] V.Ya. Pan: Computational complexity of computing polynomials over the fields of real and complex numbers. In Proc. 10th ACM STOC, pages 162–172, 1978

[408] V.Ya. Pan: Strassen's algorithm is not optimal. In Proc. 19th FOCS, pages 166–176, 1978

[409] V.Ya. Pan: Field extension and trilinear aggregating, uniting and cancelling for the acceleration of matrix multiplication. In Proc. of the 20th Ann. IEEE Symp. on Foundations of Comp. Sc., pages 28–38, 1979

[410] V.Ya. Pan: New fast algorithms for matrix operations. SIAM J. Comp. 9:321–342, 1980

[411] V.Ya. Pan: New combinations of methods for the acceleration of matrix multiplication. Comput. Math. Appl. 7:73–125, 1981

[412] V.Ya. Pan: How can we speed up matrix multiplication. SIAM Review 26:393–415, 1984

[413] V.Ya. Pan: How to multiply matrices faster. Number 179 in LNCS. Springer, 1984

[414] C.H. Papadimitriou and D.E. Knuth: Duality in Addition Chains. Bulletin of the EACTS 13:2–3, 1981

[415] M.S. Paterson and L.J. Stockmeyer: On the number of nonscalar multiplications necessary to evaluate polynomials. SIAM J. Comp. 2:60–66, 1973

[416] P. Penfield Jr., R. Spencer, and S. Duinker: Tellegen's Theorem and Electrical Networks. M.I.T. Press, Cambridge, MA 1970

[417] R.S. Pierce: Associative Algebras. Number 88 in GTM. Springer, 1982

[418] M. Pinsker: On the complexity of concentrators. In 7th International Teletrafic Conference, pages 318/1–318/4, 1973

[419] N. Pippenger: Superconcentrators. SIAM J. Comp. 6:298–304, 1978

[420] N. Pippenger: On simultaneous resource bounds. In Proc. 20th FOCS, pages 307–311, 1979

[421] N. Pippenger: On the evaluation of powers and monomials. SIAM J. Comp. 9:230–250, 1980

[422] N. Pippenger: Communication networks. In J.v. Leeuwen, editor, Handbook of Theoretical Computer Science, volume A, chapter 15, pages 805–833. Elsevier Science Publishers B.V., 1990

[423] M. Plotkin: Binary codes with specified minimum distances. IEEE Trans. Inform. Theory 6:445–450, 1960. This paper had already appeared as a research report in 1951

[424] G. Pólya: Aufgabe 424. Arch. Math. Phys. 20:271, 1913

[425] F.P. Preparata and M.I. Shamos: Computational Geometry, An Introduction. Texts and Monographs in Computer Science. Springer, 1985

[426] R.L. Probert: On the additive complexity of matrix multiplication. SIAM J. Comp. 5:187–203, 1976

[427] J.S. Provan: The complexity of reliability computations in planar and acyclic graphs. SIAM J. Comp. 15(3):694–702, 1986

[428] P. Purdom Jr: A transitive closure algorithm. BIT 10:76–94, 1970

[429] M.O. Rabin: Proving simultaneous positivity of linear forms. J. Comp. Syst. Sci. 6:639–650, 1972

[430] M.O. Rabin and S. Winograd: Fast evaluation of polynomials by rational preparation. Comm. Pure and Appl. Math. 15:433–458, 1972

[431] C.M. Rader: Discrete Fourier transform when the number of data points is prime. Proc. IEEE 56:1107–1108, 1968

[432] P. Ramanan: Obtaining lower bounds using artificial components. Inf. Proc. Letters 24:243–246, 1987

[433] E.M. Reingold: On the optimality of some set algorithms. J. ACM 19:649—659, 1972

[434] E.M. Reingold and I. Stocks: Simple proofs of lower bounds for polynomial evaluation. In R. Miller and J. Thatcher, editors, Complexity of Computer Computations, pages 21–30. Plenum Press, 1972

[435] J. Renegar: On the worst-case arithmetic complexity of approximating zeros of polynomials. J. Compl. 3:90–113, 1987

[436] J. Renegar: On the worst-case arithmetic complexity of approximating zeros of systems of polynomials. SIAM J. Comp. 18:350–370, 1989

[437] L. Revah: On the number of multiplications/divisions evaluating a polynomial with auxiliary functions. SIAM J. Comp. 4:381–392, 1975

[438] J.-J. Risler: Additive complexity and zeros of real polynomials. SIAM J. Comp. 14:178–183, 1985

[439] P. Ritzmann: Ein numerischer Algorithmus zur Komposition von Potenzreihen und Komplexitätsschranken für die Nullstellenberechnung von Polynomen. PhD thesis, Zürich University 1984

[440] P. Ritzmann: A fast numerical algorithm for the composition of power series with complex coefficients. Theoret. Comp. Sc. 44:1–16, 1986

[441] D. Rockmore: Fast Fourier analysis for abelian group extensions. Adv. Appl. Math. 11:164–204, 1990

[442] J.M. Rojas: A convex geometric approach to counting the roots of a polynomial system. Theoret. Comp. Sc. 133:105–140, 1994

[443] F. Romani: Some properties of disjoint sums of tensors related to matrix multiplication. SIAM J. Comp. 11:263–267, 1982

[444] R.A. Rueppel: Analysis and Design of Stream Ciphers. Springer, Berlin 1986

[445] H. J. Ryser: Combinatorial Mathematics. Number 14 in Carus Math. Monographs. Math. Assoc. America, New York 1963

[446] G.E. Sacks: Multiple error correction by means of parity checks. IEEE Trans. Inform. Theory 4:145–147, 1958

[447] J.E. Savage: An algorithm for the computation of linear forms. SIAM J. Comp. 3:150–158, 1974

[448] C. Scheiderer: Stability index of real varieties. Invent. Math. 97:467–483, 1989

[449] C.P. Schnorr: Improved lower bounds on the number of multiplications/divisions which are necessary to evaluate polynomials. Theoret. Comp. Sc. 7:251–261, 1978

[450] C.P. Schnorr: An extension of Strassen's degree bound. SIAM J. Comp. 10:371–382, 1981

[451] C.P. Schnorr: How many polynomials can be approximated faster than they can be evaluated. Inf. Proc. Letters 12:76–78, 1981

[452] C.P. Schnorr and J.P. van de Wiele: On the additive complexity of polynomials. Theoret. Comp. Sc. 10:1–18, 1980

[453] A. Scholz: Aufgabe 253. Jber. d. Dt. Math.Verein 92, class II 47:41–42, 1937

[454] A. Schönhage: Multiplikation großer Zahlen. Computing 1:182–196, 1966

[455] A. Schönhage: Schnelle Berechnung von Kettenbruchentwicklungen. Act. Inf. 1:139–144, 1971

[456] A. Schönhage: Unitäre Transformation großer Matrizen. Num. Math. 20:409–417, 1973

[457] A. Schönhage: A lower bound for the length of addition chains. Theoret. Comp. Sc. 1:1–12, 1975

[458] A. Schönhage: An elementary proof of Strassen's degree bound. Theoret. Comp. Sc. 3:267–272, 1976

[459] A. Schönhage: Schnelle Multiplikation von Polynomen über Körpern der Charakteristik 2. Act. Inf. 7:395–398, 1977

[460] A. Schönhage: Partial and total matrix multiplication. SIAM J. Comp. 10:434–455, 1981

[461] A. Schönhage: The fundamental theorem of algebra in terms of computational complexity. Technical report, Univ. Tübingen 1982

[462] A. Schönhage: Equation solving in terms of computational complexity. In Proc. of the International Congress of Mathematicians, volume 1, pages 131–153, 1986

[463] A. Schönhage: Probabilistic computation of integer polynomial GCDs. J. Algorithms 9:365–371, 1988

[464] A. Schönhage: Bivariate polynomial multiplication patterns. In Proceedings of AAECC 11, number 948 in LNCS, pages 70–81. Springer, 1995

[465] A. Schönhage, A. Grotefeld, and E. Vetter: Fast Algorithms. BI-Wissenschaftsverlag, Mannheim 1994

[466] A. Schönhage and V. Strassen: Schnelle Multiplikation großer Zahlen. Computing 7:281–292, 1971

[467] P. Schuster: Interpolation und Kettenbruchentwicklung. Die Komplexität einiger Berechnungsaufgaben. PhD thesis, Univ. Zürich 1980

[468] J.T. Schwartz: Fast probabilistic algorithms for verification of polynomial identities. J. ACM 27:701–717, 1980

[469] J.P. Serre: Sur le nombres des points rationelles d'une courbe algébrique sur uns corps finis. C.R. Acad. Sci. Paris 296:397–402, 1983

[470] J.P. Serre: Linear Representations of Finite Groups. GTM. Springer, 1986

[471] I.R. Shafarevich: Basic Algebraic Geometry. Springer, 1974

[472] S. Shokranian and M.A. Shokrollahi: Coding Theory and Bilinear Complexity. Number 21 in Scientific Series of the International Bureau. KFA Jülich 1993

[473] M.A. Shokrollahi: On the rank of certain finite fields. Comp. Compl. 1:157–181, 1991

[474] M.A. Shokrollahi: Beiträge zur Codierungs- und Komplexitätstheorie mittels algebraischer Funktionenkörper. Bayreuth. Math. Schriften 39:1–236, 1992

[475] M.A. Shokrollahi: Efficient randomized generation of algorithms for multiplication in certain finite fields. Comp. Compl. 2:67–96, 1992

[476] M.A. Shokrollahi: Optimal algorithms for multiplication in certain finite fields using elliptic curves. SIAM J. Comp. 21:1193–1198, 1992

[477] V. Shoup and R. Smolensky: Lower bounds for polynomial evaluation and interpolation problems. volume 32 of FOCS, pages 378–383, 1991

[478] I.E. Shparlinski, M.A. Tsfasman, and S.G. Vlădut: Curves with many points and multiplication in finite fields. In Proceedings, Luminy 1991, volume 1518 of Lecture Notes in Mathematics, pages 145–169. Springer, 1992

[479] M. Shub: Some remarks on Bézout's theorem and complexity theory. In From Topology to Computation: Proceedings of the SMALEFEST, pages 443–455. Springer, 1993

[480] M. Shub and S. Smale: Complexity of Bézout's theorem V: Polynomial time. Theoret. Comp. Sc. 133:141–164, 1994

[481] M. Shub and S. Smale: On the intractability of Hilbert's Nullstellensatz and an algebraic version of "NP ≠ P?". Duke Math. J. 81:47–54, 1995

[482] M. Sieveking: An algorithm for division of power series. Computing 10:153–156, 1972

[483] J. Simon: On some central problems in computational complexity. PhD thesis, Dept. of Computer Science, Cornell University, Ithaca, NY 1975

[484] C.C. Sims: Computation with Finitely Presented Groups. Cambridge University Press, 1994

[485] S. Smale: Algorithms for solving equations. In Proc. of the International Congress of Mathematicians, volume 1, pages 172–195, 1986

[486] S. Smale: On the topology of root finding. J. Compl. 3:81–89, 1987

[487] J.M. Steele and A.C. Yao: Lower bounds of algebraic decision trees. J. Algorithms 3:1–8, 1982

[488] H. Stichtenoth: Algebraic Function Fields and Codes. Universitext. Springer, 1993

[489] H.J. Stoß: The complexity of evaluating interpolation polynomials. Theoret. Comp. Sc. 41:319–323, 1985

[490] H.J. Stoß: Lower bounds for the complexity of polynomials. Theoret. Comp. Sc. 64:15–23, 1989

[491] H.J. Stoß: On the representation of rational functions of bounded complexity. Theoret. Comp. Sc. 64:1–13, 1989

[492] D.R. Stoutemeyer: Polynomial remainder sequences: greatest common divisors revisited. In N. Inada and T. Soma, editors, Proceedings 2nd RIKEN Int. Symp. Symbol. Algebraic Comput. by Computers, volume 2, pages 1–12, Philadelphia, 1985. World Sci. Publ. Ser. Comput. Sci

[493] V. Strassen: Gaussian elimination is not optimal. Num. Math. 13:354–356, 1969

[494] V. Strassen: Berechnung und Programm. I. Act. Inf. 1:320–335, 1972

[495] V. Strassen: Evaluation of rational functions. In R.E. Miller and J.E. Thatcher, editors, Complexity of Computer Computations, pages 1–10. Plenum Press, New York 1972

[496] V. Strassen: Berechnung und Programm. II. Act. Inf. 2:64–79, 1973

[497] V. Strassen: Die Berechnungskomplexität von elementarsymmetrischen Funktionen und von Interpolationskoeffizienten. Num. Math. 20:238–251, 1973

[498] V. Strassen: Vermeidung von Divisionen. Crelles J. Reine Angew. Math. 264:184–202, 1973

[499] V. Strassen: Polynomials with rational coefficients which are hard to compute. SIAM J. Comp. 3:128–149, 1974

[500] V. Strassen: Die Berechnungskomplexität der symbolischen Differentiation von Interpolationspolynomen. Theoret. Comp. Sc. 1:21–25, 1975

[501] V. Strassen: Computational complexity over finite fields. SIAM J. Comp. 5:324–331, 1976

[502] V. Strassen: The computational complexity of continued fractions. In ACM Symposium on Symbolic and Algebraic Computation, pages 51–67, 1981

[503] V. Strassen: Berechnungskomplexität I+II. Lectures at the department of mathematics of Zürich University, 1983

[504] V. Strassen: The computational complexity of continued fractions. SIAM J. Comp. 12:1–27, 1983

[505] V. Strassen: Rank and optimal computation of generic tensors. Lin. Alg. Appl. 52:645–685, 1983

[506] V. Strassen: Algebraische Berechnungskomplexität. In Perspectives in Mathematics, Anniversary of Oberwolfach 1984. Birkhäuser, 1984

[507] V. Strassen: The work of L.G. Valiant. In Proc. International Congress of Mathematicians, Berkeley, 1986

[508] V. Strassen: Relative bilinear complexity and matrix multiplication. Crelles J. Reine Angew. Math. 375/376:406–443, 1987

[509] V. Strassen: The asymptotic spectrum of tensors. Crelles J. Reine Angew. Math. 384:102–152, 1988

[510] V. Strassen: Algebraic complexity theory. In J. van Leeuwen, editor, Handbook of Theoretical Computer Science, volume A, chapter 11, pages 634–672. Elsevier Science Publishers B. V., Amsterdam 1990

[511] V. Strassen: Degeneration and complexity of bilinear maps: some asymptotic spectra. Crelles J. Reine Angew. Math. 413:127–180, 1991

[512] V. Strassen: Algebra and complexity. In First European Congress in Mathematics, Paris 1992, Progress in Mathematics, pages 429–446. Birkhäuser, 1994

[513] E.G. Straus: Addition chains of vectors. Amer. Math. Monthly 71:806–808, 1964

[514] G. Szegö: Zu Aufgabe 424. Archiv Math. Phys. 21:291–292, 1913

[515] L. Teichert: Die Komplexität von Bilinearformpaaren über beliebigen Körpern. PhD thesis, Technische Universität Clausthal 1986

[516] R. Thom: Sur l'homologie des variétés algébriques réelles. In S.S. Cairns, editor, Differential and Combinatorial Topology, pages 255–265. Princeton Univ. Press, 1965

[517] H.C.A. van Tilborg: An Introduction to Cryptology. Kluwer Academic Publishers, Boston 1988

[518] V. Tobler: Spezialisierung und Degeneration von Tensoren. PhD thesis, Konstanz University 1991

[519] R. Tolmieri: A non-quadratic algorithm for a system of quadratic forms. J. Algorithms 3:31–40, 1982

[520] A.L. Toom: The complexity of a scheme of functional elements realizing the multiplication of integers. Soviet Math. Dokl. 4:714–716, 1963

[521] M.A. Tsfasman, S.G. Vlădut, and Th. Zink: Modular curves, Shimura curves, and Goppa codes better than the Varshamov-Gilbert bound. Math. Nachr. 109:21–28, 1982

[522] M.A. Tsfasman and S.G. Vlădut: Algebraic-Geometric Codes. Mathematics and its Applications. Kluwer Academic Publishers, Dordrecht 1991

[523] A.M. Turing: On computable numbers, with an application to the Entscheidungsproblem. Proc. London Math. Soc. 42(2):230–265, 1936

[524] L.G. Valiant: General context-free recognition in less than cubic time. J. Comp. Syst. Sci. 10:308–315, 1975

[525] L.G. Valiant: On non-linear lower bounds in computational complexity. In Proc. 7th ACM STOC, pages 45–53, 1975

[526] L.G. Valiant: Completeness classes in algebra. In Proc. 11th ACM STOC, pages 249–261, 1979

[527] L.G. Valiant: The complexity of computing the permanent. Theoret. Comp. Sc. 8:189–201, 1979

[528] L.G. Valiant: The complexity of enumeration and reliability problems. SIAM J. Comp. 8:410–421, 1979

[529] L.G. Valiant: Reducibility by algebraic projections. In Logic and Algorithmic: an International Symposium held in honor of Ernst Specker, pages 365–380. Monogr. No. 30 de l'Enseign. Math., 1982

[530] L.G. Valiant, S. Skyum, S. Berkowitz, and C. Rackoff: Fast parallel computation of polynomials using few processors. SIAM J. Comp. 12(4):641–644, 1983

[531] J.-P. van de Wiele: Complexité additive et zéros des polynômes à coefficients réels et complexes. Technical Report 292, INRIA, Rocquencourt 1978

[532] V.N. Vapnik and A.Ya. Chervonenkis: On the uniform convergence of relative frequencies of events to their probabilities. Theory of Probability and its Applications 16(2):264–280, 1971

[533] T.M. Vari: On the number of multiplications required to compute quadratic functions. Technical Report TR 72-120, Dept. of Computer Science, Cornell University 1972

[534] R.R. Varshamov: Estimate of the number of signals in error-correcting codes. Dokl. Akad. Nauk USSSR 117:739–741, 1957

[535] V.A. Vassiliev: Cohomology of braid groups and complexity of algorithms. Funct. Anal. Appl. 22:15–24, 1988

[536] V.A. Vassiliev: Topological complexity of algorithms of approximate solving the systems of polynomial equations. Algebra Anal. 1:198–213, 1989

[537] V.A. Vassiliev: Complements of Discriminants of Smooth Maps: Topology and Applications. Number 98 in Transl. of Math. Monographs. AMS, 1992

[538] W. Vogel: On Results on Bézout's Theorem. Number 74 in Lecture Notes. TATA Institute of Fundamental Research, Bombay 1974

[539] B.L. van der Waerden: Einführung in die algebraische Geometrie. Volume 51 of Grundlehren der mathematischen Wissenschaften. Springer, Berlin 1939

[540] B.L. van der Waerden: Modern Algebra. Frederick Ungar Publishing Co., New York 1948

[541] A. Waksman: On Winograd's algorithm for inner products. IEEE Trans. Comp. C-19:360–361, 1970

[542] P.S. Wang: An improved multivariate polynomial factoring algorithm. Math. Comp. 32:1215–1231, 1978

[543] P.S. Wang: The EEZ-GCD algorithm. ACM SIGSAM Bulletin 14:50–60, 1980

[544] E. Waring: Phil. Trans. Roy. Soc. London 69:59–67, 1779

[545] A. Weil: Foundations of Algebraic Geometry. Amer. Math. Soc., Colloquium Publ. 29, 1946

[546] A. Weil: Sur le courbes algébriques et les variétés qui s'en dédusient. Hermann, Paris 1948

[547] L.R. Welch and R.A. Sholtz: Continued fractions and Berlekamp's algorithm. IEEE Trans. Inform. Theory 25:18–27, 1979

[548] D.T. Whiteside: The mathematical papers of Isaac Newton, volume 2. Cambridge University Press, 1968

[549] D. Wiedemann: Solving sparse linear equations over finite fields. IEEE Trans. Inform. Theory 32:54–62, 1985

[550] S. Winograd: On the number of multiplications required to compute certain functions. Proc. Nat. Acad. Sc. 58:1840–1842, 1967

[551] S. Winograd: A new algorithm for inner products. IEEE Trans. Comp. C-17:693–694, 1968

[552] S. Winograd: On the number of multiplications necessary to compute certain functions. Comm. Pure and Appl. Math. 23:165–179, 1970

[553] S. Winograd: On multiplication of 2 × 2 matrices. Lin. Alg. Appl. 4:381–388, 1971

[554] S. Winograd: Some remarks on fast multiplication of polynomials. In IEEE Symposium on Complexity of Sequential and parallel numerical algorithms, 1973

[555] S. Winograd: A new algorithm for computing the discrete Fourier transform. In Proceedings of Tenths Asilomar Conference on Circuits, Systems and Computers, 1976

[556] S. Winograd: On computing the discrete Fourier transform. Proc. Nat. Acad. Sc. 73:1005–1006, 1976

[557] S. Winograd: Some bilinear forms whose multiplicative complexity depends on the fields of constants. Math. Syst. Theory 10:169–180, 1977

[558] S. Winograd: On computing the discrete Fourier transform. Math. Comp. 32:175–199, 1978

[559] S. Winograd: On multiplication in algebraic extension fields. Theoret. Comp. Sc. 8:359–377, 1979

[560] S. Winograd: On the multiplicative complexity of the discrete Fourier transform. Adv. Math. 32:83–117, 1979

[561] S. Winograd: Arithmetic complexity of computations. SIAM CBMS-NSF regional conference series in applied math, 33, 1980

[562] S. Winograd: On multiplication of polynomials modulo a polynomial. SIAM J. Comp. 9:225–229, 1980

[563] A.C. Yao: On the evaluation of powers. SIAM J. Comp. 5:100–103, 1976

[564] A.C. Yao: A lower bound to finding convex hulls. J. ACM 28:780–787, 1981

[565] A.C. Yao: Lower bounds for algebraic computation trees with integer inputs. In Proc. 30th FOCS, pages 308–313, 1989

[566] A.C. Yao: Algebraic decision trees and Euler characteristic. In Proc. 33rd FOCS, 1992

[567] A.C. Yao: Decision tree complexity and Betti numbers. In Proc. 26th ACM STOC, 1994

[568] A.C. Yao and R.L. Rivest: On the polyhedral decision problem. SIAM J. Comp. 9:343–347, 1980

[569] F.F. Yao: Computational geometry. In J. van Leeuwen, editor, Handbook of Theoretical Computer Science, volume A, chapter 7, pages 343–389. Elsevier Science Publishers B. V., 1990

[570] D.Y.Y. Yun: Uniform bounds for a class of algebraic mappings. SIAM J. Comp. 8:348–356, 1979

[571] O. Zariski and P. Samuel: Commutative Algebra, volume 1. Van Nostrand, 1958

[572] O. Zariski and P. Samuel: Commutative Algebra, volume 2. Van Nostrand, 1960

[573] R.E. Zippel: Probabilistic algorithms for sparse polynomials In E.W. Ng, editor, Symbolic and Algebraic Computation, volume 72 of LNCS, pages 216–226. Springer, 1979

List of Notation

$\mathbb{N}$	set of non-negative integers
$\mathbb{N}'$	$\mathbb{N} \setminus \{0\}$
$\mathbb{Z}$	ring of integers
$\mathbb{Q}, \mathbb{R}, \mathbb{C}$	fields of rational, real, complex numbers
$\mathbb{F}_q$	finite field with q elements
$\langle v_1, \ldots, v_m \rangle_k$	k-linear hull of the elements $v_1, \ldots, v_m$ belonging to a vector space over k
V^*	dual of the vector space V
$\dim_k V$	dimension of the vector space V over k
$\mathrm{Hom}_k(V, W)$	space of all morphisms $V \to W$ of k-spaces
$\mathrm{End}_k(V)$	ring of the k-endomorphisms of V
$\mathrm{Aut}_k(V)$	group of all k-automorphisms of V
$\mathrm{rk}(\phi), \mathrm{rk}\,\phi$	rank of the vector space morphism ϕ
$\mathrm{cork}(\phi), \mathrm{cork}\,\phi$	co-rank of the vector space morphism ϕ
$\ker(\phi), \ker\phi$	kernel of the vector space morphism ϕ
$\mathrm{im}(\phi), \mathrm{im}\,\phi$	image of the mapping ϕ
id_V	identity mapping on V
$\mathrm{tr.deg}_k K$	degree of transcendency of the field extension $k \subseteq K$
$\mathrm{Gal}(K/k)$	Galois group of the field extension $k \subseteq K$
$\mathrm{Aut}(K/k)$	group of field automorphisms of K which fix the elements of the subfield k of K
$\bar{k}$	algebraic closure of the field k
$[K:k]$	degree of the field extension $k \subseteq K$
$k \to A$	k-algebra A
$\mathrm{char}\,R$	characteristic of the ring R
$R^\times$	group of units of the ring R
$\mathrm{GL}(n, R)$	group of invertible $n \times n$-matrices over the ring R
$\mathrm{SL}(n, R)$	group of $n \times n$-matrices over R with determinant one
$k[X_1, \ldots, X_n]$	polynomial ring in the indeterminates $X_1, \ldots, X_n$ over the field k
$k(X_1, \ldots, X_n)$	field of rational functions in the indeterminates $X_1, \ldots, X_n$ over the field k

$k[X_1, \ldots, X_n]_d$	localization of the polynomial ring with respect to the set of denominators $\{d^i \mid i \in \mathbb{N}\}$, where $d \in k[X_1, \ldots, X_n]$
$k[[X_1, \ldots, X_n]]$	ring of formal power series in the indeterminates $X_1, \ldots, X_n$ over the field k
$k((X_1, \ldots, X_n))$	ring of formal Laurent series in the indeterminates $X_1, \ldots, X_n$ over the field k
$\mathbb{P}^n(k)$	n-dimensional projective space over k
$\deg f$	degree of the polynomial f
$\operatorname{grad} f$	gradient of f
$[\partial_j f_i(\xi)]_{1 \le i, j \le n}$	Jacobian matrix of $f_1, \ldots, f_n$ at ξ
$Z(f_1, \ldots, f_r)$	zeroset of $f_1, \ldots, f_r$
$d_x \phi$	differential of the morphism ϕ in x
$k^{m \times n}$	vector space of the $m \times n$-matrices over k
I_n	$n \times n$ identity matrix
$\operatorname{Tr}(A)$	trace of the matrix A
$\det(A), \det A$	determinant of the matrix A
$A^\top$	transpose of the matrix A
$f = O(g)$	means there exist positive constants c and N such that for all $n \ge N$ we have $f(n) \le cg(n)$
$O(g)$	stands for a function f satisfying $f = O(g)$
$f = o(g)$	means $\lim_{n \to \infty} f(n)/g(n) = 0$
$o(g)$	stands for a function f satisfying $f = o(g)$
$f = \Theta(g)$	means $f = O(g)$ and $g = O(f)$
$f = \Omega(g)$	means there exist positive constants c and N such that for all $n \ge N$ we have $f(n) \ge cg(n)$
$f \lesssim g$	means $\limsup_{n \to \infty} f(n)/g(n) \le 1$
$f \sim g$	means $f \lesssim g$ and $g \lesssim f$
δ_{ij}	Kronecker delta, i.e., one if $i = j$ and zero otherwise
$\log, \log_2$	logarithm to the base 2
$\ln$	natural logarithm
$\underline{n}$	$\{1, \ldots, n\}$
$|M|$	cardinality of the set M
$\operatorname{conv}(M)$	convex hull of M
$1_S, \chi_S$	indicator function of a subset S
$\sqcup$	disjoint union
$\wedge$	logical and
$\vee$	logical or
DFT_N	DFT matrix, 8
VP	Valiant's class of p-computable families, 19
VNP	Valiant's class of p-definable families, 20
DET	family of generic determinants, 20
PER	family of generic permanents, 20
$\lfloor a/b \rfloor$	quotient of a and b, 31

$a \bmod b$	remainder of a and b, 31
$M(n)$	upper bound for complexity to multiply two univariate polynomials of degree $\leq n$, 34
$\mathcal{H}(d_1, \ldots, d_t)$	Entropy (non-normalized), 38
(T, π)	product tree, 40
$m(d_1, \ldots, d_t)$	minimum weighted external path length, 40
$\mathcal{H}(T, \pi; p_1, \ldots, p_t)$	set of all intermediate results when computing $p_1 \cdots p_t$ via the product tree (T, π), 40
LOG	formal power series for $\ln(1 + x)$, 49
EXP	formal power series for $\exp(x)$, 49
$GCD(a, b)$	greatest common divisor of a and b, 62
$s(A_1, A_2, \ell), s(d_1, \ldots, d_t; \ell)$	functions related to a controlled Euclidean descent, 64
$A \mid \ell$	significant part of length $\ell + 1$ of A, 64
$\Gamma(A, B, \ell)$	prefix of the Euclidean representation of (A, B), 67
$D(d_1, \ldots, d_t)$	set of pairs of polynomials having degree pattern $(d_1, \ldots, d_t)$, 71
$\mathrm{vert}(P)$	set of vertices of polytope P, 80
$\mathrm{relint}(P)$	relative interior of polytope P, 80
$\mathrm{conv}(M)$	convex hull of M, 81
KS_n	real knapsack problem, 83
$\Pi_R(A)$	restriction of range R to A, 84
$VCdim(\mathcal{S})$	VC-dimension of range space $\mathcal{S}$, 84
$\Phi_d(m)$	number of cells of an arrangement of m hyperplanes in $\mathbb{R}^d$ in general position, 85
$(\mathcal{H}_n^+, \mathcal{R}_n)$	certain range space, 86
$R_{A,\varepsilon}$	set of ranges which contain a significant fraction of the points in A, 87
$E(Z)$	expectation of random variable Z, 88
$\mathrm{Var}(Z)$	variance of random variable Z, 88
$\mathrm{lc}(A)$	leading coefficient of A, 93
Ω	set of operations, 105
$L_A^{\mathrm{tot}}(F|I)$	total complexity of F modulo I, 108
$L_A^+(F|I)$	additive complexity of F modulo I, 108
$L_A(F|I), L_A^*(F|I)$	multiplicative complexity of F modulo I, 108
P	set of relations, 115
(φ, π)	collection, 117
$C(\varphi, \pi), C^*(\varphi, \pi)$	multiplicative complexity of collection (φ, π), 118
$C^{*, \leq}(\pi)$	multiplicative branching complexity of partition π, 118
$C^{\mathrm{tot}}(\pi)$	total complexity of partition π, 118
$L^*[F], L^+[F], L^{\mathrm{tot}}[F]$	complexities with preconditioning, 126
$\mathrm{Coeff}(F), \mathrm{Coeff}_{K/k}(F)$	coefficient field, 130
$L^*(F), L^+(F), L^{\mathrm{tot}}(F)$	complexities with preconditioning, 131
$\mathrm{def}\,\sigma$	domain of definition of a substitution, 145

$\delta_{k \to K(a)}^{p}(F)$	degree of linearization, 148
$[Y_i, Y_{i-1}, \ldots, Y_0]$	Euler brackets, 151
$C_d(F)$	set of Taylor polynomials up to degree d of elements in F, 162
∂_i	differential operator, 164
$k[X]_S$	localization of $k[X]$ at S, 165
$\deg Y$	degree of a set Y of rational functions, 172
$S = k[X_0, \ldots, X_n]$	homogeneous polynomial ring, 178
$H(I; t)$	Hilbert function, 178
$h(I; t)$	Hilbert polynomial, 179
$\mathrm{DEG}(I)$	degree of a homogeneous ideal, 179
$\deg V$	degree of the variety V, 180
$\deg(f_1, \ldots, f_r)$	degree of a sequence of rational functions, 183
$C(p_1, \ldots, p_t)$	set of coefficients of polynomials $p_1, \ldots, p_t$, 188
$C'_{r,n}$	complexity class, 208
$\pi(n)$	number of primes $\leq n$, 210
φ	Euler totient function, 210
$C_{r,n}$	complexity class, 211
$\overline{C}_{r,n}$	closed complexity class, 211
$\mathrm{ht}(f)$	height of a polynomial f, 212
$wt(f)$	weight of a polynomial f, 212
$B_{s,a}$	K-subspace of $K[Z_i|i][[X]]$, 212
$J(n, m)$	set of pairs of polynomials of degrees n, m, 249
$J(d_1, \ldots, d_t)$	set of sequences of polynomials of degree pattern $(d_1, \ldots, d_t)$, 249
φ_d	map $D(d_1, \ldots, d_t) \to J(d_1, \ldots, d_t)$ induced by Euclidean algorithm, 249
$S_{\ell,i}(A, B)$	subresultant coefficient, 254
$\mathrm{SRes}_\ell(A, B)$	ℓth subresultant of A, B, 254
$\mathrm{Res}(A, B)$	resultant of A, B, 254
$b_0(X)$	number of connected components of X, 265
$\mathrm{dist}(D, E)$	distance of D and E, 270
$C^{*, \leq}(W)$	multiplicative branching complexity of $W \subseteq \mathbb{R}^n$, 272
B_n	set of Boolean functions $\{0, 1\}^n \to \{0, 1\}$, 276
T_n	set of threshold functions of n variables, 277
$C_d(W)$	(degree d) decision complexity, 279
$w(W, U)$	generic width of W in U, 281
$\mathcal{N}(U)$	ring of Nash functions on U, 282
$w^{\mathcal{N}}(W, U)$	generic Nash width of W in U, 282
$\mathcal{R}(\phi)$	set of regular values of a differentiable map ϕ, 293
$L^+(f)$	additive complexity, 296
$\Lambda(F)$	variant of the linear complexity of F, 306
$L_\infty(f_1, \ldots, f_m)$	linear complexity of the linear forms $f_1, \ldots, f_m$, 307

P/poly	class of languages recognizable in nonuniform polynomial time, 544
NP	class of languages verifiable in polynomial time, 546
$A_1 \leq_p A_2$	A_1 is polynomially reducible to A_2, 546
$\equiv_p$	p-equivalence of languages, 546
SAT	satisfiability problem, 546
$f \preceq_p g, f \preceq_p^t g$	f is a p-projection of g, 547
HC	family of Hamiltonian cycle polynomials, 547
$\text{val}(\varphi)$	polynomial represented by the expression φ, 549
$E(f)$	expression size of f, 549
VP$_e$	Valiant's class of p-expressible families, 550
VNP$_e$	equals **VNP**, 550
VQP$_e$	class of qp-expressible families, 562
VQP	class of qp-computable families, 562
VNC^d	Valiant's analogue of Nick's class, 568
#L	counting problem associated with $L \in$ **NP**, 574
#**P**	class of counting problems computable in polynomial time by a counting Turing machine, 574

Index

Grundlehren der mathematischen Wissenschaften
A Series of Comprehensive Studies in Mathematics

Springer
and the
environment

At Springer we firmly believe that an international science publisher has a special obligation to the environment, and our corporate policies consistently reflect this conviction.
We also expect our business partners – paper mills, printers, packaging manufacturers, etc. – to commit themselves to using materials and production processes that do not harm the environment. The paper in this book is made from low- or no-chlorine pulp and is acid free, in conformance with international standards for paper permanency.

Springer